Molecular Physics and Hypersonic Flows

NATO ASI Series

Advanced Science Institutes Series

A Series presenting the results of activities sponsored by the NATO Science Committee, which aims at the dissemination of advanced scientific and technological knowledge, with a view to strengthening links between scientific communities.

The Series is published by an international board of publishers in conjunction with the NATO Scientific Affairs Division

A Life Sciences	Plenum Publishing Corporation
B Physics	London and New York
C Mathematical and Physical Sciences	Kluwer Academic Publishers
D Behavioural and Social Sciences	Dordrecht, Boston and London
E Applied Sciences	
F Computer and Systems Sciences	Springer-Verlag
G Ecological Sciences	Berlin, Heidelberg, New York, London,
H Cell Biology	Paris and Tokyo
I Global Environmental Change	

PARTNERSHIP SUB-SERIES

1. **Disarmament Technologies**	Kluwer Academic Publishers
2. **Environment**	Springer-Verlag / Kluwer Academic Publishers
3. **High Technology**	Kluwer Academic Publishers
4. **Science and Technology Policy**	Kluwer Academic Publishers
5. **Computer Networking**	Kluwer Academic Publishers

The Partnership Sub-Series incorporates activities undertaken in collaboration with NATO's Cooperation Partners, the countries of the CIS and Central and Eastern Europe, in Priority Areas of concern to those countries.

NATO-PCO-DATA BASE

The electronic index to the NATO ASI Series provides full bibliographical references (with keywords and/or abstracts) to more than 50000 contributions from international scientists published in all sections of the NATO ASI Series.
Access to the NATO-PCO-DATA BASE is possible in two ways:

– via online FILE 128 (NATO-PCO-DATA BASE) hosted by ESRIN,
Via Galileo Galilei, I-00044 Frascati, Italy.

– via CD-ROM "NATO-PCO-DATA BASE" with user-friendly retrieval software in English, French and German (© WTV GmbH and DATAWARE Technologies Inc. 1989).

The CD-ROM can be ordered through any member of the Board of Publishers or through NATO-PCO, Overijse, Belgium.

Series C: Mathematical and Physical Sciences – Vol. 482

Molecular Physics and Hypersonic Flows

edited by

Mario Capitelli

Department of Chemistry
and
Centro di Studio per la Chimica dei Plasmi del CNR,
University of Bari,
Bari, Italy

Kluwer Academic Publishers

Dordrecht / Boston / London

Published in cooperation with NATO Scientific Affairs Division

Proceedings of the NATO Advanced Study Institute on
Molecular Physics and Hypersonic Flows
Maratea, Italy
May 21–June 3, 1995

A C.I.P. Catalogue record for this book is available from the Library of Congress.

ISBN 0-7923-4055-8

Published by Kluwer Academic Publishers,
P.O. Box 17, 3300 AA Dordrecht, The Netherlands.

Kluwer Academic Publishers incorporates the publishing programmes of
D. Reidel, Martinus Nijhoff, Dr W. Junk and MTP Press.

Sold and distributed in the U.S.A. and Canada
by Kluwer Academic Publishers,
101 Philip Drive, Norwell, MA 02061, U.S.A.

In all other countries, sold and distributed
by Kluwer Academic Publishers Group,
P.O. Box 322, 3300 AH Dordrecht, The Netherlands.

Printed on acid-free paper

Contents

PREFACE

The NATO Advanced Research Institute on Molecular Physics and Hypersonic Flows was held at Acquafredda di Maratea during 21 May-3 June 1995.

The scope of this NATO Institute was to establish a communication channel between the fluid dynamics and molecular physics communities emphasizing the role played by elementary processes in hypersonic flows. In particular the Institute was primarily dedicated to fill the gap between microscopic and macroscopic treatments of the source terms to be inserted in the fluid dynamics codes.

The present book tries to reproduce these lines. In particular several initial contributions describe the molecular dynamics of elementary processes both in gas phase and in the interaction with the surfaces by using quantum mechanical and phenomenological approaches.

A second group of contributions describes thermodynamics and transport properties of air components with special attention to the transport of internal energy.

Then a series of papers is dedicated to the flow of partially ionized gases from both experimental and theoretical points of view. Follow chapters treating modern computational techniques for 3-D hypersonic flows.

A series of papers is then dedicated to the description of non-equilibrium vibrational kinetics and to the coupling of vibration-dissociation processes in affecting hypersonic flows. Special emphasis is given to the interfacing of non-equilibrium models with computational fluid dynamics methods. Finally the last chapters deal with the application of direct Monte Carlo method in describing rarefied flows.

The organizing committee gratefully acknowledges the generous financial support provided by the NATO Science Committee as well as by Azienda Autonoma di Soggiorno e Turismo di Maratea, by CNR (Centro di Studio per la Chimica dei Plasmi and Comitato per la Chimica), by University of Bari and by CIRA.

The editor wishes to thank all the contributors for their co-operation in submitting the final version of their manuscripts on time. Special thanks are due to Dr. Fabrizio Esposito for his invaluable work in collecting and organizing the manuscripts.

M. Capitelli

RELEVANCE OF AEROTHERMOCHEMISTRY
FOR HYPERSONIC TECHNOLOGY

G.S.R. SARMA
Von Karman Institute for Fluid Dynamics
Rhode-St-Genèse, Belgium

Permanent address: Institute for Fluid Mechanics
DLR, Göttingen, Germany

1. Introduction

With the advent of jet propulsion it became necessary to broaden the field of aerodynamics to include problems which before were treated mostly by physical chemists. **von Kármán (1958)**

After a lull of over a decade there has been, since the late 1980's, a resurgence of interest in hypersonics with special emphasis on future hypersonic transport systems and associated technologies[37, 48]. The cited references contain a wealth of detailed information on recent re-entry capsule technology[37] and on the fundamentals of aerothermochemistry and their application to external and internal flows relevant to current and future hypersonic technology concepts[48]. (The term "Aerothermochemistry" was coined by von Kármán in the 1950's to denote this multidisciplinary field of study shown to be pertinent to the then emerging aerospace era[56]. He also credited the term "Aerothermodynamics" to G. A. Crocco, father of the famous gasdynamicist L. Crocco.) There are now competing efforts from several advanced industrial countries to put forward conceptual designs and feasibility studies for a viable hypersonic transport technology (Fig. 1 [29]). Current thinking on future programs is based on the knowhow and technology since the sixties and on the progress achieved since then in propulsion, materials and avionics, and last but not least also in CFD, inspiring a fairly confident outlook for the ambitious R & D needed for the envisaged hypersonic vehicles of the next century. In order to make

1

M. Capitelli (ed.), Molecular Physics and Hypersonic Flows, 1–20.

2

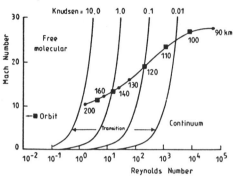

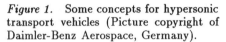

Figure 1. Some concepts for hypersonic transport vehicles (Picture copyright of Daimler-Benz Aerospace, Germany).

Figure 2. Ma-Re-Kn domain for a Space Shuttle trajectory. (Reference length for Re and Kn is that of the Shuttle.)

these technologies efficient, economical and environmentally sound, an essential aspect of the R & D is the vital need for a better understanding and control of the complex physico-chemical processes occurring in such configurations in the external flow fields, wakes and the propulsion systems. The relevant problems to be dealt with involve flows of multi- component, chemically reacting, high-temperature gases in complicated configurations. The gas dynamic problems become all the more challenging since the vehicle flight-trajectories traverse a wide range of Mach (Ma: vehicle speed/ sound speed), Knudsen (Kn: molecular mean free path/reference length), and Reynolds (Re: inertial/viscous forces) numbers (Fig. 2 after [42]). Thus the vehicle experiences flow régimes going from subsonic to hypersonic, continuum to free-molecular, and laminar to turbulent. Furthermore, the high-temperature chemical phenomena introduce several widely varying Damköhler numbers (Da: flow time/chemical reaction times) which are associated with the chemical species production terms and add numerical stiffness to the governing equations to be solved [57]. The thermochemical régimes go from frozen (: $Da \ll 1$) to equilbrium ($Da \gg 1$) states. Herein the full range of molecular physical phenomena encompassing: rarefaction, ionization, radiation, relaxation and non-equilibrium (thermal and compositional) in general, manifest themselves in macroscopically significant forms, occurring over wide Ma-Kn-Re-Da ranges. Thus effective experimental simulation in ground facilities is severely limited. Flight simulation is limited by high costs. Often variations in theoretical modelling and CFD approaches together with validation by dedicated experiments are the only means for progress in this challenging field. It must also be noted that in spite of the rapidly increasing power of the supercomputers and CFD techniques, fluid dynamicsts are nevertheless subject to the 'rate-limiting steps' of better physico-chemical modelling, associated defining parameters and

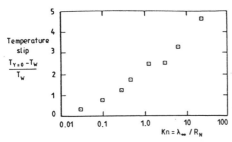

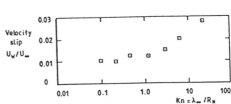

Figure 3. Temperature jump at the Shuttle stagnation point at different Knudsen numbers.

Figure 4. Velocity slip on the Shuttle at x = 1.5 m for U_∞ = 7.5 km/s at different Knudsen numbers.

related experimental data, whose inadequacy and uncertainty are now well recognized [21, 35, 48].

Some of the important phenomena in the present context will be discussed by means of a few illustrative examples touching aspects like radiation, ionization and wake flows. The main emphasis will be on indicating the effects of fluid dynamical interest from a practical point of view. The present paper is thus meant to provide briefly the background and motivation for the detailed discussions of individual phenomena and methodologies for their analysis presented in the other papers.

2. Some Special Effects in Hypersonics

2.1. RAREFACTION

Due to the low densities of air at the upper altitudes ($\sim$ 100 km) the flow around the hypersonic vehicle has, in contrast to the lower continuum régime (with no-slip, i. e., no difference between gas and wall in the velocity and temperautre), a 'slip' near the wall and the boundary conditions on velocity and temperature have to account for this. We see from Figs. 3, 4 that the slip effect can be significant as the Knudsen number ($\sim$ altitude) increases [36]. At even higher altitudes ($\sim$ 200 km) there is only a free-molecular flow around the vehicle [42]. The slip-conditions shown above are typical but the details are still an area of active research especially for nonequilbrium flows [3, 13, 18]. In the following we shall focus more on examples from the continuum flow régime.

2.2. DENSITY CHANGE

A significant difference between the nonreacting (chemically frozen) and reacting gas at high temperatures induced by the high viscous dissipation during hypersonic flight can be seen in the density change behind the bow-shock ahead of a blunt body [23]. In the frozen case ($Da \ll 1$) the

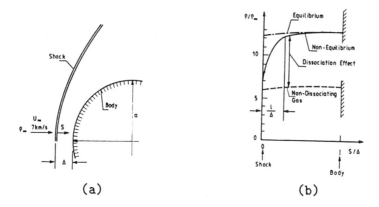

(a)   (b)

Figure 5.   Schematic sketch showing (a) bow-shock ahead of a blunt body, and (b) density profile along the stagnation streamline, in a dissociating gas flow.

temperatures are much higher than in the reactive case and the densities are smaller. If the gas reacts very fast ($Da \gg 1$) it reaches compositional equilbrium immediately behind the shock using up much of the energy for molecular processes like dissociation, vibration, rotation and ionization. Consequently the temperatures are lower than in the non-reacting case and densities are higher. Density ratio for the equilibrium case (Fig. 5b) is double that of the frozen case. Thus the dissociation effect is very significant. In general of course these molecular processes take a finite time and the steady flow quantities can attain their equilibrium values only over a 'relaxation distance', as seen in the continuous curve, Fig. 5(b).

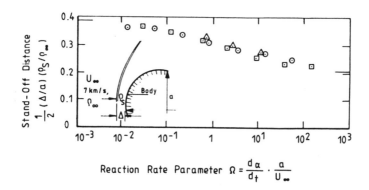

Figure 6.   Correlation of stand-off distance on spheres with reaction rate parameter Ω. ⊙: degree of dissociation $\alpha_\infty = 0.07$; ▫: $\alpha_\infty = 0.42$; △: $\alpha_\infty = 0.65$.

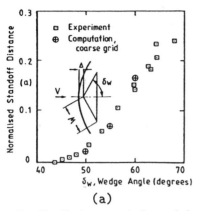 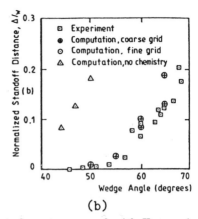

(a)   (b)

Figure 7. Candler's computations of shock-detachment compared with Hornung's experiments on flow past a wedge of: (a) Nitrogen, 55 X 50 grid; (b) Carbon dioxide 55 X 60 and 110 X 100 grid. Note triangles correspond to frozen chemistry and vibration.

2.3. BOW-SHOCK STAND-OFF DISTANCE

Stand-off distance Δ varies significantly with the dissociation reaction.Δ is an important parameter determining the stagnation point heat flux to the body and also the local pressure distributions. Computed results (Fig. 6) of Hornung [22] for the shock stand-off distance Δ ahead of a sphere at different free-stream dissociation reaction rates Ω show the good correlation between different results under proper scaling of the data. The horizontal lines at the two ends denote the asymptotic results for the equilibrium case ($\Omega \gg 1$) and the frozen case ($\Omega \ll 1$). The result here also illustrates the principle of the 'binary-scaling' law for a comparison between experiments (flight or windtunnel), viz., that the product of density and characteristic length must be held constant for given free stream velocity and temperature, in order to properly model binary-collision dominated relaxation processes such as dissociation here. Full Navier-Stokes solutions of Candler [10] also confirm the decrease of Δ with Ω shown in Fig. 6.

2.4. SHOCK DETACHMENT AND STAND-OFF DISTANCE

A remarkable event depending strongly on dissociation effects can be demonstrated using a simple wedge. The variation of stand-off distance Δ and detachment angle as a function of a wedge angle δ_w[23] for different gases show a good agreement between experiment and theory predicting a gradual and slower increase in Δ with wedge angle δ_w for a dissociating gas compared to a non-dissociating gas. The difference between dissociating nitrogen and the frozen gas was shown to be a large effect. Hornung's results also compare well with full Navier-Stokes calculations of Candler (Fig. 7). For nitrogen the coarser grid is seen to be adequate while a grid refinement

6

Figure 8. Computed and observed interferograms for flow past a cylinder (left to right): Models of Park, Macheret, and CVDV vs. experiment for N_2 at $U_\infty = 5.07$ km/s, $\rho_\infty = 0.0402 kgm^{-3}$, $T_\infty = 2260$ K.

is required for the CO_2 case: Fig. 7(b). In fact the experimental CO_2 data [23] also show more scatter than N_2. Computations ignoring chemistry and vibrational excitation (triangles in Fig. 7 b) are far from the other values and follow a steeper rising curve (as expected for the frozen gas [23]).

Recent Navier-Stokes computations of Candler [48] illustrate strikingly the power and sophistication of current CFD and post-processing techniques. The computed results are post-processed to interferograms that can be directly compared with their laboratory counterparts (Fig. 8). This example also illustrates the kind of CFD-validation (of different physico-chemical models for vibration-dissociation here) through dedicated principle-experiments, as mentioned earlier.

2.5. PITCHING MOMENT

Apollo and Space Shuttle experienced a pitching moment (taken about a spanwise axis) quite different from those of pre-flight predictions, necessitating sizeable in-flight trimming corrections (Fig. 9) [37, 40]. This significant anomaly has been re-investigated at NASA and elsewhere, and the current concensus is that the high-temperature chemistry, which was not well simulated in preflight testing, is the major cause of this discrepancy as indicated in (Fig. 10) [34]. But the proper quantitative identification of the various contributing factors is still a matter of debate. For instance the sign of viscous contribution to C_m (nose-up or nose-down) is a matter of contention [4, 28, 34]. It is important to note that not only heat transfer and skin friction but even a pressure dependent aerodynamic quantity such

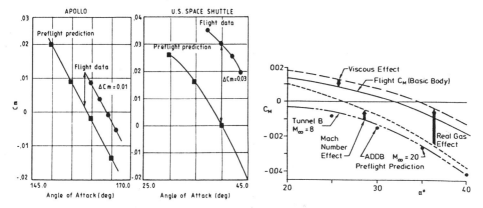

Figure 9. Pitching moment C_m for Apollo and Shuttle at different angles of attack

Figure 10. Buildup of Shuttle C_m on the basis of Maus et al. methodology.

as the pitching moment can be affected by high-temperature chemistry.

2.6. INFLUENCE OF THERMOCHEMICAL STATE

We see in Fig. 11 the influence of the thermal and chemical equilbrium, non-equilibrium, as against the frozen case, on a "hyperboloid-flare" test configuration defined for Hermes [7]. The influence of the thermochemical state is quite large on temperature and velocity profiles, with noticeable effects on heat flux and skin friction, especially around the ramp region.

3. Radiation

It is known that [14] radiation and ionization can contribute significantly to the thermal loads to hypersonic vehicles entering at very high velocities. Aerospace transfer vehicles (ASTV) at 10 km/s (Lunar mission) at 80 km altitude have a 1% ionization, but the radiative and convective heat fluxes in the bow-shock layer are comparable. At 12 to 17 km/s the (Mars return) radiative part is 90% of the total heat load and ionization level is 30%.

3.1. RADIATION AND CHEMISTRY

The stagnation region ahead of a hypersonic vehicle attains temperatures at which radiation from the ionized species enhances the heat flux to the body. The species and the energy bands of this radiative flux are also in non-equilbrium in the shock layer. Apollo return at Mach ~ 36 induces temperatures $\sim$11000 K at the nose-tip [40]. At these temperatures non-ionized species also emit radiation (certain ablation carbon products even at lower temperatures [27]).

8

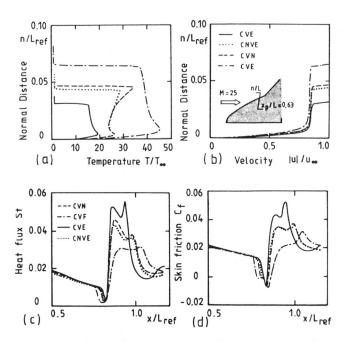

Figure 11. Influence of thermochemical state on (a) temperature, (b) velocity, (c) heat flux, St, and (d) skin friction, C_f on hyperboloid-flare ($\sim$ Hermes model, altitude 77 km) symmetry line. CVE: Chemical & thermal equilibrium; CNVE: Chemical non-equilibrium & thermal equilibrium; CVN: Chemical & thermal non-equilbrium; CVF: Frozen chemistry & vibrational modes.

Radiative heating of the nose region of 0.2-scale model of Apollo in Fire II flight experiments was analyzed by Park [4, 38, 40] using the two-temperature model for the non-equilibrium region. Representative values of T_{tr} for the translation-rotation modes of the heavier species, and T_v for the vibrational and electronic modes are shown. T_{tr} decreases monotonically from 62 000 K towards the body, whereas T_v has a broad peak at $\sim$ 13 000 K, which also corresponds to the calculated high radiative emission region, which is away from shock and wall as seen in Fig. 12a,b. Comparison of Fire II flight data and Park's non-equilibrium model calculations shows very good agreement for altitudes up to 81 km (Fig. 12c). (The disagreement above 81 km is attributed to the finite shock structure, the mean shock-thickness being no longer thin compared to stand-off distance at high altitudes.) Fire II results [53] clearly show (Fig 12d) that radiative and convective forms of heat transfer are indeed comparable. (For the Galileo probe during Jovian entry the heat transfer is almost fully radiative.)

We may mention that the radiative heat flux to the body is $\sim R_n U_\infty^{8.5} \rho_\infty^{1.6}$,

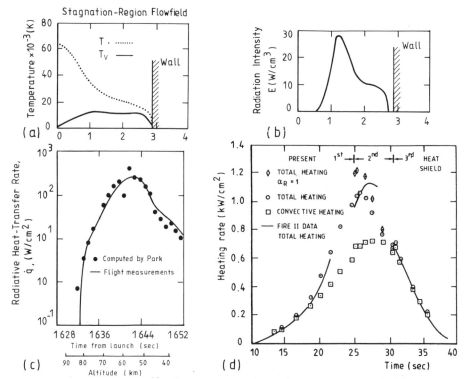

Figure 12. Stagnation point heat transfer for Fire II (73.72 km altitude) (a) Two temperatures, (b) Radiation intensity along stagnation line in the shock layer (c) comparison of theory & flight data, and (d) radiative & convective parts of entry heating.

while convective heating is $\sim R_n^{-0.5} U_\infty^3 \rho_\infty^{0.5}$ and both of them are comparable for a 0.3m radius sphere at 12.2 km/s [4]. Note also that there are opposite design criteria for radiative and convective heat shielding optimization, the flatter nose (larger R_n) for the latter and the blunter nose for the former mode of heat transfer. Also, the composition of the planetary atmosphere affects the radiation. During entry to Mars ($CO_2 - N_2$) and to Titan ($N_2 - CH_4 - Ar$) their atmospheres can raise forebody heat fluxes due to strongly radiating CN produced. Radiative transfer modelling is still in general quite inadequate as shown by comparisons of CFD based on updated radiation models for β and γ systems of NO and flight experiments [12, 31].

4. Ionization

4.1. RADIATION AND IONIZATION BEHIND BOW-SHOCK

In air electrons are first produced by the associative ionization reaction $N + O \rightarrow NO^+ + e^-$. The NO^+ ion so generated transfers its charge to other neutral species through several charge-exchange reactions. When the resulting density of electrons reaches a threshold value they collide in electron-impact reactions with N and O to produce further ions and electrons, resulting in an 'avalanche' generation of electrons [40]. At relatively low speeds (below $\sim$ 6 km/s) only the associative ionization takes place. At speeds between 6 km/s and 10 km/s the charge-exchange reactions e. g. $N^+ + N_2 \rightarrow N_2^+ + N$ increase the ionization. At still higher speeds the electron-impact reactions, e. g. $O + e^- \rightarrow O^+ + 2e^-$ lead to the electron avalanche process.

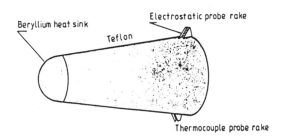

Figure 13. Schematic of RAM-C II (Radio attenuation measurement) flight model; Nose-radius = 15 cm, cone semi-angle 9 deg, length 130 cm.

4.2. ELECTRON DENSITY AND HEAT TRANSFER

For Aeroassisted Space/Orbital Transfer Vehicles (ASTV, AOTV) the flight conditions create hypersonic flows with ionization under chemical and thermal non-equilibrium conditions, affecting bow shock stand-off distance, forces and moments. Knowledge of ionization effects is also important for radio-communication with the vehicle. Radiation from charged species also alters the heat flux to the vehicle significantly. Candler and MacCormack [9] computed such flows involving five species: N_2, O_2, NO, N, O (using the QSS: Quasi Steady State assumption, viz., the reaction $N + O \rightleftharpoons NO^+ + e^-$ is in equilibrium) and seven species including the ionization reaction for NO explicitly. Energy exchange between translational and vibrational modes and electrons: $Q_{T-V_s}, Q_{V-V_s}, Q_{e-V_s}$, and Q_{T-e} are also modelled. Coupling of translational-rotational to other internal states for evaluation of chemical reaction rate coefficients is based on Park's model [38] taking T(effective) as the geometric mean $\sqrt{T_{tr}T_i}$ between T_{tr} and the other temperatures T_i. Full N-S computations were compared with experimental data on RAM (Radio Attenuation Measurement)-C II flights. The models had a beryllium heat-sink nose-cap and teflon coated afterbody (hence no additional ablation product species).

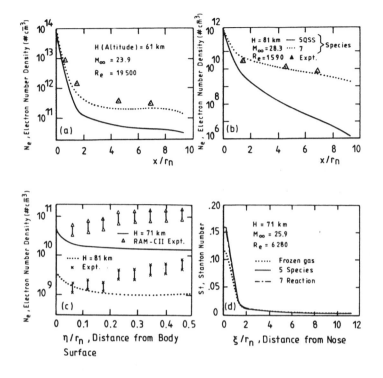

Figure 14. Effects of ionization: computation & flight results on RAM-C II.

4.3. COMPARISON OF COMPUTATIONS AND FLIGHT DATA

There is good agreement between the computed and measured electron density ditributions along the body at altitudes 81 km (non-equilibrium conditions) and 61km ($\sim$ equlibrium). QSS (5 species) works better at 61 km. However, its nose region trends are similar to the 7-species computations. For heat transfer (Stanton number, St) the two models give nearly the same results (Fig. 14d) and demonstrate the significant ($\sim$ 50%) increase in heat flux in the nose region due to the active chemistry (b. c. $T_w = 1500K$, non-catalytic wall, modelling flight experiment.)

It may be mentioned that a lot of modelling goes also into the transport coefficients and the choice of reaction schemes. Here the simplest sophistication has been shown to give good agreement, thereby "validating" the CFD and C-V-e models. Furthermore, the vibrational-chemical coupling is still an area of active research and there are several competing models of different levels of complexity in development [2, 8, 12, 26, 32, 45].

4.4. TRANSPORT COEFFICIENT MODELLING

We note that heat transfer rates do depend significantly (even at much lower speeds than those considered above) on transport coefficient modelling [25, 41, 47]. Fig. 15 shows that "lumped parameter" approaches (using constant Prandtl, Lewis numbers) can be far from the exact values of practically relevant quantities such as heat flux and skin friction. The discrepancies are found to increase with the outer flow velocity [41].

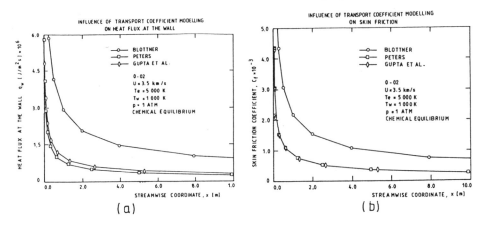

Figure 15. Influence of transport coefficient modelling in a flat plate boundary layer flow on (a) heat flux, q_w and (b) skin friction, C_f.

4.5. HYPERSONIC WAKE CHEMISTRY

4.5.1. *Radiation from the wake*

The flow in the base region immediately behind a hypersonic vehicle tends to be in non-equilibrium and this has important consequences [40]: (a) Radiative heating of the base region that can be a significant percentage of the very high forward stagnation point heat transfer. Jupiter entry studies showed that this can necessitate a thermal shield for the base as well [39, 40]. The static temperature in the wake-neck region is about the same as in the forward stagnation region and the wake recirculation region is as hot as the former. (b) The pressure acting on the base is affected by the non-equilbrium chemistry there and any outside asymmetries lead to pitching moment contributions and consequent control problems. (c) In the neck region ionization leads to significant electron densities and affects radio communication and also allows detection by microwaves. Flight measurements on spherical models established the non-equilibrium ionization phenomena in this region. (Attempts to correlate data with a single temperature model for the NO ionization reaction were unsuccessful.) (d) Optical radiation in the wake is a familiar sight observed in the luminosity of the tracks of meteors. Experiments on spherical models established the qualitative and quantitative aspects of the chemiluminiscence, radiative emission from the reactions: $NO + O \rightarrow NO_2^* \rightarrow NO_2 + h\nu$. These processes occur in the less hot regions, downstream of the neck region. Such a luminosity is a visual 'giveaway' of the vehicle.

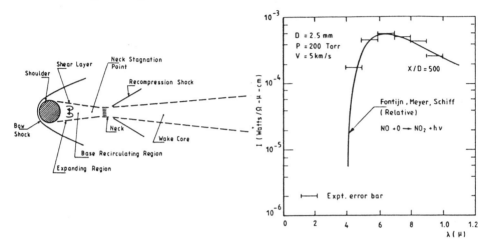

Figure 16.   Schematic representation of hypersonic wake region behind a sphere.

Figure 17.   Chemiluminescent radiation from hypersonic wake behind a sphere.

4.5.2. Luminescence in hypersonic wake flow

The following is an example [43] of a study of chemiluminescence serving to identify the responsible chemical reaction. Some possible chemiluminescent reactions are: Rayleigh-Lewis nitrogen afterglow, $N + N + M \rightarrow N_2 + M + h\nu$; blue nitric oxide afterglow, $N + O + M \rightarrow NO + M + h\nu$; and NO_2 continuum, $NO + O + M \rightarrow NO_2 + M + h\nu$ etc. These are merely the overall reactions but it is the intermediate steps that bring about chemiluminescent radiation through transitions between excited molecular states. Through ballistic experiments on spheres it was established that the dominant reaction responsible for the afterglow in the wake flow is the recombination of NO_2. In Fig. 17 a theoretical model for chemiluminescence from NO_2 and experimental data of Reis are compared. Within the experimental error-bars the spectral distribution in the wake flow shown in Fig. 17 agrees very well with that of spectroscopic model studies on NO_2 luminescence. This allows one to model the luminescence as a the result of binary recombination of NO and O to form NO_2 in an electronically excited state NO_2^* from which it decays to NO_2 ground-state emitting radiation in the visible spectrum. This phenomenon occurs at lower temperatures than at the neck

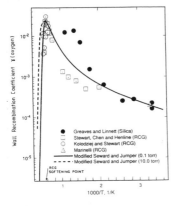

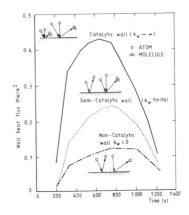

Figure 18. Comparison of new model with measured data for oxygen recombination.

Figure 19. Effect of wall catalyticity on Shuttle stagnation-point heat flux.

region and extends far downstream (several hundreds of body diameters) of it and is useful even as a diagnostic tool for studying the turbulent mixing phenomena occurring in this region. In this region and further downstream negative ion chemistry plays an important role in determining the electron density as well [52]. (The UV-radiation from the bow-shock region, in contrast, decays in intensity immediately behind the body.) This is obviously a promising area of common interest for molecular and flow physics.

5. Wall catalyticity Modelling

5.1. FINITE RATE CATALYTICITY STUDIES ON SHUTTLE

Scott [49, 50] was perhaps the first to use temperature dependent finite-rate catalyticity coefficients $\gamma(T)$ determined from arc-jet experiments for N-N and O-O recombination on high temperature reusable surface insulation (HRSI) materials like Silica, reaction cured glass (RCG). He explored the theoretical and experimental aspects of the Shuttle surface heat transfer and its increasing temperature with the successive flights, which was surmised then as probably due to surface contamination, which may increase catalyticity and/or decrease emissivity. Shuttle experienced much higher heat loads on its orbital maneuvering system than preflight predictions. Methodologies with varying degrees of predictive success have been devised over the years. The crux of the problem to date is the proper modelling of the wall catalytic recombination kinetics [8, 15, 24, 58].

5.2. A MODELLING STUDY

Modelling of catalytic recombination of oxygen and nitrogen on silica-based materials is a very important aspect in the aerothermal design for thermal

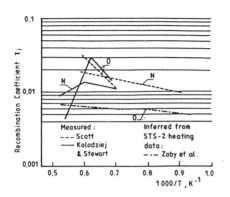

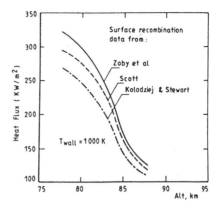

Figure 20. Comparison of different measured and correlated data for catalytic recombination coefficient $\gamma_i(T)$.

Figure 21. AOTV stagnation-point convective heat flux computed on the basis of different sets of input data for $\gamma_i(T)$.

protection of the nose region of reentry vehicles. Here there are still considerable uncertainties and ongoing studies[8, 15, 58] to clarify the associated reaction mechanisms and their quantitative description for practical estimation of aerothermal loads (which are enhanced by the exothermic recombination of O-atoms or N-atoms due to wall catalysis). This is an important area where the research interests of both surface chemistry and flow physics could be served well. The results of a recent study, wherein the heterogeneous reaction between the wall and O-atoms is envisioned as an adsorption of an O-atom at a wall site followed by a recombination via collision with a gas-phase O-atom [24, 51], are shown in Fig 18.

The significant effect of wall catalyticity on Shuttle stagnation-point heat flux during (STS-2) re-entry [16] is shown in Fig. 19. Here $k_w = \gamma\sqrt{k_B T_w / 2\pi m}$ is the catalytic speed, k_B, m , and T_w being respectively the Boltzmann constant, particle mass, and the wall temperature.

5.3. MODELLING FOR AOTV FLIGHT STUDIES

Fig. 20 shows typical uncertainties in experimental data on $\gamma_i(T)$ and the consequent nontrivial variations demonstrated in a recent detailed computational study on AOTV reentry stagnation-point heat loads [19] are depicted in Fig. 21.

6. Engine Chemistry

Engine chemistry is far more complex, involving far more reactions and species (especially for hydrocarbon fuels), than that of external air-flows and only very basic model-studies are feasible even with the current power

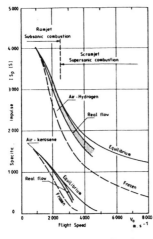

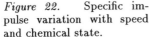

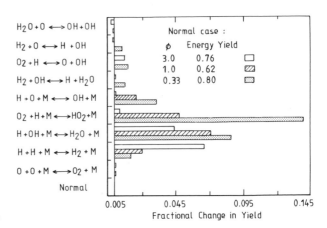

Figure 22. Specific impulse variation with speed and chemical state.

Figure 23. Sensitivity of energy-yield at nozzle-end to reaction kinetic data.

of CFD and test facilities [5, 6, 21, 48]. At the flight Mach numbers envisaged high-energy fuels like hydrogen and supersonic combustion ramjet engines (SCRAMJETS) are considered good candidates for realizing the challenging propulsion concepts under development.

6.1. FUEL CHEMISTRY

An important performance characteristic of an aerospace engine is its specific impulse I_{sp}, the ratio of thrust to fuel-weight rate, in seconds. I_{sp} depends strongly on the fuel-chemistry and flight Mach number, M. Since it decreases steadily with M, it is essential to utilize high specific-energy fuels like liquid hydrogen for hypersonic propulsion systems. It is estimated that hydrocarbon fuels give less than a third of the hydrogen performance for turbo/ram/scramjet engine cycles in their respective optimal operating ranges [4, 21]. I_{sp} depends, in addition, on the thermochemical state of the fuel in the thrust-producing nozzle. Fig. 22 [33] shows that frozen states reduce and equilibrium ones enhance engine performance. It may be noted that even a few per cent gain in I_{sp} is crucial at hypersonic speeds at which the difference between thrust and drag decreases quite rapidly.

6.2. NOZZLE STUDIES

Scramjet studies using different reaction mechanisms for the complex chemically reacting internal flows show that it is often worthwhile to carry out sensitivity studies in this context [1, 20, 44, 46] to identify viable chemistry models and get the basic performance trends. In this spirit a 22- step hydrogen/air reaction scheme (with eleven species) was employed [20] to study the sensitivity of energy-yield of a scramjet nozzle at Mach 15 flight with respect to the individual reactions and their kinetic coefficients. The results of changing the important reaction rate coefficients by a factor of five (which is a rather modest range in this context, but may be tenable in this case since some of the, a posteriori, important reactions are quite fast) showed (Fig. 23) that the three-body recombination reactions are affected the most and then the bimolecular reactions, as far as heat released is concerned. Consideration of finite-rate chemistry is essential for scramjets with shorter residence times unlike in conventional engines wherein equilibrium can possibly be attained. It may also be noted that different reactions have different sensitivities with respect to the equivalence ratio. For instance in a fuel-rich ($\phi = 3$) mixture the hydrogen recombination: $H + H + M \rightarrow H_2 + M$ induces a 7% change versus a 14% change due to $O_2 + H + M \rightarrow HO_2 + M$ under fuel-lean conditions ($\phi = 0.33$), as seen in Fig. 23.

7. Conclusion

From the brief preview of some of the important physico-chemical phenomena relevant to hypersonic flows presented above it is clear that there are many challenging tasks that call for a collaboration and cross-fertilization of ideas and techniques between hypersonics and molecular physics. Fundamental studies on underlying physico-chemical mechanisms, their viable modelling for incorporation into efficient CFD and validation by dedicated principle-experiments, and also the constant updating of the currently uncertain data on transport properties and on gaseous and gas/solid reaction rate coefficients, are essential, and are likely to occupy us for quite sometime. The ultimate technological challenge is of course the 'all-in-one' optimal design of air-breathing hypersonic transport systems plying between a conventional runway and orbit, necessitating a full integration of propulsion and aerodynamics to cover wider ranges of flow parameters and flight conditions far more stringent than those faced by the Shuttle [4, 48, 54]. With the current impetus of interest and recent activities and with proper pooling of resources, one can reasonably hope that a viable hypersonic transport technology may be realized in the next few decades.

8. Acknowledgments

Sincere thanks are due to Professors Claudio Bruno (Univ. Rome), Mario Capitelli (Univ. Bari), and Mario Carbonaro (V.K.I.) for their kind encouragement and support towards the author's participation at this ASI.

18

References

1. Ahuja, J. K., et al., "Hypersonic shock-induced combustion in a hydrogen-air system", AIAA J., Vol. 33, No. 1, 1995, pp. 173 - 176.
2. Armenise, I., et al., "On the coupling of non-equilbrium vibrational kinetics and dissociation- recombination processes in the boundary layer surrounding an hypersonic reentry vehicle", In: *Aerothermodynamics for Space Vehicles*, Ed., J. J. Hunt, ESA-SP-367, pp. 287-297, ESA, Paris, 1995.
3. Bergemann, F., "Gaskinetische Simulation von kontinuumsnahen Hyperschall Strömungen unter Berücksichtigung von Wandkatalyse", DLR- FB 94- 30, 1994.
4. Bertin, J. J., *"Hypersonic Aerothermodynamics"*, AIAA Education Series, AIAA, Washington,D.C., 1994.
5. Billig, F. S., "Current problems in nonequilbrium gasdynamics: Scramjet engines", in AIAA Professional Studies Series, *"Fundamentals of Nonequilibrium Gasdynamics"*, June 10 - 11, 1989, Buffalo, N. Y., pp. 9-1 to 9-42.
6. Billig, F. S., "Thermochemistry issues in the design of high-speed propulsion systems", In: *Aerothermochemistry for Hypersonic Technology*, VKI-LS 1995-04, 1995.
7. Brenner, G., "Numerische Simulation von Wechselwirkungen zwischen Stößen und Grenzschichten in chemisch reagierenden Hyperschall Strömungen", DLR- FB 94-04, 1994.
8. Bruno, C., "Physico-chemical input data", In: *Aerothermochemistry for Hypersonic Technology*, VKI-LS 1995-04, 1995.
9. Candler, G. V., and MacCormack, R. W.,"The computation of hypersonic ionized flows in chemical and thermal nonequilbrium", AIAA Paper No. 88-0511, 1988.
10. Candler, G. V., "On the computation of shock shapes in nonequilbrium hypersonic flows", AIAA Paper No. 89-0312, 1989.
11. Candler, G. V., "A perspective on computational aerothermodynamics: current modelling difficulties", In: *Aerothermodynamics for Space Vehicles*, Ed., J. J. Hunt, ESA-SP-367, pp. 183-187, ESA, Paris, 1995.
12. Candler, G. V., "Chemistry of external flows", In: *Aerothermochemistry for Hypersonic Technology*, VKI-LS 1995-04, 1995.
13. Daiß, A., et al., "New slip model for the calculation of air flows in chemical and thermal nonequilibrium", In: *Aerothermodynamics for Space Vehicles*, Ed., J. J. Hunt, ESA-SP-367, pp. 155-162, ESA, Paris, 1995.
14. Deiwert, G. S., "Real gas: CFD prediction methodology, flow physics for entry capsule mission scenarios", In: *Re-entry Capsule Aerothermodynamics*, AGARD-FDP-VKI-LS 1995-06, 1995.
15. Deutschmann, O., et al., "Modelling of surface reactions in hypersonic re-entry flow fields", In: *Aerothermodynamics for Space Vehicles*, Ed., J. J. Hunt, ESA-SP-367, pp. 305-310, ESA, Paris, 1995.
16. Eldem, C., "Couches Limites Hypersoniques avec Effets de Dissociation", Doctoral Thesis, L'Ecole Nationale Superiure de L'Aeronautique et de L'Espace, Thèse EN-SAE No. 9- 1987, Toulouse, France.
17. Green, M. J., and Davy, W. C., "Galileo probe forebody thermal protection", In: *Thermophysics of Atmospheric Entry*, Ed., T. E. Horton, Progr. Astron. Aeron., Vol. 82, pp. 328-353, AIAA, New York, 1982.
18. Gupta, R. N., et al.,"Slip-boundary equations for multicomponent nonequilibrium airflow", NASA Tech. Paper 2452, 1985.
19. Gupta, R. N.,"Stagnation flowfield analysis for an aeroassisted vehicle", AIAA J. Spacecrafts and Rockets, Vol. 30, No. 1, 1993, pp. 14-21.
20. Harradine, D., et al., "Hydrogen/air combustion calculations: The chemical basis of efficiency in hypersonic flows", AIAA J., Vol. 28, No. 10, 1990, pp. 1740-1744.
21. Heiser, W. H., et al., *"Hypersonic Airbreathing Propulsion"*, AIAA Education Series, AIAA, Washington, D.C., 1994.
22. Hornung, H. G., "Non-equilibrium dissociating nitrogen flow over spheres and cir-

cular cylinders", J. Fluid Mech., Vol. 53, Pt. 1, 1972, pp. 149-176.

23. Hornung, H. G., "High enthalpy facilities", 3rd Joint Europe/ US Short Course in Hypersonics, RWTH, Aachen, Germany, 1990.

24. Jumper, E. J., and Seward, W. A., "Model for oxygen recombination on reaction-cured glass", AIAA J. Thermophysics and Heat Transfer, Vol. 8, No. 3, 1994, pp. 460-465.

25. Klomfaß, A., et al., "Modelling of transport phenomena for the stagnation point heat transfer problem", AIAA Paper No. 93-5047, 1993.

26. Knab, O., et al., "Theory and validation of the physically consistent coupled vibration-chemistry-vibration model", AIAA J. Thermophysics and Heat Transfer, Vol. 9, No. 2, 1995, pp. 219-226.

27. Komurasaki, K., et al., "Radiation from an ablative shock layer around a hypersonic flight model", In: *Aerothermodynamics for Space Vehicles*, Ed., J. J. Hunt, ESA-SP-367, pp. 195-199, ESA, Paris, 1995.

28. Koppenwallner, G., "Low Reynolds number influence on aerodynamic performance of hypersonic lifting vehicles", AGARD- CP- 428, 1987, pp. 11- 1 to 11- 14.

29. Kuczera, H., et al., "The German Hypersonics Technology Programme", 43rd IAF-Congress Paper No. IAF-92-0867, August 1992.

30. Labracherie, M., et al., "CN emission measurements in Titan radiative environment", In: *Aerothermodynamics for Space Vehicles*, Ed., J. J. Hunt, ESA-SP-367, pp. 183-187, ESA, Paris, 1995.

31. Levin, D. A., et al., "Examination of theory for bow shock ultraviolet rocket experiments-I", AIAA J. Thermophysics and Heat Transfer, Vol. 8, No. 3, 1994, pp. 447-459.

32. Méolans, J. G., et al., "Vibration-dissociation coupling in high temperature nonequilbrium flows", In: *Aerothermodynamics for Space Vehicles*, Ed., J. J. Hunt, ESA-SP-367, pp. 293-297, ESA, Paris, 1995.

33. Marguet, R., "Statoréactuers" In: *Aerodynamic Problems of Hypersonic Vehicles*, AGARD-LS-42, Vol. 2, Partie 3, pp. 61 - 94, Ed., R. C. Pankhurst, 1972.

34. Maus, J. R., et al., "Hypersonic Mach number and real gas effects on Space Shuttle orbiter aerodynamics", AIAA J. Spacecrafts and Rockets, Vol. 21, No. 2, 1984, pp. 136-141.

35. Mehta, U. B., "Some Aspects of uncertainty in computational fluid dynamics results", ASME J. Fluids. Engg., Vol. 113, Dec. 1991, pp. 538-543.

36. Moss, J. N., and Bird, G. A., "Direct simulation of transitional flow for hypersonic re-entry conditions", In: *Thermal Design of Aeroassisted Orbital Transfer Vehicles*, Ed., H. F. Nelson, Progr. Astron. Aeron., Vol. 96, pp. 338-360, AIAA, New York, 1985.

37. Muylaert, J., "Capsule aerothermodynamics: missions, critical issues, overview and course roadmap", In: *Re-entry Capsule Aerothermodynamics*, AGARD-FDP-VKI-LS 1995-06, 1995.

38. Park, C., "Assessment of two-temperature kinetic model for ionizing air", AIAA Paper No. 87-1574, 1987.

39. Park, C., "Modelling of radiative heating in base region of Jovian entry probe", In: *Entry Heating and Thermal Protection*, Ed., W. B. Olstad, Progr. Astron. Aeron., Vol. 69, pp. 124-147,AIAA, New York, 1980.

40. Park, C., *"Nonequilibrium Hypersonic Aerothermodynamics"*, Wiley-Interscience, John Wiley & Sons, Inc., New York, 1990.

41. Pignataro, S. "Chemically reacting boundary layer flow of a diatomic gas", VKI Diploma Project Report 1994-11, June 1994.

42. Rault, D. F. G., "Aerodynamics of the Shuttle orbiter at high altitudes", AIAA J. Spacecrafts and Rockets, Vol. 31, No. 6, 1994, pp. 944-952.

43. Reis, V. H., "Chemiluminescent radiation from the far wake of hypersonic spheres", AIAA J., Vol. 5, No. 11, 1967, pp. 1928-1933.

44. Rhie, C. M., et al., "Numerical analysis of reacting flows using finite rate chemistry

models", AIAA J. Propulsion and Power, Vol. 9, No.1, 1993, pp. 119 - 126.

45. Riedel, U., et al., "A detailed description of chemical and thermal nonequilbrium in hypersonic flows based on elementary processes", Int. J. Mod. Phys. C, Vol. 5, No. 2, 1994, pp. 229-231.

46. Sangiovanni, J. J., et al., "Role of hydrogen/air chemistry in nozzle performance for a hypersonic propulsion system", AIAA J. Propulsion and Power, Vol. 9, No. 1, 1993, pp. 134 - 138.

47. Sarma, G. S. R., "Some parameter studies on hypersonic Couette flow", International J. of Mod. Phys. C, Vol. 5, No.2 (1994), pp. 237-239.

48. Sarma, G. S. R., "General Introduction", In: *Aerothermochemistry for Hypersonic Technology,* VKI-LS 1995-04, 1995.

49. Scott, C. D., "Space Shuttle laminar heating with finite-rate catalytic recombination", In: *Thermophysics of Atmospheric Entry,* Ed., T. E. Horton, Progr. Astron. Aeron., Vol. 82, pp. 273-289, AIAA, New York, 1982.

50. Scott, C. D., "A review of nonequilibrium effects and surface catalysis", In: *Thermophysical Aspects of Reentry Flows,* Eds., J. N. Moss and C. D. Scott, Progr. Astron. Aeron., Vol. 103, pp. 865-889, AIAA, New York, 1986.

51. Seward, W. A., and Jumper, E. J., "Model for oxygen recombination on silicon dioxide surfaces", AIAA J. Thermophysics and Heat Transfer, Vol. 5, No. 3, 1991, pp. 284-291.

52. Sutton, E. A., "Chemistry of electrons in pure-air hypersonic wakes", AIAA J., Vol. 6, No. 7, Oct. 1968, pp. 1873 - 1882.

53. Sutton, K., "Air radiation revisited", In: *Thermal Design of Aeroassisted Orbital Transfer Vehicles,* Ed., H. F. Nelson, Progr. Astron. Aeron., Vol. 96, pp. 419-441, AIAA, New York, 1985.

54. Tauber, M. E., et al., "Aerothermodynamics of transatmospheric vehicles", AIAA Paper No. 86-1257, 1986.

55. Von Kármán, Th., "Aerothermodynamic problems of combustion", Section G, Ch. 1, p.574, In: *High Speed Aerodynamics and Jet Propulsion, Vol. III: "Fundamentals of Gasdynamics",* Ed., H. W. Emmons, Princeton University Press, Princeton, N.J.,1958.

56. Von Kármán, Th., "Aerothermodynamics and combustion theory", L'Aerotecnica, Vol. 33, No. 1, pp. 80-86, 1953; followed by five other basic contributions to combustion theory and applications, reprinted in: *Collected Works of Theodore Von Kármán,* Vol. V (1952 -1963), pp. 67 - 79; 80 - 177, V.K.I., Rhode Saint Genèse, 1975.

57. Warnatz, J., "Different levels of air dissociation chemistry and its coupling with flow models", 2nd USAF-GAMNI-SMAI joint Europe/USA course in hypersonics", Univ. Texas at Austin, 1989.

58. Willey, R. J., "Comparison of kinetic models for atom recombination on high-temperature reusable surface insulation", AIAA J. Thermophysics and Heat Transfer, Vol. 7, No. 1, 1993, pp. 55-62.

VIBRATIONAL ENERGY EXCHANGES BETWEEN DIATOMIC MOLECULES OF RELEVANCE TO ATMOSPHERIC CHEMISTRY

M.CACCIATORE

CNR- Centro Studio Chimica dei Plasmi, Dipartimento di Chimica
Università di Bari, via Orabona N°4, 70126 Bari, Italy

1. Abstract

This work concerns itself with the vibration to translation and vibration to vibration energy exchanges in collisions between atmospheric molecules. The relation between results obtained within simple analytical theories, frequently used in vibrational kinetic modelling, and results obtained from more exact three-dimensional numerical calculations is presented and discussed.

2. Introduction

One of the fundamental simplifications assumed in the kinetic modelling of atmospheric processes is the LTE assumption, according to which the population distributions of the roto-vibro-electronic molecular states are described by local Boltzmann distributions [1]. Nevertheless, for many physical situations of interest in the chemistry of the upper atmosphere, the LTE approximation could be not valid. Chemical processes occurring under shuttle re-entry conditions, in ionosphere, in stellar and in interstellar media, are examples where the LTE assumption can easily fail, particularly when the vibrational degrees of freedom of diatomic molecules, such as O_2, N_2, NO, H_2 ..., are involved. Under non-equilibrium conditions, the vibrational distribution densities are no longer Boltzmann-like [2].

The determination of the vibrational population distributions for each active molecules is obviously the main object of the chemical kinetics[3]. The solution of the vibrational master equation written for each vibrational level requires the knowledge of a large number of state-to-state rate coefficients relevant to the several collisional processes that involve vibrational energy exchanges. One of the most difficult problems to solve in the context of non-equilibrium vibrational kinetics is the 'collisional' problem, that is the search for sufficiently accurate state-to-state cross-sections or rate constants for gas-phase (and gas-surface) processes.

Vibration to vibration (V-V) and vibration to translation (V-T) energy exchanges in collisions between atom/molecule with molecules are among the most effective inelastic processes for the establishment of non-equilibrium vibrational

21

M. Capitelli (ed.), Molecular Physics and Hypersonic Flows, 21–34.
© 1996 *Kluwer Academic Publishers. Printed in the Netherlands.*

distributions, under conditions where strong chemical and radiative perturbations are absent. Such distributions are generally characterised by the existence of three regions, corresponding, respectively, to the group of the first low-lying vibrational levels, the plateau region, and the region of the highly excited levels up to the dissociation continuum [2]. The vibrational energy exchanges in these regions occur through different V-V and V-T energy transfer processes, from near resonant to off resonant transitions in the full range of the vibrational manifold.

Thus non-resonant V-V processes: $AB(v') + AB(v=0) \rightarrow AB(v'-\Delta v) + AB(v=1)$, are those most important for the population of the low-lying v-levels, while in the plateau region the distribution is largely determined by resonant and near-resonant V-V process: $AB(v) + AB(v) \rightarrow AB(v-\Delta v) + AB(v+\Delta v)$. As a consequence of the molecular aharmonicity and the detailed balancing principle, the V-V exchanges are responsible for the pumping up of the vibrational quanta. Whilst the population of the highly vibrationally excited levels is maily determined by the direct V-T quenching processes: $AB(v) + AB(v=0) \rightarrow AB(v-\Delta v) + AB(v=0)$. Obviously, this is quite a simplified picture, and in real systems a large number of different elementary processes are involved [3]. It is important to realise that, until recently, the experimental determinations of V-V and V-T rate constants has been largely concentrate on the relaxation of the first few vibrational levels.

In this paper we concentrate on the theoretical determination of rate constants for V-V and V-T collisions involving molecules in specific v-state.. The literature on the determination of rate constants for vibrational energy exchanges in diatomics is extremely large [4]. Nevertheless, as we will see later in some detail, exact and near-exact quantum mechanical determinations for generating complete set of state-selected rates, are not yet feasible from a computational point of view. We therefore focus on those analytical and (mainly semiclassical) numerical approximate theoretical methods which are the most useful and most often applied for the solution of the collisional problem in the context of the kinetic modelling.

3. SSH and SSH-based approximations

Due to its simplicity for both the absolute evaluation of the rate constants and the vibrational quantum number dependence, the SSH theory [5] has been very popular for the evaluation of large sets of V-T rate constants. Although formulated in the early 5o's, this theory is still used for quantitative comparison with experimental determinations in the absence of any other better theoretical estimations. The theory is based on simple assumptions. The collision $A + BC(v=0) \rightarrow A + BC(v=1)$ is assumed to be collinear (1D), consequently the molecular rotations are not considered in the dynamics, with both the x translational and the q vibrational coordinate quantized. The atom-molecule interaction potential is assumed to be purely repulsive, $V(x) = C \cdot \exp(-x/l)$, where l is the length of the interaction. By further assuming a first order expansion of the interaction potential, that is in the limit of small vibrational perturbations, the vibrational quantum amplitude $a_{1,2}$ for the transition $(0,1)$ can be factorized in two terms: $a_{1,0} \sim H_{1,0}(x) H_{1,0}(q)$, where $H_{1,0}(x)$ is the translational integral. In the harmonic oscillator approximation the vibrational term $H_{1,0}(q) = \langle \Psi_{BC}(v=0)|q|\Psi_{BC}(v=1) \rangle$ is solved analytically with selection rule $\Delta v = \pm 1$. Therefore, only one-quantum transitions are allowed in the collision. This result establishes one of the most stringent limits for the applicability of the SSH theory, as well as for any other first order perturbation theory. The vibrational factor is generally replaced with the Morse matrix elements [6], and this is a way to partially take account for the vibrational anharmonicity. The distorted wave approximation [7] provides an analytical solution to the scattering equations for the x coordinate. The rate constant is then obtained by averaging the transition probability over the Maxwell-Boltzmann velocity distribution. The final result is the well known SSH formula:

$$K_{1,0}(T) = \frac{1}{2}\sqrt{\frac{2}{3}}\pi(\frac{\theta_1}{\theta})(\frac{\theta_1}{T})^{\frac{1}{3}}\exp[-\frac{3}{2}(\frac{\theta_1}{T})^{\frac{1}{3}} + (\frac{\theta}{2T})]]$$ (1)

$$\theta_1 = \frac{4\pi^2\mu\omega^2 l^2}{K} \quad ; \quad \theta = \frac{h\omega}{2\pi K}$$ (2)

where ω is the transition frequency, k the Boltzmann constant and μ is the reduced mass. This expression is valid in the adiabatic limit $\theta_1/T >> 1$.

From this expression, $\log K_{1,0}$ scales linearly with $T^{-1/3}$. The same behaviour is predicted by the classical Landau-Teller expression [8] which is similar to the SSH formula. One dimensional semiclassical first order perturbation models also give the same result[9]. According to the SSH formula, the absolute determination of K requires the determination of the parameter l. Historically, the intermolecular potentials for many systems have been determined from swarm scattering data and are expressed in terms of the Lennard-Jones potential which also include the long range attractive forces with potential well depth ε [10]. This explains why the most commonly used procedure for the evaluation of l is to fit the repulsive exponential to the known L-J potentials, although any other available potential can be assumed as reference potential. More often, l is considered as a semiempirical parameter. The parameter l is very sensitive to the slope of the repulsive potential, but the well depth ε of the reference potential is also important since the deeper the well depth the steeper is the repulsion. This introduces some flexibility into the SSH expression, and leads to way to take into account the attractive forces not explicitly considered in the theory , which is to change ε until the l value best fits the experimental or ab initio rate constant. This procedure has been applied by Calvert al. for the V-T relaxation in N_2-N_2 [11], and it has been brought forward in the discussion on the vibrational relaxation in NO [12-14] and atom-diatom ion systems[15].

The real test for the SSH theory is the comparison between the absolute rate constant predicted from the theory and the experimental determinations made at different temperatures. The vibrational relaxation of O_2, CO, N_2 is a good test to prove the validity of the SSH theory. The interaction potential for these systems is basically repulsive (with weak well depth). Since the moment of inertia is high, the rotational frequency is short and the vibrational and rotational motions may be decoupled in the range of the thermal collisions. Indeed, the linear dependence of the $K_{1,0}$ rate constant in the plot $\log K/ T^{-1/3}$ has been observed for atom/molecule-molecules system involving CO, N_2, O_2, H_2 in quite a large range of translational temperatures, from T=300-400K up to 3000K [16,17], that is in temperature range of interest in the re-entry conditions. This is a success for such elementary theory. The predicted relative variation of $K_{1,0}$ with ω is also qualitatively correct. Nevertheless, in the linear region the SSH $K_{1,0}$ rate constants can be overestimated by a factor 5-10 with respect to the experimental data. A serious limit for the SSH applicability is the regime of lower temperatures, T<200K, which is of relevance in the chemistry of interstellar space. In this regime significant deviations from the L-T linearity are observed, and the SSH rates are underestimated. These deviations are a consequence of the role played by the attractive forces and the roto-vibrational coupling not considered in the collisional model.

The failure of the SSH theory in predicting accurate $K_{1,0}$ rate constants in the high T regime is also a consequence of the 1st order perturbation approximation. The inclusion of the multiquantum exchanges would in fact decreases the (1,0) transition probability [18]. In order to get a better agreement between theory and experiment, 1D two-state analytical model has been developed for the V-T [18] and V-V [19] rate constant. It comes out that also the frequently used REG [19] expression is rather inadequate. An accurate multistate analytical approach for resonant V-V transitions has been developed by Zelechow et al. [20]. Nevertheless, these multistate models have not been applied as extensively as the SSH theory.

The attractive forces, particularly important for interactions between polar molecules, have been introduced in analytical models developed for V-T [21] and near-resonant V-V [22] exchanges. In these models the rotations are also considered.

In the attempt to get the correct temperature dependence of the rate constants in a wide range of T, an analytical approach has been developed by Dmitrieva et al. [23]. by matching the SSH expression to the Shin theory [21] that predicts a reliable T-dependence of the rate constant at low T. The V-T relaxation time for the transition transition (1,0) in N_2-N_2 collisions computed according to this model is reported in the L-T plot of Fig.1, taken from Ref.[23], together with the numerical semiclassical value and the experimental results. The modified SSH values are accurate in the full range of temperatures.

The SSH result is also used as the basis for the vibrational quantum number dependence of the rate constants. In this case the SSH expression for the V-T and V-V rate constants is given in the most useful form:

$$K_{v,v-1}^{0,0}(T) = K_{1,0}^{0,0} \times \frac{v}{(1-v x_e)} F(y_{v,v-1}) \tag{3}$$

$$K_{v,v-1}^{v',v'+1}(T) = K_{1,0}^{0,1} \times \frac{v(v'+1)}{(1-v x_e)[1-(v'+1) x_e]} F(y_{v,v-1}^{v',v'+1}) \tag{4}$$

$$y_{v,v-1} = (1-2x_e v)(\frac{1}{2})^{\frac{3}{2}} (\frac{\theta_1}{T})^{\frac{1}{2}} \quad ; \quad y_{v,v-1}^{v',v'+1} = x_e |v-v'-1|(\frac{\theta_1}{T})^{\frac{1}{2}} \tag{5}$$

where the adiabaticity factor F modified by Keck and Carrier [25] has been introduced; x_e is the vibrational anharmonicity factor. The REG value [19] is frequently used for $K_{1,0}^{0,1}$.

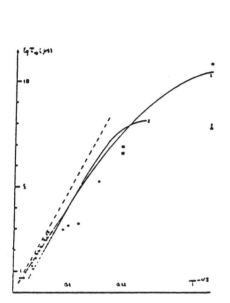

Fig.1 Relaxation time for the V-T transitions :
$N_2(v=0)+N_2(v=1)-->N_2(v=0)+N_2(v=1)$.
Analytical values(line1), experimental data (points),
Numerical values (line 2) (Taken from Ref.23)

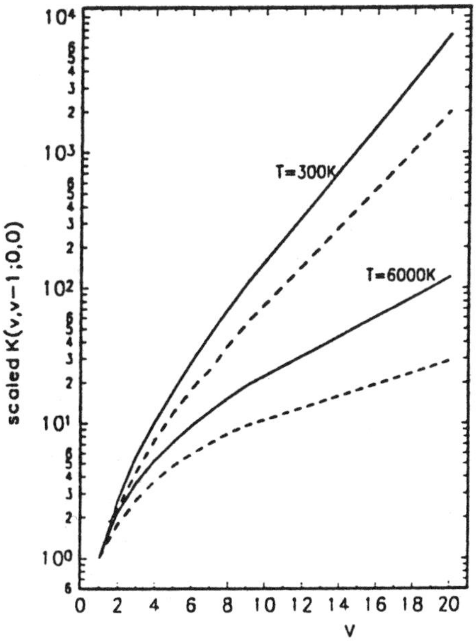

Fig.2 VT scaled rate constants for the transitions:
$N_2(v=0)+N_2(v)-->N_2(v=0)+N_2(v-1)$.
analytical SSH scaling (full lines), numerical scaling
(dotted lines) (adapted from Ref. 24)

In Fig.2 the SSH scaled rate constants $K(0,0|v,v-1)/K(0,0|1,0)$ for N_2-N_2 are reported as a function of v, together with the numerical ab initio rates computed by Billing et al.[24], at T=300K and T=6000K. The analytical rates deviate significantly from the numerical rate constant at low and high translational temperature. Furthermore, the SSH values are overestimated by a factor of 5 in the most important region of the vibrational ladder, that is in the region of highly excited levels. These deviations can be understood when considering the importance of both the attractive forces and the competition from other processes (V-V and multiquantum energy exchanges) not included in the SSH. A better agreement in the full range of T and v between the analytical rates and the reference rates is obtained from the multistate expansion theory [20] improved for non-resonant V-V transitions. Thus the transition probability at a given collisional velocity v and for the transition AB(v) + AB(v') ---> AB(v-1) + AB(v'+1) is given by [23]:

$$P_{v,v-1}^{v',v'+1}(v) = 2v(v'+1)\cos^{2(v+v'-1)}(\rho/2)\cdot\sin^2(\rho/2)\cdot\sec h^2(\pi\Delta E\cdot l/vh) \quad (6)$$

$$\rho = \frac{m_A}{m_B} \cdot \frac{vh}{2\pi l} |\Delta E_{v,v-1} \Delta E_{v',v'+1}|^{-1/2} \qquad (7)$$

The rate constants for the V-V relaxation in N_2-N_2 have been calculated by numerical integration of P over the thermal velocity distributions. These rates are reported in Fig.3 together with the reference numerical values [24].

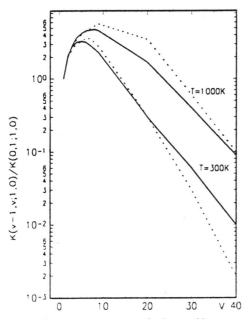

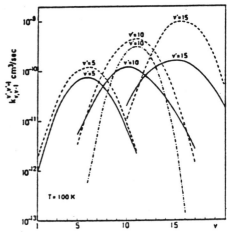

Fig.3 V-V scaled rate contants for the transitions:
$N_2(v=0)+N_2(v=1) \rightarrow N_2(v=0)+N_2(v=1)$.
From eq 6. (line1), numerical scaling (dotted line)
(Adapted from Ref. 23)

Fig.4 V-V rate constants for the processes:
$CO(v)+CO(v') \rightarrow CO(v-1)+CO(v'-1)$ at T=100K
Analytical SSH+SB scaling: dashed lines; SB scaling:
dashed-dotted; reference rates: full line.

In order to account for both the long range attractive V_{LR} forces and the short range repulsive forces V_{SR}, it has been suggested [26] to add V-V transition probabilities from the SSH theory to those obtained from the S-B theory [22]. This procedure is still applied, but it can be the cause of errors. This is shown in Fig.4 where the V-V rate constants K(v,v-1|v',v'+1) in CO-CO obtained according to the above mentioned approximation are reported together with the semiclassical rate constants [27]. The agreement is quite good at low v'. At higher v' values there is a significant disagreement: the maximum of the analytical rates is higher and the position is also shifted toward higher v. Furthermore, the analytical curves are symmetric, whereas the reference rates are asymmetric.

4. 3D numerical determination of state-selected rate constants for V-T and V-V energy transfers involving N_2, CO, O_2, H_2

The first serious obstacle for the accurate numerical determination of V-V and V-T rate constants is the need to know the interaction potential $V(R, r_i, \{\theta\})$ as a function of the intermolecular distance R between the centre of mass of the colliding molecules, the intramolecular distances r_i and the orientational angles $\{\theta\}$. At short collisional distances, the interaction potential is generally dominated by the repulsive forces, while the long range potential V_{LR} is determined by the attractive multipole interactions, including the Van der Waals dispersion forces. The range of the coordinates where the potential should be known with the highest possible accuracy depends not only on the range of collisional energies involved, but also on the type of collisional process. It comes out that the V-T processes in atom/molecule-molecule collisions are dominated by the short range potential in a broad range of collisional temperatures, above and below the thermal temperatures. The short range potential is also critical for the accurate determination of the V-V rate constants [28], although the attractive forces have a significant influence on the dynamics of resonant and near-resonant transitions.

The short range potential is that which is generally not very well known. This is also true for those collisional systems involving atoms and molecules with relatively simple electronic structures, such as $He-CO/H_2$, CO-CO, N_2-N_2, etc. When an accurate interaction potential is not available, then a semiempirical potential is constructed by assuming an adequate parametric functional form. The parameters are then varied until the computed rate constant matches with the experimental values [29].

Apart from the uncertainty in the interaction potential, a further difficulty for the exact quantum mechanical determination of the rate constant is the large number of vibrational and rotational states to include in the dynamics. In fact, the exact quantum mechanical calculations would require the solution of the Close Coupled Shrodinger equations for the transition amplitudes $a(v_f, j_f|v_i, j_i)$ through the expansion of the total scattering wave function over a (minimum) basis set which include all the energetically accessible v and j states [30]. When all the energetically accessible rotational states are included in the basis set, and they are coupled to the vibrational levels through the centrifugal term, then the number of coupled states quickly becomes large and the numerical solution is no longer computationally feasible. On the other hand, the inclusion of the molecular rotations is important for two main reasons. First because the energy mismatch $\Delta E_{v,v'}$ for the vibrational transition (v,v') can be lowered when the rotational states are included, so that the rate constants $K(v,v')$ can be in someway enhanced, secondly because the Coriolis coupling term between the rotational and vibrational molecular motions, could also be important in promoting the roto-vibrational V/R-T and V-V energy exchanges [31]. Since the molecular rotations are important, and a three-dimensional (3D) collisional model is the necessary, the problem is how they can be included in the dynamics and still have a computational tractable model. In the near-exact Coupled State (CS) approximation [32], the number of coupled equations is reduced, by leaving out the coupling between the projections of the total angular momentum in the centrifugal term $|J-j|^2$, where J

28

and j is the total and the rotational angular momentum. Nevertheless, CS calculations are still computationally expensive, particularly in the high energy (high temperature) collisional regime (with today's computer facility). In fact, very accurate V-T relaxation rate constants K(v,v-1|0,0)], for the first few vibrational levels v and in a restricted range of kinetic temperatures, have been calculated for a limited number of atom-diatom/triatom [33] and atom-diatomic ion systems [34]. This approximation has not been applied for V-V exchanges in diatom-diatom collisions.

Further approximate treatments of the rotations have led to the IOSA [35] approximation, where the rotational channels are completely decoupled and the scattering equations are solved at a fixed orientational angle. This approximation gives fairly good results for the V-T and V-V energy exchanges involving polyatomic.

A way to drastically reduce the number of coupled states in the wave function expansion is to treat the rotational motion classically. This approximation is assumed in the three-dimensional semiclassical coupled state method [36], according to which the translational and rotational motions of the colliding molecules are treated in the classical limit, while the molecular vibrations are treated quantum mechanically. Thus, the (rotationally average) cross-section is defined as:

$$\sigma_{i,i'}^{j,j'}(U,T_0) = (\frac{h}{2\pi})^6 \frac{\pi}{8\mu I_1 I_2}(KT_0)^{-3} \int_0^{l_{max}} dl \int_0^{j_1^{max}} dj_1 \int_0^{j_2^{max}} dj_2 (2j_2+1)(2j_1+1)(2l+1)$$

$$x\frac{1}{N}\Sigma|a_{i,i'}^{j,j'}|^2 \qquad (8)$$

where j_i are the rotational angular moments, I_i are the moments of inertia. N is the total number of classical trajectory for a given translational energy U [36]. The vibrational quantum amplitudes a(i,i'|j,j') are obtained by solving the classical Hamilton equations of motion for the three-dimensional rigid rotors in an effective potential $H_{eff}=<\Psi|V_{int}|\Psi>$, $\Psi(t)$ is the total vibrational wave function while V_{int} is the full interaction potential, together with the time dependent Schrodinger equations for the Morse vibrators. The Schrodinger equations are solved numerically in the Coupled State approximation, by expanding $\Psi(t)$ in a set of product Morse wave functions of the two oscillators. The rate constant is obtained by numerical integration of the cross section over a Bolzmann distribution of the symmetrized kinetic energy.

It has been shown in a number of papers that this method is quite accurate, and computationally practicable, for the generation of large set of V-V and V-T/R state-to-state rate constants for atom/diatom-diatom collisions in a large interval of the translational temperatures. In these studies some important features concerning the dynamics of the vibrationally excited states in N_2 [24,37], CO[27,38,39], O_2[40], H_2[28,31,41], O_2^+[50], have been pointed out. The collisional data obtained within the semiclassical CS model have been extensively used in the vibrational kinetic modelling in CO, N_2, H_2 [42]. The dynamics of the vibrationally excited states have different features to that of the ground state, so that a question would be as to whether the interaction potential which correctly describe the $K_{1,0}$ relaxation rate also correctly describes the relaxation of the vibrationally excited levels. In ref. [38] it has been shown that two different potential energy surfaces that give the same $K_{1,0}$ rate for the He-CO system, give rise to different $K_{v,v-1}$ values for the relaxation of the vibrationally excited states. Such rather unexpected result could put into question the applicability of analytical scaling relations for V-V and V-T rate constants.

Another important question concerns the relative importance of multiquantum transitions. The first experimental evidence of the importance of multiquantum exchanges between vibrationally excited levels in CO was given by Brechignac [43], and subsequently confirmed theoretically in a semiclassical study on the V-T and V-V relaxation in CO-CO [44]. Rate constants for multiquantum V-V exchanges in collisions:

$$CO(v) + CO(v) \rightarrow CO(v-m) + CO(v+m) , m=1,2,3. \tag{9}$$

are reported in Fig.5 (taken from Ref.44) as a function of v at T=200K. The single quantum transitions have a fast increase in the interval v=1,5 than the rate levels off with a maximum near v=28. One reason for this behaviour is the increasing competition from the multiquantum transitions, thus the ratio K(m=2)/K(m=1) increases from 0.24 to 0.45 when going from v=10 to v=20. At around v=30 even the quadrupole transitions can contribute. Multiquantum transitions are important for virtually all the molecular systems studied: N_2-N_2,N_2-CO,N_2-O_2.

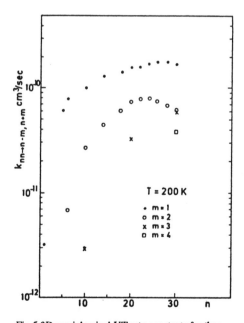

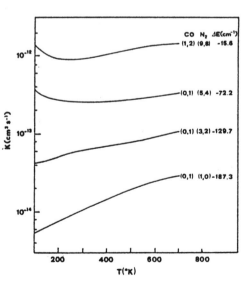

Fig.5 3D semiclassical VT rate constants for the: multiquantum transitions:
CO(n) + CO(n) --> CO(n-m) + CO(n+m)
(Figure taken from Ref.44.)

Fig.6 Semiclassical V-V rate constants: for the dissymmetric transitions:
CO(v)+N_2(v')-->CO(v+1)+N_2(v'-1).
(From Ref. 37)

The vibrational relaxation in CO-CO and CO-N_2 has been extensively studied, tables of the V-V and V-T rates can be found on the relative papers [37,27]. Vibrational energy exchanges between different diatomics, as in N_2-C0 and N_2-O_2, are dominated by the near-resonant dissymmetric transitions.

As it is shown in Fig.6, the near-resonant dissymmetric transition:

$$N_2(v=5) + CO(v=0) \text{ ---> } N_2(v=4) + CO(v=1) \quad \Delta E= -72.2cm^{-1} \tag{10}$$

is in fact the most efficient one for the pumping up of the v=0 level in CO.

Due to the different vibrational energy spacing, the multiquantum dissymmetric transitions can also be important in the dynamics of the vibrational energy exchanges in collisions between molecules of different masses [37]. This aspect of the vibrational relaxation dynamics has been considered for the N_2/CO and N_2/O_2 systems [37].

The vibrational relaxation of oxygen is important for a variety of chemical processes in the upper atmosphere. Recently, the study on the relaxation of vibrationally excited O_2 molecules has been stimulated by the importance of O_2 in v>>1 in the ozone formation reactions. According to the reaction scheme proposed by Slanger et al.[45], the photodissociation of vibrationally excited O_2 could be a further root for ozone formation, provided that O_2(v>>1) are sufficiently populated. The photodissociated oxygen atoms can , in fact, recombine via the reaction: $O(^3P)$ + $O_2(v=0)$ + M --> O_3 + M., thus increasing the O_3 formation. In the attempt to explain the kinetics of O_3 formation [46], the vibrational relaxation of O_2 has been modelled by assuming simple SSH v-scaling for the V-V rate constants in O_2-O_2. However we have shown that both approximations can be easily incorrect. Recently, the vibrational relaxation in O_2 has been very accurately studied both experimentally [45,47] and theoretically [40] in a broad range of v, v=(4,27). The experimental and theoretical deactivation rate constants are reported in Fig.7. There is substantial agreement between the two sets of experimental determinations, apart from the values at v=19, and the position of the minimum is also well defined in both cases. The agreement between theory and experiment is also good up to v=27, from there on up the experimental rates show a steeper increase. The semiclassical deactivation rates reported in Fig.7 are given as sum of two individual contributions to the vibrational quenching of each v-level, that is the V-T rates for the process:

$$O_2(v) + O_2(v=0) \text{ -----> } O_2(v-1) + O_2(v=0) \tag{11}$$

and the V-V deactivation rates for the process:

$$O_2(v) + O_2(v=0) \text{ -----> } O_2(v-1) + O_2(v=1) \tag{12}$$

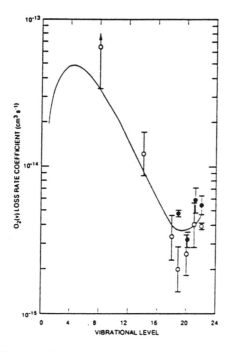

Fig.7 Total quenching rate constant for O,(v) in collisions with O,.
Experimental data: full circle (Ref.47), circle (Ref.45.).
Theory : line (Ref.40) (Figure taken from Ref.47)

An important result in the analysis made by Billing and Kolesnick [40] is that the V-V quenching processes are dominant in the interval of v between v=0,16, so that the total deactivation rates determined in the experiments in this interval are the V-V rates, basically. At higher v, the energy mismatches for the V-T processes decrease, so that the relevant V-T rates increase faster than the corresponding V-V ones. The minimum in the vibrational quenching rates is therefore due to the prevailing of the V-T rate constants for processes (11) with respect to the V-V relaxation mechanism. A further significant results pointed out in these studies [40,45,47] is that the vibrationally excited levels in O_2 are quenched more efficiently than suspected on the basis of a linear vibrational quantum number scaling of the REG rate constant [19]. As a consequence, the efficiency of the dissociation mechanism via V-V relaxation in O_2 could come into question, as well as the importance of the photodissociation of vibrationally excited O_2 invoked in the ozone formation kinetics.

The relaxation of vibrationally excited H_2 molecules has also attracted considerable attention, not only because of the fundamental questions concerning the collision dynamics, but also because of the practical importance of H_2 in the field of fuel cells, nuclear reactors, and in interstellar chemistry. Experimental determinations of the V-V and V-T rate constants are very difficult to carry out, due to the difficulty in preparing and probing vibrationally excited molecules with no permanent dipole moment, so that only very few experimental determinations exist for the V-V rate constants for H_2 in $v=2$ [48]. On the contrary, a great deal of theoretical work has been done on the vibrational (and rotational) relaxation of hydrogen in collisions with atoms (H,He,..) and molecules (H_2). H_2 is in fact a prototype molecule for electronic structure and molecular dynamics calculations. In particular, the interaction potential for the H_2-H_2 system has been computed under different ab initio schemes, and molecular dynamics models for the vibrational energy exchanges have been applied with different degrees of accuracy. Extensive calculations of state-selected rate constants for V-V and V-T exchanges in H_2-H_2 collisions have been performed within the semiclassical collisonal method [31,41,49] and a highly accurate interaction potential. It has been confirmed from these studies that the H_2-H_2 interaction is dominated by the short range repulsive forces, and, as a consequence, the vibrational relaxation in H_2 is controlled by the V-T energy exchanges. The V-V and V-T rate constants for the processes (13)(14) are reported in Fig.8 as a function of v at T=300K.

$$H_2(v) + H_2(v=1) \longrightarrow H_2(v-1) + H_2(v=0) \tag{13}$$

$$H_2(v) + H_2(v=0) \longrightarrow H_2(v-1) + H_2(v=0) \tag{14}$$

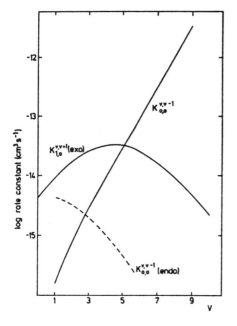

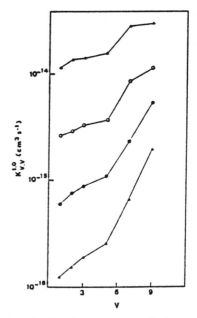

Fig.8 V-V and V-T rate constants inH_2-H_2 collisions for processes (13) and (14) at T=300K

Fig.9 Semiclassical rate constants for the processes: $H_2(v=1) + H_2(v) \to H_2(v=0) + H_2(v)$ T=300K (full triangle); 500K (full circle) 700K (circle); 1000K (triangle)

The exothermic V-V rates show a broad maximum at about v=5, then the rate constants decrease monotonically as v increases. The V-V endothermic curve crosses the V-T rates at around v=3, so that the V-T exchanges dominate the vibrational quenching in H_2 for almost the full vibrational ladder. The V-V processes play a minor role, except for the first few low-lying vibrational levels in the low temperature regime. The position of the crossing point is of great importance in determining the shape of the vibrational population distribution obtained from the kinetics modelling. A further interesting property of the vibrational relaxation in H_2 is shown in Fig.9, where the deactivation rate constants out of v=1 level are reported for the processes:

$$H_2(v=1) + H_2(v) \rightarrow H_2(v=0) + H_2(v) \tag{15}$$

The energy mismatch for these transitions does not depend on the vibrational level v, but nevertheless the results show that the vibrationally excited H_2 molecules are much efficient quenchers than H_2 in the ground level. This behaviour is a consequence of the large molecular anharmonicity [49]. Finally, it is worthwile noting that the V-V rates, as well as the V-T rate constants, are very sensitive to the details of the interaction potential, so that the accuracy of the potential interaction assumed in the scattering calculations is of fundamental importance [28].

5. Concluding remarks

In this article we have reviewed collisional models that are frequently used for calculating state-to-state rate constants for V-V and V-T energy exchanges in diatomic molecules. The attention has been focused on those aspects that can be of importance in the field of non-equilibrium vibrational kinetics. The body of material presented shows that simple analytical 1st order theories have some important limitations, so that they can easily fail in predicting both the absolute values of the rate constants and the vibrational quantum number dependence. Thus, to overcome these limitations the molecular rotations and the multiquantum vibrational exchanges must be correctly included in the collision dynamics. With today's computers, fully quantum CS calculations of V-T rate constants are feasible for atom-molecule systems with the diatom in the low vibrational levels. However, the calculation of rate constants at the high temperatures involved in the re-entry conditions would be computationally very expensive due to the rapid proliferation of the energetically accessible roto-vibrational states.

The semiclassical CS method can be very accurate for the generation of complete sets of rate constants for the full range of the vibrational ladder in a large range of collisional temperatures. Collisional data for the V-V and V-T relaxation in O_2, N_2, CO, H_2,, NO^+, O_2^+ obtained by this method are now available in literature. The results obtained in these studies have revealed some important features of the dynamics of the vibrationally excited states that can be of valuable importance in the kinetic modelling of non-equilibrium chemical phenomena occurring in the upper atmosphere. The reliability of these data lies not within the limits of the method, but rather on the accuracy of the intermolecular potential assumed in the dynamics.

References

1. Brasser, G., and Solomon, S., (1984), *Aeronomy in the middle atmosphere*, Reidel, Dordrecht
2. Treanor, C.E.,Rich,J .W.,and Rehm, R.G., (1968), *J.Chem.Phys.* **48**, 1798

34

3. Capitelli,M.,(ed.), (1986) *'Nonequilibrium vibrational kinetics'*, Springer, Berlin
4. Miller, W.H., (1976) *'Dynamics of molecular collisions'*, Plenum, New York
5. Schwartz, R.N., Slawsky, Z.I. and Herzfeld, K.F., (1952), *J.Chem.Phys.***20**,1591;
   Schwartz, R.N., and Herzfeld, K.F., (1954), *J.Chem.Phys.* **22**,767
6. DeLeon, R.L., and Rich, J.W., (1986), *Chem.Phys.* **107**,283
7. Jackson, J.M., and Mott., N.F., (1932), *Proc.Roy.Soc.* **A137**,703
8. Landau, L., Teller, E., (1936) *Phys.Z.Sow.* **10**,34
9. Rapp, A., and Kassal, T., (1969) *Chem.Rev.* **69**,61
10. Hirschfelder, J.O., Curtiss, C.F., and Bird, R.B. (1954), *Molecular theory of gases and liquids*, Wiley, New York
11. Calvert, J.B., and Amme, R.C., (1966) *J.Chem.Phys.* **45**,4710
12. Yang, X., Kim, E.H., Wodtke, A.M., (1992) *J.Chem.Phys.* **96**,5111
13. Wyson,I.J., (1994), *J.Chem.Phys.* **101**,2800
14. Saupe,S., Adamovich, I., Grassi, M.J., Rich, J.W. (1993), *Chem.Phys.* **174**,219
15. Tanner, J., and Maricq, M.M., (1987) *Chem.Phys.Lett.* **138**,495
16. Miller, J., and Millikan, R.C., (1970) *J.Chem.Phys.* **53**,3384
    Millikan, R.C., and White, D.R., (1963) *J.Chem.Phys.* **39**,98
17. Lambert, J.D., (1977) *Vibrational and rotational relaxation in gases*, Clarendon Press, Oxford
18. Rapp, D., and Sharp, T.E., (1963) *J.Chem.Phys.* **38**,2641
19. Rapp, D. and Englander-Golden, P., (1964) *J.Chem.Phys.* **40**,573
20. Zelechow, A., Rapp, D., and Sharp, T., (1968) *J.Chem.Phys.***49**,286
21. Shin, H.K, (1972) *J.Chem.Phys.* **57**,1363
22. Sharma, R.D., and Brau, C.A., (1969) *J.Chem.Phys.* **50**,924
23. Dmitrieva, I.K., Pogrebnya, S.K., Porshnev, P.I., (1990) *Chem.Phys.* **142**,25
24. Billing, G.D., and Fisher, E.R, (1979) *Chem.Phys.* **43**,395
25. Keck, J., and Carrier, G., (1965) *J.Chem.Phys.* **43**,2284
26. Rockwood, S.D., Brau, J.E., Proctor, W.A., and Canavan, G.H., (1973), *IEEE J.Quant.Electron.* **QE9**,120
27. Cacciatore, M., and Billing, G.D., (1981) *Chem.Phys.* **58**,395
28. Kolesnick, R.E., and Billing, G.D., (1993) *Chem.Phys.* **170**, 201
29. Billing, G.D., and Cacciatore, M., (1982) *Chem.Phys.Lett.* **86**,20
30. Arthurs, A.M., and Dalgarno, A., (1960) *Proc.R.Soc.* **A256**,540
31. Cacciatore, M., Caporusso, R., and Billing, G.D., (1992) *Chem.Phys.Lett.* **197**,92
32. Pack,R.T., (1974) *J.Chem.Phys.* **60**,633-639; Schatz, G.C., and Kuppermann, A.,(1976) *J.Chem.Phys.* **65**, 4668; McGuire, P., and Kouri, D.J., (1974) *J.Chem.Phys.* **60**,2488
33. Banks,A.J., Clary, D.C., (1987) *J.Chem.Phys.***86**,802
34. Pogrebnya, S.K., Kliesch, A., Clary, D.C. and Cacciatore, M., (1996) 'Vibrational relaxation in NO+-He: accurate quantum mechanical study', *Int.J.Mass Spectrom. Ion Proc.*, in press
35. Gianturco, F.A., (1979) *The transfer of molecular energies by collisions*, Springer, Heidelberg;
36. Billing, G.D., (1984) *Comp. Phys.Rep.* **1**,237
37. Cacciatore, M., Capitelli, M., and Billing, G.D., (1984) *Chem.Phys.* **89**,17
38. Cacciatore, M., Capitelli, M., and Billing, G.D., (1983) *Chem.Phys.* **82**,1
39. Cacciatore, M., Billing, G.D., (1985) *Chem.Phys.Lett.* **121**,99
40. Billing, G.D., and Kolesnick, R.E., (1992) *Chem.Phys.Lett.* **200**,382
41. Cacciatore, M. and Billing, G.D., (1992) *J.Phys.Chem.* **96**,217
42. Capitelli, M., Celiberto, R., and Cacciatore, M., (1994), *'Needs for cross-sections in plasma chemistry'*, in Adv.At.Mol.and Opt.Phys., M.Inokuti ed.,**33**,322 Academic Press, New York
43. Brechignac, Ph., (1978), *Chem.Phys.* 34,119; (1981)*Chem.Phys.* **62**,239
44. Billing, G.D., and Cacciatore, M., (1983) *Chem.Phys.Lett.* **94**,218
45. Park, H., and Slanger, T.G., (1994) *J.Chem.Phys.* **100**,287
46. Touomi, R., (1992) *J.Atm.Chem.* **15**,69
47. Price, J.M., Mack, J.A., Rogaski, C.A., and Wodtke, A.M., (1993) *Chem.Phys.* **175**,83
48. Rohlfing, E.A.,Rabitz,H.,Gelfand,J.,Miles,R.B., (1984) *J.Chem.Phys.* 81,820
49. Cacciatore, M., Capitelli, M., and Billing, G.D., (1989) *Chen.Phys.Lett.* **157**,305
50. Zenevich, V.A., Pogrebnya, S.K., Lindinger, W. and Cacciatore, M., (1993) *Int.J.Mass Spectrom. Ion Proc.*, **129**,101

REACTIVE VIBRATIONAL DEEXCITATION: THE N + N$_2$ AND O + O$_2$ REACTIONS

A.LAGANÀ, A. RIGANELLI AND G. OCHOA DE ASPURU
Dipartimento di Chimica, Università di Perugia,
Perugia, Italy

E. GARCIA
Departamento de Quimica Fisica, Universidad del Pais Vasco,
Vitoria, Spain

AND

M.T. MARTINEZ
Departamento de Maquinas y Motores Termicos, Universidad
del Pais Vasco,
Bilbao, Spain

1. Introduction

In recent years it has become increasingly apparent that the characteristics of the gaseous flux around objects flying at hypersonic speed are sensitive to the nature and the efficiency of the involved elementary chemical reactions.[1] [2] [3] This situation is common to the modelling of all complex gas phase processes (*e.g.* lasers, cold plasmas, ionic sources, etc.[4]). Due to the non-equilibrium nature of these systems the reactive properties relevant to their description are the state-to-state detailed quantum cross sections. Detailed cross sections, in fact, individually depend on the different energetic modes and, therefore, can account for the actual energetic distribution of the components of the gaseous system.

However, despite the impressive evolution of memory capacity and cpu speed of modern computers, it is still extremely difficult to compute the extended set of detailed cross sections needed for chemical modelling even when dealing with atom diatom A + BC reactions. For this reason, on the theorist's side significant efforts have been paid both to develop suitable theoretical treatments and to parallelize related computational procedures. At the same time, on the modeller' s side, target have been degraded to

35

M. Capitelli (ed.), Molecular Physics and Hypersonic Flows, 35–52.
© 1996 *Kluwer Academic Publishers. Printed in the Netherlands.*

less accurate approaches and less detailed quantities (as an example rate coefficients instead of cross sections). Rate coefficients, in fact, are quantities averaged over translational, rotational and sometimes even vibrational energy according to a given (usually Boltzmann) distribution. Therefore, their calculation may not need the determination of detailed state to state properties and may be, to a certain extent, simplified.

In this paper, however, we always refer to the molecular nature of these quantities. Therefore, we relate their value to the state to state scattering S matrix or, equivalently, to the state to state cross sections. In this scheme, vibrational state (v) to state (v') atom diatom rate coefficients are rigorously expressed as

$$k_{v,v'}(T_{tr}, T_{rot}) = \frac{\sum_j g(2j+1)e^{-\varepsilon_j/kT_{rot}}}{(k^3 T_{tr}^3 \pi \mu/8)^{1/2} Q_R(T_{rot})} \int_0^\infty dE_{tr} E_{tr} e^{-E_{tr}/kT_{tr}} \sigma_{vj,v'}(E_{tr})$$

$$= \left(\frac{8}{k^3 T_{tr}^3 \pi \mu}\right)^{1/2} \int_0^\infty dE_{tr} E_{tr} e^{-E_{tr}/kT_{tr}} \sigma_{v,v'}(E_{tr}) \qquad (1)$$

where for reactions like $N + N_2$ g is either 2 (for even rotational levels) or 1 (for odd rotational levels). In the above equation, μ is the reduced mass of the reactants ($\mu = m_A m_{BC}/M$ with $M = m_A + m_B + m_C$ and m_I being the mass of the Ith atom), k is the Boltzmann's constant, T_{tr} and T_{rot} are the translational and rotational temperatures respectively, Q_R is the rotational partition function, E_{tr} is the translational energy, ε_j the energy of the reactant rotational state j, and $\sigma_{vj,v'}$ is the degeneracy averaged ·detailed reactive cross section $\sigma_{vj,v'j'}$ summed over the product rotational states j' (product quantities are primed). By summing over j the degeneracy averaged $\sigma_{vj,v'}$ cross section one obtains the vibrational state to state $\sigma_{v,v'}$. By further averaging over v' one generates the state specific cross section and the related rate coefficient

$$k_v(T_{tr}, T_{rot}) = \sum_{v'} k_{v,v'}(T_{tr}, T_{rot}). \qquad (2)$$

The chapter is articulated as follows: In section 2 we illustrate the different theoretical approaches commonly used to calculate the reactive properties of elementary atom diatom reactions. In section 3 we analyze some of the progresses made in obtaining efficient parallel implementations of related computer codes. In section 4 we report some applications to the N + N$_2$ and O + O$_2$ prototype reactions and discuss the main features of the calculated properties.

2. Theoretical approaches

There are different levels of accuracy at which the evaluation of the efficiency of atom diatom chemical reactions can be performed. These can be classified as model, quasiclassical, reduced dimensionality quantum and three dimensional accurate quantum treatments. In the followings we shall illustrate their main features.

2.1. A MODEL APPROACH

To illustrate model approaches we shall refer to the intuitive picture of the reaction as a break up following the collision of a molecule with an impinging body. Its simplest formulation is the hard sphere model.[5] In this case, the probability $P_v(E_{tr})$ (to introduce the dependence from the reactant vibrational state v we make the size d_v of the hard sphere potential depend on v) of breaking the target molecule is formulated as a step function:

$$
\begin{aligned}
P_v(E_{tr}) \quad &= 0 \qquad \text{if } b > d_v \\
&= 1 \qquad \text{if } b \leq d_v \text{ and } E_{tr} \geq E_a
\end{aligned}
$$

where E_{tr} is the collision energy and b is the impact parameter of the collision. In other words, the reaction probability is 1 only when the impact parameter b is smaller than the average size d_v of the potential and the collision energy E_{tr} is larger than a given activation value E_a. This formulation leads to the following expression for the cross section $\sigma_v(E_{tr})$

$$
\sigma_v(E_{tr}) = \pi d_v^2 \tag{3}
$$

that when is inserted into the rate coefficient expression of Equation 1 leads to the following formulation of the rate coefficient

$$
k_v(T) = A T^n e^{-E_a/kT} \tag{4}
$$

with A being equal to $(8\pi k^3/\mu)^{1/2} d_v^2$ and n equal to $3/2$. To make the formulation of the cross section more flexible (and more realistic), the dependence of $P_v(E_{tr})$ from E_{tr} and b can be made smoother than that of the step function. Further improvements can be introduced by allowing $P_v(E_{tr})$ depend also on the collision angle Θ (steric effect) and the initial rotational state j. As a result, the formulation of the rate coefficient becomes more complex. To retain the simplicity of Equation 3 when fitting calculated or measured data, A, n and E_a need to be considered as empirical adjustable parameters.

Such a formulation can also be improved by introducing quantum corrections. To introduce also the dependence from the final vibrational state v' the reaction probability can be formulated in terms of the global state specific component $P_v(E)$ and a Franck-Condon overlap integral $S_{vv'}(s)$ between reactant and product diatomic wavefunctions displaced by a quantity s

$$P_{vv'}(E) = P_v(E)|S_{vv'}(s)|. \tag{5}$$

This formulation was found to give a realistic modelling of the state to state properties when the displacement s is made vary linearly with E, v and v'.[6]

2.2. THE QUASICLASSICAL TRAJECTORY METHOD

The quasiclassical trajectory (QCT) approach does not introduce dynamical approximations. The limitation of the QCT method is the neglecting of the quantum nature of atomic and molecular collisions. At the same time, however, it shows the advantage of being conceptually simple and giving an easy-to-understand picture of the reactive processes.

The detailed QCT cross section $\sigma_{vj,v'j'}$ is formulated as a five dimensional integral. When using a Monte Carlo technique this is usually approximated as

$$\sigma_{vj,v'j'} = \frac{\pi b_{max}^2}{N} \sum_{i=1}^{N} f_{vj,v'j'}(\xi_1, \xi_2, \xi_3, \xi_4, \xi_5) \tag{6}$$

where N is the number of integrated trajectories and b_{max} is the maximum value of the impact parameter leading to reactive encounters. The function $f_{vj,v'j'}(\xi_1, \xi_2, \xi_3, \xi_4, \xi_5)$ is a Boolean function. Its value is 1 only when, after integrating the motion equations (trajectory) starting from a given set of initial values of the five ξ variables and a given vibrotational energy corresponding to that of the quantum vj reactant state, the final outcome can be assigned to the $v'j'$ quantum state of the products. The ξ variables are related to the initial geometry and velocity parameters and have the advantage of reducing the integral to a unit cube. The 12 motion equations to be integrated relate the time derivative of the projection of position and momentum vectors ($\vec{q}_I$ and $\vec{p}_I$) of the generic atom I on the generic axis w (q_{wI} and p_{wI}) to the partial derivatives of the Hamiltonian (H)

$$\frac{dq_{wI}}{dt} = \frac{\partial H}{\partial p_{wI}}, \tag{7}$$

$$\frac{dp_{wI}}{dt} = -\frac{\partial H}{\partial q_{wI}}.$$

The initial value of the ξ variables is selected randomly so as to generate a uniform distribution in the related action angle coordinates. Assignment of continuous internal classical energies to discrete quantum states is made using a closeness criterion.

2.3. THE REDUCED DIMENSIONALITY QUANTUM METHOD

The reduced dimensionality quantum methods have the advantage over QCT ones, of considering the quantum nature of molecular processes. However, the price paid for this is the introduction of artificial dynamical constraints. The simplest reduced dimensionality treatment which keeps a three dimensional nature is the infinite order sudden (RIOS)[7] method. The RIOS approach reduces the dimensionality of the atom-diatom reactive scattering problem by decoupling rotational motions. As a result, in a time independent approach the treatment reduces to the integration of a set of fixed collision angle Θ two mathematical dimension differential equations of the type:

$$\left[\frac{\partial^2}{\partial R_\lambda^2} + \frac{\partial^2}{\partial r_\lambda^2} - \frac{A_l}{R_\lambda^2} - \frac{B_j}{r_\lambda^2} - \frac{2\mu}{\hbar^2}\left(V(R_\lambda, r_\lambda; \Theta_\lambda) - E\right)\right] \Xi(R_\lambda, r_\lambda; \Theta_\lambda) = 0$$

(8)

where λ indicates the type of atom-diatom arrangement ($\lambda = \alpha$ for reactants, $\lambda = \beta$ for products), E is the total energy, R_λ and r_λ are the mass scaled Jacobi coordinates, μ is the reduced mass of the system and $A_l = \hbar^{-2}l(l+1)$ and $B_j = \hbar^{-2}j(j+1)$ are the coefficients of the decoupled orbital and rotational terms of the Hamiltonian with l and j being the related quantum numbers.

Equation 8 can be solved using a CC technique. To this end, for each arrangement channel, R_λ is segmented into many small sectors. Within each sector i the global fixed Θ_λ wavefunction $\Xi(R_\lambda, r_\lambda; \Theta_\lambda)$ is expanded in terms of the eigenfunctions calculated by solving the one dimensional bound state equation in r at R_λ^i (the midpoint value of R_λ for sector i). By truncating the expansion to the first N_v terms, substituting the expansion into Equation 8 and averaging over r_λ, one obtains a set of N_v coupled equations of the type:

$$\left[\frac{d^2}{dR_\lambda^2} - \mathbf{D}_\lambda^l\right] \psi_v^l(R_\lambda; \Theta_\lambda) = 0$$

(9)

with $\mathbf{D}_\lambda^l$ being the coupling matrix and $\psi_v^l(R_\lambda; \Theta_\lambda)$ the coefficients of the expansion of $\Xi(R_\lambda, r_\lambda; \Theta_\lambda)$. After integrating Equation 9 through the different sectors to the asymptotes and imposing the appropriate boundary conditions one can estimate the detailed state v to state v' fixed Θ S matrix elements ($S_{lv,v'}(\Theta, E)$) and, from these, the state (v) to state (v') reactive cross section $\sigma_{v,v'}(E)$.

$$\sigma_{v,v'}(E) = \frac{\pi}{k_v^2} \sum_l (2l+1) \int_{-1}^{1} |S_{lv,v'}(\Theta, E)|^2 \, d\cos\Theta.$$

(10)

2.4. THE ACCURATE THREE DIMENSIONAL QUANTUM METHOD

The most rigorous way of calculating reactive cross sections is to make use of three dimensional quantum techniques. This can be performed using both time dependent and time independent formulations of the Schödinger equation for the nuclear motion. Up to now, most of the work has been carried out using time independent techniques. Related formalism has mainly been developed using hyperspherical coordinates.[8]. In this formalism the Hamiltonian reads as:

$$H = T_\rho + T_h + T_r + T_c + V(\rho, \theta, \chi) \tag{11}$$

where ρ is the hyperradius and θ and χ are the hyperangles (the subscripts of the different terms of the Hamiltonian stand for hyperradius, hyperangles, rotational, and Coriolis, respectively). The individual terms are given by

$$T_\rho = -\frac{\hbar^2}{2\mu\rho^5} \frac{\partial}{\partial\rho} \rho^5 \frac{\partial}{\partial\rho}, \tag{12}$$

$$T_h = -\frac{\hbar^2}{2\mu\rho^2} \left(\frac{4}{\sin 2\theta} \frac{\partial}{\partial\theta} \sin 2\theta \frac{\partial}{\partial\theta} + \frac{1}{\sin^2\theta} \frac{\partial^2}{\partial\chi^2} \right), \tag{13}$$

$$T_r = A J_x^2 + B J_y^2 + C J_z^2, \tag{14}$$

and

$$T_c = -\frac{i\hbar \cos\theta}{\mu\rho^2 \sin^2\theta} J_y \frac{\partial}{\partial\chi}, \tag{15}$$

with $A^{-1} = \mu\rho^2(1+\sin\theta)$, $B^{-1} = 2\mu\rho^2 \sin^2\theta$ and $C^{-1} = \mu\rho^2(1-\sin\theta)$ and $\mu = (m_A m_B m_C/M)^{1/2}.$.

To solve the scattering problem using a CC technique, the wavefunction $\Psi_{t\lambda}^{Jpn}$ for a given value of the total angular momentum $\mathbf{J}$ is expanded in products of Wigner rotation functions $D_{\Lambda M}^J$ of the three Euler angles (α, β and γ) and fixed ρ surface functions $\Phi_{t\lambda}^{Jp}$ of the two internal hyperangles (θ and χ). The coefficients ψ of the expansion of the global wavefunction are unknown functions of the hyperradius ρ. To carry out the numerical integration the ρ interval is divided into several small sectors. For each sector i, the appropriate set of $\Phi_{t\lambda}^{Jp}(\theta, \chi; \rho_i)$ surface functions is calculated at the sector midpoint ρ_i.

The related numerical procedure is called APH3D and, by analogy with the RIOS program, is divided into two main steps. The first of them (ABM) is devoted to the calculation of the surface functions $\Phi_{t\lambda}^{Jp}$ by integrating the two dimensional bound state equation

$$\left[T_h + \frac{15\hbar^2}{8\mu\rho_i^2} + \hbar^2 G_J + F\hbar^2 \Lambda^2 + V(\rho_i, \theta, \chi) - \mathcal{E}_{t\lambda}^{Jp}(\rho_i) \right] \Phi_{t\lambda}^{Jp}(\theta, \chi; \rho_i) = 0, \tag{16}$$

where $G_J = \frac{1}{2}J(J+1)(A+B)$ and $F = C - (A+B)/2$. The second computational step (LOGDER) is devoted to the integration, sector by sector, from small ρ values to the asymptote, of the following set of coupled differential equations

$$\left[\frac{\partial^2}{\partial \rho^2} + \frac{2\mu E}{\hbar^2} \right] \psi_{t\Lambda}^{Jpn}(\rho) = \frac{2\mu}{\hbar^2} \sum_{t'} < \Phi_{t\Lambda}^{Jp}(\theta, \chi; \rho_i) | H_i | \Phi_{t'\Lambda}^{Jp}(\theta, \chi; \rho_i) > \psi_{t'\Lambda}^{Jpn}(\rho),$$

(17)

where the internal Hamiltonian H_i reads

$$H_i = T_h + T_r + T_c + \frac{15\hbar^2}{8\mu\rho^2} + V(\rho, \theta, \chi).$$

(18)

Once the integration is performed, there are other minor steps to perform in order to map the solution into the space of the Jacobi coordinates, to impose asymptotic boundary conditions and to evaluate the fixed J **S** matrix elements and related probabilities.

3. The parallelization of the reactive scattering programs

With the only exception of the model treatments, the numerical evaluation of the reactive properties of atom diatom reactions is a heavy computational task. For this reason significant effort was paid to implement related procedures on parallel computers.

3.1. THE QCT CODE

The QCT calculation can be structured as a set of independent processes (worker) each dedicated to the integration of a subset of trajectories. Synchronization points to coordinate the process are necessary only for inputting the data common to all trajectories and assembling the results of individual processes. In a parallel organization two conditions have to be met:

1. Guarantee the reproducibility of the parallel calculations by imposing a biunivocal correspondence between the sequential position of a trajectory and the sequence of the pseudo-random numbers used to define the initial conditions for the calculation of each trajectory.
2. Select a correct method to distribute the workload among the various workers taking into account that the elaboration time for a single trajectory can only be determined at run-time.

To guarantee the reproducibility of the results the pseudorandom sequence is generated by the repeated execution of an algorithm that, given an integer number (seed), provides a pseudo-random (real) number in the interval 0-1 and a new integer number to be used as the next seed. To let each process worker independently reproduce the same initial conditions for the same trajectory it is necessary to use a technique that defines a correspondence between the index of the trajectory and its initial conditions. The technique involved requires the use of a vector that memorizes a set of sequentially generated seeds, each of which is the first seed used to calculate each trajectory. In this way, there is a unique correspondence between the sequential position of a trajectory and its initial conditions.

The most appropriate parallel programming model was found to be the task-farm with the dynamic assignment of the load.[10] According to this model the parallel program is structured in two components: a farmer program and a worker program, the latter being duplicated on all the used processors. The farmer process dynamically assigns the work (the integration of one or more trajectories) to the workers. New work is assigned each time a worker has completed its task.

An optimation of the model was obtained by placing the generation and validation of initial conditions inside the worker process. Only the generation of the first seed of each trajectory was kept at the master level to ensure determinism. Once the seed vector (of dimension N) has been constructed, it is distributed to the worker processors. From this vector, following a message of the master, the worker takes the element corresponding to the trajectory to be run. Starting from this seed, a pseudorandom substring and related initial conditions are generated and validated as a part of the in-node work. This allows a dramatic reduction of the workload of the master and of the amount of transmitted data.

A similar modification was introduced to improve the return of the results. More in detail, communications were drastically reduced by performing an in-node update of the statistical indexes. Statistical indexes are collected by the farmer at the end of the worker activity to assemble a global statistical analysis.

In this way, it has been possible to obtain a speedup of 119 when running 4096 trajectories on an nCUBE 2 128 nodes. This well compares with the single program multiple data (SPMD)[10] model used on the Cray T3D. In fact, on the Cray T3D a speedup of 58.5, 106 and 180 was measured when using 64, 128 and 256 processors, respectively. As a result, parallelized versions of the code used for production runs and integrating batches of 100,000 trajectories per job take, on the average, 7 hours on the nCUBE 2 and 1 hour on the Cray T3D.

3.2. THE RIOS CODE

The parallelization of the RIOS code can be carried out at different levels. The level that was considered in the first instance[9] is the distribution on different processors of the single l propagation of the scattering wavefunction for fixed angle and fixed energy calculations using a task farm model. The task farm model assigns to a master node (or to a host machine) the role of managing the distribution to the worker nodes of the different single l propagations of the scattering wavefunction. The choice of the task farm model was mainly motivated by the difficulty of adopting an SPMD scheme. In fact, the need for checking the convergency of the calculation with l does not allow an *a priori* determination of the number of computational tasks that need to be assigned to the individual nodes. On the contrary, the use of a task farm model allows a dynamical distribution of the load among the available processors. This keeps the processors busy all the time since a new single l propagation (for the same angle and the same energy) is assigned as soon as a processor has ended its work. Then, when the convergency is reached, the master stops the current work and new single l propagations for a different energy are issued.

This model gives a good scaling of the parallel performance on machines with a limited number of processors.[9] However, it is not suited for implementation on highly parallel architectures since the number of single l calculations may be smaller than the number of processors and the work lost when reaching convergency with l may be large.

A first attempt to improve the scalability of the code on large machines is the adoption of a two level parallel model. A way of constructing a two level parallelization model was to distribute different fixed angle calculations. In fact, since fixed angle calculations are also independent computational tasks (the RIOS method builds the 3D **S** matrix by averaging the independent fixed angle ones), when distributing fixed angle calculations the nodes can be clustered in subsets. Therefore, the two level task farm model consists of a master assigning different fixed angle calculations to the different clusters and a cluster submaster (as in the first parallelization model described above) distributing single l propagations.

A single level parallel model looping more rapidly on the energy than on the l value, can also be adopted. That means that, for example, after distributing the first single l propagation for the first energy, the first single l propagation for the second energy is distributed next instead of the second l propagation for the first energy.

In this way, the time lost in unproductive single l calculations reduces significantly. In fact, the energy loop has no convergency check and the number of energies to be calculated is given as an input data. Therefore, a

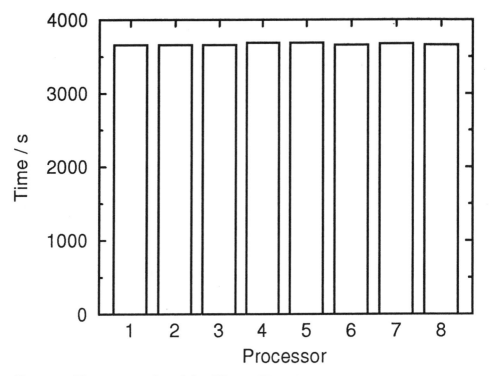

Figure 1. Time consumption of the different SP2 nodes when using the parallelization model III.

limited number of nodes (if not none) will still be carrying out calculations for the same energy when convergency with l is reached. The loss of single l propagation calculations, however, increases when approaching the end of the calculation. In this case, in fact, when the large majority of fixed energy calculations is completed, several nodes will be contemporarily running a single l propagation for the same energy. This means that more single l propagations will be stopped when convergency for that energy is reached. An example of the good load balance obtained when using this model is given in Fig. 1

3.3. THE ABM CODE

As already mentioned, the APH3D computational procedure is articulated into several programs. Among them, only ABM and LOGDER deserve to be considered for parallelization. In this paper we shall discuss the main aspects of the parallelization of ABM.

Since, as already mentioned, the fixed ρ calculation of eigenvalues and eigenfunctions of the two hyperangles is a two dimensional bound state problem, ABM is more memory demanding and time consuming than the corresponding section of RIOS. In ABM, the external loop runs over the different sectors and has to deal with the surface functions for all the Λ projections (inner loop). A difficulty of the parallelization at sector level is the order dependency associated with the calculation of overlap integrals between surface functions of adjacent sectors (each sector needs the surface functions of the previous one). Such a dependency can be avoided by duplicating the surface functions calculation when the adjacent sector is assigned to a different node. This penalizes a task farm scheme of parallelization since it requires a duplication of all surface function calculations. On the contrary the SPMD model seems to be more convenient model. In fact, when using the SPMD scheme a subset of sector calculations is assigned to every node and the determination of the surface functions has to be repeated only for the first sector of the subset. .

Performances measured when using the SPMD model on a Cray T3D[11] indicate a strong load imbalance. Most of this load imbalance, however, is not due to duplicated surface function calculations. It is instead due to the fact that small ρ value calculations are more time consuming. For example, in a 64 node run a small load imbalance on the first 38 nodes is caused by the fact that when using 64 nodes the first 38 nodes get 4 sectors each (for a total of 152) while the remaining 26 nodes get only 3 sectors each (one extra calculation out of four). There is also a larger load imbalance (affecting only the first 12 nodes). The large one is due to the fact that for the first 48 sectors ($\rho < 6 \, a_o$) the calculation of the surface functions is more time consuming. This explains also why in the case of the 32 node run there is only one type of load imbalance. In that case, in fact, the number of nodes carrying out the extra work and the most time demanding work is the same (48).

4. The calculated state to state properties

As already mentioned, the calculations have been performed for $N + N_2$ and $O + O_2$ reactions. For these systems both QCT and RIOS techniques have been used.

4.1. THE N + N$_2$ REACTION

For the reaction:

$$N(^4S_u) + N_2(^1\Sigma_g^+, v) \rightarrow N(^4S_u) + N_2(^1\Sigma_g^+, v'). \qquad (19)$$

quasiclassical calculations were carried out using a LEPS potential energy surface[12] (PES) having a collinear barrier of 1.56 eV. The quasiclassical estimate[13] of the rate coefficients are in excellent agreement with measured data[14] despite the empirical nature of the arguments used for the construction of the PES.

A detailed analysis of the quasiclassical results is given in ref. [13] where rate coefficients calculated for rotational and translational temperatures ranging from 500 to 4000 K are discussed. Here, we summarize only the main conclusions. At all temperatures rate coefficients were found to be appreciable only for fairly vibrationally excited reactants implying a moderately active role of vibration in promoting reactivity. On the contrary, even a modest rise in energy supplied as an increase of the system temperature was quite effective in enhancing reactivity.

In the above mentioned calculations, a temperature increase affects both rotational and collision energies. To figure out which of the two types of temperature is more effective in promoting reactivity, T_{rot} and T_{tr} were varied independently from 500 K to 4000 K.

The effect on $k_v(T_{tr}, T_{rot})$ of varying the rotational temperature was found to be little especially at high translational temperature. On the contrary, the effect of increasing the translational temperature of the system showed to be definitely much larger. More detailed quantities are the state-to-state rate coefficients $k_{v,v'}(T_{tr}, T_{rot})$ or their partial summations (like the vibrational deexcitation rate coefficients $k_v^d(T_{tr}, T_{rot}) = \sum_{v'<v} k_{v,v'}(T_{tr}, T_{rot})$ or vibrational excitation rate coefficients $k_v^e(T_{tr}, T_{rot}) = \sum_{v'>v} k_{v,v'}(T_{tr}, T_{rot})$). Reactive and non reactive vibrational deexcitations are always the dominant process (with the former invariably more efficient than the latter). An increase of the rotational temperature does not alter significantly the situation since it almost equally affects both reactive and non reactive $k_v^d(T_{tr}, T_{rot})$ values. Processes mostly affected by an increase of the rotational temperature are the reactive vibrational excitations $k_v^e(T_{tr}, T_{rot})$ and, to a less extent, reactive adiabatic processes. However, although the effect is large for both of them, at high v values reactive excitation $k_v^e(T_{tr}, T_{rot})$ increases so fast that the two curves tend to cross with the location of the crossing point depending on the value of T_{rot}. At high reactant vibrational states and rotational temperatures, reactive excitation becomes so efficient to compete with non reactive deexcitation.

RIOS calculations have been carried out for total energies extending up to 9 eV and reactant vibrational states as high as 25. The state specific reactive cross section σ_v summed over all product states at all values of v rises from zero at threshold to a plateau as E_{tr} increases. The rise at threshold becomes sharper at higher v values and the threshold energy lowers. However, only a fraction of the energy supplied as vibration is effective in lowering the threshold.

The absolute value of the RIOS state-to-state rate coefficients agree, on the average, with quasiclassical results. However, a plot of these rate coefficients as a function of the final vibrational number v' singles out interesting differences between the two sets of results: the quasiclassical treatment seems to lead to a more pronounced vibrational deexcitation than the RIOS one. On the contrary, RIOS calculations lead to excitation rate coefficients higher than quasiclassical ones. This reflects also on the prediction of the most populated product state. This is always lower than the reactant one for quasiclassical calculations while it is the same for RIOS calculations. The quasiclassical behaviour can be associated with the stronger tendency of classical treatments to redistribute the energy over all possible degrees of freedom. On the contrary, the RIOS behaviour can be associated with the insensitivity of Infinite Order Sudden treatments to rotations and the consequent propensity to channel the collision energy into product vibration.

As already mentioned, despite these differences, the value of the rate coefficients calculated using the two methods are in quite good agreement. Both quasiclassical ($7.0 \times 10^{-13} cm^3 molecule^{-1} s^{-1}$) and RIOS ($5.6 \times 10^{-13} cm^3 molecule^{-1} s^{-1}$) estimates of the rate coefficient at the temperature of the experiment (T=3400 K) agree with the measured ($5 \times 10^{-13} cm^3 molecule^{-1} s^{-1}$) value.[14]

4.2. THE N + N$_2$ REACTION ON MODIFIED POTENTIALS

Recently we generalized the definition of the Rotating Bond Order (ROBO) model potential[15] to the description of the interaction of systems with more than one channel open to reaction (LAGROBO[16]).

The LAGROBO model was first applied to the fitting of the potential energy surface of the H + H$_2$ and N + N$_2$ reactions. For H + H$_2$ the LAGROBO model was used for fitting *ab initio* data. In contrast, for N + N$_2$ it was used to fit a LEPS potential previously employed to perform state to state rate coefficient calculations. This was motivated by the fact that we needed a model potential easier to modify to shape the window to reaction than the LEPS. In fact, recent *ab initio* calculations[17] seem to indicate that the transition state of the N + N$_2$ reaction is not collinear. To be able to play with the height of the barrier, with its location and with the shape of the window to reaction we took advantage of the flexibility of the LAGROBO model potential.

In addition to the LAGROBO PES fitted to the LEPS (L0), we also generated three new LAGROBO surfaces (L1, L2 and L3) having a bent transition state. A preliminary dynamical check of the suitability of L0 to describe the reactive behaviour of the system was performed by comparing the calculated thermal ($T = 3400$ K) rate coefficient with the experimental information.[14] The value of the rate coefficient calculated on the LEPS PES (7.1×10^{-13} cm^{3}molecule^{-1}s^{-1}), and on L0 (7.0×10^{-13} cm^{3}molecule^{-1}s^{-1}) agree reasonably well with the experiment (5.0×10^{-13} cm^{3}molecule^{-1}s^{-1}). The agreement was found to be even more satisfactory when using the RIOS approach (5.6×10^{-13} cm^{3}molecule^{-1}s^{-1} on the LEPS and 5.4×10^{-13} cm^{3}molecule^{-1}s^{-1} on L0).

A more detailed analysis was performed by comparing the state-specific RIOS reactive cross section $\sigma_v(E_{tr})$. Cross section values calculated on L0 are smaller than those calculated on the LEPS at energies about the threshold. On the contrary, at larger energies, when the cross section becomes almost constant the L0 results are larger than those obtained on the LEPS. When the reactant vibrational energy increases the difference increases and the collision energy at which the two values tend to coincide becomes larger (about 4 eV at v=5 and 5.5 eV at v=10). This reflects on the value of the state-specific rate coefficients $k_v(T)$ as well as on that of state-to-state rate coefficients $k_{vv'}(T)$ which are lower than those calculated on the LEPS.

As far as the modified LAGROBO PESs is concerned, the height of the barrier is about 80 kcal/mol for collinear geometries to reach a minimum of 35.7 kcal/mol (L3) and 32.3 kcal/mol (for L1 and L2 respectively). The transition state energy was slightly lowered to the end of investigating whether this compensates for the displacement of the barrier to smaller values of NN̂N and for the possible related effect of a less efficient funnelling of collision energy into vibration of the target molecule. At smaller NN̂N values the barrier rises again to reach the same height of the collinear barrier at 60°. It is also worth pointing out that, except for L2 at small values of the NN̂N angle the internuclear distances at the barrier are the same as those of the LEPS.

A first indication of the dynamical effect produced by the changes made is given by the value of the thermal rate coefficient calculated on the modified PESs. At $T = 3400$ the thermal rate coefficient calculated on L1, L2 and L3 are 10.44×10^{-13} cm^{3}molecule^{-1}s^{-1}, 10.37×10^{-13} cm^{3}molecule^{-1}s^{-1} and 14.77×10^{-13} cm^3 molecule^{-1}s^{-1} when using the QCT method and 7.72×10^{-13} cm^{3}molecule^{-1}s^{-1}, 7.50×10^{-13} cm^{3}molecule^{-1}s^{-1} and 11.16×10^{-13} cm^{3}molecule^{-1}s^{-1} when using the RIOS method.

As far as the cross sections are concerned, values calculated on L1 have about the same threshold as those calculated on L0 but are larger at higher collision energy. Cross sections calculated on L2 and L3 are smaller than those calculated on L1. The difference increases with E_{tr} especially at higher v values.

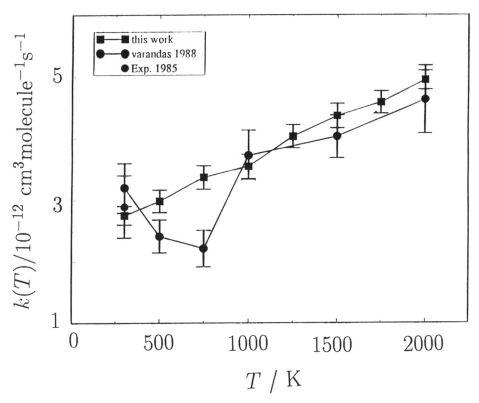

Figure 2. Quasiclassical rate coefficient for thermalized reactants plotted as a function of the temperature.

4.3. THE O + O$_2$ REACTION

As far as the reaction

$$O(^3P) + O_2(^3\Sigma_g^-, v) \rightarrow O(^3P) + O_2(^3\Sigma_g^-, v') \qquad (20)$$

is concerned, extended quasiclassical calculations have been carried out on a Double Many Body Expansion potential energy surface[18] for reactant vibrational levels ranging from 5 to 30 in steps of 5 and a temperature $T = T_{tr} = T_{rot}$ ranging from 200 K to 4000 K. Quasiclassical results (see figure 2) well compare with the experiment[19] ($2.9 \pm 0.5 \times 10^{-12}$cm^{3}molecule^{-1}s^{-1} versus $2.8 \pm 0.2 \times 10^{-12}$cm^{3}molecule^{-1}s^{-1}) in reproducing the value of the thermal rate coefficient at $T = 300$ K. Our calculations not only give a value of the thermal rate closer to the experiment than that obtained by Varandas but they also give a smoother temperature dependence.

To analyze the effect of both the reactant vibrational energy and temperature (*i.e.* of the distribution of translational and rotational energy)

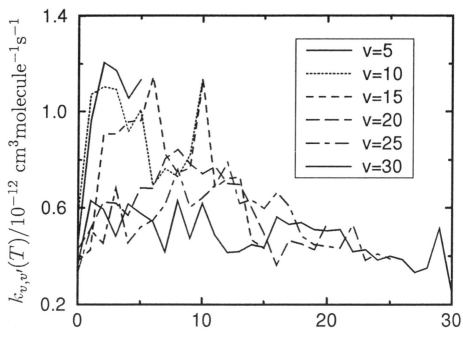

Figure 3. Quasiclassical rate coefficients calculated a T = 1000 K plotted as a function of the vibrational state jump $n = v - v'$.

we have compared the dependence of the rate coefficient on T at different distributions of the reactant internal energy. In Fig. 2 the rate coefficient of thermalized reactants are plotted as a function of T. The plot shows an almost linear increase with the temperature. Such an increase becomes small for higher vibrationally excited O_2 molecules. To single out the different contributions to the vibrational disequilibrium we have also compared $k_v^d(T)$, $k_v^e(T)$ and $k_{v,v}(T)$ values for both reactive and non reactive events. Reactive vibrational deexcitation was found to be the most efficient process on the reactive PES. A substantial contribution comes also from non reactive deexcitation (such a contribution might become the most important one if non reactive surfaces are included in the calculation).

An important outcome of the quasiclassical calculations can be evidenced by plotting the state to state rate coefficients as a function of the difference between initial and final product vibrational number $n = v - v'$. Quasiclassical calculations for $N + N_2$ (see previous section) as well as several model treatments[4] predict single vibrational quantum jumps to be the most important deexcitation mechanism. Multiquantum deexcitations are significantly less important and are mainly funneled by cascade effects. Plots of the state to state rate coefficients for the $O + O_2$ reaction as a function of n are given in Fig. 3. A feature evidenced by these plots is the great importance for this reaction of multiquantum deexcitation mechanisms. Deexcitations to vibrational states from 5 to 10 quanta lower are not unlikely and are sometimes the most efficient processes.

RIOS calculations substantially confirm quasiclassical ones. The RIOS thermal rate coefficient is $9.2 \times 10^{-13} cm^3 molecule^{-1} s^{-1}$. This value is about 1/3 the value obtained from quasiclassical calculations. The same type of deviation applies to other dynamical and kinetic properties.

5. Conclusions

To investigate the effect on vibrational deexcitation mechanisms of increasing the temperature of the system and varying the reactant vibrational state, we have performed massive quasiclassical trajectory and quantum RIOS calculations of state specific and state-to-state rate coefficients for the $N + N_2$ and $O + O_2$ reactions.

Both quasiclassical and quantum state specific rate coefficients reproduce reasonably well the available experimental information. The same type of agreement was found also when considering more detailed quantities. An important difference between calculated properties of the two reactions is the fact that for $N + N_2$ single quantum deexcitation is the most efficient process, whereas for $O + O_2$ the most efficient processes are multiquantum jumps.

6. Acknowledgments

Financial support from Agenzia Spaziale Italiana (ASI) and from NATO (grant No CRG 920051) is acknowledged. A fellowship from the Basque Government is also acknowledged.

References

1. Park, C.: (1995) Review of finite rate chemistry models for air dissociation and ionization. Nato ASI *Molecular Physics and hypersonic flows*, Maratea, 62.
2. Candler, G.V.: (1995) Interfacing nonequilibrium models with computational fluid dynamics methods. Nato ASI *Molecular Physics and hypersonic flows*, Maratea 72.
3. Frühauf, H.H.: (1995) Computation of high temperature nonequilibrium flows. *Nato ASI Molecular Physics and hypersonic flows*, Maratea 7.
4. (1986) *Non equilibrium vibrational kinetics*, Capitelli, M., Ed., Springer-Verlag, Berlin; (1989) *Nonequilibrium processes in partially ionized gases*, Capitelli, M. and Bardsley, J.N., Eds., Plenum, New York.

5. Johnston, H.S.: (1966) *Gas phase reaction theory*, Ronald Press, New York.
6. Laganà, A.: (1987) On the Franck-Condon behavior of the H + Cl_2 reaction, *J. Chem. Phys.*, **86**, 5523 - 5533; Laganà, A., Garcia, E., and Alvariño, J.M.: (1990) A modelling of accurate reduced-dimensionality quantum probabilities for H(D,T) + Cl_2 reactions, *Il nuovo Cimento*, **12 D**, 1539 - 1551; Laganà, A., Paniagua, M., and Alvariño, J.M.: (1990) Accurate and model collinear reactive probabilities of the Mg + FH reaction, *Chem. Phys. Letters*, **168**, 441 - 447.
7. Garcia, E., Gervasi, O., and Laganà, A.: (1989) Approximate Quantum Techniques for Atom Diatom Reactions, *Supercomputer Algorithms for Reactivity, Dynamics and Kinetics of Small Molecules*, Laganà, A., Ed., Kluwer, Dordrecht, 271 - 294; Laganà, A., Gervasi, O., Baraglia, R., and Laforenza, D.: (1989) Vector and Parallel Restructuring for Approximate Quantum Reactive Scattering Computer Codes, *High Performance Computing*, Delahyes, J.L., and Gelenbe, E., Eds, North Holland, Amsterdam, 287 - 298.
8. Parker, G.A., Pack, R.T and Laganà, A.: (1993) Accurate 3D quantum reactive probabilities of Li + FH. it Chem. Phys. Letters, **202**, 75-81 ; Parker, G.A., Laganà, A., Crocchianti, S., Pack, R.T.: (1995) A detailed three dimensional quantum study of the Li + FH reaction. *J. Chem. Phys.*, **102**, 1238-1250.
9. Baraglia, R., Laforenza, D., Perego, R., Laganà, A., Gervasi, O., Fruscione, M., Stofella, P.: (1991) Porting of reduced dimensionality quantum reactive scattering code on a meiko computing surface, *CNR Report 8/20*.
10. Fox, G.C., Johnson, M., Lyzenga, G., Otto, S., Salmon, J., Walker, D.: (1988) Solving problems on concurrent processors, Prentice Hall, Englewood Cliff .
11. Baraglia, R., Laforenza, D., Laganà, A.: (1995) Parallelization strategies for a reduced dimensionality calculation of quantum reactive scattering cross sections on a hypercube machine. *Lecture Notes in Computer Science*, **919**, 554-561.
12. Laganà, A., Garcia, E., and Ciccarelli, L.: (1987) Deactivation of vibrationally excited nitrogen molecules by collision with nitrogen atoms, *J. Phys. Chem.*, **91**, 312 - 314.
13. Laganà, A., Garcia, E.: (1994) Temperature dependence of N + N_2 rate coefficients, *J. Phys. Chem.*, **98**, 502 - 507.
14. Wright, A.N., Winkler, C.A.: (1968) *Active nitrogen*, Academic Press, New York.
15. Laganà, A.: (1992) A rotating bond order formulation of the atom-diatom potential energy surface, *J. Chem. Phys.*, **95**, 2216 - 2217; Laganà, A., Ferraro, G., Garcia, E., Gervasi, O., and Ottavi, A.: (1992) Potential energy representations in the bond order space, *Chem. Phys.*, **168**, 341 - 348.
16. Garcia, E., and Laganà, A.: (1995) The largest angle generalization of the rotating bond order potential: the H + H_2 and N + N_2 reactions, *J. Chem. Phys* in the press.
17. Petrongolo, C.: (1988) MRD-CI ground state geometry and vertical spectrum of N_2, *J. Mol. Struct. Theochem*, **175**, 215 - 220; Petrongolo, C.: (1989) MRD-CI quartet potential surfaces for the collinear reactions $N(^4S_u) + N_2(^1\Sigma_g^+, A^3\Sigma_g^+$ and $B^3\Pi_g)$, *J. Mol. Struct. Theochem*, **202**, 135 - 142;
18. Varandas, A.J.C., Pais, A.A.C.C.: (1988) A realistic double many-body (DMBE) potential energy surface for the ground state O_3 from a multiproperty fit to ab initio calculations, experimental spectroscopy, inelastic scattering, and kinetic isotope thermal rate data, *Mol. Phys.*, **65**, 843 - 860.
19. Anderson, S.M., Klein, F.S., Kaufman, F.: (1985) Kinetics of the isotope exchange reaction of ^{18}O with NO and O_2 at 298 K, *J. Chem. Phys.*, **83**, 1648 - 1656.

REACTIVE CROSS SECTIONS INVOLVING ATOMIC NITROGEN AND GROUND AND VIBRATIONALLY EXCITED MOLECULAR OXYGEN AND NITRIC OXIDE

M. GILIBERT, M. GONZÁLEZ, R. SAYÓS, A. AGUILAR, X. GIMÉNEZ
AND J. HIJAZO
Departament de Química Física
Facultat de Química
Universitat de Barcelona
C/ Martí i Franquès, 1
08028 Barcelona
SPAIN

ABSTRACT. The reactions of atomic nitrogen with molecular oxygen and nitric oxide play an important role in the chemistry of the upper atmosphere, and also to model accurately the conditions existing around a reentering body. The $N + O_2$ system is also an important role of chemiluminiscence in the upper atmosphere. This contribution briefly reviews some of the most powerful theoretical methodologies available for the calculation of detailed and averaged dynamical quantities and presents an overview of the most relevant results obtained in our group about these systems in the last years applying those methodologies, with special emphasis in considering diatomic molecules both in the low and highly excited vibrational levels. Finally, a new improved analytical ground potential energy surface for the reaction of nitrogen with molecular oxygen is presented and some preliminary results discussed.

1. Introduction

The distribution of elements in the atmosphere among the different radical and molecular species (O_2, N_2, O_3, $O...$) is to a great extent determined by photochemical processes. As a consequence, ions constitute the main feature of the thermosphere. Photoionization of N_2, O_2 and O in this region of the atmosphere yields the corresponding N_2^+, O_2^+ and O^+ ions which may undergo subsequent ion-molecule reactions (k between 10^9 and 10^{11}) followed by fast dissociative and radiant recombinations with electrons:

$$NO^+ + e \rightarrow N + O \qquad k \approx 10^{14} \text{ l/mol s}$$
$$O_2^+ + e \rightarrow 2O \qquad k \approx 10^{14} \text{ l/mol s}$$
$$O^+ + e \rightarrow O + h\nu \qquad k \approx 3 \ 10^9 \text{ l/mol s}$$

The translationally hot N atoms produced are afterwards replaced by O atoms according to the reactions:

M. Capitelli (ed.), Molecular Physics and Hypersonic Flows, 53–84.
© 1996 *Kluwer Academic Publishers. Printed in the Netherlands.*

$$N + O_2 \rightarrow NO + O \qquad \Delta H_0° = -134 \text{ kJ/mol} \qquad (1)$$
$$N + NO \rightarrow N_2 + O \qquad \Delta H_0° = -314 \text{ kJ/mol} \qquad (2)$$

giving nitric oxide in highly excited rovibrational states [1-3]. An important fact to be stressed is that, above 100 km., the temperature of the termosphere increases with altitude and the concentration of gas molecules is low, resulting in no thermal equilibrium at high altitudes.

In a reentering body at hypersonic velocity in the upper atmosphere, the gas is compressed inside the shock layer and the nozzle, giving rise to a strong temperature increase. Subsequent expansion as it flows downstream along the body results in nonequilibrium conditions and a strong coupling between vibration, dissociation and exchange processes. In the shock layer, the rotational and translational modes of the molecules may be equilibrated, but the vibrational and electronic degrees of freedom are definitely not equilibrated [3]. In this context, thermochemical relaxation of nitric oxide is governed by the Zel'dovich mechanism [4] which comprises reactions (1) and (2) and their reverses as well as:

$$N + OH \rightarrow H + NO \qquad \Delta H_0° = -204 \text{ kJ/mol}$$

This mechanism is also important to explain the chemistry occuring in the postflame regions of the air-breathing combustors as well as in air pollution studies. From a purely chemical point of view, reaction (2) is important also as a titrant for N atoms [5] and as a means of providing ground state molecular nitrogen [6].

Under the flight conditions expected for a reentering body the electronic temperature may be greater than 10 000 K, which implies that reactions (1) and (2) occuring in excited electronic states should also be considered. An accurate dynamical treatment of those reactions requires however a prior calculation of the corresponding electronic Potential Energy Surfaces (PES), so far not available. In this contribution we will focus on reactions (1) and (2) occuring in their ground potential energy surfaces.

On the other hand, it is worth noting that vibrational excitation of the reactant molecules may bring about important changes in the reactivity of the system and that vibrationally excited product molecules are found in a variety of processes. Thus, product molecules in high vibrational levels are observed in hidrogen [7] and nitrogen [8] plasmas, resulting in non-thermal vibrational distributions.

Phenomenologically, the influence of reactants vibrational excitation can be analysed assuming an Arrhenius-like representation for the rate constant of an elementary reaction in a given vibrational state v [9], leading to:

$$k_v = A_v e^{-(E^0 - \alpha \varepsilon_{vib,BC})/RT} \qquad (3)$$

with α the fraction of vibrational energy effectively used to reduce the reaction barrier E^0 and A_v the preexponential factor for vibrational state v. Vibrational excitation may change threshold energy as well as the excitation function shape, causing the reaction rate constant to vary significantly with v. In general, however, changes in threshold energy ($\alpha > 0$) are compensated by the Boltzman factor in the rate constant expression and it is the enhancement of the preexponential factor that may result in a non-Arrhenian behaviour of the rate constant [10]. Another aspect underlining the importance of vibrational excitation is the slow rate of vibrational energy transfer as compared to that of rotational and translational ones. While thermal distributions are easily attained for those two degrees of freedom, the equilibrium vibrational distribution may not be maintained if part of the reactive flux occurs via excited states. In this case, the rate coefficient becomes time-dependent and approaches a steady-state, the approach being determined by the way reaction is induced. In any case it is found that the coefficient for nonequilibrium kinetics is always smaller than the one appropriate for equilibrium conditions, and the deviation depends on the relative rates of vibrational relaxation and reaction [10].

In such nonequilibrium conditions, accurate information on the reaction outcome may then be obtained by means of dynamical methods. Thus, the present paper begins reviewing the necessary steps towards determining reaction rate constants and other dynamical quantities. These are obtained from the study of individual collision events by means of solving the equations of motion, ultimately based on fundamental laws. It then follows its detailed application to the important reactions (1) and (2) and the main consequences arising thereof. Finally, some conclusions are drawn.

2. Theoretical methodologies in Reaction Dynamics

2.1 INTRODUCTION

The dynamical description of a reactive system requires setting up the Schrödinger equation (SE) for all nuclei and electrons of the system. The next usual step consists in separating the nuclear and electronic motion through expansion of the total wave function in terms of electronic and nuclear basis sets. After substitution in the full Schrödinger equation, a set of coupled equations equations is obtained, with different electronic eigenvalues being coupled through nuclear motion. Neglecting these terms leads to a set of fully decoupled equations and nuclear motion can then be described on each electronic PES separately. Figure 1 summarizes the procedure and the different methods available for the dynamical study of a system.

2.2 POTENTIAL ENERGY SURFACES

Solving the Electronic Schrödinger equation by ab initio or semiempirical methods for all possible nuclear configurations yields the electronic PES. Dynamical methods employed to describe nuclear motion require in most cases the possibility of evaluating the electronic energy for any possible configuration of the nuclei of the system with an error of less than 1 kcal/mol. Even the most exhaustive PES calculations are limited to a few hundred of points for given configurations of the nuclei of the system and hence some kind of analytical representation of the relevant PES is required. One such analytical functional form must in principle be able to [11] describe correctly the asymptotic regions, match as well as possible the experimental data available in the strong interaction region, reflect the symmetry of the problem and behave reasonably in unexplored regions, while connecting smoothly the asymptotes with the strong interaction region.

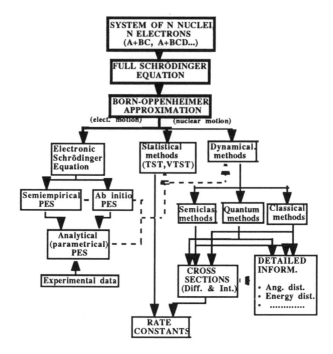

Figure 1. Schematic representation of the procedure and methods available for dynamical description of a system.

Many body expansions, first introduced by Murrell et al. [12] are possible choices for analytical functions fulfilling the abovementioned requirements. Following Murrell's scheme, the potential is written as:

$$V = \sum_{i=1}^{N} V_A^{(i)} + \sum_{i=1}^{\binom{N}{2}} V_{AB}^{(i)} (R_i) + \sum_{i=1}^{\binom{N}{3}} V_{ABC}^{(i)} (R_l, R_j, R_k)....+V_{ABC..N}^{(i)}(R_l,...) \qquad (4)$$

where each term represents the contribution of the 1,2...N-body interaction. The interaction terms must be well-behaved, that is, for instance the three-body term $V_{ABC}^{(i)} (R_l, R_j, R_k)$ must tend to zero for sufficiently large R_l, R_j or R_k. Expression (4) is useful only if simple expressions can be found for the different interaction terms. In our studies, the Sorbie-Murrell N-body expansion [13] has been employed. For a triatomic system (N=3), the diatomic terms are fitted to Extended-Rydberg curves employing the available ab initio or spectroscopic data:

$$V_{AB}^{(i)} = V_{ER(i)} = - De_{(i)} (1 + \sum_{j=1}^{3} a_j \rho_i^j) exp(-a_1 \rho_i), \qquad \rho_i = R_i - R_i^{eq} \qquad (5)$$

with R_i^{eq} the equilibrium bond distance of the diatomic molecule i, $De_{(i)}$ its dissociation energy and a_j undetermined coefficients which have to be optimized to reproduce the curve of the diatomic molecule i. The triatomic term is taken as:

$$V^{(j)} = P(R_1, R_2, R_3) T(R_1, R_2, R_3) \qquad (6)$$

where:

$$T(R_1, R_2, R_3) = \prod_{i=1}^{3} (1 - tanh(\frac{\gamma_i \rho_i}{2})) \qquad (7)$$

and:

$$P(R_1, R_2, R_3) = V^0 (1 + \sum_{i=1}^{3} c_i \rho_i + \sum_{i=1}^{3} \sum_{j \geq i}^{3} c_{ij} \rho_i \rho_j + \sum_{i=1}^{3} \sum_{j \geq i}^{3} \sum_{k \geq j}^{3} c_{ijk} \rho_i \rho_j \rho_k) \qquad (8)$$

with $\rho_i = R_i - R_i^\circ$ the displacements from a reference structure R_i° and γ_i and c_i, c_{ij}, c_{ijk} adjustable parameters. The range function (7) ensures that the triatomic term will become zero whenever any of the three displacements from the reference structure becomes sufficiently large. In case of symmetric systems (such as NO_2), the ρ_i coordinates are substituted by symmetry-adapted coordinates S_i built by applying an orthogonal transformation matrix to ρ_i. It is worth noting that, even though Sorbie-Murrell expansions furnish a very flexible functional form to reproduce all accidents on a PES while giving a correct and accurate asymptotic behaviour, they are usually difficult to fit and may give rise to spurious accidents in unexplored regions of the surface.

2.3 TRANSITION STATE THEORY

Once the analytical PES of the system has been obtained, it can be used to obtain dynamical properties. One possible choice (see fig. 1) is to use statistical methods such as the Transition State Theory (TST) [14] or the more refined Variational Transition State Theory (VTST) [15]. In both approaches the basic underlying idea is to consider a dividing surface in phase space separating reactants from products function only of the coordinates. It is assumed that there exists equilibrium between the supermolecules in the configuration corresponding to the dividing surface and those of reactants. All molecules crossing that surface in the forward direction are considered to lead to reaction (hence recrossing is not allowed). In TST the dividing surface is placed at the saddle point, whereas in VTST its location is optimized variationally to obtain the minimum reactive flux. Quantum corrections are subsequently introduced in both approaches to account for purely quantum effects such as tunneling. Even though these methods are quite cheap and require in general only limited information about the PES, they provide usually only rate coefficients. Thus, detailed dynamical properties such as cross sections or product vibrational distributions are not obtained with these approaches.

2.4 CLASSICAL TRAJECTORY METHODS

The so called dynamical methods are based upon solving the nuclear SE either classically, quantum mechanically or combining both quantum and classical mechanics. In the classical methods, the nuclei are treated as classical particles moving under an electronic potential V. This approximation is most valid [16] in the limit of high energies, for heavy nuclei and when dealing with sufficiently averaged dynamical properties (such as rate constants). For a generic A + BC system, the classical hamiltonian is a function of 9 coordinates and conjugate momenta, which after transformation to Jacobi generalized coordinates Q_i and momenta P_i and subsequent elimination of the center of mass motion give rise to 12 differential equations plus energy and angular momentum conservation [16]:

$$\dot{P}_i = -\frac{\partial H}{\partial Q_i} = -\frac{\partial V}{\partial Q_i} = \sum_{K=1}^{3} \left(\frac{\partial V}{\partial R_K} \right) \left(\frac{\partial R_K}{\partial Q_i} \right) \tag{9}$$

$$\dot{Q}_i = \frac{\partial H}{\partial P_i} = \frac{\partial T}{\partial P_i} \qquad i=1...6 \tag{10}$$

The integration of equations (9) and (10) for a given set of initial conditions defines a trajectory, which may be either reactive or non-reactive. Dynamical properties are evaluated employing a crude Monte Carlo approach, following which the integral of a given property depending on a multiparametrical function $f(\vec{\beta})$:

$$I = \int_0^1 \int_0^1 ....\int_0^1 f(\vec{\beta}) \, d\vec{\beta} \tag{11}$$

can be replaced by N evaluations (where N is the total number of trajectories run) of the function f with the elements of $\vec{\beta}$ chosen randomly from a uniform distribution:

$$I \approx \varepsilon(I) = \frac{1}{N} \sum_{i=1}^{N} f(\beta_1^{[i]}, \beta_2^{[i]}, ... \beta_k^{[i]}) \tag{12}$$

and the error in I corresponding to one mean standard deviation is:

$$\Delta\varepsilon(I) \approx \frac{N_R}{N} \left(\frac{N - N_R}{N_R \, N} \right)^{\frac{1}{2}} \tag{13}$$

where N_R is the number of reactive trajectories.

The set of 6 parameters determining a trajectory is shown in figure 2. The main drawback of the Quasiclassical Trajectory (QCT) method is the impossibility of accounting for purely quantum effects, such as tunneling and zero point energy at the saddle point. In some systems this may bring about an important source of error. Besides, quantization of the rovibrational energy of products has to be introduced after performing trajectory analysis by using a generalized momenta [18] or histogrammic binning method [17] according to which classical vibrational and rotational energies are assigned to quantum product vibrational and rotational states.

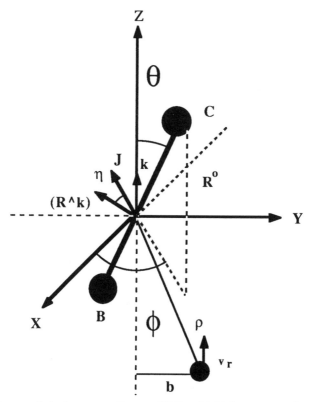

Figure 2 Reference framework in the center of mass of BC and initial parameters determining a quasiclassical trajectory. All symbols bear their usual meaning [17].

In spite of its drawbacks the QCT method provides an easy way of following the detailed reaction mechanism as well as of obtaining the most relevant dynamical properties, such as Integral (σ_R) and state-to-state($\sigma_{vj \to v'j'}$) cross sections, angle-dependent (σ^{γ}_v) and differential ($d\sigma_v/d\Omega$) cross sections, opacity functions (P_R), vibrational distributions in products ($P(v')$) and energy distributions in products. The detailed expressions for these properties within the QCT method can be found elsewhere [16]. Integrating cross sections over the relevant relative translational energy distribution at each temperature and summing over the corresponding rovibrational distributions yields rate coefficients:

$$k(T) = (8k_B T/\pi\mu)^{1/2} (k_B T)^{-2} \sum_v \sum_j g_j \frac{(2j+1)}{Q_{vj}} e^{-\varepsilon_{BC(v,j)}/k_B T} \int_0^\infty E_T \sigma_R(E_T,v,j) e^{-E_T/k_B T} \, dE_T$$

(14)

where g_j is the nuclear spin degeneracy factor, Q_{vj} is the internal rovibrational partition function, $e_{BC(v,j)}$ is the rovibrational energy of the diatomic molecule BC in the (v,j) internal state, $(2j+1)$ is the rotational degeneracy and $\sigma_R(E_T,v,j)$ is the reactive cross section for relative energy E_T with the BC molecule in the internal state (v,j).

2.5 APPROXIMATE QUANTUM METHODS

Another kind of dynamical methods is that based on the quantum description of the nuclear motion. To solve the corresponding nuclear SE, the total wave function is expanded in terms of eigenfunctions of the squared total angular momentum $\hat{J}^2$ and its projection on the Z axis $\hat{J}_z$. Subsequent substitution in the SE gives a set of coupled equations know as the Close Coupling (CC) equations [19]. Although rapidly evolving, at the present stage the solution of such a system of equations is impractical for moderately involved, realistic systems like those of the present study. Therefore there has been considerable effort to develop realistic approximate methods that allow these problems to be tractable. A scheme of the approximate 3D quantum reactive scattering methods is shown in figure 3.

On top of the hierarchy of approximate methods stands the Centrifugal Sudden (CS), j_z-conserving approximation [20]. In this approach it is assumed that the collision is dominated by the electrostatic potential and the molecular rotation, so that the orbital angular momentum operator $\hat{l}^2$ appearing in the nuclear SE can be substituted by a mean value $l(l+1)$. This implies that the Coriolis interaction term in the hamiltonian becomes diagonal.

Another approach, complementary to the former, is the Energy Sudden (ES) l_z-conserving approximation [21]. In this case it is assumed that the vibrational period is much shorter than the rotational one, so that the BC molecule remains frozen during collision (sudden rotational approximation). This results in the rotational angular momentum operator in the hamiltonian to be replaced by a mean value $j(j+1)$.

The R-IOSA approximation implies simultaneous application of the CS and ES approximations [22]. In this case, the R-IOSA Schrödinger equation becomes:

$$\left[\frac{\partial^2}{\partial R_\lambda^2} + \frac{\partial^2}{\partial r_\lambda^2} - \frac{\bar{l}_\lambda(\bar{l}_\lambda+1)}{R_\lambda^2} - \frac{\bar{j}_\lambda(\bar{j}_\lambda+1)}{r_\lambda^2} \right] \psi_{\overline{jl}}^\lambda(R_\lambda,r_\lambda,\gamma_\lambda) =$$

$$\frac{2\mu}{\hbar^2} \left[V^\lambda(R_\lambda,r_\lambda,\gamma_\lambda) - E \right] \psi_{\overline{jl}}^\lambda(R_\lambda,r_\lambda,\gamma_\lambda) \tag{15}$$

so that now all eigenfunctions of j and l in arrangement channel λ become decoupled and the wave function depends only parametrically on the collision angle γ_λ (so it is constant in every calculation). In this method it is necessary to define a matching surface between arangement channel λ and product channel v, as well as the way their internal and orbital angular momenta transform into each other.

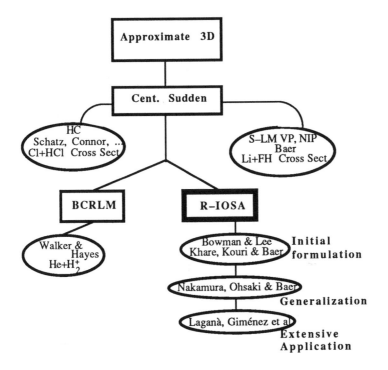

Figure 3 Some approximate methods (recent implementations) in quantum reactive scattering. Legenda: HC: Hyperspherical coordinates; S–LM VP: S–matrix Lagrange Multiplier Variational Principle; NIP: Negative Imaginary Potential; BCRLM: Bending Corrected Rotating Linear Model; R–IOSA: Reactive Infinite Order Sudden Approximation. Attached to each method are the main authors and systems studied.

In this method, provided it is properly used, it is possible to obtain all relevant dynamical magnitudes once the S matrix has been obtained by means of the usual assymptotic analysis. The R-IOSA method, employed in some of the calculations reported in this contribution, furnishes thus a cheap way of observing quantum effects in heavy and exothermic systems, while being a good approximation in the case of isotropic potentials, high energies or heavy atom exchange. Note however that because of the nature of the method, no reorientation or l-j coupling can be accounted for within this approximation.

3. The N + O₂ system

3.1 INTRODUCTION

The reaction of ground state nitrogen atoms and molecular oxygen:

$$N(^4S) + O_2(X^3\Sigma_g^-) \rightarrow NO(X^2\Pi) + O(^3P) \quad \Delta H^\circ_{298} = -31.82 \text{ kcal/mol}$$

may populate NO vibrational levels up to v' = 7 at room temperature and so, besides the considerations put forward in section 1, provides also a good opportunity of studying a system with high vibrational excitation of products.

The available experimental information about this systems consists both in rate constant measurements [23-30] and product vibrational distributions [31-35]. As for the rate constants, a summary of the most relevant results obtained with different methods is depicted in table 1.

TABLE 1. Rate constants for reaction (1) fitted to to the expression $k(T)=AT^m exp(-B/T)$ cm^3/mol s.

$k(T)$ (cm^3/mol s)	E_a (kcal/mol)	T (K)	A (cm^3/mol sK^m)	m	B (K)	Ref.
		1170-1530	3.8(13)		4022	23
	6.2	394-516	2.0(12)		3117	24
5.0(7)		350				25
	7.9		1.4(13)		3972	26
			3.0(11)	1/2	3590	26
5.3(7)		297				27
		300	8.3(12)		3569	28
		300-3000	6.4(9)	1	3150	29

Most rate coefficient determinations give a reasonable Arrhenian behaviour over an extended T range, resulting in an activation energy between 6 and 8 kcal/mol.

As for the vibrational distribution in products, the LIF and Infrared Chemiluminiscence techniques all point towards declining P(v') shapes and fractions of 0.079 and 0.22 of the available energy (E_{av}) in vibration respectively. By contrast the, in principle, more accurate Multiphoton Resonant Ionization (MPI) technique seems to indicate an alternating P(v') distribution, with even levels more populated than odd ones, and a fraction of 0.34 of E_{av} in vibration.

3.2 POTENTIAL ENERGY SURFACE

Ground state of reactants $N(^4S)+O_2(X^3\Sigma_g^-)$ and products $NO(X^2\Pi)+O(^3P)$ become correlated in C_{2v} symmetry through two surfaces of B_2 symmetry: $^2B_2$ and $^4B_2$. On descending to C_s symmetry, both surfaces become of A' symmetry, and so does the deep $^2A_1$ minimum corresponding to ground state NO_2. Thus, the ground state of the stable NO_2 molecule can be accessible on the doublet surface through C_s configurations [36]. A thorough description of this reaction will also have to take into account reactivity occuring on both the doublet and quadruplet surfaces.

The ab initio information available for the description of the surfaces of this system is rather scarce. Except for a DIM calculation [37] most other studies [38-40] indicate that reaction occurs mainly on the lowest $^2A'$ PES. The best set of calculations available at the time our PES was built [40] pointed out towards a barrier of 18 kcal/mol on the $^4A'$ surface and about 10.2 kcal/mol on the $^2A'$ one. To match the experimental information available, all points on the lower PES were rescaled by a constant shift so that the barrier at the saddle point was 7.5 kcal/mol. The resulting minimum energy reaction path is depicted in figure 4.

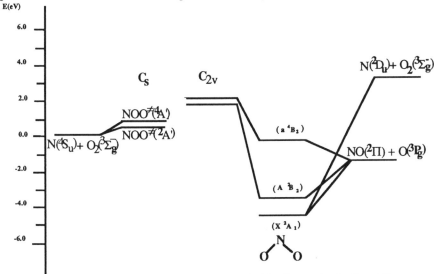

Figure 4. Minimum energy reaction path diagram for the $N + O_2$ reaction in C_s and C_{2v} approaches.

The calculations also indicate that the insertion of N into O_2 to give NO_2 must overcome a high barrier (ca. 2 eV) and that the saddle point for both the $^2A'$ and $^4A'$ PES is bent. According to the barrier heights indicated above, only the $^2A'$ PES is of relevance below 2500 K (or 0.7 eV relative energy). In this contribution we present the results obtained on this lower PES, even though work is in progress to consider the quadruplet PES as well. The fitted Sorbie-Murrell PES [41] shows a nice reproduction of the ab initio data, even though, as it will be shown below, it is still a bit too repulsive in the long range part and includes a deep $D_{\infty h}$ ONO minimum, accessible over a large barrier to insertion.

The bending potential about the saddle point for the fitted PES is presented in figure 5, including the vibrational energy levels of the O_2 molecule. As the NOO angle closes up or opens from its saddle point value (114°), the potential energy rises steeply. This fact has important consequences in what respects the reaction window at each energy.

3.3 REACTION DYNAMICS WITH O_2 IN THE GROUND AND LOW EXCITED VIBRATIONAL LEVELS

3.3.1 Reaction mechanism

The analysis of selected trajectories point towards a direct mechanism over the whole energy range and rovibrational conditions explored [41,42,44]. This means that usually O_2 vibration is not influenced until N comes close to NO equilibrium distance (sudden approach of N to O_2). Only about 3 % of the trajectories at the highest E_T considered evolve through indirect or complex mechanisms. At high energies a small number of trajectories explore the

64

$D_{\infty h}$ minimum following an indirect mechanism. It is also found that most trajectories pass through configurations close to that of the saddle point.

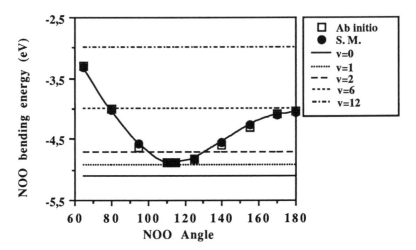

Figure 5. Ab initio and fitted bending potential on the analytical PES with R_{NO} and R_{OO} fixed at the ab initio saddle point values. The straight lines show the energies of different vibrational levels of the asymptotic O_2 molecule.

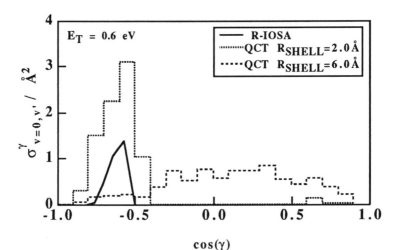

Figure 6 Jacobi fixed-angle partial cross-section. The R-IOSA results are superimposed on the QCT ones, drawn at different R_{SHELL} distances (see text).

To ascertain to what extent the shape of the bending potential at the saddle point influences reactivity, the angle-dependent QCT reaction cross section has been plotted for several distances, hereafter called R_{SHELL}, between N and the center of mass of O_2, together with the corresponding R-IOSA magnitude. As seen in figure 6, there is a clear difference between the fixed-angle cross section at the beginning of each trajectory (R_{SHELL} = 6.0 Å) and that at R_{SHELL} = 2.0 Å (the distance between N and the center of mass of O_2 at the saddle point): whereas the initial angular distribution leading to reaction is spread over a wide angular range, it becomes quite constrained on approaching the saddle point configuration, so that the angular window open for reaction corresponds roughly to the range of NOO angles for which the bending potential energy lies below the relative translational energy E_T. The results from the R-IOSA calculation are very similar qualitatively to those obtained from trajectories. This confirms the applicability of this method for this system. The changes in shape, from a broad distribution at the beginning of each trajectory to a constrained one are due to the combined effect of the anisotropy of the potential (which tends to shift the system to configurations liable to lead to reaction) and the geometrical changes occuring during the evolution of the system to products. It is also found that the shape of the fixed angle cross section, σ^γ_v, changes very little with rotational excitation of O_2 or on considering low or moderate vibrational excitation of reactants.

3.3.2 The effect of translation, rotation and low vibrational excitation on reactivity
The effect of translational energy on reaction cross section is shown in figure 7. From the QCT and R-IOSA results, a threshold energy of 0.52 eV is found [41-44], above the theoretical one of 0.34 eV. This is attributed to the excessively repulsive character of the present PES in the long R_{NO} distance range.

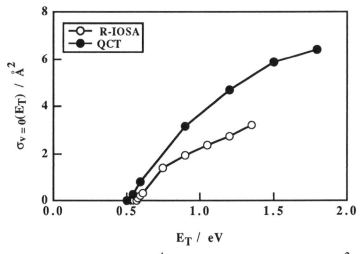

Figure 7 Integral cross sections for the N(^{4}S) + O_2 system reacting on the ground ^{2}A' PES with O_2 in the most populated rovibrational level at 300 K.

The qualitative form of both QCT and R-IOSA excitation functions is very similar and is typical for reactions presenting a barrier to reaction. The lower reactivity in the case of the R-IOSA calculation is attributed to the impossibility for this method to account for reorientation happening along the collision process. On the other hand, the plateau attained with growing E_T can be explained by taking into consideration the opening of the reaction window with energy: at low E_T only a very limited set of NOO configurations lead to reaction. With growing E_T, the reaction window expands both to NOO configurations more open and more bent than those of the saddle point, hence the rise in σ_R. However (see fig. 5), the bending potential rises in a less steep way on the side of open NOO angles, so that up from a certain energy all configurations from that of the saddle point up to 180 ° are open for reaction. From that energy on, the reaction window can only expand in the more repulsive, bent NOO angle range, causing the rise of σ_R with E_T to stall [41,44]. A similar behaviour of σ_R with E_T has also been observed in other studies on this system [2,45]

It is possible to define the average increment $\Delta\sigma_R/\Delta E_T$ with fixed v and j to quantify the enhancing effect of translational energy on reactivity [42]. For this system, values ranging from 5.9 to 4.2 Å^2/eV are found, depending on the initial rovibrational state. The analogous average for rotation ($\Delta\sigma_R/\Delta E_{rot}$ with E_T and v fixed), with successive increments of four j gives values of -15.62, -3.19, 2.79 and 1.64 Å^2/eV. For vibration ($\Delta\sigma_R/\Delta E_{vib}$ with E_T and j fixed), increasing v from 0 to 1 and 2 results in averages of 0.4 and 0.7 Å^2/eV. Thus it can be seen that translational energy is much more effective than low vibrational energy and rotational energy in promoting reactivity. This is consequent with the shape of the potential presented in figure 5: at low v values, the contribution of vibrational energy to the total accessible energy is still very low, and also with the early character of the saddle point, which implies that translational energy is more effective than internal one in promoting reactivity. In fact, vibrational enhancement implies for low v, motion perpendicular to the coordinate leading the system to products and causes it to explore the repulsive walls of the potential.

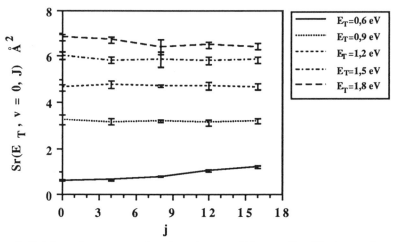

Figure 8 Total reaction cross section dependence on J for reaction (1) at several translational energies.

As it can be seen in figure 8, the system exhibits a remarkably isotropic behaviour with rotational excitation of the reactant molecule [42]. Even though the effect of j on cross section may be important in absolute terms at the lowest E_T considered, the curves for higher relative energies run much more smoothly, with only minor rising or decreasing trends. If the rotational period of the O_2 molecule is compared with the period it takes for N to reach the strong interaction zone, it is seen that except for the highest J and lowest E_T

used in our calcultions, reaction takes place almost frozen rotationally. Therefore, rotation should have a hindering effect if any, something which is not oberved in the present calculations. We interpret this fact as being due to the anisotropy of the potential being able to shift the system down to configurations leading to reaction even if it starts with rather unfavorable configurations.

Low vibrational excitation of the reactant molecule causes only a minor effect on σ_R, consequently with the early character of the barrier [46]. This is shown in figure 9. However, at the lowest E_T explored, vibrational energy (0.29 and 0.49 eV for v=1, 2 respectively) constitutes a non-negligible part of the total E_{av}, so that the enhancing effect is more important in these cases.

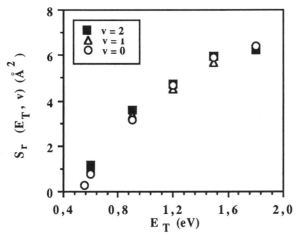

Figure 9 Total reaction cross section for reaction (1) at several initial vibrational states.

3.3.3 Energy disposal in products

The way E_{av} is distributed among the different energy modes of products has been analysed in terms of the average fractions of E_{av} appearing as vibration, rotation and translation, denoted by $\langle f_v' \rangle$, $\langle f_r' \rangle$ and $\langle f_T' \rangle$. respectively, and also the average increments produced for each type of energy between products $\langle E_i' \rangle$ and reactants E_i, $\delta E_i = \langle E_i' \rangle - E_i$ [42]. It is found that product molecules are produced usually with high translational excitation, with $\langle f_T' \rangle$ ranging between 0.70 and 0.52 depending on the initial rovibrational state considered. This fraction does not show a clear-cut behaviour with rotational or vibrational excitation of reactants. As for $\langle f_v' \rangle$, the behaviour with rotational excitation is not clear, while showing a slight decrease followed by an increase with E_T (moving usually between 0.22 and 0.14) and an overall increment with v. Perhaps the most interesting effect is that concerning $\langle f_r' \rangle$, since it shows a clear rise with E_T (between 0.10 and 0.15 units). The behaviour of this last indicator with rovibrational excitation of reactants is not clear.

The moderately exothermic nature of this reaction causes the effect of the different energy modes to be somewhat eclipsed when the analysis is performed in terms of fractions. In figure 10 we present the results for the increments in the different energy modes as a function of the translational energy.

The most interesting feature found in this analysis is the strong decrease in δE_T with translational energy (from 1.13-0.77 eV at 0.60 eV to 0.05-0.13 eV at 1.80 eV) so that reaction becomes almost translationally adiabatic. Parallely, there is a strong rise in δE_r (which becomes 2 to 3 times greater at 1.8 eV than at 0.60 eV) and a not so steep increase in δE_v. It is also worth noting that there is a remarkable decrease of δE_v with v at fixed E_T and j, so that reaction becomes almost vibrationally adiabatic for v=2, at the same time that there is a clear rise of δE_T with v, especially from v=0 to v=1. The progressive increase of energy being channeled into internal modes of products is consequent with the reaction mechanism outlined above: at low E_T a limited range of NOO configurations lead to reaction and the mechanism is direct.

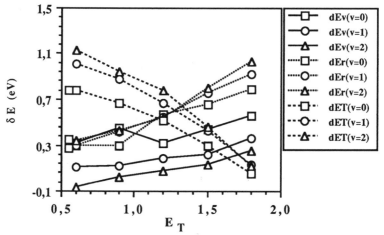

Figure 10 Increments in the different energy modes as a function of translational energy for reaction (1).

At higher energies, an increasing number of open NOO angle configurations, leading to vibrational excitation of products, and also of bent NOO angle configurations, resulting in rotational excitation of products, are open for reaction, so that internal energy of the product NO molecules is increased at the expense of the translational one. Besides, the energy of the v=2 rovibrational level of O_2 is close to the barrier for reaction (see fig. 5), which could contribute to coupling between the stretching of O_2 and that of the NOO transition state, bringing the system over the barrier to products.

3.3.4 Product vibrational distributions

An interesting behaviour is found when dealing with the vibrational distribution obtained from reaction (1). It can be seen that the distribution produced at low E_T either from trajectories [41,42] or from the R-IOSA [43] method presents a strongly nonstatistical behaviour, with a fraction of energy in product vibration of ca. 0.20 (to be compared with experimental values of 0.22 [34] and 0.34 [35]). This behaviour tends to disappear at high E_T, where a distribution close to the statistical one is found. Because of the high reaction energy threshold and the Boltzmann distribution at 300 K, the experimental vibrational distribution can be very well represented by the QCT and R-IOSA results at 0.60 eV. The different theoretical and experimental results are given in figure 11.

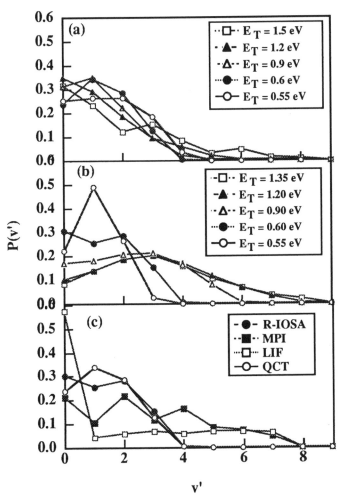

Figure 11 Product vibrational distributions for the NO molecules produced in the $N(^4S) + O_2$ reaction at several initial translational energies with O_2 in the most populated rovibrational level at 300 K. a) QCT results, b) R-IOSA results, c) Comparison between theoretical (E_T=0.60 eV) and experimental results.

There is an important difference between quantum and QCT data in the low translational energy range: whereas the quantum results show an alternating shape with v'=2 more populated than v'=1, the classical data show only an inversion peaked at v'=1. At higher energies the differences are maintained, since R-IOSA data tend to give vibrational inversions at even larger v' while QCT distributions become roughly decreasing at higher E_T. It is also found that the inversion obtained in the QCT data at low E_T smooths out when excited vibrational or rotational states of O_2 are considered. From figure 11(c) it is seen that there is a semiquantitative agreement between the MPI P(v') distribution and that arising from the R-IOSA calculations.

Note also that the analysis of the QCT vibrotational and translational product energy distributions indicates that NO molecules are produced rather translationally hot and rotationally cool, in agreement with the data outlined above.

3.4 REACTION DYNAMICS WITH O_2 IN HIGHLY EXCITED VIBRATIONAL LEVELS

3.4.1 Derivation of a new $N + O_2$ potential energy surface

Even though the PES used for the calculation reported above is capable of giving a reasonable description of the system, it still shows some weaknesses, such as the impossibility of reproducing the P(v') distribution or the energy threshold being too high.

There has been recently considerable work devoted to improving the PES of this system, either by carrying out better ab initio points or by reoptimizing the previously existing PES. Apart from the work of Clementi et al. [47], for which only VTST data are so far available, it is worth mentioning the reoptimization of the data of Ref. 40 carried out by Duff et al. [2]. These authors carried out a fitting of the $^4A'$ PES also. However, their rate constants are still too low as compared to the experimental ones. We have also very recently carried out a reoptimization of our PES by following a scheme in which the barrier of the $^2A'$ PES was lowered and the resulting Sorbie-Murrell PES combined with the $^4A'$ one of Duff et al. [2] to give rate constants. These were afterwards compared to the experimental data available. We were finally able to come to an optimized PES which, combined with the quadruplet one, was able to reproduce quantitatively the experimental rate constant dependence on T (c.f. table 1) over the whole temperature range. The optimal Sorbie-Murrell coefficients are given in table 2.

TABLE 2. Three-body parameters of the analytical $^2A'$ PES:V_o in eV, c_i in Å^{-1}, c_{ij} in Å^{-2}, c_{ijk} in Å^{-3}, and g_i in Å^{-1}. Reference structure and symmetry coordinates like in Ref. 41

V_o	1.678	c_{111}	-0.499	c_{1111}	0.229	c_{2222}	21.447
c_1	-0.780	c_{112}	4.443	c_{1112}	0.811	c_{2223}	0.000
c_2	3.640	c_{113}	0.000	c_{1113}	0.000	c_{2233}	6.088
c_3	0.000	c_{122}	-1.059	c_{1122}	15.424	c_{2333}	0.000
c_{11}	1.419	c_{123}	0.000	c_{1123}	0.000	c_{3333}	3.787
c_{12}	-2.596	c_{133}	0.856	c_{1133}	-2.720	γ_1	2.539
c_{13}	0.000	c_{222}	42.974	c_{1222}	10.261	γ_2	7.308
c_{22}	15.266	c_{223}	0.000	c_{1223}	0.000	γ_3	0.0
c_{23}	0.000	c_{233}	0.534	c_{1233}	7.707		
c_{33}	-0.304	c_{333}	0.000	c_{1333}	0.000		

3.4.2 Preliminary dynamical results

Even though this PES reproduces the expected threshold energy, it is seen to lead to $\langle f'_v \rangle$ values between 0.44 and 0.64, quite below the ones found experimentally and in our previous calculations, while giving rise to translationally cool products. A plot of the effect of v on reactivity for several initial E_T is given in figure 12. It is seen that vibrational energy shows an enhancing effect which is more evident for low E_T. Even though the reaction occurs with an early barrier, increasing vibrational energy up to v=18 implies that most of the available energy for reaction comes from vibration, and hence the similarity of σ_R at high E_T and v.

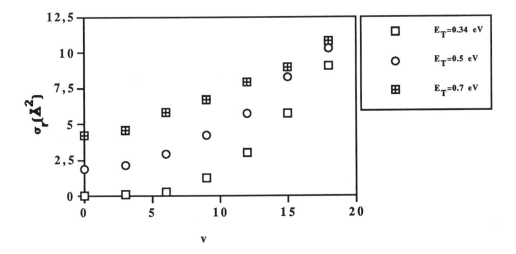

Figure 12 Total QCT reactive cross sections as a function of v for different E_T on the reoptimized $^2A'$ PES.

The analysis of the enhancing effect of the different energy modes on reactivity, performed in terms of $\Delta\sigma_R/\Delta E_i$ yields results similar to those using the previous PES: translational energy is the most effective mode, in agreement with the early character of the saddle point. As deduced from fig. 12, $\Delta\sigma_R/\Delta E_T$ is important for low v and decreases as v is increased. It is worth noting however that it lies always above $\Delta\sigma_R/\Delta E_v$, even for the highest v considered. Rotation has only a moderate, enhancing effect for the energy (0.34 eV) for which calculations have so far been performed. Work is in progress [48] on this system using this new PES and will be reported in the near future.

4. The N + NO system

4.1 INTRODUCTION

The reaction of ground state nitrogen atoms and nitric oxide:

$$N(^4S) + NO(X^2\Pi) \to N_2(X^1\Sigma_g^+) + O(^3P) \quad \Delta H°_{298} = -74.95 \text{ kcal/mol}$$

despite its intrinsic interest, has been subject of relatively few experimental and theoretical studies. The experimental rate coefficients available from the literature [49-57] are shown in table 3.

TABLE 3. Rate coefficients for reaction (2) fitted to to the expression $k(T)=A\exp(-B/T)$ cm^3/mol s.

$k(T)$ (cm^3/mol s)	E_a (kcal/mol)	T (K)	A (cm^3/mol sK^m)	B (K)	Ref.
<5(13)		300			49
		476-755	3.0(13)	200	50
1.0(13)		300			51
1.3(13)		300			52
	0.82	298-670	4.9(13)	410	53
2.0(13)		196-400			54
1.3(13)	No Ea	1600-2285			55
2.1(13)		196-3150	6.4(9)	3150	56
1.6(13)		300-5000			29

Almost all experimental data available seem to indicate that this reaction is fast and proceeds with no or very little activation energy over the whole energy range, the highest experimental value being 0.82 kcal/mol [50]. The other only experimental information available about this system consists in estimates of the fraction of exothermicity going into product vibration. The values given in the literature are 0.28 ± 0.07 [57] and 0.25 ± 0.03 [58] respectively.

4.2 POTENTIAL ENERGY SURFACE

In $C_{\infty V}$ both ground states of reactants $N(^4S) + NO(^2\Pi)$ and products $N_2(^1\Sigma_g^+) + O(^3P)$ are correlated through a $d^3\Pi$ surface which, on descending to C_s symmetry, gives rise to two $^3A'$ and $^3A''$ PESs. In contrast to what happens in the case of $N + O_2$, the ground state of nitrous oxide $(X^1\Sigma^+)$ correlates with $N(^2D) + NO(^2\Pi)$ and $N_2(^1\Sigma_g^+) + O(^1D)$ on a singlet PES an can only connect with reactants or products through nonadiabatic transitions from this singlet surface to the fundamental triple one. Assuming these type of processes to be of little inportance at low temperatures and default of any detailed information about them, it is reasonable to consider that reactivity can be well described by taking into account the $^3A''$ and $^3A'$ PESs only, even though the adequate weighing factor will have to be taken into account when giving global results.

Structural information about these surfaces is very scarce. To the best of our knowledge, the only calculation concerning this system is that of Ref. 40, pointing towards a bent saddle point in both cases and barriers of only 0.5 kcal/mol and 14.7 kcal/mol for the lower $^3A''$ and upper $^3A'$ PESs respectively. The very low barrier on the lower PES is in good agreement with experimental observations, and given the difference in barriers between both surfaces, it is reasonable to assume that at room and moderately high temperatures only the

[3]A" PES need be considered. These data, together with the N_2O states calculations, were employed to build the minimum energy reaction path shown in figure 13.

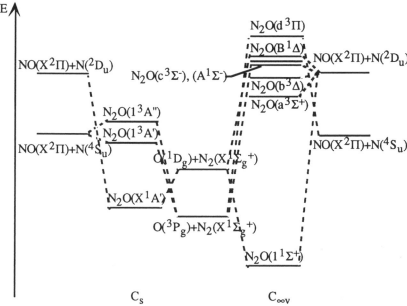

Figure 13. Minimum energy reaction path diagram for the N + NO reaction in C_s and $C_{\infty V}$ symmetries.

The fitted Sorbie-Murrell PES [59] shows no barrier to reaction. The saddle point is placed 0.54 kcal/mol below the ab initio estimate. As in the case of $N + O_2$, the surface shows a strong repulsive behaviour as it goes away from the minimum energy reaction path. As in the case of the $N + O_2$ reaction, the repulsive behaviour of the bending potential about the saddle point determines the way reaction takes place [59-61, 44]. Figure 14 gives a polar contour plot of the PES for this system. It can be seen how only a narrow scope of NNO angles allow N approach to NO at low E_T to lead eventually to reaction.

4.3 DYNAMICAL RESULTS

4.3.1 Reaction mechanism

As in the case of reaction (1) the reaction mechanism observed for this system shows a marked direct and sudden character [60], irrespective of the initial relative translational energy and rovibrational state of NO. In fact no indirect trajectories have been observed over the whole translational and rovibrational energy range studied. At most, monitoring trajectories with strong initial rotational states of reactants has led to the conclusion that NO rotation is hindered along the trajectory, due to the strong anisotropic behaviour of this PES. The NO bond does not break until the N atom has come close to almost within N_2 equilibrium bond range that the NO bond breaks, implying repulsive energy release and translational excitation of products. The analysis of the QCT angle-dependent cross section shows also a behaviour similar to that for the other atmospheric system: even though the distribution of initial NNO Jacobi angles is scattered over a relatively wide angular range, the reaction window is drastically reduced on approaching the saddle point configuration, so that finally all reactive trajectories are concentrated in the range of NNO angular configurations with energy below E_T. The agreement with R-IOSA calculations is nice, especially at low relative energies. However, monitoring the NNO angle at R_{SHELL} above that of the saddle point indicates that, at low translational energies, it is still possible for the system to reorient itself on its way down to products. This possibility does not exist within the R-IOSA framework, causing important differences between quantum and classical calculations.

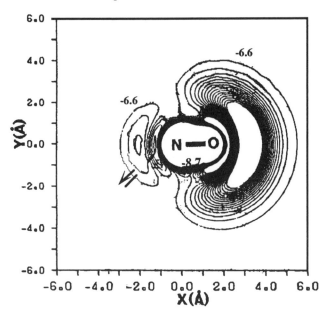

Figure 14 Equipotential potar contour plot for the $^3A''$ PES with NO fixed at its equilibrium bond distance. Contours are spaced by 0.3 eV. The equipotential contours increase in energy for collinear approach of N to the O or N side, but they decrease for angular approach to NO.

4.3.2 The effect of translation, rotation and vibrational excitation on reactivity

The effect of relative translational energy on reactivity for the most populated rovibrational level at 300 K is shown in figure 15. No energy threshold for reaction is found in either the QCT [59,60] or R-IOSA [61] calculation. On the contrary, reactivity is found even at the lowest E_T explored (0.01 and 0.0038 eV for QCT and R-IOSA respectively). The shape of the excitation function is remarkable, since as stated above this reaction proceeds with no barrier to reaction on the minimum energy reaction path. So the rising shape of σ_R, leading to a plateau at high E_T is caused by the shape of the bending potential around the saddle point: at low E_T only a limited number of NNO configurations lead to reaction, so that few trajectories can evolve to products. On increasing translational energy a greater amount of NNO configurations may lead to reaction since the angular window is expanded. This goes on until all NNO configurations with NNO angles more open than that of the saddle point are open for reaction. Up from that point as in the case of NOO, the reaction window can only expand in the more repulsive bent NNO range, causing the σ_R rise to stall.

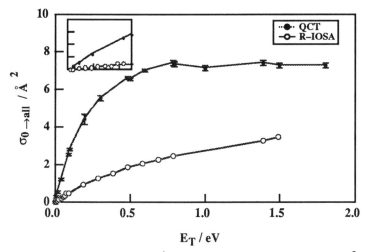

Figure 15 Integral cross sections for the N(^{4}S) + NO system reacting on the ground ^{3}A" PES with NO in the most populated rovibrational level at 300 K. The box shows the behaviour in the zone of the threshold.

As in the case of N + O$_2$, the R-IOSA curve falls quite below the QCT one. Again this effect is probably due to the impossibility for the R-IOSA method to lead to reorientation along the collision process. In general, it is also worth noting that this approximation considers only extreme internal angular momentum $\leftrightarrow$ orbital angular momentum transformations between reactants and products, which may introduce as well a partial source of error. As seen in figure 15, the QCT excitation function rises much more steeply than the R-IOSA one, especially for the lowest energies considered.

The semiquantitative analysis of the effectivity of the different energy modes in promoting reactivity has been carried out also in terms of the $\Delta\sigma_R/\Delta E_i$ averages. For the translational energy enhancement, $\Delta\sigma_R/\Delta E_T$ values between 17.08 and 12.02 Å^2/eV are found, depending on the initial rovibrational level considered. These data indicate a moderate enhancing effect of E$_T$, quite above that for rotational and vibrational excitation when NO is in low v states.

The behaviour with rotational excitation of reactants is rather striking, especially if the results for N + O$_2$ are taken into consideration. As j increases, there is a dramatic drop in reactivity for low E$_T$, with cross sections eventually becoming zero. This behaviour smooths out at high E$_T$ and it can be attributed to several factors: on the one hand rotational excitation of reactants usually means that it is much more likely for the system to shift away from the configurations that would lead it to reaction. At higher E$_T$, when a great number of configurations may lead to reaction this effect will in general be much less pronounced. This unfavourable effect may be compensated by if the PES is anisotropic enough to shift the system back to the preferred orientation required for reaction. A greater amount of rotational energy implies also that there is more energy available for reaction and may also therefore favor reactivity. The way σ_R will behave with j depends on which one of these factors is dominant . In the present case the orientation effect seems to dominate, to a great extent because of the narrow angular cone of acceptance for reaction, and cross section falls down with j at low transaltional energies.

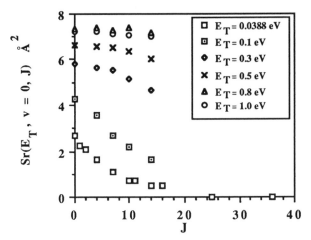

Figure 16 Evolution of the QCT total reaction cross section with j for E_T = 0.0388, 0.3, 0.5, 0.8 and 1.0 eV with NO in the v = 0 level.

As for translational energy, the average QCT $\Delta\sigma_R/\Delta E_{rot}$ has been evaluated to quantify the extent to which rotation affects reactivity. Average values of -1040, -236, -165, -21 and -13.7 Å^2/eV have been obtained for the passage from j = 0 to 1, 1 to 2, 2 to 4, 4 to 7, 7 to 10 and 10 to 14 respectively. When compared with $\Delta\sigma_R/\Delta E_T$, these results confirm our previous statement in the sense that rotation greatly hinders reactivity. This effect is maintained even when considering very high initial vibrational levels [62]. Since the available energy at high vibrational states implies that probably almost all NNO configurations are open for reaction, the rotational effect can hardly be attributed to orientational effects. Thus, reactivity is probably reduced as a result of vibration-rotation coupling in the high v regime.

The effect of low and high vibrational excitation on reactivity [60] is shown in figure 17. The results for low vibrational excitation are depicted in figure 17b, while those for high vibrational excitation are displayed in figure 17a. Low vibrational excitation does not cause a large effect on reactivity, the excitation function showing a behaviour very similar to that for the ground vibrational level. The associated enhancing effect $\Delta\sigma_R/\Delta E_v$ takes accordingly low values (1.94 and 1.07 Å^2/eV, to be compared with the much higher values for translational and rotational excitation). This behaviour is typical for systems with an early barrier showing weak vibration-translation coupling, as outlined above in the case of N + O$_2$. The increase in σ_R may be partly due to the increase in the maximum impact parameter b_{max} with vibrational excitation of reactants. Also as in the case of N + O$_2$, the enhancing effect is much more evident at low E_T since at these energies vibrational energy represents a much bigger contribution to the available energy than the translational one.

The effect of high translational excitation on reactivity is shown in figure 17a. The behaviour with vibrational excitation changes from that of a typical system with a barrier to reaction to that of a system showing no barrier and a very isotropic PES over the whole energy range.

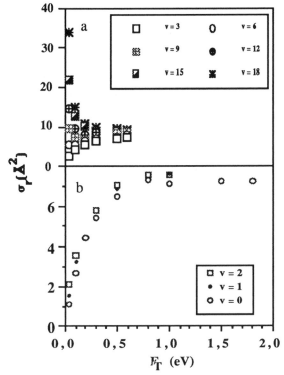

Figure 17 Excitation function for NO at several vibrational levels: the upper graph shows results for low vibrational levels, the lower graph presents data at high vibrational levels.

This evolution with growing v is very similar to that observed in the case of N + O_2, and, as in that case, the main reason for this behaviour is the fact that most excited rovibrational levels lie very high up in energy compared to the potential energy corresponding to the bending potential (see figure 5). Even though vibrational energy is much less effective than translational energy in overcoming the barrier to products because of its early nature, its contribution to E_{av} is overwhelming in the high v range as compared to E_T. In this case the reaction mechanism is no longer dominated by the fall in energy after the saddle point and the opening of the reaction window with E_T, but by the amount of energy in vibration. Almost all NNO configurations are open for reaction, and increasing translational energy causes the system to behave much like a hard sphere model, for which increasing E_T brings about a decrease in σ_R.

The preliminary analysis of the average enhancements for the different kinds of energy [62] with NO in high initial vibrational levels confirm our previous observations. For low v values, the $\Delta\sigma_R/\Delta E_T$ ratio is greater than the $\Delta\sigma_R/\Delta E_v$ one. On increasing v, the translational average decreases until it becomes negative for v = 12 (values of 8.36, 4.25, 1.46, -11.23, -23.57 and -43.69 $Å^2$/eV for j = 7 and v = 3, 6, 9, 12, 15 and 18 respectively), in agreement with what is observed in figure 17a. The vibrational average decreases also with growing E_T, going from 10.64 $Å^2$/eV at j = 7 and E_T = 0.0388 eV to 0.63 $Å^2$/eV for j = 7 and $E_T = 0.60$ eV. The behaviour of this indicator is consequent with the early character of the barrier to reaction: since at high E_T translational energy is high enough for a wide range of NNO configurations to be open for reaction, the enhancing effect of vibrational energy is felt much less than at low energies where the ability to overcome the barrier is much more linked to the amount of energy in vibration because of the small translational contribution. The rotational average $\Delta\sigma_R/\Delta E_{rot}$ has also been studied for a variety of energies and v conditions. It is always found that rotation has a hindering effect on cross section, and that this effect is much more evident at low E_T. It is also found that, roughly, rotation inhibits reaction more at high v (thus, for $E_T = 0.0388$ eV, $\Delta\sigma_R/\Delta E_{rot}$ is -85.4, -163.9, -164.2, -248.2, -419.2 and -536.7 $Å^2$/eV for v = 3, 6, 9, 12, 15 and 18 respectively) , confirming our previous considerations about vibration-rotation coupling preventing reactivity.

4.3.3 Product vibrational distributions
The product vibrational distributions obtained for reaction (2) within the QCT and R-IOSA methods are presented in figure 18

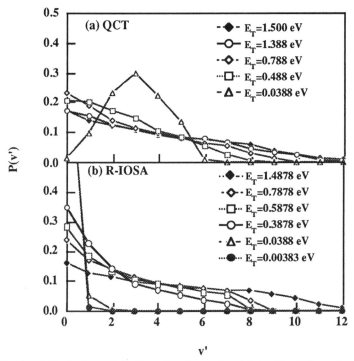

Figure 18 Vibrational distributions of the N_2 molecules produced in the N + NO reaction at several translational energies with NO in the v = 0, j = 7 rovibrational state. Figure 18a shows the QCT results and figure 18b the R-IOSA ones. The R-IOSA distributions for $E_T = 0.00388$ and 0.0388 eV show probabilities close to 1 for v' = 0.

The most striking difference between both kinds of distributions is the strong vibrational adiabaticity found for the R-IOSA one and the important vibrational inversion arising in the QCT results at low E_T. At high energies both types of calculations tend to converge, as it should be expected from the equivalence between quantum and classical mechanics at high energies. There is a strong correlation between NNO attack angle and the product vibrational distribution, with NNO open angles giving rise to strong vibrational inversions [60,61] and NNO bent angles resulting in low vibrational inversions. As stated above, there is the possibility for trajectories to reorientate after the saddle point, so that the final vibrational distribution is also determined by the angular window on falling to products. However, contrarily to QCT, the nature of the R-IOSA method precludes such possibility. The change from a decreasing product vibrational distribution to an inversion occurs over a relatively small NNO angular range. At low E_T only NNO angles near the saddle point contribute to reactivity within the R-IOSA framework (tunneling does not cause any important contribution). Those angles are not found to lead to vibrational inversion, hence the decreasing vibrational inversion found at low E_T. Consequent with this behaviour, the fraction $\langle f_v' \rangle$ is 0.30 and 0.05 for the QCT and R-IOSA calculations at 0.0388 eV respectively, to be compared with experimental values of 0.28 or 0.25.

4.3.4 Energy distribution in products
The analysis of the energy distribution in products has been carried out in terms of the fraction of energy $\langle f_i' \rangle$ in the different modes of products for all conditions explored and also by using the $\delta E_i = \langle E_i' \rangle - E_i$ increment.

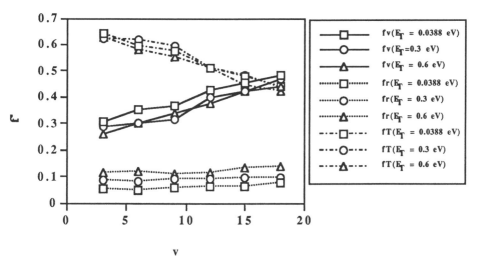

Figure 19 Fractions of the different energy modes as a function of initial vibrational state for reaction (2).

The overall differences in fractions of energy are, as in the case of $N + O_2$, hidden partly by the great exothermicity of the reaction, especially when dealing with low vibrotational levels. The most interesting feature is that $\langle f'_R \rangle$ increases clearly with E_T at fixed j and low v (almost twice between 0.0388 and 1.0 eV), while the other fractions do not show any marked tendency. When dealing with increments a clearer picture emerges, with δE_T showing a sharp decrease at E_T above 0.5 eV and δE_R becoming twice or three times bigger at E_T= 1.8 eV than at E_T = 0.0388 eV. As in the case of N + O_2, this behaviour can be rationalized by taking into account that at low E_T, the high exothermicity and sudden N + NO approach imply repulsive energy release,meaning translational excitation. Higher energies allow more open NNO configurations (leading to vibrational excitation) but also more bent NNO configurations (leading to rotational excitation) to be open for reaction, and hence internal energy contents increase.

The behaviour of the fractions of energy in products for high initial vibrational levels is similar to that observed for low initial v. The $\langle f'_v \rangle$ fraction ranges between 0.25 and 0.50, with the highest vibrational contents found for high initial v. Similarly, $\langle f'_R \rangle$ moves between 0.05 and 0.12 depending on the relative energy and initial vibrotational level considered. Translational excitation of products $\langle f'_T \rangle$ continues to be the most salient feature of the energy distribution, falling around 0.44 to 0.64.

Figure 19 presents the evolution of the different fractions of energy with v for several initial relative energies. The most relevant feature of the behaviour of $\langle f'_v \rangle$ is its increment with the initial vibrational level at fixed E_T and j, sometimes by more than 0.20. The behaviour of this fraction with j at fixed E_T and v gives almost a flat line, and only a very sligth decrease is observed when considering translational excitation at fixed v and j. As in the case of low vibrational excitation, $\langle f'_R \rangle$ increases considerably as E_T goes up (almost twice) for all initial v considered, while vibrational or rotational excitation of reactants don't cause a clear-cut behaviour to appear (at most a slightly upward behaviour with v and a small decrease with j at high v). Consequently with what has been outlined before, $\langle f'_T \rangle$ decreases clearly with vibrational excitation of reactants (cf. fig. 19), while the effect of this fraction with E_T and j is in most cases fluctuating.

4.3.5 Rate constant evaluation

The rate coefficients for reaction (2) occurring on the $^3A''$ PES are given in table 4 together with the experimental data. The most important result observed in this table is the very good agreement found for QCT, VTST and experimental determinations, whereas those from quantum calculations fall 24 to 7 times below the experimental, classical and statistical ones, in agreement with what has been outlined above.

It is expected that, once the upper $^3A'$ PES is considered, theroretical rate constants will be in even better agreement with experimental values.

TABLE 4. Comparison of rate constants for reaction (2) obtained with several methods.

T (K)	K(QCT) (cm³/mol s)	K(R-IOSA) (cm³/mol s)	K(VTST/SCSAG) (cm³/mol s)	K(Ref. 53) (cm³/mol s)	K(Ref. 29) (cm³/mol s)
300	1.0(13)	4.1(11)	7.3(12)	1.2(13)	1.6(13)
500	1.9(13)	1.5(12)	1.2(13)	2.1(13)	1.6(13)
700	2.1(13)	3.2(12)	1.7(13)	2.7(13)	1.6(13)

5. Acknowledgments

This work was carried out under grants PB91–0553 and PB92-0756 of the Spanish DGICYT. One of us (Jesús Hijazo) is grateful to the Spanish Ministry of Education and Science for a predoctoral grant.

6. References

1. Sharma, R.D., Sun, Y. and Dalgarno, A. (1993) Higly rotationally excited nitric oxide in the terrestrial thermosphere, *Geophys. Res. Lett.* **20**, 2043-2045, and references therein.
2. Duff, J.W., Bien, F. and Paulsen, D. E. (1994) Classical dynamics of the $N(4A) + O_2(X^3\Sigma_g^-) \rightarrow NO(X^2\Pi) + O(^3P)$ reaction, *Geophys. Res. Lett.* **21**, 2043-2046
3. Jaffe, R.L. (1986) Thermophysical Aspects of Re-entry Flows, in J.N. Moss and C . D. Scott (eds.), *Progress in Astronautics and Aeronautics*, AIAA, New York, 1986 Vol. 96, pp. 123-151
4. Zeldovich, Ya. B. (1946), Oxidation of N in combustion and explosion *Acta Physiochim URSS* **21**, 577
5. Schiff, H.I. (1964), Reactions involving N and O *Ann. Geophys.* **20**, 115
6. Heicklen, J., Cohen, N. (1968), Role of nitric oxide in photochemistry *Adv. Photochem.* **5**, 157
7. Gorse, C., Capitelli, M., Bretagne J. and Bacal,M. (1985) Vibrational Excitation and Negative-ion production in magnetic multicusp hydrogen discharges *Chem. Phys.* **93**, 1.
8. Capitelli, M.; Gorse C.; Ricard, A. (1986) *Top. Curr. Phys.* **39**, 5. Laganà, A.; García, E. and Ciccarelli, L. (1987) Deactivation of vibrationally excited nitrogen molecules by collision with nitrogen atoms *J. Phys. Chem.* **91**, 312.
9. Birely, J. H. and Lyman J.L. (1975) Effect of reagent vibrational energy on measured reaction rate constants *J. Photochem.* **4**, 269.
10. Zellner, R. (1984) Bimolecular Reaction Rate Coefficients, in W. C. Gardiner, Jr. (ed.), *Combustion Chemistry*, Springer Verlag, New York, pp.127-169, and references therein.
11. Kuntz, P. J. (1976) Features of Potential Energy Surfaces and their effect on Collisions, in W. H. Miller (ed.), *Dynamics of Molecular Collisions*, Plenum Press, New York, pp. 53-116.
12. Murrell J. N.; Carter, S.; Farantos, S. C.; Huxley, P. and Varandas A. J. C. (1984) *Molecular Potential Energy Functions*, John Wiley & Sons, New York.
13. Sorbie, K. S.; Murrell, J. N. (1975) Analytical potentials for triatomic molecules from spectroscopic data *Mol. Phys.* **29**, 1378.
14. Eyring, H. (1935) The activated complex in chemical reactions *J. Chem. Phys.* **3**, 107. Evans M. G. and Polanyi M., (1935) Calculation of reaction velocities *Trans. Faraday Soc.* **31**, 875.
15. Garrett, B. C. and D. G. Truhlar D. G. (1979) Generalized Transition State Theory. Classical mechanical theory and applications to collinear reactions of hydrogen molecules, *J. Phys. Chem.* **83**, 1052; (erratum (1979) **83**, 3058; (erratum) (1983) **87**, 4553.
16. Truhlar, D. G. and Muckerman, J. T. (1979) Reactive Cross Sections: Quasi and Semiclassical methods, in R. B. Bersntein (ed.) *Atom-Molecule Collision Theory*, Plenum Press, New York.
17. Karplus, M.; Porter, R. N. and Sharma, R. D. (1965) Exchange Reactions with Activation Energy. I. Simple Barrier Potential for $(H,H_2)^+$ *J. Chem. Phys.* **43**, 3259-3287.
18. Truhlar, D. G. and Blais, N. C. (1977) Legendre moment method for calculating differential scattering cross sections from classical trajectories with Monte Carlo initial conditions *J. Chem. Phys.* **67**, 1532-1538, and references therein.

19. Schatz, G. C. and Kupperman, A. (1976) Quantum mechanical reactive scattering for three-dimensional atom plus diatom systems. I. Theory *J. Chem. Phys.* **65**, 4642-4667. For the most recent collection of 3D methods, see: Bowman, J. M. (ed.) (1994) *Advances in Molecular Vibrations and Collision Dynamics.* Vols 2A and 2B. JAI Press, Greenwich.

20. Elkowitz, A.B. and Wyatt R.E. (1976) J_z-conserving approximation for the hidrogen exchange reaction *Mol. Phys.* **31**, 189. Kuppermann A.; Schatz, G. C.; Dwyer R.E. (1977) Angular momentum decoupling approximations in the quantum dynamics of reactive systems *Chem. Phys. Lett.* **45**, 71.

21. Secrest, D. (1975) Theory of angular momentum decoupling approximations for rotational transitions in scattering *J. Chem. Phys.* **62**, 710. Khare, V. (1978) On the l_z-conserving energy sudden approximation for atom–diatom scattering *J. Chem. Phys.* **68**, 4631.

22. Khare V.; Kouri, D. J. and Baer M. (1979) Infinite order sudden approximation for reactive scattering. I. Basic l–labelled theory *J. Chem. Phys.* **71**, 1188. Jellinek, J. (1985) Approximate Treatments of Reactive Scattering: Infinite Order Sudden Approximation, in M. Baer (ed.) *Theory of Chemical Reaction Dynamics*, Vol II., CRC Press, Boca Raton, pp. 2-119. Nakamura, H.; Ohsaki, A. M. Baer M. (1986) New implementation to approximate quantum mechanical treatment of atom–diatom chemical reactions *J. Phys. Chem.* **90**, 6176.

23. Kaufman, F. and Kelso, J. R. (1955) Thermal Decomposition of Nitric Oxide *J. Chem. Phys.* **23**, 1702-1707.

24. Kistiakowsky, G. B. and Volpi, G. G. (1957), Reactions of Nitrogen Atoms. I. Oxygen and Oxides of Nitrogen *J. Chem. Phys.* **27**, 1141-1149.

25. C. B. Kretschmer, H. L. Petersen, (1963) kinetics of three–body atom recombination *J. Chem. Phys.* **39**, 1772.

26. Wilson, W. E. (1966) Rate Constant for the Reaction N + O_2 → NO + O *J. Chem. Phys.* **46**, 2017-2018.

27. Barnett, A. J.; Marston, G. and Wayne, R. P. (1987) Kinetics and Chemiluminiscence in the Reaction of N Atoms with O_2 and O_3 *J. Chem. Soc. Faraday Trans.* 2 **83**, 1453-1463.

28. Clyne, M. A. A. and Thrush, B. A. (1961) Rates of the reactions with O_2 and NO *Proc. Roy. Soc. A* **261**, 259.

29. Baulch, D. L.; Drysdale D. D. and Haine, D. G. (1973) *Evaluated Kinetic Data for High Temperature Reactions*, Vol. 2. Butterworths, London.

30. Basevich, V. Ya. (1987) Detailed kinetic mechanisms for the combustion of homogeneous gaseous mixtures with oxigen–containing oxidizing agents, *Prog. Energy Combust. Sci.* **13**, 199. Hanson, R. K. and Salinian S. (1984) Survey of Rate Constants in the N/H/O System, in W. C. Gardiner, (ed.), *Combustion Chemistry*, Springer Verlag, New York, pp. 352-461. Miller, J. A. and Bowman, C. T. (1989) Mechanism and modeling of nitrogen chemistry in combustion *Prog. Energy Combust. Sci.* **15**, 287.

31. Hushfar F.; Rogers, J. W. and Stair Jr., A. T. (1971) Infrared Chemiluminiscence of the Reaction N + O_2 → NO + O *Appl. Optics* **10**, 1843-1847. (1972) **11**, 1656-1657.

32. Whitson Jr., M. E.; Darnton, L. A. and McNeal, R. J. (1976) Vibrational Energy distribution in the NO produced by the reaction of N(^{4}S) with O_2 *Chem. Phys. Lett.* **41** 552-556.

33. Rahbee, A. and Gibson, J. J. (1981) Rate constants for formation of NO in vibrational levels v = 2 through 7 from the reaction N(^{4}S) + O_2 → NO$^{\neq}$ + O *J. Chem. Phys.* **74** 5143-5148.

34. Herm, R. R.; Sullivan, B. J. and Whitson Jr., M. E. (1983) Nitric oxide vibrational excitation from the N(^{4}S) + O_2 reaction J. Chem. Phys. **79**, 2221-2230.

35. Winkler, I. C.; Stachnik, R. A.; Steinfeld, J. I. and Miller S. M. (1986) Determination of NO(v=0-7) product distribution from the N(^{4}S) + O_2 reaction using two-photon ionization *J. Chem. Phys.* **85**, 890-899.

36. Jackels, C.F. and Davidson, E. R. (1976) The two lowest ^{2}A' states of NO_2 *J. Chem. Phys.* **64**, 2908-2917.

37. Wilson Jr., C. W. (1975) Diatomics-in-molecules potentials for N(^{4}S) collisions with $O_2(^3\Sigma_g^-)$ *J. Chem. Phys.* **62**, 4843-4847.

38. Benioff, P. A.; Das, G. and Wahl, A. C. (1977) Ab initio calculations of the minimum energy path in the doublet surface for the reaction N(^{4}S) + $O_2(^3\Sigma_g^-)$ → NO($^2\Pi_u$) + O(^{3}P) *J. Chem. Phys.* **67**, 2449-2462 .

39. Das, G. and Benioff, P. A. (1980) A study of the minimum energy path of the reaction N(^{4}S) + $O_2(^3\Sigma_g^-)$ → NO($^2\Pi_u$) + O(^{3}P) *Chem. Phys. Lett.* **75**, 519-524.

40. Walch, S. P. and Jaffe, R. L. (1987) Calculated potential surfaces for the rections O + N_2 → NO + N and N + O_2 → NO + O *J. Chem. Phys.* **86**, 6946-6956.

41. Gilibert, M.; Aguilar, A.; González, M. and Sayós, R. (1993) Quasiclassical trajectory study of the $N(^4S_u) + O_2(^3\Sigma_g^-) \rightarrow NO(^2\Pi_u) + O(^3P_g)$ atmospheric reaction on the $^2A'$ ground potential energy surface employing an analytical Sorbie-Murrell potential *Chem. Phys.* **172**, 99-115.

42. Gilibert, M.; Aguilar, A.; González, M. and Sayós, R. (1993) A quasiclassical trajectory study of the effect of the initial rovibrational level and relative translational energy on the dynamics of the $N(^4S_u)$ + $O_2(^3\Sigma_g^-) \rightarrow NO(^2\Pi_u) + O(^3P_g)$ atmospheric reaction on the $^2A'$ ground potential energy surface *Chem. Phys.* **178**, 287-303.

43. Gilibert, M.; Giménez X.; González, M.; Sayós, R. and Aguilar, A. (1995) A comparison between experimental, quantum and quasiclassical properties for the $N(^4S) + O_2(^3\Sigma_g^-) \rightarrow NO(^2\Pi) + O(^3P)$ reaction *Chem. Phys.* **191**, 1-15.

44. Sayós, R.; Aguilar, A.; Gilibert, M. and González, M. (1993) Orientational dependence of the $N(^4S)$ + $NO(X^2\Pi)$ and $N(^4S) + O_2(X^3\Sigma_g^-)$ Reactions: Comparison of the Angle-dependent Line-of-centres Model with Quasiclassical Trajectories *J. Chem. Soc. Faraday Trans.* **89**, 3223-3234.

45. Jaffe, R. L.; Pattengill, M. D. and Schwenke, D. W. (1989) Classical Trajectory studies of gas phase reaction dynamics and kinetics using ab initio potential energy surfaces, in A. Laganà (ed.), *Supercomputer Algorithms for Reactivity. Dynamics and Kinetics of Small Molecules*, Kluwer Academic Publishers, Dordrecht, pp. 367-382.

46. Polanyi, J. C. (1972) Some concepts in Reaction Dynamics *Acc. Chem. Res.* **5**, 161-168.

47. Suzzi Valli, G.; Orrú, R.; Clementi, E.; Laganà, A. and Crocchianti, S. (1995) Rate coefficients for the N + O_2 reaction computed on an ab initio potential energy surface, *J. Chem. Phys.* **102**, 2825-2832.

48. Gilibert, M.; Giménez, X.; González, M.; Sayós, R.; Hijazo, J. and Aguilar, A. (1995) A theoretical study of the N + O_2 atmospheric reaction on a new optimized potential energy surface, to be submitted.

49. Kistiakowsky, G. B. and Volpi, G. G. (1958) Reactions of nitrogen atoms. II. H_2, CO, NH_3, NO and NO_2 *J. Chem. Phys.* **28**, 665.

50. Clyne, M. A. A. and Thrush, B. A. (1961) Rates of the reactions of nitroghen atoms with oxigen and nitric oxide *Nature* **189**, 56.

51. Herron, J. T. (1961) Rate of the reaction NO+N *J. Chem. Phys.* **35**, 1138.

52. Phillips, L. F. and Schiff, H. I. (1962) Mass Spectrometric Studies of Atom Reactions. I Reactions in the Atomic Nitrogen-Ozone System *J. Chem. Phys.* **36**, 1509-1516.

53. Clyne, M. A. A. and McDermid, I. S. (1975) Mass Spectrometric Determinations of the Rates of Elementary Reactions of NO and of NO_2 with Ground State N^4S Atoms *J. Chem. Soc. Faraday Trans.* **71**, 2189-2202.

54. Lee, J. H.; Michael, J. V.; Payne W. A. and Stief, L. J. (1978) Absolute Rate of the reaction of $N(^4S)$ with NO from 196-400 K with DF-RF and FP-RF techniques *J. Chem. Phys.* **69**, 3069-3076.

55. Koshi, M.; Yoshimura, M.; Fukuda, K.; Matsui, H.; Saito, K.; Watanabe, M.; Imamura, A. and Chen, C. (1990) Reactions of nitrogen ($N(^4S)$) atoms with nitric oxide and hydrogen, *J. Chem. Phys.* **93**, 8703-8708.

56. Michael, J. V. and Lim, K. P. (1992) Rate Constants for the N_2O reaction system: Thermal decomposition of N_2O; N + NO $\rightarrow$ N_2 + O; and implications for O + N_2 $\rightarrow$ NO + N, *J. Chem. Phys.* **97**, 3228-3234.

57. Morgan, J. E.; Phillips, L. F. and Schiff, H. I. (1962) Studies of Vibrationally Excited Nitrogen Using Mass Spectrometric and Calorimetric-Probe Techniques, *Discuss. Faraday Soc* **33**, 118-127. Morgan, J. E. and Schiff H. I. (1963) The study of vibrationally excited N_2 molecules with the aid of an isothermal calorimeter, *Can. J. Chem.* **41**, 903-912.

58. Black, G.; Sharpless, R. L. and Slanger, T. G. (1973) Measurements of vibrationally excited molecules by Raman scattering. I The yield of vibrationally excited nitrogen in the reaction N + NO $\rightarrow$ N_2 + O, *J. Chem. Phys.* **58**, 4792-4797.

59. Gilibert, M.; Aguilar, A.; González, M.; Mota, F. and Sayós, R. (1992) Dynamics of the $N(^4S_u)$ + $NO(X^2\Pi) \rightarrow N_2(X^1\Sigma_g^+) + O(^3P_g)$ atmospheric reaction on the $^3A''$ ground potential energy surface. I. Analytical potential energy surface and preliminary quasiclassical trajectory calculations, *J. Chem. Phys.* **97**, 5542-5552.

60. Gilibert, M.; Aguilar, A.; González, M. and Sayós, R. (1993) Dynamics of the $N(^4S_u) + NO(X^2\Pi) \rightarrow$ $N_2(X^1\Sigma_g^+) + O(^3P_g)$ atmospheric reaction on the $^3A''$ ground potential energy surface. II. The effect of reagent translational, vibrational and rotational energies, *J. Chem. Phys.* **99**, 1719-1733.

61. Aguilar, A.; Gilibert, M.; Giménez, X.; González, M. and Sayós, R. (1995) Dynamics of the $N(^4S)$ + $NO(X^2\Pi) \rightarrow N_2(X^1\Sigma_g^+) + O(^3P)$ atmospheric reaction on the $^3A''$ ground potential energy surface. III.

Quantum dynamical study and comparison with quasiclassical and experimental results J. Chem. Phys. (in press).

62. Gilibert, M.; González, M.; Giménez, X.; Sayós, R.; Hijazo, J. and Aguilar, A. (1995) Dynamics of the $N(^4S) + NO(X^2\Pi) \rightarrow N_2(X^1\Sigma_g^+) + O(^3P)$ atmospheric reaction on the $^3A''$ ground potential energy surface. IV. The behaviour with very high rovibrational excitation (manuscript in preparation).

VIBRATIONAL RELAXATION, NONEQUILIBRIUM CHEMICAL REACTIONS, AND KINETICS OF NO FORMATION BEHIND STRONG SHOCK WAVES

I.V. ADAMOVICH
Department of Mechanical Engineering, The Ohio State University, Columbus, OH 43210, USA

S.O. MACHERET
Department of Mechanical and Aerospace Engineering, Princeton University, Princeton, NJ 08544, USA

J.W. RICH
Department of Mechanical Engineering, The Ohio State University, Columbus, OH 43210, USA

C.E. TREANOR
CTSA, Inc., Williamsville, NY 14221, USA

A.A. FRIDMAN
Department of Mechanical Engineering, University of Illinois at Chicago, Chicago, IL 60607, USA

Abstract

The paper addresses development of new theoretical models for the rates of vibrational energy exchange and chemical reactions under conditions of extreme nonequilibrium, and their application to modeling calculations. These models yield analytic expressions for the specific rate constants that can be incorporated into existing high enthalpy flow codes. The models are validated by comparison with the state-of-the-art calculations and available experiments, including recent shock tube study of NO production kinetics behind shock waves. The models are also applied to simulation of NO formation behind strong bow shocks.

1. Introduction

Nonequilibrium chemical processes and radiation coupled to vibrational relaxation are of crucial importance in determining gas composition, temperature,

M. Capitelli (ed.), Molecular Physics and Hypersonic Flows, 85–104.
© *1996 Kluwer Academic Publishers. Printed in the Netherlands.*

pressure, and radiative heat flux in bow shocks accompanying reentry spacecraft as well as in expansion flows of interest in rocket plume detection [1-3]. There is a need to have a reliable model of vibrational relaxation, nonequilibrium chemistry, and radiation for reentry problems. Powerful computer codes for nonequilibrium shock or expansion flows, such as NEQAIR, LAURA, LORAN, and NOZNT [3-6], have been developed in recent years. Also, DSMC methods are being increasingly used for nonequilibrium shock modeling (see [7] and references therein). However, these codes require an input of a number of kinetic rates, and their predictive capability critically depends on the accuracy of these data.

The present paper discusses development and validation of new theoretical models for the rates of vibrational energy exchange (Section 2) and chemical reactions (Section 3) under conditions of extreme nonequilibrium. These models yield simple analytic expressions for the rate constants that can be easily incorporated into existing flow codes. Section 4 presents recent shock tube experiments and modeling calculations studying NO production kinetics behind the shock up to very high shock velocities. Finally, Section 5 gives a summary of the recommended rate models.

2. Nonperturbative Vibrational Energy Transfer Rate Model

The vibrational relaxation model, discussed in the present paper, provides the rates of vibration-translation (V-T) processes,

$$AB(i) + M \leftrightarrows AB(f) + M \tag{2.1}$$

and vibration-vibration-translation (V-V-T) processes,

$$AB(i_1) + CD(i_2) \leftrightarrows AB(f_1) + CD(f_2) \tag{2.2}$$

In Eqs. (2.1,2.2) AB, CD and M represent diatomic molecules and an atom, respectively, and i_1, i_2, f_1 and f_2 are vibrational quantum numbers. The rates of Eqs. (2.1,2.2) are not based on the first-order perturbation theory (FOPT), but evaluated according to the forced harmonic oscillator (FHO) theory [8-11]. This theory provides the V-T and V-V-T probabilities in collinear atom-diatom and diatom-diatom collisions, based on the analytic solution of Schrodinger's equation. It assumes that V-T and V-V-T transitions occur as a series of virtual one-quantum steps, during a single collision.

Summarizing the results [8-10], one obtains the following expressions for the V-T transition probabilities of Eq. (1) in atom-diatom collisions:

$$P_{VT}(i \to f, \varepsilon) = i! f! \varepsilon^{i+f} \exp(-\varepsilon) \left| \sum_{r=0}^{n} \frac{(-1)^r}{r!(i-r)!(f-r)!} \frac{1}{\varepsilon^r} \right|^2 \tag{2.3}$$

and for V-V-T probabilities of Eq. (2.2) in diatom-diatom collisions:

$$P_{VVT}(i_1, i_2 \to f_1, f_2, \varepsilon, \rho) =$$

$$\left| \sum_{r=o}^{n} C_{r+1,i_2+1}^{(i_1+i_2)} \cdot C_{r+1,f_2+1}^{(f_1+f_2)} \cdot e^{-i(f_1+f_2-r)\rho} \cdot P_{VT}^{1/2}(i_1+i_2-r \to f_1+f_2-r, 2\varepsilon) \right|^2 \quad (2.4)$$

In Eq. (2.3), $n = \min(i,f)$. In Eq. (2.4), $n = \min(i_1+i_2, f_1+f_2)$, and $C_{ij}^{(k)}$ is the transformation matrix [10]. The parameters ε and ρ in Eqs. (2.3, 2.4) are simply related to the two-state (FOPT) transition probabilities $P_{FOPT}(1 \to 0)$ and $P_{FOPT}(1,0 \to 0,1)$. For a repulsive exponential intermolecular potential $V(r) \sim \exp(-\alpha r)$ one has [12]

$$\varepsilon = P_{FOPT}(1 \to 0) \cong S_{VT} \frac{8\pi^2 \omega (m^2/\mu)\gamma^2}{\alpha^2 \hbar} e^{\frac{-2\pi\omega}{\alpha v}} \quad (2.5)$$

$$\rho = [4 P_{FOPT}(1,0 \to 0,1)]^{1/2} = \left[S_{VV} \frac{4m^2 \gamma_1^2 \gamma_2^2 \alpha^2 v^2}{\mu_1 \mu_2 \omega_1 \omega_2} \right]^{1/2} \quad (2.6)$$

In Eqs. (2.5, 2.6), ω, ω_1 and ω_2 are oscillator frequencies, m is the collision reduced mass, μ, μ_1 and μ_2 are oscillator reduced masses, $\gamma = \gamma_1 = m_B/(m_A + m_B)$, $\gamma_2 = m_C/(m_C + m_D)$, and v is the symmetrized collision velocity. S_{VT} and S_{VV} are "steric" factors, $0 < S < 1$, which will be discussed later. The probabilities (2.3-2.6) have been compared with quantum calculations for collinear geometry and show very good agreement at low quantum states and high collision energies, including multi-quantum transitions [13].

To use the FHO probabilities for simulation of vibrational relaxation of real gases, some corrections have to be made. First, we suggest determining the "steric" factors S_{VT} and S_{VV} from the comparison of the FHO probabilities $P(1 \to 0)$ and $P(1,0 \to 0,1)$ with the results of 3-D close-coupled calculations. Second, to generalize Eq. (2.4) for non-resonance V-V transitions, we use the two-state FOPT result [14], replacing ρ given by Eq. (2.6) by the expression

$$\rho_\xi = \rho \cdot \frac{\xi}{\sinh(\xi)} , \qquad \xi = \frac{\pi^2(\omega_1 - \omega_2)}{4\alpha v} = \frac{\pi^2 \Omega}{4\alpha v} \quad (2.7)$$

Finally, to account for the anharmonicity, we use a corrected harmonic oscillator, replacing frequencies in Eqs. (2.5-2.7) by "average" frequencies

$$\omega_{1,2} = \frac{E_i^{1,2} - E_f^{1,2}}{i - f} , \qquad \begin{array}{l} \omega = (\omega_1 + \omega_2)/2 \\ \Omega = |\omega_1 - \omega_2| \end{array} \quad (2.8)$$

where $E_v = \omega_e \, v \cdot [1 - x_e(v+1)]$ is the energy of a molecule on vibrational level v.

Eqs. (2.3,2.4) can be simplified [14,15]. For the V-T probabilities of Eq. (2.3) and for purely V-T probabilities (M=i_2=f_2) of Eq. (2.4) one can use,

$$P_{VT}(i \rightarrow f, \varepsilon) = P_{VT}(i, M \rightarrow f, M, \varepsilon)e^{\varepsilon} \cong \frac{(n_s)^s}{(s!)^2} \varepsilon^s \exp\left(-\frac{2n_s}{s+1}\varepsilon\right) \qquad (2.9)$$

where

$$s = |i - f|, \qquad n_s = [\max(i,f)!/ \min(i,f)!]^{1/s} \qquad (2.10)$$

and for the purely V-V probabilities (i_1+i_2=f_1+f_2) of Eq. (2.4) one can use

$$P_{VV}(i_1, i_2 \rightarrow f_1, f_2, \rho) = \frac{(n_s^{(1)}n_s^{(2)})^s}{(s!)^2}\left(\rho_{\xi}^2/4\right)^s \exp\left(-n_s^{(1)}n_s^{(2)}\rho_{\xi}^2/2(s+1)\right) \quad (2.11)$$

After averaging the probabilities (2.9,2.11) over the one-dimensional Boltzmann distribution, one obtains thermally averaged V-T and V-V rates. For V-T processes, the result of steepest descent integration is:

$$k(i \rightarrow f, T) = Z \cdot \left(\frac{2\pi}{3+\delta}\right)^{1/2} \frac{(n_s S_{VT}\theta'/\theta)^s}{(s!)^2} C_{VT} \left(\frac{s^2\theta'}{T}\right)^{1/6} \cdot$$

$$\exp\left[-\left(\frac{s^2\theta'}{T}\right)^{1/3}\left(\frac{C_{VT}^2}{2} + \frac{1}{C_{VT}}\right) - s(1 - C_{VT})^3\right]\exp\left(\frac{\theta s}{2T}\right), \qquad (2.12)$$

where Z is the gas-kinetic collision frequency, $\theta = h\omega/2\pi k$,

$$\theta' = \frac{8\pi^2\omega^2(m^2/\mu)\gamma^2}{\alpha^2 k}, \qquad \delta = \frac{1 - C_{VT}^3}{C_{VT}^3}\frac{2\pi\omega}{\alpha v_{m0} C_{VT}}, \qquad (2.13)$$

and C_{VT} determines the velocity v_m at which the integrand reaches the maximum,

$$C_{VT} = \frac{v_m}{v_{m0}} = \left[1 - \frac{v + 2n_s/(s+1)}{s} \cdot \varepsilon\left(v_{m0}C_{VT}\right)\right]^{1/3} \qquad (2.14)$$

where v=0 for AB-M collisions and v=1 for AB-CD collisions. In Eqs. (2.13,2.14), v_{m0} is the Landau-Teller "most effective" velocity, v_{m0}=$(2\pi s k T/\alpha m)^{1/3}$. Transcendental Eq. (2.14) has a single root and can be easily solved by the Newton method. At s=1 and ε<<1, ρ<<1, $C_{VT}\cong$1, and Eq. (2.12) reduces to the result of the SSH theory [12]. A result, similar to Eq. (2.12), can be also obtained from Eq. (2.11) for far-from-resonance V-V processes ($\pi\Omega >> \alpha v$) [15]. For resonance V-V energy exchange ($\Omega = 0$), Boltzmann averaging of the probability (2.11) gives

$$k(i_1, i_2 \rightarrow i_2, i_1, T) = Z \cdot \frac{n_s^{2s}}{s!} \frac{\langle P(10 \rightarrow 01) \rangle^s}{\left(1 + \frac{2n_s^2 \langle P(10 \rightarrow 01) \rangle}{s+1}\right)^{s+1}}, \tag{2.15}$$

where

$$\langle P(10 \rightarrow 01) \rangle = \frac{S_{VV} \alpha^2 kT}{2\omega^2 m} \tag{2.16}$$

is the thermally averaged FOPT probability of Eq. (2.6) [12]. Eq. (2.15) gives the result of the SSH theory, if $s=1$, $<P(10\rightarrow01)><<1$. The gap between the far-from-resonance and the exact resonance V-V rates can be filled by multiplying the resonance rate (2.15) by the energy defect function $G(\lambda)$, similar to that suggested in [16]

$$G(\lambda) = \frac{1}{2} \frac{3 - \exp(\lambda)}{\exp(\lambda)} \tag{2.17}$$

where λ is proportional to the vibrational energy defect ΔE,

$$\lambda = -\frac{2}{3} \cdot 2^{-\frac{3}{2}} \cdot \sqrt{\theta' / T} \frac{|\Delta E|}{sk\theta} = -\frac{2}{3} \cdot 2^{-\frac{3}{2}} \cdot \sqrt{\theta' / T} \frac{\Omega}{\omega}, \tag{2.18}$$

providing very good accuracy both for single-quantum and multi-quantum transitions, compared to the numerically averaged probabilities of Eq. (2.4).

The comparison of the FHO and the FOPT probabilities (see [15]) proves that both V-T and V-V multi-quantum processes principally occur via a sequential FHO mechanism, rather than via a straightforward FOPT mechanism.

Let us outline the area of applicability of the FHO model. Since the FHO theory is based on the simple collision model, it does not account for strong vibration-rotation coupling, important, for example, in H_2 and in HBr [17,18]. Also, the model cannot predict V-V-T rates in electronically nonadiabatic collisions, as occurs in NO-NO [19]. Further, the FHO model, in its present form, cannot be applied to transitions induced by long-range attractive forces, which determine the V-V rates for CO-CO and CO-N_2 at low temperatures [20,21]. Finally, the FHO model is expected to work well for relatively heavy homonuclear diatomics, such as N_2 and O_2. Note that the single-quantum V-V rates, based on the FHO model [22], have been previously compared with close-coupled calculations [26] for N_2-N_2, showing good agreement. However, the rates [22] have not been presented in a closed analytic form.

To validate the use of the corrected FHO model, we compared it to the experiments [23-25] and to the 3-D semiclassical calculations for N_2-N_2, O_2-O_2, and N_2

-O₂ [26-28], considered to be the most reliable data available at this time. In our calculations, we used an intermolecular potential with the exponential repulsion parameter α=4.0 A , chosen to fit the potentials used in [26-28]. The corrective "steric factors" S_{VT} and S_{VV}, originating from the non-collinear nature of collisions, have been obtained from matching the two FHO probabilities at low collision velocities,

$$P(1,0 \to 0,0) \cong S_{VT} \cdot \varepsilon , \quad P(1,0 \to 0,1) \cong S_{VV} \cdot \rho^2 / 4 \qquad (2.19)$$

to the results of [26-28]. These factors do not reflect the dynamics of a real collision and are purely phenomenological. It has been found that S_{VT}=1/2, S_{VV}=1/27 are temperature independent and the same for all three systems, which is the only justification for their use. These values, used in all subsequent calculations, are the only "fitting" parameters in the present model.

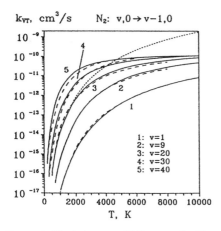

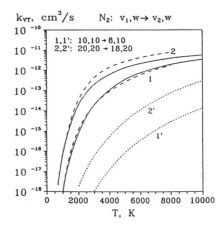

Figure 1. Single-quantum V-T rates for N₂-N₂. Solid lines, FHO model; dashed lines, calculations [26]; dotted line, FOPT, $k_{VT}(20 \to 19)$

Figure 2. Double-quantum V-T rates for N₂-N₂. Solid lines, FHO model; dashed lines, calculations [26]; dotted line, FOPT, $k_{VT}(10 \to 8)$ and $k_{VT}(20 \to 18)$

Figs. 1-3 compare the FHO model rates, calculated by Eqs. (2.12, 2.15-2.18) to the results of [26-28], and to the experimental data. One can see remarkably good agreement, both for the temperature and quantum number dependence of single-quantum and double-quantum V-T and V-V rates, in the temperature range 200<T<8000 K, and for the vibrational quantum numbers 0<v<40. The good agreement between the FHO theory and the semiclassical model [26-28] results from the fact that both models account for the coupling of many vibrational states during a collision. This effect is of crucial importance in both high-energy single-quantum transitions and multi-quantum transitions, as discussed in [29]. As a result, we obtain simple analytic formulae (2.12, 2.15-2.18) for nonperturbative V-T and V-V rates that are applicable up to very high temperatures, at high vibrational quantum numbers, and also for multi- quantum vibrational energy transfer.

3. Nonequilibrium Chemical Reaction Rate Model

In this Section, we present analytic rates of dissociation and bimolecular exchange reactions, based on the classical impulsive model [30-32]. Chemical reactions are considered to occur mostly through an optimum configuration, that is, a set of collision parameters minimizing the energy barrier. This configuration defines the threshold kinetic energy for the reaction, or the exponential factor of the rate coefficient. The probability of finding the colliding system near the optimum configuration determines the preexponential factor of the rate. Consider the classical atom-diatom or diatom-diatom impulsive collision. By applying energy and momentum conservation equations, an expression can be derived for the translational energy, E_t, required to make the after-collision vibrational energy greater than D (the chemical bond energy):

$$E_t = G_{E_v,E_r,E_v',E_r'}(\gamma_1,...,\gamma_n) \tag{3.1}$$

Here γ_i (i=1,...,n) is a set of angles and oscillation phase(s) at the moment of collision, whereas E_v, E_r, E_v', E_r' are vibrational and rotational energies of the dissociating molecule and, in the case of diatom-diatom collision, its collision partner (E_v', E_r'), prior to the collision.

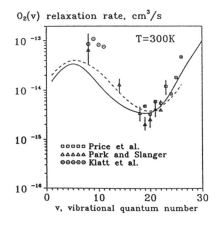

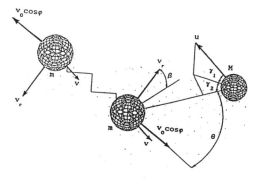

Figure 3. Relaxation rate of oxygen. Solid lines, FHO model; dashed lines, calculations [27]; symbols, experiments [23-25]

Figure 4. Collision geometry for diatom-atom collisions.

An optimum configuration is determined by the set of equations:

$$\partial G / \partial \gamma_i = 0, \quad i = 1,...,n \tag{3.2}$$

Solution of Eq. (3.2) yields the optimum set $\gamma_i = \gamma_{i,0}$ and the threshold function

$$E_t = F(E_v, E_r, E_v', E_r'),$$
(3.3)

so that the dissociation probability is non-zero only if $E_t > F(E_v, E_r, E_v', E_r')$. To get the probability of finding the system in the vicinity of the threshold (3.3), we expand the G-function in a Taylor series, taking into account Eq. (3.2):

$$\Delta E = E_t - F(E_v, E_r, E_v', E_r') \cong \frac{1}{2}\sum_{i,j}\frac{\partial^2 G}{\partial\gamma_i\partial\gamma_j}\Big|_{\gamma_i=\gamma_{i,0}} \cdot \Delta\gamma_i\Delta\gamma_j = \frac{1}{2}\sum_{i,j}G_{i,j}\Delta\gamma_i\Delta\gamma_j \quad (3.4)$$

With the right choice of γ_i, cross-derivatives would vanish, so that,

$$\Delta E = \sum_i G_{ii}\Delta\gamma_i^2$$
(3.5)

Having analytic expressions for G_{ii} as functions of E_v, E_r, E_v', E_r', we can find the dissociation probability in terms of these energies and ΔE, which is the ratio of volumes of the ellipsoid and the parallelepiped in the phase space of γ_i [32]

$$P = \frac{2(2\pi\Delta E)^{n/2}}{n\pi^n\Gamma(n/2)\sqrt{G_{11}...G_{nn}}}$$
(3.6)

Eq. (3.6) may be written in the form

$$P = \Phi\cdot(E_t - F)^s$$
(3.7)

where F is the threshold function (3.3), Φ is a function of vibrational and rotational energies, and $s=n/2$. Note that the rather common assumption $F=D-E_v$, might not be true, at least for low vibrational energies.

Integration of the probability (3.7) over a Maxwellian distribution function of the translational energy gives the "microscopic" rate coefficient dependent on temperature and vibrational/rotational energy (through the F and Φ functions),

$$k(E_v, E_r, E_v', E_r') = Z\cdot\Gamma(s+2)\cdot\left(1+F/T(s+1)\right)\cdot\Phi T^s\exp(-F/T)$$
(3.8)

where Z is the gas kinetic frequency. Integrating Eq. (3.8) over vibrational and rotational energies with appropriate distribution functions, assumed in this paper to be Boltzmann with different temperatures, T_v and T_r, would give the nonequilibrium two- or three-temperature rate coefficient. This procedure is implemented analytically for both diatom-atom and diatom-diatom collisions. Details of calculations, as well as handling of singularities, or "resonances", when some of $G_{ii}=0$ in Eq. (3.6), can be found in [32]. This method is well suited for hypersonic applications, since it is based on the assumption of high collision energies, and because it emphasizes the importance of reactions from low vibrational levels.

3.1. DISSOCIATION

The 3D diatom-atom center-of-mass collision geometry is shown in Fig. 4. In applying the general method described above, a difficulty arises, since in a non-collinear collision, a part of the pre-collision rotational energy is converted, together with translational energy, into vibrational energy (TRV process). Such a TRV process makes the optimum configuration non-collinear, and the derivations become analytically intractable. Although TRV processes deserve more detailed further investigation, at this time the TRV effect will be neglected, the optimum configuration being collinear, with the proper correction taken later for the change of the dissociation energy due to rotation (see [32]).

Neglecting the TRV effect, one obtains:

$$E = G = \frac{2(1-\sqrt{\alpha})}{(1+\sqrt{\alpha})\cos^2\gamma_1\cos^2\gamma_2} \left(\frac{\sqrt{D - E_v\sin^2\varphi} + \sqrt{E_v}\cos\varphi}{(1-\sqrt{\alpha})\cos\theta} - \sqrt{E_v}\cos\varphi\cos\theta \right)^2 \quad (3.9)$$

where $\alpha = [m/(m+M)]^2$. A minimum of this function gives the optimum configuration,

$$\gamma_1 = \gamma_2 = \theta = 0 \,;$$

$$\cos\varphi = \begin{cases} -1, & \text{if } E_v \leq \alpha D \\ -\left(\dfrac{\alpha}{1-\alpha}\dfrac{D-E_v}{E_v}\right)^{1/2}, & \text{if } E_v > \alpha D \end{cases} \quad (3.10)$$

and the threshold function is

$$E_t = F(E_v) = \begin{cases} \dfrac{(\sqrt{D} - \sqrt{\alpha E_v})^2}{1-\alpha}, & \text{if } E_v \leq \alpha D \\ D - E_v, & \text{if } E_v > \alpha D \end{cases} \quad (3.11)$$

The physical meaning of these optimum conditions is the following. Obviously, with no rotational effects, the best chance for translation-vibration energy transfer is provided by the collinear geometry (all angles = 0). In this collinear geometry, translational energy can be completely transferred into the vibrational mode in the situation when the molecule's atom immediately participating in the collision is not moving at the instant this orientation occurs. Therefore, translational velocity of this atom should be equal by magnitude and directed oppositely to its vibrational velocity, imposing a condition on the oscillator phase. However, at low vibrational energy, even the amplitude vibrational velocity is too small to be able to compensate for the translational velocity. In this range of E_v, $\cos\phi$ stays at (-1), and the translational energy cannot be completely used for dissociation, resulting in a threshold higher than $(D-E_v)$, as reflected in Eq. (3.11).

The probability of finding the system in the vicinity of the optimum configuration, according to the general method, is

$$P = \frac{(3/8)\cdot(E_t - F)^{3/2}}{F\cdot\left(4(\sqrt{D} - \sqrt{\alpha E_v})(\sqrt{D} - (2 - \sqrt{\alpha})\sqrt{E_v})/(1-\alpha)\right)^{1/2}}, \qquad E_v \leq \alpha D$$

$$P \cong \frac{\Delta E^2}{(D - E_v)^{3/2}\cdot\sqrt{E_v - \alpha D}}, \qquad E_v > \alpha D$$
(3.12)

where T is the "average" temperature,

$$T_a = \alpha T_v + (1-\alpha)T. \qquad (3.13)$$

This probability may be integrated over the Boltzmann distribution with temperature T_v. The rationale for this is that at small T_v/T, when populations at high levels are strongly non-Boltzmann, dissociation proceeds mostly from the low vibrational levels; and when T_v/T is not small (late stages of relaxation behind shocks), the vibrational distribution is quite close to the quasi-Boltzmann one [33]. The results for the dissociation from the low levels $(E_v \leq \alpha D)$ is

$$k_l(T,T_v) = \frac{9\sqrt{\pi(1-\alpha)}}{64} Z\left(\frac{T}{D}\right)^{1/2}\left(1 + \frac{5(1-\alpha)T}{2D}\right)\exp\left(-\frac{D}{T_a}\right) \qquad (3.14)$$

The contribution to the nonequilibrium dissociation from the high vibrational levels should be roughly proportional to the nonequilibrium-to-equilibrium population ratio near the dissociation limit (see [32]), that is,

$$k_h(T,T_v) = AT^n \frac{1 - \exp(-h\nu/T_v)}{1 - \exp(-h\nu/T)} H(T)\cdot\exp\left(-\frac{D}{T_v}\right) \qquad (3.15)$$

where A(T) is the Arrhenius preexponential factor of the equilibrium dissociation rate constant, and $h\nu$ is the molecule's vibrational quantum. The total nonequilibrium rate coefficient is approximately the sum of (3.14) and (3.15),

$$k(T_v,T) = k_l + k_h, \qquad (3.16)$$

and parameter H(T) in Eq. (3.15) is chosen so that at $T_v=T$ the sum (3.16) would give the Arrhenius rate constant

$$k = AT^n \exp(-D/T) \qquad (3.17)$$

When $T_v \ll T$, dissociation proceeds mostly from low vibrational levels, so that $k_l \gg k_h$.

Similar calculations, omitted here, can be done for diatom-diatom collisions (see [32]). Then one obtains the three-temperature rate constant for dissociation from low vibrational levels

$$k_l(T_v, T_r, T) = \frac{2(1-\alpha)}{\pi^2 \alpha^{3/4}} Z \frac{T}{D} \cdot \frac{T_k^3}{T^{5/2} T_r^{1/2}} \cdot \left(1 + \frac{7(1-\alpha)(1+\sqrt{\alpha})T_k^2}{2DT} \right) \cdot \exp\left(-\frac{D}{T_a} \right), \qquad (3.18)$$

where

$$T_k = \left(T + \sqrt{\alpha} T_r \right) / \left(1 + \sqrt{\alpha} \right), \qquad (3.19)$$

which we call the "kinetic" temperature, and T_a is the "average" temperature,

$$T_a = \alpha T_v + (1-\alpha)T_k . \qquad (3.20)$$

The rate coefficient for dissociation from high vibrational levels is determined, as before, by Eq. (3.15) and total rate coefficient by Eq. (3.16).

3.2. BIMOLECULAR REACTIONS

The exponential factor of the rates of nonequilibrium bimolecular reactions

$$AB(v) + C \rightarrow A + BC \qquad (3.21)$$

have been evaluated by the classical impulsive model in [30]. The state-specific rate is given by the equation

$$k(v \rightarrow, T) = A(T) \cdot \exp(-F(E_v)/T), \qquad (3.22)$$

where E_v is the vibrational energy of a molecules, $F(E_v)$ is the threshold function, and $A(T)$ is a factor obtained from normalization on the thermal reaction rate. The threshold energy for the endoergic exchange reaction (3.22) is

$$F(E_v) = \begin{cases} W + \dfrac{1}{1-\alpha}\left[(E_a - W)^{1/2} - (\alpha E_v)^{1/2} \right]^2 , & if \quad E_v \le (E_a - W)/\alpha \\ W , & if \quad E_v > (E_a - W)/\alpha \end{cases} \qquad (3.23)$$

In Eq. (3.23), E is the reaction activation energy,

$$\alpha = \frac{m_B(m_A + m_B + m_C)}{(m_A + m_B)(m_B + m_C)} , \qquad (3.24)$$

and W can be found from the relation

$$\beta = \alpha \cdot (1 - W / E_a), \tag{3.25}$$

where β is the energy fraction that goes into vibrations in the reverse exoergic reaction. For reaction (3.21) this parameter is measured to be $\beta=0.25\pm0.03$ [34]. The

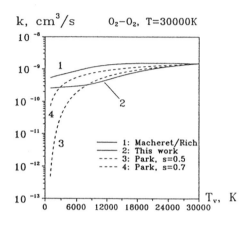

Figure 5. Oxygen dissociation rate by different models: Macheret/Rich [31], Park [1], and this work.

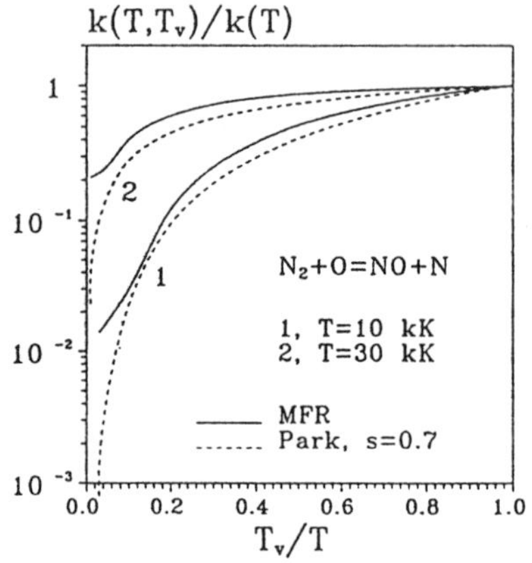

Figure 6. Nonequilibrium rate of reaction (4.5) by the Macheret-Fridman-Rich [30] and the Park [1] models

two-temperature rate constant can be obtained by averaging the specific rate (3.22) over the Boltzmann distribution with temperature T_v,

$$k(T) = A(T) \cdot \exp\left[-\frac{E_a - W}{\alpha T_v + (1-\alpha)T} - \frac{W}{T} \right] \tag{3.26}$$

The two- or three-temperature rates obtained in this Section can be directly applied to nonequilibrium hypersonic calculations. The rates of Eqs. (3.18,3.15) and of Eq. (3.26) are shown in Figs. 5,6 as functions of vibrational temperature T_v.

4. Kinetics of NO Formation behind Strong Shock Waves

4.1. EXPERIMENTAL

The time-history of infrared (IR) radiation behind a normal shock was measured using a pressure-driven shock tube with a 7.62 cm inside diameter. The driver section of the shock tube is 1.5 m long, and was operated at pressures of up to 260 atm of hydrogen. The routine double-diaphragm technique provided excellent run-to-run reproducibility in wave speed (~1%) and radiation records. Initial test gas pressures were measured to better than 1%. Scientific grade (99.999%) O_2 and N_2 were used, with H_2O and CO_2 at less than 1 ppm, and were premixed to provide the O_2/N_2 mixtures used in the experiments.

An InSb detector was used in the radiometer to measure the NO IR radiation behind the shock wave. The combined filter-detector bandpass function between 5-5.5 μm was determined by separate bench experiments. A standard blackbody source through a scanning monochromator was calibrated and then used to establish the wavelength dependence of the radiometer. As described in [35], the observed in-band infrared radiation is related to the NO concentration and the gas translational temperature through the relation

$$S = 1.29 \cdot 10^{-20} \cdot L \cdot n_{NO} \cdot \left[1 - 3.21 \cdot 10^{-4} \left(T - 3000 \right) \right] \quad W/cm^2 \cdot sr \qquad (4.1)$$

with n_{NO} in cm^{-3}, L=7.62 cm and T in K. This equation is consistent with the equilibrium values of NO radiation measured in the shock tube and with the known band strength of the molecule. It was used to relate the measurements of non-equilibrium radiation to the non-equilibrium NO molecular density.

The optical system used in the experiment was designed to produce a sharp definition of the region behind the shock contributing to the radiation received by the detector. The physical apertures (the InSb detector and a razor-blade slit) are both external to the shock tube, and each provides a 3 mm image on a shock tube window. The external placement precludes any reflections from the apertures. With shock speeds of 3 to 4 mm/μs, this results in a temporal resolution of 1 μs.

4.2. KINETIC MODELING

To simulate NO formation kinetics behind normal shocks, we use master equation modeling. The kinetic model incorporates (i) 1-D gas dynamics equations for nonequilibrium reacting gases [11]; (ii) chemical kinetics equations for reacting species; and (iii) master equations for vibrational levels populations of each diatomic species. The system of equations incorporates chemistry-vibration coupling terms and is self-consistent. The equations (ii) and (iii) can be found in [36].

It is assumed that the mixture components take part in dissociation reactions

$$N_2(v) + M \leftrightarrows N + N + M \qquad (4.2)$$

$$O_2(v) + M \leftrightarrows O + O + M \qquad (4.3)$$

$$NO(v) + M \leftrightarrows N + O + M \qquad (4.4)$$

and in Zel'dovich mechanism bimolecular reactions

$$N_2(v) + O \leftrightarrows NO + N \qquad (4.5)$$

$$O_2 + N \leftrightarrows NO(v) + O \qquad (4.6)$$

The notation AB(v) indicates that the effect of vibrational excitation of molecule AB on the rate of an endoergic reaction is taken into account. The rates of reactions (4.2-4.6) in thermal equilibrium are taken the same as in [37], where they were incorporated from [38,39]. The nonequilibrium rates of reactions (4.2-4.6) are evaluated by the Macheret-Fridman-Rich (MFR) model [30-31], described in Section 3. The rates of V-T and V-V processes, used in master equation, are evaluated by the FHO model [11,15], described in Section 2.

To compare this master equation model with kinetic models, commonly used in nonequilibrium hypersonic calculations, we also ran a code based on simplified kinetics [37]. It incorporates one equation for the vibrational energy of each diatomic species. The rates of nonequilibrium chemical reactions (forward reactions (4.2-4.5)) are calculated according to the Park model [1],

$$k(T, T_v) = k_{eq}(T^s, T_v^{1-s}), \quad s = 0.7 \qquad (4.7)$$

where k_{eq} is the equilibrium rate constant at $T = T_v$. This model will be called "Model I" in the present paper, while the master equation model will be referred to as "Model II". Model I has been validated by comparing the predicted and the experimental [35] NO radiation profiles behind the shock for shock velocities u_s=3-4 km/s,

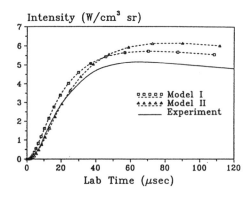

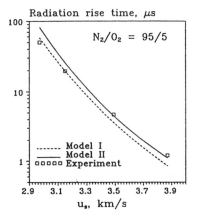

Figure 7. NO IR radiation profiles behind the normal shock. $N_2:O_2$=60:40, u_s=3.06 km/s, P_0 = 2.25 torr

Figure 8. NO IR radiation rise time (time to the half-maximum) as a function of the shock velocity

showing good agreement [37]. However, Model I is based on experimental relaxation data, available only for relatively low (T<10 kK) temperatures, and on the empirical nonequilibrium reaction rate model (4.7). Therefore, it has no predictive capability at u_s>4 km/s. On the other hand, Model II, that incorporates nonperturbative relaxation rate model and nonempirical impulsive reaction rate model, can be used up to very high temperatures.

To validate the use of Model II with kinetic rates described in Sections 2,3 for strong shocks (u_s=7-9 km/s), we first compared its predictions with experiments at lower shock velocities, u_s=3-4 km/s [35]. The NO IR signal intensity was obtained from the calculated NO concentration according to Eq. (4.1). Fig. 7 shows typical experimental IR signal behind the shock, for the low shock velocity u_s =3.05 km/s. The conditions before the shock are T_0=300 K, P_0=2.25 torr, which corresponds to a 40 km altitude. Also shown in Fig. 7 are results of calculations by Model I and Model II, which both agree well with the experiment. One can see from Fig. 7 that NO radiation is delayed by an incubation time τ_{inc}(NO). The incubation time reaches τ_{inc}(NO)~20 ms at u_s=2.97 km/s in an N_2-O_2=95%/5% mixture. At these temperatures (T=4-5 kK for $u_s \cong 3$ km/s), the N_2 vibrational relaxation time τ_{VT} (N_2)~30 ms is much longer than both $\tau_{VT}(O_2)$~1 ms and the O_2 dissociation incubation time, $\tau_{inc}(O_2)$~1 ms. This proves that the bottleneck for the NO production at u_s~3 km/s is reaction (4.5) due to vibration-chemistry coupling, rather than dissociation of O_2 (4.2). The same effect, although not seen in the experiments at higher shock velocities due to the limited time resolution has been observed in all normal shock calculations for u_s =3-4 km/s.

Calculations made for 16 experimental conditions (u_s=3-4 km/s) in three N_2-O_2 mixtures (95%/5%, 78%/22%, and 60%/40%) all show good agreement between experimental and predicted radiation signals. For example, Fig. 8 shows the experimental and calculated radiation rise time (time to signal half-maximum), as functions of shock velocity. Also, the calculations did not reveal any substantial difference between the predictions of Model I and Model II, which were quite close for all considered experimental conditions. This is expected for two reasons. First, nonequilibrium chemical reaction rates (4.2-4.6) by the MFR model can be approximated fairly well by the Park formula (4.7), except for the case of extreme disequilibrium T_v/T<0.1 (see Figs. 5,6). Second, for the temperatures T<8 kK, the effect of multi-quantum relaxation on the vibrational energy distribution function (VDF) of N_2 and O_2 is not very dramatic, so that the VDF, calculated by the state-to-state kinetic model (Model II), is Boltzmann-like. A considerable deviation from a Boltzmann distribution occurs only at T_v /T<<1 for the high vibrational levels, when their populations are very low. Finally, since at these temperatures most of the NO behind the normal shock is produced at $T_v(O_2)/T \geq T_v(N_2)/T \geq 0.4$, both models should therefore give close results, as they actually do.

To compare the predictions of Model I and Model II at very high shock velocities, we ran the normal shock code for air at u_s=9 km/s, T_0=220 K, P_0=40 mtorr (these conditions correspond to a 70 km altitude). We note that in this case the first-order vibrational relaxation model used in Model I and based on the low-temperature experimental data, have no predictive capability and may be not applicable at all. However, the predictions of NO production by the two models again turn out to be very close.

The reasons for this behavior are as follows. First, molecular dissociation at these high temperatures (T≥20 kK in the region of the most intensive NO production) is very rapid. The dissociation rates in this region only weakly depend on the vibrational temperature, while the rates for N_2 and O_2 dissociation are comparable (see [32]). Dissociation becomes the major source of both N and O atoms, so that the Zel'dovich mechanism reactions (4.5,4.6) are no longer coupled in a chain. Under these conditions, NO is produced mainly in the second Zel'dovich reaction (4.6), which has low activation energy and therefore is not vibrationally stimulated. Second, at these high temperatures the rate of another NO-producing reaction, the vibrationally induced reaction (4.5), also weakly depends on the vibrational temperature or on the VDF (see Fig. 6). Obviously, when the gas temperature is comparable to the reaction activation energy (E_a =38 kK for reaction (4.5)), a considerable part of all N_2 molecules can react with O atoms regardless of how much vibrational energy they have. Thus, NO production behind strong shocks becomes weakly coupled to the vibrational energy of the gas.

A considerable difference between the two model predictions does occur in the first stage of relaxation when $T_v/T \ll 1$. The Model II NO production rate in this region exceeds the prediction of Model I by an order of magnitude. This effect is due to the unphysical behavior of the Park model reaction rates (4.7), incorporated in Model I, at $T_v/T \to 0$, when they approach zero (see Figs. 5,6). However, this effect is strongly overshadowed by the much greater amount of NO produced in the less nonequilibrium stage (at higher T_v/T ratio), where the prediction of both models are getting much closer.

The effect of the reaction rate behavior at the extreme vibrational disequilibrium (when $T_v/T \ll 1$), first studied by Boyd et al. [40], may be very dramatic for low density bow shocks, where there is no clearly defined shock wave. To separately estimate the sensitivity of the NO production rate to the chemical reaction and vibrational relaxation rate models, we used the results of the DSMC bow shock calculations [40]. In the following calculations, the gas velocity, temperature, and pressure along the stagnation streamline (see Figs. 9,10) were used as input data for the overlay modeling of vibrational relaxation and chemical reactions in the gas flow on this streamline.

The input flowfield data and the results of the calculations for air at the two altitudes of 88 and 100 km are shown in Figs. 9,10, respectively. One can see that in both cases there exists a region of extreme vibrational disequilibrium, $T_v/T \le 0.1$, while the stagnation point temperature is much less than the maximum gas temperature behind the shock, $T_{st}/T_{max} \sim 0.1$. In these calculations, we used four different kinetic models: (i) Model II; (ii) Model I (both described in the previous Section); (iii) Model I with s=0.5 instead of s=0.7 in Eq. (4.7); and (iv) Model I with the MFR reaction rates instead of the Park rates of Eq. (4.7).

At a 88 km altitude, the use of the MFR rate model leads to a one and a two orders of magnitude increase in NO concentration, compared to the Park model (4.7) with s=0.7 and s=0.5, respectively. This result is in qualitative agreement with the 3-D

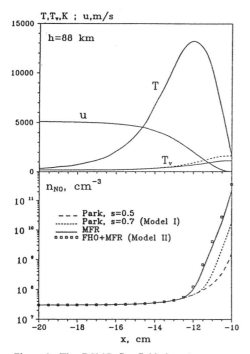

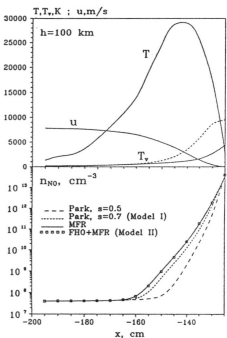

Figure 9. The DSMC flowfield data (upper part), and the NO concentration (lower part) on the stagnation streamline. h=88 km, u_s= 5 km/s, P_0 =1.9 mtorr.

Figure 10. The DSMC flowfield data (upper part), and the NO concentration (lower part) on the stagnation streamline. h=100 km, u_s= 8 km/s, P_0 =1.6 mtorr.

DSMC calculations of Boyd et al. [40], who use the MFR model for dissociation reactions only, and predict an order of magnitude increase compared to the Park model with s=0.5. The much faster NO production, predicted by the MFR model, occurs due to two equally important factors. First is the greater rate of oxygen dissociation at $T_v/T \to 0$ (see [40]), discussed in [40] and resulting in the earlier triggering of the Zel'dovich mechanism chain reactions (4.5,4.6), so that they occur at a higher translational temperature. Second is the higher rate of the key NO production reaction (4.5) in the region of extreme disequilibrium. One can also see that the NO production rate is weakly sensitive (within a factor of two) to the vibrational relaxation model (curves "MFR" and "FHO+MFR" in Fig. 9). This happens because the calculated vibrational temperature is so low that chemical reactions (4.2-4.6) proceed mainly from the ground vibrational level.

The main difference of NO production kinetics at a 100 km altitude is that there is already enough photodissociated oxygen in the incoming flow (~4% of the mixture [40]). Thus, delayed oxygen dissociation is no longer a trigger for the Zel'dovich mechanism reactions. Calculations at this altitude, shown in Fig. 10, therefore demonstrate only the effect of the variation of the reaction (5) nonequilibrium rate. One can see that the predictions of all models are closer than at an 88 km altitude, giving about an order of magnitude difference between the Park, s=0.5 model and the MFR model. The predicted NO concentrations near the stagnation point are very close for all models. NO production here is compensated by its decomposition in the reverse reaction (4.5,4.6), so that NO mole fraction reaches a quasi-equilibrium value and is not rate-sensitive. We note that the overlay approximation in the stagnation point region is no longer valid, since NO model fraction here reaches several percent. Finally, the effect of the vibrational kinetics rate model on NO concentration is again negligibly small, due to the low vibrational temperature in the region of the most intensive NO production.

5. Summary

In the present paper, we analyzed, corrected and validated the nonperturbative semiclassical analytic V-T and V-V-T rate models, previously suggested for the forced harmonic oscillator (FHO) [8-10]. The model developed shows remarkably good agreement with 3-D close-coupled calculations and provides analytic V-T and V-V rates for single-quantum and multi-quantum processes, including high temperatures. We also developed a theoretical method for calculation of nonequilibrium reaction rates, based on the classical impulsive model, that provides analytic two- and three- temperature rate constants.

We applied these rate models to analysis of NO production kinetics behind strong normal and bow shocks. The self-consistent state-to-state kinetic model, described here (Model II), was validated by comparison with shock tube experiments, for the shock velocities u_s=3-4 km/s. The results of the calculations illustrate, however, that a simplified kinetic model (Model I) and the Model II predict similar NO production rates behind normal shocks up to very high shock velocities u_s=9 km/s. This justifies the use of Model I in normal shock calculations.

NO production kinetics behind bow shocks is simulated by an overlay modeling, with the results of the DSMC flowfield calculations [40] used as inputs. The results of the overlay calculations show that correct modeling of vibration-chemistry coupling in extreme vibrational disequilibrium is one of the most crucial points in prediction of NO radiation behind the strong bow shocks. We also note that a reliable prediction of the NO ultraviolet radiation behind the shock requires additional studies of mechanisms of electronically excited NO formation, which are far from being understood at this time.

6. Acknowledgements

The work reported here was supported by NASA SBIR Research Grant NAS1-20140. We would like to thank Prof. Iain Boyd who promptly made available to us the results of the DSMC bow-shock calculations [40]. We would also like to express our gratitude to I.K. Dmitrieva, S.K. Pogrebnya, and P.I. Porshnev for drawing our attention to their paper [22], where they applied the results of the FHO theory to evaluate the single-quantum V-V rates.

References

1. C. Park, "Nonequilibrium Hypersonic Aerodynamics", Wiley, New York, 1990
2. G.A. Tirsky, in: Ann. Rev. Fluid Mech., vol.25, 1993, p.151
3. C. Park, in: "Thermal Design of Aeroassisted Orbital Transfer Vehicles", ed. H.F. Nelson, Progress in Astronautics and Aeronautics, vol.96, AIAA, New York, 1985, p.511
4. P.A. Gnoffo, AIAA Paper 89-1972, 1989
5. L.C. Hartung, Journal of Thermophysics and Heat Transfer, vol. 6, 1992, p.618
6. C. Park and S.H. Lee, AIAA Paper 93-2862, 1993
7. I.D. Boyd and T. Goksen, J. of Thermophysics and Heat Transfer, v.7, 1993, p.406
8. E.H. Kerner, Can. J. Phys., v.36, 1958, p.371
9. C.E. Treanor, J. Chem. Phys., v.43, 1965, p.532
10. A. Zelechow, D. Rapp and T.E. Sharp, J. Chem. Phys., v.49, 1968, p.286
11. I.V. Adamovich, S.O. Macheret, J.W. Rich and C.E. Treanor, AIAA Journal, vol.33, No.6, 1995, p.1064
12. D. Rapp and T. Kassal, Chem. Rev., v.69, 1969, p.61
13. W.R. Gentry, in: "Atom-Molecule Collision Theory", ed. by R.B. Bernstein, Plenum Press, New York, 1979, p.391
14. E.E. Nikitin and A.I. Osipov, "Vibrational Relaxation in Gases", in "Kinetics and Catalysis" series, v.4, VINITI, Moscow, 1977
15. I.V.Adamovich, S.O.Macheret, J.W.Rich and C.E.Treanor, AIAA Paper 95-2060, 1995
16. J. Keck and G. Carrier, J. Chem. Phys., v.43, 1965, p.2284
17. G.D. Billing and E.R. Fisher, Chem. Phys., v.18, 1976, p.225
18. L.L. Poulsen and G.D. Billing, Chem. Phys., v.36, 1979, p.271
19. E.E. Nikitin, "Theory of Elementary Atomic and Molecular Processes in Gases", Clarendon Press, Oxford, 1974
20. G.D. Billing, Chem. Phys., v.50, 1980, p.165
21. M. Cacciatore and G.D. Billing, Chem. Phys., v.58, 1981, p.395
22. I.K. Dmitrieva, S.K. Pogrebnya, and P.I. Porshnev, Chem.Phys., v.142, 1990, p.25
23. J.M.Price, J.A.Mack, C.A.Rogaski and A.M.Wodtke, Chem. Phys., v.175, 1993, p.83
24. H. Park and T.G. Slanger, J. Chem. Phys., v.100, 1994, p.287
25. M. Klatt, I.W.M. Smith, R.P. Tuckett and G.N. Ward, Chem. Phys. Lett., v.224, 1994, p.253
26. G.D. Billing and E.R. Fisher, Chem. Phys., v.43, 1979, p.395
27. G.D. Billing and R.E. Kolesnick, Chem. Phys. Lett., v.200, 1992, p.382
28. G.D. Billing, Chem. Phys., v.179, 1994, p.463
29. G.D. Billing, in:"Nonequilibrium Vibrational Kinetics", ed. by M. Capitelli, Springer-Verlag, Berlin, 1986, p.85

104

30. S.O. Macheret, A.A. Fridman and A.A. El'kin, Khimicheskaya Fizika (Sov. Chem. Phys.), vol.9, 1990, p.174

31. S.O. Macheret and J.W. Rich, Chem. Phys., vol.174, 1993. p.25

32. S.O. Macheret, A.A. Fridman, I.V. Adamovich, J.W. Rich and C.E. Treanor, AIAA Paper 94-1994, June 1994

33. I.V. Adamovich, S.O. Macheret, J.W. Rich and C.E. Treanor, AIAA Journal, vol.33, No. 6, 1995, p.1070

34. G. Black, R.L. Sharpless and T.G. Slanger, J. Chem. Phys., vol.58, p.4792, 1973

35. W.H. Wurster, C.E. Treanor and M.J. Williams, "Non-equilibrium Radiation from Shock-Heated Air", Final Report. U.S. Army Research Office, Contract No. DAAL03-88K-0174, July 1991, Calspan UB Research Center, Buffalo, NY

36. J.F.Clark and M.McChesney, "Dynamics of Real Gases", Butterworths, London, 1976

37. C.E. Treanor and M.J. Williams, "Kinetics of Nitric Oxide Formation Behind 3 to 4 km/s Shock Waves", Final Report. U.S. Army Research Office, Contract No. DAAL03-92K-0003, February 1993, Calspan UB Research Center, Buffalo, NY

38. M. Camac, R. Feinberg and J.D. Teare, "The Production of Nitric Oxide in Shock-Heated Air", Avco Research Report 245, December 1966

39. J.P. Monat, R.K. Hanson and C.H. Kruger, "Shock Tube Determination of the Rate Coefficients for the Reaction $N_2 + O \rightarrow NO + N$", Proceedings of the 17th International Symposium on Combustion", 1978, p.543

40. I.D.Boyd, G.V.Candler and D.A.Levin, 'Dissociation Modeling in Low Density Hypersonic Flows of Air", to be published in Physics of Fluids, July 1995

NON-EQUILIBRIUM CHEMISTRY MODELS FOR SHOCK-HEATED GASES

PHILIP L. VARGHESE
Department of Aerospace Engineering & Engineering Mechanics
The University of Texas at Austin, Austin, Texas 78712, USA

DAVID A. GONZALES
Department of Aeronautics & Astronautics
Massachusetts Institute of Technology, Cambridge, Massachusetts 02139, USA

Abstract

We demonstrate the utility of Information Theory for the efficient computation of state-specific reaction rates needed in hypersonic flow computations. We summarize our recent work on vibrational state-specific inelastic rates and dissociation cross-sections. The reaction rate matrix we generate is used to compute the spatially homogenous relaxation process of a diatomic gas dilute in an inert bath gas when the temperature is suddenly raised to a high value. Linear algebraic methods are used to solve the master equation efficiently. We find that almost the entire relaxation process, including vibrational state populations, can be described accurately using data related to the translational temperature only.

1. Introduction

Non-equilibrium flows are of fundamental importance in hypersonic flows. The design and development of advanced propulsion systems, high speed aircraft, and spacecraft depend critically on an accurate description of flows where chemical reactions occur while the gas is also relaxing thermally, i.e. when the rotation-vibration distributions are non-Boltzmann. Non-equilibrium reaction rates are very different from thermal equilibrium rates. Most present non-equilibrium rate expressions are highly empirical and cannot be extrapolated with confidence. However, a complete description of non-equilibrium by tracking individual vibration-rotation states is completely unfeasible. First, most state-specific rates are unknown, and it would be extremely expensive to use chemical dynamics calculations to generate the very large matrix of rates needed. Second, even if these rates were available, it is impossible to track many thousands of individual quantum states when computing any flow of engineering interest. What one needs are good engineering models that will describe non-equilibrium situations accurately but with a minimum number of parameters. We believe that Information Theory can assist in the solution of these problems. This paper summarizes our recent work in this area [1-3].

M. Capitelli (ed.), Molecular Physics and Hypersonic Flows, 105–114.
© *1996 Kluwer Academic Publishers. Printed in the Netherlands.*

2. Information theory

The application of Information Theory (IT) to computation of chemical reaction rates was pioneered by Levine and Bernstein [e.g. 4, 5]. Despite some successes the method is not widely used in the chemistry community because, in general, it does not yield accurate absolute reaction rates. However, as we show below, the method does provide good relative rates and hence may be used for scaling purposes. Additionally, the absolute predictions are substantially better than simple first order methods. Thus the method is useful for preliminary estimates when better data (obtained by quasi-classical or quantum trajectory calculations) are unavailable. The essential features of the IT method are outlined below; further details are given by Procaccia and Levine [5].

(1) It provides an efficient way to generate a large matrix of state-specific cross-sections or rate coefficients; both inelastic and dissociation processes can be computed.

(2) Although the absolute cross-sections/rates are not very accurate, the scaling relations of the rates with internal state quantum number and collision energy or temperature are good. Thus, if a few absolute values of state-specific cross-sections or rates are measured experimentally, or computed by trajectory calculations, then the entire rate matrix may be filled out with the help of IT. The method works best when filling out a large matrix when sparse but accurate data are available. It may be regarded as complementary to the trajectory calculations and helps provide maximal utilization of such computations.

(3) The theory provides a rational approach to describing non-equilibrium systems with the minimum information, i.e. the smallest number of parameters. We hope to exploit this feature in future work aimed at developing compact but accurate descriptions of non-equilibrium reaction rates for use in hypersonic flow computations.

The fundamental postulate in this approach is that the probability of a particular outcome in a collision process would be determined purely by statistics in the absence of bias. This *a-priori* probability of a particular outcome of a collision is just the ratio of the number of states representing this outcome to the total number of states that are energetically accessible. In fact, the actual outcomes of collisions are *not* purely statistical: some outcomes are more likely and others less likely than predicted from the *a-priori* probability. This bias is a measure of the information content in the distribution of actual outcomes. The departure from *a-priori* statistical expectation is measured quantitatively using the "surprisal". We illustrate the approach by examining state-specific vibrational energy transfer rates.

2.1 STATE-TO-STATE INELASTIC RATES

Consider the vibrationally inelastic process of a diatom A_2 with an inert collision partner M:

$$A_2(v) + M \rightarrow A_2(v') + M$$

At a defined temperature of the bath gas and assuming that the rotational temperature is the same as the bath gas temperature, the surprisal $I(v,v'; T)$ for the state specific process $(v \rightarrow v')$ is given by

$$I(v,v'; T) \equiv -\ln[k(v \rightarrow v'; T)/k^\circ(v \rightarrow v'; T)] \tag{1}$$

where $k°(v \to v'; T)$ is the "prior" rotationally-averaged rate computed from the density of vibrational states v' under the specified conditions, and $k(v \to v'; T)$ is the actual rotationally-averaged rate coefficient. Similar expressions can be written for rovibrational state-specific rates, and even velocity dependent cross-sections [5]. As written, the equation permits one to *analyze* known rate data. In order to *synthesize* reaction rates, a model for the surprisal is needed. Experimentally, a linear surprisal model has been found to be reasonable in many (though certainly not all) cases

$$I(v,v'; T) = \lambda_o + \lambda \frac{|E_v - E_{v'}|}{\kappa T}; \qquad (2)$$

here λ_o is a normalizing factor related to elastic rates, and λ is the surprisal parameter. E_v is the energy of vibrational state v, and κ is Boltzmann's constant.

Using this model for I gives

$$k(v \to v'; T) = A(T) \, k°(v \to v'; T) \, exp\left(-\lambda \frac{|E_v - E_{v'}|}{kT}\right), \qquad (3)$$

where

$$A(T) = exp(-\lambda_o) = \frac{k(v \to v; T)}{k°(v \to v; T)}. \qquad (4)$$

The surprisal parameter λ is obtained by the further assumption that the individual inelastic rates obey a "sum-rule" [5]

$$\sum_v (E_{v'} - E_v) \, k(v \to v'; T) = \frac{V(\infty) - E_v}{(p\tau)/\kappa T}, \qquad (5)$$

where $V(\infty)$ is the average vibrational energy at equilibrium, p is the pressure, and τ is the vibrational relaxation time. The sum-rule is a necessary and sufficient condition that the mean vibrational energy relaxes exponentially (in the absence of dissociation) [5]. This is close to the experimentally observed behavior. If the vibrational relaxation time is measured experimentally, or calculated by the formula of Millikan and White [6] (for example), then Eq. 5 provides the means to determine λ from a macroscopic observable [2, 5].

2.2 COMPARISON WITH QCT RESULTS

The results of the information theoretic formulation may be compared to direct quasi-classical (QCT) calculations for p-H_2 by Duff, et al. [7]. Figure 1 shows vibrational energy transfer rates from $v = 4$ of p-H_2 dilute in Ar at 4500 K; all rates are normalized by the elastic collision frequency Z. The vibrational relaxation time for H_2 in Ar was obtained from [8]. All the information theoretic results were scaled by a factor of 40 so that $k_{IT}(1 \to 0,T) = k_{QCT}(1 \to 0,T)$. Using this single scaling factor IT provided a good match of the inelastic rates for all other states, as shown for $v = 4$ on Fig. 1. Note also that both single and multi-quantum rate predictions are reasonable. In contrast, Fig. 1 shows that inelastic rates computed by the Keck and Carrier [9] implementation of SSH theory [10], similarly scaled to match the $1 \to 0$ rate, do not scale well at high v, and predict multi-quantum inelastic rates that are too low by many orders of magnitude. Additional details and other computations are given in [2].

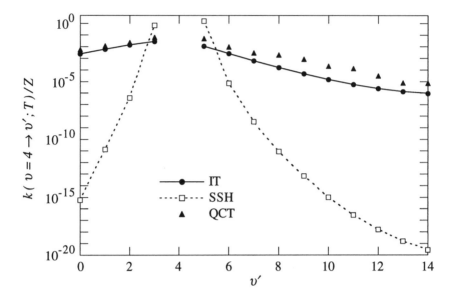

Figure 1 Normalized vibrational energy transfer rates from $v = 4$ for $p\text{-H}_2\text{-Ar}$ collisions at 4500 K.

2.3. ROVIBRATIONAL STATE-SPECIFIC DISSOCIATION MODEL

Information theory can also be used to predict state-specific dissociation cross-sections. We have extended the work of Kafri and Levine [11] and Kiefer and Hajduk [12] to allow for the contribution of rotational energy to the probability of dissociation; details are provided in [1]. The result of the information theoretic synthesis is a rovibrational state-specific dissociation rate that is assumed to follow a surprisal of the form

$$k(vN\rightarrow c;T) = B(v,N;T)\, k^{o}(vN\rightarrow c;T)\, exp\left(-\lambda_D \frac{D_o - E_{vN}}{kT}\right) \tag{6}$$

Here B is a scaling term determined by the model assumptions [1] for the relative efficiency of the contribution of vibrational and rotational energy to dissociation probability, k^{o} is the prior rate obtained from statistical mechanics, and λ_D is the surprisal parameter for dissociation that must be determined. An analytical expression for $k(vN\rightarrow c;T)$ can be obtained by suitable approximations [1]. The vibrational state-specific dissociation rate $k(v\rightarrow c;\ T)$ is obtained by averaging over the appropriate rotational distribution:

$$k(v\rightarrow c\ ;\ T) = \sum_{N} Y_{N}(v\,|\,T)\, k(vN\rightarrow c\ ;\ T) \tag{7}$$

where $Y_N(v\,|\,T)$ is the fractional population of rotational state N in vibrational state v. In our work [1-3] we have approximated Y_N by a Boltzmann distribution characterized by the bath gas temperature T. The surprisal parameter is chosen so that the temperature dependence of the steady-state dissociation rate matches experimental shock-tube data. The steady-state rate is defined by

$$k_d^{ss}(T) = \sum_{\upsilon} P_{\upsilon}^{ss} k(\upsilon \to c; T), \tag{8}$$

where P_{υ}^{ss} is the fractional population in level υ at steady-state. The steady-state distribution is determined by solving the master equation as described below. In practice we have found $\lambda_D \approx 1$ for several diatomics (H_2, N_2, O_2, CO) colliding with Ar.

The performance of the model may be assessed by direct comparison of synthesized rovibrational state-specific dissociation rates with the QCT results of Haug, *et al.* for H_2 in Ar [13]. Haug, *et al.* specify an effective cross-section defined by

$$\sigma(\upsilon N \to c; T) \equiv k(\upsilon N \to c; T)/\langle v; T \rangle \tag{9}$$

where $\langle v; T \rangle$ is the average relative thermal velocity. The state-specific rate coefficients computed using IT were converted to effective cross-sections using Eq. 9 and are plotted in Fig. 2 with the corresponding QCT values. The figure shows that the IT results [1] agree well with the QCT calculations [13].

3. Solutions of the Master Equation

The matrix of vibrational state-specific inelastic and dissociative rates computed by IT may be used to solve the master equation describing the time-dependence of the vibrational state population. The master equations are a set of stiff ordinary differential equations that are expensive to solve. If the diatom is dilute in a bath gas, the problem is linear provided that atomic recombination reactions can be neglected. This is true at early times, i.e. before a significant fraction of the diatomic molecules have dissociated. Under these conditions, it is well known that the master equations can be written as a linear system and solved efficiently using eigenvalue methods. The solution of this linear system predicts that the population evolves to a steady-state distribution, where the fractional population in individual levels remains constant. The final equilibrium distribution is only obtained when recombination reactions are included. This non-linear problem can also be solved approximately by linearizing about the final equilibrium distribution. The two approaches are outlined in the following sections.

3.1 EVOLUTION TO STEADY-STATE

The solution of the linear master equation neglecting atomic recombination is straightforward. There are some numerical problems associated with expanding the initial population distribution in the eigenvlaues for very low initial temperatures (e.g. 300 K) because of the very wide range of magnitudes involved. A solution to these problems has been developed [14,15]. In general, the solutions of the master equation show a very rapid evolution to steady-state determined entirely by the final temperature; the initial conditions only influence the population evolution at early times. Thus the early transient behavior is not important for most applications.

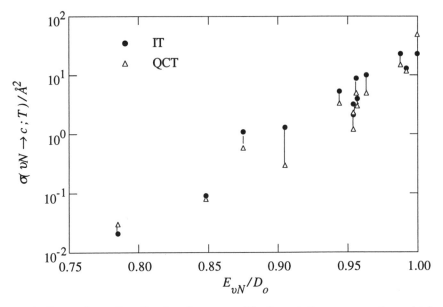

Figure 2 Comparison of rovibrational state-specific dissociation cross-sections obtained by IT [1] with those from QCT [17].

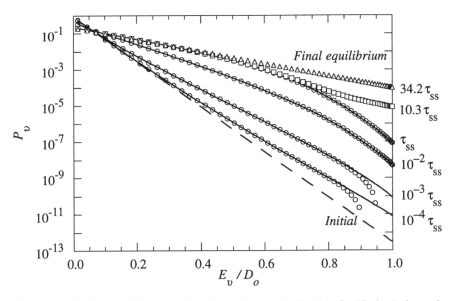

Figure 3 Evolution of the vibrational population distribution for N_2 in Ar heated suddenly from 4000 K to 15000 K. The dashed line is the initial Boltzmann distribution, and the solid lines are linear master equation calculations neglecting recombination. The symbols are the linearized master equation calculations that include atomic recombination.

Fig. 3 shows the time-dependent vibrational population distributions of 5% N_2 dilute in Ar when the gas initially at 4000K is suddenly heated to 15000 K. The dashed line is the initial Boltzmann distribution at 4000 K. The solid lines are the vibrational population distributions computed from the linear eigenvalue problem neglecting recombination. The symbols are the solution of the linearized problem including recombination discussed in the next section. τ_{ss} is the steady-state time. The rapid initial rise of the high-lying states at very early times arises because the rates of multi-quantum V-T processes dominate dissociation from these levels [14]. As steady-state is approached, the population of these states is depleted because of dissociation.

The fractional contributions f_v^{ss} of individual vibrational states to the overall steady-state dissociation rate k_d^{ss} can be computed from the state-specific rates and the steady-state vibrational distribution:

$$f_v^{ss} \equiv \frac{P_v^{ss} k(v \to c)}{\sum_v P_v^{ss} k(v \to c)} = \frac{P_v^{ss} k(v \to c)}{k_d^{ss}}. \tag{10}$$

Typically we find that the dominant contribution to the steady-state dissociation comes from states near the middle of the vibrational manifold [1, 2]. This reflects the balance between increasing dissociation rate and decreasing steady-state population with increasing v. These results, based on information theoretic rates, are in good agreement with QCT calculations for dissociation of H_2 [13] and Br_2 [16].

3.2 APPROACH TO FINAL EQUILIBRIUM

Once steady-state is established, the population of atoms begins to grow, and the final equilibrium distribution is only established when recombination processes are included. This process is non-linear in the atom concentration. However, the master equation may be linearized about the final equilibrium state to obtain an eigenvalue problem. Details of the procedure are given in [3]. The smallest eigenvalue for this problem is zero, so that an equilibrium distribution is obtained as $t \to \infty$. This is in contrast to the linear problem described above where the distribution does not evolve beyond a steady-state. The symbols on Fig. 3 show the time-dependent N_2 vibrational population distributions computed both for times prior to steady-state and for times after steady state. The distribution is at the final equilibrium at the longest time shown ($t=34.2\tau_{ss}$). It is evident that the two computations agree very well except for the highest lying states (fractional populations $< 10^{-8}$) at very early times ($t \le 10^{-3}\tau_{ss}$). Indeed, it is remarkable that the distributions for times as early as $t = 10^{-4}\tau_{ss}$ computed by linearizing about the *final* state agree with the distributions computed for the pure dissociation problem. This indicates that the vibrational population distribution for nearly the entire coupled vibrational relaxation and dissociation process can be accurately described knowing a *single* parameter T_f, even though the distributions are non-Boltzmann. This observation could guide the search for simple models for non-equilibrium dissociation rates.

The mean vibrational energy changes during the relaxation to equilibrium. The linear relaxation rate (LRR) equation is commonly used to simulate vibrational energy relaxation:

$$\frac{dV(t)}{dt} = \frac{V^e - V(t)}{\tau_{\text{vib}}} \tag{11}$$

where V^e is the equilibrium vibrational energy and τ_{vib} is the vibrational relaxation time [6]. This is often referred to as the Landau-Teller model and predicts a simple exponential relaxation for the conditions of the simulation shown in Fig. 3. In contrast, the vibrational relaxation obtained by solution of the master equation exhibits a two-step relaxation. There is an initial rise to steady-state, a nearly constant value during steady state, and a final rise to equilibrium [3]. Similar results were shown by Schwenke [17].

Park has proposed a modification of the LRR to reduce the rate of vibrational relaxation [18]. The model assumes a diffusion of vibrational energy through the vibrational manifold and reduces the rate of vibrational relaxation from that computed by the simple LRR model. Figure 4 compares the variation of dV/dt computed from the different models during the relaxation process. For convenience, we use the dimensionless variables $v \equiv V/V^e$, $\tau \equiv t/\tau_{\text{vib}}$ and plot $dv/d\tau$. The vibrational relaxation *rate* computed from the solution of the master equation is high initially, plateaus at a low value during the steady-state period, accelerates again at the end of this period, and then rapidly decays as the final equilibrium is established [3]. If the instantaneous values of $V(t)$ computed from the master equation are used in the LRR model or in the Park model one obtains the other curves shown in Fig. 4. The Park model reduces the LRR relaxation rate as expected. However, the figure shows that the model does not reduce the relaxation rate enough during the steady-state period, and reduces it too much during the approach to equilibrium.

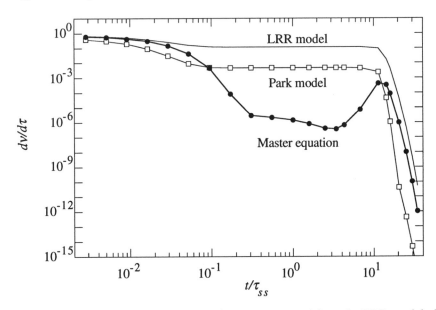

Figure 4 Normalized vibrational relaxation rate computed from the LRR model, the Park model, and directly from the linearized master equation solution.

4. Summary and Conclusions

We have illustrated how Information Theory may be used to generate a large number of state-specific inelastic and dissociation rates. The theory provides a good scaling procedure, and may be used to supplement quasi-classical or quantum trajectory calculations of individual state-specific inelastic and dissociative cross-sections such as described by Billing [19, 20], Laganà [21], and Gilibert [22]. (See also the papers in these Proceedings.) Even in the absence of scale factors provided by a few accurate calculations the information theoretic method is much more reliable than SSH-type models. Our model for rovibrational state-specific dissociation indicates that the dominant contribution to dissociation come from highly-lying rotational states with moderate vibrational excitation. This is in agreement with QCT calculations.

At least for the case of a diatomic dilute in an inert bath gas, the non-equilibrium relaxation and dissociation processes can be very accurately modeled knowing the translational temperature only. This could provide the basis for an accurate *single* parameter model for non-equilibrium dissociation rate. However, similar results must first be demonstrated for the relaxation of non-dilute mixtures of diatoms. We plan to extend information theoretic methods to diatom-diatom inelastic and reactive collisions and synthesize the very large state-specific rate matrices involved.

Acknowledgements

This work was supported by NASA Minority Graduate Engineering Grant NGT-70032 and the Texas Advanced Research Program.

5. References

1. Gonzales, D.A. and Varghese, P.L. (1993) A Simple Model for State-Specific Diatomic Dissociation, *J. Physical Chemistry* **97**, 7612–7622.
2. Gonzales, D.A. and Varghese, P.L. (1994) Evaluation of Simple Rate Expressions for Vibration-Dissociation Coupling, *J. Thermophysics and Heat Transfer* **8**, 236–243.
3. Gonzales, D.A. and Varghese, P.L. (1995) Vibrational Relaxation Models for Dilute Shock-Heated Gases, *J. Chemical Physics* **195**, 83–91.
4. Levine, R.D. and Bernstein, R.B. (1976) Thermodynamic Approach to Collision Processes, in W.H. Miller (ed.), *Dynamics of Molecular Collisions, Part B*, Plenum Press, New York, pp. 323–364.
5. Procaccia, I. and Levine, R.D. (1975) Vibrational Energy Transfer in Molecular Collisions: An Information Theoretic Analysis and Synthesis, *J. Chemical Physics* **63**, 4261–4279 (see also references therein)
6. Millikan, R.C. and White, D.R. (1963) Systematics of Vibrational Relaxation, *J. Chemical Physics* **39**, 3209–3213.

114

7. Duff, J.W., Blais, N.C., and Truhlar, D.G. (1979) Monte Carlo Trajectory Study of Ar+H_2 Collisions: Thermally Averaged Vibrational Transition Rates at 4500 K, *J. Chemical Physics* **71**, 4304–4320.

8. Dove, J.E. and Teitelbaum, H. (1974) The vibrational relaxation of H_2. I. Experimental measurements of the rate of relaxation by H_2, He, Ne, Ar, and Kr, *Chemical Physics* **6**, 431-444.

9. Keck, J. and Carrier, G. (1965) Diffusion Theory of Nonequilibrium Dissociation and Recombination," *J. Chemical Physics* **43**, 2284–2298.

10. Schwartz, R.N., Slawsky, Z.J., and Herzfeld, K.F. (1952) Calculation of Vibrational Relaxation Times in Gases, *J. Chemical Physics* **20**, 1591–1599.

11. Kafri, A. and Levine, R.D. (1976) Comment on the dynamics of dissociation of diatomic molecules: Mass and temperature effects, *J. Chemical Physics* **64**, 5320–5321.

12. Kiefer, J.H. and Hajduk, J.C. (1979) A Vibrational Bias Mechanism for Diatomic Dissociation: Induction Times and Steady Ratres for O_2, H_2 and D_2 Dilute in Ar, *Chemical Physics* **38**, 329–340.

13. Haug, K., Truhlar, D.G., and Blais, N.C. (1987) Monte Carlo Trajectory and Master Equation Simulation of the Nonequilibrium Dissociation Rate Coefficient for Ar+$H_2 \rightarrow$Ar+2H at 4500 K, *J. Chemical Physics* **86**, 2697–2716.

14. McElwain, D.L.S. and Pritchard, H.O. (1971) The Temperature Coefficients of Diatomic Dissociation and Recombination Reactions, *Thirteenth Symposium (International) on Combustion,* Combustion Institute, Pittsburgh, pp. 37-49.

15. Pritchard, H.O. (1975) Network Effects in the Dissociation and Recombination of a Diatomic Gas, *Reaction Kinetics* **1**, 243–290.

16. Itoh, H.; Koshi, M.; Asaba, T.; Matsui, H. (1985) Vibrational Non-equilibrium Dissociation of Br_2 in Collisions with Ar and Br Atoms,.*J. Chemical Physics* **82**, 4911–4915

17. Schwenke, D.W. (1990), A theoretical prediction of hydrogen molecule dissociation-recombination rates including an accurate treatment of internal state nonequilibrium effects, *J. Chemical Physics* **92**, 7267–7282.

18. Park, C. (1989) Assessment of Two-Temperature Kinetic Model for Ionizing Air, *J. Thermophysics Heat Transfer* **3**, 233–244.

19. Billing, G.D. (1986) Vibration-vibration and Vibration-Translation Energy Transfer, Including Multiquantum Transitions in Atom-Diatom and Diatom-Diatom Collisions, in M. Capitelli (ed.) *Nonequilibrium Vibrational Kinetics*, Springer-Verlag, Berlin, pp. 85-112.

20. Billing, G.D. (1994) Classical path method in inelastic and reactive scattering, *International Reviews in Physical Chemistry* **13**, 309-335.

21. Laganà, A. and Garcia, E. (1994) Temperature Dependence of N + N_2 Rate Coefficients, *J. Physical Chemistry* **98**, 502–507.

22. Gilibert, M., Giménez, X., González, M., Sayós, R. and Aguilar, A. (1995) A comparison between experimental, quantum and quasiclassical properties for the $N(^4S_u) + O_2(X \ ^3\Sigma_g^-) \rightarrow NO(X \ ^2\Pi) + O(^3P_g)$ reaction, *Chemical Physics* **191**,1–15.

ELECTRON - AIR MOLECULE COLLISIONS IN HYPERSONIC FLOWS

WINIFRED M. HUO AND HELMAR T. THÜMMEL
NASA Ames Research Center, Moffett Field, CA 94035-1000, U.S.A.

1. Introduction

At high vehicle velocity, the flow field surrounding a space vehicle becomes partially ionized, with the amount of ionization determined by the vehicle velocity. For a transatmospheric flight at an altitude of 80 km and a velocity of 10 km/sec, a maximum of 1% ionization is expected, whereas up to 30% of the atoms and molecules in the flow field will be ionized in a Mars return mission with a velocity of 12 - 17 km/sec. Even when the flow field is only slightly ionized, electron collisions still have an important role in determining the internal energy and state distribution of the atoms and molecules in the flow field. This in turn contributes to the flow characteristic and the amount of radiative heat load on the vehicle.

Electron-molecule collisions differ from atom-molecule and molecule-molecule collisions in two aspects. First, the mass of an electron is more than four orders of magnitude smaller than the reduced mass of N_2. Thus its average speed, and hence its average collisional frequency, is more than 100 times larger. Even in the slightly ionized regime with only 1% electrons, the frequency of electron-molecule collisions is equal to or larger than that of molecule-molecule collisions, an important consideration in the low density part of the atmosphere where the reaction probability is frequently controlled by the collisional frequency. Second, the interaction potential between a charge particle (electron) and a neutral particle is longer range than neutral-neutral interactions. Hence electron-molecule collision cross sections tend to be larger. Also, low-energy electron collisions are particularly effective in spin changing excitations.

Three electron-molecule collisional processes relevant to hypersonic flows will be considered in this chapter: (1) vibrational excitation/de-excitation of an air molecule by electron impact, (2) electronic excitation/de-excitation,

M. Capitelli (ed.), Molecular Physics and Hypersonic Flows, 115–137.

and (3) dissociative recombination in electron-diatomic ion collisions. A review of available data will be given. The emphasis is on tailoring the molecular physics to the condition of hypersonic flows. For example, the high rotational and vibrational temperatures in a hypersonic flow field mean that most experimental data measured at room temperature are not applicable. Also, the average electron temperature is expected to be between 10,000 and 20,000 K. Thus only data for low energy electrons will be discussed.

The lack of experimental data at the high rotational and vibrational temperatures appropriate to the description of a hypersonic flow means that the necessary database for modeling must come from theoretical calculations. In the last decade, significant progress has been made in improving the computational techniques [1] so that currently cross section data determined from theory can be as accurate as the corresponding experimental values in the region where experimental data are available. In particular, *ab initio* methods, which calculate cross sections from first principles without relying on experimental data or adjustable parameters, have proven to be highly successful. In the following, a brief summary of three *ab initio* methods currently in use will be given before the rates and cross section data relevant to the modeling of hypersonic flows are presented. For readers who are interested only in the cross section data, Sec. 2 can be skipped.

2. *Ab Initio* Methods for the Calculation of Electron-Molecule Collision Cross Sections

Due to the light electronic mass, e-molecule collisions must be treated quantum mechanically. The Schrödinger equation for the e-molecule system is given by

$$(H - E)\Psi(\tau, \mathbf{R}) = 0, \tag{1}$$

where H is the Hamiltonian of the e + molecule system, E the total energy, and Ψ the corresponding wave function. Also, τ the spatial and spin coordinates of the electrons and $\mathbf{R}$ the nuclear coordinates. The total Hamiltonian H consists of the molecular Hamiltonian, H_{mol}, the kinetic energy operator of the continuum electron, T_e, and the Coulomb potential between the continuum electron and the molecule, V.

$$
\begin{aligned}
H &= H_{mol} + T_e + V, \\
H_{mol} &= -\frac{1}{2}\sum_{i=1}^{N}\nabla_{r_i}^2 - \sum_{k=1}^{M}\nabla_{R_k}^2 + \sum_{i>j}^{N}\frac{1}{|\mathbf{r}_i - \mathbf{r}_j|} \\
&\quad - \sum_{i=1}^{N}\sum_{k=1}^{M}\frac{Z_k}{|\mathbf{r}_i - \mathbf{R}_k|} + \sum_{k>l}^{M}\frac{Z_k Z_l}{|\mathbf{R}_k - \mathbf{R}_l|},
\end{aligned}
$$

$$T_e = -\frac{1}{2}\nabla^2_{\mathbf{r}_{N+1}}$$

$$V = \sum_{i=1}^{N}\frac{1}{|\mathbf{r}_{N+1}-\mathbf{r}_i|} - \sum_{k=1}^{M}\frac{Z_k}{|\mathbf{r}_{N+1}-\mathbf{R}_k|}. \qquad (2)$$

Here the label $N+1$ denotes the continuum electron. Note that due to the indistinguishability of electrons, the label $N+1$ is purely arbitrary.

Making use of the large difference between the electronic and nuclear masses, the solution of Eq. (1) is generally separated into two steps. First, electron dynamics is solved assuming the nuclei are fixed. This solution is repeated for a range of relevant nuclei geometries. Nuclear dynamics calculation is then carried out based on the data from the electron dynamics study.

Virtually all fixed-nuclei calculations have been carried out in a coordinate system which rotates with the molecule and with the origin at its center of mass, the so called molecular frame or body frame. Note that the molecule is rotationless in the this frame. Thus for a diatomic molecule the only relevant nuclear coordinate in the molecular frame is the internuclear distance R. Note also that molecular rotation can only be described in the laboratory frame, but vibrational motion can be treated using the molecular frame or laboratory frame. Furthermore, at electron energies above the rotational threshold, the adiabatic nuclei rotational approximation (ANR) is generally valid [5]. In ANR, it is assumed that the characteristic collision time is short in comparison with the rotational period so that rotational motion can be uncoupled from electronic and vibrational motion. Thus the major effect of rotation is to broaden the elastic or inelastic vibronic cross sections, but the area under the cross section curve, and hence the vibronic excitation rates, are little changed by including rotation. This approximation will be employed in the discussions below. Another effect of rotational motion, the shift of vibrational levels due to the change in the centrifugal potential, will be discussed in Sec. 3.

Three *ab initio* methods frequently used in the study of electron collision of air molecules and their ions are: the R-matrix method, Schwinger variational method, and complex Kohn variational method. So far, these methods provide the most accurate cross sections available in the literature. They are described below.

2.1. THE R-MATRIX METHOD

The R-matrix method [2], [3] takes advantage of the natural separation of electron-molecule interactions and divide the interaction space into two regions separated by a sphere of radius s. Here s is taken to be large enough to envelop the charge distribution of the target states of interest. Thus, when the scattering electron is in the internal region ($r < s$) it lies within the molecular charge cloud and electron exchange and correlation effects must be included in the treatment. In this region the wave function of the target molecule plus scattering electron resembles that of a negative ion bound state, and the variational R-matrix method is analogous to bound state molecular configuration-interaction calculations [4]. However, unlike bound state calculations, the one-particle basis is augmented with functions representing the continuum electron. In the external region the scattering electron moves in the long-range multipole potential of the target. General solutions of the external Schrödinger equation with proper scattering boundary conditions can be obtained analytically. These solutions are matched with the solution at the inner region via the R-matrix and the cross sections are determined.

2.1.1. Fixed-Nuclei R-Matrix

Let R_o represent a given internuclear distance. In the molecular frame the wave function of the $N + 1$-electron system in the inner region is expressed as

$$\Psi(\tau, R_o) = \sum_a \mathcal{A}\left\{\Phi_a(\tau_1 \ldots \tau_N, R_o) f_a(\tau_{N+1}, R_o)\right\}$$
$$+ \sum_d \Theta_d(\tau_1 \ldots \tau_{N+1}, R_o); \ r_i \leq s, i = 1, \ldots N + 1. \ (3)$$

Here Φ_a represents the ath target state, f_a represents the scattering electron, and Θ_d is a multicentered square-integrable function which vanishes at $r_i = s$, i=1, ... N+1. The antisymmetrizer $\mathcal{A}$ permutes the electronic coordinates τ_{N+1} with $\tau_i, i = 1, \ldots N$ in order to account for the indistinguishability of electrons.

The Schrödinger equation in the inner region is rewritten as

$$(H + L - E)\Psi = L\Psi. \tag{4}$$

where the Bloch operator L is introduced so that $H + L$ is a Hermitian operator.

$$\left.(L - \sum_i \frac{1}{2}\nabla^2_{\mathbf{r}_i})\right|_s = \sum_i \frac{1}{2}(\nabla^{\mathcal{L}}_{\mathbf{r}_i} \bullet \nabla^{\mathcal{R}}_{\mathbf{r}_i})\Big|_s. \tag{5}$$

Here s is the value of $\mathbf{r}_i$ at the surface of the R-matrix sphere. The superscripts $\mathcal{L}$ and $\mathcal{R}$ denote the gradient to be taken on the function to the left or right of the operator.

The R-matrix corresponds to a logarithmic derivative matrix which connects the wave function of the continuum electron in the inner and outer regions. The variationally stable form of the R-matrix [6], [7] is given by

$$\mathcal{R}_{ij} = \frac{1}{2}\psi_i(\tau, R_o) (A^{-1})_{ij} \psi_j^\dagger(\tau, R_o), \qquad (6)$$
$$A = H + L - E,$$

with ψ_i, ψ_j the $N + 1$-electron basis functions used to construct the wave function Ψ in Eq. (3). Also the superscript $\dagger$ denotes a Hermitian conjugate.

The wave function in the external region $r_{N+1} \geq s$ is analogous to the first term in Eq. (3).

$$\Psi(\tau, R_o) = \sum_a \left\{ \Phi_a(\tau_1 \ldots \tau_N, R_o) f_a(\tau_{N+1}, R_o) \right\};$$
$$r_{N+1} \geq s; \quad r_i \leq s, i = 1, \ldots N. \qquad (7)$$

Recall that s is chosen such that the molecular charge density vanishes outside $r_i = s, i = 1, \ldots N$. Because r_{N+1} is defined in a different region of space from the molecular electrons, exchange contribution vanishes. The absence of the nonlocal exchange interaction allows us to integrate the interaction potential, V, over the target wave function and obtain a local, one-electron potential for the continuum electron in this region.

$$(T_e + V_{aa} - E + E_a)f_a(\tau_{N+1}, R_o) = -\sum_b V_{ab}f_b(\tau_{N+1}, R_o); \qquad (8)$$

with

$$V_{ab} = \int d\tau_1 \ldots d\tau_N \Phi_a^\dagger(\tau_1 \ldots \tau_N, R_o)V\Phi_b(\tau_1 \ldots \tau_N, R_o). \qquad (9)$$

The wave function of the continuum electron and its derivative at the surface is related by the R-matrix

$$f_a(s, R_o) = \sum_b \hat{\mathcal{R}}_{ab} \nabla_{\mathbf{r}_{N+1}} f_b(r_{N+1}, R_o) \cdot \mathbf{n} \Big|_s, \qquad (10)$$

$$\hat{\mathcal{R}}_{ab} = \langle \Phi_a(\tau_1 \ldots \tau_N, R_o)\sigma_a(N+1)|\mathcal{R}|\Phi_b(\tau_1 \ldots \tau_N, R_o)\sigma_b(N+1)\rangle.$$

Here $\mathbf{n}$ is an unit vector outward normal to the surface, σ_a and σ_b the spin function of the continuum electron, and $\mathcal{R}$ is the R-operator whose matrix representation is given in Eq. (6). Eq. (10) connects the wave function Ψ in the internal and external regions.

Alternately, a non-variational form of the R-matrix can be used, employing basis functions with fixed boundary conditions. It is then necessary to introduce the Buttle correction [8] to render accurate results. This approach is used in the UK package of R-matrix codes [2], [9].

2.1.2. *Inclusion of Vibration, Dissociative Attachment, and Dissociative Recombination*

The wave function of the electron + diatomic target in the molecular frame, including nuclear vibration and dissociation attachment or dissociative recombination, can be written as,

$$\Psi(\tau, R) = \sum_k \mathcal{A}\{\Phi_k(\tau_1 \ldots \tau_N, R)f_k(\tau_{N+1}, R)\}\chi_{v_k}(R)$$
$$+ \sum_l \Psi_l(\tau_1 \ldots \tau_N, \tau_{N+1}, R)\chi_{\epsilon_l}(R). \quad (11)$$

where the first term on the right-hand-side of Eq. (11) represents the elastic channel and vibronic excitation channels, including direct dissociation by electron impact. In the second term, the additional electron is now bound, and the nuclear motion is in the dissociative continuum, denoted by the continuum energy ϵ_l. For a neutral target, this is a dissociative attachment process. If the target is a positive ion, it is called dissociative recombination.

The R-matrix equation for nuclear motion is obtained by adding the operator L to both sides of the Schrödinger equation, Eq. (1), and projecting out the electronic functions,

$$\sum_{ij}(E_{ij}(R) + H_{nucl} + L^P - E)\chi_j(R)$$
$$= \langle \mathcal{A}\Phi_i(\tau_1, \ldots \tau_N, R)f_i(\tau_{N+1}, R)|L|\Psi(\tau, R)\rangle_\tau, \quad (12)$$

Here H_{nucl} is the nuclear Hamiltonian and the subscript τ on the left-hand-side of Eq. (12) indicates integration over the electronic coordinates. Also, $E_{ij}(R)$ and L^P are respectively the electronic energy and Bloch operator projected onto the electronic wave function.

The UK molecular R-matrix package specializes in diatomic targets. Calculations on H_2^+, H_2, HeH^+, N_2, O_2, CO, NO, HF, and HCl have been reported. The excitation processes studied include rovibrational and electronic excitations and dissociative recombination.

2.2. THE SCHWINGER VARIATIONAL METHOD

Instead of working with the Schrödinger equation, the Schwinger variational method deals with the corresponding integral equation, the Lippmann-Schwinger equation. It is given by,

$$\Psi^{(+)} = S + G^{(+)}V\Psi^{(+)}, \quad (13)$$

where the Green's function, $G^{(+)}$ is given by

$$G^{(+)} = (E - H_o + i\delta)^{-1},$$
$$H_o = H - V. \quad (14)$$

with δ infinitely small. Also S is the solution of the corresponding homogeneous equation and the superscript $(+)$ denotes that the wave function satisfies the boundary condition of an incoming plane wave and outgoing spherical waves. Because the correct boundary condition is automatically incorporated through the use of Green's function, Schwinger calculations need not employ basis functions which satisfy correct boundary conditions. Thus a common practice in Schwinger calculations is to employ a square-integrable basis. If a Gaussian basis is used, all one- and two-electron integrals are evaluated analytically, without resorting to numerical integration techniques. The Schwinger variational method also converges one iteration faster than variational methods based on the Schrödinger equation approach.

2.2.1. The Schwinger Multichannel Method in the Fixed-Nuclei Approximation

The Schwinger multichannel (SMC) method [11] is a hybrid method which uses the Lippmann-Schwinger equation to describe the energetically accessible channels (open channels) and the Schrödinger equation for the energetically inaccessible channels (closed channels). Using a projection operator defined by the open channel target functions

$$P = \sum_{m \in open} |\Phi_m(\tau_1 \ldots \tau_N, R_o)\rangle\langle\Phi_m(\tau_1 \ldots \tau_N, R_o)|, \qquad (15)$$

A fixed value R_o is used here for the internuclear distance. The projected Lippmann-Schwinger equation is given by

$$P\Psi^{(+)} = S + G_P^{(+)}V\Psi^{(+)}, \qquad (16)$$

with the projected Green's function $G_P^{(+)} = PG^{(+)}P$. The closed channel part of the wave function is determined by the Schröd- inger equation with the open channel contribution projected out,

$$(1 - aP)\hat{H}\Psi^{(+)} = 0. \qquad (17)$$

Here $\hat{H} = E - H$. The SMC equation is obtained by combining Eqs. (16) and (17),

$$A^{(+)}\Psi^{(+)} = VS. \qquad (18)$$

The operator $A^{(+)}$ is given by

$$A^{(+)} = \left\{ \frac{1}{2}(PV + VP) - VG_P^{(+)}V + \frac{1}{a}\hat{H} - \frac{1}{2}(P\hat{H} + \hat{H}P) \right\}, \qquad (19)$$

where a is a projection parameter.

A variationally stable expression of the T-matrix, which is related to the collision amplitude, is given by

$$T_{ij} = \sum_{mn} \langle S_i | V | \zeta_m \rangle D_{mn} \langle \zeta_n | V | S_j \rangle, \tag{20}$$

where ζ is the basis function used, and D is the inverse matrix of of $A^{(+)}$,

$$(D^{-1})_{mn} = \langle \zeta_m | A^{(+)} | \zeta_n \rangle. \tag{21}$$

The collision cross section can then be calculated from the T-matrix.

A number of different methods can be employed, in conjunction with the fixed-nuclei SMC results, to treat nuclear dynamics. In the following, the projection operator method [12], which has been applied successfully to study resonant enhanced vibrational excitations of N_2 under hypersonic conditions, will be described.

2.2.2. The Projection Operator Method for Resonant Enhanced Vibrational Transitions

Resonance enhanced scattering results from the incoming electron being temporarily trapped inside the molecule, forming a transient negative ion. There are two types of resonances: a shape resonance where the electron is temporarily trapped due to the presence of a potential barrier, and a Feshbach resonance where the negative ion of an excited state of the target is stable with respect to the parent neutral state, but unstable with respect to the ground state of the target. The occurrence of a resonance greatly enhances the collision cross section, and hence resonances play an important role in e-molecule collisions. Shape resonances are most commonly encountered in low-energy e-molecule collisions. For example, the low-energy shape resonance of $^2\Pi_g$ symmetry in e-N_2 collisions, centered at ≈ 2.3 eV in the total cross section, plays an important role in the vibrational excitation of N_2 by electron impact.

The optical potential treatment of nuclear dynamics [12] makes use of the Feshbach projection operator to partition the interaction potential for electron scattering into two parts, the direct (non-resonant) term and the resonant contribution. The Feshbach projection operator, $\hat{P}$, [13] is defined in the $N+1$-electron space, i.e., the space of the target + scattering electron. It is to be distinguished from the SMC projection operator in Eq. (15), which is defined in the target electron (N-electron) space alone.

$$\hat{P} = \sum_r |\Psi_r(\tau_1 \ldots \tau_N, \tau_{N+1}, R)\rangle \langle \Psi_r(\tau_1 \ldots \tau_N, \tau_{N+1}, R)|. \tag{22}$$

where Ψ_r is the part of the wave function which describes the resonance. The operator $\hat{Q}$ which defines the non-resonant space is the complement of

$\hat{P}$

$$\hat{Q} = 1 - \hat{P} \tag{23}$$

The effective Hamiltonian acting in the resonant part of configuration space alone is

$$H_{eff} = H_{\hat{P}\hat{P}} + H_{\hat{P}\hat{Q}} \frac{1}{E - H_{\hat{Q}\hat{Q}}} H_{\hat{Q}\hat{P}}. \tag{24}$$

Based on the Lippmann-Schwinger equation, the resonant T-matrix can be expressed as

$$T_{res} = V_{\hat{P}\hat{Q}} \frac{1}{E - H_{\hat{Q}\hat{Q}} - \Delta + \frac{i}{2}\Gamma} V_{\hat{Q}\hat{P}}. \tag{25}$$

The width function Γ, potential $V_{\hat{Q}\hat{P}}$ and shift Δ are determined from fixed-nuclei calculations such as SMC calculations.

The fixed-nuclei SMC method was the first *ab initio* method developed to treat general polyatomic molecules without relying on one-center expansion of the target wave function. It has been applied to the study of H_2, N_2, O_2, CO, and NO, as well as a large number of polyatomic systems. The processes studied include vibrational and electronic excitations, spin polarization, and direct dissociation. The projection operator method has also been employed to study nuclear dynamics in low-energy resonant scattering of many diatomic and polyatomic systems.

2.3. THE COMPLEX KOHN VARIATIONAL METHOD

Like the R-matrix method, the complex Kohn method [14] is based on the Schrödinger equation. However, instead of separating the wave function into regions, the Kohn formulation is based on global functions. The complex Kohn method has been applied to a number diatomic molecules such as H_2, Li_2, F_2, Cl_2, HBr, N_2^+, and CO^+ as well many polyatomic molecules and the processes studied include vibrational and electronic excitation, direct dissociation, and dissociative recombination. Because the *ab initio* data presented below are not based on the complex Kohn method, readers are referred to recent review articles by Rescigno, et al. [14], [15] for a description of the Kohn formulation.

3. Electron Impact Vibrational Excitations

Due to the disparity between electronic and molecular masses, electron impact vibrational excitation cross sections are expected to be small. However, when a resonance is involved, the formation of a transient negative ion allows strong coupling between the continuum electron and the nuclear motion. Under these circumstances, the vibrational excitation cross section is greatly enhanced [17].

124

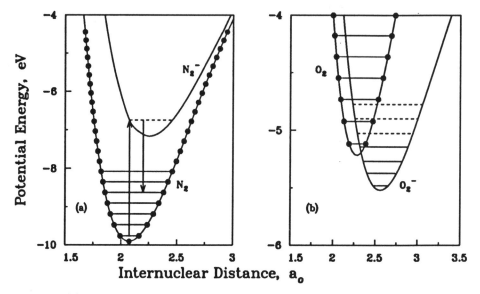

Figure 1. (a) Potential energy curves of N_2 and N_2^-. The lower curve is the RKR potential energy curve for the electronic ground state of N_2. The upper curve is a schematic curve for the transient N_2^- ion. (b) Potential energy curves of O_2 and O_2^-. The O_2 curve is the RKR potential energy curve for the electronic ground state. The O_2^- curve is a Morse curve based on experimental spectroscopic constants.

For the purpose of flow field modeling, non-resonant contribution to electron-impact vibrational excitation is so small that it can be neglected. Thus this section only considers the resonant process. Low-energy resonances have been observed in electron collisions with many diatomic and polyatomic molecules. The cases of N_2, O_2, and NO are considered separately below.

3.1. RESONANT ENHANCED VIBRATIONAL EXCITATION IN N_2

A prominent feature in low-energy e-N_2 collisions is a shape resonance of $^2\Pi_g$ symmetry. This is one of the best studied resonances, and enhancement in the cross section due to this resonance has been observed in experimental measurements of the total, elastic, vibrational, and ro-vibrational excitation cross sections.

Figure 1a illustrates the resonant enhanced excitation mechanism for N_2. The lower curve is the RKR potential energy curve of the ground electronic state of N_2 [18] where the low-lying vibrational energy levels are also shown. The upper curve is a schematic curve for the transient N_2^- ion. As discussed in Sec. 2.2.2, the actual potential energy curve of N_2^-, due to the contributions of the width and shift functions in Eq. (25), is complex

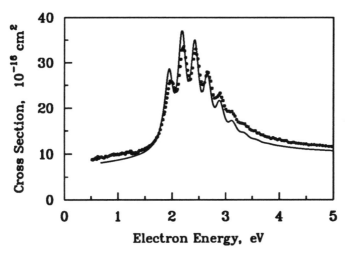

Figure 2. Total cross section of e-N_2 collisions. The full curve is from the *ab initio* calculation of Huo et al. [19], and the experimental data points are from Kennerly [20].

and dependent on the electron energy. Let the molecule initially be at the ground vibrational level v_o. When the incident electron energy matches the energy difference between the v_o and a vibrational levels of N_2^-, the electron may be temporarily attached to the molecule and a transient negative ion is formed, indicated by the arrow pointing up in Fig. 1a. Due to the finite lifetime of N_2^-, the electron will eventually detach. However, the molecule may return to a vibrational level different from v_o, as indicated by the down arrow in the figure, resulting in vibrational excitation.

A wealth of *ab initio* studies has also been reported for this system. Fig. 2 compares the total cross sections calculated using the SMC method [19] with the experimental data of Kennerly [20]. A distinctive feature of this resonance is the series of peaks in the cross section curve, corresponding to temporary electron attachment to different vibrational levels of the ion. A recent *ab initio* calculation using the noniterative partial differential equation technique [21] obtained similar overall agreement with Kennerly's data whereas the R-matrix calculation of Gillan et al. [9] gives somewhat larger deviations.

From the preceding discussion of the resonant excitation mechanism, it is obvious that excitations involving large Δv are possible. Indeed this is the case, as observed experimentally and from *ab initio* calculations. Table 1 presents the vibrational excitation cross section at the maximum of the cross section curve for $v = 0 \rightarrow 1$ - 17. The data are from the experiment of Allan [23], normalized using the measured cross section of Jung et al. [24], and the of *ab initio* calculations using the SMC method [25] and the R-matrix approach [26]. It is seen that the $\Delta v = 1$ - 5 cross sections are sizable,

TABLE 1. Electron-impact $0 \rightarrow v$ excitation of N_2. Position and magnitude of the maximum in the cross section curve.

Transition	Position (eV)			Cross section (cm^2)		
	Expt[a]	SMC[b]	R-matrix[c]	Expt[a]	SMC[b]	R-matrix[c]
$0 \rightarrow 1$	1.95	1.96[d]	2.12	5.6 (-16)	5.75(-16)	6.94 (-16)
$0 \rightarrow 2$	2.00	2.00[d]	2.32	3.7 (-16)	3.30(-16)	3.39 (-16)
$0 \rightarrow 3$	2.15	2.18	2.40	3.1 (-16)	2.90(-16)	3.09 (-16)
$0 \rightarrow 4$	2.22	2.22	2.62	2.1 (-16)	1.79(-16)	1.39 (-16)
$0 \rightarrow 5$	2.39	2.42	2.68	1.3 (-16)	1.23(-16)	8.9 (-17)
$0 \rightarrow 6$	2.48	2.48[d]	2.93	7.1 (-17)	5.24(-17)	3.7 (-17)
$0 \rightarrow 7$	2.64	2.66	2.99	3.8 (-17)	3.04(-17)	1.3 (-17)
$0 \rightarrow 8$	2.82	2.87	3.25	1.6 (-17)	1.31(-17)	5.0 (-18)
$0 \rightarrow 9$	2.95	3.06	3.50	6.1 (-18)	4.57(-18)	1.5 (-18)
$0 \rightarrow 10$	3.09	3.16	3.75	2.2 (-18)	1.44(-18)	3.5 (-19)
$0 \rightarrow 11$	3.30	3.35	3.97	6.3 (-19)	4.67(-19)	6.8 (-20)
$0 \rightarrow 12$	3.87	3.55	4.15	1.4 (-19)	1.31(-19)	1.2 (-20)
$0 \rightarrow 13$	4.02	3.75	4.85	4.5 (-20)	3.23(-20)	1.9 (-21)
$0 \rightarrow 14$	4.16	3.96	5.08	1.3 (-20)	7.06(-21)	4.7 (-22)
$0 \rightarrow 15$	4.32	4.16	5.30	3.6 (-21)	1.37(-21)	9.7 (-23)
$0 \rightarrow 16$	4.49	4.71	5.52	9.1 (-22)	3.62(-22)	1.8 (-23)
$0 \rightarrow 17$	4.66	4.88	5.72	2.4 (-22)	0.96(-22)	2.7 (-24)

[a]Experimental data from Allan [23].
[b]SMC calculation from Huo et al. [25].
[c]R-matrix calculation from Morgan [26].
[d]First maximum. Calculated second maximum is slightly higher.

all larger than 1 Å^2 (10^{-16} cm^2), indicating that electron impact excitation has a significant role in determining the vibrational energy distribution in a hypersonic flow.

The cross section data presented above were measured at room temperature. Since only levels with low rotational quantum numbers, J, were populated, the *ab initio* calculations neglected rotational effects in the dynamical treatment (the ANR approximation, see Sec. 2). The centrifugal potential for the vibrational motion, $J(J+1)h/(8\pi^2 c\mu R^2)$, with h Planck's constant, c the velocity of light, and μ the reduced mass of the molecule, was also neglected. From the general agreement between theory and experiment shown in Fig. 2 and Table 1, such approximation appeared to be justified. However, under hypersonic flow conditions, the rotational temperature is high, above 10,000 K. Under these circumstances, the rotational quantum number is large and the centrifugal potential can no longer be neglected. Figure 3a shows the electron impact cross section for the $v = 0 \rightarrow 1$ tran-

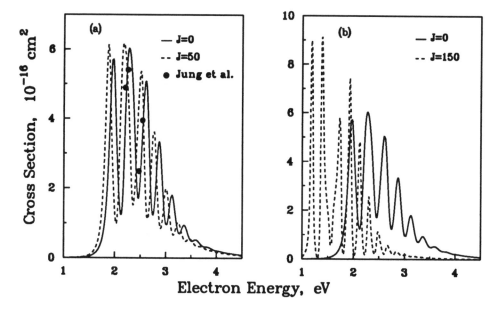

Figure 3. Electron impact cross section for the $v = 0 \rightarrow 1$ transition in N_2. (a) The full and dashed curves are from SMC calculations at $J = 0$ and 50 [27], and the data points are from the 500 K experiment of Jung et al. [24]. (b) The full and dashed curves are from SMC calculations at $J = 0$ and 150 [27].

sition in N_2, calculated at $J = 0$ and 50 using the SMC method [27]. The experimental data of Jung et al. [24], measured at about 500 K, also are displayed. The $J = 0$ result is in good agreement with experiment, but at $J = 50$, the peaks in the cross section curve shift to lower electron energies by ≈ 0.1 eV. This is due to the fact that the neutral and ion vibrational energy levels shift differently under the centrifugal potential. Figure 3b compares the cross sections calculated at $J = 0$ and 150 [27]. The peak positions of the $J = 150$ curve now shift down significantly from the $J = 0$ curve, by ≈ 0.9 eV. The peaks are much sharper, and the relative peak heights are different. These results indicate that room temperature data, either from experiment or *ab initio* theory, are not suitable for modeling hypersonics flows. Note that the SMC calculation takes rotational effects into account by using of the correct centrifugal potential, but the ANR approximation is still employed to separate rotational motion from vibronic motion. While the ANR approximation may still be valid at $J = 50$, it is likely to break down when J is as high as 150. This aspect needs to be studied in the future.

Using the SMC cross sections determined over a range of electron energies and at $J = 50$, the vibrational excitation/de-excitation rate coefficients,

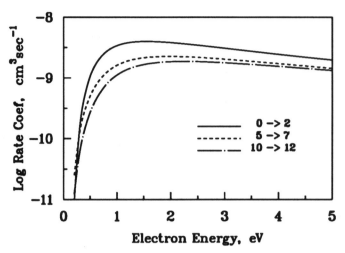

Figure 4. N$_2$ vibrational excitation rate coefficients by electron impact for the transitions $v = 0 \rightarrow 2$, $5 \rightarrow 7$, and $10 \rightarrow 12$, assuming $J = 50$ [27].

k_{if}, of N$_2$ by electron impact have been calculated assuming a Maxwellian electron velocity distribution.

$$k_{if} = C_o T_e^{-1.5} \int E\sigma_{if}(E)e^{-E/T_e} dE. \tag{26}$$

Here $C_o = 6.6971 \times 10^{-7}$, T_e the electron temperature in eV, E the electron energy, and σ_{if} the cross section in units of cm^2. Also, k_{if} from Eq. (26) is in cm^{3}s^{-1}. Tabulation of k_{if}, for initial vibrational quantum number v_i = 0 - 12 and Δv= 1 - 5 and for electron temperatures 0.1 - 5 eV (1 eV = 11605.4 K), are given in Ref. [27]. An example of the temperature and v_i dependence of k_{if} are presented in Fig. 4 for the $0 \rightarrow 2$, $5 \rightarrow 7$, and $10 \rightarrow 11$ transitions. All three rate coefficients are strongly temperature dependent. The sharp increase at the low temperature end indicates the onset of the resonance. The maximum rate is reached at a temperature between 1.5 and 2 eV. The position of this maximum is dependent on v and J, relating to the v, J-dependence of the resonance. Also, the rate coefficient decreases with increasing v_i; its maximum value decreases by a factor of 2 as v_i increases from 0 to 10.

3.2. RESONANT ENHANCED VIBRATIONAL EXCITATION IN O$_2$

The transient negative ion responsible for the shape resonance in low energy electron collisions of O$_2$ is different from its counterpart in N$_2$ because the lowest four vibrational levels of O$_2^-$ are stable. This is shown in Fig. 1b where the RKR potential energy curve of the ground electronic state of O$_2$ [28] is plotted together with the Morse curve of the ground state of O$_2^-$. The

TABLE 2. The positions and widths of low energy resonances in e-O_2 collisions and the associated vibrational levels of O_2^-. The data are from the R-matrix calculation of
Higgins et al. [29].

Vibrational level of O_2^-	Position (eV)	Width (meV)
4	0.0892	0.122
5	0.2230	0.896
6	0.3495	2.17
7	0.4776	3.32
8	0.6030	4.99
9	0.7219	7.02

parameters of the Morse curve are taken from experimental data [28]. Only
for v=4 and above is O_2^- unstable. Those levels are shown by dashed lines to
indicate their finite lifetime. Because the progression of unstable O_2^- levels
begins at energies just slightly above the ground vibrational level of the
neutral molecule, O_2 shape resonance appears at much lower energies than
N_2. Table 2 presents the positions and widths of the low energy resonances
in O_2 due to the the $v' = 4$ - 9 levels of O_2^-, determined by a recent R-
matrix calculation [29]. The R-matrix value for the position of the first
resonance is 0.0892 eV, to be compared with the measured value of 0.076
eV [30] and 0.091 eV [31]. According to Wigner's threshold law, the width
of a resonance Γ near threshold can be expressed as [32], [33], [34]

$$\Gamma = cE^{(2l+1)/2}, \qquad (27)$$

where l is the angular momentum quantum number of the partial wave responsible for the resonance and c is a constant. For O_2, the shape resonance
is due to the attachment of a d wave electron, and hence l=2. Since the
resonance occurs at such low energy, it should be extremely narrow. Table
2 confirms this. The lowest resonance, corresponding to $v' = 4$, only has a
width of 0.122 meV. The width of the $v' = 6$ level is calculated to be 2.17
meV, falling within the range of 2 - 3 meV estimated by the high resolution
experiment of Field et al. [35]. The width of the broadest resonance listed
there, located at 0.72 eV and associated with the $v' = 9$ level of O_2^-, is
7 meV. Because the width is inversely proportional to the lifetime of the
transient negative ion, this implies that the associated O_2^- ion is long lived.
The N_2 resonance, on the other hand, is much broader, with a fixed-nuclei
width of 0.41 eV at its equilibrium internuclear distance.

The width of the resonance also has relevance to the vibrational excitation rate. Because the rate coefficient is obtained by integrating the cross

section over electron energy (see Eq. (26)) narrow resonances contribute little to the rate coefficient. Linder and Schmidt gave the energy integrated cross section of the $v = 0 \rightarrow 1$ excitation of O_2 as 4.4×10^{-18} cm^{2}eV [30], whereas the corresponding transition for N_2 has a value of 4.7×10^{-16} cm^{2}eV [36]. A comparison of these two values shows that the resonant enhanced vibrational excitation rate coefficient in e - O_2 collision will be much smaller than the N_2 case, Thus, contrary to N_2, the low energy resonance in e - O_2 collision does not appear to play an important role in the vibrational energy distribution of O_2 under hypersonic flow conditions.

3.3. RESONANT ENHANCED VIBRATIONAL EXCITATION IN NO

The two lowest electronic states of NO$^-$, the ground $^3\Sigma^-$ state and the first excited $^1\Delta$ state, both contribute to the shape resonances observed in the low energy vibrational excitation in e-NO collisions [37], [38]. The $v'=0$ vibrational level of the $^3\Sigma^-$ state of NO$^-$ is stable, lying below the ground vibrational level of the neutral molecule. Analogous to the O_2 case, the $^3\Sigma^-$ state of NO$^-$ contributes to a series of narrow resonance structures in the cross section, beginning at ≈ 0.26 eV [39]. While the widths are somewhat larger than the O_2 case, this resonance is still too narrow to be important in hypersonic modeling. On the other hand, the resonance width associated with the $^1\Delta$ state of NO$^-$ is significantly larger. The R-matrix calculation of Tennyson and Noble [40] gives a fixed-nuclei width of 0.39 eV at the equilibrium internuclear distance, R_e, of the ground state of NO, with the center of resonance at 1.4 eV. The width is comparable to that of N_2, but occurring at a lower energy. The contribution of the $^1\Delta$ resonance is reflected in Reinhardt's estimate for the energy integrated cross section of the $v = 0 \rightarrow 1$ transition in NO, 0.39×10^{-16} cm^{2}eV [41], approximately a factor of 10 larger than the corresponding value for O_2, but also a factor of 10 smaller than the N_2 value.

Detailed study of the resonant enhanced vibrational excitation of NO, especially under hypersonic flow conditions, remains to be done. A recently initiated R-matrix study on this system [42] should be useful towards this direction.

4. Electronic Excitations by Electron Collisions

Because the selection rules for electron impact excitation is not as strict as photon impact or heavy particle collision, electrons can transfer energy efficiently to many different molecular electronic states. The status of electron collision data of N_2 and O_2, including electronic excitation, ionization, dissociation, as well as ro-vibrational excitation, has recently been reviewed by Itikawa [43],

Analogous to vibrational excitations discussed in Sec. 3, most cross section data for electronic excitations important to hypersonic modeling are unavailable. So far, all experimental studies on electronic excitations of air molecules start with the lowest vibrational level of the ground electronic state. Theoretical studies either use the same initial conditions or employ a fixed-nuclei approximation with calculations carried out at the R_e of the ground electronic state. On the other hand, the electron temperature under hypersonic conditions is $\approx$ 1 - 2 eV, significantly lower than the excitation energy of 6.17 eV for the lowest excited state of N_2, the $A^3\Sigma_u^+$ state. The excitation energy of O_2 to the two lowest lying excited states, $a^1\Delta_g$ and $b^1\Sigma_g^+$, are 0.98 and 1.63 eV. These two states arise from different spin multiplets of the same electronic configuration as the ground $X^3\Sigma_g^-$ state. On the other hand, excitations to the $B^3\Sigma_u^-$ state, the upper state of the Schumann-Runge band, is significantly higher, 6.12 eV. Large excitation energies require electrons at the tail of the Maxwellian distribution, thus diminishing the excitation probability. However, the vibrational temperature in a hypersonic flow is quite high, also $\approx$ 1 - 2 eV. Excitation from a hot vibrational level requires electrons with lower energies. Furthermore, the R_e of the ground state tends to be smaller than the excited states, so that excitations from a vibrationally hot level, with a larger R_e, tends to have a more favorable Franck-Condon factor. Based on these arguments, electronic excitation cross sections from a vibrationally hot level of the ground electronic state is a likely excitation pathway in a hypersonic flow field. So far, such data is not available.

Electronic energy transfer among excited electronic states due to collision with electrons is another aspect which has been largely neglected. Since the transition energy involved is significantly lower, such process may be more favorable. They are considered below.

4.1. ELECTRONIC EXCITATION OF N_2

The most comprehensive set of measurements on the electronic excitations of N_2, starting from the ground vibrational level of the $X^1\Sigma_g^+$ state, comes from the work of Cartwright et al. [45]. It should be noted that a renormalization factor is required and the renormalized data has been tabulated in a review article by Trajmar et al. [46].

Figure 5a presents the SMC cross sections for the $X^1\Sigma_g^+ \rightarrow B^3\Pi_g$ and $A^3\Sigma_u^+ \rightarrow B^3\Pi_g$ transitions [44]. The calculations used the fixed-nuclei approximation at the experimental R_e of the $X^1\Sigma_g^+$ state. Note that the data are plotted against scattered electron energy instead of incident electron energy so that cross sections for the two transitions can be represented using the same energy scale. The SMC result for the first transition are in

132

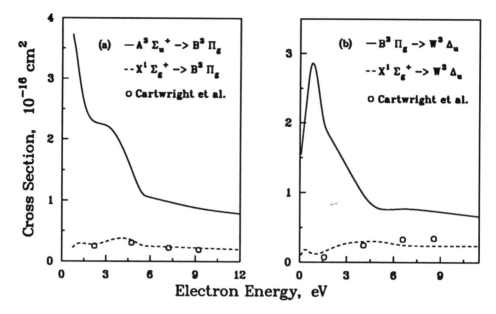

Figure 5. Electronic excitation in e - N_2 collisions. (a) The SMC cross sections [44] for the $A^3\Sigma_u^+ \to B^3\Pi_g$ and the $X^1\Sigma_g^+ \to B^3\Pi_g$ transitions are shown by full and dashed curves. The experimental data points are from Cartwright et al. [45], [46]. (b) The corresponding results for the $B^3\Pi_g \to W^3\Delta_u$ and the $X^1\Sigma_g^+ \to W^3\Delta_u$ transitions.

general agreement with the experimental data of Cartwright et al. [45], [46]. The second transition, starting from an electronic excited state, is found to be significantly stronger than the transition starting from the ground electronic state. This is particularly so near the threshold, where the cross sections differ by an order of magnitude. Similar results are found in Fig. 5b for the $X^1\Sigma_g^+ \to W^3\Delta_u$ and $B^3\Pi_g \to W^3\Delta_u$ transitions.

The trend of larger excited to excited cross sections continue for other transitions in N_2 as well. Experimental measurements of metastable atoms also found relatively large excitation cross sections, especially near the excitation threshold [47]. Probably one reason for the larger cross section comes from the fact that the charge distribution of an excited state is more extended, rendering it a better scatterer. In the two examples presented here, the ground $\to$ excited transitions are spin forbidden, and the excited $\to$ excited transitions are spin and dipole allowed. However, similar increase in the excited $\to$ excited cross sections has been observed even when both transitions are spin allowed. In Figs. 5a and 5b, the sharp increase of the excited $\to$ excited cross section near the threshold is related to the dipole allowed nature of the transition. Also, both calculations are fixed nuclei calculations, and the vibrational overlap has not been taken into account. While the vibrational overalp is case dependent, the underlying trend of larger cross sections for excited → excited transitions considered here should continue to hold.

Excited → excited transitions also have lower excitation energies, Thus for electron impact excitations in a hypersonic flow field, the Maxwellian factor is much more favorable. If the initial state in the excited → excited transition is sufficiently long lived, the excitation energy will be 'scrambled' efficiently by additional collisions with electrons. This effect has significant consequences in modeling the radiative heat load.

4.2. ELECTRONIC EXCITATION OF O_2

As pointed out earlier in this section, the excitation energy of the two lowest lying excited states of O_2, the $a^1\Delta_g$ and $b^1\Sigma_g^+$ states, falls inside the estimated electron temperature range of ≈ 1 - 2 eV in a hypersonic flow field. Thus these two states are expected to have significant populations. A recent nine-state R-matrix calculation of Noble and Burke [48] finds good agreement in the $X^3\Sigma_g^- \rightarrow a^1\Delta_g$ and $X^3\Sigma_g^- \rightarrow b^1\Sigma_g^+$ excitation cross sections with experiment [30], [49], [50], [51], especially in the low energy region important to flow field modeling.

In view of the fact that significant population is expected for the $a^1\Delta_g$ and $b^1\Sigma_g^+$ states and that both are relatively long lived, a ladder climbing process by electron collision of these two states is a possibility. Data for such study is not yet available.

5. Dissociative Recombination of Molecular Ions

The dissociation energy of air molecules is significantly higher than the average electron energy in the hypersonic flow field, making direct dissociation by electron impact unlikely. Furthermore, molecular dissociation by electron- and photon-impact generally proceeds via predissociation, i.e., by first exciting the ground electronic state to a predissociative state lying above the ground state dissociation limit. This makes direct dissociation by electron impact even more unfavorable energetically. Under these circumstances, dissociative recombination, DR, of molecular ions with electrons

$$e + AB^+ \rightarrow A + B \tag{28}$$

becomes an important dissociation pathway.

Figure 6 presents two DR mechanisms important for air molecules. The direct process is shown in Fig. 6a where the electron recombines with the ion forming a neutral molecule in a repulsive state which crosses the ion curve. In the indirect mechanism shown in Fig. 6b the electron is first captured to

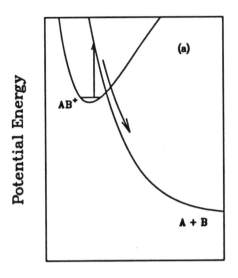
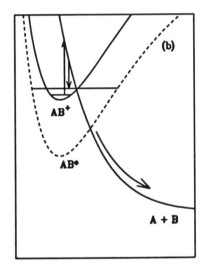

Internuclear Distance

Figure 6. Schematic diagrams of two DR mechanisms; (a) direct mechanism, and (b) indirect mechanism.

form a Rydberg state of the neutral molecule, denoted by AB*, which then crosses the repulsive curve. Because DR proceeds through curve crossing mechanisms [52], DR cross sections, and hence DR rates, are sensitive to the initial v and J levels of the ion. In general, more than one repulsive neutral curve may cross the ion curve. Thus different neutral repulsive states may be responsible for the DR mechanism in different vibrational levels, resulting in vastly different DR cross sections. Also, the centrifugal potential associated with a high J quantum number shifts the vibrational levels and consequently the crossing point. Thus high J levels is expected to have an effect on the DR cross section.

Experimental measurements of the DR of N_2^+, O_2^+, and NO^+ have been reviewed by Johnsen [53] and the DR rates are tabulated there together with their temperature dependence. The dependence of the DR cross sections on the vibrational level of N_2^+ has been studied in a merged beam experiment [54], which finds that the DR cross section of N_2^+ at the $v=0$ level is approximately a factor of five smaller than vibrationally hot N_2^+. However, this result appears to be inconsistent with the microwave afterglow and flowing afterglow/Langmuir probe experiments [53].

Theoretical calculations using an electron capture width extrapolated from Rydberg data and multichannel quantum defect theory have been reported by Guberman [55], [56], [57]. His study includes the dependence on vibrational levels and electron temperature. His result for the DR of N_2^+ also does not support the merged beam experiment. So far no study on the dependence on J has been reported.

6. Summary

The role of electron - air molecule collisions in hypersonic flows is reviewed. Due to the high rotational and vibrational temperatures in the flow field, cross sections required for its modeling generally are not available from laboratory measurements. Current *ab initio* methods capable of furnishing the necessary data base are reviewed. For electron impact vibrational excitation, the cases of N_2 and O_2 are well understood. Further study of NO should be pursued. For electron impact electronic excitation, the dependence of the excitation cross sections on the initial vibrational level should be studied. Further investigations of excited $\rightarrow$ excited transitions are also worthwhile. For dissociative recombination, the dependence on the rotational and vibrational temperature of the ion and the effect on the product branching ratio need to be further investigated.

Acknowledgments

HTT acknowledges a NASA-NRC Research Associateship.

References

1. Huo, W.M. and Gianturco, F.A. (eds.) (1995) *Computational Methods for Electron-Molecule Collisions*, Plenum Press, New York.
2. Burke, P.G. and Berrington, K.A. (eds.) (1993) *Atomic and Molecular Processes: An R-matrix Approach*, Institute of Physics Publishing, Bristol.
3. A detailed description of the R-matrix method as applied to electron-molecule collisions are given in Chapters 8-14 in Ref. [1].
4. Partridge H., Stallcop, J.R., and Levin, E. (1995) Potential energy curves and collision integrals of air components, in M. Capitelli, (ed.) *Molecular Physics and Hypersonic Flows*, Kluwer Academic Publishing (Dordrecht).
5. Shimamura, I. (1980) State-to-state rotational transition cross sections from unresolved energy-loss spectra, *Chem. Phys. Lett.* **73**, 328-333.
6. Jackson J. L. (1951) A variational approach to nuclear reactions, *Phys. Rev.* **83** 301-304.
7. Oberoi, R.S. and Nesbet, R.K. (1973) Variational formulation of the R-matrix method for multichannel scattering, *Phys. Rev. A* **8**, 215-9. Oberoi, R.S. and Nesbet, R.K. (1974) Addendum to "Variational formulation of the R-matrix method for multichannel scattering", *Phys. Rev. A* **8**, 2804-5.
8. Buttle, P.J.A. (1967) Solution of coupled equations by R-matrix techniques, *Phys. Rev. A* **160**, 719-29.
9. Gillan, C.J., Tennyson, J., and Burke, P.G. (1995) The UK molecular R-matrix scattering package: a computational perspective, in Huo, W.M. and Gianturco, F.A. (eds.) *Computational Methods for Electron-Molecule Collisions*, Plenum Press, New York.
10. Morgan, L.A. (1995) Non-adiabatic effects in vibrational excitation and dissociative recombination, in Huo, W.M. and Gianturco, F.A. (eds.) *Computational Methods for Electron-Molecule Collisions*, Plenum Press, New York.
11. Takatsuka, K. and McKoy, V. (1981) Extension of the Schwinger variational principle beyond the static-exchange approximation, *Phys. Rev. A* **24**, 2473-2480; Takat-

136

suka, K. and McKoy, V. (1984) Theory of electronically inelastic scattering of electrons by molecules, *Phys. Rev. A* **30**, 1734-1740.

12. Domcke, W. (1991) Theory of resonance and threshold effects in electron-molecule collisions: the projection operator approach, *Phys. Rep.* **208** 97-188.

13. Feshbach, H. (1958) Unified theory of nuclear reactions, *Ann. Phys.* **5** 357-390; Feshbach, H. (1962) A unified theory of nuclear reactions, II, *Ann. Phys.* **19**, 287-313.

14. Rescigno, T.N., McCurdy, C.W., Orel, A.E., and Lengsfield, B.H., III (1995) The complex Kohn variational method, in Huo, W.M. and Gianturco, F.A. (eds.) *Computational Methods for Electron-Molecule Collisions*, Plenum Press, New York.

15. Rescigno, T.N., Lengsfield, B.H. III, and McCurdy, C.W. (1995) in Yarkony, D.R., (ed.) *Modern Electronic Structure Theory, Part I*, World Scientific, Singapore.

16. Abramowitz, M. and Stegun, I (1965) *Handbook of Mathematical Functions*, Dover, New York.

17. Herzenberg, A. (1984) Vibrational excitation of molecules by slow electrons, in Shimamura, I. and Takayanagi, K. (eds.) *Electron-Molecule Collisions*, Plenum Press, New York.

18. Lofthus, A. and Krupenie, P.H. (1977) The spectrum of molecular nitrogen, *J. Phys. Chem. Ref. Data* **6**, 113-307.

19. Huo, W.M., Lima, M.A.P., Gibson, T.L., and McKoy, V. (1987) Correlation effects in elastic e-N_2 scattering, *Phys. Rev. A* **36**, 1642-1648.

20. Kennerly, R.E. (1980) Absolute total electron scattering cross sections for N_2 between 0.5 and 50 eV, *Phys. Rev. A* **21**, 1876-1883.

21. Weatherford, C.A. and Temkin, A. (1994) Completion of a hybrid theory calculation of the Π_g resonance in electron-N_2 scattering, *Phys. Rev. A* **49**, 2580-2586.

22. Gillan, C.J., Nagy, O., Burke, P.G., Morgan, L.A., and Noble, C.J. (1987) *J. Phys. B* **20**, 4585.

23. Allan, M. (1985) Excitation of vibrational levels up to $v = 17$ in N_2 by electron impact in the 0-5 eV region *J. Phys. B* **18**, 4511-4517.

24. Jung, K., Antoni, Th., Müller, R., Kochum, K.-H., and Ehrhardt, H. (1982) Rotational excitation of N_2, CO, and H_2O by low-energy electron collisions, *J. Phys. B* **15**, 3535-3555.

25. Huo, W.M., Gibson, T.L., Lima, M.A.P., and McKoy, V. (1987) Schwinger multichannel study of the $^2\Pi_g$ shape resonance in N_2, *Phys. Rev. A* **36**, 1632-1641.

26. Morgan, L.A., (1986) Resonant vibrational excitation of N_2 by low-energy electron impact, *J. Phys. B* **19**, L439.

27. Huo, W.M., McKoy, V., Lima, M.A.P., and Gibson, T.L. (1986) Electron-nitrogen molecule collisions in high-temperature nonequilibrium air, in Moss J.N. and Scott, C.D. (eds.) *Thermophysical Aspects of Re-entry Flows*, AIAA, New York.

28. Krupenie, P. H. (1972) The spectrum of molecular oxygen, *J. Phys. Chem. Ref. Data* **1**, 423-534.

29. Higgins, K., Gillan, C.J., Burke, P.G., and Noble, C.J.. (1995) Low-energy electron scattering by oxygen molecules: II. vibrational excitation, *J. Phys. B* **28**, 3391-3402.

30. Linder, F. and Schmidt, H. (1971) Experimental study of low energy e-O_2 collision processes, *Z. Naturf.* **26a**, 1617-1625.

31. Land, J.E., and Raith, W. (1974) *Phys. Rev. A* **9**, 1592-1602.

32. Parlant, G. and Fiquet-Fayard, F. (1976) The O_2^- $^2\Pi_g$ resonance: theoretical analysis of electron scattering data, *J. Phys. B* **9**, 1617-1628.

33. Fiquet-Fayard, F. (1975) Angular distributions for pure resonant scattering of electrons by diatomic molecules in Hund's cases a and b,

34. Wigner, E.P. (1948) On the behavior of cross sections near thresholds, *Phys. Rev.* **73**, 1002-1009. *J. Phys. B* **8**, 2880-2897.

35. Field, D., Mrotzek, G., Knight, D.W., Lunt, S., and Ziesel, J.P. (1988) High-resolution studies of electron scattering by molecular oxygen, *J. Phys. B* **21**, 171-

188.

36. This value is determined from the SMC cross sections at $J = 0$ [27].

37. Spence, D. and Schulz, G.J. (1971) *Phys. Rev.* **3**, 1968-76.

38. Tronc, M., Huetz, A., Landau, M. Pichou, F. and Reinhardt, J. (1975) Resonant vibrational excitation of the NO ground state by electron impact in the 0.1-3 eV energy range, *J. Phys. B* **8**, 1160-9.

39. Teillet-Billy, D. and Fiquet-Fayard, F. (1977) The NO^- $^3\Sigma^-$ and $^1\Delta$ resonances: theoretical analysis of electron scattering data, *J. Phys. B* **10**, L111-117.

40. Tennyson, J. and Noble, C.J. (1986) Low-energy electron scattering by the NO molecule, *J. Phys. B* **19**, 4025-4033.

41. This value is attributed to J. Reinhardt (1976) in Ref. [39].

42. Morgan, L.A. and Gillan, C.J. (1995) Low energy electron scattering by NO, in Mitchell, J.B.A., McConkey, J.W., and Brion, C.E., *XIX ICPEAC Scientific Program and Abstracts of Contributed Papers*, Whistler, Canada (1995).

43. Itikawa, Y. (1994) Electron collisions with N_2, O_2, and O: what we know and do not know, in Inokuti, M. (ed.) *Advances in Atomic, Molecular, and Optical Physics Vol. 33, Cross Section Data*, Academic Press (Boston).

44. Huo, W. (1990) Electron collisions cross sections involving excited states, in Capitelli, M. and Bardsley, J.N., (eds.) *Nonequilibrium Processes in Partially Ionized Gases*, Plenum, New York.

45. Cartwright, D.C., Trajmar, S., Chutjian, A., and Williams, W., (1977) Electron impact excitation of the electronic states of N_2. II. Integral cross sections at incident energies from 10 to 50 eV. *Phys. Rev. A* **16**, 1041-1051.

46. Trajmar, S., Register, D.F., and Chutjian, A., (1983) Electron scattering by molecules II. experimental methods and data, *Phys. Rep.* **97**, 219-356.

47. Schappe, R.S., Schulman, M.B., Anderson, L.W., and Lin, C.C. (1994) Measurements of cross sections for electron-impact excitation into the metastable levels of argon and number densities of metastable argon atoms, *Phys. Rev. A* **50**, 444-461.

48. Nobel, C.J. and Burke, P.G. (1992) R-matrix calculations of low-energy electron scattering by oxygen molecules, *Phys. Rev. Lett.* **68**, 2011-2014.

49. Trajmar, S., Cartwright, D.C., and Williams, W. (1971) Differential and integral cross sections for the electron-impact excitation of the $a^1\Delta_g$ and $b^1\Sigma_g^+$ states of O_2. *Phys. Rev. A* **4**, 1482 - 1492.

50. Doering, J.P. (1992) Absolute differential and integral electron excitation cross sections for the O_2 $(a^1\Delta_g \leftarrow X^3\Sigma_g^-)$ transition, *J. Geophys. Res.* **97**, 12267-12270.

51. Middleton, A.G., Teubner, P.J.O., and Brunger, M.J. (1992) Experimental confirmation for resonance enhancement in the electron impact excitation cross sections of the $a^1\Delta_g$ and $b^1\Sigma_g^+$ electronic states of O_2, *Phys. Rev. Lett.* **69** 2495-2498.

52. While curve crossing is considered the DR mechanism for N_2^+, O_2^+, and NO, DR can also proceed without curve crossing. See Sarpal, B.K, Tennyson, J, and Morgan, L.A., Dissociative recombination without curve crossing: study of HeH^+, *J. Phys. B* **27**, 5943-5953. See also Guberman, S.L. (1994) Dissociative recombination without a curve crossing, *Phys. Rev. A* **49**, R4277-4280.

53. Johnsen, R. (1989) Recombination measurements of microwave discharge plasma afterglows, in Mitchell, J.B.A. and Guberman, S.L. (eds) *Dissociative Recombination: Theory, Experiment, and Applications*, World Scientific, Singapore.

54. Noren, C., Yousif, F.B., and Mitchell, J.B.A. (1989) The dissociative recombination and excitation of N_2^+, *J. Chem. Soc. Faraday Trans.* **285**, 1697.

55. Guberman, S.L. and Giusti-Suzor, A. (1991) The generation of $O(^1S)$ from the dissociative recombination of O_2^+. *J. Chem. Phys.* **95**, 2602-2613.

56. Guberman, S.L. (1991) Dissociative recombination of the ground state of N_2^+, *Geophys. Research Lett.* **18**, 1051-1054.

57. Guberman, S.L. (1988) The production of $O(^1D)$ from dissociative recombination of O_2^+, *Planet. Space Sci.* **36**, 47-52.

HETEROGENEOUS CATALYSIS:
THEORY, MODELS AND APPLICATIONS

M. BARBATO
CRS4 Research Center
Cagliari, Italy

AND

C. BRUNO
Dip. Meccanica e Aeronautica – Università di Roma I
Roma, Italy

1. Introduction

When air flows past a body moving at high Mach numbers, a bow shock forms in front of the body. If the flight Mach number is high enough, air will dissociate and form species as O, N, NO; if the Mach number corresponds to energies of order about 1 eV (about 11,000 K), ionization also takes place, and O^+, NO^+, N^+ and e^- will appear.

Past the bow shock the airflow entering the shock layer will eventually reach the body surface, which (for structural reasons) must be kept at surface (or: wall) temperatures T_w of order 300 K - 2000 K at most. At these temperatures the greatest part of the species above will tend to recombine. Recombination will occur throughout the shock layer, since the gas temperature drops going from the shock toward the wall, and a (usually) substantial percentage of atoms that have escaped gas-phase recombination will recombine on the surface of the body.

Solid surfaces may be thought as a medium where bonds present inside the bulk of the material have been severed by the fabrication process. Besides, the complex solid structure has many more degrees of freedom than molecules in the gas. It is intuitive that gas species will naturally tend to form bonds with the surface that may facilitate recombination. When this occurs, the surface is said to be catalytic.

If surface recombination occurs, the formation energy of the recombining species will initially stay within the new species formed. For instance, when two O atoms recombine, initially the total formation energy of the two O

139

M. Capitelli (ed.), Molecular Physics and Hypersonic Flows, 139–160.

atoms will tend to stay within the O_2 formed. But then the O_2 just formed would be translationally and roto-vibrationally excited so much that it would again dissociate, unless part of its energy is deposited on the surface.

Since formation energies of O, N or ionized species are large, even if the percentage deposited is small, the surface may be considerably heated by catalytic recombination. This heat flux is additional to molecular conduction and that due to sensible enthalpy transported by diffusing individual species (convective fluxes at the surface are zero, if the surface velocity is assumed zero, as is the case for non-permeable or non porous walls). Convincing estimates of the heat flux contribution due to catalysis yield as much as 40% of the total heat flux inside the stagnation region [1].

Since the inception of the manned space program, there has been therefore much interest in preventing recombination over the Thermal Protection System (TPS). In essence, one would like to coat TPS with materials as little catalytic as possible. Old work on recombination (dating back to the '50s and '60s, and performed at room temperature) indicated that glass is among the best material, a fact well known to combustion researchers investigating explosion limits [2]. Follow-on research and testing have shown that even glasses tends to increase their "catalyticity" with temperature, and by a large factor. Much work has been performed to predict this increase, since the actual composition of dissociated air during a re-entry trajectory is difficult to measure, and some sort of modeling is needed to preliminarily predict heat fluxes.

To bypass this problem, back in the early '70s, an extreme assumption became popular, i.e., the surface was assumed to be "fully catalytic", meaning that all species would recombine and form O_2 and N_2 only. In this case the heat flux contribution due to catalysis is maximum, which at first sight sounds like a conservative way of designing TPS. However, this assumption results also in overestimating Boundary Layer temperatures, therefore underestimating densities, and eventually wrong C_p predictions. Ever since, there has been an acute awareness of the need to understand finite-rate recombination.

What follows wants to be a primer of what is known to-date in this area; it is hoped that the material covered will be used as a starting point toward better understanding and modeling of surface catalysis in hypersonics.

2. Heterogeneous Catalysis

The word catalysis was first used by Berzelius in 1835: "Catalysts are substances that by their mere presence evoke chemical reactions that would not otherwise take place" [3]. Wilhelm Ostwald defined a catalyst as: "a substance that changes the velocity of a chemical reaction without itself appearing in the end products" [3]. In accord with that, the Chinese word used for a catalyst is "tsoo mei" which also mean "marriage broker".

This brief historical introduction brings to the real meaning of catalyst. A catalyst participating to a reaction does not appear in the stoichiometric equation of the reaction. His effect, for an assigned reaction, is to increase the reaction velocities for an assigned temperature or to decrease the temperature at which the reaction achieves a given rate. For simple reactions, without thermodynamically unstable intermediate product, this means that the catalyst action is to increase both the reaction velocities (forward k_f and backward k_b) in such a manner that the ratio k_f/k_b does not change. For this reason equilibrium obtained with a catalyst is the same as that ultimately arrived at when no catalyst is present. It is important to notice that the catalyst cannot initiate a reaction which is thermodynamically impossible [4]. Its action is always subject to the laws of thermodynamics.

The "speed-up" obtained using a catalyst is due to a reduction of the activation energy E necessary for a specific reaction (see Fig. 1). For example, we consider a solid catalyst: according to "transition state theory" a higher reaction velocity means a higher velocity constant [4]:

$$k_v = \frac{kT}{h} e^{-\triangle G^{\ddagger}/RT} \qquad (1)$$

where k is the Boltzmann constant, h is the Plank constant, T is the temperature, R is the gas constant and $\triangle G^{\ddagger}$ is the free energy for the activated state. For a fixed temperature, an higher k_v means a lower free energy $\triangle G^{\ddagger}$ for the catalysed reaction. But

$$\triangle G^{\ddagger} = \triangle H^{\ddagger} - T \triangle S^{\ddagger} \qquad (2)$$

and for the catalysed reaction $\triangle S^{\ddagger}$ is lower than the same quantity for the non catalysed reaction because the particle bonded on the surface of the catalyst loses some degree of freedom. This means that, to have a lower $\triangle G^{\ddagger}$ with respect to the non-catalysed reaction, the $\triangle H^{\ddagger}$ must be lower. From transition state theory we know that $\triangle H^{\ddagger} = E_a$ provided there is no change in the number of molecules in the reaction, which explains the activation energy reduction (see Fig. 1).

It is possible to specify different kinds of catalysis depending on the chemical phase of catalyst and reactants:

Homogeneous Catalysis : the catalyst is in the same phase as the reactants and no phase boundary exists;

Heterogeneous Catalysis : there is a phase boundary separating the catalyst from the reactants;

Enzymatic Catalysis : neither a homogeneous nor heterogeneous system.

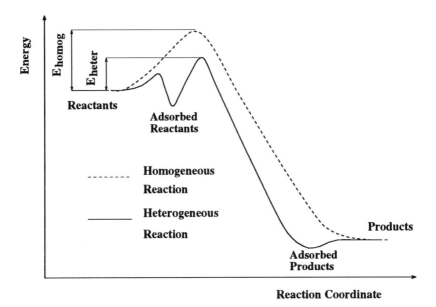

Figure 1. Energy reaction path for heterogeneous and homogeneous reaction.

A gas flow past a solid wall belongs to the second case: the reactants are the gas chemical species and the wall is the catalyst.

2 Modelling Heterogeneous Catalysis

A scheme representing the elementary steps involved in heterogeneous catalysis can be drawn as follow [3]:

1 diffusion of reactants to the surface;
2 adsorption of reactants at surface;
3 chemical reactions on the surface;
4 desorption of products from the surface;
5 diffusion of products away from the surface.

Each one of these steps has a different velocity and the slower one is the process-determining rate. Steps 1 and 5 are usually fast; exceptions exist when the catalyst is very efficient. The limit for the catalytic reaction rate is the quantity of reactants that goes to the wall, and the surface reaction rates become independent of the kinetics properties of the surface [5]. This is the case of "diffusion controlled" catalysis.

Coatings for space vehicles TPS are studied to have very low catalytic efficiency, to lower, as much as possible, the surface heat flux due to recombination effects. Therefore, for these applications, the rate determining step for the heterogeneous catalysis process has to be found among steps

2, 3 and 4. In the following we will focus our attention on these elementary steps.

2.1. ADSORPTION

Adsorption is the formation of a bond between the atom, or the molecule, and the solid surface [4].

We speak of **physisorption** (physical adsorption) when the bond between surface and gas particles is due to van der Waals forces (forces between inert atoms and molecules). The forces involved are electrostatic forces for molecules with permanent dipole moments, induced polar attractions for readily polarizable molecules, dispersion forces for non polar atoms and molecules (forces due to the fluctuation of electron density). The strength of these bonds depends upon the physical properties of the adsorbed gas, and is connected to the boiling point (or condensation) of the gas. Physisorption does not depend much on the chemical nature of the solid [4]. The particle-surface bond energy is low (10-50 kJ/mole) and the bond is important only at low temperatures ($\sim$ 100-300 K): as the temperature rises the gas is removed more or less completely [6]. In heterogeneous catalysis, the importance of this kind of adsorption lies in the fact that it can be a precursor step for **chemisorption**.

Chemisorption is associated to the solid surface properties. The surface of a solid can have properties markedly different from those of the bulk solid because on the surface there are unsaturated bonds. In fact, an atom on the surface is not in the same condition of a "bulk" atom, because it does not have its full complement of neighbors [4]. This is true either for a covalent solid (e.g. SiO_2, SiC) as for an ionic solid (e.g. $NaCl$). A simple way to understand this situation is to think about an ideal fracture of a crystal: the fracture sides are new surfaces where atoms which were before inside the bulk are now surface atoms with "dangling" bonds.

When a free gas atom is near to a solid surface there is an attractive interaction between it and the surface atoms: the atom is attracted to the surface and it can find in one of the "dangling" bonds a **site**, i.e. a physical location corresponding to a potential well (low energy state) where it remains trapped. The bond formed is a true **chemical** bond, usually covalent: there is an interpenetration of particle shells with electron sharing and the bond energy is high (40-800 kJ/mole [4]). Chemisorption occurs until the unsaturated surface valences are filled. This explains why, if multilayer physisorption is possible, no more than one layer can be adsorbed via chemisorption [7] although **new** species can form at surface (e.g. due to oxidation in metals). Physisorption and chemisorption are spontaneous ($\triangle H < 0$) but the latter can have a substantial activation energy E_A (see

Fig. 2). This means that it starts at temperatures higher than those relative to physisorption.

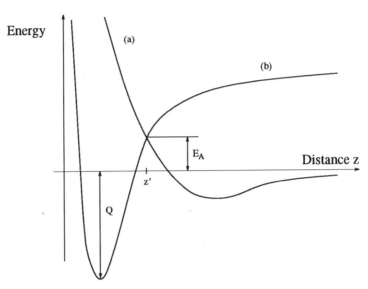

Figure 2. Potential energy curves for adsorption: (a) physisorption of a molecule; (b) chemisorption of two atoms.

Two kinds of chemisorption are possible: the first one is called **associative** or **non-dissociative** and it occurs to atoms $(O, N, H, ...)$ and to molecules (CO, OH) without molecular bond breaking. A very simple schematic representation of associative chemisorption is:

$$A + * \rightarrow A^*$$

where the "$*$" represent the **site** and the A^* the adsorbed atom (adatom). Assuming a uniform surface with uniformly distributed sites and assuming sites activity do not be function of the fraction of surface already covered, θ, the adsorption rate can be expressed as:

$$v_{ads} = s_0 v_c (1 - \theta)$$

where s_0 is the initial sticking coefficient, representing the probability of a molecule to stick on a bare surface, and v_c is the collision rate:

$$v_c = \frac{n}{4} \left(\frac{8kT}{\pi m} \right)$$

where n is the gas particle density and m the particle mass.

The second kind of chemisorption is called **dissociative** and it may occur to diatomic molecules like O_2, N_2, H_2 and CO. In this case chemisorption occurs with breaking of molecular bonds. Each atom is adsorbed from one site:

$$A_2 + 2* \rightarrow 2A^*$$

For this reaction the rate of adsorption reads:

$$v_{ads} = s_0 v_c (1 - \theta)^2 \frac{z}{2}$$

where z represents the number of neighbors sites. In fact, in this case we assume that the atoms coming from the broken molecule can adsorb only on adjacent sites.

Dissociative adsorption may occur also following another path:

$$A_2 + * \rightarrow A^* + A$$

In this case the striking molecule dissociates, an atom is adsorbed and the other goes back to the gas. Usually this process is far less probable than the previous one. In fact the energy content of the system adatom plus atom is greater than the energy associated to the system of two adatoms.

Before describing the reaction step, we need to spend a few words about the initial sticking coefficient expression. This important quantity, assuming an uniform sites activity over the surface, depends on T. For atoms impinging on a surface a general expression is [8]:

$$s_0 = P e^{-E/(kT)}$$

where P is expected to be < 1. s_0 decreases with temperature and this behavior can be understood with simple considerations: with increasing temperature gas atoms impinging on the surface have higher kinetic energy and thus higher probability to "rebound" after the collision with a site without sticking. Notice that for impinging molecules this trend can be the reverse (s_0 increases with T): in fact, if there is an activation energy associated to dissociative adsorption, at higher temperature it becomes easier to overcome this energy barrier.

2.2. REACTION

The kind of reactions interesting to hypersonics applications are **recombinations** of atoms forming O_2 and N_2 and reactions producing NO, CO, and CO_2.

In heterogeneous catalysis at least one of the reacting particles must be first chemisorbed in order to react catalytically on a solid surface [4]. Therefore, if physisorption does not play a direct role, chemisorption is a fundamental step to "prepare" atoms and molecules to react. This mechanism furnishes a reaction path with an activation energy lower than the one of the correspondent homogeneous reaction (see Fig. 1).

The first reaction mechanisms we consider is the recombination between an atom from the gas and an adsorbed one: the free gas atom strikes the adatom and reacts with it. The result is a recombined molecule. This is called the **Eley-Rideal mechanism** (E-R) and a simple scheme of it is [9]:

$$A + B^* \rightarrow AB^*$$

where the symbols A, B mean an atom. After the recombination the molecule AB^* is still adsorbed. The rate of this reaction depends by the partial pressure of the gas phase atoms and by the surface coverage. On this basis for the step rate we can write [10]:

$$v_{ER} = k\,[A]\,[B^*]$$

The second mechanism is the recombination between two adatoms: this is called the **Langmuir-Hinshelwood mechanism** (L-H) and a schematic representation is [9]:

$$A^* + B^* \rightarrow AB^* + *$$

Obviously this kind of reaction is possible if adatoms diffuse over the surface. For adatoms a way to move it is to "hop" from one potential well to another [8] overcoming a potential barrier E_m that is lower than the desorption energy E_d; in fact particles during this movement do not leave the surface completely. In literature it can be found that $E_m \sim 0.1 - 0.2$ E_d. This value depends on the surface coverage extent (it decreases when θ increases), on the surface defects and on the crystallographic orientation of the surface. The presence of the energy barrier E_m implies that the L-H mechanism is an activated process which becomes efficient at higher temperature with respect to the E-R mechanism. This is evident in catalysis involving metal surfaces were the recombination coefficient drops sharply when the L-H mechanism starts (see Fig. 3).

2.3. DESORPTION

After a E-R or a L-H recombination, there is not chemical bond between surface and molecule, which is free to leave the wall. This phenomenon

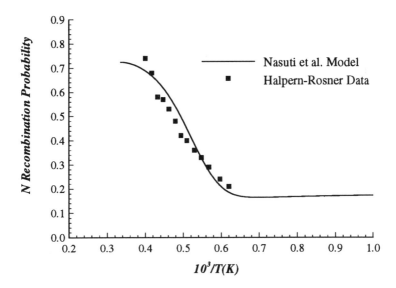

Figure 3. Nitrogen recombination over tungsten: γ_N *vs* inverse temperature. From [15].

is called **desorption**. It may happen that desorption is not "immediate" and the molecule may fall in a physisorption well before the leaving (see Fig. 1). In any case, desorption is a very fast process and many authors consider reaction and desorption like a single step. The importance of this phenomenon lies in the fact that recombination energy can be left, in part or totally, on the wall when the molecule leaves the wall. The characteristic time of desorption is strictly linked to the amount of energy left on the wall by the molecular internal energy relaxation process [11]. This means that the molecule can leave the surface in an excited state and "quench" later in the gas near the wall, or they can leave the surface in thermal equilibrium with it. To take in to account this phenomenon a **chemical energy accomodation** (CEA) coefficient β is defined as the fraction of the equilibrium dissociation energy delivered to the catalyst surface per recombination event [11]:

$$\beta = \frac{\dot{q}}{D_{AB}\dot{\Delta n}/2} \qquad (3)$$

where $\dot{q}$ is the effective energy flux to the wall, $\dot{\Delta n}$ is the flux of recombined atoms and D_{AB} is the AB molecule dissociation energy.

Experimental values of β [11] show a dependence on the interaction between surface and recombined molecules. If the bond between recombined molecule and surface is "strong" there is enough time to have a total accomodation ($\beta = 1$). On the contrary, if the bond is "weak" the molecules desorb rapidly and the energy left on the wall is smaller ($\beta < 1$).

Large savings in TPS appear, in principle, possible if β could be determined reliably and accurately, and if it could be "designed" a priori based on material properties and recombining atoms type. This is a frontier of surface catalysis still in need of exploration.

2.4. THERMAL DESORPTION

Besides the mechanisms cited, we have to consider also **thermal desorption**. In fact, adsorbed atoms vibrate with the frequency kT/h; at sufficiently high temperatures this vibrations can extract the atom from the potential well leading to desorption [12]:

$$A^* \to A + * \; .$$

Following transition state theory, the rate of desorption reads [8]:

$$v_{des} = N_a \frac{kT}{h} e^{\triangle S/k} e^{-\triangle H/(kT)} \tag{4}$$

where $\triangle S$ and $\triangle H$ are respectively entropy and enthalpy differences between two states (adsorbate-gas). Usually $\triangle H$ is supposed to be of the same order of magnitude of the adsorption energy. Due to the high values of the latter, thermal desorption becomes important at high temperatures (e.g. $T > 2000$ for N over W [11]).

2.5. RATE DETERMINING STEP

In the following we will make some considerations about the rates of heterogeneous phenomena on a generic surface. For this reason we will speak about low and high temperatures without any precise specifications. In any case, what we mean for "low" are temperatures not far for ambient, and for "high" temperatures over 1000 K. In a more detailed discussion, it would be better to distinguish between metallic and non-metallic surfaces, but the following discussion is general enough to fit both cases.

At low temperatures the surface coverage θ is near to unity; this because the atoms mobility is very low and thermal desorption is not efficient. In these conditions the E-R recombination mechanism is expected to be more effective with respect to L-H [13]. In fact, there is a very high probability, for an incoming atom, to strike an adsorbed one. In this conditions the rate of adsorption is expected to be larger than the rate of surface atoms depletion:

$$v_{ads} \gg v_{reac}, v_{des}$$

When the temperature rises, the coverage reduces both due to thermal and L-H recombination desorption. High temperature enhances adatom mobility and consequently their probability of recombination. Besides, the L-H mechanism is very efficient because it removes two adatoms at once. The E-R mechanism become less effective due to the low probability of a collision between a gas atom and an adatom, since surface coverage is low and adatoms are very mobile and difficult "targets" for striking gas atoms. Under these conditions the rate limiting step can be adsorption:

$$v_{des}, v_{reac} \gg v_{ads}$$

Besides defining a rate limiting steps, the previous discussion leads to some further comment. We can observe that the molecules formed via a E-R mechanism with high probability are, more likely to leave the surface in an excited state [13]. They would then take with them part of the recombination energy. This is not the case of L-H recombined molecules, because they would have enough time to leave all the energy excess energy to the surface, and would desorb in thermal equilibrium. Therefore the β factor would be expected to be small at low temperatures and to increase toward unity with increasing temperature.

3. Surface kinetics for reentry flows

For reentry applications, one of the problems to solve is to predict the catalytic activity of TPS surfaces. The US Shuttle TPS coatings are made of borosilicate Reacting Cured Glass (RCG: 92% SiO_2, 5% B_2O_3, 3% SiB_4), a surface that can be thought as a pure SiO_2 surface. In fact, there is experimental evidence that a RCG surface may be more similar to SiO_2 than to the borosilicate bulk [14]. This is the reason why the existing models simulating RCG catalytic activity, built by starting from the gas/surface physics seen above, are based on SiO_2 structure. These models can be used to set realistic catalytic boundary conditions in CFD codes, and in the following we will present three of the most recent.

Before going through these models we recall here some definitions starting from the Molecular Recombination Coefficient [1]:

$$\gamma = \frac{\text{flux of atoms recombining at surface}}{\text{flux of atoms impinging the surface}}$$

γ represents also the recombination probability for an atom impinging the surface. The Thermal Recombination Coefficient is defined as [1]:

$$\gamma' = \beta\gamma$$

This coefficient is a measure of the energy effectively left on the surface by heterogeneous chemistry. In fact values of β can be lower than 0.1 [14], meaning that only a small portion of recombination energy is left on the wall. If recombination rates (or probabilities) are measured via calorimetry, telling γ and γ' apart is a difficult task.

Finally we recall the expression for the surface catalytic recombination rate, usually named catalitycity [1]:

$$K_w = \gamma \sqrt{\frac{kT_w}{2\pi m}} \tag{5}$$

where the subscript $_w$ stands for "wall" and m is the atom mass.

3.1. HETEROGENEOUS CATALYSIS MODELS

A model for recombination of Oxygen atoms on SiO_2 has been presented by Jumper and Seward [16]. These authors assume the surface adsorption sites are the Si atoms of the SiO_2 crystal lattice. They assume that an O atom of the SiO_2 surface can desorb, either thermally or after a recombination with a gas atom (E-R mechanism), leaving available a Si atom that becomes a site where an oxygen atom can adsorb forming a double bond. Therefore the number of adsorption sites available on the surface is equal to the number of Si atoms on the surface (see Table 1).

To obtain the recombination coefficient γ, Jumper and Seward write the equation for the time rate variation of the surface concentration of oxygen atoms (n_s) [16]:

$$\frac{dn_s}{dt} = \dot{n}_{\text{ads}} - \dot{n}_{\text{tdes}} - \dot{n}_{\text{rec}} \tag{6}$$

where $\dot{n}_{\text{ads}}$ represents the rate at which O adsorbs at surface, $\dot{n}_{\text{tdes}}$ the rate at which the adatoms desorb without recombination, and where $\dot{n}_{\text{rec}}$ is the rate at which atoms leave the surface after recombination with a gas atom following the E-R mechanism. Jumper and Seward, assuming steady state ($\frac{dn_s}{dt} = 0$), calculate the surface coverage and the recombination coefficient that are functions of wall temperature and O partial pressure. The initial sticking coefficients are function of wall temperature too (see Table 1). This model was extended to RCG in [17] and these same authors, together with Newman and Kitchen, proposed a similar model for N recombination over SiO_2-based TPS also [18]. These models are the first that can simulate heterogeneous catalysis in Hypersonics taking into account the **local** flow and wall characteristics.

A model for air recombination over SiO_2-based materials has been presented by Nasuti et al. [15]. In this model the recombination coefficients are also obtained by writing a surface balance for incoming and leaving particles based on [11]. For steady state conditions the flux of atoms adsorbing on the surface has to be balanced by the flux of atoms desorbing due to thermal desorption, E-R recombination and L-H recombination (in the latter case two atoms desorb at one time). Therefore simultaneous E-R and L-H recombination mechanisms are accounted for. The surface considered is SiO_2 and adsorption sites are specified as potential energy wells. This model calculates the recombination coefficients for N_2, O_2 and also NO surface formation. Therefore it produces 3 independent recombination coefficients functions of wall temperature and of O and N surface partial pressures [15]. Constant values are assumed for the initial sticking coefficients. This model depends on knowledge of basic surface physics, and in principle may be extended to any kind of gas species and any surface.

The last model recently proposed is by Deutschmann et al. [19]. In this case, again for a SiO_2 surface, a more conventional kinetic approach is used. Rate constants for adsorption, thermal desorption surface reactions, including E-R and L-H mechanisms, and desorption after recombination are calculated. The rate constants for reaction and desorption are in Arrhenius-like form. Recombination coefficients (for N_2 and O_2 only) depend on the wall temperature and on adsorbed species concentration. As in Nasuti et al. [15], constant values for the initial sticking coefficients are assumed.

Whereas in the other two models the desorption step is always assumed to be collapsed in the reaction step, Deutschmann et al. assume this to be true for the E-R mechanism only. In fact, after L-H recombination, the molecule is assumed still adsorbed and a rate for the desorption step is calculated (a very low activation energy is actually used: 20 kJ/mol). In accordance with that, these authors assume complete accomodation ($\beta = 1$) for L-H recombined molecules, while they assume a constant $\beta = 0.2$ for E-R recombined molecules.

The crucial point of these models is the incorporation of physiochemical quantities, such as, for example, atoms adsorption energies, sticking coefficients, number of sites for unit area and others. There is still a large range of uncertainty in the determination of these quantities and in each model more or less complicated expressions for steric factor are used to cover these "holes". In any case the effort to produce such a models, in our opinion, has to be appreciated. In fact, these tools are valid under several flow conditions, and wall temperatures [21]. They can cover a wide range of flight and wind tunnel flow conditions with a larger flexibility with respect to experimental fits obtained for a small range of temperature. For example, a spacecraft with a complex geometry re-entering our atmosphere, undergoes different flow conditions near to the surface at different locations and in this case using a single fit could be too restrictive.

TABLE 1. Main parameters for recombination coefficient models: initial sticking coefficient s_0; surface site density N_s $(sites/m^2)$; steric factor S; desorption energy E_d $(kJ/mole)$.

authors	s_0	N_s	S	E_d
Jumper-Seward (SiO_2)	$0.05 \cdot e^{0.002T}$	$5.00 \cdot 10^{18}$	$2.24 \cdot 10^{-5} \cdot e^{0.00908T}$	339
Jumper-Seward (RCG)	$1.00 \cdot e^{0.002T}$	$2.00 \cdot 10^{18}$	$2.00 \cdot 10^{-4} \cdot e^{0.003T}$	339
Nasuti et al.	0.05	$4.50 \cdot 10^{18}$	0.1	250
Deutschmann et al.	0.10	$1.39 \cdot 10^{19}$	-	200

Oxygen recombination coefficient calculated using the three models presented above are shown in Fig. 4. These results, obtained for an O partial pressure $= 400$ Pa, are compared with the experimental data of Greaves and Linnett (SiO_2) [22] and of Kolodziej and Stewart (RCG) [23]. All the three models represent qualitatively well the high temperature behavior by reproducing the γ_O rollover for $T \sim 1600$ K shown by the experimental results of [23]. In all three this second order effect has been found to be due mainly to thermal desorption, the most important process at those temperatures. Moving toward the low temperature range, the Deutschmann et al. model underpredicts γ of order of magnitude showing its intentional high-temperature-addressed design. The other two models represent qualitatively and quantitatively well the recombination coefficient trend. An evident difference between these two models is their different curvatures. It is due to the difference in the sticking and steric factors laws chosen by the authors: Nasuti et al. assume constant values whereas Jumper and Seward adopt temperature dependent expressions (see Table 1). These factors are one critical to such a models, because by necessity, they try to include physics not otherwise modeled by elementary steps. A way to clarify this point may be to have more experimental data for γ and γ' over a wide range of temperature going from ambient to the material melting limit.

4. Heterogeneous catalysis applications to hypersonic flows

It is already some time that the CFD community is moving towards the inclusions of this kind of models in complex thermochemical nonequilibrium flow solvers [20], [21], but often "rude" boundary conditions, such as "fully catalytic wall" or "non catalytic wall", are still used. The only justification for them is that they are simple to understand and to implement in a code. Clearly this is not acceptable when real conditions are very far from these extremes. In the following some results showing the large differences due to

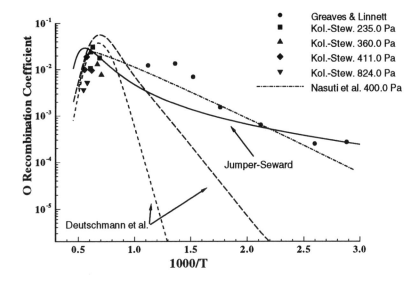

Figure 4. γ_O for TPS coating materials: comparison among several models.

the use of one or another condition are reported. A brief summary of the different wall boundary conditions used in CFD for heterogeneous catalysis and for a 5 species air is [21]:

1 Non catalytic wall:

$$\gamma_i = 0 \quad i = O, N$$

The wall is completely indifferent to the flow chemistry.

2 Fully catalytic wall:

$$\gamma_i = 1 \quad i = O, N$$

in this case the wall is a "perfect catalyst" and catalysis is "diffusion controlled." Since each atom hitting the wall recombines, the catalytic efficiency depends by the velocity of diffusion of atomic species trough the boundary layer.

3 Fully Equilibrium wall:

$$Y_i = \overline{Y}_i \quad i = O_2, N_2, NO, O, N$$

where $\overline{Y}_i$ are the equilibrium species mass fractions for the local values of temperature and pressure. This is a very strong condition. It forces the flux of atoms to the wall (usually is $Y_O = Y_N = 0$) imposing very strong mass fractions gradients at the wall. Without any doubt these boundary conditions are stressing over the limit the near wall chemistry.

4 Finite rate catalysis:

$$0 < \gamma_i < 1 \qquad i = O, \, N$$

In this case a measure of real interface physics is introduced in the CFD flow solver. This means attributing the same importance to homogeneous and heterogeneous chemistry.

4.1. SOME APPLICATIONS

Solving hypersonic flows, the additional computational effort required by Finite Rate Catalysis (FRC) models is balanced by the increase in quality of the results. This concept can be clarified by a simple example: we consider a flow past a 40 cm long blunt cone (scaled down ELECTRE reentry capsule geometry). The inflow conditions are those of the HEG [24] shock tunnel test chamber for $p_0 = 500 \, Pa$, $T_0 = 9500 \, K$ stagnation conditions [25]. The body surface is assumed to be at 1500 K. The flow is solved with the TINA code [26] and two different Boundary Conditions (BCs): (a) finite rate catalysis; (b) fully catalytic wall. The results of these calculations are shown in Figs 5 and 6. From the stagnation line results, shown in Fig. 5, we see that, besides the expected results for the wall mass fractions, i.e. $Y_{N_2}^{(b)} > Y_{N_2}^{(a)}$, $Y_{O_2}^{(b)} > Y_{O_2}^{(a)}$, the wall NO mass fraction in the fully catalytic case is higher than in the finite rate catalysis case, notwithstanding the fact that the FRC model used [15] predicts also catalytic NO formation. The reason for this result is that in case (b) the strong production of N_2 and O_2 ($K_{wO_2}^{(b)} \simeq K_{wN_2}^{(b)}$) and the diffusion of these species away from the surface due to the strong species gradients, drive the gas phase mechanism:

$$O_2 + N \longrightarrow NO + O + 32.4 \text{ kcal/mole} \qquad (7)$$

$$N_2 + O + 76.4 \text{ kcal/mole} \longrightarrow NO + N \qquad (8)$$

This mechanism pumps NO in the flow and stores energy in the gas phase; it is very effective slightly away from the surface and it quenches near to the wall, where $Y_O, Y_N \to 0$.

For case (a) instead this mechanism is not effective: in the gas phase a smaller quantity of NO is produced by the reaction (8), whereas in the zone close to the wall, due to the surface formation of O_2 and to the presence of N atoms, reaction (7) produces NO and releases energy. This latter phenomenon is similar to what was reported by Tirsky in [27], where an exchange mechanism "(7) + reverse of (8)" is presented as very effective in the gas close to the wall when $K_{wO_2} > K_{wN_2}$. Notice that in the case (a) shown here, in the zone closer to the wall, there is not any sensible contribution from either reaction (8) direct or reverse.

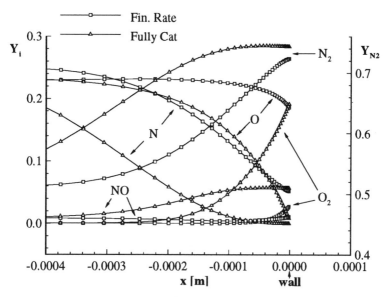

Figure 5. Comparison between two solutions with different catalytic boundary conditions at $T_w = 1500\ K$: stagnation line mass fractions.

The effects of this on the surface heat flux can be seen in Fig. 6 where the ratio q_i/q_{tot} for $i =$ "tr (translational), vib (vibrational), df (diffusive)" are plotted along the body. For case (b) the strong recombination leads to $q_{df}/q_{tot} > 0.4$ all along the body, whereas for case (a) this ratio is < 0.3. Still for case (a), the energy released by reaction (7) near to the surface yields a quite larger q_{tr}/q_{tot} with respect to case (b). This result is also partly due to the desorption of a larger amount of molecules than in case (b). In fact, assuming that the recombined molecules leave the wall in thermal equilibrium $(T = T_V = T_w)$, the BL gas is cooled by this "cold" flow of particles. Therefore, due to the larger recombination, in case (b) the wall ∇T and ∇T_V are smaller than in case (a).

An other important issue arises when we compare flow conditions for flight and wind tunnels simulations. In Fig.s 7-10 we show the results of two calculations performed for the same shape (i.e. sphere plus cone) and obeying to the binary scaling rule $(\rho_\infty L = 0.0006)$

For the two cases the amount of dissociated species reaching the wall is quite different, as shown by Fig.s 7 and 8. This leads to different amounts of recombined species. In fact the atoms depletion term can be expressed as:

$$\dot{w}_a = K_{wa}\rho_w Y_{wa} \tag{9}$$

where Y_{wa} is the atoms mass fraction at the wall.

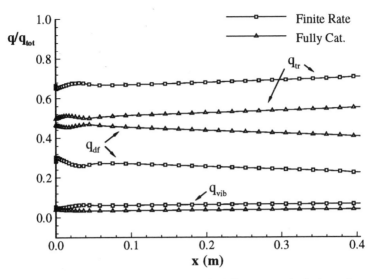

Figure 6. Comparison between solutions with different catalytic boundary conditions at $T_w = 1500\ K$: different contributions as a percent of the total wall heat flux.

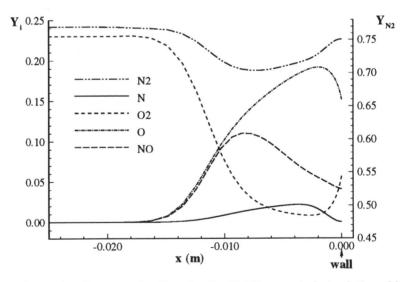

Figure 7. Stagnation line mass fractions for "in flight" numerical simulation: $Ma = 15.3$, $\rho_\infty = .00033\ kg/m^3$, $L = 2.0\ m$, $T_w = 1000\ K$.

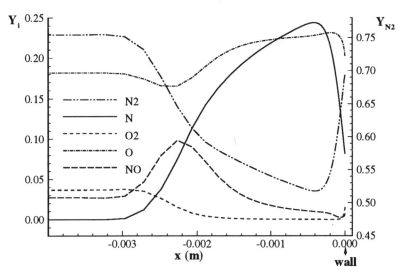

Figure 8. Stagnation line mass fractions for shock tunnel numerical simulation: $Ma = 9.7$, $\rho_\infty = .002017 \ kg/m^3$, $L = .33 \ m$, $T_w = 1000 \ K$.

Therefore the catalytic activity has different effects as it is shown by diffusive contribution to the total surface heat flux in Fig. 10. In fact, along the body the ratio q_{df}/q_{tot} is higher than 0.15 for the wind tunnel numerical simulation whereas, apart from the zone near to the stagnation point, it is lower than 0.1 for the "in flight" case. Near the stagnation zone this difference reduces, and on the stagnation point the two contributions are very close to each other. For the "in flight" numerical simulation we see a rapid decrease of the catalytic activity from the stagnation point to $x/L \simeq$ 0.16.

In cases similar to the latter presented here, the necessary "binary scaling" principle is insufficient to guarantee reactive flow similarity then, to do that, the Damköheler numbers related to heterogeneous chemistry (i.e., inside the BCs) must be the same of flight conditions. This further condition yields that, to reproduce "real flight conditions" in a wind tunnel, a coating material different from that applied to the flying object must be used.

A consequence of what presented above is that numerical simulations with realistic catalytic boundary conditions may help in the design of ground testing models. In fact the choice of the model skin has to be done on the basis of "scaling factors" for the heterogeneous chemistry. Therefore the main question is how important is the influence of catalysis on the quantities measured in a test; if this importance is high, the similarity

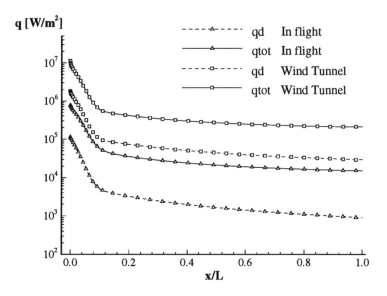

Figure 9. Wall heat fluxes for shock tunnel and "in flight" numerical simulations.

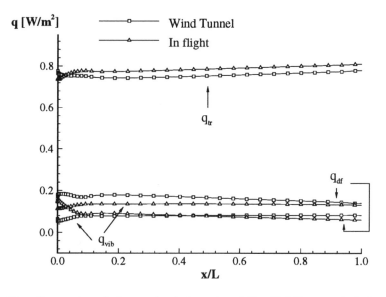

Figure 10. Comparison between shock tunnel and "in flight" numerical simulations: different contributions as a percent of the total wall heat flux.

parameters (Damköheler numbers), associated with catalytic activity must be set identical to those in flight. This conclusion should be supported by two suggested actions: (a) an experimental study of the catalytic behavior of several materials, from oxides to metals, resulting in a database containing data for a wide range of temperatures; (b) starting from these data, the definition of more accurate and more effective models for the catalytic activity of a set of materials of interest for hypersonic flight. Finally, the right coupling among numerical simulations, ground testing and in flight measurements may lead to more reliable predictions.

5. Conclusions

The brief discussion presented here has the aim of stimulate the interest in FRC models and on their use in CFD hypersonic flow solvers. This choice becomes necessary if one wants to move towards a more quantitative evaluation of thermal loads, a real estimate of gas composition near the wall and also in the entire boundary layer. Several models are now available that couple homogeneous and heterogeneous chemistry for hypersonic reactive flows. The catalytic activity of a surface is strongly dependent on surface structure, on temperature and on local flow conditions. The use of existing fits is a way to account for heterogeneous chemistry but it is also a poor substitute of that offered by FRC models. Clearly these models are still far from being perfect but, based on the availability of more experimental data, work can be done to make them more and more reliable.

6. Acknowledgement

The authors would like to acknowledge Dr. Jean Muylaert for his interest and his support. This work has been carried out with the financial support of the Sardinia Regional Government.

References

1. Scott, C.D. (1987) The Effects of Thermochemistry, Non-Equilibrium and Surface Catalysis in the Design of Hypersonic Vehicles, 1th Joint Europe-US Short Course on Hypersonics, Paris, France.
2. Lewis, B. and von Elbe, G. (1987) *Combustion, Flames and Explosions of Gases*, Academic Press, Orlando.
3. Moore, W.J. (1978) *Physical Chemistry*, Longman Group Limited, London.
4. Bond, G.C. (1974) *Heterogeneous Catalysis: Principles and Applications*, Clarendon Press, Oxford.
5. Rosner, D.E. (1966) Convective Diffusion Limitations on the Rates of Chemical Reactions at Solid Surfaces - Kinetic Implications, 11th Symposium (International) on Combustion, The Combustion Institute, Pittsburgh.
6. Kensington, A. N. (1968) *The Physics and Chemistry of Surfaces*, Dover Publications Inc., New York.
7. Tamaru, X. (1978) *Dynamic Heterogeneous Catalysis*, Academic Press, New York.

8. Boudart, M. and Djéga-Mariadassou, G. (1984) *Kinetics of Heterogeneous Catalytic Reactions*, Princeton University Press,Princeton N.J..

9. Zangwill, A. (1988) *Physics at Surfaces*, Cambridge University Press, Cambridge.

10. Atkins, P.W. (1994) *Physical Chemistry*, Oxford University Press, Fifth Edition, Oxford.

11. Halpern, B. and Rosner, D.E. (1978) Chemical Energy Accomodation at Catalyst Surfaces, *Chemical Society, Faraday Transactions I*, 74, 1833-1912.

12. Park, C. (1990) *Non-Equilibrium Aerothermodynamics*, John Wiley & Sons, New York.

13. Bruno, C. (1989) Nonequilibrium Thermochemistry and Catalysis, 2nd Joint Europe-US Short Course on Hypersonic, US Air Force Academy, Colorado Springs, CO.

14. Carleton, K.L. and Marinelli, W. J. (1992) Spacecraft Thermal Energy Accomodation from Atomic Recombination, *J. of Thermophysics and Heat Transfer*, 6, 4, 650-655.

15. Nasuti, F., Barbato, M. and Bruno, C. (1995) Material-Dependent Catalytic Recombination Modeling for Hypersonic Flows, also AIAA Paper 93-2840, revised and accepted for publication in *J. of Thermophysics and Heat Transfer*.

16. Seward, W.A. and Jumper, E.J. (1991) Model for Oxygen Recombination on Silicon-Dioxide Surface, *J. of Thermophysics and Heat Transfer*, 5, 3, 284-291.

17. Jumper, E.J. and Seward, W.A. (1994) Model for Oxygen Recombination on Reaction-Cured Glass, *J. of Thermophysics and Heat Transfer*, 8, 2, 460-465.

18. Jumper, E.J., Newman, M., Seward, W.A., and Kitchen, D.R. (1993) Recombination of Nitrogen on Silica-Based, Thermal-Protection-Tile-Like Surfaces, AIAA Paper 93-0477, 31st Aerospace Sciences Meeting & Exhibit, Reno, NV.

19. Deutschmann, O., Riedel, U., and Warnatz, J. (1994) Modeling of Nitrogen and Oxygen Recombination on Partial Catalytic Surface, *ASME J. of Heat Transfer*, June, Preprint 23.

20. Grumet, A.A., and Anderson, J.D. Jr. (1994) The effects of Surface Catalysis on the Hypersonic Shock Wave/Boundary Layer Interaction, AIAA Paper 94-2073, 6th AIAA/ASME Joint Thermophysics and Heat Transfer Conference.

21. Barbato, M., Giordano, D. and Bruno, C. (1994) Comparison Between Finite Rate and Other Catalytic Boundary Conditions For Hypersonic Flows, AIAA Paper 94-2074, 6th AIAA/ASME Joint Thermophysics and Heat Transfer Conference.

22. Greaves, J.C. and Linnett, J.W. (1955) Recombination of Oxygen Atoms on Silica from $20°C$ to $600°C$, *Transactions of the Faraday Society*, 55, 1355-1361.

23. Kolodziej, P. and Stewart, D.A. (1987) Nitrogen Recombination on High-Temperature Reusable Surface Insulation and the Analysis of its Effects on Surface Catalysis, AIAA Paper 87-1637, 22nd Thermophysics Conference, Honolulu, Hawaii.

24. Eitelberg, G. (1993) Calibration of the HEG and Its Use for Verification of Real Gas Effects in High Enthalpy Flows, AIAA Paper 93-5170, AIAA/DGLR Fifth International Aerospace Planes and Hypersonic Technologies Conference.

25. Walpot, L.M.G. (1991) Quasi One Dimensional Inviscid Nozzle Flow in Vibrational and Chemical Non-Equilibrium, Technical Report EWP-1664, ESA-Estec, Noordwijk, The Netherlands.

26. Netterfield, M.P., (1992) Validation of a Navier-Stokes Code for Thermochemical Non-Equilibrium Flows, AIAA Paper 92-2878, 27th Thermophysics Conference, Nashville, TN.

27. Tirsky, G.A. (1993) Up-to-Date Gasdynamic Models of Hypersonic Aerodynamics and Heat Transfer with Real Gas Properties, *Annual Review of Fluid Mechanics*, 25, 151-181.

RECOMBINATION OF ATOMIC SPECIES ON SURFACES

A Phenomenological Approach

C. D. SCOTT
NASA Johnson Space Center
Houston, Texas 77058
USA

1. Introduction - Defining Catalysis

In hypersonic flight and in high enthalpy test facilities it is normal for the flow to be dissociated and not in chemical equilibrium. There may be a significant amount of energy in latent form, particularly the heat of formation of atomic species. When an object or spacecraft encounters the flow the atoms may recombine on surfaces and give up part of that latent heat of dissociation to the surface. The amount of energy transferred to the surface depends on diffusion to the surface, the energy of the chemical bond, and on the probability of recombination on the surface. Other factors affecting the energy transfer is the fraction of the bond energy given to the surface per recombination event.

In aerospace reentry thermal protection system applications what is wanted are poor catalysts. That is, thermal protection materials that are desirable do not enhance recombination - indeed they impede or retard recombination. It has been fortuitous that many of the surfaces of flight vehicles have consisted mainly of silica, one of the least catalytic materials for air atom recombination. There is a tendency for metals and many metal oxides to be highly catalytic whereas some oxides to have very low catalytic activity. Many salt compounds fall in between. Carbon monoxide oxidation is an important reaction related to aeroheating during entry into the Mars or Venus atmospheres. A review of relative rates of catalytic CO oxidation on many compounds has been given by Golodets [1]. He gives the general activity pattern for CO oxidation at 150 C to be:

$MnO_2 > CoO > Co_3O_4 > MnO > CdO > Ag_2O > CuO > NiO > SnO_2 > Cu_2O > Co_2O_3 > ZnO > TiO_2 > Fe_2O_3 > ZrO_2 > Cr_2O_3 > CeO_2 > V_2O_5 > HgO > WO_3 > ThO_2 > BeO > MgO > GeO_2 > Al_2O_3 > SiO_2$.

It has been observed that there may be a strong dependence of catalytic reaction rates on surface temperature. Often times the rates have an Arrhenius dependence.

161

M. Capitelli (ed.), Molecular Physics and Hypersonic Flows, 161–180.
© *1996 Kluwer Academic Publishers. Printed in the Netherlands.*

In this paper several phenomenological models for catalytic recombination will be described. How one determines catalytic reaction rates will be explored from a theoretical point of view, but more importantly various means of measuring catalytic rates will be described, including a proposed technique that involves molecular beam techniques. Chemical mechanisms that relate to a theoretical description of catalytic reaction rates will be discussed briefly.

To predict the effect of catalytic reaction rates on aeroheating one needs to use techniques that incorporate these mechanisms into the boundary conditions, thus we will discuss the form of the boundary conditions and how they may be implemented into flow codes.

Finally, some the effects of catalysis on aeroheating for an example Mars entry vehicle will be given. A comprehensive review of various results was given in [2].

2. Phenomenological Models

In the following we will discuss catalytic atom recombination from a phenomenological rather than a chemical mechanism approach. However, we will discuss the use of mechanistic approaches to help understand the expected behavior catalytic reaction rates in Section 3. For now we will be content to deal with the subject in terms of reaction rates inferred from measurements and phenomenological models. We will first consider a simple model of gas/surface interactions then address additional features concerning energy accommodation.

2.1 BINARY INTERACTION MODEL WITH FULL ENERGY ACCOMMODATION.

The simplest model of atom recombination on a surface is one in which we presume the atoms that diffuse to a surface either specularly reflect from the surface and neither gain nor lose energy at the surface - or, the atoms stick to the surface and are fully accommodated. Of those that stick, some fraction may recombine with other adsorbed atoms, or with atoms directly impinging from the gas as shown in Fig.1. Later we will distinguish these two as the Langmuir-Hinshelwood mechanism and the Eley-Rideal mechanism, respectively. The molecules that are formed are presumed to lose all of their latent heat of dissociation to the surface. With few exceptions this model has been assumed in most of the calculations of aeroheating up to this time.

The fraction of incident atoms impinging on the surface that recombine is defined as γ;

$$\gamma_i \equiv \frac{M_i \cdot \hat{n}}{M_i^{\downarrow} \cdot \hat{n}} \tag{1}$$

where $\hat{n}$ is the surface unit normal vector, $M_i^{\downarrow}$ is the mass flux of the incident atoms and M_i is the mass flux of recombining atoms or the net mass flux of atoms to the surface. This ratio γ is often referred to as the atom *recombination coefficient* or the atom *recombination probability*. This number depends on the particular atoms and surface involved, and is a temperature dependent quantity. It is often considered to be independent of the pressure or density because recombination on surfaces is usually a first order reaction, except at high temperature. More about that will be discussed later.

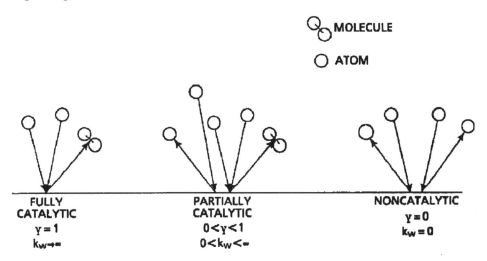

Figure 1 Schematic of atom recombination on surfaces

2.2 BINARY INTERACTION MODEL WITH PARTIAL ENERGY ACCOMMODATION

A refinement to the previous model is made by considering the chemical energy accommodation to be incomplete. Mellin and Madix [3] and Halpern and Rosner [4] have measured recombination and the energy transfer independently and have found that the energy transferred to the surface less than total. That is, they found that even though a large fraction of the atoms recombine, the energy transferred per molecule to many materials is much less than the dissociation energy. This implies that the molecules formed on the surface must leave the surface in excited states or with a velocity much higher than would be expected based on the wall temperature. A fraction β called the chemical energy accommodation coefficient, has been defined in [3]. Specific measurements of β for a number of metals is given in [4]. This ratio of chemical energy transferred to the surface compared to the available energy from recombination is

$$\beta \equiv q_c / M_i h_{Di} \qquad (2)$$

where q_c is the chemical energy flux, M_i is the net mass flux of atoms and h_{Di} is the energy of dissociation per unit mass. The disposition of dissociation energy may occur by molecules leaving the surface in excited

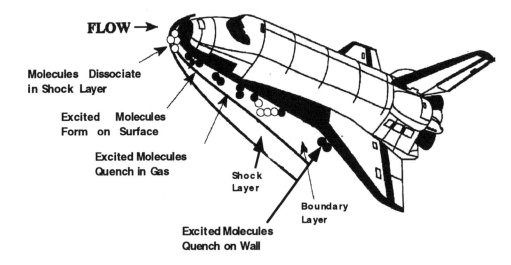

FLOW →

Molecules Dissociate
in Shock Layer

Excited   Molecules
Form   on   Surface

Excited Molecules
Quench in Gas

Shock
Layer

Boundary
Layer

Excited Molecules
Quench on Wall

Figure 2 Schematic of flow field with catalytic recombination, excited state production, and quenching.

states. They may quench in the boundary layer or by re-striking the wall and giving up that excitation energy as indicated in Fig. 2. There is some evidence that glassy materials such as the reaction cured glass (RCG) coating of the tiles that are is used on the Space Shuttle also have incomplete energy accommodation. The measurements of Breen, et al. [5] in which numbers of atoms recombining were measured and those of Scott [6], who determined the recombination coefficients from heat fluxes are compared in Fig. 3. It can be seen that the recombination based on numbers is about a factor of 5 greater than that based on heat flux. This implies that β is on the order of 0.2 for tile materials. One note of caution is that the coatings in the two cases were not exactly the same. A theory of the effects of incomplete energy accommodation and quenching on heat flux was developed by Rosner and Feng [7]. Also, there has been a number of papers that appeared in the Russian literature and have found their way into translation that address the effects of incomplete energy accommodation and quenching of excited states [8], [9], [10], and [11]. In [9] there is a review of the values taken from the literature of quenching coefficients for a number of excited states of air species on various materials at room temperature. It is conjectured that the temperature dependence is not great. However, since very little is known about β for real space vehicle surfaces, this refinement in the concept of catalytic heating effects has not been implemented in heat transfer calculations for vehicles. There have been a number of parametric investigations including [8], [9], and [11]. In most of the following we will assume that the recombination coefficient determined from heat transfer measurements are based on the simple binary interaction model with full energy accommodation.

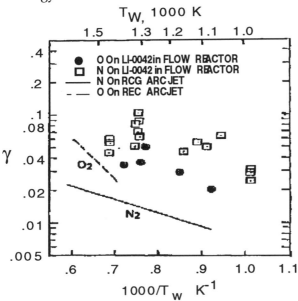

Figure 3 Comparison of recombination coefficients determined from atom counting and heat flux experiments.

3. Mechanistic Models

Mechanistic models have been developed to investigate the behaviour of recombination with the aim of eliciting the temperature and pressure dependence and eventually of being able to predict catalytic recombination rates from parameters known about the gas and surface. At present the models can only give approximate values; but parameters in the models can be estimated by fitting the models to actual measured recombination coefficients. Jumper, et al. [12] developed a model of fluorine recombination on nickel. Later

Seward and Jumper [13] extended the general technique to oxygen recombination on SiO_2. Willey [14] investigated several possible models of nitrogen and oxygen recombination on high temperature reusable surface insulation. He fit the model of Seward and Jumper [13] to some previously unpublished data, the data of Scott [6] and Kolodziej and Stewart [15] and the single point at room temperature of Marinelli [16]. He also fit the data to both the simple models for the Eley-Rideal and Langmuir-Hinshelwood mechanisms, as well as a four-parameter model that was not based on established mechanisms, but simply on an empirical fit. This latter model gave the best fit of the experimental data, but the model of Seward and Jumper [13] also gave very good results. Jumper gives a more thorough description of the mechanisms that are included in his model in another paper in this volume [17].

4. Measurement Techniques of Catalytic Reaction Rates

Several techniques have been used to measure the catalytic recombination coefficient or recombination rates on a variety of materials. There is a large body of literature on techniques near room temperature for many combinations of gas species and surface materials. A very good review of the literature of published values of recombination coefficients was published in Russian by Berkut, Kudriavtsev, and Novikov [18]. They include temperature dependent recombination coefficients for O, N, and H atom recombination on many surfaces. In this section we will confine ourselves to *air* atom recombination.

Techniques differ in both the method used and in the basic information determined. As mentioned earlier, there are recombination coefficients based on direct atom loss measurements; and there are measurements based on heat transfer measurements. All of the atom loss measurements are done in flow tube reactor experiments. The method of detection of atoms and the range of temperature differ in the experiments. Some methods of detecting and measuring species are: atom titration, side-arm oxygen atom detection, laser induced fluorescence of molecules, laser multiphoton ionization of molecules, and electron magnetic resonance spectroscopy. Usually, atoms such as nitrogen and oxygen are generated in a microwave discharge device. However, they may be generated chemically or in a high power radio-frequency discharge. The atoms are allowed to flow over a specimen made of the material of interest. The change in atom concentration with distance is a measure of the rate of atom loss on the surface due to recombination.

Heat flux measurements also have been used to determine atom recombination rates in flow reactor experiments, while others have been done in arc jets in RF induction torches, and in shock tubes. Attempts have been made to determine the oxygen recombination coefficient for the Space Shuttle tile material from flight heat flux measurements. Such inferences require accurate heating and flow field predictions as well as accurate measurements.

Whereas the measurements are believed to be reliable, in this author's opinion, the heat flux prediction techniques that involve calculating the flow field properties over a 3-dimensional body are not sufficiently accurate to obtain accurate recombination coefficients.

Evaluation of the large amount of data reveals that recombination coefficients are difficult to measure accurately, since surface conditions and possible contaminants significantly affect the measurements. Measurements by different investigators are given in Table 2 of [2] for nitrogen and oxygen recombination on room-temperature nickel. It can be seen that there is a large uncertainty in the data. Most likely, the scatter is due to surface preparation and partly due to different measurement techniques.

4.1 FLOW REACTORS

Techniques for measuring catalytic recombination coefficients fall into two basic categories: atom loss experiments and heat flux experiments. The atom loss experiments fall into two basic types: side arm flow reactor experiments and duct reactor experiments. In each the loss of atoms is measured and related by simple theory to the recombination coefficient.

4.1.1 *Side arm reactors*
One of the earliest techniques was a side arm technique developed by Smith [19] and adapted by Greaves and Linnett [20]. The technique makes use of a side arm duct coated with the material of whose recombination coefficient is to be determined. A moveable probe which is sensitive to the atoms of interest is inserted into the side arm. The probe response as function of distance will yield the recombination coefficient via a solution of the diffusion equation for a catalytic wall tube having zero convection. The conditions must be such that gas phase recombination is negligible. The solution to this equation is given in [26]. An example of the physical arrangement for a side arm reactor is shown in Fig. 4. Similar techniques have been used by other researchers who used various detector techniques, such as, Wrede-Hartek gages, afterglow emission measurements and titration, electron magnetic resonance spectroscopy, etc. Many detection techniques may be used.

4.1.2 *Straight flow reactors*

Straight flow reactors have been used in which the wall of the tube is lined with the material of interest and similar detection techniques have been used as above. Marinelli, et al. [16] used laser induced fluorescence to measure the atom and molecular nitrogen and oxygen production or loss rates on the Shuttle tile RCG coating at near room temperature. Instead of the wall being coated with the RCG, a coated rod was inserted into the flow tube as shown in Fig. 5.

168

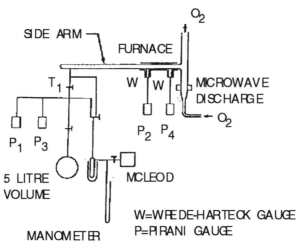

Figure 4 Side arm reactor schematic diagram

Some techniques rely on the measurement of the heat flux to determine the recombination coefficient. The results of this technique may result in data that is more applicable to aerothermodynamics applied to the reentry spacecraft because the question of incomplete energy accommodation is bypassed somewhat. At least, it is an effect that already may be accounted for approximately in the measurements. Flow reactor experiments that have provided nitrogen recombination on platinum and copper were done by Prok [21]. Myerson [22] measured the oxygen recombination coefficient on noble metals and titanium dioxide; whereas, Hartunian, et al. [23] measured oxygen recombination on silver. Most of these measurements were made for surfaces at room temperature. For temperatures of hypersonic vehicle interest heating devices need to be used in conjunction with the atom loss or heat flux measurements. In some cases the heating is provided externally, such as with a heating coil or oven or even radiatively heated with a high power cw laser. Another technique to be discussed in the next section is the arc jet, which is particularly appropriate for thermal protection materials.

4.2 ARC JET

In arc jet flows the samples for which γ is to be determined are subjected to heating and a dissociated flow environment. The energy flux catalytic atom recombination coefficient is derived from measurements of the heat flux via theory such as that of Goulard [24] or some other heating prediction such as viscous shock layer or boundary layer equation solutions. One difficulty of using arc jets is that the free stream flow has not been well characterized. Approximations have to be made to estimate the atom fractions, since the kinetics of the flow and the initial conditions are not well-known.

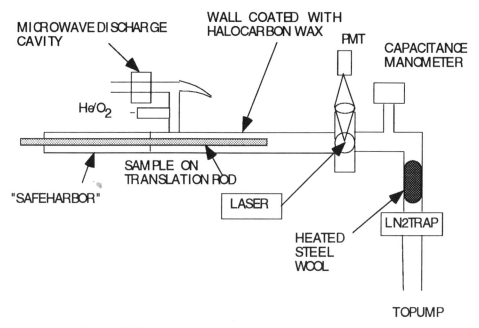

Figure 5 Flow reactor recombination experiment schematic

Determinations of energy flux recombination coefficients have been made by Pope [25], who measured nitrogen and air recombination on copper, nickel silver, and their oxides, gold platinum, tungsten, silicon monoxide, and Teflon at discrete temperatures. Anderson [26] studied nitrogen recombination on nickel, platinum, copper, gold, and silica at wall temperatures of 400, 500, and 1000 K. Scott [27] measured room temperature energy transfer recombination rates on chrome oxide, siliconized carbon/carbon, hafnium/tantalum carbide on carbon/carbon, and niobium silicide. Scott [6], [28] also measured temperature dependent energy transfer recombination coefficients on the Space Shuttle tile RCG coating and an overcoat used on the Orbiter Experiment (OEX) wall catalysis experiment of Stewart et al. [29] Subsequently, Kolodziej and Stewart [15] have extended these measurements to higher temperatures.

4.3 SHOCK TUBES

Shock tubes and shock tunnels have been used to generate dissociated flows for catalytic heating studies [30], [31] as well as for determining catalytic recombination rate coefficients [32]. In a shock tube a shock wave generated by explosive compression of gas dissociates molecules of air which then are allowed to flow over a sample of material in which a temperature sensor is installed. Frequently, these are mounted on flat plates. The heat flux is determined from the sensor and compared with theory to determine the recombination coefficients. This technique can be used for wall temperatures near ambient, but one needs some external heat source to do measurements at elevated surface temperatures. Because of these difficulties, shock tubes have not found much use in the West.

4.4 ATOMIC BEAMS

None of the techniques already discussed have addressed the issue of incomplete energy accommodation and the production and quenching of excited species. It has been one of the great disappointment that measuring excited species produced by atom recombination has proven to be very difficult. One technique attempted at Physical Sciences, Inc. is an atomic beam technique [33] in which a beam of atoms is generated in an RF discharge then impinges on a sample in an ultra high vacuum chamber. The incident flux of atoms and the resultant recombined molecules are measured using mass spectrometry and laser multiphoton ionization. Detection of the ions is attempted by a channel plate having a large number of detection elements. This technique should be able to determine the energy state of the molecules produced using a retarding potential in front of the plate. The experiment would also be used to determine the probability of formation of the excited molecules, since the energy required to ionize the molecules depends on its excited state. The samples may be "prepared" in the ultra high vacuum chamber by flowing suitable gasses over the surface of the sample prior to the measurement. The sample also may be rotated and translated to position it for various measurements. Unfortunately, this technique ran into some difficulties due to the production of photoelectrons by the laser source, therefore the technique awaits further development.

The difficulty in measuring the excited state production on surfaces lies with the sensitivity of techniques to measure excited states in a harsh environment. N_2 and O_2 are non polar molecules whose vibrational excited states do not radiate. Their electronically excited states are long lived and do not radiate strongly, therefore, other techniques such as laser induced fluorescence or laser multiphoton ionization must be used. This has so far proven to be difficult.

Molecular beams do have the potential of being useful for recombination measurements on some surfaces. In particular, they can be used to study the residence time of adsorbed atoms or molecules [34]. Fig. 6 shows the time dependent flux of reflected O_2 from a Kapton surface. A schematic diagram of the atomic beam facility at NASA Jet Propulsion Laboratory, in which these measurements were made, is given in Fig. 7. The measurements of the type given in Fig. 6 lead to the velocity distribution functions as well as the possibility of measuring the reflected fluxes to determine recombination rates. Information about the species scattered off or produced on surfaces may form the basis of models used to determine excited state production probabilities or recombination rates.

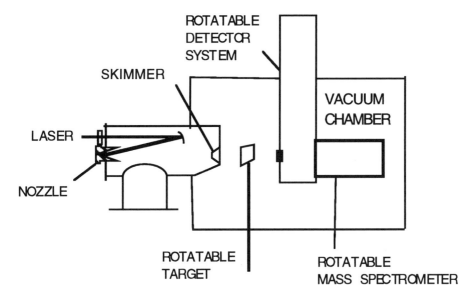

Figure 6 Example atomic beam apparatus

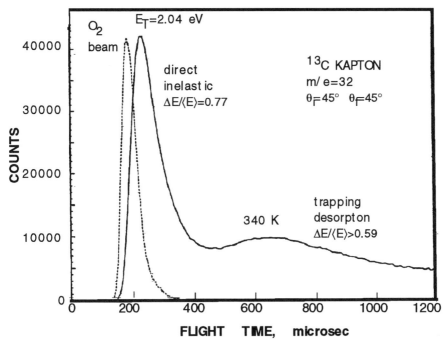

Figure 7 Velocity distribution function of molecular oxygen produced on a Kapton surface in a molecular beam facility (courtesy Minton, et al. [34]).

5. Boundary Conditions

5.1 HEAT FLUX AT A SURFACE

In a chemically reacting flow the heat transfer to a surface is composed of the usual conduction term plus the term resulting from diffusion of species to the surface. We can write the heat flux as

$$q = -\lambda \partial T/\partial y + \sum_{i=1}^{n_s} \beta_i h_{ci} M_i \qquad (3)$$

where λ is the thermal conductivity, T is the temperature, y is the coordinate normal to the surface, h_{ci} is the enthalpy of formation of the ith species, β_i is the chemical energy accommodation coefficient, and M_i is the normal mass flux of the ith species

$$M_i = \rho_i V_i = n_i m_i \sum_{j \neq i}^{n_s} D_{ij} \frac{\partial X_j}{\partial y} \qquad (4)$$

where thermal diffusion has been neglected. The expression for the diffusion velocity V_i is written in terms of the mole fraction X_j.

The problem for the aerothermodynamicist is to determine these quantities. Most authors have assumed that $\beta_i = 1$. In principle, finding the heat flux requires solving the flow equations with appropriate boundary equations and for the particular geometry of interest. The heat flux q is then calculated from the solution using equation (3).

5.2 BOUNDARY CONDITIONS FOR THE SPECIES EQUATIONS

The expression for the boundary equation for an atomic species is obtained by equating the diffusion mass flux relation (4) with the rate of recombination at the surface $\gamma_i m_i N_i^{\downarrow}$.

$$\gamma_i m_i N_i^{\downarrow} = \gamma_i n_i m_i \sqrt{\frac{kT}{2\pi m_i}} = n_i m_i \sum_{j \neq i}^{n_s} D_{ij} \frac{\partial X_j}{\partial y} \qquad (5)$$

All quantities are evaluated at the wall. Note that these mixed boundary equations are coupled in the sense that flux of one species depends on the gradient of all the other mole fractions X_j, including through the dependence of the multicomponent diffusion coefficients D_{ij} on all the X_i.

5.3 EXAMPLE FOR A SIMPLIFIED MARS ATMOSPHERE MODEL

The Martian atmosphere consists of primarily of carbon dioxide (97%), nitrogen (<3%), and argon (1%). Although there are about 16 possible molecules, atoms and ions that could be considered for studying the effects of catalysis on a surface entering the atmosphere of Mars one should be primarily concerned with the surface chemistry of the dissociation products of CO_2. Unless an entry vehicle enters at extremely high velocity the ions are negligible, therefore model chosen for illustration here consists only of CO_2, CO, O and O_2. Nitrogen is neglected because it is a minor species as far as surface reactions are concerned. This follows the model addressed in [35] in which fully recombined CO_2 was studied, i. e., an extreme case of a fully catalytic wall.

When encountering a surface atoms or molecules may be adsorbed on the surface or they may react directly with species that are already adsorbed. Adsorbed species may either react or may desorb. We will assume that the Eley-Rideal mechanism applies and that the reactions are first order.

Consider the set of elementary encounters of the species with a surface:
O-atom adsorption

$$\gamma_1 \qquad\qquad O + s \rightarrow O\text{-}s \qquad\qquad (6)$$

O-atom recombination on adsorbed O

$$\gamma_2 \qquad\qquad O + O\text{-}s \rightarrow O_2 + s \qquad\qquad (7)$$

CO recombination on adsorbed O

$$\gamma_3 \qquad\qquad CO + O\text{-}s \rightarrow CO_2 + s \qquad\qquad (8)$$

CO adsorption

$$\gamma_4 \qquad\qquad CO + s \rightarrow CO\text{-}s \qquad\qquad (9)$$

O-atom recombination on adsorbed CO

$$\gamma_5 \qquad\qquad O + CO\text{-}s \rightarrow CO_2 + s \qquad\qquad (10)$$

In equations (6)-(10) s is a surface site. The coefficients γ_r are elementary-step recombination or adsorption probabilities. We neglect other species on the surface since they will not affect the heat flux significantly.

In general, these coefficients depend on the state of the surface, its coverage, and temperature. However, in this formulation we will not attempt to determine the coverage, but assume it is a property of the surface which is incorporated into the specific γ_r. Here we denote the specific reaction by the subscript r. The incident flux of species i is given from kinetic theory by

$$N_i^{\downarrow} = -n_i \sqrt{\frac{kT_w}{2\pi m_i}} \tag{11}$$

where, n_i the number density of species i at the wall, m is the mass, and T_w is the wall temperature. For a no-slip condition, the gas temperature at the wall is the same as the wall temperature. A positive flux is away from the wall.

The desorption rates are determined by the surface coverage of the species θ_i, the bond energy, and the surface temperature, which in this study are not treated explicitly, but are phenomenological dependencies. We will denote the positive desorption rate by N_i^{des}. We now write the expressions for the net fluxes of O, CO, O_2, and CO_2.

$$N_O^{net} = N_O^{\downarrow} \gamma_1 + N_O^{\downarrow} \gamma_2 + N_O^{\downarrow} \gamma_5 + N_O^{des} \tag{12}$$

$$N_{CO}^{net} = N_{CO}^{\downarrow} \gamma_3 + N_{CO}^{\downarrow} \gamma_4 + N_{CO}^{des} \tag{13}$$

$$N_{CO_2}^{net} = -N_{CO}^{\downarrow} \gamma_3 + N_O^{\downarrow} \gamma_5 \tag{14}$$

$$N_{O_2}^{net} = -N_O^{\downarrow} \gamma_2 \tag{15}$$

We omit adsorption and desorption terms for CO_2 and O_2 by assuming that CO_2 and O_2 do not adsorb appreciably on the surface as compared to CO or O, or by assuming that the adsorption and desorption are balanced   Equations (12)-(15) can be simplified by replacing the adsorption and desorption terms with "known" quantities.   Recall that in the Eley-Rideal mechanism the adsorbed species CO and O are site-bound and may react with an impinging reactant species.   Otherwise, they may spontaneously desorb with a rate depending on bond energy and temperature. In steady state where there is no net build up or depletion on the surface the rate of adsorption is balanced by the rates of desorption and reaction; the net rate of adsorption is zero. Thus, we can write

$$0 = N_O^{\downarrow} \gamma_1 + N_O^{reac} + N_O^{des} \tag{16}$$

and
$$0 = N_{CO}^{\downarrow} \gamma_4 + N_{CO}^{reac} + N_{CO}^{des} \tag{17}$$

O-atoms on the surface may react with incoming O-atoms to form O_2 or with incoming CO to form CO_2. Then

$$-N_O^{reac} = N_O^{\downarrow} \gamma_2 + N_{CO}^{\downarrow} \gamma_3 \qquad (18)$$

The superscript ^{reac} refers to reactions involving adsorbed species. Adsorbed CO reacts with impinging O-atoms to form CO_2 in which case

$$-N_{CO}^{reac} = N_O^{\downarrow} \gamma_2 \qquad (19)$$

Combining (18)-(19) with (12)-(15) we can then write the net number fluxes as

$$N_O^{net} = 2N_O^{\downarrow} \gamma_2 + N_O^{\downarrow} \gamma_5 + N_{CO}^{\downarrow} \gamma_3 \qquad (20)$$

$$N_{CO}^{net} = N_{CO}^{\downarrow} \gamma_3 + N_O^{\downarrow} \gamma_5 \qquad (21)$$

$$N_{CO_2}^{net} = -N_{CO}^{\downarrow} \gamma_3 - N_O^{\downarrow} \gamma_5 = -N_{CO}^{net} \qquad (22)$$

$$N_{O_2}^{net} = -N_O^{\downarrow} \gamma_2 = -\frac{1}{2}(N_O^{net} + N_{CO}^{net}) \qquad (23)$$

Thus the adsorption and desorption terms have been eliminated, greatly simplifying the form of the boundary conditions. This enables us to investigate limiting cases without having to manipulate the adsorption and desorption terms. The factor 2 in (20) is a consequence of the fact that γ_2 refers to an elementary step, not the total recombination reaction. One atom of oxygen recombines from direct impingement, reaction step (7), and another from an adsorbed atom, reaction step (6). Thus the total contribution to oxygen recombination is $2\gamma_2$ times $N_O^{\downarrow}$. If we combined (6) and (7) into a single reaction then

$$O + s \rightarrow \frac{1}{2}O_2 + s \qquad (24)$$

A single recombination coefficient can be defined $\gamma_{O \rightarrow O2} = 2\gamma_2$. Then, (20) would become

$$N_O^{net} = N_O^{\downarrow} \gamma_{O \rightarrow O2} + N_O^{\downarrow} \gamma_5 + N_{CO}^{\downarrow} \gamma_3 \qquad (20')$$

and (23) becomes

$$N_{O_2}^{net} = -\frac{1}{2}N_O^{\downarrow} \gamma_{O \rightarrow O2} = -\frac{1}{2}(N_O^{net} + N_{CO}^{net}) \qquad (23')$$

The range of values permitted for γ_2, γ_3, and γ_5 depend on a number of parameters, including the fractional site coverage of each reactant and their elementary reaction probabilities. Since it is assumed that sites may be occupied by O or CO, then increasing the coverage of one species may be at the expense of the other. Thus, γ_2, γ_3, and γ_5 cannot all be unity. Suppose that all sites are occupied by O-atoms, then no CO could be adsorbed, thus

$$N_{CO}^{des} = N_{CO}^{reac} = 0$$

which implies that $\gamma_5 = 0$. For this case $\gamma_3 + \gamma_2$ could range from 0 to 1. On the other hand, if all sites are occupied by CO and no O-atoms, then

$$N_{O}^{des} = N_{O}^{reac} = 0$$

This implies that $\gamma_3 = \gamma_2 = 0$. For this case γ_5 could be in the range from 0 to 1. We can see that where multiple surface reactions are possible one cannot ignore the site densities in the boundary problem, except for extreme cases such as these. It may be possible to simply define the γ's as phenomenological parameters to be determined empirically.

6. Catalysis and Aeroheating for Planetary Atmospheres

To illustrate the use of boundary conditions such as the ones of Section 5 we will summarize some of the results of [45]. In their paper the maximum heating to the Mars Pathfinder entry vehicle, known as MESUR, was wanted for design purposes. This is because there is no recombination data on the thermal protection materials that are proposed for MESUR. One can see from equations (22)-(23) that there are several pathways one can achieve a "fully catalytic" condition. In [45] it was assumed that no O_2 was produced and that the maximum heating was associated with whichever of the CO recombination pathways would yield the greatest CO incident flux. Viscous shock layer equations were solved and the heat flux was determined using boundary conditions equivalent to (20)-(23), but written in terms of mass fluxes instead of number fluxes and given that $\gamma_2 = 0$. The heating distribution on the 70° sphere-cone is given in Fig. 8. It can be seen that the heating due to fully recombined CO_2 is very close to the heating to a equilibrium catalytic wall and that for a fully equilibrium flow field. However, the noncatalytic heating is much less for this case. These results reflect the need to determine the actual recombination mechanism and recombination rates for the CO_2 system.

Conclusions

The effects of catalytic recombination on aeroheating in nonequilibrium flow environments has been addressed. Measurement techniques for determining recombination rates were reviewed. New atomic beam techniques hold some promise for determining recombination probabilities and excited states in which these molecules leave the surface. Understanding the mechanisms associated with the reactions will help guide the experiments by indicating what species may be important and should be considered in experimental techniques. The experimental techniques that depend on solving flow equations as well as use of recombination coefficients for heating predictions require suitable boundary conditions. Boundary conditions for calculating the flow around partially catalytic bodies were addressed. An example for a simplified Mars atmosphere was formulated and some brief results were given that show that in a Martian environment knowledge of CO recombination rates is very important for accurate heating predictions.

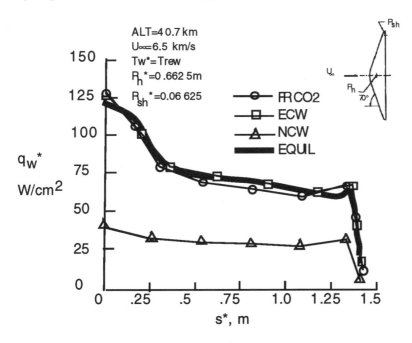

Figure 8 Calculated heat flux to MESUR spacecraft with various catalytic boundary conditions.

References

[1]Golodets, G. I. (1983) *Heterogeneous Catalytic Reactions Involving Molecular Oxygen*, Elsevier, Amsterdam.

[2]Scott, C. D. (1992) Wall Catalytic Recombination and Boundary Conditions in Nonequilibrium Flows - with Applications, in *Advances in Hypersonics*, Vol. 2, J. J. Bertin, J. Périaux, and J. Ballmann (eds.), Birkhaüser, Boston, pp. 176-250.

[3]Mellin, G. A. and Maddix, R. J. (1971) Energy Accommodation During Oxygen Atom Recombination on Metal Surfaces, *Faraday Society Transactions* **67**, pp.198-211.

[4]Halpern, B. and Rosner, D. E. (1974) Chemical Energy Accommodation at Catalyst Surfaces, *Chemical Society of London, Faraday Transactions I. Physical Chemistry* **74**, part 8, pp. 1883-1912.

[5]Breen, J. , Rosner, D. E., et al. (1973) Catalysis Study for Space Shuttle Vehicle Thermal Protection Systems, NASA CR-134124.

[6]Scott, C. D. (1981) Catalytic Recombination of Nitrogen and Oxygen on High Temperature Reusable Surface Insulation, *Progress in Astronautics and Aeronautics*, Vol. 77, edited by A. L. Crosbie, AIAA, New York, pp. 192-212.

[7]Rosner, D. E. and Feng, H. H. (1974) Energy Transfer Effects of Excited Molecule Production by Surface Catalyzed Atom Recombination, *J. of the Chemical Society, Faraday Transactions I* **70**, pp. 884-907.

[8]Berkut, V. D., Kudriavtsev, N. N., and Novikov, S. S. (1987) Heat Transfer at a Surface Flushed by Dissociated Nitrogen in the Presence of Heterogeneous Recombination of Electronically Excited Molecules, *High Temperature* **25**, pp. 256-263.

[9]Doroshenko, V. M., Kudriavtsev, N. N., and Smetanin, V. V. (1991) Equilibrium of Internal Degrees of Freedom of Molecules and Atoms in Hypersonic Flight in the Upper Atmosphere, *High Temperature* **29**, pp. 815-832.

[10]Tirsky, G. A. (1993) Up-to-date Gasdynamic Models of Hypersonic Aerodynamics and Heat Transfer with Real Gas Properties, *Annual Rev. Fluid Mech* **25**, pp. 151-181.

[11]Suslov, O. N. and Tirskiy, G. A. (1994) The Kinetics of the Recombination of Nitrogen Atoms on High-Temperature Reusable Surface Insulation in Hypersonic Thermochemical Non-equilibrium Flows, *Proceedings of the Second European Symposium on Aerothermodynamics for Space Vehicles*, held in ESTEC, Noordwijk, The Netherlands, 21-25 Nov. 1994, (ESA SP-367), pp. 413-419.

[12]Jumper, E. J., Utlee, C. J., and Dorko, E. A. (1980) A Model for Fluorine Atom Recombination on a Nickel Surface, *Journal of Physical Chemistry* **84**, pp. 41-50.

[13]Seward, W. A. and Jumper, E. J. (1991) Model for Oxygen Recombination on Silicon-Dioxide Surfaces, *J. Thermophysics and Heat Transfer* **5**, pp. 284-291.

[14]Willey, R. J, (1993) Comparison of Kinetic Models for Atom Recombination on High-Temperature Reusable Surface Insulation, *J. Thermophysics and Heat Transfer* **7**, 55-62.

[15]Kolodziej, P. and Stewart, D. A. (1987) Nitrogen Recombination on High-Temperature Reusable Surface Insulation and the Analysis of its Effect on Surface Catalysis, AIAA-87-1637.

[16]Marinelli, W. J. and Campbell, J. P. (1986) Spacecraft-Metastable Energy Transfer Studies, Physical Sciences, Inc., Andover, Mass., Final Report PSI-G565/TR-595, Contract No. NAS9-17565, 31 July 1986.

[17]Jumper, E. J. (1996) Recombination of Oxygen and Nitrogen on Silica-Based Thermal Protection Surfaces: Mechanism and Implications, in *Molecular Physics and Hypersonic Flows*, M. Capitelli (ed.), Kluwer Academic Publishers, Dordrecht, pp. 181-191.

[18]Berkut, V. D., Kudriavtsev, N.N., and Novikov, S. S. (1986) Surface Thermophysical Properties Due to the Accommodation of Chemical Energy of a Supersonic Dissociated Gas Flow, *Surveys on the Thermophysical Properties of Substances*, TFTs Moscow, IVTAN No. 2(58) pp. 3-135.

[19]Smith, (1943) *J. Chem. Phys.*, Vol. 11, pp. 110.

[20]Greaves, J. C. and Linnett, J. W. (1958) The Recombination of Oxygen Atoms at Surfaces, *Transactions of the Faraday Society* **54**, pp. 1323-1330.

[21]Prok, G. M. (1961) Effect of Surface Preparation and Gas Flow on Nitrogen Atom Surface Recombination, NASA TN D-1090.

[22]Myerson, A. L. (1965) Mechanisms of Surface Recombination form Step-Function Flows of Atomic Oxygen over Noble Metals, *J. Chem. Phys.* **42**, No. 9, pp. 3270-3276.

[23]Hartunian, W. P. Thompson, and S. Safron (1965) Measurements of Catalytic Efficiency of Silver for Oxygen Atoms and the O-O2 Diffusion Coefficient, *J. Chem. Phys.* **43**, No. 11, pp. 4003-4006.

[24]Goulard, R. J. (1958) On Catalytic Recombination Rates in Hypersonic Stagnation Heat Transfer, *Jet Propulsion* **28**, pp. 737-745.

[25]Pope, R. B. (1968) Stagnation-Point convective Heat Transfer in Frozen Boundary Layers, *AIAA J.*, **6**, pp. 619-626.

[26]Anderson, L. A. (1973) Effect of Surface Catalytic Activity on Stagnation-Point Heat Transfer Rates, *AIAA J.* **11**, pp. 649-656.

[27]Scott, C. D. (1973) Measured Catalycities of Various Candidate Space Shuttle Thermal Protection System Coatings at Low Temperatures, NASA TN D-7113.

[28]Scott, C. D. (1983) Catalytic Recombination of Nitrogen and Oxygen on Iron-Cobalt-Chromia Spinel, AIAA Paper 83-0585.

[29]Stewart, D. A., Rakich, J. V., and Lanfranco, M. J. (1982) Catalytic Surface Effects Experiment on the Space Shuttle, *Progress in Astronautics and Aeronautics* **82**, edited by T. E. Horton, AIAA, New York, pp. 248-272

[30]Vidal, R. J. and Golian, T. C (1967) Heat-Transfer Measurements with a Catalytic Flat Plate in Dissociated Oxygen, *AIAA Journal* **5**, pp. 1579-1588.

[31]East, R. A., Stalker, R. J., and Baird, J. P. (1980) Measurements of heat transfer to a flat plate in a dissociated high-enthalpy laminar air flow, *J. Fluid Mechanics* **97**, pp. 673-699.

[32]Berkut, V. D., Kovtun, V. V., Kudriavtsev, N. N., Novikov, S. S., and Sharovatov, A. I. (1986) Determination of the time-resolved probabilities of heterogeneous recombination of atoms in shock tube experiments, *Int. J. Heat Mass Transfer* **29**, pp. 1-19.

[33]Carleton, K. L. and Marinelli, W. J. (1992) Spacecraft Thermal Energy Accommodation from Atomic Recombination, *J. Thermophysics and Heat Transfer* **6** pp. 650-655.

[34]Minton, T. K. and Moore, T. A. (1993) Molecular Beam Scattering from ^{13}C-Enriched Kapton and Correlation with the EIOM-3 Carousel Experiment, in Proceedings of the 3rd LDEF Post-Retrieval Symposium, Williamsburg, VA, November 1993, NASA CP-3275, pp. 1115-1128.

[35]Gupta, R. N., Lee, K. P., and Scott, C. D. (1995) An Aerothermal Study of MESUR Pathfinder Aeroshell, submitted to the *Journal of Spacecraft and Rockets*.

RECOMBINATION OF OXYGEN AND NITROGEN ON SILICA-BASED THERMAL PROTECTION SURFACES: MECHANISM AND IMPLICATIONS

E.J. Jumper
Hessert Center for Aerospace Research
Department of Aerospace and Mechanical Engineering
University of Notre Dame
Notre Dame, Indiana 46556
USA

1. Introduction

A number of papers, see for example [1], have pointed out the importance of being able to model the catalytic recombination of oxygen and nitrogen on silica-based materials because of the role that these reactions play in re-entry heating. Because the energies of molecular oxygen and nitrogen are less than that of two atoms of oxygen and nitrogen, respectively, when these species recombine, the excess energy can be transferred directly to the surface, representing an additional heat load [2]. Silicon dioxide is the principal material making up the Space Shuttle thermal-protection-tile surface; to be more precise, the surface is a borosilicate reaction-cured glass (RCG) [3]. This paper describes a kinetic approach to modeling the recombination of oxygen and nitrogen on Reaction-Cured Glass (RCG). The paper is meant to present only a brief review of these models; more detail on these models can be found in [4-7].

1.1 HETEROGENEOUS CATALYSIS

By now there are many standard references on the proposed mechanisms for simple recombinative heterogeneous catalysis; my favorite reference is still Emmett [8]. In general, heterogeneous catalysis

M. Capitelli (ed.), Molecular Physics and Hypersonic Flows, 181–191.

mechanisms can be complex, involving physical adsorption, chemisorption, migration of adhered species, reaction between adhered species and reaction between gas-phase and adhered species. The paper by Bruno [9] attempts to lay out these various complications as they might apply to the recombination of oxygen and nitrogen on silica-based surfaces. The paper by Barbato [10] has incorporated at least two possible mechanisms for modeling recombination of oxygen and nitrogen on silica-based surfaces. In this paper, and in my work in general, I have found that a single mechanism, the Langmuir-Rideal (L-R) mechanism [8] (also referred to as the Rideal and the Eley-Rideal mechanism) is sufficient to model those recombination systems I have studied [4-7]. The rationale for why this mechanism is preferred over others is too lengthy to go into here, but discussions on this matter can be found in [4-7]; I have found that the use of the L-R mechanism alone provides a rationale basis for what is observed in the data and is sufficient to explain the data without resorting to other mechanisms.

1.2   LANGMUIR-RIDEAL MECHANISM

The L-R mechanism is essentially a two step process where in step one a gas-phase atom impinges upon and adheres to an empty surface site (activation site). The second step is that a gas-phase atom collides and reacts with a surface-adhered atom to form a gas-phase molecule. These steps are described by

$$
\begin{array}{ccc}
A \; + & --> & A \\
| & & | \\
S & & S
\end{array}
\tag{1}
$$

$$
\begin{array}{ccc}
A \; + \; A & --> & + \; A_2 \\
| \quad | & & \\
S \quad S & &
\end{array}
\tag{2}
$$

where A is the applicable atomic species and S is a surface site. It should be pointed out that although the direction of Eq. (1) is toward the direction necessary for catalysis to progress, the reverse reaction is also possible via thermal desorption. These processes are shown schematically in *Figure 1*.

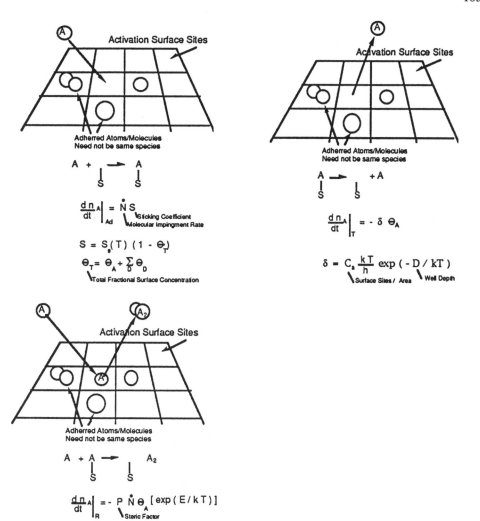

Figure 1. Schematic representation of Step one, its reverse process and the removal of atoms from the surface by recombination.

The time rate of change of the surface concentration of adhered atoms can be written as

$$\frac{dn_S}{dt} = \dot{n}_A - \dot{n}_T - \dot{n}_R \qquad (3)$$

where n_S is the surface concentration (particles/cm^2), $\dot{n}_A$ is the rate at which atoms adhere to the surface, $\dot{n}_T$ is the rate at which atoms thermally desorb, and $\dot{n}_R$ is the rate at which atoms are removed from the surface by recombinations. Each of the three processes on the right hand side of Eq. (3) have the following form:

$$\dot{n}_A = \dot{N} S_0 (1 - \Theta_T) \tag{4}$$

where S_0 is the initial sticking coefficient, Θ_T is the fraction of the surface sites occupied by A or any other atomic or molecular species, and $\dot{N}$ is the atomic impingement rate given by

$$\dot{N} = n (\bar{c} / 4) \tag{5}$$

where n is the gas-phase number density of A and $\bar{c}$ is the average kinetic velocity of an A particle given by

$$\bar{c} = \sqrt{\frac{8k_B T}{\pi m}} \tag{6}$$

where k_B is the Boltzmann constant, T is the absolute temperature and m is the mass of A. The thermal desorption rate is given by

$$\dot{n}_T = \delta \Theta_A \tag{7}$$

where,

$$\delta = Ca \frac{k_B T}{h} \exp\left(-\frac{D}{k_B T}\right) \tag{8}$$

Finally, the rate of removal by recombination is the product of the rate of impingement of A, times the probability of the impact site being occupied by a surface-adhered A, times the fraction of collisions that overcome an activation energy, times the fraction of the sufficiently energetic collisions that actually recombine (steric factor); this may be written as

$$\dot{n}_R = \dot{N}\,\Theta_A \left[\exp\left(-\frac{E}{k_BT} \right) \right] P \tag{9}$$

where E is the activation energy and P is the steric factor. Taken together, Eq. (3) may be written as

$$\frac{dn_s}{dt} = \dot{N}\,S_0\,(1 - \Theta_T) - \delta\,\Theta_A - \dot{N}\,\Theta_A \left[\exp\left(-\frac{E}{k_BT} \right) \right] P \tag{10}$$

In general, an equation of this type may be written for each recombining species (i.e., $O_{gas-phase}$ with $O_{surface-bound}$ to form O_2, $N_{gas-phase}$ with $O_{surface-bound}$ to form NO, $N_{gas-phase}$ with $N_{surface-bound}$ to form N_2 and $O_{gas-phase}$ with $N_{surface-bound}$ to form NO) and equations lacking the last term of Eq. (10) for non-recombining species (i.e., other atomic and molecular species present in the gas phase). In the case of Eq. (10), and other applicable equations, the total surface-coverage fraction would then be equal to the sum of all coverage fractions

$$\Theta_T = \Theta_A + \sum_{i=1}^{m} \Theta_i \tag{11}$$

where i is summed over all species other than A. Equation (10) and similar equations for the other surface-adhered species can be solved simultaneously, at steady state, for the surface coverage fractions and, in particular, for Θ_A. Once Θ_A is known, Eq. (9) can be used to compute the rate of recombination of A. The recombination coefficient, γ, can then be obtained from the recombination rate by

$$\gamma = 2\,(\dot{n}_R / \dot{N}) \tag{12}$$

Where the 2 accounts for the disappearance of two gas-phase atoms for each recombination, one in the reaction and a second to fill a void somewhere on the surface, thereby maintaining the steady state.

2. The Oxygen / Silica-Based Surface System

Section 1.2 discussed the L-R mechanism in general, without reference to the specifics of either the oxygen-on-silica-based or the nitrogen-on-silica-based surface systems. Since the nitrogen system follows directly from the oxygen system, oxygen will be address first. In all cases, the reader is referred to the more-detailed references [4-7]. The most fundamental and significant part of the models is the identification of the participating surface-bond partner in the recombinations.

It is now generally acknowledged that the surface of a silicon-dioxide substrate is an oxygen, double bonded to a silicon atom which is itself bonded to two substrate oxygens through single bonds. The substrate oxygens are, in turn, single bonded to substrate silicon atoms forming one of four oxygens surrounding the substrate silicon to form tetrahedrons. These tetrahedrons can swivel around the oxygen bonds to other substrate silicon-oxygen tetrahedrons, and the orientations of the substrate tetrahedrons determines the form of the silica. *Figure 2* shows a schematic of a surface for quartz (discussion of surface N in later section). Although RCG contains some B, its surface can be expected to be similar.

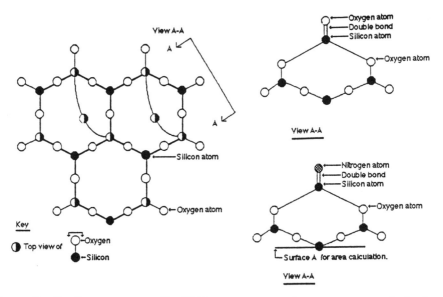

Figure 2. Surface structure for SiO_2 on quartz substructure; cut A-A shows participating surface bound O and alternate A-A shows N replacement for O.

In our work, we have identified the surface oxygen as the participating surface-bound oxygen. It should be pointed out that this identification is a fundamentally-new concept, and has been referred to by Grumet [11,12] as a new heterogeneous mechanism (the "Jumper Mechanism"), separate from the L-R mechanism. Although an overstatement on Grumet's part, it points out the importance of this participating oxygen in both simplifying the model and in understanding the ramifications of the model, over and above its ability to compute the recombination coefficient as a function of local gas/surface conditions. If one considers an oxygen and inert-gas flow over a silica surface, then in a general L-R reaction system at temperatures sufficiently high to discount physical bonding (i.e., Van der Waal type bonding), one might consider the possibility of both chemisorbed atomic and molecular oxygen; however, when the participating surface-bound oxygen is the oxygen of the surface matrix itself, bound molecular oxygen may be discounted. Thus only one equation of the Eq. (10) type need be considered, and the physiochemical parameters required are the steric factor, P, the activation energy, E, the initial sticking coefficient, S_0, the surface sites, C_a and the thermal desorption energy, D, for only the atomic oxygen. The reader is referred to [5-7] for a discussion of how the identification of the participating surface-bound species aided in determining the physiochemical parameters; these parameters for the oxygen-silica-based systems (SiO_2 and RCG) are given in TABLE 1. The curves generated by these parameters are shown in *Figures 3a* and *3b* along with data for O recombination on Silica (SiO_2) and RCG; the reader is referred to [5,6] for further details.

TABLE 1. Values/form of the physiochemical parameters for catalytic recombination of Oxygen on SiO_2 and RCG

Parameter	O on SiO_2	O on RCG
P, steric factor	2.24×10^{-5} exp (0.00908 T) (max value = 0.1)	2×10^{-4} exp (0.03 T) (max value = 0.1)
E, activation energy	1.0 Kcal / mole	1.0 Kcal / mole
S_0, initial sticking coefficient	0.05 exp (-0.002 T)	1.0 exp (-0.002 T)
Ca, surface sites	5×10^{14} sites / cm^2	2×10^{14} sites / cm^2
D, thermal desorption energy	81 Kcal / mole	81 Kcal / mole

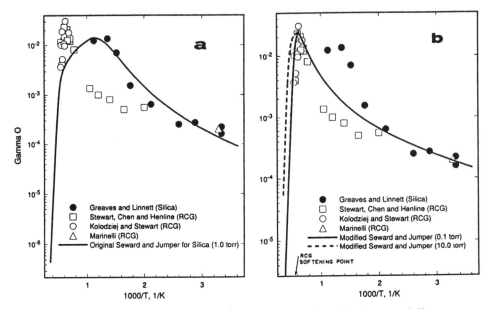

Figure 3. Comparison of the model with data [6], with the model's coefficients for a.) SiO$_2$ and b.) RCG (see TABLE 1).

3. The Nitrogen / Silica-Based Surface System

For various reasons discussed in [7], it seems reasonable to suppose that the surface bound oxygen can be removed by a recombination with a gas-phase O or combination with a gas-phase N to form NO, or through thermal desorption. Once the site is empty, the O could be replaced by N, as shown in the alternate A-A of *Figure 2*, either without competition from O in a pure nitrogen-inert gas stream or in competition with O if both atomic species are present in the gas stream. Some data exists for both the nitrogen-quartz and nitrogen-RCG systems. From these data physiochemical parameters have been extracted as given in TABLE 2; however, unlike the oxygen case, these parameters represent N on RCG only. The purpose of including the quartz data is only to argue that a curving of the data (increasing negative slope with increasing temperature) between room temperature and the temperature for maximum recombination coefficient should be present. The curve(s) generated by these parameters are shown in *Figure 4* along with data for N recombination on quartz and RCG; the reader is referred to [7] for further details.

TABLE 2. Values/form of the physiochemical parameters for catalytic recombination of Nitrogen on SiO_2 and RCG

Parameter	N on RCG
P, steric factor	1.62×10^{-4} exp (0.003 T) (max value = 0.02)
E, activation energy	0.64 Kcal / mole
S_0, initial sticking coefficient	0.95 exp (-0.002 T)
Ca, surface sites	2×10^{14} sites / cm^2
D, thermal desorption energy	79 Kcal / mole

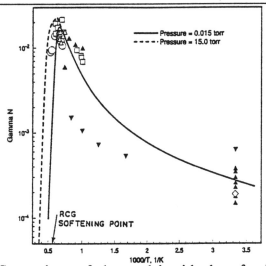

Figure 4. Comparison of the model with data for RCG (open symbols) and quartz (closed symbols) see [7] for origin of the data and discussion.

4. Form of the Steric Factor for Both N and O Systems

The functional form of the steric factor, P, in both the O and N systems represents a curve fit to the data. Rationale for why it might be temperature dependent may be found in [5-7]; however, the form of P actually represents what is unknown about the phenomena. Various discussions

190

were held with attendees present during the presentation of this paper and several suggestions offered; however, the most that can be said at this point is that P represents the functional form of what is unknown about the process and an avenue for future research.

5.   Ramifications of the Models

Figures 5a and *5b* give energy-state diagrams for molecular oxygen and nitrogen overlaid with the energy states for two $O(^3P)$ and two $N(^4S^o)$ atoms, respectively, and broken Si=O and Si=N relative to the surface-bonded Si=O and Si=N, respectively. From this information it is clear that, because the participating surface-bound atoms are bound in deep wells, a large fraction of the energy available in the recombinations goes into breaking the surface bonds. Thus, we may conclude that these fractions of the energy go directly into raising the potential of the surface; once the site is refilled, the surface lowers its potential and the this energy must be transferred directly to the surface, representing an additional "heat load." In the case of oxygen recombination, step one, Eq. (1), represents 81 Kcal/mole out of the available 117.6 Kcal/mole in the recombination (i.e., 69% of the energy). This leaves only 36.6 Kcal/mole available to produce vibrationally and electronically excited species; therefore, the only viable electronic species is singlet delta. There is, in fact, evidence that singlet-delta oxygen is preferentially produced in surface recombination of O on Pyrex, with nascent yields of up to 36% at room temperature. Similar arguments can be made for nitrogen recombination; the reader is directed to [6,7] for further discussions.

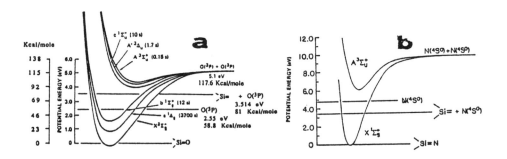

Figure 5. Energy-State Diagrams for a.) Oxygen and b.) Nitrogen.

References

1. Scott, C.D. (1983) Effect of Nonequilibrium and Catalysis on Shuttle Heat Transfer, AIAA Paper 83-1485.

2. Rosner, D.E. and Cibrian, R. (1974) Non-Equilibrium Stagnation Region Aerodynamic Heating on Hypersonic Glide Vehicles, AIAA Paper 74-755.

3. Goldstein, H.E., Leiser, D.G., and Katvala, W., (1978) Reaction Cured Borosilicate Glass Coating for Low-Density Fibrous Silica Insulation, in *Borate Glasses*, Plenum Press, New York, 623-634.

4. Jumper, E.J., Ultee, C.J. and Dorko, E.A. (1980) A Model for Fluorine Atom Recombination on a Nickel Surface, *Journal of Physical Chemistry*, **84**, 41-50.

5. Seward, W.A. and Jumper, E.J. (1991) Model for Oxygen Recombination on Silicon-Dioxide Surfaces, *Journal of Thermophysics and Heat Transfer*, **5**, 284-291.

6. Jumper, E.J. and Seward, W.A. (1994) Model for Oxygen recombination on Reaction-Cured Glass, *Journal of Thermophysics and Heat Transfer*, **8**, 460-465.

7. Jumper, E.J., Newman, M., Kitchen, D.R. and Seward, W.A. (1993) Recombination of Nitrogen on Silica-Based, Thermal-Protection-Tile-Like Surfaces, AIAA Paper 93-0477.

8. Emmett, P.H. (ed.) (1954) *Catalysis, Vol. 1*, Reinhold, New York.

9. Bruno, C. (1995) Surface Catalysis, *Proceedings of the NATO Advanced Study Institute on Molecular Physics and Hypersonic Flows*, 21 May - 3 June 1995.

10. Barbato, M. (1995) Heterogeneous Catalysis Models, *Proceedings of the NATO Advanced Study Institute on Molecular Physics and Hypersonic Flows*, 21 May - 3 June 1995.

11. Grumet, A.A., (1994)*The Effects of Surface Catalysis on the Hypersonic Shock Wave/Boundary layer Interaction*, UM-AERO-94-09, PhD Dissertation, University of Maryland.

12. Grumet, A.A., Anderson, J.D. Jr., and Lewis, M.J. (1994) A Numerical Study of the Effects of Wall Catalysis on shock Wave/Boundary layer Interaction, *Journal of Thermophysics and Heat Transfer*, **8**, 40-47.

PHENOMENOLOGICAL THEORY FOR HETEROGENEOUS RECOMBINATION OF PARTIALLY DISSOCIATED AIR ON HIGH-TEMPERATURE SURFACES

V.L.KOVALEV, O.N.SUSLOV AND G.A.TIRSKIY
Institute of Mechanics, Moscow State University,
Michurinskiy pr.1, Moscow 119899, Russia

1. Introduction

Simulation of heterogeneous catalytic recombination of partially dissociated air without chemical equilibrium in flow on the high temperature surfaces of hypersonic vechicle thermoprotection system started in the 1950-th [1],[2]. As energy of oxigen or nitrogen molecule less than one of two oxigen or nitrogen atoms respectively, the energy releasing during the atomic recombination transmitted to the surrounding gas (homogeneous recombination) or to the surface (heterogeneous recombination). In 1960-1970 calculated models of heterogeneous reactions in dissociated air were reduced to the formal consideration of two extreme boundary condition cases: unreal noncatalytic surface model and assumption of ideal catalytic one, that essentially encrease radiation equilibrium surface temperature [3]. Since 1980 systematic experimental [4],[5] and theoretical [6]–[10] investigations carried out to predict heterogeneous recombination probability (HRP) γ_O and γ_N on the thermal protection reusable insolations ("tiles"). The approximation of HRP experimental data $\gamma_i (i = O, N)$ in the Arrhenius form [4] for the materials was found to be unusable in the wide range of surface temperature; for the agreement with the experimental data an activation energy must change with the temperature. The recombination reaction rate dependence on pressure was also discovered. Moreover, the comparison Space Shuttle experiments data with the calculated heat fluxes obtained using only the temperature depending catalytic constant $\left(k_{wi} = \gamma \sqrt{kT/2\pi m_i} \right)$ does not give satisfactory agreement everywhere along the vehicle surface and in all points of glide reentry trajectory. This indicates that hetero-

M. Capitelli (ed.), Molecular Physics and Hypersonic Flows, 193–202.
© *1996 Kluwer Academic Publishers. Printed in the Netherlands.*

geneous recombination of partially dissociated air on real thermoprotection materials is multi-step process including elementary steps. On the base of ideal adsorbed layer Langmuir concept generalized and developed by M.P.Temkin [11], the structural dependence of the effective coefficient of catalytic activity via pressure, temperature and mixture composition first was obtained in references [7]–[10]. The heterogeneous reaction brutto mechanism involved atoms and molecule reactions adsorption-desorption, Eley-Rideal and Langmuir-Hinshelwood mechanisms. For the body surface covering by reaction cured glass with high content of silica, the boundary conditions for component diffusive fluxes containing unknown rate constant of elementary stages were obtained by means of interpretation of the experimental data for catalytic activity coefficient [4]. It was shown for the silicon material that HRP $\gamma_i(i = O, N)$ dependence on temperature having nonmonotone behavior. That is after reaching some sufficiently high wall temperature T_w, depending on pressure, HPR $\gamma_i(i = O, N)$ begin decrease sharply with temperature increase. This effect was first theoretically predicted and developed in references [8]–[10] and then was discovered experimentally [12]. Calculations values of the heat flux near the windward centerline have been compared with experimental data for the second, third and fifth Space Shuttle flights. It shows a good agreement of obtained results along the apparatus length and at different points of reentry trajectory. Achieved values of effective coefficients of catalytic activity depend on location along the surface. Their values have a good agreement with data derived from flight and laboratory experiments. Soon after the preparation of the report had finished, reference [13, 14] in which the rate constants of elementary stages of heterogeneous processes had estimated become available to us. The application of this rate coefficients in our heterogeneous recombination theory for dissociated air will be published.

2. Phenomenological theory for heterogeneous catalytic reactions of partially dissociated air

a. After the references [8]–[10] let us consider following set of elementary physical-chemical interaction between atoms and molecules of air dissociated in shock layer near the nonablating nonoxidating surface.

1. Adsorption-desorption of atoms

$$O + (S) \underset{\longleftarrow}{\overrightarrow{}} (O - S) \qquad N + (S) \underset{\longleftarrow}{\overrightarrow{}} (N - S) \qquad (1), (2)$$

2. Eley-Rideal recombination mechanism (and inverse reaction)

$$O + (O - S) \underset{\longleftarrow}{\overrightarrow{}} O_2 + (S) \qquad N + (N - S) \underset{\longleftarrow}{\overrightarrow{}} N_2 + (S) \qquad (3), (4)$$

3. Langmuir-Hinshelwood recombination mechanism (and inverse reaction)

$$2(O - S) \overset{\longrightarrow}{\longleftarrow} O_2 + 2(S) \qquad 2(N - S) \overset{\longrightarrow}{\longleftarrow} N_2 + 2(S) \qquad (5), (6)$$

4. Adsorption-desorption of molecules

$$O_2 + (S) \overset{\longrightarrow}{\longleftarrow} (O_2 - S) \qquad N_2 + (S) \overset{\longrightarrow}{\longleftarrow} (N_2 - S) \qquad (7), (8)$$

Here $(X - S)$, (S) designated a species X atom or molecule absorbed by the surface and free adsorption place (active centers) on the surface respectively.

Reactions involving NO are not written out in (1-8). This permission is given because of two reasons. Firstly, NO gas mass fraction in the vicinity of body surface is less then 5 %. Secondly, it is supposed, that rate constants of reaction involving NO are small.

With the aid of elementary reaction set written above we can find the expression of atom formation (vanishing) brutto-reaction at the surface.

b. According to the acting surfaces law, that is similar to acting mass law for homogeneous reactions, the specific mass formation (vanishing) rates (per unit surface area) of O and N atoms in reactions (1,2,5,6) are given by following

$$\dot{r}_1(O) = -m_O p k_1 (x_O \theta - \theta_O / p K_1), \qquad \dot{r}_2(N) = -m_N p k_2 (x_N \theta - \theta_N / p K_2)$$

$$\dot{r}_3(O) = -m_O p k_3 (x_O \theta_O - \theta x_{O_2} / K_3), \quad \dot{r}_4(N) = -m_N p k_4 (x_N \theta_N - \theta x_{N_2} / K_4) \quad (9)$$

$$1 = \theta + \theta_O + \theta_N + \theta_{O_2} + \theta_{N_2}, \qquad x_i = p_i / p, \qquad (i = O, N, O_2, N_2) \quad (10)$$

Here x_i, p_i, m_i – molar fractious, partial pressures and molecular masses of species, p – mixture pressure, $K_i = k_i / k_{-i}$ – the equilibrium constant, $k_i, k_{-i} (i = 1, \dots, 8)$ are the rate constants of forward and reverse reactions respectively, θ_i is the surface fraction of the adsorbed i –th species, θ is the surface fraction of the free adsorption centers. In addition it is assumed that all the adsorption active centers are energetic homogeneous.

Specific mass formation (vanishing) brutto rate of free and adsorbed atoms and molecules for given reaction set (1)-(8) will be equal to

$$\dot{r}(O) = \dot{r}_1(O) + \dot{r}_3(O), \qquad \dot{r}(N) = \dot{r}_2(N) + \dot{r}_4(N)$$

$$\dot{r}(O-S) = \dot{r}_1(O-S) + \dot{r}_3(O-S) + \dot{r}_5(O-S)$$

$$\dot{r}(N-S) = \dot{r}_2(N-S) + \dot{r}_4(N-S) + \dot{r}_6(N-S)$$

$$\dot{r}(O_2-S) = \dot{r}_7(O_2-S), \qquad \dot{r}(N_2-S) = \dot{r}_8(N_2-S) \qquad (11)$$

Thus being restricted to elementary reactions (1)-(8) at the surface, it is necessary to know the 16 reaction rate constants or 8 forward reaction rate constants k_i and 8 equilibrium constants K_i.

Stoichiometric reaction equation (5) is linear combination of equations (1) and (3) and equation (6) is linear combination of equations (2) and (4). It leads as a result to two expressions connecting respective equilibrium constants

$$K_5 = K_1/K_3, \qquad K_6 = K_2/K_4 \qquad (12)$$

c. If the quasisteady state of the surface is assumed, then four relations are obtained as that last four expressions in (11) is equal to zero and taking into account (7), (8) and (12) we have

$$(k_1 + 2k_5 K_3 \theta)(x_O \theta_O - x_{O_2} \theta/K_1) = (k_3 + 2k_5 K_3 \theta_O)(x_O \theta - \theta_O/pK_3)$$

$$(k_2 + 2k_6 K_4 \theta)(x_N \theta_N - x_{N_2} \theta/K_2) = (k_4 + 2k_6 K_4 \theta_N)(x_N \theta - \theta_N/pK_4)$$

$$\theta_{O_2}/\theta = pK_7 x_{O_2} = kTK_7 n_{O_2}, \qquad \theta_{N_2}/\theta = pK_8 x_{N_2} = kTK_8 n_{N_2} \qquad (13)$$

Here k, n_i are the Boltzmann constant and number density. The first two relations can be represented as quadratic equations with respect to θ_O and θ_N with coefficients depending on θ. The solution of the equations are not written out here and we can refered to [8]. The equation (13) and (10) permit us to define all surface fractions.

d. Then from the first two equations (11) granting (13) we have

$$\dot{r}(O) = -2m_O p\theta \frac{k_1 k_3 + k_1 k_5 K_3 \theta_O + k_3 k_5 K_3 \theta}{(k_1 + 2k_5 K_3 \theta)x_O + (k_3 + 2k_5 K_3 \theta_O)/pK_3} \left(x_O^2 - \frac{K_O}{p} x_{O_2} \right)$$

$$\dot{r}(N) = -2m_N p\theta \frac{k_2 k_4 + k_2 k_6 K_4 \theta_N + k_4 k_6 K_4 \theta}{(k_2 + 2k_6 K_4 \theta)x_N + (k_4 + 2k_6 K_4 \theta_N)/pK_4} \left(x_N^2 - \frac{K_N}{p} x_{N_2} \right) \quad (14)$$

Here in brackets equilibrium constant K_O, K_N of dissociation reactions in gas $O_2 \rightleftarrows 2O$, $N_2 \rightleftarrows 2N$ are appeared:

$$K_O = 1/K_1 K_3 = p_O^2/p_{O_2}, \qquad K_N = 1/K_2 K_4 = p_N^2/p_{N_2} \quad (15)$$

The expressions in brackets in (14) represent reaction "deviation" from equilibrium.

e. The final expressions (14) include plenty unknown constants. So let us suppose only reaction (1)–(4), taking account that L–H reactions have a small thermodynamical probability at wall temperature $300 \le T_w \le 2000K$ and at pressure $10^{-3} \le p \le 1atm$ as mentioned in references [15], that is

$$k_5 \ll pk_1, \qquad k_6 \ll pk_2 \quad (16)$$

Moreover, following estimation taking place in these ranges of temperature and pressure [15, 16]

$$pK_7 \ll 1, \qquad pK_8 \ll 1 \quad (17)$$

that is under the temperature range mentioned above molecule surface fractions are sufficiently small (see (13)). Granting (16), (17) from the equations written above we obtain

$$\frac{\theta_O}{\theta} = \frac{x_O + l_O x_{O_2}/K_1}{l_O x_O + (pK_3)^{-1}}, \qquad \frac{\theta_N}{\theta} = \frac{x_N + l_N x_{N_2}/K_2}{l_N x_N + (pK_4)^{-1}} \quad (18)$$

$$\theta = 1 - \theta_O - \theta_N$$

$$\dot{r}(O) = -\frac{2m_O \theta p k_1}{l_O x_O + (pK_3)^{-1}} \left(x_O^2 - \frac{K_O}{p} x_{O_2} \right),$$

$$\dot{r}(N) = -\frac{2m_N \theta p k_2}{l_N x_N + (pK_4)^{-1}} \left(x_N^2 - \frac{K_N}{p} x_{N_2} \right) \quad (19)$$

Under the condition of assumptions (16), (17) let us consider binary mixture case, i.e. gas mixture contains N_2 and N. Then

$$\frac{1}{\theta} = 1 + \frac{x_N + (l_N/K_2)x_{N_2}}{l_N x_N + 1/(pK_4)}$$

$$\dot{r}(N) = -\frac{2m_N pk_2\theta}{l_N x_N + (pK_4)^{-1}} \left(x_N^2 - \frac{K_N}{p} x_{N_2} \right) \qquad (20)$$

At the sufficiently low wall temperatures the reaction E–R go rather fast that is $K_2 \gg 1$. At the same time the rate constant of dissociation $N_2 \rightleftarrows 2N$ is small, i.e. $K_N/p \ll 1$. Under this conditions from (20) we obtain

$$\dot{r}(N) = -\frac{2\rho R_A T k_2}{l_N + 1} \left[1 - \frac{1}{1 + (1 + l_N)pK_4 x_N} \right] c_N \qquad (21)$$

where $c_N = (m_N/m)x_N$ – mass fraction of nitrogen atoms, R_A – the universal gas constant, ρ – the density of mixture.

As shown in [17] at $10^{-3} \le p \le 1 atm$, $T \le 1350K$ the function $K_4(T)$ satisfies the condition $pK_4 \gg 1$ and expression in square brackets in (21) became close to unit. At such a wall temperatures it became possible to introduce recombination coefficient that does not depend on composition and pressure and is a function on temperature. So we can identify this function with experiment.

At this temperatures we have

$$2k_2 R_A T/(1 + l_N) = k_w(N) = \gamma_N \sqrt{R_A T/2\pi m_N} \qquad (22)$$

where γ_N for reacting cured glass was represented in experiment data [4] in the form of Arrhenius law

$$\gamma_N = 0.0734 \exp(-2219/T), \qquad 1090 < T < 1670K.$$

Substituted (22) to (21) we obtain expression for $\gamma^{eff}(N)$

$$\gamma^{eff}(N) = \gamma(N)\left[1 - \frac{1}{1 + (1 + l_N)pK_4 x_N}\right], \quad k_w^{eff}(N) = \gamma_N \sqrt{R_A T/2\pi m_N} \qquad (23)$$

Expressions (18), (19) for dissociated air will be simplified for sufficiently low temperature when $K_1 \gg 1$, $K_2 \gg 1$ and $K_O/p \ll 1$, $K_N/p \ll 1$ and under the assumtions

$$l_O(= k_1/k_3) \sim l_N(= k_2/k_4), \quad l_O pK_3 \ll 1, \; l_N pK_4 \ll 1 \qquad (24)$$

from (18) we obtain ($K_3 = K_4$)

$$\dot{r}(O) = -\frac{2\rho R_A T k_1 x_O}{x_O + x_N} \left[1 - \frac{1}{1 + pK_3(x_O + x_N)}\right] c_O$$

$$\dot{r}(N) = -\frac{2\rho R_A T k_2 x_N}{x_O + x_N} \left[1 - \frac{1}{1 + pK_3(x_O + x_N)}\right] c_N \qquad (25)$$

To determine the last constant k_1 we turn to experiment [4]. For the mixture consist of atoms and molecules of   oxigen and nitrogen   from [4] when $T < 3000K$ we have

$$2k_1 R_A T x_O/(x_O + x_N) = \bar{k}_w(O) = \bar{\gamma}_O \sqrt{R_A T/2\pi m_O} \qquad (26)$$

where $\bar{\gamma}_O$ is approximated by Arrhenius expression [4]:

$$\bar{\gamma}_O = 16 \exp(-10271/T), \qquad 1400 < T < 1650 \qquad (27)$$

Substituting (26) to the former equation (25) we obtain

$$\gamma^{eff}(O) = \bar{\gamma}(O) \left[1 - \frac{1}{pE(x_O + x_N)}\right] \qquad (28)$$

where $E = K_3 = K_4 = 10\sqrt{20} \exp[Q(492/T - 0.311)]$, $\quad Q = 46kal/mol$. In what follows, describing catalytic properties of surface covering reacting cured glass, we will use expressions (19) in which coefficients k_1 and k_2 are given by (22), (26) under the condition provided (24).

In Figures 1, 2 the effective recombination coefficients $\gamma^{eff}(N)$ and $\gamma^{eff}(O)$ are represented. Curves $\gamma^{eff}(N)$ and $\gamma^{eff}(O)$ substantially depend on quantities $p_N = px_N$ and $p^* = p(x_O + x_N)$. Straight lines on the figures correspond to γ_N and γ_O from [4], numbers near curves 1,2,3,4 correspond to the values of p_N and p^* with $1, 10^{-1}, 10^{-2}, 10^{-3} atm$ respectivly.

In Figure 3 equilibrium surface temperature variations are presented (surface emissivity $\epsilon = 0.8$) in the vicinity of stagnation point during reentering the body with bluntness radius $R_0 = 1m$ along the glide trajectory of Space Shuttle [17] for the various models describing catalytic surface properties. The maximum value of heat flux to the surface J_q and equilibrium wall temperature T_w along the trajectory achieve in the high range $65 \div 75$ km. Recombination coefficient approximation dependence on temperature suggested in reference [4] leads not to substantial decrease of J_q and T_w along this trajectory (curve 2) comparatively to ideal catalytic surface (curve 1), since they take not into account coefficient dependence on real flight condition (pressure and mixture composition in the surface vicinity). Computation results, using the model of gas and catalytic surface interaction suggested in this paper (curve 3), in which atom adsorption-desorption reaction supposed fast, provide the results close to noncatalytic surface (curve 4). The maximum equilibrium temperature T_w more than by *300 K* lower comparatively with maximum for the ideal catalytic surface and by *100 K* higher than for noncatalytic surface. Maximum value of heat flux in this case is twice as less as maximum heat flux to the ideal catalytic surface.

In Figures 4 and 5 comparison for the computation results of heat flux with experimental data, obtained during second, third and fifth Space Shuttle flight at the most thermostress part of trajectory is presented. Circles denote the measurements during the second flight, triangles and rhombuses refired to the measurements in third and fifth flight respectively. A good agreement of calculated and experimental data have been observed. Spatial flow in the vicinity of spreading line at every point of trajectory was simulated by flowing near the hiperboloid with corresponding angle and effective radius of bluntness [18]. Heat flux in Figure 4 correspond to $V = 7.16 km/sek$, $H = 74,7km$; ones in Figure 5 correspond to $V = 6.57 km/sek$, $H = 70,2km$. Presented in Figure 6 deviation of effective catalytic activity coefficient $K_w(O)$, $K_w(N)$ under the condition is shown, that their amounts vary sufficiently along the surface and in dependence of free stream conditions. Solid lines correspond to $K_w(O)$ dotted lines correspond to $K_w(N)$, numbers near curves denoted trajectory points. Note, that catalytic activity effective coefficients have a good agreement with laboratory experimental data [19] and data recommended on the base of flight experiments [18, 19]. At the same time, proposed model, describing catalytic surface properties in contrast to fixed values used and temperature depended simulation [18] – [20] permit to describe satisfactory the heat transfer along the apparatus side surface and everywhere at the trajectory.

This work was supported by Russian Foundation Fundamental Resources (grant N 94–01–01363a, grant N 94–01–00634a and grant N 95–01–01261a).

References

1. Tirsky, G.A. (1993) Up–to–date gasdynamic models of hypersonic aerodynamics and heat transfer with real gas properties, *Annu. Rev. Fluid Mech*, **25**, 151–181.
2. Goulard, R. (1958) On catalytic recombination rates in hypersonic stagnation heat transfer, *Jet Propulsion* **28**, 737–745.
3. Gershbein, E.A., Peigin, S.V., and Tirskiy, G.A. (1993) Supersonic flows at low to moderate Reynolds number, in G.K.Mikhailov and V.Z.Parton(eds.), *Super- and Hypersonic Aerodynamics and Heat Transfer*, CRC Press, pp. 1-88.
4. Scott, C.D. (1985) Effect of nonequilibibrium and wall catalysis on Shuttle heat transfer, *J. of Spacecraft and Rockets*, **22**, 489–498.
5. Berkut, V.D., Kudryavtsev, N.N., and Novikov, S.S. (1992) Thermophysical properties of surface induced by chemical energy accomodation of supersonic flow of

dissociated gas, *Obzory po Teplofiz. Svoystvam Veschestv*, Inst. Visokih Temperatur AN SSSR,**2(52)**, 3–135.

6. Halpern, B., and Rosner, D.E. (1978) Chemical energy accomodation at catalytic surfaces, *J. of Chem. Soc., Faraday Trans.I*, **74**, part 8, 1883–1912.

7. Suslov, O.N. (1981) Asymptotic investigations of chemical nonequilibrium boundary layer equations, in G.A.Tirskiy (ed.), *Giperzvuk. Prostranst. Techeniya pri nalichii Fiziko–Khimich. Prevrasch.*, Inst. Mech., Moscow State University, pp. 138–213.

8. Kovalev, V.L., and Suslov, O.N. (1983) Multicomponent nonequilibrium viscous shock layer on a catalytic surface, in G.A.Tirskiy (ed.), *Giperzvuk. Techenia pri Obtekanii Tel i v Sledah.*, Inst. Mech., Moscow State University, pp. 44–62.

9. Kovalev, V.L., and Suslov, O.N. (1982) Convective heat flux to a catalytic surfaces with regard to nonequilibrium reactions and multicomponent diffusion, Inst. Mech., Moscow State Univ., Tech. Report No.2729, 1–56.

10. Alferov, V.I., Kovalev, V.L., Suslov, O.N., and Tirskiy, G.A. (1985) Viscous gas flow around a body at nonequilibrium homogeneous and heterogeneous reactions regime, in *Mech. Neodnorodnykh System*, Novosibirskiy Inst.Teor. i Prikl. Mekh. SO AN SSSR, pp. 255–280.

11. Temkin, M.I., (1970) Kinetics of complicated reactions, in *Mechanisms and Kinetics of Complicated Catalytic Reactions*, Nauka Publishers, Moscow.

12. Kolodziej, P., and Sewart, D.A. (1987) Nitrogen recombination on high–temperature reusable surface insulation and the analysis of its effect on surface catalysis, *AIAA Paper* No. 87–1637.

13. Jumper, E.J., Newman, M., Kitchen,D.R., and Seward, W.A. (1993) Recombination of nitrogen on silica–based thermal–protection tile–like surfaces, *AIAA Paper* No. 93–0477.

14. Jumper, E.J., and Seward,W.A. (1994) Model for oxygen recombination on reaction–curved glass, *J. of Thermoph. and HeatTransfer*, **8**, 460–465.

15. Wise, H. and Wood, B.J. (1967) Reactive collisions between gas and surface atoms, *Advances in atomic and molecular physics* **3**, 291–353.

16. Tompkins, F.C. (1974) Heterogeneous catalysis. Simple molecules reactions on metal surface, in T.S. Jayadevaiah and R. Vanselov (eds.), *Surface Science: Recent Progress and Perspectives*, CRC Press, Inc., Clevland.

17. Agafonov, V.P., and Kuznetsov, M.M. (1979) The simulation of nonequilibrium heat flux to the catalytic surface, *Uchionye zapiski TsAGI*, **10**, No.4.

18. Zoby, E.V., Gupta, R.N., and Simmonds, A.L. (1984) Temperature-dependent reaction-rate expression for oxigen recombination at Shattle entry conditions, *AIAA Paper* No. 84–224.

19. Baronets, P.N., Gordeev, A.N., Kolesnikov, A.F., et all. (1990-1991) The waste of Buran orbiter thermoprotection materials in induction plasmatron, in *Gagarinskie Nauchnye Chtenia po Aviatsii i Kosmonavtike*, Moscow, pp. 41–52.

20. Tong, H., Morse, H.L., and Carry, D.M. (1975) Nonequilibrium viscous–layer computation of Space Shattle TRS requirements, *J. of Spacecraft and Rockets*, **12**, 739–743.

202

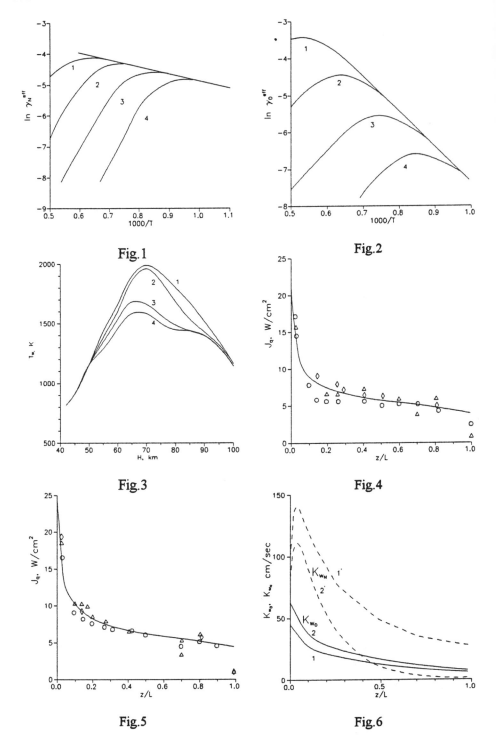

Fig.1

Fig.2

Fig.3

Fig.4

Fig.5

Fig.6

CHEMICAL REACTIONS AND THERMAL NONEQUILIBRIUM ON SILICA SURFACES

A. DAIß, H.-H. FRÜHAUF, E.W. MESSERSCHMID
Institut für Raumfahrtsysteme, Universität Stuttgart
Pfaffenwaldring 31, 70550 Stuttgart, Germany

1. Introduction

In the past many efforts have been made to understand the physics of the gasphase of re-entry flows in chemical and thermal nonequilibrium but much less is known about catalytic effects and thermal nonequilibrium at solid surfaces although important progresses have been achieved [10, 12, 13]. Up to now thermal equilibrium or adiabatic surface assumptions have served as boundary conditions for the vibrational and electron energy equations within multiple temperature Navier Stokes simulations. However, it seems that no justification for these assumptions has been given in literature. Another crucial point of surface modeling is that particle fluxes which are calculated with the well-known effusion relation become the more inaccurate the more the catalytic efficiency of a surface rises [5]. As these particle fluxes are needed within catalysis models, significant errors may be introduced in the calculation of reaction rates and surface heat fluxes.

Slip relations similar to those of Gupta et al. [5] can provide appropriate boundary conditions for both vibrational and electron energy equations. Moreover, they give a better estimate of the particle fluxes at catalytic surfaces and thus allow for a more realistic description of chemical processes at surfaces.

The catalysis models for silica/air systems available in literature [10, 12, 13] focus their attention on the recombination processes of oxygen and nitrogen which are the sources of a heat flux increase of up to a factor 2. However, under special conditions as they occur in the slip flow regime also dissociative adsorption and dissociation reactions might become important since in this case the flows next to the surface possess such a high amount of thermal energy that these processes become energetically possible.

M. Capitelli (ed.), Molecular Physics and Hypersonic Flows, 203–218.
© *1996 Kluwer Academic Publishers. Printed in the Netherlands.*

The purpose of the first part of this paper is to present validation results obtained with a recently developed slip model [4] which follows the approach of Gupta et al. [5] closely. This model can take into account chemical and thermal nonequilibrium effects and therefore provides a basis for the modeling of catalytic surface reactions and vibrational excitation of molecules appearing in surface recombination reactions. The second part is devoted to the modeling of the surface reactions with a special focus on dissociative adsorption reactions which might reduce heat fluxes considerably.

2. Slip Model for Multiple Temperature Flows

2.1. MODELING CONCEPT

The slip model used in these calculations is based on the separation of net species-, momentum and energy fluxes into contributions due to particles which approach the surface and contributions due to particles which leave the surface after some type of interaction. The difference of these two contribution can be seen as a source term which must be equal to the net fluxes at the surface.

Details about the model can be found in Ref. [4]. Unfortunately, this paper contains some typing errors so that the interested reader is asked to contact the author in order to get a corrected version.

2.2. VALIDATION OF THE SLIP RELATIONS

The present model has been implemented into the 5 temperature, 11 species Navier-Stokes code URANUS which is described in Ref. [3]. The modeling of the gasphase nonequilibrium reaction rates has been presented in Ref. [7] in detail.

2.2.1. *Flows around Hyperboloids*

In order to validate the presented slip model computations have been carried out for flows around Shuttle equivalent hyperboloids for various flow conditions ranging from a near continuum flow with a freestream Knudsen number of about 0.03 to transitional flows with Knudsen numbers of about 0.54. In Table 1 the freestream and geometrical conditions which have been considered within the computations are summarized.

The calculations were carried out on meshes 72×69 which were well suited to resolve the merged shock-boundary layer and the physical effects near the surface.

For the comparison with other computational results a non catalytic surface was prescribed and a constant surface temperature was assigned. Furthermore, full accommodation to the surface conditions was assumed for the

Altitude [km]	ρ_∞ [kg/m^3]	T_∞ [K]	Ma_∞ [−]	Angle of Attack [o]	R_N [m]	Kn_∞ [−]	T_W [K]
92.35	$2.184 \cdot 10^{-6}$	180	27.90	40.4	1.295	0.028	1040
109.75	$1.049 \cdot 10^{-7}$	242	24.02	40.7	1.309	0.539	420

TABLE 1. Freestream conditions and geometrical data for the STS-2 re-entry trajectory

scattered particles.

In Figure 1 the surface heat flux at an altitude of $92.35\,km$ calculated with the present model is compared to DSMC results obtained by Carlson et al. [2]. As can be seen good agreement is found, however it must be noted that this agreement was only obtained because the gasphase chemistry models used in the DSMC and Navier-Stokes calculations were quite similar. The use of different models would have produced deviations of up to 20% since at this flight altitude chemical reactions in the gasphase already become significant.

As a second quantity the skin friction coefficient at an altitude of $109.75\,km$ has been calculated and compared to the values computed by Moss et al. [9]. Figure 2 shows clearly that the skin friction coefficients obtained by the two methods also compare favorably.

More results for flows around Shuttle equivalent hyperboloids can be found in Ref. [4].

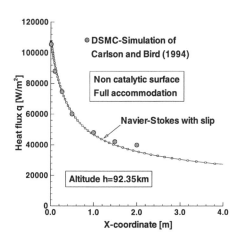

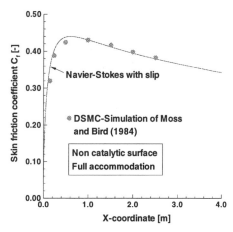

Figure 1.   Heat flux at the surface of a hyperboloid

Figure 2.   Skin friction coefficient for a flow around a hyperboloid

2.2.2. *Flow around a Sphere - Experiment of Vetter*

In order to check the validity of the presented slip relations in the continuum case ($Kn \rightarrow 0$) we calculated the flow around a sphere of $0.1m$ diameter which has been investigated experimentally by Vetter et al. [14]. The freestream conditions are: $Kn_\infty = 4 \cdot 10^{-4}$, $Ma_\infty = 12.7$, $\rho_\infty = 1.6 \cdot 10^{-3} kg/m^3$ and $T_\infty = 196K$. Since the experiment has been performed in a shock tube the mass fractions of atomic oxygen and nitric oxide are 0.5% and 4.0% respectively.

For the considered flow case the results for the translation-rotation temperature and the slip velocity obtained with the slip relations should be very close to the results computed with no-slip boundary conditions. In the figures 3 the temperatures profiles on the stagnation streamline near the surface are displayed for a non catalytic and a fully catalytic surface. The surface temperature T_W is in both cases $300K$. It becomes apparent that the translation-rotation temperature approaches the surface temperature in both cases closely. In contrast to this the vibrational temperature profiles show different behaviours. $T_{v,NO}$ depends strongly on the catalytic properties of the surface: Since the assumption was made that in the fully catalytic case all NO-molecules which collide with the surface vanish there is no flux of NO-molecules back into the gasphase and thus no accommodation of $T_{v,NO}$ occurs. T_{v,N_2} and T_{v,O_2} show also more pronounced jumps than $T_{trans/rot}$ at the surface (non and fully catalytic cases). This behaviour can be explain by the fact that at a temperature of $300K$ the vibrational modes are not fully excited (nonlinearity of the relationship between vibrational energy and temperature).

For the slip velocity very small values have been calculated indicating the transition to the no-slip boundary conditions. Also the heat fluxes and skin friction coefficients differ only slightly for the slip and no-slip conditions.

3. Catalysis Model for Silica Surfaces

At the surface of a silica/air system various chemical process may occur. Nitrogen and oxygen atoms can be adsorbed at certain "active sites" which they may leave after some time due to a thermal desorption process (Fig. 4) or migration processes on the surface. They may also be involved in recombination processes according to the Eley-Rideal-(ER)- and Langmuir-Hinshelwood-(LH)-mechanism (Fig. 5 and 6). In the Eley-Rideal-mechanism a gasphase atom strikes an occupied active site and recombines with the adsorbed atom by forming a gasphase molecule. In contrast to this, the Langmuir-Hinshelwood-mechanism describes a recombination reaction where a migrating surface atom collides with another surface atom in order to form a gasphase molecule. If the thermal energy in the gasphase next to

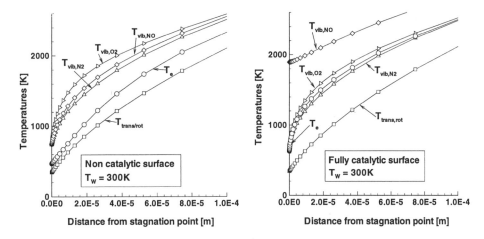

Figure 3. Temperature profiles along the stagnation streamline near the surface for a non and fully catalytic surface - Experiment of Vetter et al, $T_W = 300K$

surface is high enough also the reverse process of the ER-mechanism, the dissociative adsorption (Fig. 5), becomes important: A gasphase molecule hits the surface and produces an adsorbed atom and a gasphase atom. Since this process is endothermic it can only happen if the gasphase temperature is much higher than the usually met surface temperatures ($T_W \leq 1700K$). However, in the slip flow regime the temperature jump at the surface can be high enough for the occurence of this reaction. Also the dissociation of a molecule into gasphase atoms may take place (Fig. 7). However, this reaction mechanism is not considered in this work.

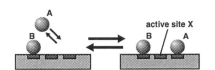

Figure 4. Adsorption and desorption reaction of an atom A on an active site X

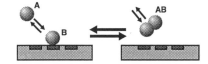

Figure 5. Eley-Rideal-mechanism and dissociative adsorption of a molecule AB

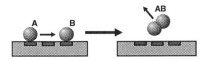

Figure 6. LH-mechanism for AB

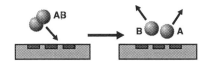

Figure 7. Dissociation reaction of AB

3.1. BASIC MODEL ASSUMPTIONS

The construction of a tractable catalysis model necessitates some simplifications. Therefore the following assumptions are made concerning the behaviour of the adsorbed atoms:

- The adatoms behave like harmonic oscillators on the surface. They possess different vibrational frequencies $\theta_\parallel$ and $\theta_\perp$ for their oscillations parallel and perpendicular to the surface.
- The partition function Q_{AX} of the adsorption complex AX can be factorized into a contribution Q_X coming from the surface and a contribution $Q_{Ad,A}$ coming from the oscillations
- The vibrational levels are Boltzmann-populated with the surface temperature T_W.

These assumptions allow us to write the partition function of the adsorption complex AX as

$$Q_{AX} = Q_X \cdot Q_{Ad,A} = Q_X \cdot \left[\frac{exp\left(-\frac{\theta_{\perp,A}}{2T_W}\right)}{1 - exp\left(-\frac{\theta_{\perp,A}}{T_W}\right)} \right] \cdot \left[\frac{exp\left(-\frac{\theta_{\parallel,A}}{2T_W}\right)}{1 - exp\left(-\frac{\theta_{\parallel,A}}{T_W}\right)} \right]^2 . \quad (1)$$

In literature typical values of $\theta_\parallel$ and $\theta_\perp$ for chemisorbed molecules are $\theta_\parallel = 100K$ and $\theta_\perp = 600K$. In our approach we use the values $\theta_{\parallel,N} = 100K$, $\theta_{\perp,N} = 600K$ for nitrogen and $\theta_{\parallel,O} = 75K$, $\theta_{\perp,O} = 600K$ for oxygen.
Also, for the gasphase particles simplification are necessary. It is important to note, that they should be consistent with the assumptions made for the modeling of the gasphase processes:

- The electronic excitation of the gasphase particles is neglected.
- For the description of the rotational motion of the gasphase molecules the infinite rigid rotator model is used.
- The vibrational motion is reproduced by the truncated harmonic oscillator model and the vibrational levels of the oscillators are Boltzmann-populated.

With the above simplifcations one gets for the partition functions of the molecules AB

$$Q_{AB} = \left[\frac{8\pi^2 I_{AB} kT}{\sigma_{AB} h^2} \right] \cdot \left[\frac{1 - exp\left(-\frac{E_{A-B}}{RT_v}\right)}{1 - exp\left(-\frac{\theta_{AB}}{T_v}\right)} \right] exp\left(-\frac{\theta_{AB}}{2T_v}\right) \cdot \left[\frac{2\pi m_{AB} kT}{h^2} \right]^{\frac{3}{2}} \quad (2)$$

and for the atoms A

$$Q_A = \left[\frac{2\pi m_A kT}{h^2} \right]^{\frac{3}{2}} . \quad (3)$$

3.2. ELEMENTARY REACTION STEPS

In this section we will briefly describe how the frequency factors and the activation energies which are required for the calculation of the rates of the various elementary reaction steps are determined.
We will mainly use transition state and collision rate theory for the evaluation of these quantities.

3.2.1. *Adsorption/Desorption reactions*

The adsorption reaction of an atom A on an active site X and its corresponding reverse reaction, the thermal desorption, can formally be written as

$$A + X \rightleftharpoons AX \tag{4}$$

where AX represents the adsorption complex adatom/active site. For the rate of adsorption it is supposed that the expression

$$\dot{\omega}_{Ad} = s_A Z_A^+ \frac{n_X}{n_X^0} \tag{5}$$

is valid [10, 13]. Herein s_A is the sticking probability of an atom A striking a vacant active site X. Z_A^+ denotes the molar flux of the atoms A impinging on the surface per area and n_X/n_X^0 is the ratio of the actual active site concentration to the active site concentration of a completely free surface. For n_X^0 Newman [11] gives the value $2 \cdot 10^{18} m^{-2}$.
The rate of desorption is expressed as a function of the rate coefficient k_{Des} and the concentration of adatoms n_{AX} on the surface

$$\dot{\omega}_{Des} = k_{Des} n_{AX}. \tag{6}$$

k_{Des} is determined by using the transition state theory (TST). In the framework of this theory k_{Des} reads

$$k_{Des} = P_{Des} \frac{kT_W}{h} \frac{Q_{AX}^\ddagger}{Q_{AX}} \cdot exp\left(-\frac{E_{A-X}}{\mathcal{R}T_W}\right) \tag{7}$$

with Q_{AX} being the partition function of the adsorption complex AX and $Q_{AX}^\ddagger$ being the partition function of the activated adsorption complex (without the reaction coordinate) which has the form

$$Q_{AX}^\ddagger = Q_X \cdot \left[\frac{exp\left(-\frac{\theta_{\parallel,A}}{2T_W}\right)}{1 - exp\left(-\frac{\theta_{\parallel,A}}{T_W}\right)}\right]^2. \tag{8}$$

Using equation (1) we get for the rate coefficient of desorption

$$k_{Des} = P_{Des} \frac{kT_W}{h} \left[\frac{1 - exp\left(-\frac{\theta_{\perp,A}}{T_W}\right)}{exp\left(-\frac{\theta_{\perp,A}}{2T_W}\right)} \right] exp\left(-\frac{E_{A-X}}{RT_W}\right). \tag{9}$$

The parameter P_{Des} is a catch-all factor which will be adjusted later.
The sticking coefficient s_A is evaluated by means of equilibrium considerations. In the case of thermodynamic equilibrium the rates of desorption and adsorption must be equal, thus $\dot{\omega}_{Ad} = \dot{\omega}_{Des}$. Furthermore, the particle flux of atoms A colliding with the surface now satisfies the effusion relation $Z_A^+ = n_A \bar{c}_A/4$, where n_A is the concentration of atoms A and $\bar{c}_A$ represents their average peculiar velocity, $\bar{c}_A = \sqrt{8kT/\pi m_A}$. Using these two relations together with the equations (5) and (6) we get

$$s_A = \frac{4n_X^0}{\bar{c}_A} \left[\frac{n_A n_X}{n_{AX}} \right]^{-1} k_{Des}. \tag{10}$$

The ratio of concentrations in (10) can be expressed by the adsorption/desorption equilibrium constant

$$K_{Ad/Des} = \frac{n_A n_X}{n_{AX}} = \frac{Q_A Q_X}{Q_{AX}} \cdot exp\left(-\frac{E_{A-X}}{RT_W}\right). \tag{11}$$

Introducing this expression into (10) together with the expressions for the partition functions gives the final form of the sticking coefficient

$$s_A = P_{Des} \frac{n_X^0 h^2}{2\pi m_A k} \cdot \frac{1}{T_W} \cdot \left[\frac{exp\left(-\frac{\theta_{\|,A}}{2T_W}\right)}{1 - exp\left(-\frac{\theta_{\|,A}}{T_W}\right)} \right]^2. \tag{12}$$

The sticking process is strongly exothermic and no bond has to be broken. As authors [6, 10] we assume that the adsorption reaction needs no activation energy. Different values have been given for the desorption energy E_{A-X} in literature [6, 10]. In our approach we use for nitrogen and oxygen the values $E_{N-X} = 300 KJ/mol$ and $E_{O-X} = 300 KJ/mol$ respectively.

3.2.2. Eley-Rideal-Mechanism/Dissociative Adsorption
The ER recombination rate $\dot{\omega}_{ER}$ of an adatom A with a gasphase atom B

$$AX + B \rightleftharpoons AB + X \tag{13}$$

is proportional to molar flux Z_B^+ of atoms B colliding with the surface and also proportional to the surface coverage n_{A-X}/n_X^0 with adatoms A. Hence, we write

$$\dot{\omega}_{ER} = k_{ER} \cdot \frac{n_{A-X}}{n_X^0} \cdot Z_B^+ \tag{14}$$

where k_{ER} is the rate coefficient. For the reverse reaction, the dissociative adsorption, the rate is proportional to the flux Z_{AB}^+ of molecules AB and proportional to the surface coverage with vacant active sites X, thus we have

$$\dot{\omega}_{DA} = k_{DA} Z_{AB}^+ \frac{n_X}{n_X^0}. \tag{15}$$

The rate coefficient k_{DA} is evaluated by means of a simple collision rate approach: It is supposed that every particle AB which possesses enough energy for surmounting the activation barrier of dissociative adsorption will react when it hits the surface. Assuming that the vibrational energy fully contributes to the activation energy and that only the translational degree of freedom perpendicular to surface provides energy for the reaction we get the minimum normal-velocity v_z^l of a reacting particle in the vibrational state l

$$v_z^l = \sqrt{max\,[\,2 \cdot (E_{DA} - \epsilon^l)/M_{AB}\,,\,0\,]} \tag{16}$$

where $\epsilon^l = \mathcal{R}\theta_{AB}l$ is the energy of the oscillator AB in the vibrational level l and E_{DA} the activation energy of dissociative adsorption. M_{AB} represents the molar mass of the molecule. For simplicity, we have supposed here that the considered surface is perpendicular to z-axis of a cartesian coordinate system. A v_z-weighted integration of the velocity distribution function f_{AB} over the velocity half space $-\infty < v_z \le -v_z^l$ and subsequent summation over all vibrational energy levels gives the flux of molecules being able to react at the surface. By multiplying the resulting expression with the surface coverage of the vacant active sites one gets

$$\dot{\omega}_{DA} = \chi P_{DA} \sum_{l=0}^{l_D-1} p^l \int_{-\infty}^{-v_z^l} \int_{-\infty}^{+\infty} \int_{-\infty}^{+\infty} v_z f_{AB}\, dv_x\, dv_y\, dv_z \cdot \frac{n_X}{n_X^0}. \tag{17}$$

with $p_l = exp(-\epsilon^l/\mathcal{R}T_v)$. The two additional factors χ and P_{DA} have been introduced. χ takes into account that a molecule AB can dissociate into $A + BX$ but also into $AX + B$. Thus we assign for χ the value 0.5 if $A \ne B$, and 1 if $A = B$. P_{DA} is a catch-all factor which includes all the uncertainties in the modeling and which has to be adjusted by means of comparison with experimental data. Evaluation of equation (17) for a Maxwellian velocity distribution function and comparison with equation (15) now yields the coefficient

$$k_{DA}(T, T_v) = \chi P_{DA} \cdot \Lambda(T, T_v) \cdot exp\left(-\frac{E_{DA}}{\mathcal{R}T}\right) \tag{18}$$

with Λ being

$$\Lambda(T, T_v) = \frac{Q_v^{DA}(T^*) + \left[Q_v^D(T_v) - Q_v^{DA}(T_v)\right] exp\left(\frac{E_{DA}}{\mathcal{R}T}\right)}{Q_v^D(T_v)} \tag{19}$$

The subscripts DA and D signify that in the vibrational partition functions Q_v of the truncated harmonic oscillators AB the truncation limits E_{DA} (dissociative adsorption) and E_{A-B} (dissociation) have to be introduced respectively. The pseudo-temperature T^* is defined by $1/T^* = 1/T_v - 1/T$. Similar to the sticking coefficient s_A the rate coefficient k_{ER} is evaluated from equilibrium considerations. Under this special condition $\dot{\omega}_{ER}$ must satisfy the equality $\dot{\omega}_{ER} = \dot{\omega}_{DA}$ and the flux of particles B can be calculated from $Z_B^+ = n_B \bar{c}_B / 4$. From the equations (14), (15) and the above relations one gets

$$k_{ER} = \frac{\bar{c}_{AB}}{\bar{c}_B} \left[\frac{n_{AX} n_B}{n_{AB} n_X} \right]^{-1} k_{DA}. \tag{20}$$

Once again the ratio of the concentrations is expressed by the equilibrium constant

$$K_{ER/DA} = \frac{n_{AX} n_B}{n_{AB} n_X} = \frac{Q_{AX} Q_B}{Q_{AB} Q_X} \cdot exp \left(\frac{\Delta H_{ER}}{\mathcal{R} T_W} \right). \tag{21}$$

Introduction of the partition functions and some algebraic manipulations finally give

$$k_{ER} = \chi P_{DA} \Lambda(T_W) \cdot \left(\frac{8\pi^2 I_{AB} k T_W}{\sigma_{AB} h^2} \right) \cdot \left[\frac{1 - exp\left(-\frac{E_{A-B}}{T_W}\right)}{1 - exp\left(-\frac{\theta_{AB}}{T_W}\right)} \right] exp\left(-\frac{\theta_{AB}}{2T_W}\right)$$

$$\times \frac{m_{AB}}{m_B} \left[\frac{exp\left(-\frac{\theta_{\parallel,A}}{2T_W}\right)}{1 - exp\left(-\frac{\theta_{\parallel,A}}{T_W}\right)} \right]^{-2} \cdot \left[\frac{exp\left(-\frac{\theta_{\perp,A}}{2T_W}\right)}{1 - exp\left(-\frac{\theta_{\perp,A}}{T_W}\right)} \right]^{-1} exp\left(-\frac{E_{ER}}{\mathcal{R} T_W}\right). \tag{22}$$

The activation energies E_{ER} and E_{DA} are still unspecified. We use the semi-empirical correlation of Hirschfelder [8] for their estimation from the desorption energy E_{A-X}. The relation of Hirschfelder reads $E_{ER} = 0.055 \cdot E_{A-X}$. The calculated activation energies E_{ER} are then corrected slighty to give best comparison with experimental data. If the activation energy E_{ER} is known one can determine the activation energy of the dissociative adsorption by $E_{DA} = -\Delta H_{ER} + E_{ER}$. Numerical values for the activation energies of nitrogen are $E_{ER,2N \to N_2} = 16.6 KJ/mol$ and $E_{DA,N_2 \to 2N} = 658.7 KJ/mol$ and for oxygen one gets $E_{ER,2O \to O_2} = 20.8 KJ/mol$ and $E_{DA,2O \to O_2} = 214.4 KJ/mol$.

3.2.3. Langmuir-Hinshelwood-Mechanism

The modeling of the Langmuir-Hinshel-mechanism has been adopted from the work of Nasuti et al. [10]. Only the determination of the activation energy for surface migration E_m which plays a major role in this model has been modified. For this activation energy we use the rough estimate $E_{m,A} = E_{A-X}/2$ [1]. Additionally, a steric factor P_{LH} has been introduced since the modeling of the surface migration process contains some simplifications. The activation energies used in our work are $E_{LH,2N \to N_2} = 150.0 KJ/mol$ and $E_{LH,2O \to O_2} = 150.0 KJ/mol$.

3.3. ENERGETIC STATE OF RECOMBINING MOLECULES

For the determination of the surface heating of a re-entry body it is not only important to know the surface reaction rates but it is also necessary to know in which state the various particles are formed. For the estimation of the states of particles being formed in the ER recombination we first consider the state of molecules which are vanishing in the reverse process. The vibrational energy flux $F^+_{v,DA}$ due to molecules vanishing in dissociative adsorption reactions is given by the expression

$$F^+_{v,DA} = \dot\omega_{DA}\, G^+_{v,DA}$$

$$= \chi P_{DA} \sum_{l=0}^{l_D-1} \epsilon^l\, p^l \int_{-\infty}^{-v_z^l} \int_{-\infty}^{+\infty} \int_{-\infty}^{+\infty} v_z f_{AB}\, dv_x dv_y dv_z \cdot \frac{n_X}{n_X^0} \quad (23)$$

and for the translational energy flux $F^+_{t,DA}$ one finds

$$F^+_{t,DA} = \dot\omega_{DA} \cdot G^+_{t,DA}$$

$$= \chi P_{DA} \sum_{l=0}^{l_D-1} p^l \int_{-\infty}^{-v_z^l} \int_{-\infty}^{+\infty} \int_{-\infty}^{+\infty} v_z \frac{M_{AB}\vec{v}^2}{2} f_{AB}\, dv_x dv_y dv_z \cdot \frac{n_X}{n_X^0}. \quad (24)$$

In the thermodynamic equilibrium the rates of the ER recombination and the dissociative adsorption must be equal, hence $\dot\omega_{ER} = \dot\omega_{DA}$. Moreover the vibrational and translational energy fluxes of the particles reacting in the two processes have to be the same in this case, thus we get $F^+_{v,DA} = F^-_{v,ER}$ and $F^+_{t,DA} = F^-_{t,ER}$. This implies that also the mean vibrational excitation and the mean translational energies are equal for the direct and the reverse process, $G^+_{t,DA} = G^-_{t,ER}$ and $G^+_{t,DA} = G^-_{t,ER}$. Using these arguments we get for the mean vibrational excitation and the mean translational energy of particles which come into being within the ER-mechanism under equilibrium conditions

$$G^-_{v,ER} = \frac{\frac{E_{DA}(E_{DA}-\mathcal{R}\theta_{AB})}{2\mathcal{R}\theta_{AB}} + \left[Q_v^D(T_W)L^D(T_W) - Q_v^{DA}(T_W)L^{DA}(T_W) \right] e^{\frac{E_{DA}}{\mathcal{R}T_W}}}{\frac{E_{DA}}{\mathcal{R}\theta_{AB}} + \left[Q_v^D(T_W) - Q_v^{DA}(T_W) \right] e^{\frac{E_{DA}}{\mathcal{R}T_W}}}$$

$$(25)$$

with

$$L^D(T_W) = \frac{\mathcal{R}\theta_{AB} \cdot e^{-\frac{\theta_{AB}}{T_W}}}{1 - e^{-\frac{\theta_{AB}}{T_W}}} - \frac{E_{A-B} \cdot e^{-\frac{E_{A-B}}{\mathcal{R}T_W}}}{1 - e^{-\frac{E_{A-B}}{\mathcal{R}T_W}}}. \quad (26)$$

For the determination of $L^{DA}(T_W)$ the dissociation energy E_{A-B} in equation (26) has to be replaced by the activation energy E_{DA} of the dissociative adsorption.

The molar translational energy of a molecule being formed in the ER recombination reads

$$G^-_{t,ER} = [2\mathcal{R}T_W + E_{DA}]$$

$$- \frac{\frac{E_{DA}(E_{DA} - \mathcal{R}\theta_{AB})}{2\mathcal{R}\theta_{AB}} + E_{DA}\left[Q^D_v(T_W) - Q^{DA}_v(T_W)\right]e^{\frac{E_{DA}}{\mathcal{R}T_W}}}{\frac{E_{DA}}{\mathcal{R}\theta_{AB}} + [Q^D_v(T_W) - Q^{DA}_v(T_W)]e^{\frac{E_{DA}}{\mathcal{R}T_W}}}. \quad (27)$$

These values are not exact under nonequilibrium conditions since the energies of the molecules AB coming into being in the recombination reaction are also affected by the energetic state of the atoms B. Since the energy of the atoms B is usually much smaller than the energy values $G^-_{v,ER}$ and $G^-_{t,ER}$ calculated for AB under equilibrium conditions, the nonequilibrium values of AB will not be strongly affected by the states of the atoms B and thus the equilibrium values can be used as a quite reasonable approximation for nonequilibrium situations.

The LH recombination reaction is an endothermic process. Therefore it is assumed that molecules which are formed within this process do not possess any excess energy and we assign the values $G^-_{v,LH} \approx e_v(T_W)$ and $G^-_{t,LH} \approx 2RT_W$ for the average vibrational and translational excitation, where $e_v(T_W)$ is the molar average vibrational energy of Boltzmann-populated truncated harmonic oscillators.

The rotational degrees of freedom are assumed to be in equilibrium with the surface conditions for all considered reactions, hence $G^-_{r,LH/ER} \approx RT_W$.

3.4. RESULTS

Extensive parameter studies have been carried out in order to adjust the coefficients P_{Des}, P_{DA} and P_{LH}. For a pure nitrogen system N/N_2 best agreement with existing experimental data has been found with $P_{Des,N} = 0.25$, $P_{ER,N} = 0.1$, $P_{LH,N} = 0.1$. In figure 8 the recombination coefficient $\gamma^I_{N \to N_2}$ which is defined by

$$\gamma^I_{N \to N_2} = \frac{Z^+_N - Z^-_N}{Z^+_N} \quad (28)$$

and the energy recombination coefficient $\gamma_{N \to N_2}$ which is the product of $\gamma^I_{N \to N_2}$ times the average energy accommodation coefficient $\beta_{N \to N_2}$ are plotted as a function of the inverse surface temperature $1/T_W$. The partial pressures of N and N_2 in the gasphase are $p_N = 200 N/m^2$ and $p_{N_2} = 0 N/m^2$. It is assumed that thermal equilibrium exists between the surface, the gasphase and the different degrees of freedom. As one can see good agreement is found between the calculated value $\gamma_{N \to N_2}$ and the measured values (which in effect are energy recombination coefficients because they have been determined by heat flux measurements). The recombination coefficients show the typical roll over at a temperature of about $1600 K$. Furthermore, one can see that γ^I and γ differ strongly at low temperatures where the ER-mechanism is dominant but lie closely together where the LH-mechanism becomes the more important process. This can be explained by the fact that in the model presented the accommodation coefficient $\beta_{LH,N \to N_2}$ of the LH-mechanism is unity whereas the ER-mechanism has a much smaller value $\beta_{ER,N \to N_2}$ of about 0.3 (Fig. 10). The values of β_{ER} shown in figures 10 and 11 are calculated from the terms $G_{v,ER}^-$ and $G_{t,ER}^-$. For a pure oxygen system the same investigations have been carried out as for nitrogen. The most favorable comparison with experimental data has been achieved by using the values $P_{Des,O} = 0.3$, $P_{ER,O} = 0.1$, $P_{LH,O} = 0.1$. In the figure 9 the comparison is shown for the recombination coefficients $\gamma_{O \to O_2}$ and $\gamma_{O \to O_2}^I$. As for the nitrogen system the agreement between the experimental data and the calculated values is quite satisfactory. Also, the other features which have been depicted for nitrogen are found for oxygen, although the differences between the two coefficients γ^I and γ is not as much pronounced in this case. This arises from the fact that the coefficient $\beta_{ER,O \to O_2}$ is about 0.6 (Fig. 11) and thus much closer to unity than the value for nitrogen.

Also, the influence of the gasphase composition on the recombination coefficient has been investigated for an O/O_2 system. In figure 12 the recombination coefficient $\gamma_{O \to O_2}$ is shown as a function of the inverse surface temperature $1/T_W$ for different partial pressure ratios $f = p_O/p_{O_2}$. Since the translational temperature of the particles impinging on the surface is chosen to be $5000 K$ and the pressure $p = p_O + p_{O_2}$ is $1000 N/m^2$ this can be seen as a typical low density case. Furthermore, the vibrational temperature is assumed to have the same value as the surface ($T_v = T_W$). One can see that the recombination coefficient defined above becomes negative for large ratios f because more atomic oxygen is formed in dissociative adsorption reactions than vanishs in recombination reactions. This means that in low density flows the catalytic surface reactions may also decrease surface heat fluxes.

It is also interesting to look at the influence of temperature T of the impinging gasphase molecules on the recombination coefficient. In figure 13 the recombination coefficient $\gamma_{O \to O_2}$ is plotted as a function of the inverse surface temperature $1/T_W$ for temperatures $T = 2000 K - 6000 K$ with a

216

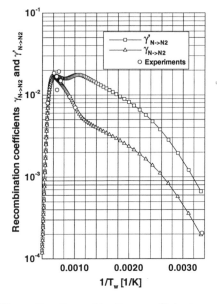

Figure 8. Recombination coefficients γ_N and γ_N^I

Figure 9. Recombination coefficients γ_O and γ_O^I

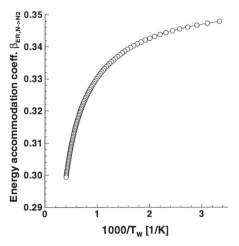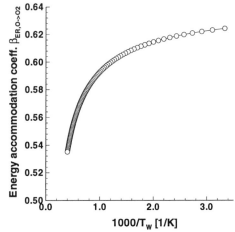

Figure 10. Energy accommodation coefficients $\beta_{ER,N \to N_2}$ of the ER-mechanism

Figure 11. Energy accommodation coefficient $\beta_{ER,O \to O_2}$ of ER-mechanism

constant ratio $f = p_O/p_{O_2} = 2$. Once again the recombination coefficient is strongly reduced by dissociative adsorption reactions if the temperature T is increased.

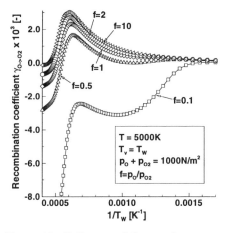 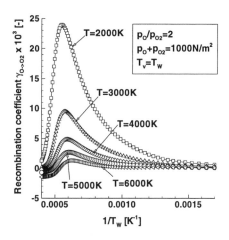

Figure 12. Influence of the gasphase composition on the recombination coefficient $\gamma_{O \to O_2}$

Figure 13. Influence of the translational temperature on the recombination coefficient $\gamma_{O \to O_2}$

4. Important Conclusions and Future Work

- Since the presented catalysis model takes into account dissociative adsorption reactions an important reduction of the catalytic efficiency of hot surfaces is found in the slip flow regime.
- Heat fluxes for low density flows may be smaller than heat fluxes calculated for a non catalytic surface
- An extensive validation of the catalysis model has to be carried out for test cases ranging from continuum to slip flows

5. Acknowledgements

This work is supported by the Deutsche Forschungsgemeinschaft DFG. One of us (A.D.) would like to thank the Landesgraduiertenförderung Baden-Württemberg and the Flughafen Frankfurt Main Stiftung for the financial support of his work.

References

1. Barrer, R.M. (1954) *British Journal of Applied Physics*, Suppl. 3, **41** and 49.
2. Carlson, A.B. and Bird, G.A. (1994) Implementation of a Vibrationally Linked Chemical Reaction Model for DSMC, NASA TM 109109.
3. Daiß, A., Schöll, E., Frühauf, H.-H. and Knab, O. (1993) Validation of the Uranus Navier-Stokes Code for High-Temperature Nonequilibrium Flows, *AIAA-93-5070*.
4. Daiß, A., Frühauf, H.-H., Messerschmid, E.W. (1994) New Slip Model for the Calculation of Air Flows in Chemical and Thermal Nonequilibrium, *Proc. of*

the Second European Symposium on Aerothermodynamics for Space Vehicles, Noordwijk, The Netherlands.

5. Gupta, R.N., Scott, C.D. and Moss, J.N. (1985) Surface-Slip Equations for Multicomponent Nonequilibrium Air Flow, NASA Technical Memorandum 85820.

6. Kim, Y.C. and Boudart, M. (1991) Recombination of O, N and H Atoms on Silica: Kinetics and Mechanism. *Langmuir* **7**, pp.2999-3005.

7. Knab, O., Frühauf, H.-H. and Messerschmid, E.W. (1995) Theory and Validation of the Physically Consistent Coupled Vibration-Chemistry-Vibration Model, *Journal of Thermophysics and Heat Transfer* **9**, No.2, pp.219-226.

8. Laidler, K.J. (1987) *Chemical Kinetics*, Third Edition, Harper & Row, Publishers, New York, p.73.

9. Moss, J.N. and Bird, G.A. (1984) Direct Simulation of Translational Flow for Hypersonic Re-entry Conditions, *AIAA-84-0223*.

10. Nasuti, F. and Bruno, C.: Material Dependent Catalytic Recombination Modeling for Hypersonic Flows, *AIAA-93-2840*.

11. Newman, M. (1987) A Model for Nitrogen Recombination on a Silicon-Dioxide Surface, Master's Thesis, Air Force Institute of Technology, Wright Patterson Air Force Base, OH.

12. Scott, C.D. (1981) Catalytic Recombination of Oxygen and Nitrogen in High Temperature Reusable Surface Insulation, *Progress in Astronautics and Aeronautics* **77**, pp.192-212.

13. Seward, W.A. and Jumper, E.J. (1993) Model for Oxygen Recombination on Silicon-Dioxide Surfaces. *Journal of Thermophysics and Heat Transfer* **5**, No.3.

14. Vetter, M., Olivier, H. and Grönig, H. (1992) Flow over Double Ellipsoid and Sphere - Experimental results, *Hypersonic Flows for Reentry Problems* **III**, Springer Verlag, pp.489-500.

SPUTTERED EXCITED ATOMS: BOUNDARY CONDITIONS

G. FALCONE[*] and F. PIPERNO
Dipartimento di Fisica, Università della Calabria, 87036-Rende (CS) and Unità INFM di Cosenza-Italy.
) Dipartimento di Fisica, Università di Bari, via G. Amendola, 173 - 70126-Bari-Italy

Abstract

Neutral and excited sputtered atoms differ from each other not only in energy spectra and sputtering yields, but also in regions of kinetic energy and angular variables. The structure of this physical region, due to its general character, can be used to obtain direct information on inelastic processes which are responsible for the formation of the excited states. In particular, the existence of a limiting angle for recoiling particles, never described in the literature, introduces a further anisotropy in the angular distribution of excited species.

1. Introduction

Ion bombardment of solid metals can produce sputtered excited atoms. The observation of Auger electrons [1-4] or photons [5,6] offers a clear evidence of the existence of excited states. In particular, the sharp peaks of the Auger spectra are the consequence of sputtered excited atoms transiting in the vacuum.

Excitation mechanisms and the subsequent transport toward the surface are the aspects that encounter an evergrowing consideration. Here we do not discuss directly any particular excitation mechanism, but present some general results which are independent from an excitation mechanism used. The advantage of such approach is to provide some boundary conditions for physical quantities, from which direct information about basic processes can be derived.

2. The one-step integral formulation

In the linear regime, where the number of particles moving in the solid is small compared with the total number of particles, the linear Boltzmann transport equation is proved adequate to describe the sputtering phenomenon. Since

M. Capitelli (ed.), Molecular Physics and Hypersonic Flows, 219–230.

Harrison's paper [7], several types of Boltzmann transport equations have been proposed. The most used has been the *integro-differential form* proposed by Sigmund [8]. The relation with the traditional Boltzmann transport equation, and the limits of this formulation when applied to the sputtering phenomenon have been discussed by Williams [9].

In order to relate the emission spectra to the depth of origin of sputtered atoms the *kernel form* of the Boltzmann transport equation was introduced [10]. This form is also known as *three step model* for emission phenomena.

Here, we shall introduce the one-step integral form of the Boltzmann transport equation [11,12]. To understand this new formulation better, we shall first discuss the *three step model*. According to this model, the emission of a particle from a bombarded solid can be described as occurring in three steps. In the first step, the incident ion produces possible candidates for the emission (*primary events*: atoms are set in motion in the ground state or in an excited state). In the second step, the atoms set in motion are transported through the solid up to the surface, whereas in the final step, the particles overcome the surface barrier.

The theoretical structure of the three step model is the following: If $D_0(E, \vec{e}; E_0, \vec{e}_0, x)$ is the *recoiling spectrum*, namely if $D_0\, dx dE_0 d^2 e_0$ is the average number of target atoms *set in motion*, by the incident radiation, at depth (x, dx) with initial energy (E_0, dE_0) in the solid angle $(\vec{e}_0, d^2 e_0)$, the double differential emission yield will be given by

$$J(E_2, \vec{e}_2) = \int dx dE_0 d^2 e_0\, K(E_0, \vec{e}_0, x; E_2, \vec{e}_2)\, D_0(E, \vec{e}; E_0, \vec{e}_0, x) \qquad (1)$$

where $K dE_2 d^2 e_2$ represents the probability that a recoiling atom $(E_0, \vec{e}_0, x)$ will be ejected from the surface with energy (E_2, dE_2) in the solid angle $(\vec{e}_2, d^2 e_2)$.

The function K, also the kernel of the integral formulation, has a meaning of an *escape probability*. It is clear that the form of the kernel cannot be known *a priori*, because particles during their motion towards the surface can lose energy, can be scattered and also disappear, like f.e., excited atoms. In some cases, the expression of the kernel is postulated and its validity is checked, *a posteriori*, by comparing theoretical and experimental results.

A typical example of this last approach is that described in Ref.10. By assuming that each recoil which is to be sputtered, slows down continuously along a straight line, the following expression of function K was obtained [10]:

$$K\left(E_0, \theta_0, \varphi_0, x; E_2, \theta_2, \varphi_2\right) = \delta\left(E_2 - E_0\left(1 - \frac{x}{R(E_0)\cos\theta_0}\right)^{\frac{1}{2m}}\right)$$
$$\times \delta\left(\varphi_2 - \varphi_0\right)\delta\left(\cos\theta_2 - \cos\theta_0\right) \qquad (2)$$

where E_0, θ_0 and φ_0 are, respectively, the energy, polar angle and azimuth with respect to the surface normal of a target atom set in motion at depth (x, dx), whereas E_2, θ_2, φ_2 are the corresponding quantities at the surface. Moreover, δ is the Dirac delta, $R\left(E_0\right)$ is the path length [13] and m is the power of the Lindhard power potential [14]. The energy spent by the ejected particle to overcome the surface barrier is omitted in eq.(2), but it can be easily introduced.

From the previous formulation [10], a distribution of the emerging depth can be derived and a *mean escape depth*, λ_e (the depth from which the escape probability is $1/e$) of sputtered atoms calculated.

Subsequently, an exponential form of the escape probability has been proposed [15], already used in the emission of electrons from borbarded solid surfaces [16-18]:

$$K\left(E_0, \theta_0, \varphi_0, x; E_2, \theta_2, \varphi_2\right) = \delta\left(E_2 - E_0\exp\left(-x/\lambda_a\cos\theta_0\right)\right)$$
$$\times \delta\left(\varphi_2 - \varphi_0\right)\delta\left(\cos\theta_2 - \cos\theta_0\right) \qquad (3)$$

In eq.(3) the quantity λ_a is the *attenuation length*, (the distance within which a particle flux is reduced to $1/e$ of its initial value). The same quantity gives a measure of the mean escape depth. Both quantities (mean escape depth and attenuation length) can be, in principle, measured by experiments and for this reason are indicated prevalently as *experimental quantities* [18].

An important aspect, often not mentioned, of the three step model is that, to obtain a deeper insight into the physical emission process, the previous two quantities (λ_e, λ_a) must be related to a *theoretical concept*, the *mean free path*, λ_X (mean distance within which $1/e$ particles have not suffered a collision of X-type, or mean distance between two consecutive collisions of X-type). The connection between the previous quantities can be achieved, in principle, because the three step model is a *kernel form* of the Boltzmann transport equation [11,12] and the collision term of the *integro-differential* Boltzmann transport equation provides the precise definition of the mean free path [19].

The three step model, although very useful to the understanding of some experiments, has been always considered a spurious approach for the difficulties of finding the form of the kernel and for the not clear, *a priori*, position of the mean free path.

The third form of Boltzmann transport equation is the *one-step integral form* of Boltzmann equation, a formulation well known to the reseachers working with neutron transport formulations [11].

The passage from the integro-differential form to the integral form of Boltzmann equation can be found in several text-books [11, 12]. In particular, the double differential emission yield can be written

$$J\left(E_2, \vec{e}_2\right) = \int dx dE_0 d^2 e_0 \, P\left(E_0, \vec{e}_0, x; E_2, \vec{e}_2\right) \, D\left(E, \vec{e}; E_0, \vec{e}_0, x\right) \qquad (4)$$

where $D \, dx dE_0 d^2 e_0$ is the average number of target atoms *moving* at depth (x, dx) with initial energy (E_0, dE_0) in the solid angle $(\vec{e}_0, d^2 e_0)$ generated by the incident radiation. The function P has the following *general* form

$$
\begin{aligned}
P\left(E_0, \theta_0, \varphi_0, x; E_2, \theta_2, \varphi_2\right) = \quad & \exp\left(-x/\lambda_X \cos\theta_0\right) \, \delta\left(E_2 - E_0\right) \\
& \times \delta\left(\varphi_2 - \varphi_0\right) \, \delta\left(\cos\theta_2 - \cos\theta_0\right)
\end{aligned}
\qquad (5)
$$

In the one-step formulation, function P, except for the pertinent form of the mean free path, is well established. In other words, the integral approach to the emission problem stresses the existence of a reference plane (determined by a mean free path, which, in turn, depends on target properties and on the emitted studied particles) below the surface of which emitted particles do not undergo X-type collisions.

The one-step integral formulation has been used, in neutron [11] and in sputtering [20] calculations, essentially in connection with the Newmann's serie expansion. Here, we present a different property of the one-step formulation, which allows to clarify, for the first time, the relation between this particular form of Boltzmann equation and a sputtering integral formulation, presented some years ago [21]. The function D in eq.(4), which describes atoms *moving* at depth x, can be always separeted in two functions. The first function, D_0, also present in eq.(1), describes atoms *set in motion* at depth x, and the second function, indicated with D_1, describes atoms moving at x, *but originateing (set in motion!) at depth different from x*:

$$D\left(E, \vec{e}; E_0, \vec{e}_0, x\right) = D_0\left(E, \vec{e}; E_0, \vec{e}_0, x\right) + D_1\left(E, \vec{e}; E_0, \vec{e}_0, x\right) \qquad (6)$$

In the sputtering, due to the surface character of the phenomenon (almost all sputtered atoms originate from the first monolayer!) the function D_1 is negligible and eq.(4) reduces itself to the sputtering formulation of Ref. [21]. In other words, while the use of the exponential form for the kernel in eq.(1) remains within the kernel approach [15], the sputtering formulation of Ref.21, is an approximate form of the one-step integral Boltzmann transport equation.

The above formulation has several new characteristics. The first one is that it offers a possibility to study separately the influence of primary and relocated events in the emission spectrum (and yield). The second and more important for the excitation species, is a new class of equations that determine the accessible region of the values for several physical parameters. The new equations do not explain the physical mechanism of phenomena, but they determine the boundary conditions for any possible formulation and, under adequate experimental conditions, one can derive from them direct information about basic processes.

3. The boundary conditions

The new and general result of the above formulation, which is the principal argument of this paper, is the restriction of physical parameters of the ejected excited particles, in comparison with the corresponding neutral ones.

In fact, independently from the explicit form of function D (thus independently from the excitation mechanism), the kinetic energy E_2 of the excited sputtered atoms must be limited to a value comprised between a minimum E_m and a maximum E_M:

$$E_m \leq E_2 \leq E_M \tag{7}$$

In particular, for sputtered neutral atoms, the following relation was found [22]:

$$E_M = B\,E \tag{8}$$

where B is the maximum fraction of the projectilés energy that can be transferred to a target atoms (to be sputtered) in a single elastic collision. Due to the boundary conditions, the kinetic energy of the emitted particles, in the plane (E, E_2), is limited by two straight lines, $E_2 = 0$ and eq.(8).

In the case of sputtered excited atoms, the minimum and the maximum values belong to the following hyperbola:

$$E_2^2 - \frac{4A}{(A+1)^2}\cos^2\alpha_2 E E_2 + \frac{2}{A+1}Q E_2 + \frac{Q^2}{(1+A)^2} = 0 \tag{9}$$

with $A = M_2/M_1$, where M_1 and M_2 are the masses of the incident and target atoms respectively. Moreover, α_2 is the recoil angle of the target particle, with respect to the incident direction in the laboratory system; the target particles are supposed at rest before collisions and Q is the excitation energy, namely the energy lost in an elementary inelastic event. Eq.(9) is the consequence of the principle of conservation of energy, linear and angular momentum, and masses, the validity of which is explicitly assumed in eq.(5).

From the same principles of conservation, it is possible to derive the following limiting angle, α_2^M, for recoiling particles in a binary inelastic collision:

$$\cos \alpha_2^M = \sqrt{\frac{E_{th}}{E}} \tag{10}$$

where

$$E_{th} = \frac{M_1 + M_2}{M_2} Q \tag{11}$$

is the threshold for the excitation process in the laboratory system.

4. Boundary conditions for direct recoil sputtering

For neutral sputtered atoms, the accessible values of energy and angular variables have been indicated in Ref.22 and are

$$E_m = 0 \quad E_M = B E - U \tag{12}$$

for the energy variables (U is the surface binding energy) and

$$0 \leq \eta_2 = \cos \alpha_2 \leq 1 \tag{13}$$

for the angular variables.

For sputtered excited atoms, to determine the boundary conditions we have to separate the case of excited recoils generated by ion-atoms collisions from the case ot atom-atom collisions. In the first case, when recoil sputtering is the predominant ejection mechanism, the boundary conditions for inelastic collisions, characterized by a specific Q value, are for the energy variables

$$E_m - U \leq E_2 \leq E_M - U \tag{14}$$

where (E_m, E_M) are the two solutions of eq.(9), and for the angular variable

$$\sqrt{\eta_m^2 - \frac{U}{E_2}(1 - \eta_m^2)} \leq \eta_2 \leq 1 \tag{15}$$

where $\eta_m = \cos \alpha_2^M$ is given in eq.(10).

It is evident that the region of accessible values for the sputtering variables depends on the excited species, and with adequate experimental measurements, we can derive direct information about the ion-atom excitation mechanism [23].

Moreover, the limiting angle enables us to introduce a new quantity [24], which is equivalent, at the microscopic scale, to the restitution coefficient [25] (a parameter used up to now to describe, at the macroscopic scale, the binary inelastic collisions):

$$\epsilon \equiv \sin \alpha_2^M \qquad (16)$$

We have shown [24] that this new quantity enables to better describe the binary inelastic collisions. We just mention tow new results. The first one is a generalization of the Lindhard power cross section [14] for inelastic processes:

$$d\sigma\left(E, E_2\right) = \epsilon^{-3m} C_m E^{-m} \left(E_2 - \hat{E}_m\right)^{-1-m} dE_2 \qquad (17)$$

where both m and C_m are the quantities used in the elastic case [26]. Moreover

$$\hat{E}_m = \frac{T_M}{4}(1-\epsilon)^2 \qquad (18)$$

with $T_M = 4AE/(A+1)^2$.

The second result is the determination, in terms of this quantity, of the excitation cross section for the Kessel-Sigmund model [27,28]:

$$\sigma_e = \pi r_c^2 \, \epsilon^2 \qquad (19)$$

where r_c is the critical distance at which, in a binary collision, an excitation takes place. For the excitation of 2p electron in Si, we have found out [23] $r_c = 0.414 \times 10^{-8} cm$.

In conclusion, we have shown that in each binary inelastic collision the region of accessible values of the energy-angular variables of the target atoms is clearly different from the elastic binary collision. These restrictions have never been mentioned or used in previous theoretical or computer simulation studies. Under particular conditions, experimental measurements on sputtering species, together with the derived expressions, allow to obtain direct information about excitation mechanisms. Finally, a new parameter, for describing binary inelastic processes, has been presented. On the basis of this new parameter several basic quantities can be obtained.

Acknowledgments

The authors wish to thank D. Aiello and L. Forlano for the numerous conversations on the subject.

5. References

1. J. J. Wrakking and A. Kroes, Auger spectra induced by Ne and Ar impact on Mg, Al and Si, *Surf. Sci.*, **84**, 153 (1979).

2. R. A. Baragiola, J. Fine and H. Raiti, Ion-induced Auger-electron emission from aluminum, *Phys. Rev.* **A25**, 1969 (1982).

3. T. D. Andrealis, J. Fine and J. A. D. Matthew, Auger electron emission from decay of collisionally excited atoms sputtered from Al and Si, *Nucl. Instrum. Meth.*, **209/210**, 495 (1983).

4. A. Bonanno, F. Xu, M. Camarca, R. Siciliano and A. Oliva, Angle-resolved Auger study of 10-keV-ion-induced Si LMM atomic lines *Phys. Rev.*, **B41**, 12590 (1990).

5. E. Veje, Emission of secondary electrons and photons from silver bombarded with Sb, Sb_2 and Sb_3, *Rad. Effect.* **58**, 35 (1981).

6. J.J. Jimenez-Rodriguez, D.S. Karpuzov and D.G. Armour, The angle of incidence dependence of ion-bombardment induced photon emission from solids, *Surf. Sci.* **136**, 155 (1984).

7. D. E. Harrison, Theory of the sputtering process, *Phys. Rev.* **102**, 1473 (1956).

8. P. Sigmund, Theory of sputtering I, *Phys. Rev.*, **164**, 383 (1969).

9. M. M . R. Williams, The role of the Boltzmann transport equation in radiation damage calculations, *Prog. Nucl. Energy*, 1, 1 (1979).

10. G. Falcone and P. Sigmund, Depth of origin of sputtered atoms, *Appl. Phys.*, **25**, 307 (1981).

11. J. Spanier and E. M. Gelbard,(1969) *Monte Carlo Principles and Neutron transport Problems*, Addison-Wesley Publ. Comp., Reading, Massachusetts.

12. V. Boffi,(1974) *La Teoria del trasporto dei Neutroni*, Pàtron Ed., Bologna.

13. J. Lindhard, M. Scharff and H. Schiott, Range concepts and Heavy ion ranges, *Mat. Fys. Medd. Dan. Vid. Selsk*, **33**, No. 14 (1963)

14. J. Lindhard, V. Nielsen and M. Scharff, Approximation method in classical scattering by screened Coulomb fields, *Mat. Fys. Medd. Dan. Vid. Selsk.* **36** N.10 (1968)

15. G. Falcone, Ejection process in collisional sputtering, *Phys. Rev.*, **B33**, 263 (1986).

16. E. S. Parilis and L. M. Kishinevskii, The theory of ion-electron emission, *Sov. Phys. Solid State*, **3**, 855 (1960).

17. M. Rosler and W Brauer, Theory of secondary electron emission, *Phys. stat. sol. (b)*, **104**, 161 (1981).

18. M. Cailler, J. P. Ganachaud and D. Roptin, Quantitative Auger electron spectroscopy, *Ad. Electronics and Phys.*, **61**, 162 (1983).

19. G. H. Wannier, (1966) *Statistical Physics*, Dover, N.Y..

20. V. V. Pletnev, D. S. Semenov and V. G. Tel'Kovsky, On the theory of binary alloy sputtering by light ions, *Radiat. Eff.*, **83**, 113 (1984).

21. G. Falcone, Theory of collisional sputtering, *Surf. Sci.*, **187**, 212 (1987).

22. G. Falcone, Sputtering transport theory: the mean energy, *Phys. Rev.*, **B38**, 6398 (1988).

23. G. Falcone and F. Piperno, Kinematic of sputtered excited atoms *Surf. Sci.*, submitted for publication

24. G. Falcone and F. Piperno, Generalized Lindhard power cross section, *Phys. Rev. B*, submitted for publication.

25. H. W. Harkness, (1964) *Elementary dynamics of particles*, Academic Press, N.Y., London.

26. P. Sigmund, Collisional theory of displaced damage, ion ranges and sputtering, *Rev. Roum. Phys.*, **17**, 823 (1972).

27. R.K. Cacak, Q.C. Kessel and M.E. Rudd, Emission of Auger electrons resulting from simmetric argon and neon ion-atom collisions, *Phys. Rev.*, **A2**, 1327 (1970).

28. P. Sigmund, (1981) *Inelastic Particle-Surface Collisions*, E. Taglauer and W. Heiland eds, Springer, Berlin.

SEMI-CLASSICAL TREATMENT OF THE DYNAMICS OF MOLECULE SURFACE INTERACTION

Gert D. Billing
Department of Chemistry
H. C. Ørsted Institute, 2100 Copenhagen Ø,
Denmark

1 Introduction

The dynamics of molecule surface scattering is a many-body problem in which not only the equations of motion of the molecule must be considered but also the excitation of various collective modes of the solid. Furthermore the molecule may undergo dissociation or being adsorbed as well as scattered inelastically from the surface. This requires that the dynamical treatment must include not just energy transfer channels but also reactive ones. The adsorption probability will obviously depend upon not only the surface temperature and other characteristics of the surface as site, steps, coverage etc.. During the last decade we have developed a semi-classical model which can in a realistical fashion incorporate all of these features. In order to describe the motion of the gas-atoms it is most convenient to use classical mechanics, whereas the modes in the solid being it phonon or electronic excitation is a many body problem, which can be handled only in the

M. Capitelli (ed.), Molecular Physics and Hypersonic Flows, 231–257.
© *1996 Kluwer Academic Publishers. Printed in the Netherlands.*

second-quantization approximation. The dynamics of the gas-atoms may however not always be well described by classical dynamics, but quantum effects as proper conservation of the zero-point vibrational energy, tunneling through barriers for dissociative chemisorption and non-adiabatic processes may make a quantum-treatment necessary.

However the most severe stumbling block is at present the lack of reliable potential energy surfaces for the motion of the gas-atoms. This is not surprising since the number of accurate surfaces for gas-phase chemical reactions is also modest - the many electron nature of the molecule-surface interaction makes such calculations even more difficult. We may therefore rely on semi-empirical expressions for which the free parameters must be fitted either to experimental data as spectroscopic data on surface bound molecules or ab initio calculations at a limited number of geometries or both.

Although detailed information on molecule surface interaction potentials is lacking ab initio calculations have been attempted [1],[2] on a few systems as e.g. the hydrogen-copper system. However the quality of most surfaces are still debated owing to the necessary approximations, which have to be introduced in the ab initio calculations. As for the barrier for dissociative chemisorption, it is now believed that it should be around 0.5 eV [3]. But the ab initio calculations have predicted values between about 0.3 eV for the gradient corrected LDA (local density approximation) calculations and about 1 eV for SCF/CI calculations. It is however important to realize that the barrier for chemical reactions is not only a matter of electronic structure calculations. It is a dynamical entity, which depends upon features as: corrugation, surface site, presense of steps, surface temperature and excitation processes of phonons, electrons and internal degree's of freedom of the molecule. Thus the barrier which is measured will depend upon an average over all these features.

Thus before a meaningful comparison between experimental and theoretical predictions of the sticking probability can be carried out it is necessary to know which dynamical processes one should include. The abovementioned model for the dynamical calculations include the following aspects: a three-dimensional calculation of the motion of the hydrogen molecule either by using classical mechanics or a mixed classical-wavepacket propagation technique, full corrugation, coupling to the surface phonons and for the classical trajectory calculations also coupling to electron-hole pair excitation. In the EDIM-method the valence-bond ideas which have led to the construction of the LEPS-surfaces [5] known from three-particle reactive scattering are extended to molecule-surface problems. The advantage of using this methodology is that the shape or topology of the potential is "correct" in the sense that the asymptotic limits as in DIM (diatomics in molecules methods) are build in correctly. The method contains a number of parameters which can be fitted to either ab-initio calculations, experimental data or both as it is done in the present case. Using the EDIM-potential we can in principle obtain the two (or more) lowest electronically adiabatic potentials. Since nonadiabatic transitions may play a role in the dissociative sticking process it is important to have a procedure for obtaining these. At present, ab initio methods are not in general capable of producing this information with sufficient accuracy. Thus semi-empirical methods have to be used and the parameters of these determined such that the potential is capable of representing or explaining available experimental data.

2   The semiclassical model

Methods, which are not either completely classical or quantum mechanical are denoted "semi-classical". In the present chapter we shall describe a semiclassical approach, which is capable of reducing the numerical problem to one, which is more manegable than that arising from a complete classical description. The method assumes a classical mechanical treatment of the motion of teh gas phase atoms but quantizes the lattice motion. If the lattice is treated within a harmonic approximation it is advantageous to introduce a quantum mechanical description. The reason being that quantum operator technique in this case gives a solvable algebra. Furthermore we avoid the averaging over oscillator phase angles which is necessary in the classical description of the oscillators. The transformation between displacement and normal mode coordinates is:

$$\eta_{\alpha\gamma} = \sum_r T_{\alpha\gamma;r} Q_r \tag{1}$$

where **T** is the matrix diagonalizing the force constant matrix. The hamiltonian for the solid is in the HO-description

$$H_c = V_0 + \sum_{r=1}^{M} H_r \tag{2}$$

where

$$H_r = \frac{1}{2}(\dot{Q}_r^2 + \omega_r^2 Q_r^2) \tag{3}$$

The $M = 3N - 6$ non-zero frequencies are as mentioned obtained by diagonalizing the second derivative matrix. We now quantizes the vibrational modes r and write the normal mode coordinates in terms of the phonon creation/annihilation operators:

$$Q_r = b_r(a_r^+ + a_r) \tag{4}$$

where $b_r = (\hbar/2\omega_r)^{\frac{1}{2}}$. The operator a_r^+ creates a quantum in mode r when operating on the wavefunction, i.e.

$$a_r^+ \mid n_r >= \sqrt{n_r + 1} \mid n_r + 1 > \tag{5}$$

and

$$a_r \mid n_r >= \sqrt{n_r} \mid n_r - 1 > \tag{6}$$

The probability for being in quantum state n_k for oscillator k is in the classical limit given by

$$P_{n_k} = z_k^{n_k}(1 - z_k) \tag{7}$$

where

$$z_k = exp(-\beta\hbar\omega_k) \tag{8}$$

and $\beta = 1/kT_s$. Transitions among the phonon quantum states may now be induced by anharmonic terms or eventually through coupling to the electronic motion. Transitions induced by gas-surface collisions are however the processes which are the concern of the present chapter. As mentioned the potential should be written as a sum of atom-atom pair potentials if we wish to extract phonon coupling from it. Thus we assume

$$V_I = \sum_{\alpha} V_{a\alpha}(R_{a\alpha}) \tag{9}$$

where $R_{a\alpha}$ is the distance from the gas phase atom a to the surface atom α, i.e.

$$R_{a\alpha} = \sqrt{\sum_{i=1,2,3} (\bar{X}_i - X_{\alpha i})^2} \tag{10}$$

The position of the atom a may be written in terms of polar coordinates:

$$\bar{X}_1 = Rsin\theta cos\phi \tag{11}$$
$$\bar{X}_2 = Rsin\theta sin\phi \tag{12}$$
$$\bar{X}_3 = Rcos\theta \tag{13}$$

The interaction potential is now expanded in normal mode coordinates

$$V_I = V_I^{(0)} + \sum_r V_r^{(1)} Q_r + \frac{1}{2} \sum_{rr'} V_{rr'}^{(2)} Q_r Q_{r'} \tag{14}$$

where $V_I^{(0)}$ is the interaction potential with the lattice atoms in their equilibrium position $X_{\alpha i}^{eq}$ and

$$V_r^{(1)} = \sum_\alpha \frac{\partial V_{a\alpha}}{\partial R_{a\alpha}} |_{eq} \frac{\partial R_{a\alpha}}{\partial Q_r} \tag{15}$$

In order to calculate $\frac{\partial R_{a\alpha}}{\partial Q_r}$ we use eqs.(1), (10) and that:

$$\eta_{\alpha\gamma} = m_\alpha^{1/2}(X_{\alpha\gamma} - X_{\alpha\gamma}^{eq}) \tag{16}$$

i.e.

$$\frac{\partial R_{a\alpha}}{\partial Q_r} = -m_\alpha^{-1/2} \sum_i T_{\alpha i;r}(\bar{X}_i - X_{\alpha i}^{eq})/R_{a\alpha}^{eq} \tag{17}$$

The terms $V_r^{(1)}$ and $V_{rr'}^{(2)}$ in (14) are those responsible for collision induced phonon excitation. These terms depend upon time through the time-dependence of the trajectory. If only the linear terms $V_r^{(1)}$ are included the oscillators are "linearly forced". From extensive studies on the linearly forced and also more general atom-molecule interactions we know that good agreement between the semiclassical description and the full quantum description of the excitation process is obtained throug the socalled "symmetrized Ehrenfest" approach. This approach is based upon the following two conjectures:
1) The classical particle motion should be governed by an average potential: $V_{eff} =< \Psi \mid V_I \mid \Psi >$ where Ψ is the total wavefunction of the

quantum system, i.e. the phonons in the present case. The averaging is carried out over the quantum variables Q_k.

2) If the quantum system goes from state k to state $k + 1$ with the probability $P_{n \to m}$ as a result of the collision with the classical particle, then one should calculate this probability using an average velocity $v_{av} = (v_n + v_m)/2$ where $v_n = \sqrt{2(E - E_n)/\mu}$, E the total energy and E_n the energy of the quantum state n.

In order to illustrate this let us consider the collision of an atom and a harmonic oscillator with a linear interaction potential, i.e.

$$V^{(1)} = f(R(t))Q \tag{18}$$

The total hamiltonian is

$$H = \hat{H}_0 + V^{(1)} + V^{(0)} + \frac{1}{2\mu}P_R^2 \tag{19}$$

where $\hat{H}_0$ is the hamiltonian for the quantum oscillator

$$\hat{H}_0 = \hbar\omega(a^+ a + \frac{1}{2}) \tag{20}$$

Asymptotically the total energy is

$$E = E_n + E_{kin} = \hbar\omega(n + \frac{1}{2}) + \frac{1}{2\mu}P_R^2(t_0) \tag{21}$$

where μ is the reduced mass of the atom-diatom system. The first conjecture now says that the relative (classical) motion is governed by the effective hamiltonian:

$$H_{eff} = < \Psi \mid H \mid \Psi > = < \Psi \mid \hat{H}_0 \mid \Psi > + < \Psi \mid V^{(1)} \mid \Psi > + V^{(0)} + \frac{1}{2\mu}P_R^2 \tag{22}$$

where the two first terms will be evaluated below.

The second conjecture says that the initial velocity should be

$$\frac{1}{\mu}P_R(t_0) = \frac{1}{2}(v_n + v_m) \tag{23}$$

for the quantum transition n to m. The classical trajectory is obtained from the effective hamiltonian and is defined by the hamiltons equations of motion, i.e.

$$\dot{R} = \frac{1}{\mu}P_R \tag{24}$$

and

$$\dot{P}_R = -\frac{\partial}{\partial R} H_{eff} \qquad (25)$$

This trajectory defines the time-dependence of the perturbation on the quantum system, i.e. $f(R(t)) = f(t)$ in eq.(18). In order to obtain the probability $P_{n \to m}$ and the wavefunction to be used in (22) we should solve the time-dependent Schrödinger equation

$$i\hbar \frac{\partial \Psi}{\partial t} = H\Psi \qquad (26)$$

Introducing the evolution operator such that

$$\Psi(t) = U(t, t_0)\Psi(t_0) \qquad (27)$$

we get in the interaction representation

$$i\hbar \frac{dU(t, t_0)}{dt} = exp(iH_0 t/\hbar)V^{(1)}(R(t))exp(-iH_0 t/\hbar)U(t, t_0) \qquad (28)$$

Introducing the quantum operators a^+ and a^- we get:

$$i\hbar \frac{dU(t, t_0)}{dt} = exp(i\hbar\omega(a^+ a + \frac{1}{2}))bf(t)(a^+ + a)exp(-i\hbar\omega(a^+ a + \frac{1}{2}))U(t, t_0) \qquad (29)$$

This equation is easily solvable. One obtains:

$$U(t, t_0) = exp(i(\beta(t, t_0) + \alpha^+(t, t_0)a^+ + \alpha^-(t, t_0)a)) \qquad (30)$$

where

$$\beta = \frac{1}{\hbar} \int_{t_0}^{t} dt' f(t')[exp(i\omega t')\alpha^-(t') - exp(-i\omega t')\alpha^+(t')] \qquad (31)$$

and

$$\alpha^\pm = -\frac{b}{\hbar} \int_{t_0}^{t} dt' f(t') exp(\pm i\omega t') \qquad (32)$$

The probability for a quantum transition is then

$$P_{nm} = |< m \mid U(t, t_0) \mid n >|^2 \qquad (33)$$

where

$$< m \mid U(t, t_0) \mid n >=$$
$$exp(i\beta - \rho/2)(i\alpha^+)^{m-n}\sqrt{n!m!} \sum_{k=0}^{n} \frac{(-1)^k}{k!} \frac{\rho^k}{(m-n+k)!(n-k)!} \qquad (34)$$

and

$$\rho = \alpha^+ \alpha^- = \mid \alpha^+ \mid^2 \tag{35}$$

From eq.(34) we see that the quantity ρ determines the magnitude of the transition probability, ρ is sometimes called the excitation strenght and for small values of ρ we have

$$P_{nm} \sim \rho^{|m-n|} \tag{36}$$

In order to determine the value one has to evaluates the integrals (32), i.e. the Fourier transform of the force $f(t)$ driving the oscillator. The integral (32) may be evaluated in a number of cases. Assuming e.g. an interaction of the form

$$V(z) = Aexp(-2\beta z) - Bexp(-\beta z) \tag{37}$$

and that the hamiltonian for the motion along the z-axis is given by

$$E_0 = \frac{1}{2}m\dot{z}^2 + V(z) = \frac{1}{2}m(v_0cos\theta)^2 \tag{38}$$

we get:

$$I = \int_{-\infty}^{\infty} dtexp(i\omega t)V(z(t)) =$$

$$\frac{\pi m}{\beta^2}sinh^{-1}(\frac{\pi\omega}{\beta v_0cos\theta})[\omega cosh(\frac{\omega\tau}{\beta v_0cos\theta}) - tan(\phi)\beta v_0cos\theta sinh(\frac{\omega\tau}{\beta v_0cos\theta})]$$

where $\hspace{9cm} (39)$

$$cot\tau = -\sqrt{E_m/E_0} \tag{40}$$

$$\phi = \tau - \frac{\pi}{2} \tag{41}$$

and the well depth $E_m = B^2/4A$. In the perturbation limit, i.e. where $\pi\omega/\beta v_0cos\theta > 1$ we obtain from eq.(39)

$$\rho = I^2 = Aexp(-\frac{2\pi\omega}{\alpha v_0cos\theta}(1 - \frac{2\phi}{\pi})) \tag{42}$$

where A is independent of v_0 the initial velocity. According to conjecture 2 above we should substitute v_0 with $\frac{1}{2}(v_n + v_m)$ when calculating the probability P_{nm}. Inserting this in eq.(42) we obtain

$$\rho = Aexp(-\frac{2\pi\omega}{\alpha v_n cos\theta}(1 - \frac{2\phi}{\pi}))exp(\mp \mid m - n \mid x) \tag{43}$$

where

$$x = \frac{\pi\omega^2\hbar}{\alpha m v_n^3 cos\theta}(1 - \frac{2\phi}{\pi}) \qquad (44)$$

and where we have used that

$$\frac{1}{2}mv_m^2 = \frac{1}{2}mv_n^2 \mp |m - n| \hbar\omega \qquad (45)$$

where the upper sign is for excitation $(m > n)$ and the lower for deexcitation $(m < n)$. We have also assumed when deriving (43) that

$$|m - n| \frac{\hbar\omega}{mv_n^2} << 1 \qquad (46)$$

This condition is fulfilled for thermal velocities and phonon frequencies less than 0.2-0.5 10^{14} sec^{-1}. Eq.(43) shows that the symmetrization decreases the excitation and increases the deexcitation probabilities as compared to the original expression (42).

It is now relatively straight forward to extend this forced oscillator model for a single oscillator to the situation in a solid where we have a distribution of oscillators.

Omitting the zero-point energy of the crystal we can write the total hamiltonian as

$$H = \frac{1}{2}\sum_{r=1}^{M}(\dot{Q}_r^2 + \omega_r^2 Q_r^2) + V_I^{(0)} + \sum_r V_r^{(1)}Q_r +$$

$$+ \sum_{rr'} V_{rr'}^{(2)}Q_r Q_{r'} + \frac{1}{2M}P_R^2 \qquad (47)$$

The evolution operator for the phonons perturbed linearly and quadratically is known from semi-classical theories for energy transfer in polyatomic molecules and may be expressed in terms of the phonon creation/annihilation operators. Thus the effective hamiltonian is obtained as:

$$H_{eff} = < \Psi | H | \Psi > = < \Psi_0 | U^+HU | \Psi_0 > \qquad (48)$$

where $U(t, t_0)$ is the evolution operator and $| \Psi_0 >$ the initial wavefunction, i.e.

$$| \Psi_0 > = \Pi_{k=1}^{M} | n_k^0 > \qquad (49)$$

The final expression for the effective hamiltonian is:

$$H_{eff} = \sum_k \hbar\omega_k\rho_k + V_I^{(0)} + \sum_k \omega_k^{-1}\epsilon_k(t) + \frac{P_R^2}{2M} + \sum_{k>l} b_k b_l V_{kl}^{(2)}\{F_{kl}^+[n_k^0 Q_{lk} +$$
$$(n_l^0 + 1)Q_{kl}^*] + F_{kl}^-[Q_{lk}^*(n_k^0 + 1) + Q_{kl}n_l^0]\} + O(Q_{kl}^2)$$
$$(50)$$

In eq.(50) ρ_k is the excitation strength for oscillator k. Initially we have $\rho_k(t_0)=0$ and finally, i.e. for $t_1 \to \infty$ where the atom has left the surface the only surviving term in (50) is:

$$E_{int} = \sum_{k=1}^{M} \hbar\omega_k\rho_k \qquad (51)$$

Thus E_{int} is the energy transfer to the solid. The quantity $\epsilon_k(t)$ is defined by:

$$\epsilon_k(t) = V_k^{(1)}(R(t)) \int_{t_0}^t dt' V_k^{(1)}(R(t'))sin(\omega_k(t' - t)) \qquad (52)$$

This term arises from the expectation value of the linear $V_r^{(1)}Q_r$ terms in (47). These terms are called phonon VT terms because they exchange one quantum of vibrational energy when forced by the translational motion. The second derivative terms in (47) contain operators as $a_r^+ a_{r'}$ and $a_r a_{r'}^+$, i.e. these terms create a quantum in one mode and destroys one in another. It has however turned out that these socalled VV terms are of minor importance for energy transfer in solids. These terms may therefore be neglected. The quantities $F_{kl}^\pm$ are defined as:

$$F_{kl}^\pm = exp(\pm i(\omega_k - \omega_l)t) \qquad (53)$$

and the functions $Q_{kl}(t)$ are do first order given by

$$Q_{kl} = Q_{lk} = \frac{1}{i\hbar} \int_{t_0}^t dt b_k b_l V_{kl}^{(2)}(R(t)) F_{kl}^-(t) \qquad (54)$$

Eq.(51) shows that $E_{int} > 0$ irrespective of the initial state. However the symmetrization principle discussed above showed that ρ_k should be replaced by

$$\rho_k \to \rho_k exp(\mp \mid m - n \mid x_k) \qquad (55)$$

where x_k is given by eq.(44) with ω replaced by ω_k. It is therefore necessary to introduce an alternative derivation which divides the energy transfer in a phonon creation and a phonon anhilation process and then introduce the correction (55). However, if we neglect the VV processes we would get the following effective hamiltonian from eq.(50)

$$H_{eff} = E_{int} + V_I^{(0)}(R) + \sum_k \omega_k^{-1} \epsilon_k(R,t) + \frac{P_R^2}{2M} \quad (56)$$

and the equations of motion are obtained using hamiltons principle, i.e.

$$\frac{dH_{eff}}{dt} = \frac{\partial H_{eff}}{\partial t} + \frac{\partial H_{eff}}{\partial R}\dot{R} + \frac{\partial H_{eff}}{\partial P_R}\dot{P}_R \quad (57)$$

yields:

$$\dot{R} = \frac{\partial H_{eff}}{\partial P_R} = \frac{P_R}{M} \quad (58)$$

and

$$\dot{P}_R = -\frac{\partial H_{eff}}{\partial R} = -\frac{\partial V_I^{(0)}}{\partial R} + \sum_k \omega_k^{-1}\frac{\partial \epsilon_k(R,t)}{\partial R} \quad (59)$$

Thus if we require $\dot{H}_{eff}=0$ we get:

$$\frac{\partial H_{eff}}{\partial t} = \sum_{k=1}^M \hbar\omega_k\dot{\rho}_k + \sum_k \omega_k^{-1}\frac{\partial \epsilon_k}{\partial t} = 0 \quad (60)$$

or

$$\frac{\partial \epsilon_k}{\partial t} = -\hbar\omega_k^2\dot{\rho}_k = -\omega_k\frac{d}{dt}\Delta E_k \quad (61)$$

where $\Delta E_k = \hbar\omega_k\rho_k$. Eq.(61) may be used to evaluate the effective potential

$$< V^{(1)} >= \sum_k \omega_k^{-1}\epsilon_k(R,t) \quad (62)$$

from ΔE_k. Thus if the symmetrization is included in ΔE_k we can use eq.(61) to calculate the "symmetrized" potential.

3 Calculation of ΔE_{int}

We assume that the surface phonons initially are in a distributed according to a distribution function $p_{n_k^0}$. For a Boltzmann distribution we would have:

$$p_{n_k^0} = (1 - z_k)z_K^{n_k^0} \quad (63)$$

where $z_k = exp(-\hbar\omega_k/kT_s)$. The energy transfer to the solid is given by:

$$\Delta E_{int} = \sum_k \sum_{n_k} \sum_{n_k^0} P_{n_k^0}(E_{n_k} - E_{n_k^0})P_{n_k^0 \to n_k} \tag{64}$$

where $P_{n_k^0 \to n_k}$ is the probability for a transition from quantum state n_k^0 to n_k in mode k. Within the HO approximation we have

$$E_{n_k} - E_{n_k^0} = \hbar\omega_k(n_k - n_k^0) \tag{65}$$

Using now that the excitation probability is:

$$P_{n_k^0 \to n_k} = exp(-\rho_k)\frac{n_k^0!}{n_k!}\rho_k^{n_k - n_k^0}[L_{n_k^0}^{n_k - n_k^0}(\rho_k)]^2 \tag{66}$$

and for de-excitation:

$$P_{n_k^0 \to n_k} = exp(-\rho_k)\frac{n_k!}{n_k^0!}\rho_k^{n_k^0 - n_k}[L_{n_k}^{n_k^0 - n_k}(\rho_k)]^2 \tag{67}$$

Inserting these expressions in (64) we get for excitation (+) and de-excitation (-)

$$\Delta E_{int}^{\pm} = \pm\sum_k \hbar\omega_k exp(\rho_k(1 + z_k)/(z_k - 1))T_k^{\pm} \tag{68}$$

where

$$T_k^{\pm} = \sum_{m=0}^{\infty} mI_m(t_k)z_k^{\mp m/2} \tag{69}$$

and

$$t_k = 2\rho_k\sqrt{z_k}/(1 - z_k) \tag{70}$$

The summation (68) and (69) may be carried out to give

$$\Delta E_{int} = \Delta E_{int}^+ + \Delta E_{int}^- = \sum_k \hbar\omega_k\rho_k \tag{71}$$

The symmetrization is now introduced by using ρ_k^+ in ΔE_{int}^+ and ρ_k^- in ΔE_{int}^-, where

$$\rho_k^{\pm} = \rho_k exp(\mp x_k m) \tag{72}$$

where m is the number of quanta in the excitation/de-excitation process. The correction is most important in the terms $T_k^{\pm}$, i.e. we use in (69)

$$t_k = 2\rho_k z_k^{1/2}exp(\mp x_k m)/(1 - z_k) \tag{73}$$

Once $\Delta E_{int}^{\pm}$ is obtained we return to eq.(61) and write (see eq.(52))

$$\epsilon_k(R,t) = V_k^{(1)}(R)\eta_k(t)\omega_k \tag{74}$$

where $\eta_k(t)$ is to be determined. From Eq.(61) we get:

$$V_k^{(1)}(R)\dot{\eta}_k(t) = -\frac{d\Delta E_k}{d\rho_k}\dot{\rho}_k \tag{75}$$

where

$$\Delta E_k = \Delta E_k^+ + \Delta E_k^- = \hbar\omega_k exp[\rho_k(1+z_k)/(1-z_k)](T_k^+ - T_k^-) \tag{76}$$

and where we by using eq.(32) and (35) get:

$$\dot{\rho}_k = \frac{1}{\hbar\omega_k}(I_{ck}sin(\omega_k t) + I_{sk}cos(\omega t))V_k^{(1)}(R) \tag{77}$$

Thus we have

$$\eta_k(t) = -\int dt'(\hbar\omega_k)^{-1}\frac{d}{d\rho_k}[\Delta E_k^+ + \Delta E_k^-] \times$$
$$\{I_{ck}(t')sin(\omega_k t') + I_{sk}(t')cos(\omega_k t')\} \tag{78}$$

where

$$I_{ck}(t) = \int_{t_0}^{t} dt'V_k^{(1)}(R(t'))cos(\omega_k t') \tag{79}$$

and

$$I_{sk}(t) = \int_{t_0}^{t} dt'V_k^{(1)}(R(t'))sin(\omega_k t') \tag{80}$$

Although the approach given above may seem complicated it has many advantageous features, they are:
1) The surface temperature is build into the potential
2) The classical trajectory is not tied to a specific quantum transition. The velocity symmetrization is included when defining the effective poential.
3) The time-consuming part, i.e. the evalutaion of the effective potential increases only linearly with the number of crystal atoms (N).
4) One avoids to treat some surface atoms in a special way (primary atoms) - all atoms are treated evenhandedly.
5) We avoid averaging over oscillator phase angles. Only the aiming point has to be chosen randomly. Thus quantities as energy accomodation converges rapidly (10-20 trajectories).
6) If necessary a phonon spectrum is also obtained i.e. information upon the most excited modes can be detected.

4 Quantum effects

As mentioned previously quantum dynamical effects may be important for estimation of e.g. the below the barrier tunneling contribution. We have previously introduced a quantum description of two degree's of freedom namely the hydrogen bonding coordinate and the z distance from the center of mass of the molecule to the surface. The remaining 4 variables (x,y), (θ, ϕ) and their conjugate momenta are treated classically. Here (x,y) denote the position of the center of mass at the surface and the polar coordinates θ and ϕ the orientation of the molecule axis in a coordinate system with z-axis normal to the surface. This gives 8 classical equations of motion, which are solved simultaneous with the time dependent Schrödinger equation for the 2 D wavepacket $\Psi(r, z, t)$, the equations for the phonon and electron-hole pair excitation. The various motions are coupled together in a self-consistent manner using an SCF type wavefunction (see e.g. [14]). In the present calculations only the lower adiabatic surface (W_{11}) is used when propagating the wave-packet. The propagation is carried out using a 2D grid representation of the wavefunction, a fast Fourier transform method for evaluating the kinetic energy terms and a Split operator technique for the time propagation [20]. Thus the hamiltonian is divided in the following fashion:

$$H = T_{cl} + T_q + H_0 + W_{11} \tag{81}$$

where

$$T_{cl} = \frac{1}{2M}(P_x^2 + P_y^2) + \frac{1}{2m <r>^2}(p_\theta^2 + \frac{1}{sin^2\theta}p_\phi^2) \tag{82}$$

is the classical part of the kinetic energy of the hydrogen molecule and $<r>$ the expectation value of the bond-length, M and m the total and the reduced mass of the diatomic molecule respectively. The momenta for the (x, y) motion are P_x and P_y and for the rotational motion p_θ and p_ϕ. The vibrational motion and the motion towards the surface are quantized, i.e.

$$T_q = -\frac{\hbar^2}{2M}\frac{\partial^2}{\partial z^2} - \frac{\hbar^2}{2m}\frac{\partial^2}{\partial r^2} \tag{83}$$

H_0 is the hamiltonian for the solid and W_{11} the lower adiabatic potential expressed e.g. by the EDIM-theory. The wavepacket is initialized as

$$\Psi_i(r, z, t = 0) = f_n(r)g(z, t = 0) \tag{84}$$

a product of a Morse-vibrational wavefunction and a Gaussian translational wavepacket:

$$g(z, t = 0) = N exp(-\frac{(z - z_{foc} - \hbar k_0 t_0/M)^2}{4\Delta z^2 - 2i\hbar t_0/M} - ik_0 z) \tag{85}$$

where z_{foc} is the focusing distance, $p_0 = \hbar k_0$ the average initial momentum, N a normalization constant and $t_0 = (z_0 - z_{foc})M/\hbar k_0$. This wavepacket will (without potential) contract untill it arrives at the focusing point and then expand. The grid necessary for the propagation can then be made smaller. If the width of the wavepacket is chosen such that the spread in momentum space $\Delta p = \frac{\hbar}{2\Delta z}$ is small compared to the average momentum p_0, the sticking probability at p_0 can be obtained simply by calculating the the the flux over a line at a value of r just before the wavepacket is absorbed by an optical potential placed near the borders of the grid. If a broad distribution in momentum space is used, then the energy resolved sticking probabilities can be obtained by projecting the scattered wavefunction onto asymptotic functions of the type

$$\phi_{nk}(r, z) = f_n(r)exp(ikz) \tag{86}$$

5   Electron-hole pair excitation

From experimental spectroscopic data and/or ab initio calculations we get information on the electronic adiabatic surface, i.e. the surface where the electrons have redistributed themselves according to the nuclear structure in the usual Born-Oppenheimer sense. We use this information to fit the lower adiabatic potential energy surface. This surface includes phenomena as charge-transfer from the metal to the molecule which then obtains an excess negative charge. We have fitted an analytical expression to this excess charge using the numbers obtained by Mullikan analyses of Madhavan and Whitten. The excess charge is about -0.15e at $z = 1$ Å, and $r \sim 0.74$ Å. It drops rapidly with z (as z^{-3}) and is therefore neglible for physisorbed molecules. Hence the coupling to electron-hole pair excitation is significant only for molecules coming close to the surface ($z < 1$ Å).

On top of the electronically adiabatic processes we imagine that a number of nonadiabatic or inelastic events can occur. These are: Nonadiabatic electronic transitions from the ground electronic state to excited states, inelastic excitation of the metal phonons, vibrational and rotational excitation in the molecule and inelastic electronic excitations in the metal, i.e. excitation of electrons from states below to states above the Fermi-level. The calculations used are based upon this separation in electronically adiabatic effects included through the potential energy surface and inelastic or non-adiabatic effects included through the dynamics. But it could of course be argued that the effect of some of these processes are hidden in the applied potential energy surface and that one should solve the nuclear and electronic dynamical motion simultaneously rather than invoke a Born-Oppenheimer approximation and then add the deviations from this afterwards. The reason for doing so, is that it is not known at present how to include the influence of electronic non-adiabadicity on the nuclear dynamics in such calculations. This and the fact that inclusion of quantum nuclear tunneling requires knowledge of the global potential energy surface are the reasons for dividing the processes in the fashion suggested here. Thus we imagine that the inelastic (non-adiabatic) electronic excitation of the metal-electrons occur through an interaction between the excess charge on the molecule and the electrons in the metal. The interaction is described by a screened Coulomb potential, i.e.

$$V(R, z, r) = \frac{e^2 \delta(z, r)}{R} exp(-k_c R) \qquad (87)$$

where r is the bond and z the distance to the surface. The excess charge on the atom/molecule is $e\delta(z, r)$. For the present system $H_2 - Cu$ this excess charge has been determined by Mullikan analyses of the ab initio data [1]. The distance between the charge on the particle and an electron in the metal is denoted by R. Introducing the position of the electron in the metal as (x, y) we expand the above potential in a Fourier serie as:

$$\frac{e^2 \delta(z, r)}{\sqrt{x^2 + y^2 + \tilde{z}^2}} exp(-k_c \sqrt{x^2 + y^2 + \tilde{z}^2}) = \sum_m \sum_n A_{nm} exp(im\tilde{x} + in\tilde{y})$$

$$(88)$$

where $\tilde{x} = x/a$, $\tilde{y} = y/a$, a the lattice constant and $\tilde{z} = z - z_{jell}$. z_{jell} is the position of the "jellium" edge above the surface. The Fourier coefficients A_{nm} can be obtained by using that the Kronecker delta function can be written as:

$$\delta_{nm} = \frac{1}{2\pi} \int_{-\infty}^{\infty} dx exp(i(m-n)x) \qquad (89)$$

We then obtain:

$$A_{nm} = \frac{e^2 \delta(z,r)}{2\pi a} \frac{1}{\sqrt{n^2 + m^2 + (ak_c)^2}} exp(-\tilde{z}\sqrt{k_c^2 + (n/a)^2 + (m/a)^2})$$
$$(90)$$

The reason for expressing the potential in this fashion is that it is now easy to evaluate the matrix-elements over the 2D Bloch-wavefunctions used as the single particle electronic wavefunctions. Within a 2-band model we have:

$$\psi_{k_x} = C(k_x - G)exp(i(k_x - G)x) + C(k)exp(ik_x x) \qquad (91)$$

and a similar expression is obtained for the y-component. Here $G = 2\pi/a$ and $C(k)$ the eigenvector components obtained by solving the Schrödinger equation for the two band model. Using the δ function normalization we get

$$< \psi_{k_x'} \mid exp(im\tilde{x}) \mid \psi_{k_x} >= \begin{cases} C(k_x' - G)C(k_x - G)\delta(\frac{m}{a} + k_x - k_x') \\ C(k_x')C(k_x)\delta(\frac{m}{a} + k_x - k_x') \\ C(k_x' - G)C(k_x)\delta(\frac{m}{a} + k_x - k_x' + G) \\ C(k_x')C(k_x - G)\delta(\frac{m}{a} + k_x - G - k_x') \end{cases} \qquad (92)$$

These matrix-elements are used in single particle electron model, i.e. for N electrons we have the hamiltonian given as:

$$\hat{F}_1 = \sum_i \epsilon_i C_i^+ C_i + \sum_{ij} < i \mid \hat{H}(\mathbf{r}) \mid j > C_i^+ C_j \qquad (93)$$

where C_i^+ and C_i are creation/anhilation operators for the electron in state $\mid i >= \mid k_x k_y >$ and ϵ_i the energy of state i. Within this second-quantized electron model it is possible to solve the timedependent Schrödinger-equation:

$$i\hbar \frac{\partial U}{\partial t} = \hat{F}_1 U \qquad (94)$$

where U is the evolution operator for the electrons. This solution, which can be carried out in operator space will define the electronic wavefunction

$$\Psi(t) = U(t, t_0)\Psi(t_0) \qquad (95)$$

where $\Psi(t_0)$ is the initial wavefunction. The wavefunction is used when defining the self-consistent field effective potential used for the dynamical motion of the incoming particles, i.e.

$$H_{eff}^e = < \Psi(t) \mid \hat{F}_1 \mid \Psi(t) > \tag{96}$$

This effective potential can also be evaluated using operator-algebraic methods [12]. The result is:

$$H_{eff}^e = \sum_k \epsilon_k n_k + \sum n_k H_{kk} + \sum_{k \neq l} V_{kl}(Q_{kl}^* n_l + Q_{lk} n_k) + \sum_{kl} \sum_i Q_{ki}^* Q_{li} n_i V_{kl} \tag{97}$$

where the last term constitute an effective potential V_{eff}^e and where the timedependent functions Q_{lk} are obtained by solving a set of coupled equations

$$i\hbar \frac{d\mathbf{R}}{dt} = \mathbf{HR} \tag{98}$$

with $\mathbf{Q=R\text{-}I}$ and $\mathbf{R}(t_0)=\mathbf{I}$ (a unit matrix). Up to this point the solution of the time dependent Schrödinger equation is general. Thus eq.(93) involves the use of one- electron wavefunctions. Here and in our previous work we have used a simple Bloch-potential model in order to obtain these wave functions. However more elaborate calculations using the Kohn-Sham equations [19] can be introduced at this point. The effective Hamiltonian (97) is used to couple the inelastic electronic transitions in the metal to the motion of the incoming molecule. The hamiltonian depends upon time through the "classical" variables $z(t)$ and $r(t)$ and the quantum mechanical excitation functions Q_{kl}. In the wave-packet treatment of the two degree's of freedom we introduce the expectation value, i.e. $z(t) = < \Psi \mid z \mid \Psi >$ and $r(t) = < \Psi \mid r \mid \Psi >$.

We notice that in a first order treatment of the electron-hole processes, i.e. creation of a particle in state k and a hole in state l the probability for this event is $\sim \mid Q_{kl} \mid^2$. However a first order theory for the excitation processes would fail to give a realistic description of the electronic friction and grossly overestimate the effect. We find that the electrons can jump back and forth between states below and states above the fermi level leaving the net excitation at the energies considered here modest compared to what a first-order theory would give.

In order to include the electronic inelastic effects on the collision it is necessary to solve the coupled equations (98). Although it is possible

to solve this set of equations including a large number of electronic
states (of the order 1000-2000) we have chosen to solve the equations
to infinite order in the time-dependent "sudden" limit. Furthermore
a situation, where states up to the Fermi-level are occupied is considered. Hence in the operator $C_k^+ C_l$ and the matrix element V_{kl}, k
denotes a state above and l a state below the Fermi level F. These
two approximations allows an analytical solution of eq.(98), and gives
a very compact expression for the effective potential coupling the motion of the incoming particles to the inelastic electronic processes in
the metal, i.e. for each atom we get [12] the following effective potential arising as an effect of the dynamical coupling between the motion
of the atom and the excitation processes in the solid induced by this
motion:

$$V_{eff}^e = 2 \sum_{k=1}^{F} j_0(2\sqrt{A_{kk}^{(2)}}) \sum_{l=F+1}^{N} V_{kl}(r(t), z(t)) \frac{1}{\hbar}$$
$$\times \int_{t_0}^{t} dt' V_{kl}(r(t'), z(t')) sin(\omega_{kl}(t' - t)) \tag{99}$$

where j_0 is a Bessel-function and

$$A_{kk}^{(2)} = \sum_{l} \mid A_{kl}^{(1)} \mid^2 \tag{100}$$

$$A_{kl}^{(1)} = \frac{1}{\hbar} \int_{t_0}^{t} dt' V_{kl}(r(t'), z(t')) exp(i\omega_{kl} t') \tag{101}$$

and $\omega_{kl} = (\epsilon_k - \epsilon_l)/\hbar$. Thus the total effective potential consists of the
term above and a similar potential arising from the coupling to the
surface phonons, i.e. the total potential in which the molecule moves
is a sum of a static (defined by the lowest eigenvalue to the secular
equation [4]) and two dynamical potentials, i.e:

$$V_{tot} = V_{static} + V_{eff}^{el} + V_{eff}^{ph} \tag{102}$$

where the last term is the effective potential arising from the coupling
to the phonons. The advantage of the present treatment of both the
phonon and the electron-hole pair excitation is that equations of motion are solved selfconsistently, i.e. the two effective potentials are
updated along the trajectory and hence the influence of the inelastic
surface processes is coupled directly to the motion of the molecule.

Table 1: Overview of systems and phenomena studied with the semi-classical molecule surface collision model

System	Phenomena	Reference
Ar+W(110)	Sticking/energy transfer	[6]
CO+Pt(111)	Sticking/energy transfer	[7]
C+O(ads)+Pt(111)	Recombination	[8]
O+CO(ads)+Pt(111)	Recombination	[9]
He+NaCl	Phonon/diffraction	[11]
H_2,D_2+Cu(111)	Dissociation, stikcing	[15]
$\hbar\omega$+NaCl	Laser evaporation	[10]
CO+Pt(111)	Role of electron-hole pair excitation	[12]
$\hbar\omega$+CO+Pt(111)	Laser induced energy transfer	[13]
N_2+Re	Dissociation, non-adiabatic processes	[14]
CO+Cu(111)	Diffusion/energy transfer	[16]
CO+Pt(111)	Diffusion/energy transfer	[17]

6 Review of results

The semi-classical model described above has been used to study a number of surface-dynamical processes (see table 1). More specifically it has been used for the calculation of energy-accomodation, sticking probabilities, diffusion constants, non-adiabatic processes, recombination and dissociation processes, laser induced processes etc. (see table 1). Usually the intermolecular potentials have been fitted to available spectroscopic information on surface bound atoms and molecules. However recently also ab/initio calculations using density functional theory have given insight in the barrier heights for dissociative chemisorption. These ab/initio calculations are however only available at a limited number of sites and geometries. It is therefore important to have a realistical analytical interpolation scheme to use for fitting these data in order to be able to solve the dynamical problem in the full dimensionality. Such a scheme is provided by the EDIM-potential mentioned previously. So far the EDIM-parameters have been estimated only for the interaction of hydrogen with surfaces at Ni and Cu.

The EDIM-surface has recently been used to calculate probabilities for dissociative chemisorption and for recombination reactions between adsorbed hydrogen and gas-atoms [21]. The tables 2 and 3 show

Table 2: Hydrogen scattered from a Cu(111) surface. The effect of phonon and electron-hole pair excitation on dissociative sticking probabilities (P_d), rotational and vibrational excitation E'_r and E'_v. The initial kinetic energy is 0.558 $\hat{\epsilon}$, the final E'_k and the initial vibrational rotational state $(v,j)=(0,0)$ and the approach angles $(\theta,\phi)=(0,0)$. The energy transfer to the phonons is E_{int} and the surface temperature is 300 K.

P_d	E'_k	E'_r	E'_v	E_{int}
0.29[a]	.0.505	0.103	0.211	0.00
0.34[b]	0.436	0.092	0.204	0.090
0.33[c]	0.433	0.089	0.203	0.095

Energies in units of $\hat{\epsilon}=100$ kJ/mol.
a) Without phonon coupling
b) Without electron hole pair excitation
c) With electron hole pair coupling

that the influence of phonon-coupling and electron coupling changes the sticking probability slightly. Since the hydrogen molecule delivers about 15 % of the kinetic energy to the phonons, the kinetic energy of the scattered (i.e. the non-dissociative) trajectories is somewhat smaller if the inelastic energy transfer processes to the solid is included. We notice that the inclusion of electron-hole pair excitation has an effect on the phonon-excitation - namely to increase it by about 5 % . The reason for this is that the energy transfer to the electrons slightly increases the life-time at the surface and thereby increases the phonon-coupling.

We also notice that the classical treatment of the vibrational/rotational degree's of freedom of hydrogen has the usual zero point vibrational energy problem - namely that this energy in principle is available to the other degree's of freedom, i.e. it is not conserved. Another aspect not accounted for is the tunneling effect. Both of these shortcomings will be dealt with if the quantum wavepacket description (section 4) is introduced. The price is however a significant increase in computer time. Though time-consuming the storage requirements are modest and the calculations can be carried out even on workstations. Table 3 shows that the phonon-excitation increases for deuterium as compared with hydrogen and that the inclusion of electron-hole pair excitation increases the phonon-coupling and the sticking probability. This is

Table 3: Energy transfer and dissociative sticking probabilities for D_2-Cu(111) collisions. The initial vibrational/rotational states are $(v, j)=(0,0)$ and approach angles $(\theta, \phi)=(0,0)$. The inital kinetic energy is $0.558\ \hat{\epsilon}$ and incident angles $(\theta, \phi)=(0,0)$.

P_d	E'_k	E'_r	E'_v	E_{int}
0.21^a	0.507	0.077	0.158	0.00
0.30^b	0.367	0.065	0.126	0.189
0.34^c	0.354	0.070	0.114	0.202

a) Without electron-hole pair and phonon-excitation.
b) Without electron-hole pair coupling.
c) With phonon and electron-hole pair coupling.

due to the stronger coupling, which in turns increases the life-time at the surface and hence the phonon-interaction. The strong interaction also makes part of the zero-point vibrational energy available for the dissociation process, i.e. the dissociative sticking probability increases.

Thus the problems with conserving the zero point energy for the reflected trajectories indicate that the zero-point energy is available for the other degree's of freedom in a non-physical way, i.e. also for surpassing the barrier. In order to confirm this conjecture one should perform a quantum mechanical calculation, in which at least the motion in r and z is quantized. This calculation is possible using the semi-classical methodology given above. We notice that the quantum sticking is a factor of about 3-4 smaller than the corresponding classical sticking probability. If quantum tunneling was the only factor missing from the classical calculations, then one would have expected the opposite trend. However, the failure of properly conserving the zero-point vibrational energy of hydrogen makes an additional $0.266\ \hat{\epsilon}$ of energy available for surpassing the barrier. This and the possibility for "above barrier quantum reflection" makes us conclude that the calibration of the barrier by comparison between experiment and classical trajectory calculations may not be meaningful. We notice (see table 4), that the quantum-sticking probability is a factor of about 5 smaller than the classical at $0.558\ \hat{\epsilon}$ and that the difference decreases with energy. This demonstrates that classical trajectory calculations at low energies are questionable unless something is done to preserve the zero-

Table 4: Comparison of dissociative sticking of hydrogen colliding with a Cu(001) surface obtained quantum mechanically and classically. The initial vibrational/rotational states are (v,j)=(0,0), approach angles (θ, ϕ)=(0,0) and surface temperature 300 K. N is the number of trajectories.

E_k	P_d(classical)	P_d(quantum)	N	classical/quantum
0.558	0.30	0.062	29	4.8
0.635	0.41	0.12	44	3.4

Energies in units of 1 $\hat{\epsilon}$=100 kJ/mol.

Table 5: Probabilities for recombination (P_r), i.e. for the process H+H$_{ads}$ →H$_2$ on a Cu(001) surface. The surface temperature is 300 K. E_{int} is the energy transfer to the surface phonons and P_a the adsorption probability. The energies are in units of 100 kJ/mol. N_t the number of trajectories.

Kinetic energy	P_r	E_{int}	P_a	N_t
0.045	0.011	1.07	0.22	1415
0.124	0.014	1.47	0.21	1388
0.496	0.036	0.51	0.28	1218
1.98	0.051	0.75	0.096	2700
4.46	0.047	0.61	0.010	1397

point energy (a quantity which itself changes when the molecule approaches the surface). The 2-D quantum wavepacket calculations are straightforward but unfortunately much more time-consuming (about 100 times) than the corresponding classical calculations. Thus each trajectory takes about 400 minutes CPU-time on a Convex 220 computer. The quantum calculations give, due to their proper treatment of the molecule vibrational motion lower sticking probabilitites than the corresponding classical calculations. This seam to indicate that the easy to apply classical treatment of the dynamics of the gas phase molecule must be abandoned near the dissociation threshold.

The calibration of the barrier was based upon inspection of potential contour plots. However the experimentally determined barrier is a "dynamical barrier" and our quantum calculations seem indicate that one in order to obtain a dynamical barrier of about 0.5 eV in agreement withe experimental data [3] should use a static potential barrier lower than this value.

Table 5 shows the probabilities for recombination between an incom-

Table 6: Recombination rates for the reaction H+H(ads)→ H_2 as a function of gas-temperature, the surface temperature is 300 K. The last column gives the contribution to the rates for molecules formed in vibrational states 0,1 and 2.

Temperature	Rateconstant cm^3/sec	v=0,1,2 fraction
200 K	$1.3 \ 10^{-13}$	0.85
300 K	$3.0 \ 10^{-13}$	0.82
400 K	$5.0 \ 10^{-13}$	0.78
500 K	$6.9 \ 10^{-13}$	0.78

ing and an adsorbed atom. The reaction probability is small from about 1 % at low to about 5 % at higher energies. The reason for this difference is the longer interaction time and thereby coupling to the surface phonons at lower collision energies. The trajectories are said to lead to adsorption if the energy transfer to the surface is larger than the available energy, i.e. the binding energy of a hydrogen molecule +the intial kinetic energy. From the reactive trajectories it is possible to calculate the final state distribution and thereby the rate constants for forming a hydrogen-molecule in a specific quantum vibrational state. Table 6 gives some preliminary values for the recombination rates on a Cu(001) surface. Although the reaction has no barrier we see that the rateconstant increases modestly with temperature. The reason for this is that the competing process - namely the adsorption decreases with energy (see table 5) and hence the recombination rate will increase slightly. The final vibrational state distribution is not in equilibrium with the surface temperature. This is also not expected since the molecules are formed by a direct mechanism. But the main part of the molecules are formed in the low-lying vibrational states (see table 6).

7 Conclusion

We have seen that even for hydrogen the inclusion of the coupling to the surface phonons is essential and especially for recombination reaction one would get a completely unrealistical picture if the phonon processes are neglected.

Contrary to the phonon processes, the treatment of the electronic processes is more problematic. We have divided the electronic processes (electronic friction) in adiabatic and non-adiabatic ones. The adiabatic are included when defining the potential energy surfaces and non-adiabatic effects such as "Landau-Zener" transitions in the gasphase and inelastic electron-hole pair excitation of the metal electrons from levels below the Fermi-level are added as "additional" electronic frictional effects. However some of these effects might be included if potential parameters are fitted to experimental data, where the processes have so to speak occured. But since the inelastic electronic effects are more important at higher collision energies and the experimental information as e.g. spectroscopic information is coming from low temperature (thermal) data - this problem is not expected to be significant. The main conclusion is however that unless care is taken to include all the relevant dynamical processes the comparison between theory and experiment has little meaning - and especially will the calibration of the potential barrier by comparing with experimental data be of little value if inadequate dynamics has been used for the convolution of the potential energy surface to experimental observables.

References

[1] See e.g. P. Madhavan and J. L. Whitten,"Theoretical studies of the chemisorption of hydrogen on copper", **J. Chem. Phys.** 77:2673(1982).

[2] J. A. White and D. M. Bird, **Chem. Phys. Lett.** 213:422(1993); G. Wiesenekker, G. J. Kroes, E. J. Baerends and R. C. Mowrey, "Dissociation of H_2 on Cu(100): Dynamics on a new two-dimensional potential energy surface", **J. Chem. Phys.** 102:1995,3873

[3] C. T. Rettner, D. J. Auerbach, and H. A. Michelsen, Phys. Rev. Lett. 68:1164(1992); C. T. Rettner, H. A. Michelsen and D. J. Auerbach, "From Quantum-state-specific Dynamics to Reaction Rates: The Dominant Role of Translational Energy in Promoting the Dissociation of D_2 on Cu(111) under Equilibrium Conditions", **Faraday Discuss.** 96:17(1993).

[4] T. N. Truong, D. G. Truhlar and B. C. Garrett, "Embedded Diatomics-in-Molecules: A Method To Include Delocalized Elec-

tronic Interactions in the Treatment of Covalent Chemical Reactions at Metal Surfaces", **J. Phys. Chem.**:93,8227(1989).

[5] London, F. Z. Elektrochem. **1929**, 35,552; Eyring, H. Polanyi, M. Z. Physik. Chem. **1931**, B12,279; Sato, S. J. Chem. Phys. **1955**, 23,592.

[6] G. D. Billing, "On a semiclassical approach to energy transfer by atom/molecule-surface collisions", **Chem. Phys.** 70:223(1982)

[7] G. D. Billing, "Inelastic scattering and chemisorption of CO on a Pt(111) surface", **Chem. Phys.** 86:349(1984).

[8] G. D. Billing and M. Cacciatore, "Semiclassical calculation of the reaction probability for the process $C + O \rightarrow CO$ pn a $Pt(111)$ surface", **Chem. Phys. Lett.** 113:23(1985)

[9] G. D. Billing and M. Cacciatore, "Semiclassical calculation of the probability for formation of CO_2 on a $Pt(111)$ surface", **Chem. Phys.** 103:137(1986).

[10] G. D. Billing, "Laser fragmentation of solids", **Chem. Phys.** 115:229(1987).

[11] G. D. Billing, "Influence of Phonon Inelasticity upon Atom-solid Scattering Intensities", **Surf. Sci.** 203:257(1988).

[12] G. D. Billing, "Electron-hole pair versus phonon excitation in molecule-surface collisions", **Chem. Phys.** 116:269(1987).

[13] G. D. Billing, "Laser-assisted molecule-surface collisions", **Surf. Sci.** 195:187(1988)

[14] N. E. Henriksen, G. D. Billing and F. Y. Hansen, "Dissociative chemisorption of N_2 on Rhenium: Dynamics at high impact energies", **Surf. Sci.** 227:224(1990); G. D. Billing, A. Guldberg, N. E. Henriksen and F. Y. Hansen, "Dissociative chemisorption of N_2 on rhenium: dynamics at low impact energies", **Chem. Phys.** 147:1(1990)

[15] M. Cacciatore and G. D. Billing, "Dynamical relaxation of $H_2(v,j)$ on a copper surface", **Surf. Sci.** 232:35(1990); G. D. Billing and M. Cacciatore, "Semi-classical Multi-dimensional study of the inelastic and reactive interaction of $D_2(v,j)$ with a

non-rigid $Cu(111)$ surfcae", **Faraday Disc.** 96:33(1993); Lichang Wang, Qingfeng Ge and G. D. Billing, "Molecular dynamics study of H_2 diffusion on a Cu(111) surface", **Surf. Sci.** 301:353(1994).

[16] Qingfeng Ge, Lichang Wang and G. D. Billing, "Inelastic scattering and chemisorption of CO on a Cu(111) surface", **Surf. Sci.**, 277:237(1992).

[17] Lichang Wang, Qingfeng Ge and G. D. Billing, "Study of the surface diffusion of CO on Pt(111) by MD simulation", **Surf. Sci. Lett.** 304:L413(1994)

[18] G.D. Billing, "The dynamics of Molecule-Surface Interaction", **Comp. Phys. Rep.**, 12:383(1990).

[19] W. Kohn and L. Sham, **Phys. Rev. A** 140:1133(1965).

[20] M. D. Feit, J. A. Fleck Jr. and A. J. Steiger, **Comput. Phys.** 47:412(1982).

[21] G. D. Billing, "Semiclassical formulation of molecule surface scattering using an EDIM potential", **J. Phys. Chem.** (in press).

THERMODYNAMIC EQUILIBRIUM OF MULTI-TEMPERATURE GAS MIXTURES

D. GIORDANO

European Space Research & Technology Center
P. O. Box 299, 2200 AG Noordwijk, The Netherlands

Abstract. This theoretical paper deals with the thermodynamic equilibrium of multi-temperature gas mixtures. After a brief review of basic notions and methods relative to the chemical equilibrium of gas mixtures in complete thermal equilibrium, the fundamental relations energy and entropy describing the thermodynamics of gas mixtures in disequilibrium with respect to energy and mass exchanges are formally introduced. Possible constraints arising from the subset of state parameters associated with the molecular degrees of freedom are recognized and the importance of the role they play in determining the thermodynamic equilibrium of the system is discussed. In particular, the influence that a partially constrained thermal equilibrium exercises on the equations governing the associated chemical equilibrium is made evident by performing the equilibrium analysis for the cases of entropic and energetic freezing of the molecular degrees of freedom; the analysis shows that the minimization/maximization of energy/entropy may not necessarily lead to the vanishing of the chemical reaction affinities and that, therefore, the uniqueness of the chemical equilibrium equations is lost when the thermal equilibrium is partially constrained. The equivalence of the fundamental relations energy and entropy to determine the equilibrium conditions is shown to be maintained also in multi-temperature circumstances; the problem of the lack of such equivalence for the Helmholtz and Gibbs potentials is briefly mentioned. As an application, the chemical equilibrium of a two-temperature partially ionized gas is considered with the purpose to show how the uncertainty associated with the two-temperature Saha equation can be resolved in the framework of the theory proposed in this work.

M. Capitelli (ed.), Molecular Physics and Hypersonic Flows, 259–280.
© 1996 *Kluwer Academic Publishers. Printed in the Netherlands.*

List of Symbols

Symbols without Running Subscripts

F	Helmholtz potential
f	number of frozen chemical reactions
G	Gibbs potential
H	enthalpy
h	Planck constant
k	Boltzmann constant
$\mathcal{L}$	Lagrangian function
$\mathcal{M}_A, \mathcal{M}_{A^+}$	atom, ion molecular mass
$\mathcal{M}_e$	electron mass
m	total mass
N_A, N_{A^+}	atom, ion number density
N_A°	low temperature atom number density
N_{e^-}	electron number density
n	number of components
p	thermodynamic pressure
q_A, q_{A^+}	atom, ion electronic partition function
R_G	universal gas constant, 8.3144 J/K
r	number of independent chemical reactions
S	entropy
T	temperature (thermal equilibrium)
T_e	electron temperature
T_h	heavy species temperature
U	energy
V	volume
θ	arbitrary reference temperature
$\lambda_S, \lambda_S', \lambda_U$	Lagrange multipliers
φ, Φ, Ψ	entropic potentials

Symbols with Running Subscripts

$\mathcal{A}_k$	chemical affinity
$\overline{\mathcal{A}}_k$	no standard name[39]
ℓ_ϵ	number of independent degrees of freedom
$\overline{\ell}_\epsilon$	number of (entropically/energetically) frozen degrees of freedom
M_ϵ	molar mass
$\mathcal{M}_\epsilon$	molecular mass
m_ϵ	mass
m_ϵ°	initial mass
N_ϵ	number density
$p_{\epsilon\delta}$	partial pressure

$q_{\epsilon\delta}$	partition function
$S_{\epsilon\delta}$	entropy
$T_{\epsilon\delta}$	temperature
$U_{\epsilon\delta}$	energy
$\varepsilon_{\epsilon\delta,i}$	quantum state energy
μ_ϵ	chemical potential
$\mu_{\epsilon\delta}$	contribution to chemical potential
$\nu_{\kappa\epsilon}$	stoichiometric coefficient
ξ_κ	progress variable

Running Subscripts

δ	degrees of freedom
ϵ	components
κ	chemical reactions
i	quantum states

Mathematical Operators

$(dZ)_{X,Y}$	differential of Z with X, Y constant
$\displaystyle\sum_{\epsilon,\delta}^{\mathrm{nf}}$	summation extended to non-frozen degrees of freedom
$\displaystyle\sum_{\epsilon,\delta}^{\mathrm{f}}$	summation extended to frozen degrees of freedom
$\displaystyle\sum_{\kappa}^{\mathrm{nf}}$	summation extended to non-frozen chemical reactions

1. Introduction

The concept of thermodynamic equilibrium has always been a central subject in any theorization[1-4] of thermodynamics. In the case of gas mixtures in complete thermal equilibrium, and when electric and magnetic phenomena are absent, thermodynamic equilibrium becomes synonymous with chemical equilibrium; as such, the subject has been exhaustively dealt with in the literature.[1-17] On the contrary, the state of the art relative to the thermodynamic equilibrium of multi-temperature gas mixtures, comprehensively reviewed by Eddy and Cho,[18] appears rather unsatisfactory. The treatment proposed by Napolitano[4] seems to be the only example dealing with the subject in a rigorous and consistent manner; unfortunately, such a theory appears incomplete because it misses the recognition of the influence that a partially constrained thermal equilibrium can have on the equations governing the associated chemical equilibrium.

The purpose of this theoretical paper is to briefly review the basic notions relative to the chemical equilibrium of the gas mixtures in complete thermal equilibrium, to revise the concept of thermodynamic equilibrium for multi-temperature gas mixtures and to complete Napolitano's theory with regard to the chemical equilibrium of such gas mixtures in situations of partially constrained thermal equilibrium.

2. Gas Mixtures in Complete Thermal Equilibrium

When the gas mixture is in complete thermal equilibrium,[1-4] there is no need to specify how energy and entropy are distributed among the molecular degrees of freedom. One needs only to know the functional dependence between their total amounts

$$U = U\,(S, V, \text{ all } m_\epsilon) \qquad \text{or} \qquad S = S\,(U, V, \text{ all } m_\epsilon)$$

and a single temperature

$$T = \left(\frac{\partial U}{\partial S}\right)_{V,\text{ all } m_\epsilon} \qquad \text{or} \qquad \frac{1}{T} = \left(\frac{\partial S}{\partial U}\right)_{V,\text{ all } m_\epsilon}$$

is sufficient to characterize the thermal behaviour of the mixture.

The conditions of chemical equilibrium, i.e. the equilibrium values of the component masses, are found from the minimization/maximization[19] of the fundamental relation energy/entropy or of the corresponding thermodynamic potentials. In this regard, the equivalence of energy and entropy is guaranteed from the following relation

$$(dU)_{S,V} = -T(dS)_{U,V} = \sum_{\epsilon=1}^{n} \mu_\epsilon dm_\epsilon$$

Similar relations exist for the energetic potentials

$$(dU)_{S,V} = (dF)_{T,V} = (dH)_{S,p} = (dG)_{T,p}$$

and for the entropic potentials[20]

$$(dS)_{U,V} = (d\varphi)_{1/T,V} = (d\Phi)_{U,p/T} = (d\Psi)_{1/T,p/T}$$

Any fundamental relation can, therefore, be used for the purpose of finding the chemical equilibrium composition provided that the appropriate couple of state parameters is held constant. The analysis leads invariably to the condition

$$\sum_{\epsilon=1}^{n} \mu_\epsilon dm_\epsilon = 0 \qquad (1)$$

governing the equilibration of the mass exchanges. The further elaboration of the condition (1) to compute the equilibrium composition is accomplished according to two different, although supposed necessarily equivalent, classes of methods.

The chemical affinity methods are based on the introduction of the r independent chemical reactions which can take place in the gas mixture; then, the variations[21] of the component masses are determined by the progress variables ξ_κ of the reactions

$$m_\epsilon = m_\epsilon^\circ + \sum_{\kappa=1}^r \xi_\kappa \nu_{\kappa\epsilon} M_\epsilon \tag{2}$$

The initial masses m_ϵ° must be considered assigned in (2). These linear variations ensure the conservation of the total mass and the conservation of the atomic species. The condition (1), when account is taken of (2), leads to the vanishing of the chemical reaction affinities

$$\sum_{\epsilon=1}^n \nu_{\kappa\epsilon} M_\epsilon \mu_\epsilon = 0 \tag{3}$$

The r equations (3), in turn, originate the well known law of mass action[22] and its equilibrium constants.

The other class of methods is based on the *a priori* imposition of the atomic species conservation. However, this idea applies only when the chemical equilibrium is unconstrained, i.e. when there is no restriction affecting the variations of the progress variables; in such a case, the knowledge of the chemical reactions is not even required. The use of the atomic species conservation as an *a priori* constraint represents the core of what is known in the literature as the free energy[23] minimization method.[8] Liu and Vinokur[17,24] have shown how to unify the two classes of methods.

3. Multi-Temperature Gas Mixtures

In a multi-temperature situation,[4] the molecular degrees of freedom are in disequilibrium with respect to the exchanges of energy and entropy. In this case, the energies $U_{\epsilon\delta}$ distributed over and the entropies $S_{\epsilon\delta}$ associated with the molecular degrees of freedom of each component must be introduced and accounted for separately. They are related as

$$U_{\epsilon\delta} = U_{\epsilon\delta}(S_{\epsilon\delta}, V, m_\epsilon) \tag{4}$$

or

$$S_{\epsilon\delta} = S_{\epsilon\delta}(U_{\epsilon\delta}, V, m_\epsilon) \tag{5}$$

The first-order partial derivatives of the functions (4) and (5) define the contributions, associated with the degrees of freedom, to the state equations. In the energetic formulation, the following definitions of temperature, partial pressure and partial chemical potential are in force

$$T_{\epsilon\delta} = \left(\frac{\partial U_{\epsilon\delta}}{\partial S_{\epsilon\delta}}\right)_{V,m_\epsilon}$$

$$-p_{\epsilon\delta} = \left(\frac{\partial U_{\epsilon\delta}}{\partial V}\right)_{S_{\epsilon\delta},m_\epsilon}$$

$$\mu_{\epsilon\delta} = \left(\frac{\partial U_{\epsilon\delta}}{\partial m_\epsilon}\right)_{S_{\epsilon\delta},V}$$

In the alternative, and equivalent, entropic formulation, the corresponding definitions read

$$\frac{1}{T_{\epsilon\delta}} = \left(\frac{\partial S_{\epsilon\delta}}{\partial U_{\epsilon\delta}}\right)_{V,m_\epsilon}$$

$$\frac{p_{\epsilon\delta}}{T_{\epsilon\delta}} = \left(\frac{\partial S_{\epsilon\delta}}{\partial V}\right)_{U_{\epsilon\delta},m_\epsilon}$$

$$-\frac{\mu_{\epsilon\delta}}{T_{\epsilon\delta}} = \left(\frac{\partial S_{\epsilon\delta}}{\partial m_\epsilon}\right)_{U_{\epsilon\delta},V}$$

Accordingly, the differentials of the functions (4) and (5) read respectively

$$dU_{\epsilon\delta} = T_{\epsilon\delta}dS_{\epsilon\delta} - p_{\epsilon\delta}dV + \mu_{\epsilon\delta}dm_\epsilon \tag{6}$$

$$dS_{\epsilon\delta} = \frac{1}{T_{\epsilon\delta}}dU_{\epsilon\delta} + \frac{p_{\epsilon\delta}}{T_{\epsilon\delta}}dV - \frac{\mu_{\epsilon\delta}}{T_{\epsilon\delta}}dm_\epsilon \tag{7}$$

The fundamental relations of the gas mixture are obtained by adding up the contributions (4) and (5). They read formally

$$U = \sum_{\epsilon,\delta} U_{\epsilon\delta}(S_{\epsilon\delta}, V, m_\epsilon) = U \text{ (all } S_{\epsilon\delta}, V, \text{ all } m_\epsilon) \tag{8}$$

$$S = \sum_{\epsilon,\delta} S_{\epsilon\delta}(U_{\epsilon\delta}, V, m_\epsilon) = S \text{ (all } U_{\epsilon\delta}, V, \text{ all } m_\epsilon) \tag{9}$$

The short notation $\displaystyle\sum_{\epsilon,\delta}$ represents the double summation $\displaystyle\sum_{\epsilon=1}^{n}\sum_{\delta=1}^{\ell_\epsilon}$; these summation operators are not permutable, in general, because the number ℓ_ϵ of independent degrees of freedom may change from component to component. The differentials of the functions (8) and (9) read respectively

$$dU = \sum_{\epsilon,\delta} dU_{\epsilon\delta} = \sum_{\epsilon,\delta} T_{\epsilon\delta}dS_{\epsilon\delta} - \left(\sum_{\epsilon,\delta} p_{\epsilon\delta}\right)dV + \sum_{\epsilon=1}^{n}\left(\sum_{\delta=1}^{\ell_\epsilon}\mu_{\epsilon\delta}\right)dm_\epsilon \tag{10}$$

$$dS = \sum_{\epsilon,\delta} dS_{\epsilon\delta} = \sum_{\epsilon,\delta} \frac{1}{T_{\epsilon\delta}} dU_{\epsilon\delta} + \left(\sum_{\epsilon,\delta} \frac{p_{\epsilon\delta}}{T_{\epsilon\delta}}\right) dV - \sum_{\epsilon=1}^{n} \left(\sum_{\delta=1}^{\ell_\epsilon} \frac{\mu_{\epsilon\delta}}{T_{\epsilon\delta}}\right) dm_\epsilon \quad (11)$$

The inspection of the right-hand sides of (10) and (11) indicates the state equations of the gas mixture. In particular, the temperatures $T_{\epsilon\delta}$ keep unaltered their role of state equations; the sum of the partial pressures yields the total pressure of the gas mixture

$$p = -\left(\frac{\partial U}{\partial V}\right)_{\text{all } S_{\epsilon\delta}, \text{all } m_\epsilon} = \sum_{\epsilon,\delta} p_{\epsilon\delta}$$

and the sum of the partial chemical potentials, for a given component, yields the chemical potential of that component

$$\mu_\epsilon = \left(\frac{\partial U}{\partial m_\epsilon}\right)_{\text{all } S_{\epsilon\delta}, V, \text{all } m_{\nu \neq \epsilon}} = \sum_{\delta=1}^{\ell_\epsilon} \mu_{\epsilon\delta}$$

but there is a lack of standard nomenclature and notation relative to the state equations

$$\left(\frac{\partial S}{\partial V}\right)_{\text{all } U_{\epsilon\delta}, \text{all } m_\epsilon} = \sum_{\epsilon,\delta} \frac{p_{\epsilon\delta}}{T_{\epsilon\delta}}$$

$$\left(\frac{\partial S}{\partial m_\epsilon}\right)_{\text{all } U_{\epsilon\delta}, V, \text{all } m_{\nu \neq \epsilon}} = -\sum_{\delta=1}^{\ell_\epsilon} \frac{\mu_{\epsilon\delta}}{T_{\epsilon\delta}}$$

The sets of independent state parameters which the functions (8) and (9) depend on indicate clearly that the concept of thermodynamic equilibrium broadens in a multi-temperature situation because it includes also the equilibration of the exchanges of energy and entropy among the degrees of freedom. In this regard, a major role is played by the constraints which are imposed on the system. In the case of a single temperature, one has to deal only with the component masses and common constraints are the chemical inertia of a component ($m_\epsilon = const$) or the freezing of a chemical reaction ($\xi_\kappa = const$). In the case of multiple temperatures, additional constraints may arise from the subset of state parameters associated with the degrees of freedom; for example, a degree of freedom may be entropically frozen ($S_{\epsilon\delta} = const$) or energetically frozen ($U_{\epsilon\delta} = const$) but many other kinds of constraints may be active. Accordingly, one speaks of partially constrained thermal equilibrium.

The existence of a partially constrained thermal equilibrium has a profound impact on the associated, possibly partially constrained too, chemical equilibrium because, depending on which additional constraints are acting on the gas mixture, the minimization/maximization of energy/entropy may not necessarily lead to mathematical conditions governing the equilibration of the mass exchanges similar to (1) or (3). To the best of the present author's knowledge, there seems to be no clear understanding of this fact in the literature. Some authors[18,25] affirm that the vanishing of the reaction affinities

$$\sum_{\epsilon=1}^{n} \nu_{\kappa\epsilon} M_{\epsilon} \left(\sum_{\delta=1}^{\ell_{\epsilon}} \mu_{\epsilon\delta} \right) = 0 \tag{12}$$

yields the correct chemical equilibrium equations for multi-temperature gas mixtures. Other authors,[26-30] instead, reject the equations (12) and sustain the unconditional correctness of the form

$$\sum_{\epsilon=1}^{n} \nu_{\kappa\epsilon} M_{\epsilon} \left(\sum_{\delta=1}^{\ell_{\epsilon}} \frac{\mu_{\epsilon\delta}}{T_{\epsilon\delta}} \right) = 0$$

Furthermore, disagreements[26,27,31,32] manifestly exist also concerning which thermodynamic function should be used to determine the equilibrium conditions. The existing confusion is determined mainly by the tendency to disregard the primary importance of the kind of constraints imposed on the gas mixture and the role of the entropy as a fundamental relation equivalent to the energy. An outstanding example of how such a confusion may affect the applications is given by the uncertainty afflicting the two-temperature Saha equation.[18,25,27-30,33-35] With concern to this confusing situation, it will appear evident in the following sections that: (a) there is no generalized form of the chemical equilibrium equations for multi-temperature gas mixtures but the applicable form of the equations depends on the kind of active constraints; (b) energy and entropy are perfectly equivalent for the purpose of finding the conditions of thermodynamic equilibrium.

The role of the entropy is recognized in the treatment of Napolitano[4] but the influence of a partially constrained thermal equilibrium on the equilibration of the mass exchanges appears to have been overlooked because of a seeming inconsistency in the identification[36] of the state equations associated with the progress variables in the entropic formulation.

4. Basic Statement of Equilibrium

The concept of thermodynamic equilibrium is founded on the following, basic statement:[37] the equilibrium conditions, i.e. the equilibrium values of the independent state parameters, of an isolated system[38] are those that minimize the energy U or, equivalently, maximize the entropy S compatibly with the additional constraints, if any, imposed on the system.

The isolation conditions in the energetic scheme, i.e. the one based on the fundamental relation (8), are expressed by the constancy of total entropy, volume and total mass

$$S = \sum_{\epsilon,\delta} S_{\epsilon\delta} = const \tag{13}$$

$$V = const \tag{14}$$

$$m = \sum_{\epsilon=1}^{n} m_{\epsilon} = const \tag{15}$$

In the entropic scheme, which is based on the fundamental relation (9), the condition (13) is replaced with the equivalent one expressing the constancy of the total energy

$$U = \sum_{\epsilon,\delta} U_{\epsilon\delta} = const \tag{16}$$

The additional constraints express restrictions affecting the variations of some of the independent state parameters; as such, they appear in the form of mathematical relations which have to be taken into due account in the procedures to obtain the equilibrium conditions. The great number of state parameters associated with the degrees of freedom gives rise to a variety of constraints which makes the equilibrium analysis somewhat dependent on the particular application but, in principle, every problem of thermodynamic equilibrium can be solved from the mathematical exploitation of the equilibrium statement, the isolation conditions and the additional constraints.

In the following sections, the cases of entropic and energetic freezing of the molecular degrees of freedom are dealt with. The reason behind this choice is to show, in an evident manner, the influence that a partially constrained thermal equilibrium exercises on the equations governing the associated chemical equilibrium. In order to confer a certain generality to the analysis, it is assumed, in both cases, that $f (\leq r)$ chemical reactions are frozen and that, therefore, their progress variables are constrained to be constant

$$\xi_\kappa = const \tag{17}$$

(κ of frozen reactions only)

5. Equilibrium Analysis

5.1. THERMODYNAMIC EQUILIBRIUM WITH ENTROPICALLY FROZEN DEGREES OF FREEDOM

When the exchange of the entropy associated with a molecular degree of freedom is restrained, the degree of freedom is entropically frozen and its entropy must necessarily be a constant. If each component has $\bar{\ell}_\epsilon$ ($\leq \ell_\epsilon$) entropically frozen molecular degrees of freedom, the corresponding constraints read

$$S_{\epsilon\delta} = const \tag{18}$$

(all ϵ; δ of frozen degrees of freedom only)

There are $\sum_{\epsilon=1}^{n} \bar{\ell}_\epsilon$ constraints (18) in total and their nature suggests that it is convenient to perform the equilibrium analysis via the energetic formulation because, in that scheme, the entropies $S_{\epsilon\delta}$ assume the role of independent state parameters. Thus, in order to determine the thermodynamic equilibrium conditions which settle under the assumed circumstances, the fundamental relation (8) is to be minimized compatibly with the expressions (2) governing the variations of the component masses, the isolation conditions (13), (14), and the additional constraints (17), (18); the isolation condition (15) is not required because it is identically satisfied when the masses vary according to (2).

The minimization can be accomplished, mathematically, by the Lagrange multiplier method. In the present case, the Lagrangian function reads

$$\mathcal{L} = \sum_{\epsilon,\delta} U_{\epsilon\delta}(S_{\epsilon\delta}, V, m_\epsilon) + \lambda_S \left(S - \sum_{\epsilon,\delta} S_{\epsilon\delta} \right)$$

and, when account is taken of (2), (10), (14), (17) and (18), its differential reduces to

$$d\mathcal{L} = \sum_{\epsilon,\delta}^{nf} (T_{\epsilon\delta} - \lambda_S)\, dS_{\epsilon\delta} + \sum_{\kappa}^{nf} \left[\sum_{\epsilon=1}^{n} \nu_{\kappa\epsilon} M_\epsilon \left(\sum_{\delta=1}^{\ell_\epsilon} \mu_{\epsilon\delta} \right) \right] d\xi_\kappa \tag{19}$$

The conditions of thermodynamic equilibrium are those that make the differential (19) vanish; the imposition of such a mathematical condition generates the $\sum_{\epsilon=1}^{n}(\ell_\epsilon - \bar{\ell}_\epsilon)$ equations

$$T_{\epsilon\delta} - \lambda_S = 0 \tag{20}$$

(all ϵ; δ of non-frozen degrees of freedom only) and the $r - f$ equations

$$\sum_{\epsilon=1}^{n} \nu_{\kappa\epsilon} M_\epsilon \left(\sum_{\delta=1}^{\ell_\epsilon} \mu_{\epsilon\delta} \right) = 0 \tag{21}$$

(κ of non-frozen reactions only)

The $\left(\sum_{\epsilon=1}^{n} \ell_\epsilon + n + r + 1 \right)$ equations (2), (13), (17), (18), (20), (21) consti-

tute a system for the unknown equilibrium values of the $\left(\sum_{\epsilon=1}^{n} \ell_\epsilon \right)$ entropies

$S_{\epsilon\delta}$, the n component masses m_ϵ, the r progress variables ξ_κ and the Lagrange multiplier λ_S.

The equations (20) govern the equilibration of the exchanges of energy
and entropy among the non-frozen degrees of freedom. They show that,
at equilibrium, the temperatures of those degrees of freedom must equal-
ize to the common value λ_S; one refers to this situation by saying that
the non-frozen degrees of freedom are in mutual thermal equilibrium at
the temperature λ_S. It is extremely important to realize that nothing can
be said concerning the temperatures of the frozen degrees of freedom; in
general, they must be expected to differ.

The equations (21) govern the equilibration of the mass exchanges. They
are a clear extension of the equations (3) to the multi-temperature case and
indicate that the chemical equilibrium driven by an entropically constrained
thermal equilibrium is characterized by the vanishing of the affinities of the
non-frozen chemical reactions.

The equilibrium analyses usually found in the literature[18,25] concentrate
on the particular case obtained from the present one when all the degrees
of freedom are entropically frozen ($\bar{\ell}_\epsilon = \ell_\epsilon$, all ϵ). Other authors, however,
prefer to minimize the Helmholtz potential

$$F = \sum_{\epsilon,\delta} F_{\epsilon\delta}(T_{\epsilon\delta}, V, m_\epsilon) = F \left(\text{all } T_{\epsilon\delta}, V, \text{ all } m_\epsilon \right) \tag{22}$$

or the Gibbs potential

$$G = \sum_{\epsilon,\delta} G_{\epsilon\delta}(T_{\epsilon\delta}, p_{\epsilon\delta}, m_\epsilon) = G \left(\text{all } T_{\epsilon\delta}, p, \text{ all } m_\epsilon \right) \tag{23}$$

keeping constant all the temperatures and the volume or the pressure, respectively. The equivalence of the functions U, F and G is guaranteed by
the relations

$$(dU)_{\text{all } S_{\epsilon\delta}, V} = (dF)_{\text{all } T_{\epsilon\delta}, V} = (dG)_{\text{all } T_{\epsilon\delta}, p}$$

but it is important to keep in mind that such relations hold only in the particular situation $\bar{l}_\epsilon = l_\epsilon$, all ϵ.

5.2. THERMODYNAMIC EQUILIBRIUM WITH ENERGETICALLY FROZEN DEGREES OF FREEDOM

Another interesting case to investigate is when the restraint is applied to the exchange of the energy distributed over a degree of freedom; in such a case, the degree of freedom is energetically frozen and it is its energy to be necessarily constant this time. If each component has $\bar{l}_\epsilon$ ($\leq l_\epsilon$) energetically frozen molecular degrees of freedom, the corresponding constraints read

$$U_{\epsilon\delta} = const \tag{24}$$

(all ϵ; δ of frozen degrees of freedom only)

Again, there are $\sum\limits_{\epsilon=1}^{n} \bar{l}_\epsilon$ constraints (24) in total; in this case, however, it is convenient to adopt the entropic formulation. Thus, the fundamental relation (9) is to be maximized compatibly with the expressions (2) governing the variations of the component masses, the isolation conditions (14), (16), and the additional constraints (17), (24). The Lagrangian function reads

$$\mathcal{L} = \sum_{\epsilon,\delta} S_{\epsilon\delta}(U_{\epsilon\delta}, V, m_\epsilon) + \lambda_U \left(U - \sum_{\epsilon,\delta} U_{\epsilon\delta} \right)$$

When account is taken of (2), (11), (14), (17) and (24), its differential reduces to

$$d\mathcal{L} = \sum_{\epsilon,\delta}^{nf} \left(\frac{1}{T_{\epsilon\delta}} - \lambda_U \right) dU_{\epsilon\delta} - \sum_{\kappa}^{nf} \left[\sum_{\epsilon=1}^{n} \nu_{\kappa\epsilon} M_\epsilon \left(\sum_{\delta=1}^{l_\epsilon} \frac{\mu_{\epsilon\delta}}{T_{\epsilon\delta}} \right) \right] d\xi_\kappa \tag{25}$$

and the imposition of its vanishing generates the $\sum\limits_{\epsilon=1}^{n}(l_\epsilon - \bar{l}_\epsilon)$ equations

$$\frac{1}{T_{\epsilon\delta}} - \lambda_U = 0 \tag{26}$$

(all ϵ; δ of non-frozen degrees of freedom only)

and the $r - f$ equations

$$\sum_{\epsilon=1}^{n} \nu_{\kappa\epsilon} M_\epsilon \left(\sum_{\delta=1}^{l_\epsilon} \frac{\mu_{\epsilon\delta}}{T_{\epsilon\delta}} \right) = 0 \tag{27}$$

(κ of non-frozen reactions only)

The $\left(\sum_{\epsilon=1}^{n}\ell_\epsilon + n + r + 1\right)$ equations (2), (16), (17), (24), (26), (27) consti-

tute a system for the unknown equilibrium values of the $\left(\sum_{\epsilon=1}^{n}\ell_\epsilon\right)$ energies
$U_{\epsilon\delta}$, the n component masses m_ϵ, the r progress variables ξ_κ and the Lagrange multiplier λ_U.

The equations (26) have the same mathematical form of the equations (20) and, again, they affirm the mutual thermal equilibrium at the temperature $1/\lambda_U$ of the non-frozen degrees of freedom. The equations (27), however, are different from the equations (21); the former cannot be reduced to the structure of the latter because the summation in brackets on the left-hand side of (27) includes also the energetically frozen degrees of freedom whose corresponding temperatures cannot be factorized out of it. Consequently, it is the quantity

$$\overline{\mathcal{A}}_k = \sum_{\epsilon=1}^{n} \nu_{\kappa\epsilon} M_\epsilon \left(\sum_{\delta=1}^{\ell_\epsilon} \frac{\mu_{\epsilon\delta}}{T_{\epsilon\delta}}\right) \tag{28}$$

that vanishes rather than the affinity

$$\mathcal{A}_k = \sum_{\epsilon=1}^{n} \nu_{\kappa\epsilon} M_\epsilon \left(\sum_{\delta=1}^{\ell_\epsilon} \mu_{\epsilon\delta}\right) \tag{29}$$

when the equilibrium of the corresponding non-frozen chemical reaction is driven by an energetically constrained thermal equilibrium.[39]

Chemical equilibrium equations in the form (27) are not new in the literature. Van de Sanden and coworkers[26,27] obtained similar equations for the particular case when all the degrees of freedom are energetically frozen $(\bar{\ell}_\epsilon = \ell_\epsilon$, all $\epsilon)$; Morro and Romeo[28–30] also derived equations similar to (27) but on the grounds of a more fluid dynamics nature. However, these authors support the erroneous idea that the form (21) of the chemical equilibrium equations must be rejected and replaced unconditionally with the form (27).

5.3. NON-UNIQUENESS OF THE CHEMICAL EQUILIBRIUM EQUATIONS

The expressions (21) and (27) show clearly how the uniqueness of the equations governing the chemical equilibrium, to which one has become so much used when dealing with single-temperature gas mixtures, is lost when a partially constrained thermal equilibrium prevails. The applicable form of the equations depends on the kind of constraints affecting the exchanges of energy and entropy among the degrees of freedom. The different forms come to coincide only when the thermal equilibrium of the gas mixture is uncon-strained ($\bar{\ell}_\epsilon = 0$) and, therefore, complete; in this case, all the degrees of freedom share the same temperature $T = \lambda_S = 1/\lambda_U$, the expressions (28), (29) become dependent on each other via the relation

$$\overline{\mathcal{A}}_k = \frac{A_k}{T}$$

and the vanishing of one is equivalent to the vanishing of the other.

5.4. EQUIVALENCE OF THE FUNDAMENTAL RELATIONS ENERGY AND ENTROPY

The choice made in Sec. 5.1 and Sec. 5.2 concerning which fundamental relation to use according to which constraints are prescribed is determined only by circumstances of convenience suggested by the nature of the con-straints (18) and (24). In reality, the fundamental relations (8), (9) are perfectly equivalent; there is no reason whatsoever that imposes the use of one rather than the other. In order to prove this equivalence, the equi-librium equations (26), (27), determined by the energetic freezing of some degrees of freedom, will be reobtained here by minimizing the fundamental relation (8). The minimization must be compatible with the expressions (2) governing the variations of the component masses, the isolation conditions (13), (14), and the additional constraints (17), (24).

The Lagrangian function reads

$$\mathcal{L} = \sum_{\epsilon,\delta} U_{\epsilon\delta}(S_{\epsilon\delta}, V, m_\epsilon) + \lambda'_S \left(S - \sum_{\epsilon,\delta} S_{\epsilon\delta} \right)$$

When account is taken of (24), its differential becomes

$$d\mathcal{L} = \sum_{\epsilon,\delta}^{\mathrm{nf}} dU_{\epsilon\delta} - \lambda'_S \sum_{\epsilon,\delta} dS_{\epsilon\delta} \tag{30}$$

The summation of the last term on the right-hand side of (30) can be expanded explicitly in terms of the frozen and non-frozen degrees of freedom to yield

$$d\mathcal{L} = \sum_{\epsilon,\delta}^{\mathrm{nf}} dU_{\epsilon\delta} - \lambda'_S \sum_{\epsilon,\delta}^{\mathrm{nf}} dS_{\epsilon\delta} - \lambda'_S \sum_{\epsilon,\delta}^{\mathrm{f}} dS_{\epsilon\delta} \tag{31}$$

The differentials $dU_{\epsilon\delta}$ relative to the energetically non-frozen degrees of freedom are obtained from (6) after imposing the isolation condition (14); in this way, the first two terms on the right-hand side of (31) rearrange to yield

$$d\mathcal{L} = \sum_{\epsilon,\delta}^{\mathrm{nf}} (T_{\epsilon\delta} - \lambda'_S) dS_{\epsilon\delta} + \sum_{\epsilon,\delta}^{\mathrm{nf}} \mu_{\epsilon\delta} dm_\epsilon - \lambda'_S \sum_{\epsilon,\delta}^{\mathrm{f}} dS_{\epsilon\delta} \tag{32}$$

The differentials $dS_{\epsilon\delta}$ appearing in the last term on the right-hand side of (32) are relative to the energetically frozen degrees of freedom and are obtained from (7) after imposing the isolation condition (14) and the constraints (24); they read

$$dS_{\epsilon\delta} = -\frac{\mu_{\epsilon\delta}}{T_{\epsilon\delta}}dm_\epsilon$$

(all ϵ; δ of frozen degrees of freedom only)

Thus, the last term on the right-hand side of (32) becomes

$$\sum_{\epsilon,\delta}^{f}dS_{\epsilon\delta} = -\sum_{\epsilon,\delta}^{f}\frac{\mu_{\epsilon\delta}}{T_{\epsilon\delta}}dm_\epsilon = -\sum_{\epsilon,\delta}\frac{\mu_{\epsilon\delta}}{T_{\epsilon\delta}}dm_\epsilon + \sum_{\epsilon,\delta}^{nf}\frac{\mu_{\epsilon\delta}}{T_{\epsilon\delta}}dm_\epsilon \quad (33)$$

The substitution of (33) into (32) leads to

$$d\mathcal{L} = \sum_{\epsilon,\delta}^{nf}(T_{\epsilon\delta} - \lambda'_S)dS_{\epsilon\delta} + \sum_{\epsilon,\delta}^{nf}\mu_{\epsilon\delta}\left(1 - \frac{\lambda'_S}{T_{\epsilon\delta}}\right)dm_\epsilon + \lambda'_S\sum_{\epsilon,\delta}\frac{\mu_{\epsilon\delta}}{T_{\epsilon\delta}}dm_\epsilon \quad (34)$$

After introducing the expressions (2) and the associated constraints (17), the differential (34) reduces to the final, required form

$$d\mathcal{L} = \sum_{\epsilon,\delta}^{nf}(T_{\epsilon\delta} - \lambda'_S)dS_{\epsilon\delta} + \sum_{\kappa}^{nf}\left[\sum_{\epsilon,\delta}^{nf}\nu_{\kappa\epsilon}M_\epsilon\mu_{\epsilon\delta}\left(1 - \frac{\lambda'_S}{T_{\epsilon\delta}}\right)\right]d\xi_\kappa + $$
$$+ \lambda'_S\sum_{\kappa}^{nf}\left[\sum_{\epsilon=1}^{n}\nu_{\kappa\epsilon}M_\epsilon\left(\sum_{\delta=1}^{\ell_\epsilon}\frac{\mu_{\epsilon\delta}}{T_{\epsilon\delta}}\right)\right]d\xi_\kappa \quad (35)$$

The imposition of the vanishing of the differential (35) returns the thermal equilibrium equations (26) ($\lambda'_S = 1/\lambda_U$) and the chemical equilibrium equations (27). In an analogous manner, it is possible to obtain the equilibrium equations (20) and (21) by appropriately maximizing the fundamental relation (9) rather than minimizing the function (8).

In circumstances of partially constrained thermal equilibrium, the Helmholtz and Gibbs potentials (22), (23) do not possess the characteristic of being equivalent, to the extent of finding the conditions of thermodynamic equilibrium, to energy and entropy for arbitrary constraints affecting the state parameters associated with the degrees of freedom. The lack of equivalence stems from the fact that, given the ample variety of such parameters and corresponding Legendre transformations, there is not a unique way to define Helmholtz- and Gibbs-like thermodynamic potentials for multitemperature gas mixtures. Interested readers are recommended to consult the account given by Napolitano[4,40] for a thorough understanding of this matter.

5.5. CHEMICAL EQUILIBRIUM CONSTANTS

A direct consequence of the chemical equilibrium equations (21) and (27) is the existence of different chemical equilibrium constants as functions of the various temperatures.

From statistical thermodynamics, the chemical potential contributions $\mu_{\epsilon\delta}$ are obtained in terms of the partition functions

$$q_{\epsilon\delta} = \sum_i \exp\left(-\frac{\varepsilon_{\epsilon\delta,i}}{kT_{\epsilon\delta}}\right) \tag{36}$$

For the translational ($\delta = 1$) degrees of freedom, the sum-over-states (36) can be obtained in closed form and the corresponding contribution to the chemical potential reads

$$\mu_{\epsilon 1} = -\frac{R_G}{M_\epsilon} T_{\epsilon 1} \ln\left[\left(\frac{2\pi k M_\epsilon T_{\epsilon 1}}{h^2}\right)^{3/2} \frac{1}{N_\epsilon}\right] \tag{37}$$

For the internal ($\delta > 1$) degrees of freedom, the expression of the contribution to the chemical potential directly in terms of the partition function

$$\mu_{\epsilon\delta} = -\frac{R_G}{M_\epsilon} T_{\epsilon\delta} \ln q_{\epsilon\delta} \tag{38}$$

$$(\delta > 1)$$

can be used. The partition function $q_{\epsilon\delta}$ associated with the generic internal degree of freedom depends only on the temperature $T_{\epsilon\delta}$. The substitution of (37) and (38) into (21) leads to the equilibrium equation

$$\prod_{\epsilon=1}^n N_\epsilon^{\nu_{\kappa\epsilon} T_{\epsilon 1}/\theta} = \prod_{\epsilon=1}^n \left[\left(\frac{2\pi k M_\epsilon T_{\epsilon 1}}{h^2}\right)^{3/2} \prod_{\delta=2}^{\ell_\epsilon} q_{\epsilon\delta}^{T_{\epsilon\delta}/T_{\epsilon 1}}\right]^{\nu_{\kappa\epsilon} T_{\epsilon 1}/\theta} \tag{39}$$

(κ of non-frozen reactions only)

in which θ is an arbitrary, but appropriately chosen, reference temperature to make the exponents non-dimensional. The espression (39) is in formal agreement with those[41] proposed by Eddy and Cho[18] and by Potapov;[25] the exponential term containing the zero-point energies does not appear explicitly in (39) because the quantum state energies $\varepsilon_{\epsilon\delta,i}$ on the right-hand side of (36) are not referred to the ground state value $\varepsilon_{\epsilon\delta,0}$. The substitution of (37) and (38) into (27) yields, instead, the equilibrium equation

$$\prod_{\epsilon=1}^n N_\epsilon^{\nu_{\kappa\epsilon}} = \prod_{\epsilon=1}^n \left[\left(\frac{2\pi k M_\epsilon T_{\epsilon 1}}{h^2}\right)^{3/2} \prod_{\delta=2}^{\ell_\epsilon} q_{\epsilon\delta}\right]^{\nu_{\kappa\epsilon}} \tag{40}$$

(κ of non-frozen reactions only)

The most important feature of (40), as compared to (39), is the absence of the temperature ratios in the exponents on both sides of the equation.

The right-hand sides of (39) and (40) give, by definition, the multi-temperature expressions of the chemical equilibrium constants corresponding to the number densities N_ϵ.

6. Chemical Equilibrium of a Partially Ionized Gas

6.1. INTRODUCTORY CONSIDERATIONS

The problem of the chemical equilibrium of a partially ionized gas, as arising from plasma physics and hypersonics applications, has been devoted attention for decades in the pertinent literature.[35] In this case, the gas mixture is composed by atoms A, ions A^+ and electrons e^- ($n = 3$) subjected to the chemical reaction ($r = 1$)

$$A \rightleftharpoons A^+ + e^- \tag{41}$$

The problem consists in determining the equilibrium composition compatibly with the existence of different temperatures. Various authors have attempted the solution via the minimization of the potentials (22), (23) keeping constant the temperatures[42] and the volume or the pressure, respectively; such an approach leads to a multi-temperature Saha equation whose mathematical structure is expectedly consistent with the equilibrium equation (39). Morro and Romeo,[28-30] however, have examined the problem in a fluid dynamics context. Their treatment is based on the analysis of the entropy production term in the associated balance equation; the two-temperature[43] Saha equation they obtain is determined from the imposition of the vanishing of the chemical reaction rates[44] and shows a mathematical structure consistent with the equilibrium equation (40). The equation proposed by Morro and Romeo differs, therefore, from the one derived by other authors in a pure thermodynamic context mainly because the composition parameters of atoms and ions appear powered to unity rather than to appropriate temperature ratios.

The existence of mathematically non-equivalent two-temperature Saha equations has generated, obviously, some uncertainty.[45] This uncertainty, however, is illusory: it disappears when one realizes that the different equations are determined by different constraints and that, therefore, they are not in conflict with each other because they describe different situations of partially constrained thermodynamic equilibrium. The purpose of the following sections is to show how the uncertainty can be resolved in the framework of the theory proposed in this paper. For the sake of simplicity, only two-temperature chemical equilibrium situations will be considered.

6.2. IONIZATION EQUILIBRIUM WITH ENTROPIC FREEZING

The thermodynamic circumstances that lead to an ionization equilibrium governed by the two-temperature Saha equation as used, for example, by Kannappan and Bose[34] arise when the entropy exchanges are allowed among the translational and electronic degrees of freedom of the heavy species A, A^+ ($\epsilon = 1, 2$; $\ell_\epsilon = 2$) but are restrained for the translational degree of freedom of the electrons e^- ($\epsilon = 3$; $\ell_3 = 1$). Thus, the translational degree of freedom of the electrons is entropically frozen

$$S_{31} = const$$

From (20), the degrees of freedom of the heavy species must be in mutual thermal equilibrium at the temperature $\lambda_S = T_h$ which, however, differs from the translational temperature $T_{31} = T_e$ of the electrons. From (21),

TABLE 1. Stoichiometric coefficients of reaction (41)

	A	A^+	e^-
ϵ	1	2	3
$\nu_{1\epsilon}$	-1	1	1

the chemical equilibrium is governed by the vanishing of the affinity of the reaction (41). When account is taken of the thermal equilibrium equations (20) specialized for the present plasma and the stoichiometric coefficients ($\kappa = 1$) listed in Table 1, the particularization of the expression (39) leads, with the arbitrary reference temperature being set to the electron translational temperature, i.e. $\theta = T_e$, to the two-temperature Saha equation

$$N_{e^-} \left(\frac{N_{A^+}}{N_A} \right)^{T_h/T_e} = 2 \left[\left(\frac{\mathcal{M}_{A^+}}{\mathcal{M}_A} \right)^{3/2} \frac{q_{A^+}}{q_A} \right]^{T_h/T_e} \left(\frac{2\pi k \mathcal{M}_e T_e}{h^2} \right)^{3/2} \tag{42}$$

In (42), q_A, q_{A^+} represent the electronic partition functions of the heavy species and the factor 2 arises from the quantum degeneracy due to the spin of the electrons. The ionization equilibrium equation (42) is in substantial agreement with the one used by Kannappan and Bose.[46]

The other two equations required for the determination of the equilibrium composition derive from the conservation of the atomic species; in the subject case, they read

$$N_A + N_{A^+} = N_A^\circ \tag{43}$$

$$N_{A^+} = N_{e^-} \tag{44}$$

The equation (44) expresses also the conservation of the electric charge across the reaction (41).

6.3. IONIZATION EQUILIBRIUM WITH ENERGETIC FREEZING

The ionization equilibrium governed by the two-temperature Saha equation similar to the one proposed by Morro and Romeo[28-30] corresponds to the thermodynamic situation in which the energy exchanges are allowed among the translational and electronic degrees of freedom of the heavy species but are restrained for the translational degree of freedom of the electrons. In this case, therefore, the energetic freezing of the translational degree of freedom of the electrons prevails

$$U_{31} = const$$

Thus, taking into account the thermal equilibrium equations (26) ($1/\lambda_U = T_h$) specialized for the present plasma and the stoichiometric coefficients of Table 1, the corresponding ionization equilibrium equation is obtained from the particularization of the expression (40); this leads to the two-temperature Saha equation[47]

$$\frac{N_{e^-} N_{A^+}}{N_A} = 2 \left(\frac{\mathcal{M}_{A^+}}{\mathcal{M}_A}\right)^{3/2} \frac{q_{A^+}}{q_A} \left(\frac{2\pi k \mathcal{M}_e T_e}{h^2}\right)^{3/2} \tag{45}$$

The equilibrium composition is determined by the set of equations (43), (44) and (45).

Acknowledgments

I wish to thank Prof. Mario Capitelli (Centro di Studio per la Chimica dei Plasmi del CNR, Dip. Chimica, Universitá di Bari, Italy) for his willingness to discuss the subject dealt with in this work and for his invaluable help and advice.

References and Notes

1. H. B. Callen (1963) *Thermodynamics*, John Wiley & Sons, New York.
2. L. Tisza (1966) *Generalized Thermodynamics*, The M.I.T. Press, Cambridge, MA.
3. H. S. Robertson (1993) *Statistical Thermophysics*, P T R Prentice Hall, Englewood Cliffs (NJ).
4. L. G. Napolitano (1971) *Thermodynamique des Systèmes Composites en Équilibre ou Hors d'Équilibre*, Gauthier-Villars Éditeurs, Paris.
5. C. F. Hansen (1976) Molecular physics of equilibrium gases, NASA-SP-3096.

6. K. Denbigh (1955) *The Principles of Chemical Equilibrium*, Cambridge University Press, Cambridge.

7. V. I. Golovitchev (private communication).

8. W. B. White, S. M. Johnson, and G. B. Dantzig (1958) Chemical equilibrium in complex mixtures, *J. Chem. Phys.* **28**, 751.

9. F. J. Zeleznik and S. Gordon (1960) An analytical investigation of three general methods of calculating chemical-equilibrium compositions, NACA-TN-D-473.

10. F. J. Zeleznik and S. Gordon (1966) Equilibrium computations for multi-component plasmas, *Can. J. Phys.* **44**, 877.

11. F. J. Zeleznik and S. Gordon (1968) Calculation of complex chemical equilibria, *Ind. Eng. Chem.* **60**, 27.

12. Various authors (1970) Kinetics and thermodynamics in high-temperature gases, NASA-SP-239.

13. M. Capitelli and E. Molinari (1970) Problems of determination of high temperature thermodynamic properties of rare gases with application to mixtures, *J. Plasma Phys.* **4**, 335.

14. F. Van Zeggeren and S. H. Storey (1970) *The Computing of Chemical Equilibria*, Cambridge University Press, Cambridge.

15. R. Holub and P. Vonka (1976) *The Chemical Equilibrium of Gaseous Systems*, Reidel Publishing Co., Dordrecht.

16. W. R. Smith and R. W. Missen (1982) *Chemical Reaction Equilibrium Analysis: Theory and Algorithms*, John Wiley & Sons, New York.

17. Y. Liu and M. Vinokur, "Equilibrium gaseous flow computations. I. Accurate and efficient calculation of equilibrium gas properties", AIAA Paper No. 89-1736, 24th Thermophysics Conference, Buffalo, NY, June 12–14, 1989.

18. T. L. Eddy and K. Y. Cho (1991) A multitemperature model for plasmas in chemical nonequilibrium, in K. Etemadi and J. Mostaghimi (eds.), *HTD, Vol. 161: Heat Transfer in Thermal Plasma Processing*, ASME, New York, pp. 195–210.

19. There are instances in the literature where the conditions of chemical equilibrium are found from the imposition of the vanishing of the entropy production; see Sec. 7 at p. 75 of W. G. Vincenti and C. H. Kruger jr. (1965) *Introduction to Physical Gasdynamics*, John Wiley & Sons, New York. However, this method is not rigorously proper because the vanishing of the entropy production, when the conditions of chemical equilibrium prevail, is a result of the theory and not information available *a priori*.

20. The entropic potentials are known in the literature as Massieu functions. See, for example, Sec. 5.4 at p. 101 of Ref. 1 or Sec. 3.2.2 at p. 58 of Ref. 4.

21. See Sec. 4.4.1 at p. 99 and Sec. 4.4.4 at p. 109 of Ref. 4. See also the section "Conditions d'équilibre chimique" at p. 283 in vol. 1 of L. Sédov (1975) *Mécanique des Milieux Continus*, Editions Mir (Traduction française), Moscou.

22. The term "law of mass action" is also used to refer directly to the equations (3).

23. The term "free energy" is also used to refer to the Gibbs potential.

24. See Sec. 3.3 at p. 18 of Ref. 17.

25. A. V. Potapov (1966) Chemical equilibrium of multitemperature systems, *High Temp. (USSR)* 4, 48.

26. M. C. M. Van de Sanden (1991) *The Expanding Plasma Jet: Experiments and Model*, Thesis, Technische Universiteit Eindhoven, Eindhoven.

27. M. C. M. Van de Sanden, P. P. J. M. Schram, A. G. Peeters, J. A. M. van der Mullen, and G. M. W. Kroesen (1989) Thermodynamic generalization of the Saha equation for a two-temperature plasma, *Physical Review A* 40, 5273.

28. A. Morro and M. Romeo (1988) The law of mass action for fluid mixtures with several temperatures and velocities, *J. Non-Equilib. Thermodyn.* 13, 339.

29. A. Morro and M. Romeo (1988) Thermodynamic derivation of Saha's equation for a multi-temperature plasma, *J. Plasma Phys.* 39, 41.

30. A. Morro and M. Romeo (1986) On the law of mass action in fluid mixtures with several temperatures, *Il Nuovo Cimento* 7 D, 539.

31. K. Chen and T. L. Eddy (1995) Investigation of chemical affinity for reacting flows of non-local thermal equilibrium gases, *J. Thermophys. Heat Transfer* 9, 41.

32. A. E. Mertogul and H. Krier (1994) Two-temperature modeling of laser sustained hydrogen plasmas, *J. Thermophys. Heat Transfer* 8, 781.

33. D. Giordano and M. Capitelli (1995) Two-temperature Saha equation: a misunderstood problem, *J. Thermophys. Heat Transfer* 9, No. 4, 803.

34. D. Kannappan and T. K. Bose (1977) Transport properties of a two-temperature argon plasma, *Phys. Fluids* 20, 1668.

35. See references from 6. to 23. in the list given in Ref. 18.

36. See the introduction to ch. 5 at p. 123 of Ref. 4. Using Napolitano's notation, when the state parameter η_r in the entropic formulation is a progress variable the identification of its associated state equation B_r with the ratio of the reaction affinity A_r to the temperature T relative to the degrees of freedom in mutual equilibrium, i.e. $B_r = A_r/T$, does not appear to be correct.

37. See Sec. 1.15 at p. 3 of Ref. 4.

38. A system is defined isolated if it exchanges neither energy nor mass

with its environment. Upon a basic postulate, an isolated system cannot exchange any other physical property with its environment. See Secs. 1.5–1.6 at p. 3 of Ref. 4.

39. In order for the terminology to conform to this more general situation, it seems appropriate to extend the definition of affinity. Given that the chemical potential $\mu_\epsilon = \sum_{\delta=1}^{\ell_\epsilon} \mu_{\epsilon\delta}$ and the quantity $\sum_{\delta=1}^{\ell_\epsilon} \mu_{\epsilon\delta}/T_{\epsilon\delta}$ are state equations [see (10), (11)] in, respectively, the energetic and entropic schemes, it appears consistent to define the expression (28) as entropic affinity and to rename the expression (29) as energetic affinity. In this way, the equations (21), (27) indicate, respectively, that the chemical equilibrium driven by an entropically/energetically constrained thermal equilibrium is characterized by the vanishing of the energetic/entropic affinities of the non-frozen chemical reactions.

40. See Sec. 5.4 at p. 130 of Ref. 4.

41. See Eq. (106) at p. 201 of Ref. 18 and Eq. (21) at p. 50 of Ref. 25.

42. The choice of the degrees of freedom in mutual thermal equilibrium is not unique but varies from author to author. See the paragraph following Eq. (107) at p. 201 of Ref. 18.

43. Morro and Romeo assume, implicitly, the mutual thermal equilibrium of the degrees of freedom of the heavy species A and A$^+$; thus, and using their notation, the gas mixture they consider is characterized by the equilibrium temperature of the heavy species θ_h and the translational temperature θ_e of the electrons.

44. See Sec. 4 of Ref. 29.

45. See, for example, the paragraph following Eq. (14) at p. 783 of Ref. 32.

46. See Eq. (1) at p. 1668 of Ref. 34. However, the equation used by Kannappan and Bose, as it appears in their paper, shows a slight inaccuracy because the ratio of the electron translational temperature to the heavy species temperature, $1/\theta$ in the notation of the reference, is missing in the exponent of the internal partition functions. The same inaccuracy appears also in the corresponding equation reported by Morro and Romeo [See Eq. (28) at p. 47 of Ref. 29] for comparison with their own equation. Kannappan and Bose refer to S. Veis, "The Saha equation and lowering of the ionization energy for a two-temperature plasma" [in *Proceedings of the Czechoslovak Conference on Electronics and Vacuum Physics* (Prague, 7-12 Oct. 1968), edited by L. Paty, p. 105–110, Karlova Universita, 1968] for the derivation of the two-temperature Saha equation they used.

47. See Eq. (26) at p. 47 or Eq. (29) at p. 48 of Ref. 29.

THERMODYNAMIC APPROACH OF A FLOW WITH THERMAL NON-EQUILIBRIUM (translation-electron-vibration)

M. DUDECK
Université de Paris 6
Laboratoire d'Aérothermique - CNRS
4 ter route des Gardes, 92190, Meudon, France

1. Introduction

A fluid description of a high enthalpy and rarefied flow under non-equilibrium conditions is fruitful, because it allows to determine macroscopic parameters that are essential for the definition of ballistic vehicles or planetary probes.

A macroscopic approach is presented by taking into account a non-equilibrium situation, with different temperatures for the translation, the vibration of each molecular species and for the electrons. Based upon these temperatures the different equations for a fluid description are introduced.

Fundamental thermodynamic relations [1,2] are presented (Gibbs, Euler and Gibbs-Duhem) extended to complex mixtures and generalized to the case of several vibrational temperatures, a single translational temperature for heavy species (rotation will be always considered in balance with translation) and an electronic temperature. From these equation, one deduces the second differential of internal energy and entropy [3,4] to obtain the conditions of stability in the case of the description of a gas with several local temperatures.

From the assumption of a local Gibbs relation (quasi local equilibrium), expressed with the center of mass velocity field, for a nonequilibrium gas, one determines the balance equation for entropy [2,5] after having recalled the usual expressions for the balance equations of each species, global mass, internal energies and vibrational energies.

The entropy balance equation allows to precise the expression of the entropy flux and the rate of entropy production by unit of volume. This entropy production rate that contains contributions of the different irreversible phenomena is commented.

For a near equilibrium evolution of the gas, linear relationships are deduced from the entropy production for the different heat flow terms and for the terms of scalar energy exchanges.

2. Fundamental Relations for Internal Energy and Entropy

Consider a gas whose properties are uniform but that is characterized by different temperatures. One admits therefore that this medium can be represented by a set of subsystems, each subsystem being in local equilibrium conditions but in possible non-euilibrium with the other subsystems. Expressions of internal energy and the entropy

M. Capitelli (ed.), Molecular Physics and Hypersonic Flows, 281–292.

will be established in three cases: T(translation)-T_{vj}(vibration of the species j), T(translation)-T_e(electron), T(translation)-T_{vj}(vibration of the species j)-T_e(electron).

2.1 TRANSLATIONAL-ROTATIONAL ENERGY AND VIBRATIONAL ENERGIES

In the first case, one admits only one translational temperature for all species (heavy species and electrons) and different vibration temperatures for each molecule. The gas is therefore represented by a set of subsystems in mutual non-equilibrium, each being characterized by a temperature, $T \neq T_{v1} \neq T_{v2} \neq T_{v3}$.... The mole number of each species is constant and the global system is closed.

The existence of an entropy for each subsystem being admitted, total internal energy depends on the entropy of the different subsystems (S_{tr}, S_{v1}, S_{v2},...), the total volume V as well as of the mole number of each subsystem (N_1, N_2,...)

$$U = U (S_{tr}, S_{v1}, S_{v2},..., V, N_1, N_2,...). \tag{1}$$

This expression, that generalizes the usual relations associated to a simple system can be considered as a postulate, the internal energy being a function of all extensive variables of the gas [1]. Additivity is assumed for the entropy function which allows to formulate the pressure as the partial derivative of U to volume V at constant entropies. Under the assumption that there is no coupling between the entropies of subsystems and that additivity applies to the internal energy, temperatures T_{vj} can be introduced in the generalized Gibbs, Euler and Gibbs-Duhem relations,

$$dU = T \ dS_{tr} + \sum_j T_{vj} dS_{vj} - p \ dV = T \ dS + \sum_j (T_{vj} - T) \ dS_{vj} - p \ dV \tag{2}$$

$$U = T \ S + \sum_j (T_{vj} - T) \ S_{vj} - p \ V + \sum_i \mu_i \ N_i \tag{3}$$

$$0 = S \ dT + \sum_j S_{vj} \ d(T_{vj} - T) - V \ dp + \sum_i N_i \ d\mu_i \tag{4}$$

where S is the total entropy and μ_i is the chemical potential of the component i (molecule or atom of the gas, i=1,...n), defined as the partial derivative of internal energy to the mole numbers at constant total entropy and volume. The subscript i corresponds to the external (translational) and rotational energy of each species in the gas. We assume the rotational temperature to be equal to the translational temperature allowing to treat rotation together with translation. The subscript j corresponds to the vibrational internal energy and sums over all molecular species.

* according the Gibbs relation, $T_{vj} - T = (\dfrac{\partial U}{\partial S_{vj}})_{S_{tr}, S_{vi \neq j}, V}$

* per unit of mass

$$u = T \ s + \sum_j (T_{vj} - T) \ Y_j \ s_{vj} - p \ v + g \tag{5}$$

$$0 = s \ dT + \sum_j Y_j \ s_{vj} \ d(T_{vj} - T) - v \ dp + dg \tag{6}$$

$$du = T \ ds + \sum_j (T_{vj} - T) \ Y_j \ ds_{vj} - p \ dv \tag{7}$$

u, s, g, v are respectively , the internal energy, the entropy, the Gibbs function and the volume, Y_j is the mass fraction of the species j and s_{vj} is the entropy. All expressions are per unit of mass of j. The relation (7) is used to identify on which variables the internal energy u depends. An expression for a small, finite variation of the internal energy can now be formulated with constant mass fractions Y_j :

$$\delta u = T \ \delta s \ + \sum_j (T_{vj} - T) Y_j \ \delta s_{vj} - p \ \delta v + \frac{1}{2} \delta^2 u . \tag{8}$$

Here $\delta^2 u$ is the second derivative of the internal energy per unit of mass. It can be shown that the second law of thermodynamics imposes that for any real system this term must be positive (the so-called condition of stability) and consequently this term cannot be left out in the formulation. To this second derivative, one associates the quadratic expression whose the matrix (so-called *matrix of stability*) is

$$\begin{pmatrix} u_{ss} & u_{ss_{v1}} & u_{ss_{v2}} & u_{sv} \\ u_{s_{v1}s} & u_{s_{v1}s_{v1}} & u_{s_{v1}s_{v2}} & u_{s_{v1}v} \\ u_{s_{v2}s} & u_{s_{v2}s_{v1}} & u_{s_{v2}s_{v2}} & u_{s_{v2}v} \\ . & . & . & . \\ . & . & . & . \\ u_{vs} & u_{vs_{v1}} & u_{vs_{v2}} & . & u_{vv} \end{pmatrix} = \begin{pmatrix} \dfrac{T}{c_v} & 0 & 0 & -\dfrac{l_v}{c_v} \\ 0 & Y_1\dfrac{T_{v1}}{c_{v1}} & 0 & 0 \\ 0 & 0 & Y_2\dfrac{T_{v2}}{c_{v2}} & 0 \\ . & . & . & . \\ -\dfrac{l_v}{c_v} & 0 & 0 & . & -\gamma\dfrac{l_v}{l_p} \end{pmatrix}$$

with $\tag{9}$

$$du = c_v dT + \sum_j c_{v_j} dT_{vj} + (l_v - p) dv \tag{10}$$

and c_v is the heat capacity for translation-rotational energy and c_{vj} is the heat capacity for the vibrational energy for the molecule species j. We use the notation $u_{xy} = \dfrac{\partial^2 u}{\partial x \partial y}$.

In consequence of the condition of stability, the diagonal elements must be positive. This implies $u_{ss} > 0$, giving $c_v > 0$ and $u_{s_j s_j} > 0$ inducing $c_{vj} > 0$ and $u_{vv} > 0$. The physical interpretation of this is that pressure and volume vary in opposite sense when the entropies are constant.

One can proceed the same manner with the entropy function. The entropy is supposed to be an extensive function of the internal energy function of all subsystems, the volume and the mole numbers as

$$S = S \ (U_{tr}, U_{v1}, U_{v2},..., V, N_1, N_2,...). \tag{11}$$

From relation (3) the following expression for the total entropy S is deduced :

$$S = \frac{U}{T} - \sum_j \frac{T_{vj} - T}{T} S_{vj} + \frac{p}{T} V - \frac{\sum_i N_i \ d\mu_i}{T} . \tag{12}$$

284

In order to reformulate this equation in terms of internal energy U_{vj}, seperate descriptions for the translational-rotational energy and for the vibrational internal energy have to be introduced.

Translation -rotation

The internal energy U_{tr} for translation-rotation is assumed to be a function of total translational-rotational entropy S_{tr}, global volume V and mass M_i.

$$U_{tr} = U_{tr}(S_{tr}, V, M_1, ..., M_n) \tag{13}$$

Then,
$$dU_{tr} = T \, dS_{tr} - p \, dV \tag{14}$$

where p is the total pressure. The Euler and Gibbs-Duhem relations are :

$$U_{tr} = T S_{tr} - p V + \sum_i g_i M_i = T S_{tr} - p V + \sum_i \mu_i N_i \tag{15}$$

$$0 = S_{tr} \, dT - V \, dp + M \, dg_{tr} \tag{16}$$

Vibrations

The vibrational energy U_{vj} is assumed to be a function of the vibrational entropy S_{vj} only.

$$dU_{vj} = T_{vj} \, dS_{vj} \quad (17) \quad U_{vj} = T_{vj} S_{vj} + g_{vj} M_j = T_{vj} S_{vj} + \mu_{vj} N_j \tag{18}$$

$$0 = S_{vj} \, dT_{vj} + M_j \, dg_{vj} \tag{19}$$

From (2) and (3), one deduces dS as a function of dU, dS_{vj}, dV and S as a function of U, S_{vj}, V. Using from (17) and (18) for the expressions of dS_{vj} and S_{vj}, the expressions of dS and S are :

$$dS = \frac{dU}{T} - \sum_j (\frac{1}{T} - \frac{1}{T_{vj}}) \, dU_{vj} + \frac{p}{T} \, dV \tag{20}$$

$$S = \frac{U}{T} - \sum_j (\frac{1}{T} - \frac{1}{T_{vj}}) U_{vj} + \frac{p}{T} V + \sum_j [(\frac{1}{T} - \frac{1}{T_{vj}}) \mu_{vj} - \mu_i] N_j - \frac{1}{T} \sum_{i \neq j} \mu_i N_i. \tag{21}$$

Per unit of mass, this writes :

$$ds = \frac{du}{T} - \sum_j (\frac{1}{T} - \frac{1}{T_{vj}}) Y_j \, du_{vj} + \frac{p}{T} \, dv \tag{22}$$

$$s = \frac{u}{T} - \sum_j (\frac{1}{T} - \frac{1}{T_{vj}}) Y_j \, u_{vj} + \frac{p}{T} v + \sum_j (\frac{1}{T} - \frac{1}{T_{vj}}) g_{vj} Y_j - \frac{1}{T} \sum_i g_i Y_i \tag{23}$$

Analogous to the precedent expression of a small variation of internal energy, a small variation of the entropy is

$$\delta s = \frac{\delta u}{T} - \sum_j (\frac{1}{T} - \frac{1}{T_{vj}}) Y_j \, \delta u_{vj} + \frac{p}{T} \, \delta v + \frac{1}{2} \delta^2 s \tag{24}$$

with the following matrix associated to the quadradratic form of the second derivative $\delta^2 s$:

$$\begin{pmatrix} s_{uu} & s_{uu_{v1}} & s_{uu_{v2}} & s_{uv} \\ s_{u_{v1}u} & s_{u_{v1}u_{v1}} & s_{u_{v1}u_{v2}} & s_{u_{v1}v} \\ s_{u_{v2}u} & s_{u_{v2}u_{v1}} & s_{u_{v2}u_{v2}} & s_{u_{v2}v} \\ \cdot & \cdot & & \cdot \\ \cdot & \cdot & \cdot & \cdot \\ \cdot & & \cdot & \cdot \\ s_{vu} & s_{vu_{v1}} & s_{vu_{v2}} & \cdot & s_{vv} \end{pmatrix}$$

$$= \begin{pmatrix} -\dfrac{1}{c_v T^2} & 0 & 0 & \dfrac{\partial}{\partial v}(\dfrac{1}{T})_{u_{vi},u} \\ 0 & -\dfrac{Y_1}{c_{v1}T_{v1}^2} & 0 & 0 \\ 0 & 0 & -\dfrac{Y_2}{c_{v2}T_{v2}^2} & 0 \\ \cdot & \cdot & \cdot & \cdot \\ \dfrac{\partial}{\partial v}(\dfrac{1}{T})_{u_{vi},u} & 0 & 0 & \cdot & \dfrac{\partial}{\partial v}(\dfrac{p}{T})_{u_{vi},u} \end{pmatrix}$$

Note that when the non-equilibrium terms are put to zero, all established formulae reduce to the classical thermodynamical relations, the latter thus being a special case of the here represented more general treatment.

2.2 TRANSLATIONAL ENERGY FOR HEAVY SPECIES AND ELECTRONS

Now, consider the case of an atomic gas, characterized by only two temperatures ; one for the electrons, T_e, and the other T for the heavy species. Only two subsystems are therefore introduced, each subsystem containing a constant molar number. The method is similar to the previous presented method. The total internal energy of the gas is in the form $U= U (S_{tr}, S_e, V, N_1, N_2,...)$. For the internal energy, one deduced the relations

$$du=T\ ds+(T_e-T)\ Y_e\ ds_e-p\ dv \tag{25}$$

$$u=T\ s+(T_e-T)\ Y_e\ s_e-p\ v+g \tag{26}$$

$$\delta u=T\ \delta s+(T_e-T)\ Y_e\ \delta s_e-p\ \delta v+\frac{1}{2}\delta^2 u. \tag{27}$$

In order to describe the properties of the entropy, the following relations for each subsystem are introduced.

Heavy components (translation)

$$dU_{tr}=T\ dS_{tr}-p_1\ dV \tag{28} \qquad\qquad U_{tr}=T\ S_{tr}-p_1\ V+ \sum_{i\neq \acute{e}l} g_i\ M_i \tag{29}$$

Electrons

$$dU_e=T_e\ dS_e-p_e\ dV \tag{30} \qquad\qquad U_e=T_e\ S_e-p_e\ V+g_e\ M_e \tag{31}$$

One obtains for the entropy per unit of mass

$$ds=\frac{du}{T}+(\frac{1}{T}-\frac{1}{T_e})\,Y_e du_e+[(\frac{1}{T}-\frac{1}{T_e})\,p_e+\frac{p}{T}]\,dv \tag{32}$$

$$s=\frac{u}{T}-(\frac{1}{T}-\frac{1}{T_e})Y_e u_e+[-(\frac{1}{T}-\frac{1}{T_e})p_e+\frac{p}{T}]\,v+(\frac{1}{T}-\frac{1}{T_e})\,Y_e\,g_e-\frac{g}{T}. \tag{33}$$

The matrices of stability associated to the development of δu and δs are

$$\begin{pmatrix} u_{ss} & u_{se} & u_{sv} \\ u_{es} & u_{ee} & u_{ev} \\ u_{vs} & u_{ve} & u_{vv} \end{pmatrix} = \begin{pmatrix} \dfrac{T}{c_v} & 0 & -\dfrac{l_{vtr}}{c_v} \\ 0 & \dfrac{T_e}{c_{ve}} & -\dfrac{l_{ve}}{c_{ve}} \\ -\dfrac{l_{vtr}}{c_v} & -\dfrac{l_{ve}}{c_{ve}} & -\dfrac{c_{p_{tr}}}{c_{v_{tr}}}\dfrac{l_{v_{tr}}}{l_{p_{tr}}} \end{pmatrix} \tag{34}$$

and

$$\begin{pmatrix} s_{uu} & s_{ue} & s_{uv} \\ s_{eu} & s_{ee} & s_{ev} \\ s_{vu} & s_{ve} & s_{vv} \end{pmatrix} = \begin{pmatrix} -\dfrac{1}{c_v T^2} & 0 & (\dfrac{\partial}{\partial v}\dfrac{1}{T})_{u,u_e} \\ 0 & -\dfrac{1}{c_{ve} T_e^2} & 0 \\ (\dfrac{\partial}{\partial v}\dfrac{1}{T})_{u,u_e} & 0 & s_{vv} \end{pmatrix}, \tag{35}$$

and can be used to analyse the conditions of stability.

2.3 SYSTEM WITH ONE TRANSLATIONAL TEMPERATURE, ONE ELECTRON TEMPERATURE AND SEVERAL VIBRATIONAL TEMPERATURES

Let us now consider a gas with one translational temperature T for heavy species, one temperature for electrons and different vibrational temperatures for the each molecular species. For reason of simplicity, the derivation of the equations is first established with one vibrational temperature. Extension to systems with more vibrational temperatures is then achieved by simply replacing all terms with T_v by equivalent terms with T_{vj}, with summation over j. The principle of additivity allows this approach. Starting from the internal energy for the three-temperature case, $U = U (S, S_e, S_v, V, N_1, N_2,....)$ we write:

$$du=T\,ds+(T_e-T)\,Y_e ds_e+(T_v-T)Y_v\,s_v-p\,dv \tag{36}$$

$$u=T\,s+(T_e-T)\,Y_e s_e+(T_v-T)\,Y_v s_v-p\,v+\sum_i g_i\,Y_i\,. \tag{37}$$

Using the properties of subsystems, one can write for the function entropy

$$ds=\frac{du}{T}-(\frac{1}{T}-\frac{1}{T_e})\,Y_e du_e-(\frac{1}{T}-\frac{1}{T_v})\,Y_v du_v+[(\frac{1}{T}-\frac{1}{T_e})\,p_e+\frac{p}{T}]\,dv \tag{38}$$

$$s = \frac{u}{T} - (\frac{1}{T} - \frac{1}{T_e}) Y_e u_e - (\frac{1}{T} - \frac{1}{T_v}) Y_v u_v + [-(\frac{1}{T} - \frac{1}{T_e}) p_e + \frac{p}{T}] \, v$$

$$+ (\frac{1}{T} - \frac{1}{T_e}) \, Y_e g_e + (\frac{1}{T} - \frac{1}{T_v}) \, Y_v g_v - \frac{g}{T}.$$

(39)

Generalization of these equations to multiple vibrational temperatures and chemical reactions now is straightforward. The global system is closed and the total entropy is:

$$ds = \frac{du}{T} - (\frac{1}{T} - \frac{1}{T_e}) \, dY_e u_e - \sum_j (\frac{1}{T} - \frac{1}{T_{vj}}) \, dY_j u_{vj} + [-(\frac{1}{T} - \frac{1}{T_e}) \frac{p_e}{T_e} + \frac{p}{T}] dv \quad (40)$$

$$- \sum_j (\frac{\mu_1}{T} + \frac{\mu_{vj}}{T_{vj}}) \, dY_j - \sum_{i \neq j, i \neq \text{él.}} \frac{\mu_i \, dY_i}{T} - \frac{\mu_e \, dY_e}{T_e}$$

3. Quasi local equilibrium

In the case of quasi local equilibrium it is assumed that the last relation is locally valid. In a flowing system this implies that for a mass element, following it's center of mass motion, we find the following relation by multiplying equation (40) by the total density ρ of the gas:

$$\rho \frac{ds}{dt} = \frac{1}{T_e} \rho \frac{dY_e u_e}{dt} + \sum_j \frac{1}{T_{vj}} \rho \frac{dY_j u_{vj}}{dt}$$

$$+ \frac{1}{T} \sum_{i \neq j, i \neq \text{él.}} \rho \frac{dY_i u_i}{dt} + (\frac{p_e}{T_e} + \frac{\sum\limits_{j \neq \text{él.}} p_j}{T}) \rho \frac{dv}{dt} \quad (41)$$

$$- \frac{g_e}{T_e} \rho \frac{dY_e}{dt} - \sum_j (\frac{g_{vj}}{T_{vj}} + \frac{g_j}{T}) \rho \frac{dY_j}{dt} + \sum_{i \neq j, i \neq \text{él.}} \frac{g_i}{T} \rho \frac{dY_i}{dt}$$

These expressions can only be used, on the condition that the gradients are not too large; in practice, this is considered to be true. when the theory appears to be adequate to describe a studied system. The derivation of the entropy balance equation for a non-equilibrium gas as presented in the following is based upon this expression. The right hand side of equation (41) will be reorganized into a group of flux terms and a group of production terms. For this operation the balance equations for internal energy, electron energy, vibrational energy, mass of each species and total mass will be used.

4. Balance equations

The macroscopic balance equations are recalled in their usual form.

4.1 MASS
The mass balance of component j is written as

$$\rho\frac{dY_j}{dt}+\mathrm{div}\,\vec{J}_j=\dot{w}_j \tag{42}$$

where $\vec{J}_j$ is the diffusion flow of the species j, $\vec{J}_j=\rho_j(\vec{V}_j-\vec{V})$ and $\dot{w}_j$ is the production rate by unit of volume due to chemical processes. The derivative is calculated by using the velocity of the centre of mass velocity, $\vec{V}=\sum Y_j\,\vec{V}_j$. In an ionized gas, $\vec{J}_j$ is gouverned by ambipolair diffusion [9], [12]. In our presentation, the set of equations will be closed after the analysis of the entropy production.

The total mass balance equation is $\rho\dfrac{dv}{dt}-\mathrm{div}\,\vec{V}=0$. $\tag{43}$

4.2 MOMENTUM

The momentum balance equation of species j can be written as

$$\rho\frac{d\,Y_j\vec{V}_j}{dt}=-\mathrm{div}(\vec{V}_j\otimes\rho_j(\vec{V}_j-\vec{V})+\bar{\bar{P}}_j)+\dot{w}_{\vec{V}_j} \tag{44}$$

where ρ_j is the specific density of the species j, $\bar{\bar{P}}_j$ is the pressure tensor of component j and $\dot{w}_{\vec{V}_j}$ is the momentum production of species j. Taking into account that the ambipolair field reduces the electronic diffusion, several authors ([12], for example) simplify this equation by neglecting the electronic viscosity. A source term for each species j representing the effect of friction can be introduced [7]. In general, the momentum equation is not solved for each different species. The total momentum equation, however, is necessary to close the set of equations [9],[10],[14],[12].

4.3 TRANSLATIONAL ENERGY

The translational energy balance equation for of species j is

$$\rho\frac{dY_i u_i}{dt}=-\mathrm{div}(\vec{q}_i+u_i\vec{J}_i)-p_i\,\mathrm{div}\vec{V}_i-\bar{\bar{\Pi}}_i:\mathrm{grad}\ \vec{V}_i+\vec{f}_i.\vec{J}_i+\dot{Q}_i \tag{45}$$

This equation is used in [14] to describe a non-equilibrium argon plasma flow, without viscosity and external forces and by admitting that $\vec{V}_e\approx\vec{V}$. This last assumption is also used in [11] to describe a nitrogen plasma.

4.4 VIBRATIONAL ENERGY

The balance equation for the vibrational energy of the component j is

$$\rho\frac{dY_j u_{vj}}{dt}=-\mathrm{div}(\vec{q}_{vj}+u_{vj}\vec{J}_i)+\dot{Q}_{vj} \tag{46}$$

where $\vec{q}_{vj}$ is the vibrational heat flow and the scalars $\dot{Q}_{vj}$ are the vibrational energy exchange terms : vibrational, translational and chemical processes involving the vibration of the component j. For example, Coquel F.,[12] uses equation (46) without the flux term and without the exchange term, but keeps the convective term with the centre-of-mass velocity. Sleziona P.C.,[11] uses all the terms with the same assumption for the velocity.

4. Entropy Flux and Entropy Production

With the help of the different balance equations, the balance equation for entropy is written as $\rho \dfrac{ds}{dt} = - \text{div} \vec{J}_s + \dot{w}_s$, where the diffusion entropy flow is

$$
\vec{J}_s = -\frac{\vec{q}_e + u_e \vec{J}_e}{T_e} - \sum_j \frac{\vec{q}_{vj}}{T_{vj}} - \sum_{i \neq j, i \neq \acute{e}l.} \frac{\vec{q}_i + u_i \vec{J}_i}{T}
$$

$$
- \frac{g_e}{T_e} \vec{J}_e - \sum_j (\frac{g_{vj}}{T_{vj}} + \frac{g_j}{T}) \vec{J}_j + \sum_{i \neq j, i \neq \acute{e}l.} \frac{g_i}{T} \vec{J}_i
$$

(47)

The entropy production shows terms as $-p_i \text{div} \vec{V}_i$ due to the compressibility of subsystems. The sum of these terms appearing in the balance of energy must give $-p \text{div} \vec{V}$ in the equation for the global energy that conduces to neglecte the diffusion terms. If the terms due to diffusion effects are neglected, the entropy production is then

$$
\dot{w}_s = (\vec{q}_e + u_e \vec{J}_e). \overrightarrow{\text{grad}} (\frac{1}{T_e}) + \sum_j \vec{q}_{vj}. \overrightarrow{\text{grad}} (\frac{1}{T_{vj}}) + \sum_{i \neq j, i \neq \acute{e}l.} (\vec{q}_i + u_i \vec{J}_i). \overrightarrow{\text{grad}} (\frac{1}{T})
$$

$$
- \frac{\overline{\overline{\Pi}}_e : \overline{\text{grad}} \, V_e}{T_e} - \sum_{i \neq j, i \neq \acute{e}l.} [\frac{\overline{\overline{\Pi}}_i : \overline{\text{grad}} \, V_i}{T}] + \frac{\vec{f}_e . \vec{J}_e}{T_e} + \sum_{i \neq j, i \neq \acute{e}l.} [\frac{\vec{f}_i . \vec{J}_i}{T}]
$$

$$
+ \frac{\dot{Q}_e}{T_e} + \sum_j \frac{\dot{Q}_{vj}}{T_{vj}} + \sum_{i \neq j, i \neq \acute{e}l.} [\frac{\dot{Q}_i}{T}]
$$

$$
- \vec{J}_e. \overrightarrow{\text{grad}} (\frac{g_e}{T_e}) - \sum \vec{J}_j. \overrightarrow{\text{grad}} (\frac{g_{vj}}{T_{vj}} + \frac{g_j}{T}) - \sum_{i \neq j, i \neq \acute{e}l.} \vec{J}_i. \overrightarrow{\text{grad}} (\frac{g_i}{T})
$$

$$
+ \frac{g_e \dot{w}_e}{T_e} + \sum_j (\frac{g_{vj}}{T_{vj}} + \frac{g_j}{T}) \dot{w}_j + \sum_{i \neq j, i \neq \acute{e}l.} \frac{g_i \dot{w}_i}{T}
$$

(48)

This expression can be reduced to the usual form for a local thermal equilibrium.

5. Consequences

Some consequences using the expression of the entropy production are presented.

5.1 HEAT CONDUCTION

Using the notations for the heat flow $\vec{q}_e^* = \vec{q}_e + u_e\vec{J}_e$, $\vec{q}_i = \vec{q}_i + u_i\vec{J}_i$. For equilibrium conditions, heat flow and thermal gradients are zero, the different temperatures are uniform in the gas. For an evolution near equilibrium conditions, one can admit phenomenological relations introducing direct interactions as

$$
\begin{bmatrix} \vec{q}_e^* \\ \vec{q}_{v1} \\ . \\ \vec{q}_{vn} \\ \vec{q} \end{bmatrix} = - \begin{bmatrix} \lambda_e & & & & \\ & \lambda_{v1} & & & \\ & & . & & \\ & & & \lambda_{vn} & \\ & & & & \lambda \end{bmatrix} . \begin{bmatrix} \overrightarrow{\text{grad } T_e} \\ \overrightarrow{\text{grad } T_{v1}} \\ . \\ \overrightarrow{\text{grad } T_{vn}} \\ \overrightarrow{\text{grad } T} \end{bmatrix} \tag{49}
$$

where heat conductivity coefficients have to be positive according to the second law of thermodynamics. For the heat tranfer terms, the entropy production is

$$
\dot{w}_s = \frac{\lambda_e}{T_e^2} . (\overrightarrow{\text{grad } T_e})^2 + \sum_j \frac{\lambda_{vj}}{T_{vj}^2} . (\overrightarrow{\text{grad } T_{vj}})^2 + \sum_{i \neq j, i \neq \acute{e}l.} \frac{\lambda_i}{T_i^2} . (\overrightarrow{\text{grad } T_i})^2 \tag{50}
$$

Coupling effects can be introduced by completing the transfer matrix

$$
\begin{bmatrix} \vec{q}_e^* \\ \vec{q}_{v1} \\ . \\ \vec{q}_{vn} \\ \vec{q} \end{bmatrix} = - \begin{bmatrix} \lambda_e & \lambda_{ev_1} & . & \lambda_{ev_n} & \lambda_{e\,tr} \\ \lambda_{v_1 e} & \lambda_{v_1 v_1} & . & \lambda_{v_1 v_n} & \lambda_{v_1\,tr} \\ . & . & . & . & . \\ . & \lambda_{v_n v_1} & . & \lambda_{vn} & . \\ \lambda_{tr\,e} & . & . & . & \lambda \end{bmatrix} . \begin{bmatrix} \overrightarrow{\text{grad } T_e} \\ \overrightarrow{\text{grad } T_{v1}} \\ . \\ \overrightarrow{\text{grad } T_{vn}} \\ \overrightarrow{\text{grad } T} \end{bmatrix} \tag{51}
$$

5.2 SCALAR ENERGY EXCHANGES

Now the term of source in the entropy production due to scalar exchanges of energy is considered. In the entropy production, this term can be written as

$$
\dot{Q}_e(\frac{1}{T_e} - \frac{1}{T}) + \sum_j \dot{Q}_{vj}(\frac{1}{T_{vj}} - \frac{1}{T}) \tag{52}
$$

using the global conservation of energy. For equilibrium conditions ($T_e=T=T_{vj}$), the different terms of exchange cancel and for a situation near equilibrium, one can admit phenomenological relations as

$$
\begin{bmatrix} \dot{Q}_e \\ \dot{Q}_{v1} \\ \cdot \\ \dot{Q}_{vn} \end{bmatrix} = \begin{bmatrix} \Lambda_e & \Lambda_{ev_1} & \cdot & \Lambda_{ev_n} \\ \Lambda_{v_1e} & \Lambda_{v_1} & & \Lambda_{v_1v_n} \\ \cdot & & \cdot & \\ \Lambda_{v_ne} & \Lambda_{v_nv_1} & \cdot & \Lambda_{v_n} \end{bmatrix} \cdot \begin{bmatrix} T - T_e \\ T - T_{v1} \\ \cdot \\ T - T_{v1} \end{bmatrix}
\tag{53}
$$

with the following relation for the entropy production

$$
\dot{w}_s = \Lambda_e \frac{(T-T_e)^2}{TT_e} + \sum_j \Lambda_{ev_j} \frac{(T-T_e)(T-T_{vj})}{TT_e} + \sum_j \Lambda_{v_je} \frac{(T-T_e)(T-T_{vj})}{TT_{vj}}
$$
$$
+ \sum_{i,j} \Lambda_{v_ivj} \frac{(T-T_{vi})(T-T_{vj})}{TT_{vj}}
\tag{54}
$$

This expression is positive according to the second law of thermodynamics.

6. Conclusions

The presented description corresponds to an extension of the usual notions of macroscopic thermodynamics in the presence a local thermal nonequilibrium. Fundamental equations of Gibbs, Euler and Gibbs-Duhem have been generalized to flowing gas systems with one electron temperature, one translational temperature for heavy particles and multiple vibrational temperatures for the molecules. Matrices associated to the second derivative of the internal energy and of the entropy have been established, they serve to define criteria of thermodynamical stability.

The balance equation for the entropy is deduced from the balance equation for the mass of each species, for the total mass, for the momentum of species, for the translational and for the vibrational energy. The entropy flux and the entropy production are identified and phenomenological relations are proposed for heat flow and scalar heat exchanges.

References

1. Napolitano, L. (1971) *Thermodynamique des Systèmes Composites*, Mémoire des Sciences Physiques, Gauthiers-Villars, Paris.
2. Prud'homme, R. and Flament, C. (1991) Rapport du Laboratoire d'Aérothermique RL. 91-2, janvier 1991.
3. Glansdorff, P. and Prigogine, I., *Struture, Stabilité et Fluctuations*, Masson, France, 1971.
4. Emanuel, G. (1987) *Advanced Classical Thermodynamics*, AIAA Education Series.
5. De Groot, S.R. and Mazur, P. (1969) *Non Equilibrium Thermodynamics*, North-Holland Publ.Comp.
6. Benoy, D.A. (1993) Thesis, University of Eindhoven, North-Holland.

7. Maire, P.H. (1994) Ionisation dans un Ecoulement Aérodynamique, Rapport CEA-N-2774.

8. Park, C. and Seung-Ho Lee (1993) Validation of Multi-temperature Nozzle Flow Code NOZNT, AIAA paper 93-2862.

9. Beulens, J.J., Milojevic, D., Schram, D.C. and Vallinga, P.M. (1991) A two-dimensional non equilibrium model of cascaded arc plasma flows, *Phys.Fluids B* **3** (9), 2548-2557.

10. Canupp, P.W., Candler, G.V., Perkins, J.N. and Erickson, W.D. (1993) Analysis of hypersonic nozzles including vibrational nonequilibrium and intermolecular force effects, AIAA Journal **11**, No.7.

11. Sleziona, P.C., Gogel, T.H.and Messerschmid, E.W.(1992) Non-equilibrium Flow in a Hypersonic Expansion Nozzle, Report for Hermès R&Q Program.

12. Coquel, F. and Marmignon, C. (1995) A Ror-type linearization for the Euler equations for weakly ionized multi-component and multi-temperature gas, 26th AIAA Plasmadynamics and Lasers Conference, June 19-22, San Diego, USA.

13. Buuron, A. (1995) private communication.

THERMODYNAMIC PROPERTIES OF HIGH-TEMPERATURE AIR COMPONENTS.

M. CAPITELLI, G. COLONNA, C. GORSE
Centro di Studio per la Chimica dei Plasmi del CNR
Dip. Chimica, Università di Bari
Via Orabona 4, 70126 Bari, Italy

AND

D. GIORDANO
European Space Research & Technology Center
P.O. Box 299, 2200 AG Noordwijk, The Netherlands

Abstract. The paper gives a synthetic description of the basic methods to calculate the high-temperature thermodynamic properties of air species. Emphasis is given to the strong dependence of the electronic partition function and the thermodynamic properties of atomic species on the number of electronic levels taken into account. The specific heat of the diatomic species presents a maximum followed by a steep decrease towards zero. This effect, which is usually not reported in standard textbooks of statistical thermodynamics, is a direct consequence of the finite number of energy levels accounted for in the internal partition function. A similar effect exists also for the atoms.

1. Introduction

In the last three decades, many papers have been published on the thermodynamic properties of high temperature gases, with particular reference to air species because of their importance in the aerothermodynamic analysis of the hypersonic flow surrounding a space vehicle during its reentry into the Earth's atmosphere.

Recently, new data relating to the thermodynamic properties of some air molecular species have been presented.[1-3] These data have been ob-

293

M. Capitelli (ed.), Molecular Physics and Hypersonic Flows, 293–301.

tained from sophisticated approaches to the calculation of the rotovibrational energies of the diatomic molecules. In general, these new data are in satisfactory agreement with those provided by older methods based on analytical expressions of the rotovibrational energy. However, at sufficiently high temperatures, an unexpected effect is found: the internal constant pressure specific heat of the diatomic molecules vanishes[1] instead of converging to the expected value of two times the universal gas constant given in standard textbooks. This peculiarity, which is due to the use of a finite number of rotovibrational energy levels in the calculation of the internal partition function and its derivatives, was already found[4] many years ago for molecular hydrogen. On the other hand, the new methods are important in that they make it easier to understand the problem of separation of the different contributions (electronic, rotational, vibrational) to the internal partition function, a problem which has to be solved if one must construct thermodynamic models in thermal non-equilibrium. In a situation of thermal disequilibrium, different temperatures characterise the independent degrees of freedom of the molecule and it becomes necessary to separate the associated contributions to the basic thermodynamic properties. This problem was dealt with in the past by introducing simple models for the different energy modes of a molecule but it becomes more complicated when a rigorous quantum-mechanical approach is followed; strictly speaking, a rigorous quantum-mechanical approach works against the separation of the electronic, rotational and vibrational contributions.

The separation between translational and electronic degrees of freedom raises no problem for the atomic species and one may believe that the calculation of the thermodynamic properties is straightforward. This impression is, however, illusory. In fact, the problems[5,6] met many years ago with respect to the calculation of the electronic partition functions of monoatomic species have not yet been solved: the electronic partition function of an isolated monoatomic species is not bounded and a suitable cut-off criterion must be used to truncate it. Unfortunately, there is no universal cut-off criterion; the existing cut-off criteria yield partition functions, and their derivatives, that depend on either the electron density or the gas pressure. This means that the thermodynamic properties of single species depend not only on the temperature but also on the pressure. Additional problems arise from the determination of the electronic energy levels of the atomic species: Moore's well-known tables[7] help to some extent because they provide energies and statistical weights for many observed levels. Nonetheless, supplementing these tables with semiempirical laws is still the only way to obtain the missing energy levels. Exact quantum-mechanical calculations for high-lying levels, i.e. near the continuum limit, are still a prohibitive task.

The purpose of this paper is to give a synthetic description of the basic methods to calculate the high-temperature thermodynamic properties of air species (N, N^+, N^{2+}, N^{3+}, N^{4+}, O, O^+, O^-, O^{2+}, O^{3+}, O^{4+}, N_2, N_2^+, O_2, O_2^+, O_2^-, NO, NO^+). A more detailed presentation is given in a recent report.[8]

2. The Partition Function

The partition function Q is the basic construct of statistical thermodynamics required to calculate the thermodynamic properties of single species. In general, the partition function can be factorised in the form of a product of the translational Q_{tr} and the internal Q_{int} contributions

$$Q = Q_{tr} Q_{int}$$

The translational partition function Q_{tr} is given in analytical form as

$$Q_{tr} = \left(\frac{2\pi m k T}{h^2} \right)^{3/2} V$$

where m is the mass of the molecule, k is the Boltzmann constant, T is the temperature, h is the Planck constant, and V is the volume.

The internal partition function is defined as

$$Q_{int} = \sum_n g_n \exp\left(-\frac{E_n}{kT}\right)$$

where g_n and E_n represent the statistical weight and the energy of the nth internal quantum level of the molecule under consideration, the sum running over all the accessible levels. The calculation of Q_{int} is straightforward when the characteristics g_n and E_n are known. Despite the apparent simplicity, however, the existing values of Q_{int} can differ by orders of magnitude, especially for the atomic species.

3. The Internal Partition Function of Atomic Species

In the case of atomic species, the internal partition function is the sum over the infinite electronic levels

$$Q_{int} = \sum_n^\infty g_n \exp\left(-\frac{E_n}{kT}\right) \tag{1}$$

In principle, the electronic energies E_n and the corresponding statistical weights g_n can be calculated by quantum mechanics. However, semi-empirical methods are still used for practical calculations because of the mathematical difficulties inherent in the quantum mechanical approach when dealing with the near-continuum electronic states. Another difficulty deriving from the use of (1) is the lack of appropriate cut-off criteria to avoid the divergence of the partition function. In a recent calculation,[9] the internal partition function of air atomic species has been obtained by summing up to the last energy level arising from the limitation

$$E_n \leq E_i - \Delta E_i \tag{2}$$

In (2), E_i is the ionisation potential of the atomic species and ΔE_i is the corresponding lowering of the ionisation potential. The latter quantity has been considered as a parameter. The recent calculation reported in Refs. 8 and 9 shows the strong dependence of the internal partition function of air atomic species on the assumed ionisation potential lowering.

4. Thermodynamic properties

The thermodynamic properties of the single species follow in a straightforward manner when the internal partition function and its logarithmic derivatives are known. The internal contribution E_{int} to the thermodynamic energy and the internal constant pressure specific heat $C_{p,int}$ are obtained, respectively, from

$$E_{int} = RT^2 \left(\frac{\partial \ln Q_{int}}{\partial T} \right)_V$$

$$C_{p,int} = R \left[2T \left(\frac{\partial \ln Q_{int}}{\partial T} \right)_V + T^2 \left(\frac{\partial^2 \ln Q_{int}}{\partial T^2} \right)_V \right]$$

The global quantities are obtained by adding up the translational contributions; they read

$$E = E_{int} + \frac{3}{2} RT$$

$$C_p = C_{p,int} + \frac{5}{2} R$$

The internal contribution to the thermodynamic energy and the internal constant pressure specific heat depend strongly on the assumed ionisation potential lowering (see Refs. 8 and 9). It is worth noticing that the internal specific heat, after reaching a maximum, strongly decreases and practically vanishes at high temperature. This effect, which is a consequence of considering a limited number of levels in the partition function and its derivatives, is similar to the one occurring in the molecules when a finite number of rotovibrational states is considered.

5. The Internal Partition Function of Molecular Species

The starting point for calculating the internal partition function of molecular species is to solve the Schrödinger equation of a representative molecule to obtain the energy levels corresponding to the independent molecular degrees of freedom. The Schrödinger equation is solved in the well known Born-Oppenheimer approximation, which separates the motion of the electrons from that of the nuclei. The solution of the Schrödinger equation for the electrons yields the approximate wave functions of the infinite electronic states as well as their degeneracies. The energy corresponding to the nth electronic state is expressed as a function of the internuclear distance r and constitutes the potential energy $V_n(r)$ seen by the internal motion of the nuclei, governed by the nuclear Schrödinger equation.

The internal partition function of diatomic molecules is calculated from the expression

$$Q_{int} = \frac{1}{\sigma} \sum_{n}^{n_{max}} g_n \exp\left[-\frac{E_{el}(n)}{kT}\right] \sum_{v}^{v_{max}(n)} \exp\left[-\frac{E_{vib}(n,v)}{kT}\right] \cdot$$
$$\cdot \sum_{J}^{J_{max}(v)} (2J+1) \exp\left[-\frac{E_{rot}(n,v,J)}{kT}\right] \tag{3}$$

In (3), σ is a symmetry factor that equals one or two for, respectively, heteronuclear and homonuclear diatomic molecules, $E_{el}(n)$ is the electronic excitation energy, $E_{vib}(n,v)$ is the vibrational energy and $E_{rot}(n,v,J)$ is the rotational energy. Details to determine the different energy contributions and the methods to obtain the maximum vibrational and rotational quantum numbers $v_{max}(n)$, $J_{max}(v)$ are given in recent reports.[8,9]

When the internal partition function is known, one can determine the thermodynamic properties of the molecular species in the manner described in Section 4.

Examples of results obtained from (3) are shown in Figs. 1-2. In particular, the former figure shows the internal partition function of molecular nitrogen calculated by Jaffe[1] and by Capitelli et al.;[8,9] the agreement between the two calculations is remarkable, despite the different sets of rotovibrational energy used by the authors. From the latter figure, the same agreement exists also for the internal constant pressure specific heat. In this case, however, one should note the tendency of the specific heat to vanish at high temperature; this behaviour is due to the finite, though great, number of internal energy levels accounted for in the summation of the internal partition function.

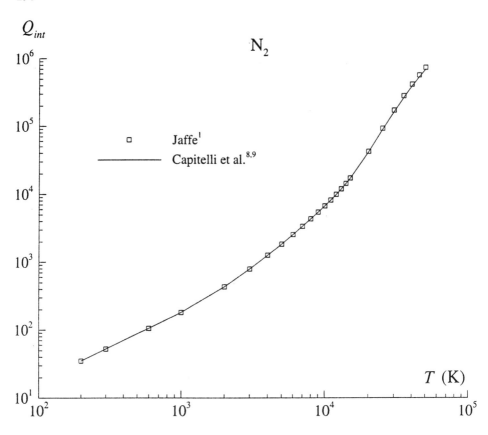

Figure 1. Internal partition function of N_2.

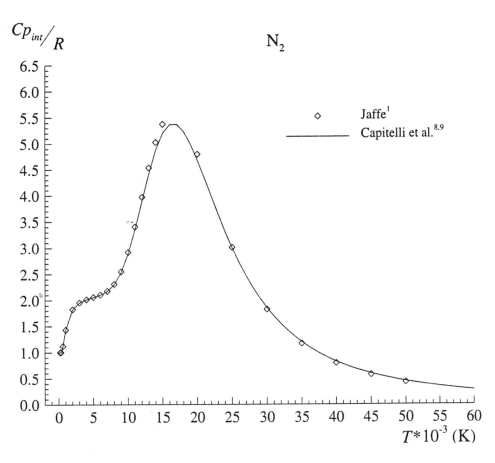

Figure 2. Internal constant pressure specific heat of N_2.

6. Conclusions

In this paper we have briefly examined the basic methods to calculate the high-temperature thermodynamic properties of air species, with particular reference to recently published results.[8,9] The impossibility of calculating the internal partition functions of the atomic species as a function of temperature only has been emphasised. The situation described in Refs. 8 and 9 emphasizes the dependence of the electronic partition function on the ionisation potential lowering needed to truncate the internal partition function and its derivatives; this latter parameter, in turn, depends[8,9] either on the pressure or on the number density of the charged particles. No problem, on the contrary, exists concerning the separation of the internal partition function from the translational one. In this case a two-temperature approach can be used to account for thermal non-equilibrium, even though one must be careful to use the appropriate transport properties of the species. We need, in fact, to remember that a large contribution to the internal energy of the atomic species comes from the excited electronic states near the ionisation continuum. These states possess transport collision cross sections much greater than the corresponding cross sections of the ground state. Consequently, use of thermodynamic properties which account for contributions from the excited electronic states to the internal energy should be accompanied by consistent sets of transport cross sections.[10]

As regards the diatomic molecules, recent calculations[1-3] based on the solution of the vibrational Schrödinger equation in the field of an effective potential have shown the accurate way to obtain the internal partition function and the subsequent thermodynamic properties. The obtained results, however, are in excellent agreement with older methods based on analytical expressions of the rotovibrational energies. All these studies show how difficult it is to separate the different internal contributions, even in the framework of the Born-Oppenheimer approximation. The construction of thermodynamic models based on different temperatures associated with the internal structure of the molecules (rotational, vibrational and electronic) appears open to question.[8] The calculation of the internal partition function of diatomic molecules may seem not to be affected by the divergence problem existing for the atomic species: this point is, however, illusory. In fact, while a well-founded cut-off criterion has been applied to the rotovibrational energies of each electronic state, a practical cut-off criterion has been applied for the number of electronic states. The approach is, in fact, to insert only few electronic states in the summation to determine the internal partition function. The justification could be that possible crossings between very excited bound electronic states and repulsive electronic states dissociate the molecule. This point needs further study.

As a conclusion we can say that caution should be exercised when the thermodynamic properties of high temperature gases are used in fluid dynamics codes. Apparently, the difficulties discussed for both molecules and atoms occur in temperature ranges in which entropic effects favour the dissociation of the molecule or the ionisation of the atom. It is to be hoped that this point, which is true for equilibrium situations, may also hold for nonequilibrium cases, thus eliminating some of the problems discussed above.

References

1. Jaffe, R. L. (1987) The Calculation of High-Temperature Equilibrium and Non-Equilibrium Specific Heat Data for N_2, O_2 and NO, *AIAA-87-1633*

2. Liu Y., and Vinokur, M. (1989) Equilibrium Gas Flow Computations. I. Accurate and Efficient Calculation of Equilibrium Gas Properties, *AIAA-89-1736*

3. Liu, Y., Shakib, F., and Vinokur, M. (1990) A Comparison of Internal Energy Calculation Methods for Diatomic Molecules, Phys. Fluids A **2**, 1884–1903

4. Capitelli, M., and Ficocelli Varracchio, E. (1977) Thermodynamic Properties of Ar-H_2 Plasmas, Rev. int. Htes Temp. et Refract. **14**, 195–200

5. Capitelli, M., and Molinari, E. (1970) Problems of Determination of High Temperature Thermodynamic Properties of Rare Gases with Application to Mixtures, J. Plasma Phys. **4**, 335–355

6. Capitelli, M., Ficocelli Varracchio, E., and Molinari, E. (1971) Electronic Excitation and Thermodynamic Properties of High Temperature Gases, Z. Naturforsch. **26a**, 672–683

7. Moore, C. (1971) Atomic Energy Levels as Derived from the Analyses of Optical Spectra, *NSRDS-NBS-35-VOL-1*

8. Capitelli, M., Colonna, G., Gorse, C. and D. Giordano (1994) Survey of Methods of Calculating High-Temperature Thermodynamic Properties of Air Species, *ESA STR-236*

9. Giordano, D., Capitelli, M., Colonna, G. and Gorse, C. (1994) Tables of Internal Partition Functions and Thermodynamic Properties of Air Species from 50 K to 100000 K, *ESA STR-237*

10. Capitelli, M., Celiberto, R., Gorse, C., and Giordano, D. (1995) Old and New Problems Related to High-Temperature Transport Properties, in M. Capitelli (ed), *Molecular Physics and Hypersonic Flow*, Kluwer Academic Publisher, Dordrecht

OLD AND NEW PROBLEMS RELATED TO HIGH TEMPE-
RATURE TRANSPORT PROPERTIES.

M. CAPITELLI, R. CELIBERTO, C. GORSE
Centro di Studio per la Chimica dei Plasmi del CNR
Dip. Chimica, Università di Bari
Via Orabona 4, 70126 Bari, Italy

AND

D. GIORDANO
European Space Research & Technology Center
P.O. Box 299, 2200 AG Noordwijk, The Netherlands

Abstract. The methods for the calculation of the transport properties of high temperature gases are reviewed from the points of view of both kinetic theory and transport cross sections. The transport cross sections (collision integrals) of high temperature air components are then discussed by comparing old and new calculations particularly emphasizing atom-atom, atom-molecule and atom-ion interactions. Special consideration is dedicated to the knowledge of transport cross sections of electronically excited states.

1. Introduction

The transport properties of high temperature air is a subject which has been widely studied in the past in connection with the reentry problem as well as with the use of thermal plasmas in some technological applications. Pionering work in this field was made by the groups of Hirschfelder and of Mason [1-2], which developed not only the kinetic theory necessary to calculate the transport coefficients but also methods of quantum chemistry to estimate the potential energy curves between the different species, this last aspect being the basis of any serious calculations of transport properties. Other pionering work was made by Devoto [3] who extended the Chapman-Enskog method to higher order approximations and showed the slow convergence of the method to calculate the properties of free electrons.

M. Capitelli (ed.), Molecular Physics and Hypersonic Flows, 303–321.

Apparently, all the ingredients for calculating the transport properties of high temperature gases exist and, therefore, one may believe that the problem is completely solved. However, the present fluid dynamics codes need accurate data for transport coefficients but workers in the field still prefer to use well known and, often, too old tabulations without any criticism on the accuracy of the adopted values. An example is the recent report by Gupta et al. [4] who present a review of transport cross sections (collision integrals) of the most important air components in the temperature range 300-30000 K. This report contains a curve fit of the relevant collision integrals as well as of the binary diffusion coefficients ready to be inserted in the kinetic equations. These data, of course, have a wide popularity in the fluid dynamics community because of their easy use. A careful reading of the report, however, shows that the reported data are based on calculations which can be considered obsolete when compared with tabulations existing before the year that their report was published. Moreover, recent calculations made by Levin et al. [5-6] make the results of Gupta's report open to question. It should be, also, pointed out that the data given by Gupta et al. are largely based on the revised tables by Yos [7] and on the tabulations by Yun and Mason [8]. These results, even though considered milestones in the literature pertinent to the transport properties, have been updated by different authors. The aim of this report is to review critically old and new efforts made to obtain transport cross sections (collision integrals) and kinetic equations used to calculate the transport properties (heat conductivity, viscosity) of a multicomponent gas. We will divide the paper into two parts: the first part deals with the kinetic formulation of transport properties, the second one with a comparison of the existing transport cross sections.

2. Transport Coefficients

The calculation of the total thermal conductivity of a v-component gas is based on the expression of the heat flux vector $\mathbf{q}$ [1]

$$\mathbf{q} = -\lambda'\nabla T + \sum_i n_i h_i \mathbf{v}_i - nkT \sum_i D_i^T \mathbf{d}_i/m_i n_i - \lambda_{int}\nabla T \qquad (1)$$

Here $\mathbf{v}_i$ and $\mathbf{d}_i$ are the diffusion velocity and the diffusion force of the ith species, respectively, while λ' and λ_{int} are the translational and internal thermal conductivities. The quantities n_i, h_i, T, D_i^T and m_i are, in order, the number density, specific enthalpy, temperature, multicomponent thermal diffusion and mass of the ith species. The equation (1) can be rearranged under equilibrium conditions as follows

$$\mathbf{q} = -(\lambda' + \lambda_{int} + \lambda_R)\nabla T \qquad (2)$$

with

$$\lambda = \lambda' - \rho k/n \sum_{i,j} E_{ij} D_i^T D_j^T / n_i m_i m_j \tag{3}$$

Here, ρ is the mass density and E_{ij} is defined as an element of the inverse of the matrix whose general element is $D_{ij} m_j$ (D_{ij} is the multicomponent diffusion coefficient). The thermal conductivity λ defined by (3) represents the true translational one corresponding to a mixture which has come to a steady state composition in the temperature gradient. According to Chapman and Enskog's fourth approximation, λ can be calculated as

$$\lambda_4 = -75/8k(2\pi kT)^{1/2} \begin{bmatrix} q_{ij}^{11} & q_{ij}^{12} & q_{ij}^{13} & n_i \\ q_{ij}^{21} & q_{ij}^{22} & q_{ij}^{23} & 0 \\ q_{ij}^{31} & q_{ij}^{32} & q_{ij}^{33} & 0 \\ q_{ij}^{41} & n_i/m_j^{1/2} & 0 & 0 \end{bmatrix} \cdot [q]^{-1} \tag{4}$$

where $[q]$ represents the same matrix without the last row and the last column. To obtain the third approximation λ_3, all q^{mp} blocks (each block is a matrix of order v) with $m, p = 3$ are deleted in (4), while, in this formalism, the Chapman-Enskog first approximation vanishes. The problem of convergence of (4) has been studied in detail for argon plasmas (A, A^+, e) by Devoto [9]. It turns out that λ_3 is two times greater than λ_2 when the plasma ionization degree is greater than 50%, while λ_4 and λ_3 are practically coincident. Similar results have been obtained by Capitelli [10] for atmospheric nitrogen plasmas in equilibrium. The primary cause of the slow convergence of the Chapman-Enskog theory lies in the fact that the mass of the electrons is very different from that of the heavy components; a secondary cause is represented by the different behaviour of the cross sections for the various interactions. Simplified expressions for λ have been obtained by considering that the distribution functions of the heavy components are not affected by electron-heavy particle collisions. As a consequence, λ can be split into two contributions: the first (λ_E) is due to the electrons and the second (λ_H) is due to the heavy particles [11]

$$\lambda = \lambda_E + \lambda_H \tag{5}$$

It should be noted that λ_H is unaffected by the properties of the electrons, while λ_E depends on the collisions between electrons and all other species. One of the causes of the slow convergence of the Chapman-Enskog method disappears in λ_H, so that it is possible to use a different level of approximation for the two terms of (5). It turns out that the third approximation to λ_E and the second approximation to λ_H give the same results as those obtained with the third approximation as given by (4) (see ref.11).

The second approximation to λ_H has the form [1]

$$(\lambda_H)_2 = 4 \begin{bmatrix} L_{11} & L_{1v} & x_1 \\ L_{v1} & L_{vv} & x_v \\ x_1 & x_v & 0 \end{bmatrix} \cdot \begin{bmatrix} L_{11} & L_{1v} \\ L_{v1} & L_{vv} \end{bmatrix}^{-1} \tag{6}$$

where x_i is the molar component of ith component and the L_{ij}'s are functions of the collision integrals between heavy components, temperature, molecular weights and composition (see Hirschfelder et al.[1]). Attempts to simplify eq.6 have been discussed in ref.12.

The reactive thermal conductivity λ_R, which represents the contribution due to chemical reactions occurring in the plasma, can be calculated with a good approximation by means of the general Butler and Brokaw's expression [13]

$$\lambda_R = -1/RT^2 \begin{bmatrix} A_{11} & A_{1\mu} & \Delta H_1 \\ A_{\mu 1} & A_{\mu\mu} & \Delta H_\mu \\ \Delta H_1 & \Delta H_\mu & 0 \end{bmatrix} \cdot \begin{bmatrix} A_{11} & A_{1\mu} \\ A_{\mu 1} & A_{\mu\mu} \end{bmatrix}^{-1} \tag{7}$$

Elements $A_{1\mu}$ (μ is the number of independent chemical reactions) are given by

$$A_{ij} = A_{ji} = \sum_{k,l}(RT/D_{kl})x_k x_l \left[n_k^{(i)}/x_k - n_l^{(i)}/x_l \right] \left[n_k^{(j)}/x_k - n_l^{(j)}/x_l \right] \tag{8}$$

where D_{kl} represents a binary diffusion coefficient, $n_k^{(i)}$ is the stoichiometric coefficient of species k in the ith reaction while ΔH_i represents the enthalpy difference in the ith chemical reaction. Finally, the calculation of λ_{int} can be performed via the extension of the Eucken approximation

$$\lambda_{int} = \sum_i (\lambda_{int})_i/(1 + \sum_j D_{ii}x_j/D_{ij}x_i) \tag{9}$$

$$(\lambda_{int})_i = pD_{ii}/RT \cdot (C_{p,i} - 5/2R) \tag{10}$$

where $C_{p,i}$ is the specific heat at constant pressure and D_{ii} is the self-diffusion coefficient of the ith species.

3. Viscosity

The calculation of the viscosity of a v-component gas can be calculated by expressions similar to those described for the translational thermal conductivity, even though in this case the first approximation of the Chapman-Enskog method does not vanish. As in the case of λ, η can be expressed as a sum of the two contributions due to electrons and heavy particles

$$\eta = \eta_H + \eta_E \simeq \eta_H \tag{11}$$

The calculation of η can be performed with the first approximation of the Chapman-Enskog method which assumes a form very similar to $(\lambda_H)_2$. Simplified expressions have been proposed to calculate η. The most popular were proposed by Sutherland [14] and by Buddemberg and Wilke [15]. The last give very poor results for partially ionized gases.

4. Collision Integrals

Accurate calculations of transport coefficients can be obtained when the collision integrals (transport cross sections) of the different interactions are known [1]. These quantities can be obtained by performing three integrations. First, one determines the classical deflection angle $\Theta(b, E)$ as a function of impact parameter b and relative energy E

$$\Theta(b, E) = \pi - 2b \int_0^\infty dr/r^2 F(r, b, E)^{1/2} \tag{12}$$

A further averaging over the impact parameter b yields the relevant cross section

$$0Q^{(l)}(E) = 2\pi - \{1 - [1 + (-1)^l]/2(1 + l)\}^{-1} \int_0^\infty (1 - \cos^l \Theta) b \, db \tag{13}$$

and the above quantities can in turn be employed for a further energy averaging that produces the collision integrals as a function of temperature

$$\Omega^{(l,s)}(T) = [(s + 1)!(kT)^{s+2})]^{-1} \int_0^\infty Q^{(l)}(E) E^{s+1} \exp(-E/KT) dE \tag{14}$$

The problem of calculating $\Omega^{(l,s)}$, therefore, reduces to the knowledge of $V(r)$. Apparently, we have all the ingredients for calculating the transport properties of high temperature gases. Difficulties, however, arise because of the lack of informations (both theoretical and experimental) about $V(r)$. Fortunately, the collision integrals are the result of three average procedures so that their values do not dramatically depend on $V(r)$.

5. Main Interactions

In a high temperature environment, different species can survive depending on the considered temperature range. As an example, let us consider the species in an atmospheric nitrogen plasma. Taking into account the following equilibria

$$
\begin{array}{llll}
N_2 & \rightleftharpoons & 2N & 5000 < T < 10000K \\
N & \rightleftharpoons & N^+ + e & 10000 < T < 20000K \\
N^+ & \rightleftharpoons & N^{++} + e & 20000 < T < 30000K \\
N^{++} & \rightleftharpoons & N^{+++} + e & 30000 < T < 35000K
\end{array}
$$

We understand that we are passing from a low temperature system, where molecules and atoms coexist, to an high temperature system, where electrons, atoms and parent ions survive, and, finally, reach a temperature range in which only ions and electrons exist. We will examine the different interactions, emphasizing the situation for nitrogen and oxygen systems. Note that equilibria similar to those discussed for nitrogen hold for oxygen plasmas, while in the case of air one should also consider the NO species. Other species such as molecular positive ions as well as molecular and atomic negative ions are not considered in the present review due to their small concentrations under thermal plasma conditions.

5.1. DISSOCIATION RANGE

In this case, we must know the interaction potentials of molecule-molecule, atom molecule and atom-atom systems. Despite the apparent simplicity of the systems under consideration we still have problems in their characterization. Ab initio potentials for all internuclear distances are still far from being completely known so that one usually builds up the interaction potential by using the available theoretical and experimental data. Theoretically the semiempirical method developed by Mason and coworkers [8], who obtained the different potentials for atom-molecule and molecule-molecule interaction as the sums of the interactions between the different atoms averaged over all molecular interactions, is still used. As an example for N_2–N_2 interaction Yun and Mason were able to obtain the following potential which can be used in the different temperature ranges

$$
\phi(r) = \epsilon(1 - 6/\alpha)^{-1}\{6/\alpha \exp[\alpha(1 - r/r_m)] - (r_m/r)^6\} \tag{15}
$$

$$
\phi(r) = k/r^s \tag{16}
$$

The first potential was used in the low and intermediate temperature range, the second one for the high temperature range (for the parameters see Refs. [8,16]). Collision integrals for the N_2–N_2 interaction have been tabulated by Yun and Mason and are still used today (see Ref. [4]). Table 1 shows a comparison of these collision integrals diffusion type with the corresponding values calculated by Gorse [17], who used different experimental potentials

TABLE 1. A comparison of collision integrals diffusion type for the interaction N_2–N_2

$T \cdot 10^{-3} K$	Ref. [19]	Ref. [17]	Ref. [17]	Ref. [8]
5	5.966	5.933	6.387	6.599
6	5.635	5.648	6.102	6.187
7	5.364	5.418	5.867	5.863
8	5.134	5.225	5.667	5.595
9	4.936	5.062	5.493	5.373
10	4.762	4.920	5.341	5.199
11	4.607	4.795	5.204	5.069
12	4.468	4.683	5.081	4.907
13	4.342	4.583	4.969	4.810
14	4.227	4.492	4.866	4.727
15	4.121	4.409	4.772	4.630

[18] and by Capitelli and Devoto [19]. We can see that the agreement is rather satisfactory in the temperature range $5000 - -15000 K$ Similar considerations apply to the N–N_2 interaction (see Refs. [20], [21]). It should be noted that the weakness of the N–N_2 interaction is the lack of any attractive component which could be important at low gas temperature (see Ref. [22] for the O–O_2 case).

Let us now consider the N–N interaction. In this case several potentials arise in the adiabatic representation. In fact the two $N(^4S)$ atoms can interact along four potential curves, spectroscopically denoted by $^1\Sigma$, $^3\Sigma$, $^5\Sigma$ and $^7\Sigma$. The valence bond theory suggests that the potential with the higher spin multiplicity (e.g. $^7\Sigma$) should be repulsive, while the one with the lower spin multiplicity (e.g. $^1\Sigma$) should be attractive. One of the first calculations for this case has been performed by Yun and Mason, who used two completely attractive potentials of the form

$$\phi(r) = -k/r^s \qquad (17)$$

for $^1\Sigma$ and $^3\Sigma$ potentials, the exp-6 potential for the $^5\Sigma$ potential and an exponential repulsive potential for $^7\Sigma$ state (see Ref. [8] for the relevant parameters). Capitelli and Devoto [19] used a Morse function

$$\phi(r) = \phi_0\{\exp[-(2C/\sigma)(r - r_e)] - 2\exp[-(C/\sigma)(r - r_e)]\} \qquad (18)$$

for the $^1\Sigma$, $^3\Sigma$ and $^5\Sigma$ potentials and the exponential repulsive potential for $^7\Sigma$ (see Ref. [19] for the relevant parameters). Subsequently, Capitelli et al. [23] calculated the N-N collision integrals by using a new $^7\Sigma$ potential obtained by the Heitler-London quantum mechanical approach [24]. More recently, Levin et al. [5] recalculated the four N-N potentials by using a

TABLE 2. A comparison of collision integrals diffusion type for $O(^3P)-O(^3P)$ and $N(^4S)-N(^4S)$ interactions calculated by different authors as a function of temperature

$T \cdot 10^{-3} K$	$O(^3P)-O(^3P)$			$N(^4S)-N(^4S)$			
	Ref. [25]	Ref. [6]	Ref. [8]	Ref. [23]	Ref. [6]	Ref. [8]	Ref. [19]
2	4.689	4.837	5.271	5.53	5.145	5.106	
4	3.977	4.003	4.386	4.57	4.388	4.166	
6	3.576	3.567	3.905	4.10	3.936	3.683	4.082
8	3.300	3.270	3.575	3.80	3.614	3.365	3.721
10	3.093	3.046	3.335	3.56	3.366	3.131	3.440
12	2.928	2.866	3.148	3.36	3.165	2.949	3.207
14	2.791	2.717	2.996	3.19	2.996	2.801	3.010
16	2.676	2.591		3.04	2.853		2.839
18	2.579	2.481		2.92	2.728		2.702
20	2.494	2.385		2.81	2.619		2.579

new set of ab initio potentials modified to take into account experimental information.

The collision integrals of multi-potential interaction is an average of the different potentials with statistical weights p_n equal to the spin multiplicity for the Σ states and two times the spin multiplicity for Π, Δ, ... states; it reads

$$< \Omega^{(l,s)} > = \sum_n p_n \Omega_n^{(l,s)} / \sum_n p_n \qquad (19)$$

Table 2 shows a comparison of diffusion type collision integrals for N-N interaction. Once again, we can see that the data do not differ by as much as 12%.

The insensitivity of collision integrals to the progress made in the potentials is essentially due to two factors. The first is due to the averaging procedure described by equations (12)–(14); the second is due to some compensation effects arising in the further averaging when the interaction occurs on several potential curves. This last point has been discussed by Capitelli and Ficocelli [25] several years ago for O-O interaction and confirmed in the results reported in Table 2. The interactions O_2-O_2, $O-O_2$ and O-O follows the behaviour described for nitrogen species. In particular, the interaction between oxygen atoms occurs along 18 potential curves so that compensation effects should be more effective in decreasing the differences between the collision integrals calculated by the different authors. This is indeed the case as can be noted by comparing the calculations of Levin et al. and of Capitelli and Ficocelli. In this case, the differences in the collision integrals do not exceed 5% while up to 12% is present when this comparison is made with the old results of Yun and Mason (see Table 2).

5.2. PARTIAL IONIZATION RANGE

In this temperature range N, N^+ and electrons are present so that we should take into account N–N, N–N^+, e–N , e–e and e–N^+ interactions. For the ion-atom interaction, we must distinguish the collision integrals diffusion type from the corresponding one relative to viscosity. The first is dominated by the charge exchange process while the second (viscosity type) is dominated by valence forces, even though, at low temperature, polarizability forces should be taken into account. Let us first consider viscosity type collision integrals for N–N^+. We should treat this interaction as in the atom-atom case. The interaction of ground state atoms $N(^4S)/N^+(^3P)$ occurs along 12 potential curves ($^{2,4,6}\Sigma_{g,u}$; $^{2,4,6}\Pi_{g,u}$ states). Collision integrals were calculated by Capitelli and Devoto [19] by using Morse and repulsive potentials for the different interactions. The parameters of the different potentials were estimated by the theoretical and experimental data available at that time. More recently Levin et al. [6] calculated a complete set of N–N^+ potentials by using the ab initio CASSCF method. Moreover, they have implemented the wells of the bound states with some experimental information. A comparison between the results of Ref. [19] and Ref. [6] in the temperature range 5,000-20,000 K shows differences up to 15%. More serious (up to 35%) are the differences with the Yos data which are still used in the recent work of Gupta. Let us now consider the collision integrals diffusion type. At high temperature the diffusion cross section Q^1 is dominated by charge transfer and is equal [26]

$$Q^1 = 2S_{ex} \tag{20}$$

where S_{ex} is the cross section relative to the process

$$A + A^+ \rightleftharpoons A^+ + A$$

Values of S_{ex} can be obtained either experimentaly or theoretically. ¿From the experimental point of view, the technique of crossed beams as used by Belyaev et al.[27] seems to give reliable results (see also Ref. [28] for O–O^+). On the other hand, numerous theoretical results exist for the resonant charge transfer cross sections involving N–N^+ and O–O^+ systems. The theoretical problem to calculate Q_{ex} reduces to the calculation of the gerade-ungerade differences of the different couples of potentials arising in the interaction, followed by an averaging procedure of the type of (19). Different attempts to calculate ΔV_{gu} have been undertaken. The first authors [26] used experimental ΔV_{gu} for the pairs $^2\Sigma_{g,u}$, $^2\Pi_{g,u}$ in combination with the following relations for the other pairs [29]

$$(\Delta V_{gu})_{sextet} = 3/2(\Delta V_{gu})_{quartet} = 3(\Delta V_{gu})_{doublet} \tag{21}$$

TABLE 3. A comparison of collision integrals diffusion type for $N(^4S)$–$N^+(^3P)$ calculated by different authors

$T \cdot 10^{-3} K$	Ref. [31]	Ref. [5]	Ref. [19]	Ref. [27]	Ref. [30]	Ref. [23]
10	16.3	26.2	13.44	25.86	25.73	19.90
12	16.0	25.62	13.19	25.15	25.23	19.60
14	15.8	25.13	12.97	24.57	24.81	19.35
16	15.6	24.7	12.79	24.06	24.45	19.13
18	15.3	24.32	12.63	23.62	24.14	18.95
20	15.2	23.00	12.48	23.23	23.86	18.78

The collision integrals calculated in this way appear to be too low compared with the experimental results based on Belyaev's experimental charge transfer cross sections. The first ab initio calculation of ΔV values for both N–N$^+$ and O–O$^+$ interactions (and relative charge exchange cross sections) was performed by Yos [30], using the Heitler and London's method; these results, which have never been published in a journal, appear to be in very good agreement with the experimental results. This kind of calculation was repeated by Capitelli et al.[31] for the Σ_{gu} states of N–N$^+$ interaction. Their results confirm that the old estimations of ΔV_{gu} based on experimental potentials give charge transfer cross sections (and collision integrals diffusion type) that are too low with respect to the experimental values. More recently Stallcop et al. improved both the potential energy curves and the quantum mechanical treatment of the charge transfer cross sections for both N–N$^+$ and O–O$^+$ systems. The new results are in excellent agreement with the experimental data and not too far from those calculated by Capitelli et al.[23] the latter based on potentials of Ref. [31]. The situation is well illustrated in Table 3 where the collision integrals diffusion type for the interaction N–N$^+$, calculated by different authors, have been reported. It appears a good agreement exists between different sets of collision integrals diffusion type so that the recent calculation of Levin et al.can be used as a guide for correcting collision integrals diffusion type of N–N$^+$ in different electronic states. It should be also noted that the collision integrals diffusion type for N–N$^+$ obtained by Capitelli [32], based on the ab initio potentials of Ref. [33], are still too low when compared with the experimental values. A similar situation holds for O–O$^+$ interaction, as can be appreciated from Table 4 where the calculations of Capitelli [10], based on the ab initio potentials of Ref. [34] have been compared with the results of Levin et al. based on an other set of ab initio potential curves. It is interesting to note that the differences between the collision integrals diffusion type for O–O$^+$ calculated in Refs. [10] and [5] are approximately the same as the corresponding difference in the collision integrals for N–N$^+$ calculated in Ref.

TABLE 4. A comparison of collision integrals diffusion type for $O(^3P)$–$O^+(^4S)$ and $N(^4S)$–$N^+(^3P)$ calculated by different authors

$T \cdot 10^{-3} K$	$O(^3P)$–$O^+(^4S)$			$N(^4S)$–$N^+(^3P)$		
	Ref. [10]	Ref. [5]	Δ %	Ref. [32]	Ref. [5]	Δ %
10	13.0	21.45	39.4	16.3	26.2	37.8
12	12.8	21.07	39.3	16.0	25.62	37.5
14	12.6	20.74	39.2	15.8	25.13	37.1
16	12.5	20.45	38.9	15.6	24.7	36.8
18	12.3	20.19	39.2	15.3	24.32	37.1
20	12.2	19.95	38.9	15.2	23.93	36.5

[32] and Ref. [5] (see Table 3). This behaviour can be probably ascribed to the differences in potential curves between the results of Refs. [33-34] and those of Ref. [5]. The last point we want to analyze is the fact that charge transfer dominates the collision integrals at high temperature while polarization can be important at low temperature. As an example at 300 K the polarization model gives a value of $\Omega^{11} = 26.1 Å^2$ to be compared with a value of 40.51 $Å^2$ given in Ref. [5]. This means that the collision integrals diffusion type for N–N$^+$ should also include at low temperature the contribution coming from polarization forces. The problem has been recently examined by Murphy and Arundell [35], who proposed, for the calculation of collision integrals diffusion type of neutral-ion interaction, the following formula

$$\Omega^{11}(T) = [\Omega_{in}^{11}(T)^2 + \Omega_{el}^{11}(T)^2]^{1/2} \qquad (22)$$

i.e. a combination of collision integrals for inelastic and elastic interactions, the inelastic contribution in the present case being that one coming from the exchange process and the elastic one coming from polarization. At 300 K, therefore, the collision integral diffusion type for N–N$^+$ should be 48.2 $Å^2$. The exchange process does not enter for symmetry reasons in the evaluation of collision integrals viscosity type [26] so that their calculation can be done as previously outlined. A question should arise about the role of polarization in the collision integrals viscosity type. To a good approximation, we can say that the theoretical and experimental potentials used contain to a given extent the polarization forces. One check which can be done is if the calculated collision integrals converges asymptotically at low temperature to the corresponding values calculated by using the polarizability model (see Ref.[10]).

Let us now consider the charged-charged interaction. In this case we can use the screened Coulomb potential written as

$$V(r) = (e^2/r)\exp(-r/d) \qquad (23)$$

where d is the Debye length. Collision integrals for this interaction are known either in tabular form [36] or approximated with closed forms of the type [37]

$$\Omega^{(l,s)} = f(l,s)b_0^2[\ln \Lambda - 0(1)\text{terms}] \qquad (24)$$

where $\Lambda = 2d/b_0$ is the ratio between the Debye length and the average closest impact parameter b_0 and f(l,s) is an implicit function ot the type of collision integrals. To the predominant $\ln \Lambda$ term, these formulae give the same results as those derived by Spitzer and Harm [38], using a different approach. Other formulations have been proposed to calculate the terms after $\ln \Lambda$. The results obtained by more sophisticated theories range in between those obtained including and neglecting the $O(1)$ terms. Use of (24) with and without the $O(1)$ terms yields, therefore, lower and upper limits for $\Omega^{(l,s)}$. This property has been used to calculate the thermal conductivity with the second approximation in combination with the collision integrals which neglect the $0(1)$ terms [39]. A similar approach has been followed by Yos [7] for reproducing the behaviour of the thermal conductivity of a completely ionized gas as compared with the Spitzer and Harm formulae.

5.3. TOTAL IONIZATION RANGE

In this case the shielded Coulomb potential must be used as previously discussed.

6. Electron-molecule and electron-atom transport cross sections

The transport cross sections for these interactions can be directly calculated from the differential elastic cross section $q(E, \theta)$ at the electron energy E and the scattering angle θ. In turn, the differential elastic cross sections can be calculated by using the quantum scattering theory or measured by using electron-beam scattering and/or swarm techniques. A lot of experiments have beem performed for electron-molecular air components while the experimental situation is still very bad for electron-atom differential cross section. The situation for e–N_2, e–O_2, e–O cross sections has been recently reviewed by Itikawa and coworkers [40-43]. Collision integrals for e-(air neutral components) have been calculated by Capitelli and Devoto [19] and by Devoto [44]. A comparison of these values with the most recent ones calculated by Murphy and Arundell shows a very satisfactory agreement, while large differences can be noted when comparing with Gupta *et al.*'s results. The sensitivity of the thermal conductivity of electrons to a variation of electron-neutral cross sections has been discussed in Ref. [10].

7. Collision Integrals of electronically excited states

The presence of electronically excited states is usually excluded from the calculation of the internal contribution to the thermal conductivity and to the viscosity. The main reason of this neglect lies in the poor knowledge of the collision integrals diffusion and viscosity type of the electronic states. Inspection, however, of the internal specific heat $c_{p,int}$ of the different atoms shows a large contribution of excited states which in general at high temperature can exceed the corresponding vibrorotational contribution present in the molecules. Inspection of Ref. [45] shows that for atomic nitrogen the internal specific heat , based on electronic partition function which includes a complete set of electronic levels , can reach values ten times higher than the translational contribution so that insertion of this specific heat in an Eucken treatment of λ_{int} could strongly increase this quantity when the diffusion coefficient of electronically excited states is taken equal to that one of the ground state. This last assumption is, however, an extreme simplification, especially taking into account that a large contribution to $c_{p,int}$ comes from high-lying excited states, i.e. from states obtained by promoting electrons to principal quantum numbers different from the ground state. Of course, excited states with the same principal quantum number as the ground state also contribute to $c_{p,int}$. We will denote these states as low-lying excited states, while the other ones will be denoted as high-lying excited states. To better understand this point, let us consider the nitrogen atom. The ground state is formed by the following configuration

$$1s^2 \, 2s^2 \, 2p^3 \qquad {}^4S \qquad \text{energy} = 0$$

A different rearrangement of electrons, without changing the principal quantum number, gives the two low lying excited states

$$1s^2 \, 2s^2 \, 2p^3 \qquad {}^2D \qquad \text{energy} = 2.38 \text{ eV}$$

$$1s^2 \, 2s^2 \, 2p^3 \qquad {}^2P \qquad \text{energy} = 3.57 \text{ eV}$$

High lying excited states are then obtained by promoting a $2p$ electron on the nl orbital ($n > 2$). To calculate the collision integrals diffusion and viscosity type of the different states one should consider these states as "true" atoms and apply the methods developed for atoms in the ground state. To this end we must know the different potential curves arising in the interaction of atoms in these states with themselves, with the atoms in the ground state and with other species. In general, an enormous number of potential curves arises in the interaction of these states. However, despite the enormous progress made by quantum chemistry in the last 20 years, the knowledge of these potential curves is still scanty for low lying excited states and practically absent for high lying excited states. In particular, the situation for collision integrals of low lying excited states of atomic nitrogen and oxygen interacting with themselves and with parent ions in the ground and excited states is still that one developed by one of the present authors many years ago[10] , with the only exception of the pionering work of Nyeland and Mason [46] on the interaction $N(^2D)$-$N(^4S)$, $N(^2P)$-$N(^4S)$ and with the recent paper of Levin et al. on the collision integrals of oxygen atoms including low lying excited states.

Collision integrals of high lying excited states of air components are completely unknown. An estimation of their values could be attempted by noting that the behaviour of high lying excited states should become more and more hydrogen-like with increasing principal quantum number n. Therefore their values should be more and more close to the corresponding collision integrals for atomic hydrogen in the excited states. The problem is, therefore, shifted to the knowledge of collision integrals of H(n)–H(1s) and H(n)–H$^+$ species. In this case the situation is a little better since many years ago Capitelli and Lamanna [47] calculated collision integrals diffusion and viscosity type for these interactions.

Less work is known on the collision integrals diffusion and viscosity type for the interaction H(n)–H(n). To our knowledge, the only data available are those relative to the interaction H(n=2)–H(n=2) which have been calculated in Ref. [47] by using the potential curves of Hirschfelder and Linder [48]. These data can be used to evaluate the corresponding collision integrals for high lying excited states of air atomic components. It should be also noted that the analogy between H, N and O systems can be understood by comparing viscosity and diffusion collision integrals of the three systems. Coming back to how the electronically excited states affect the internal thermal conductivity, we can refer to the evaluation made by Capitelli [10] for equilibrium nitrogen and oxygen plasmas. In this case the internal-reactive contribution was evaluated using (7) including ionization reactions for ground and excited states. The results of these calculations have been reported in Ref. [10] where it has be shown that the inclusion of suitable cross sections for low lying excited states can change λ_R up to 20% for nitrogen and up to 10% for oxygen. These differences should increase when considering non-equilibrium situations and electronically excited atoms and ions not belonging to the same principal quantum number. It should be also noted that in the report of Gupta et al. the transport of electronic energy is made with collision integrals diffusion type equal to N–N$^+$ or O–O$^+$ species in their ground states.

8. Nonequilibrium situations

The results reported in the previous sections consider, implicitely, situations in which only one temperature characterizes the translational and internal states of components. This situation does not hold for hypersonic flows where the heavy components are characterized by a translational temperature T_h different from the corresponding one T_e of the electrons. Moreover, the internal degrees of freedom are usually characterized by corresponding temperatures T_r, T_v, $T_e l$ (rotational, vibrational and electronic temperatures). Under many situations $T_e l = T_e$ and $T_r = T_h$, while the following inequalities hold

$$T_e > T_v > T_h \qquad (25)$$

This situation should be taken into account in the relevant equations. Differences in T_e and T_h are easily accounted for in the translational thermal conductivity; in this case, the decoupling made in (5) can be still applied with the obvious caution to calculate the collision integrals for e-(heavy particle) interaction at the T_e while the collision integrals between heavy particles should be calculated at T_h. Also the transport of internal energy should be adjusted to the new situation. Once again equations similar to (9)–(10) should be used paying attention to insert the contributions of rotational and vibrational states with appropriate diffusion coefficients calculated at the relevant temperatures. The approach to be followed for the transport of electronic energy is different since each excited state should be considered as a separate species with his own cross section.

We conclude this section by noting that a more elegant treatment of the transport of internal (vibrational and rotational) energy can be found in the contributions reported elsewhere by Kustova and Nagnibeda [57], Brun [61] and Nyeland [62]. Particularly interesting is the case when strong nonequilibrium vibrational distributions do exist [57]. In this case an increase of the internal thermal conductivity up to 40

9. Conclusions

In the present paper, we have presented the different efforts made in the last 30 years for obtaining reliable information of transport cross sections and transport properties of high temperature air components. The conclusion of this paper is that accurate values of transport coefficients can be today obtained due to the progress made by quantum chemistry in calculating the potential surfaces which are the primary ingredient of transport cross sections. To obtain these tables one can still follow the procedures outlined by Hirschfelder's and Mason's groups, i.e. use of classical scattering and of averaged (angular) potentials and neglect of inelastic processes. A recent effort in this direction has been recently performed by Murphy [50] The transport of internal energy due to electronically excited states should be done by using appropriate cross sections. The last can be estimated by scaling the corresponding values for $H(n)-H$, $H(n)-H(n)$ and $H(n)-H^+$ interactions. The present paper also warns against the use of simplified expressions for the transport coefficients, especially in the partial ionization regime. The Chapman-Enskog procedure converges poorly for the calculation of the contribution of electrons to the thermal conductivity, while simplified expressions for the calculation of the thermal conductivity and viscosity of heavy components can give poor results.

Acknowledgments

This work has been partially supported by ASI (Agenzia Spaziale Italiana)

References

1. Hirschfelder, J., Curtiss, C. F., and Bird, R. B. (1966) Molecular theory of gases and liquids, John Wiley and Sons, third printing
2. Mason, E. A., and Marrero, T. R. (1970) The diffusion of atoms and molecules, Advances in Atom. and Molec. Physics **6**, 155-232
3. Devoto, R. S. (1966) Transport properties of ionized monatomic gases, Phys. Fluids **9**, 1230-1240
4. Gupta, R. N., Yos, J. M., Thompson, R. A., and Lee, K. P. (1990) A review of reaction and thermodynamic properties for an 11-species air model for chemical and thermal non-equilibrium calculations to 30000 K, NASA Reference Publication 1232
5. Stallcop, J. R., Partridge, H., and Levin, E., (1991) Resonance charge transfer, transport cross sections and collision integrals for $N^+(^3P)$-$N(^4S)$ and $O^+(^4S)$-$O(^3P)$ interactions, J. Chem. Phys. **95**, 6429-6439 (see also AIAA paper 87-1632)
6. Levin, E., Partridge, H., and Stallcop, J. R. (1990) Collision integrals and high temperature transport properties for N-N, O-O and N-O, J. Thermoph. and Heat Transfer 4, 469-477
7. Yos, J.M. (1967) Revised transport properties for high temperature air and its components, AVCO-RAD
8. Yun, K. S., and Mason, E. A.(1962) Collision integrals for the transport properties of dissociating air at high temperature, Phys. Fluids **5**, 380-386
9. Devoto, R. S. (1967) Transport coefficients of partially ionized argon, Phys. Fluids **10**, 354-365
10. Capitelli, M. (1977) Transport coefficients of partially ionized gases, J. de Phys. (Paris) C3-38, 227-237

11. Devoto, R. S. (1967) Simplified expressions for the transport properties of ionized monatomic gases Phys. Fluids **10**, 2105-2112

12. Capitelli, M. (1972) Simplified expressions for the calculation of the contribution of the heavy components to the transport coefficients of partially ionized gases, Z. fur Naturforschung **27a**, 809-812

13. Butler, J. N., and Brokaw, R.S. (1957) Thermal conductivity of gas mixtures in chemical equilibrium, J. Chem. Phys. **26**, 1636-1643

14. Sutherland, W. (1895) The viscosity of mixed gases, Phil. Magaz. **40**, 421-431

15. Buddemberg, J. W., and Wilke, C. R. (1949) Calculation of gas mixture viscosities, Ind. Eng. Chem. **41**, 1345-1347

16. Yun, K. S., Weisman, S., and Mason, E. A. (1962) High-temperature transport properties of dissociating nitrogen and dissociating oxygen, Phys. Fluids **5**, 672-678

17. Gorse, C. (1975) Contribution au calcul des propriffitffis de transport des plasmas des mffilanges argon-hydrogen et argon-azote, Thfflse of 3fflme cycle 75-10, Universitffi de Limoges

18. Leonas, V. B. (1973) Studies of short range intermolecular forces Sov. Phys. Uspekhi **15**, 266-281

19. Capitelli, M., and Devoto, R. S. (1973) Transport coefficients of high-temperature nitrogen, Phys. Fluids **16**, 1835-1841

20. Cubley, S. J. (1970) M. S. thesis, Division of Engineering, Brown University; Cubley, S. J., and Mason, E. A. (1975) Atom-molecule and molecule-molecule potentials and transport collision integrals for high temperature air species, Phys. Fluids **18**, 1109-1111

21. Capitelli, M., Celiberto, ?, Gorse, C., and Giordano, D. (1966) Transport properties of high-temperature air components: a review, Plasma Chem. and Plasma Process., in press.

22. Brunetti, B., Liuti, G., Luzzatti, E., Pirani, F., and Vecchiocattivi, F. (1981) Study of the interaction of atomic and molecular oxygen with O_2 and N_2 by scattering data, J. Chem. Phys. **74**, 6734-6741

23. Capitelli, M., Gorse, C., and Fauchais, P. (1977) Transport coefficients of high temperature N_2-H_2 mixtures, J. Phys. (Paris) **38**, 653-657

24. Capitelli, M., Lamanna, U. T., Guidotti, C., and Arrighini, G. P.(1983) Comment on spin-polarized atomic nitrogen and the $^7\Sigma_u^+$ state of N_2, J. Chem. Phys. **79**, 5210-5211

25. Capitelli, M., and Ficocelli, E. (1972) Collision integrals of oxygen atoms in different electronic states, J. Phys. B: At. Mol. Phys. **5**, 2066-2073; Capitelli, M. and Ficocelli, E. (1973) Collision integrals of carbon-oxygen atoms in different electronic states, J. Phys. B: At. Mol. Phys. **6**, 1819-1823

320

26. Knof, H., Mason, E. A., and Vanderslice, J. T. (1964) Interaction energies, charge exchange cross sections, and diffusion cross sections for N–N$^+$ and O–O$^+$ collisions, J. Chem. Phys. **40**, 3548-3553

27. Belyaev, Y. N., Brezhnev, B. G., and Erastov, E. M. (1968) Resonant charge transfer of low energy carbon and nitrogen ions, Sov. Phys. JETP **27**, 924-926

28. Aubreton,J., Bonnefoi, C., and Mexmain, J. M. (1986) Calcul de proprietes thermodynamiques et des coefficients de transport dans un plasma Ar-O$_2$ en non-equilibre thermodynamique, Rev. Phys. Appl. **21**, 365-376

29. Stallcop, J. R. (1971) N$_2^+$ potential-energy curves, J. Chem. Phys. **54**, 2602-2605

30. Yos, J. M. (1965) Theoretical and experimental studies of high temperature gas transport properties, AVCO-RAD-TR-65-7 (section III)

31. Capitelli, M., Guidotti, C., Lamanna, U. T., and Arrighini, P. (1976) The gerade-ungerade splitting of N$_2^+$ potentials: effects on the resonant charge transfer cross sections of nitrogen atoms, Chem. Phys. **19**, 269-278

32. Capitelli, M. (1975) Charge transfer from low-lying excited states: effects on reactive thermal conductivity, J. Plasma Phys. **14**, 365-371; Capitelli, M. (1972) Cut-off criteria of electronic partition functions and transport properties of thermal plasmas, J. Plasma Phys. **7**, 99-106.

33. Andersen, A., and Thulstrup, E. W. (1973) Configuration interaction studies of the low-lying quartet states of N$_2^+$, J. Phys. B: At. Mol. Phys. **B6**, L211-L213

34. Beebe, N. H. F., Thulstrup, E. W., and Andersen, A. J. (1976) Configuration interaction calculations of low lying electronic states of O$_2$, O$_2^+$ and O$_2^{++}$, J. Chem. Phys. **64**, 2080-2093

35. Murphy, A. B. and Arundell, C. J. (1994) Transport coefficients of argon, nitrogen, oxygen, argon-nitrogen and argon-oxygen plasmas, Plasma Chem. Plasma process. **14**, 451-490

36. Hahn, H. S., Mason, E. A., and Smith, F. J. (1971) Quantum transport cross sections for ionized gases, Phys. Fluids **14**, 278-287

37. Liboff, R. L. (1959) Transport coefficients determined using the shielded Coulomb potential, Phys. Fluids **2**, 40-46

38. Spitzer, L., and Harm, R. (1953) Transport phenomena in a completely ionized gas, Phys. Rev. **87**, 977-985

39. Capitelli, M. (1970) Problems of determination of transport properties of argon-nitrogen mixture at one atmosphere between 5000 and 15000 K, Ing. Chim. Ital. **6**, 94-103

40. Itikawa, M., Ichimura, A., Onda, K., Sakimoto, K., and Takayanagi, K. (1989) Cross sections for collisions of electrons and photons with oxygen molecules, J. Phys. and Chem. Ref. Data **18**, 23-42

41. Itikawa, M., Hayashi, Y., Ichimura, A., Onda, K., Sakimoto, K., and Takayanagi, K. (1986) Cross sections for collisions of electrons and photons with nitrogen molecules, J. Phys. and Chem. Ref. Data **15**, 985-1010

42. Itikawa, Y., and Ichimura A. (1990) Cross sections for collision of electrons and photons with atomic oxygen, J. Phys. and Chem. Ref. Data **19**, 637-651

43. Itikawa, Y., (1994) Electron collisions with N_2, O_2 and O: what we do know and do not know, Adv. Atom. Molec. and Optic. Physics **33**, 253-274

44. Devoto, R. S. (1976) Electron transport properties in high-temperature air, Phys. Fluids **19**, 22-24

45. Capitelli, M., Colonna, G., Gorse, C., and Giordano, D. (1996) Thermodynamic properties of high temperature air components, in M. Capitelli (ed), *Molecular Physics and Hypersonic Flows*, Kluwer Academic Publishers, Dordrecht. [see also ESA STR-236 (1995)]

46. Nyeland, C., and Mason, E. A. (1967) Adiabatic excitation transfer in gases: effects on transport, Phys. Fluids **10**, 985-991

47. Capitelli, M., and Lamanna, U. T. (1974) Collision integrals of electronically excited states and transport coefficients of thermal plasmas, J. Plasma Phys. **12**, 71-79; Capitelli, M. (1974) The influence of excited states on the reactive thermal conductivity of an LTE Hydrogen plasma, Z. Naturforsch. **29a**, 953-954; Capitelli, M., Guidotti, C., and Lamanna, U. T. (1974) Potential energy curves and excitation transfer cross sections of excited hydrogen atoms, J. Phys. B: At. Mol. Phys. **B7**, 1683-1691;

48. Linder, B., and Hirschfelder, J. O. (1958) Energy of interaction between two excited hydrogen atoms in either $2s$ or $2p$ states, J. Chem. Phys. **28**, 197-207 c Publishers, Dordrecht.

49. Murphy, A. B. (1995) Transport coefficients of air, argon-air, nitrogen-air and oxygen-air plasmas, Plasma Chem. and Plasma Process. **15**, 279-308

POTENTIAL ENERGIES AND COLLISION INTEGRALS FOR THE INTERACTIONS OF AIR COMPONENTS

I. Ab initio Calculation of Potential Energies and Neutral Interactions

HARRY PARTRIDGE AND JAMES R. STALLCOP
Computational Chemistry Br., Space Technology Div.,
NASA Ames Research Center, 230-3,
Moffett Field, California, 94035-1000, U.S.A.

AND

EUGENE LEVIN
Thermosciences Institute[†],
NASA Ames Research Center, 230-3,
Moffett Field, California, 94035-1000, U.S.A.

1. Introduction

Transport data can be calculated from collision integrals which are thermal averages of the transport cross sections. The limiting factor in this approach has been the lack of accurate potential energy surfaces describing the pairwise interactions of the gas constituents. The accuracy achievable by *ab initio* calculations has improved dramatically in the last few years[1]. Here and in the following paper[2] we present an overview of the results of our extensive theoretical investigations of the interactions among the components of Earth's atmosphere. The primary objective of these studies is the determination of accurate interaction energies and the application of these interaction energies to establish a definitive data base that leads to reliable calculations of the transport properties for gases/plasmas formed from these components.

We have performed calculations for interaction partners that are the atoms/ions or molecules of the major constituents nitrogen and oxygen, light constituents hydrogen and helium which are important for studies of the upper atmosphere, and some minor constituents (e.g., argon and water). The interactions involving hydrogen are important for aeronautical

[†]Support by contract NAS2-14031 from NASA to Eloret Corporation is gratefully acknowledged.

M. Capitelli (ed.), Molecular Physics and Hypersonic Flows, 323–338.

applications, e.g., design and development of a future aircraft such as the National Aerospace plane (NASP) and studies of the preliminary heating of the fuel and combustion processes for air-breathing, hydrogen-burning, supersonic-combustion ram-jet engines (e.g., SCRAM).

We have determined accurate potential energy curves and surfaces for a large number of interactions and used these results to calculate transport cross sections and collision integrals. The collision integrals are our primary contribution to a definitive transport data base, however our reports have also included transport cross sections to enable the results to be applied to severe nonequilibrium conditions. We have reported transport data[3],[4],[5],[6],[7],[8] for all the interactions of the atoms and ions of nitrogen and oxygen (N–N, N–O, O–O, N^+–N, N^+–O, O^+–N, O^+–O, N^+–N^+, N^+–O^+, and O^+–O^+). For the case of interactions involving hydrogen, we have calculated the collision integrals for the pairs: H–H, H–N[9], H–H_2, H–N_2[10], H–O_2, H–H_2O, H_2–H_2, and H_2–N_2. Other work completed or in progress is described below and in paper II.

The *ab initio* quantum mechanical calculation of interaction energies is described in Sec. 2. The calculation of transport coefficients from the collision integrals is outlined in Sec. 3. Our potential energy curves and calculated transport data for atomic interactions are examined in Sec. 4. Our potential energy surfaces for the interaction of a hydrogen atom with H_2[11], N_2[10], and O_2[12] molecules and their applications are reviewed in Sec. 5. The construction of the atom–molecule potential energy surfaces, the calculation of transport cross sections from the potential energies, and applications to collisions involving ions are described in paper II.

2. Quantum Chemistry: Method for Calculating the Potential Energy Curves

Considerable progress has been made in *ab initio* quantum mechanical methods; the computed results have reached a level of accuracy where for small systems they can often compete with high-resolution spectroscopy[1]. A review of the principle methods is beyond the scope of this work, but summaries and reviews are available[13]. In this section we wish to emphasize certain aspects of the calculations which the non-specialist needs to consider in evaluating a potential.

Most of the accurate methods available employ a double basis set expansion. The first expansion, referred to as the one-particle basis set, consists of a linear combination (contraction) of atomic centered Gaussian type functions. The simplest approximation to the electronic Schrödinger equation is the self-consistent-field (SCF) approximation, where each electron is assumed to move in the average field of the other electrons. The second basis set expansion accounts for the electron correlation or the instantaneous

TABLE 1. Calculated values[a] of D_e(kcal/mol) for N_2.

Basis	D_e	
cc-pVTZ	218.6	$[4s\ 3p\ 2d\ 1f]$
cc-pVQZ	224.9	$[5s\ 4p\ 3d\ 2f\ 1g]$
cc-pV5Z	227.0	$[6s\ 5p\ 4d\ 3f\ 2g\ 1h]$
cc-pV6Z	227.9	$[7s\ 6p\ 5d\ 4f\ 3g\ 2g\ 1i]$
Expt.	228.4	

[a] Includes N $1s$ correlation contribution of 0.8 kcal/mol.

interactions between the electrons. This basis is commonly referred to as the n-particle basis set where n is the number of electrons in the system. The expansion generally consists of combinations of antisymmetrized n-fold products of the molecular orbitals (called configurations). For a given one-particle basis set the inclusion of all n-fold products in the second basis yields the full configuration-interaction (FCI) wavefunction. In the limit of an infinite one-particle basis set, a FCI calculation gives the exact solution of the electronic Schrödinger equation. The computational effort for a FCI calculation scales factorially with the number of electrons and orbitals and rapidly becomes intractable. FCI benchmark calculations have proven useful, however, for determining the appropriate level to obtain accurate results; the FCI gives the "exact" result with a given one-particle basis set and thus provides an unambiguous standard with which to compare other correlation approaches.

FCI benchmark studies[14] have demonstrated that high-level correlation methods such as a complete-active-space self-consistent-field (CASSCF) multireference configuration-interaction (MRCI) treatment[15] and coupled-cluster singles and doubles approach[16] (CCSD) supplemented with a perturbative estimate for the triple excitations[17] accurately reproduce the correlation energy. This suggests that the major limitation in most high-level ab initio calculations comes from the use of incomplete basis sets. Even when employing modern basis sets such as the atomic natural orbital[18] (ANO) and correlation consistent sets, cc-pVnZ[19], the basis set incompleteness is the major source of error. This is demonstrated in Table 1, where we have listed the CASSCF/MRCI results of Ref. [20] for the dissociation energy D_e of N_2.

It is important to realize the limits on accuracy dictated by the choice of basis set. For a computational chemist, the choice of basis set is probably the single most important decision affecting the accuracy of a calculation. For a dynamist, understanding the basis set limitations is probably the most important criterion in understanding the limitations of a given potential energy surface. The benchmark calculations are important as they demonstrate the accuracy at a given level and give confidence that one has the correct potential energy surface and hence, can calculate accurate transport properties.

3. Reduced Transport Coefficients and Applications

According to Chapman-Enskog theory[21], the transport properties of dilute monatomic gases can be expressed in terms of the mean collision integrals,

$$\sigma^2 \bar{\Omega}_{n,s}(T) = \frac{F(n,s)}{2(\kappa T)^{s+2}} \int_0^\infty e^{-E/\kappa T} E^{s+1} \bar{Q}_n(E) dE \qquad (1)$$

where κ is the Boltzmann constant, T is the kinetic temperature, and the scale factor $F(n,s)$ is defined in Ref [3]. The mean transport cross section, $\bar{Q}_n(E)$, is the average of the cross sections for all states weighted by their degeneracy factors.

For convenience, we present our results describing the contribution to transport properties in the form of reduced transport coefficients. For example, the binary (self for a pure gas) diffusion coefficient D in units of cm^2/s can be obtained from our reduced coefficient D^* defined by

$$pD^*(T) = 10^4 T^{-3/2} pD(T) = 26.287(2\mu)^{-1/2}/\sigma^2 \bar{\Omega}_{1,1}(T); \qquad (2)$$

similarly, the viscosity coefficient η in units of gm/cm s is obtained from

$$\eta^*(T) = 10^6 T^{-1/2} \eta(T) = 26.696(2\mu)^{1/2}/\sigma^2 \bar{\Omega}_{2,2}(T) \qquad (3)$$

where T is in K, the pressure p is in atmospheres, the reduced mass μ for the collision partners is in amu, and $\sigma^2 \bar{\Omega}$ is in $\text{Å}^2 = 10^{-16} \ cm^2$ as in our published tables.

The transport properties of a pure (single species) gas can be readily obtained from the tabulated values of $\sigma^2 \bar{\Omega}$; for example, the diffusion and viscosity coefficients can be calculated from Rels. (2) and (3). The corresponding relations for a binary (two species) mixture are slightly more complex [22]. The transport coefficients for a multi-component gas, in general, can be calculated from the relative concentrations of the various species and $\sigma^2 \bar{\Omega}$ using the formulation of Refs. [22] or [23].

4. Atom-Atom Interactions

As in previous calculations of transport cross sections, the potential energy wells of bound states have been constructed from available experimental data (e.g., measured dissociation energies, RKR data from spectroscopic measurements, *etc*). Our treatment of the the regions outside of the well, however, differs considerably from previous calculations. Our potential energies for this region have been determined from the results of accurate *ab initio* calculations rather than from the extrapolated results of an analytical potential function fitted to the data for the potential well. In addition, the potential energies $V(r)$ at large separation distances r for our calculations of the transport cross sections are determined from the proper long-range expansions for the interaction energies. This approach yields more reliable transport data at high temperatures and allows an accurate determination of low-temperature (e.g., ≤ 1000 K) data required for certain nonequilibrium studies.

The values of the dispersion coefficients for the long-range expansions are obtained primarily from available measured data or accurate quantum mechanical calculations; in some cases, we have used our *ab initio* results for large r to obtain unknown values. We point out that the quadrupole-quadrupole electrostatic interaction term, which falls off as r^{-5}, has been included for O–O long-range interactions.

Accurate *ab initio* calculations of the potential energy curves for the weakly-bound or unbound states are essential for an accurate determination of transport data for the interaction of nitrogen and oxygen atoms. For example, we have found[3] that the contributions from the bound states (with potential wells that can be obtained from experimental measurements) to the viscosity integral for O–O, N–N, and N–O interactions are only 15%, 25%, and 30%, respectively.

Typical atom–atom potential energy curves are shown in Fig. 1; these curves for the interaction of ground state nitrogen atoms illustrate the results of the approach described above. The major contributions to the N–N transport cross sections come from collisions for the high-spin states (because their statistical weighting factors are large). In earlier transport calculations the largest contribution, which comes from the septet state collisions, was estimated from exponential repulsive energy curves based on semiempirical models of the interaction. In contrast, the potential energy curve for our work (shown in Fig. 3 of Paper II) was obtained from an accurate calculation of Partridge *et al.*[24] that covers both the repulsive region and the attractive van der Waals region at large r. The *ab initio* results reproduce the long-range behavior obtained using measured dispersion data; the calculated values of the well depth and r_e are 2.6 meV and $7.52a_0$, respectively. Recently, the inner potential energy well of the quintet state has been experimentally characterized[25]; the measured data provides confirmation that the calculated potential energy curve[26] for this state is accurate.

The diffusion and viscosity coefficients for the interactions of nitrogen

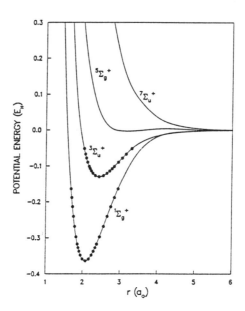

Figure 1. $V(r)$ for the states of N_2 corresponding to the interaction of nitrogen atoms in their ground states; the potential energy wells of the lower curves were constructed from experimental data ($\bullet$) obtained from spectroscopic and other measurements.

and oxygen are shown in Fig. 2. Note, that the N–O results (actually unknown before our calculations) lie below the corresponding curves for the like interactions; as a result the transport properties exhibit an unexpected behavior. For example, the viscosity coefficient for mixture of nitrogen and oxygen exhibits a minimum for a certain relative concentration.

5. Atom–Molecule Interactions

5.1. H–H$_2$

The repulsive potential energy surface for H_3 is the best characterized polyatomic potential energy and is considered to be accurate to one kcal/mol. The potential energy surface is of considerable interest for the theoretical determination of reactive scattering cross sections and has become a benchmark system to resolve differences between theory and experiment.

The determination of the transport properties requires the potential energy at large H–H$_2$ separation distances r. The potential for the region for r greater than about $4a_0$ had not been accurately characterized by previous theoretical studies. Furthermore, the values of the interaction energies deduced from scattering measurements are not in agreement[11], [27]. Therefore, we have performed[11] *ab initio* calculations employing very large basis

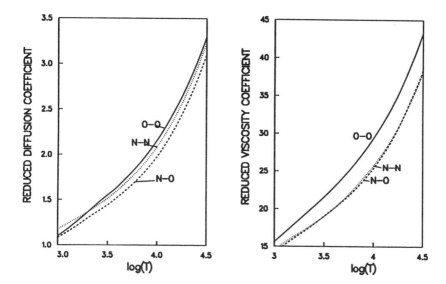

Figure 2. $D^*(T)$ and $\eta^*(T)$ for the interactions of nitrogen and oygen atoms; T is in Kelvin.

sets to narrow the uncertainty in the potential energy and thus permit a definitive determination of transport data.

To facilitate scattering calculations, our results[11] for the H–H_2 potential energy surface are provided in the form of a table of coefficients $V_{2n}(r, r_{H_2})$ for an expansion in Legendre polynomials

$$V(r, r_{H_2}, \gamma) = \sum_{n=0} V_{2n}(r, r_{H_2}) P_{2n}(\cos \gamma) \tag{4}$$

where r_{H_2} is the internuclear separation distance for the H_2 molecule and γ is the angle between the H_2 internuclear axis and the line joining the atom to the center-of-mass of the H_2 molecule.

In general, the calculation of high-energy transport cross sections that include the inelastic contributions from rotational and vibrational transitions requires the complete potential energy surface described above. One expects, however, that a rigid-rotor approximation (with r_{H_2} fixed at the expectation value $1.449a_0$ of the lowest vibrational state) is sufficient for calculating certain transport cross sections such as those needed for determining diffusion or viscosity coefficients. One further expects that slightly more accurate low-energy cross sections can be determined from an improved rigid-rotor estimate $V_M(r, \gamma)$ for the van der Waals region that was predicted[11] from calculations with larger basis sets. The values of $V_M(r, \gamma)$

TABLE 2. Comparison of parameters for H–H$_2$ potential energy.

Source	V(r)	r_0(Å)	r_m(Å)	V_m(meV)
	Scattering Experiments[a]			
σ(10–600)	BMSV[b]	3.14	3.55	2.34
σ^*(10–120)	$\exp(\alpha,6)^c$	2.98	3.35	2.40
$d\sigma^*/d\omega$(36)	BMD[d]	2.99	3.46	1.67
	MBMD[d]	2.99	3.47	1.82
σ(1–270)	$\exp(\alpha,6)^e$	2.97	3.40	2.06
	BMD[e]	2.98	3.42	2.04
	BMD(ρ)[e]	2.99	3.43	2.02
mean[f]		3.01	3.44±0.1	2.05±0.4
	Ab initio Calculations			
Eq. (4)	$V_0(r)^g$	3.05	3.50	1.82
paper II	$\bar{V}_M(r)^h$	3.01	3.45±0.02	2.04±0.08

[a]Measured cross sections (* indicates relative values only) are listed in first column; the collision energies are specified in meV within the parenthesis.
[b]Gengenbach et al. (1975) fit to D–H$_2$ data.
[c]Best fit of Hishinuma (1976) to H–H$_2$ data.
[d]D–H$_2$ results from crossed beam measurements of Torello and Dondi (1979); MBMD also fits data of Gengenbach et al. (1975)
[e]H–D$_2$ results of Hishinuma (1981).
[f]Average of scattering results ± largest deviations.
[g]Results of Table IV of Ref. 11 for $r_{H_2} = 1.449a_0$.
[h]Uncertainties determined from upper and lower bounds listed in Table V of Ref. 11.

can be obtained from Rels. (2–4) of paper II, using the repulsive potential parameters

$$\ln A(\gamma) = 1.619 + 0.233P_2(\cos\gamma) \qquad (5a)$$

$$\alpha(\gamma) = 1.669 + 0.018P_2(\cos\gamma) \qquad (5b)$$

and the data for the dispersion coefficients found in Table 3.

The potential well parameters determined from the spherical averages $V_0(r)$ and $\bar{V}_M(r)$ of the ab initio results for $V(r, 1.449, \gamma)$ (i.e., set of coefficients of the leading term on the r.h.s. of Rel. (4) for $r_{H_2} = 1.449a_0$) and $V_M(r, \gamma)$, respectively, are compared with the corresponding results obtained from fits to published data from scattering measurements in Table 2. Following the analysis of Tang and Toennies[27], we take the average values of the parameters from measurements to represent the experimental data; consistent with their approach, we have listed the results[28], [29] from the

TABLE 3. H atom–diatomic-molecule dispersion coefficients.

	H–H_2	H–N_2	H–O_2
$\bar{C}_6(a_0^6 E_h)$	8.7882[a]	21.07[b]	19.16[b]
$\bar{C}_8(a_0^8 E_h)$	161.98[a]	549[c]	352[c]
$\bar{C}_{10}(a_0^{10} E_h)$	3999.1[d]	16090[c]	8678[c]
Γ_6	0.1053[a]	0.114[e]	0.211[e]
Γ_8	0.2625[a]		1.15[f]
Γ_{10}	0.278[d]		

[a] Vibrationally-averaged results of Bishop and Pipin (1993).
[b] Ziess, G.D. and Meath, E.A. (1977) *Mol. Phys.* **33**, 1155–1176.
[c] Determined from Aufbau and related methods.
[d] Calculated by Meyer; reported by Tang and Toennies (1988).
[e] Langhoff, P.W., *et al.* (1971) *J. Chem. Phys.* **55**, 2126–2145.
[f] Deduced from *ab initio* results for large r.

best potential fits to the high-energy experiments. The potential energy in the region of the van der Waals minumum is not very sensitive to data from the high-energy experiments, but, on the other hand, is sensitive to the data from the measurements of the differential scattering cross sections $d\sigma/d\omega$ from the crossed beam experiment[30] and to the integral scattering cross sections σ from the low-energy beam experiment[31]. The values for $\bar{V}_M(r)$ agree well with the corresponding averages from the scattering data; in addition, the uncertainties in r_m and V_m from our work are considerably smaller than the corresponding uncertainties inferred from the spread in the values from the scattering data.

The potential functions of Ref. [31] not only provide good fits to the low-energy scattering data, but also can account for the data[30] of the crossed beam experiment; hence, one would select a potential function of Ref. [31] from the scattering results of Table 2 to obtain the best experimental estimate of the spherically-averaged potential energy in the van der Waals region at large r. We compare $V_0(r)$ and $\bar{V}_M(r)$ with the best fit potential BMD(ρ) from Ref. [31] in Fig. 3; again, the agreement of $\bar{V}_M(r)$ and the scattering result is quite satisfactory considering the uncertainties in each of the compared quantities (e.g., the uncertainties listed in the last row of Table 2 vs. the spread of the parameters of Table 2 or the values listed in Table II of Ref. [31] for the three measured potentials).

Our H–H_2 potential energy surface[11] has been used in a recent close-coupling calculation of diffusion and viscosity cross sections by Clark and McCourt[32]. The values of the reduced transport coefficients that have been calculated from the cross sections of Table 1 of Ref. [32] are compared

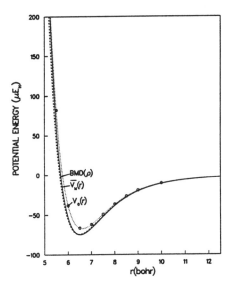

Figure 3. Spherical average of the potential energy surface for the interaction of a H atom with a H_2 molecule.

with the corresponding measured results in Fig. 4. We have selected the results of the two most recent diffusion measurements (i.e., by Lede and Villermaux[33] and Blyth *et al.*[34]) known to us; the selection of the results of the viscosity measurements by Browning and Fox[35] was based on the review by Marrero and Mason[36]. The theoretical results agree well with the measured data; i.e., to within about 2%.

5.2. H–N$_2$ AND H–O$_2$

The rigid-rotor potential energy surface for H–N$_2$ was constructed using the results of the *ab initio* calculations of Walch[37] for small r ($\leq 5a_0$) and additional calculations for $r_{N_2} = 1.2a_0$ at larger r. The *ab initio* results for small r are expected to be sufficiently accurate for the calculation of transport cross sections, but more accurate calculations would be required to determine the interaction energy in the van der Waals region at larger r. We chose an alternate approach to determine the potential for this region. As discussed in paper II, the *ab initio* results were joined to the results from the dispersion energy expansion using the potential function $V_M(r, \gamma)$ from Rels. (2–5) of paper II; the data for the dispersion coefficients is listed in Table 3.

The resulting potential for the van der Waals region is shown in Fig. 5. The uniform spacing of the repulsive walls with respect to γ is consistent

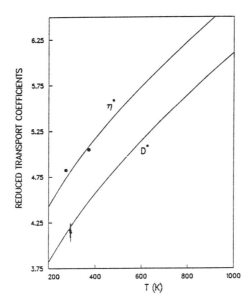

Figure 4. $D^*(T)$ and $\eta^*(T)$ for H–H$_2$ interactions. The theoretical values (solid lines) were obtained from the results of Clark and McCourt (1995); the data points with error bars (vertical lines) were obtained from the diffusion measurements of Lede *et al.* (1976) at 293 K and Blyth *et al.* (1987) at 295 K. The experimental data for viscosity is represented by the measurements of Browning *et al.* (1964). T is in Kelvin.

with equal differences in the $P_2(\cos\gamma)$ values; a similar uniform behavior is exhibited by the *ab initio* results for H–H$_2$ interaction energy (e.g., see Fig. 5 of Ref. [11]).

Our calculated value 1.2 cm^2/s for the H–N$_2$ diffusion coefficient at a pressure of one atm and room temperature agrees well with the measured results of Chery *et al.*[38] and Blyth *et al.*[34] 1.2±0.1 and 1.3±0.1 cm^2/s, respectively, for these conditions.

The ground state of O$_2$ is X$^3\Sigma_g^-$ and the H–O$_2$ collisions can interact along either the X$^2 A'$ or $^4 A'$ states. The potential energy curves $V(r,\gamma)$ derived from the *ab initio* calculation described in Ref. [12] are shown in Fig. 6, for the doublet and quartet states. Note the more complex behavior exhibited by the potential energy of the quartet state compared to that of the H–H$_2$ (see Fig. 5 of Ref. [11]) and H–N$_2$ (Fig. 5) potentials. Note further from the dispersion data of Table 3 that Γ_6 for H–O$_2$ is about a factor 2 higher than the corresponding values for H–H$_2$ or H–N$_2$; i.e., the long-range interaction energy for H–O$_2$ is more anisotropic than that for the latter cases.

The coefficient of P_4, for the polynomial expansion (Rel. 3 of paper II)

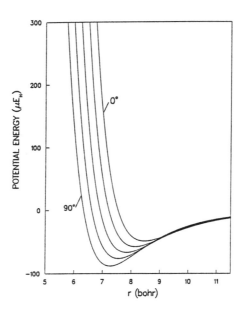

Figure 5. $V(r, \gamma)$ for the H–N$_2$ interactions; potential energy curves are shown for the following values of γ: 0°, 30°, 45°, 60°, and 90°. The repulsive wall recedes monotonically toward smaller r for increasing values of γ.

for $C_8(\gamma)$, is very small 0.009[39] for H–H$_2$ interactions. Likewise, our values of $V_4(r, \gamma)$ found in Table IV of Ref. [11] are very small for large r. Similarly, the *ab initio* results for the H–N$_2$ repulsive potential discussed above indicate that the contribution from P_4 is small. In contrast to the H–H$_2$ and H–N$_2$ interactions decribed above, one expects that higher-order polynomials would be needed to describe the more anisotropic quartet H–O$_2$ potential energy surface. For example, an application of $V_M(r, \gamma)$ using Eqs. (2–5) of paper II would require higher-order polynomials for the expansions of the potential parameters. Since the higher-order dispersion coefficients were unavailable, calculations with large basis sets were used to determine the interaction energy surface at large r.

Recently, we have used the quartet and doublet potential energy surfaces described above to calculate spin-flip cross sections for H–O$_2$ collisions. Our calculated value 16.4Å^2 for the thermal averaged cross section at room temperature is in agreement with the corresponding result 16.9±1.7Å^2 obtained from the stored atomic-beam spectroscopy measurements of Anderle *et al.*[40].

The reduced transport coefficients calculated for H–O$_2$ interactions are compared in Fig. 7 with the corresponding results for H–N$_2$ obtained from the collision integrals of Table IV of Ref. [10]. Since the mass affects are

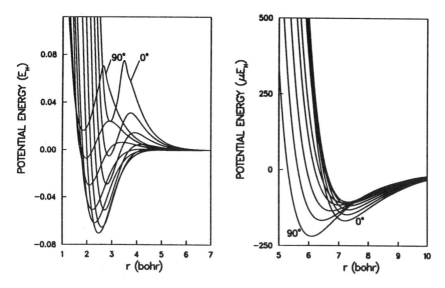

Figure 6. $V(r, \gamma)$ for H–O_2 interactions. Potential energy curves are shown for the following values of γ: $0°$, $15°$, $22.5°$, $30°$, $37.5°$, $45°$, $54.735°$, $60°$, $67.5°$, $75°$, and $90°$. The potential energy well for the doublet state (l.h.s plots) falls from the value for $0°$ to the minimum shown for $45°$ with increasing values of $\gamma < 45°$; it then rises to value shown for $90°$ with increasing values of $\gamma > 45°$. The behavior of the potential energy well for the quartet state (r.h.s. plots) is inverted compared to that described above for the doublet state; i.e., it first rises from the value for $0°$ to a maximum and then falls with increasing γ.

small (e.g., using Eqs. (2) and (3) we find that the differences are only about 1% and 0.4% for D^* and η^*, respectively) the differences between the results for the two molecular collision partners can only be explained by the differences in the complex interaction energies (compare the curves shown in Fig. 1 of Ref. [10] with those of Fig. 6).

6. Concluding Remarks

In conclusion we point out that the accuracy of our potential energy curves for atom–atom or ion–atom interactions at small values of r, as discussed in Sec. 2, can be established directly by comparison with accurate experimental data (e.g., RKR curves from spectroscopic measurements). Some comparisons with experimental results are mentioned in Sec. 4 for larger r. The agreement[4], [6], [7], of our theoretical resonance charge exchange cross sections with the results of beam measurements validates the accuracy of our calculated electron-exchange contribution to the potential

336

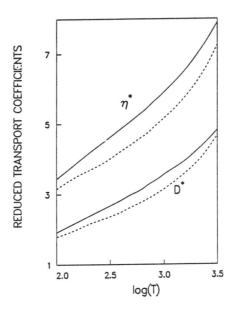

Figure 7. $D^*(T)$ and $\eta^*(T)$ for the interactions of a H atom with N_2 (dashed lines) and O_2 (solid lines) molecules; T is in Kelvin.

energy curves at large r.

The good agreement of the theoretical results with the measured scattering data described in Sec. 5 provides credibilty for our atom–molecule potential energy surfaces at large r. Since *ab initio* calculations are less difficult at smaller r (i.e., accuracy can be achieved with smaller basis sets; see Sec. 2); we expect that the transport data calculated from the potential surfaces is accurate at high temperatures also. Although our present atom–molecule transport calculations are focused on properties that are primarily governed by elastic scattering; we point out, however, that accurate potential energy surfaces, such as described in Sec. 5, are essential for an accurate calculation of the inelastic scattering contributions that are required, for example, for determining thermal conductivity.

Acknowledgements. The authors wish to acknowledge the valuable contributions of other members (and coauthors) of the Ames computational chemistry group: Drs. Charlie Bauschlicher, Steve Langhoff, Dave Schwenke, and Steve Walch. In particular, we wish to thank Dr. Peter Toennies for calling our attention to his studies of the H–H_2 interaction and scattering measurements overlooked in our prior work. We also want to thank Dr. Mario Capitelli for favorable comments on our work and point out that his published work shows that he recognized very early on that quantum structure calculations could provide an important contribution to the determination of transport properties.

References

1. Langhoff, S. R., (ed.), *Quantum Mechanical Electronic Structure Calculations with Chemical Accuracy*, Kluwer Academic Publishing, Dordrecht (1995).
2. Stallcop, J.R., Partridge, H., and Levin, E. (1995) Potential energies and collision integrals for the interactions of air components. II. Scattering calculations and interactions involving ions, in M. Capitelli (ed.), *Molecular Physics and Hypersonic Flows*, Kluwer Academic Publishing, Dordrecht.
3. Levin, E., Partridge, H., and Stallcop, J.R. (1990) Collision integrals and high temperature transport properties for N–N, O–O, and N–O, *J. Thermophysics and Heat Transfer* **4**, 469–477.
4. Stallcop, J.R., Partridge, H., and Levin, E. (1991) Resonance charge transfer, transport cross sections, and collision integrals for $N^+(^3P)$–$N(^4S^\circ)$ and $O^+(^4S^\circ)$–$O(^3P)$ interactions, *J. Chemical Physics* **95**, 6429–6439.
5. Partridge, H., Stallcop, J.R., and Levin, E. (1991) Transport cross sections and collision integrals for $N(^4S^\circ)$–$O^+(^4S^\circ)$ and $N^+(^3P)$–$O(^3P)$ interactions, *Chemical Physics Letters* **184**, 505–512.
6. Stallcop, J.R. and Partridge, H. (1985) N^+–N long-range interaction energies and resonance charge exchange, *Physical Review A* **32**, 639–642.
7. Partridge, H. and Stallcop, J.R. (1986) N^+–N and O^+–O interaction energies, dipole transition moments, and transport cross sections, in J.N. Moss and C.D. Scott (eds.), *Progress in Astronautics and Aeronautics: Thermophysical Aspects of Reentry Flows* **103**, pp. 243–260.
8. Stallcop, J.R., Partridge, H., and Levin, E. (1992) Collision integrals for the interaction of the ions of nitrogen and oxygen in a plasma at high temperatures and pressures, *Physics of Fluids B* **4**, 386–391.
9. Stallcop, J.R., Bauschlicher, C.W., Partridge, H., Langhoff, S.R., and Levin, E. (1993) Theoretical study of hydrogen and nitrogen interactions: N–H transport cross sections and collision integrals, *J. Chemical Physics* **97**, 5578–5585.
10. Stallcop, J.R., Partridge, H., Walch, S.P., and Levin, E. (1992) H–N_2 interaction energies, transport cross sections, and collision integrals, *J. Chemical Physics* **97**, 3431–3438.
11. Partridge, H., Bauschlicher, C.W., Stallcop, J.R., and Levin, E. (1993) *Ab initio* potential energy surface for H–H_2, *J. Chemical Physics* **99**, 5951–5960.
12. Stallcop, J.R., Partridge, H., and Levin, E. (in press) *Ab initio* potential energy surfaces and electron spin-exchange cross sections for H–O_2 interactions, *Physical Review A*.
13. Lawley, K.P. (Ed.) (1987) Ab initio methods in quantum chemistry, *Advances in Chemical Physics* **67**, Wiley-Interscience, New York; *ibid* **69**, Wiley-Interscience, New York.
14. Bauschlicher, C. W., Langhoff, S.R., and Taylor, P.R. (1990) *Advances in Chemical Physics* **77**, 103.
15. Saunders, V.R. and van Lenthe, J.H. (1983) *Molecular Physics* **48**, 923.
16. Bartlett, R.J. (1981) *Annual Reviews Physical Chemistry* **32**, 359.
17. Raghavachari, K., Trucks, G.W., Pople, J.A., and Head-Gordon, M. (1989) *Chemical Physics Letters* **157**, 479.
18. Almlöf, J. and Taylor, P.R. (1987) *J. Chemical Physics* **186**, 4070.
19. Dunning, T.H. (1989) *J. Chemical Physics* **90**, 1007; Kendall, R.A., Dunning, T.H., and Harrison, R.J. (1992) *J. Chemical Physics* **96**, 6796.
20. Bauschlicher, C.W. and Partridge, H. (1994) How large is the effect of $1s$ correlation on the D_e, ω_e, and r_e of N_2?, *J. Chemical Physics* **100**, 4329–4335.

21. Chapman, S. and Cowling, T.G. (1970) *The Mathematical Theory of NonUniform Gases*, 3rd ed., Cambridge University Press, Cambridge.

22. Hirschfelder, J.O., Curtiss, C.F., and Bird, R.B. (1964) *Molecular Theory of Gases and Liquids*, Wiley-Interscience, New York.

23. Maitland, G.C., Rigby, M., Smith, E.B., and Wakeham, W.A. (1981) *Intermolecular Forces. Their Origin and Determination*, Oxford University Press, Oxford.

24. Partridge, H., Langhoff, S.R., and Bauschlicher, C.W. (1986) Theoretical study of the $^7\Sigma_u^+$ state of N_2, *J. Chemical Physics* **84** 6901–6906.

25. Huber, K.P. and Vervloet, M. (1988) Rotational analysis of the Herman infrared bands of nitrogen, *J. Chemical Physics* **89**, 5957–5959.

26. Partridge, H., Langhoff, S.R., Bauschlicher, C.W., and Schwenke, D.W. (1988) Theoretical study of the $A'\,^5\Sigma_g^+$ and $C''\,^5\Pi_u$ states of N_2: Implications for the N_2 afterglow, *J. Chemical Physics* **88**, 3174–3186.

27. Tang, T. and Toennies, J.P. (1988) A model for the potential energy surface of H–H_2 in the intermediate- and long-range region, *Chemical Physics Letters* **151**, 301–307. **227**, 669–675.

28. Gengenbach, R., Hahn, Ch., and Toennies, J.P. (1975) Molecular beam measurements of the D–H_2 potential and recalibration of the reactive cross section, *J. Chemical Physics* **62**, 3620–3630.

29. Hishinuma, N. (1976) Determination of the H–H_2 potential from intergal cross section measurements at thermal energy, *J. Physical Society (Japan)* **41**, 1733–1738.

30. Torello, F. and Dondi, M.G. (1979) Experimental determination of the isotropic part of the D–H_2 potential surface, *J. Chemical Physics* **70**, 1564–1565.

31. Hishinuma, N (1981) Determination of the H–D_2 spherically averaged potential and the H–Ne potential from absolute integral cross section measurements, *J. Chemical Physics* **75**, 4960–4969.

32. Clark, G.B. and McCourt, F.R.W. (1995) Accurate calculation of diffusion and shear viscosity coefficients for H_2–H mixtures, *Chemical Physics Letters* **236**, 229–234.

33. Lede, J. and Villermaux, J. (1976) Measurement of atom diffusivity by a method independent of the wall effects, *Chemical Physics Letters* **43**, 283–286.

34. Blyth, G., Clifford, A.A., Gray. P., and Waddicore, J.I. (1987) Direct measurement of the diffusion coefficients of hydrogen atoms in six gases, *J. Chemical Society Faraday Transactions* **83**, 751.

35. Browning, R. and Fox, J.W. (1964) The coefficient of viscosity of atomic hydrogen and the coefficient of mutual diffusion for atomic and molecular hydrogen, *Proceedings Royal Society (London) A* **278**, 274–286.

36. Marrero, T.R. and Mason, E.A. (1972) Gaseous diffusion coefficients, *J. Physical Chemical Reference Data* **1**, 3–118.

37. Walch, S.P. (1990) Theoretical characterization of the potential energy surface for H+N_2 → HN_2. II. Computed points to define a global potential, *J. Chemical Physics* **93**, 2384–2392.

38. Chery, D. and Villermaux, J. (1972) *J. Chimie Physique* **69**, 452.

39. Bishop, D.M. and Pipin, J. (1993) Vibrational effects for the dispersion-energy and dispersion-polarizability coefficients for the interactions between H, He, and H_2, *J. Chemical Physics* **522**, 522–524.

40. Anderle, A., Bassi, D., Ianotta, S., Marchetti, S., and Scholes, G. (1981) Measurement of the spin-exchange cross section in the collision of H atoms with O_2 and NO by means of stored atomic-beam spectroscopy, *Physical Review A* **23**, 34–38.

POTENTIAL ENERGIES AND COLLISION INTEGRALS
FOR THE INTERACTIONS OF AIR COMPONENTS

II. Scattering Calculations and Interactions Involving Ions

JAMES R. STALLCOP AND HARRY PARTRIDGE
Computational Chemistry Br., Space Technology Div.,
NASA Ames Research Center, 230-3,
Moffett Field, California, 94035-1000, U.S.A.

AND

EUGENE LEVIN
Thermosciences Institute[†],
NASA Ames Research Center, 230-3,
Moffett Field, California, 94035-1000, U.S.A.

1. Introduction

The calculation of interaction energies by *ab initio* methods is described in the preceeding paper [1]. The construction of potential energy surfaces from the *ab initio* results and the scattering calculations to determine transport cross sections are described below.

The major contribution to the transport cross sections for nitrogen and oxygen atoms (Sec. 4 of paper I) comes from collisons for the high-spin states. The construction of the potential energies for high-spin state and atom-molecule interactions (Sec. 5 of paper I) is described in Sec. 2. We have developed [2], [3], [4], [5] scattering codes to calculate transport cross sections from accurate spline-fits to the potential energy data for a wide range of collision energies E (usually around 10^{-4} to about 1 E_h). The methods for the scattering calculations are outlined in Sec. 3. A discussion of our results [3], [6], [7], [8] for ion–atom interactions and an outline of our determination[9] of corrections to the ion–ion shielded Coulomb collision integrals, arising from the use of realistic interaction energies from *ab initio* calculations, is contained in Secs. 4 and 5, respectively. Combining rela-

[†]Support by contract NAS2-14031 from NASA to Eloret Corporation is gratefully acknowledged.

M. Capitelli (ed.), Molecular Physics and Hypersonic Flows, 339–349.

tions for potential energies are of interest for predicting unknown transport properties; some results [10] of our studies are summarized in Sec. 6.

2. Construction of Repulsive Potential Energy Curves and Surfaces

The determination of the potential energy surface at large r by the *ab initio* methods of Sec. 2 of paper I is generally very difficult; it requires large basis sets, much computational effort, and thus may not be feasible. In this case, the *ab initio* results for the repulsive potentials at small r are joined to interaction energies from the long-range expansion using potential energy functions based on the formulation of Tang and Toennies [11].

The atom-atom potential energy for the van der Waals region is represented by the sum [11]

$$V_{TT}(r) = V_{SR}(r) + V_{DLR}(r); \tag{1a}$$

where the short-range repulsive energy V_{SR} (from first-order Pauli electron repulsion) and the damped long-range attractive energy V_{DLR} (from the second-order dispersion interaction) have the forms

$$V_{SR}(r) = A \exp(-r/\rho) \text{ and } V_{DLR}(r) = -\sum_{n=3} f_{2n}(r/\rho)\frac{C_{2n}}{r^{2n}}, \tag{1b, 2a}$$

respectively. The damping function f_{2n} is related to an incomplete gamma function of order $2n+1$

$$f_{2n}(x) = 1 - \exp(-x)\sum_{k=0}^{2n} \frac{x^k}{k!}. \tag{2b}$$

Our application [10] of V_{TT} to combining relations is outlined in Sec. 6 below.

We have modified the potential function of Eqs. (1–2) to construct the potential energy surfaces for atom–molecule interactions. For the case of a diatomic molecule with like atoms, the dispersion coefficients can be represented by an expansion in Legendre polynomials

$$C_{2n}(\gamma) = \bar{C}_{2n}[1 + \Gamma_{2n}P_2(\cos\gamma) + \cdots \tag{3}$$

the angle γ specifies the orientation of the atom with respect to the molecular axis (as described in paper I). A suitable modification to represent the potential energy surface is

$$V_M(r,\gamma) = V_{SR}(r,\gamma) + V_{DLR}(r,\gamma); \tag{4}$$

analogous to Eq. (1b) the repulsive energy has the form

$$\ln V_{SR}(r, \gamma) = \ln A(\gamma) + \alpha(\gamma)r \tag{5a}$$

where the strength and the decay parameter $\alpha = 1/\rho$ are obtained from the expansions

$$\ln A(\gamma) = \bar{a} + a_2 P_2(\cos \gamma) + \cdots \tag{5b}$$

$$\alpha(\gamma) = \bar{\alpha}_0 + \alpha_2 P_2(\cos \gamma) + \cdots . \tag{5c}$$

Similarly, the attractive energy $V_{DLR}(r, \gamma)$ is obtained from Eqs. (2) using the the expansions on the right-hand side of Rels. (3) and (5c) for C_{2n} and ρ^{-1}, respectively.

We have found [12] that only the first two terms in the above polynomial expansions are required to obtain a good fit to our *ab inito* results for the H–H$_2$ potential energy surface. We have applied this potential function to construct an improved potential energy surface [12] as described in Sec. 5 of paper I.

We have used $V_M(r, \gamma)$ to determine [4] the potential energy surface for H–N$_2$ interactions (see Sec. 5 of paper I) for the calculation of transport properties. Similarly, in more recent work, we have adapted the form $V_M(r, \theta)$ to treat ion-molecule interactions; this has allowed us to extend our *ab initio* results for N$^+$–N$_2$ interactions to the region of r where the interaction energy can be accurately determined from the electrostatic and dispersion energies of a long-range expansion, and thus, calculate low-energy transport cross sections.

3. Scattering Calculations

We use a quantum mechanical formulation to determine the scattering; the transport cross sections are calculated from the scattering phase shifts. In this approach, all the physics of the interaction energy is contained in the phase shifts; hence, accurate calculations of the phase shifts are critical to the determintion of transport data.

3.1. SCATTERING PHASE SHIFTS

At low E the scattering phase shifts η_l are obtained from a direct numerical solution of the Schrödinger equations for the partial radial wave functions $G_l(r)$

$$\left[\frac{d^2}{dr^2} + [k(r)]^2 \right] G_l(r) = 0 \tag{6}$$

where the local wave number is obtained from

$$k(r) = k \left[1 - \frac{V(r)}{E} - \frac{l(l+1)}{(kr)^2} \right]^{1/2} . \tag{7}$$

The quantity k provides the asymptotic value (i.e., $k(r) \to k$ as $r \to \infty$). The values of the phase shifts are extracted from the asymptotic behavior of the wave functions at large r; i.e.,

$$G_l(r) \to \sin(kr - l\pi/2 + \eta_l); \quad \text{as } r \to \infty. \tag{8}$$

At higher energies, the scattering phase shifts are computed from the JWKB approximation or (when E lies in the region of a potential energy barrier) a uniform semiclassical approximation [3], [13] formulated by Stallcop. In the latter case, the phase shifts are obtained from

$$\eta_l \approx \eta_l^o + \rho + \tan^{-1}[P(I+\tau)\tan(\phi+\rho)] \tag{9}$$

$$\rho = \frac{1}{2}\arg\left[\Gamma\left(\frac{1}{2}+i\frac{I}{\pi}\right)\right] - \frac{I}{2\pi}\ln\left|\frac{I}{e\pi}\right| \tag{10}$$

$$\tau = \frac{1}{2}\ln\left[\Gamma\left(\frac{1}{2}+\frac{\phi}{\pi}\right)/\sqrt{2\pi}\right] - \frac{\phi}{2\pi}\ln\left|\frac{\phi}{e\pi}\right|. \tag{11}$$

The barrier penetration function P has the form

$$P(x) = \frac{\sqrt{1+e^{-2x}}-1}{\sqrt{1+e^{-2x}}+1} \tag{12}$$

and $\Gamma(x)$ is the gamma function for argument x. The phase integrals are defined by the relations

$$\eta_l^o = Re\left[\int_{r_3}^{\infty}[k(r)-k]dr - kr_3\right] + \frac{1}{2}(l+1/2)\pi, \tag{13}$$

$$\phi = Re\left[\int_{r_1}^{r_2}k(r)dr\right], \quad \text{and} \quad I = -i\int_{r_2}^{r_3}k(r)dr; \tag{14a, b}$$

where, $k(r)$ is obtained from Eq. (7) using the Langer approximation, $l(l+1)$ is replaced by $(l+1/2)^2$. The integration limits r_1 and r_2 correspond to the inner and outer, respectively, classical turning points of the potential well while the limit r_3 is the outermost turning point for the potential barrier when the energy is below the barrier maximum. A complete description for their determination and the details of the integrations for the phase integrals are contained in Refs. [3] and [13].

The above approximation accounts for resonance effects (as is evident from the form of the last term of Rel. (9) above) associated with the metastable and virtual energy levels of the potential energy well lying below and above the barrier maximum, respectively. When E is far above the maximum of the potential energy barrier or below the potential energy well $\eta_l \to \eta_l^o$; i.e., the phase shift reduces to the JWKB approximation.

The semiclassical scattering approach described above allows the determination of transport cross sections for like atom–ion (e.g., N–N$^+$) collisions where resonance charge exchange is an important [3] scattering mechanism for diffusion processes. The semiclassical approximation, in addition to taking quantal effects into account, also avoids the tedious summation of the (unrealistic) decreasing orbiting scattering contributions encountered in a classical description of the collision for energies near a barrier maximum.

A full quantum mechanical treatment of the scattering is required for low energies, especially when a colliding partner is light (e.g., hydrogen) to obtain accurate cross sections. We have compared [5] the transport cross sections from the semiclassical method with the corresponding results from the exact calculation for collisions involving hydrogen. Both calculations yield considerable resonance structure in the cross sections; we find roughly that the semiclassical results are shifted slightly in energy compared to the exact results. The integration over a Maxwellian velocity distribution for the collision integrals averages the structure and differences in the cross sections; we found that the corrections to the semiclassical hydrogen diffusion and viscosity collision integrals from a full quantum mechanical treatment of the scattering rise to about 8% and 4%, respectively, as the temperature decreases to 100K.

The exact calculation of phase shifts becomes impractical at high collision energies because of increasing computational effort, especially for heavier collision partners. Fortunately, however, the results of the semiclassical approximation become accurate for such scattering conditions. We have compared [5] the results of the two scattering methods for increasing values of the reduced mass of the collision partners. ¿From this study we conclude that the error in the collision integrals from the semiclassical method is not significant for collision partners as heavy as the atoms (or ions) of the major constituents of air (nitrogen and oxygen); the error is less than that introduced by the uncertainty in the potential energy data.

3.2. TRANSPORT CROSS SECTIONS

The transport cross sections can be calculated from the formulation of Ref. [10] when certain symmetry affects are not significant. At low energies, for example, the effects of nuclear symmetry must be taken into account. The calculation of the transport cross sections for like atom–ion collisions also requires a different treatment; we have reported [3] a formulation of these cross sections that also includes the effects of nuclear symmetry.

The accurate calculation of transport cross sections for collisions involving molecules is, in general, considerably more difficult than that described above for atoms or ions. A close-coupling calculation, for example, can provide the contribution from inelastic scattering, but it is limited to low energies because of the computational effort required.

We use the sudden approximations applied by Parker and Pack [14] to formulate cross sections in the body-fixed frame for atom-molecule collisions. Thus, for fixed orientations of the molecule, we can use the methods described above to calculate the phase shifts for a central field potential, combine the results to obtain cross sections as a function of the angle(s) that specify the orientation, and then integrate over the angle(s) to determine the complete cross section. The centrifugal sudden approximation is expected to be accurate for small values of the impact parameters; for large values, one might follow the approach of Stallcop [15] that requires only the energy sudden approximation.

Sudden approximations have been found[16] to be accurate for the collisions of a light atom with a heavy molecule (e.g., He–N_2). We have compared [17] our scattering results for H–H_2 interactions with that of close-coupling calculations [16] for the same potential energy surface. We found [17] that our values of the diffusion and viscosity coefficients agree with the corresponding values shown in Fig. 5 of paper I to within about 0.5%. The high-energy approximations (neglect or extrapolation of certain matrix elements) of Ref. [16] preclude a meaningful comparison at higher temperatures.

4. Ion-atom Interactions

An understanding of the physics of the interactions is necessary for guiding an efficient and accurate calculation of transport cross sections or for predicting transport properties from limited potential energy data. For example, the calculation of the high-energy resonance charge exchange cross sections $Q_{ex}(E)$ for like atom–ion collisions (and likewise, for the diffusion cross section $Q_1 \approx 2Q_{ex}$) requires careful attention to the determination of gerade–ungerade splitting energies $\Delta E(r)$ at large r.

¿From a hydrogenic model of the electron exchange interaction, one expects that $\Delta E(r)$ should have roughly an exponential behavior at large r. We found that the results of molecular calculations with basis sets constructed from Gaussian–type orbitals allow one to assess convergence readily and reliably. The $\Delta E(r)$ calculated from smaller basis sets exhibit the expected exponential behavior on a semilog plot at smaller r, but at larger r fall off much too rapidly; i.e., the "building up" process to obtain $\Delta E(r)$ requires the contributions from many orbitals with various ranges. To reduce computation effort, an exponential extrapolation may be applied to extend calculated values of $\Delta E(r)$ to larger r since the gerade and ungerade states have the same long-range interaction energies.

For N$^+$–N interactions and O$^+$–O interactions, we have found that the ratios of the gerade-ungerade splitting energies $\Delta E_\Lambda^{2S+1}(r)$ predicted by the asymptotic valence bond relations developed by Stallcop [18]

$$\Delta E_\Lambda^{2S+1} \rightarrow -\frac{1}{3}(-1)^{2S+1}(2S+1)W_\Lambda(r), \quad \text{as } r \rightarrow \infty \qquad (15)$$

where $W_\Lambda(r)$ contains the one electron exchange terms, agree well [3], [7] with the corresponding results obtained from the results of the large *ab initio* molecular calculations at large r. Hence, the application of valence bond type relations and fairly accurate (to within a few pecent [3]) analytical approximations [6] for Q_{ex}, based on exponential fits to $\Delta E(r)$ at large r, allow reasonably accurate diffusion cross sections to be obtained with a relatively small computational effort. This appoach, for example, is satisfactory for determining the contribution from excited states.

At low energies, $Q_{ex}(E)$ is much larger than the corresponding result predicted from $\Delta E(r)$; here the scattering is governed by the long-range forces. If the long-range interaction energy has the form $V(r) \rightarrow C_\nu/r^\nu$, we have found [19] that the exchange cross section can be estimated from

$$Q_{ex} \rightarrow \frac{\nu}{4}\left(\frac{\nu}{2} - 1\right)^{\frac{2}{\nu} - 1}\left(\frac{C_\nu}{E}\right)^{\frac{2}{\nu}}. \qquad (16)$$

The leading long-range term for N$^+$–N interactions arises from charge polarization ($\nu = 4$); in this case, we find, as is well known, that Q_{ex} falls off as $E^{-1/2}$. For O$^+$–O interactions, the leading terms come from the charge-quadrupole interaction ($\nu = 3$), we find that Q_{ex} falls as $E^{-2/3}$ i.e., more rapidly compared to that from charge induced-dipole forces.

Our calculated N$^+$–N and O$^+$–O charge exchange cross sections are in agreement [3] with the results of high-energy beam measurements. We have calculated [3], [8] the collision integrals for all the ion–atom interactions of nitrogen and oxygen and tabulated the results for a broad range of temperatures (100K to 100,000K). The reduced diffusion and viscosity coefficients that have been determined from these tabulations and Eqs. (2) and (3) of paper I are shown in Figs. 1 and 2, respectively.

Note that the like atom-ion diffusion coefficients lie far below the results for the collisions of unlike pairs (e.g., N–O$^+$) at high temperatures; this can be explained by the resonance charge exchange mechanism discussed above. The corresponding differences for the viscosity coefficient are small; in addition, note that the N–O$^+$ curve lies above the O–O$^+$ curve.

We have determined analytical approximations [20] to the collision integrals for the ion-atom interactions decribed above and also for atom-atom interactions [2] examined in Sec. 4 of paper I to facilitate transport applications.

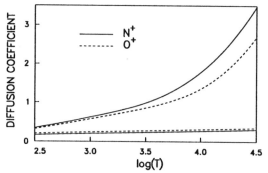

Figure 1. $D^*(T)$ for unlike (upper curves) and like (lower curves) atom collision partners of N^+ and O^+; T is in Kelvin.

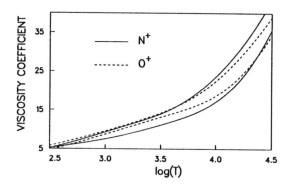

Figure 2. $\eta^*(T)$ for unlike (upper and lower curves) and like (lower and upper curves) atom collision partner of the ions (N^+ and O^+, respectively); T is in Kelvin.

5. Ion-ion Interactions

The calculation of the ion–ion transport cross sections by the methods described above is not practical (because of the long-range behavior of the Coulomb interaction) when the shielding parameter (Debye cut-off) λ is large. We have, however, developed an approach [9] that significantly reduces the computational effort required to obtain transport cross sections Q_n for realistic interaction potentials (e.g., constructed from our *ab initio* potential energies at small r). We calculate only the difference

$$\Delta Q_n(\lambda, E) = Q_n(\lambda, E) - Q_n^C(\lambda, E) \tag{17}$$

where Q_n^C is the shielded Coulomb cross section. Using the semiclassical approximations to the phase shifts described above one need only perform the integration to a value r_c that is sufficiently large such that the charge-induced dipole interaction energy (charge-quadrupole interaction energy, if N^+ is one of the interaction partners) is negligibly small compared to the Coulomb interaction. The phase shifts for the various molecular states can then be obtained from the sum of this contribution from smaller r and the contribution from the Coulomb tail for $r \geq r_c$. Moreover, the phase shifts for large l where the corresponding impact parameter $b = (l + 1/2)/k$ is greater than r_c need not be calculated; note these contributions cancel out in the summations of Eq. (18) for ΔQ_n.

Since the Coulomb collision integrals $\Omega_{n,s}^C(\lambda, T)$ have been tabulated, the ion–ion collision integrals $\Omega_{n,s}(\lambda, T)$ can be obtained from the difference

$$\Delta\Omega_{n,s}(\lambda, T) = \Omega_{n,s}(\lambda, T) - \Omega_{n,s}^C(\lambda, T). \tag{18}$$

We have calculated $\Delta\Omega$ for the diffusion and viscosity integrals for all the interactions of the ions of nitrogen and oxygen [9]; note, however, that this approach can also be readily used to determine the collision integrals for higher-order terms or other ion–ion interactions.

We have determined analytical fits [9] to $\Delta\Omega$ from our calculated results and also to Ω^C from tabulated results in the literature to facilitate application of the collision integrals described above. We have found [9] that our calculated corrections $\Delta\Omega$ to the shielded Coulomb collision integrals can be significant at higher temperatures and pressures.

6. Combining Relations for Predicting Interaction Energies

The second-order dispersion contribution to the potential energy can be obtained with the help of combining relations that have been developed for the dispersion coefficients. The first-order contribution V_{SR} for applications to combining relations can be obtained from $V(r)$ using Eqs. (1–2) and the values of C_{2n}.

The accurate potential energy data from our work provides a realistic foundation for testing combining rules for $V_{SR}(r)$. Based on the physics of the electron interactions, Smith [21] has proposed the following combining relations for obtaining the parameters of a repulsive potential for unlike interactions a–b from those for like interactions

$$\rho(a, b) = \frac{1}{2}[\rho(a, a) + \rho(b, b)] \tag{19}$$

$$\rho(a, b) \ln\left(\frac{A(a, b)}{\rho(a, b)}\right) = \frac{1}{2}\left[\rho(a, a) \ln\left(\frac{A(a, a)}{\rho(a, a)}\right) + \rho(b, b) \ln\left(\frac{A(b, b)}{\rho(b, b)}\right)\right]. \tag{20}$$

We have applied [10] these relations, to predict the values of $A(N, H)$ and $\rho(N, H)$ from the values of the corresponding parameters obtained from accurate *ab initio* H-H and N-N potential energies for the highest-spin states.

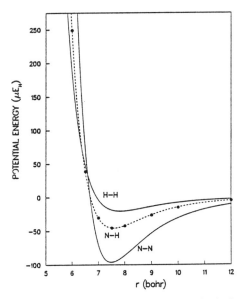

Figure 3. Van der Waals potential energy curves for the $^3\Sigma_u^+$, $^5\Sigma^-$, and $^7\Sigma_u^+$ states of H$_2$, NH, and N$_2$, respectively.

The predicted N–H potential energy curve (dashed curve) calculated using the above predicted values and Eqs. (1–2) is compared with the data ($\bullet$) from our accurate *ab initio* calculation [10] in Fig. 3; good agreement is also found [10] for smaller r, i.e., to about $4.5a_0$ where the departure of the *ab initio* results from an exponential can be attributed [10] to an avoided curve crossing.

We have found the transport coefficients that were determined from the spherically averaged potential energy agree well with the corresponding results obtained from the complete potential energy surface for both H–H$_2$ [17] and H–N$_2$ (see Fig. 3 of Ref. [4]) interactions. Consequently, one expects that an extension of the above combining rule to include the average potential energies for interactions involving the molecules of hydrogen and nitrogen could provide meaningful interaction energies for the calculation of transport properties.

References

1. Partridge, H., Stallcop, J.R., and Levin, E. (1995) Potential energies and collision integrals for the iteractions of air components. I. *Ab initio* calculation of potential energies and neutral interactions, in M. Capitelli (ed.), *Molecular Physics and Hypersonic Flows* Kluwer Academic Publishers, Dordrecht.
2. Levin, E., Partridge, H., and Stallcop, J.R. (1990) Collision integrals and high tem-

perature transport properties for N–N, O–O, and N–O, *J. Thermophysics and Heat Transfer* **4**, 469–477.

3. Stallcop, J.R., Partridge, H., and Levin, E. (1991) Resonance charge transfer, transport cross sections, and collision integrals for $N^+(^3P)$–$N(^4S^o)$ and $O^+(^4S^o)$–$O(^3P)$ interactions, *J. Chemical Physics* **95**, 6429–6439.

4. Stallcop, J.R., Partridge, H., Walsch, S.P., and Levin, E. (1992) H–N_2 interaction energies, transport cross sections, and collision integrals, *J. Chemical Physics* **97**, 3431–3438.

5. Levin, E., Schwenke, D.W., Stallcop, J.R., and Partridge, H. (1994) Comparison of semiclassical and quantum mechanical methods for the determination of transport cross sections, *Chemical Physics Letters* **227**, 669–675.

6. Stallcop, J.R. and Partridge, H. (1985) N^+–N long-range interaction energies and resonance charge exchange, *Physical Review A* **32**, 639–642.

7. Partridge, H. and Stallcop, J.R. (1986) N^+–N and O^+–O interaction energies, dipole transition moments, and transport cross sections, in J.N. Moss and C.D. Scott (eds.), *Progress in Astronautics and Aeronautics: Thermophysical Aspects of Reentry Flows* **103**, pp. 243–260.

8. Partridge, H., Stallcop, J.R., and Levin, E. (1991) Transport cross sections and collision integrals for $N(^4S^o)$–$O^+(^4S^o)$ and $N^+(^3P)$–$O(^3P)$ interactions, *Chemical Physics Letters* **184**, 505–512.

9. Stallcop, J.R., Partridge, H., and Levin, E. (1992) Collision integrals for the interaction of the ions of nitrogen and oxygen in a plasma at high temperatures and pressures, *Physics of Fluids B* **4**, 386–391.

10. Stallcop, J.R., Bauschlicher, C.W., Partridge, H., Langhoff, S.R., and Levin, E. (1993) Theoretical study of hydrogen and nitrogen interactions: N–H transport cross sections and collision integrals, *J. Chemical Physics* **97**, 5578–5585.

11. Tang, K.T. and Toennies, J.P. (1984) An improved simple model for the van der Waals potential based on universal damping functions for the dispersion coefficients, *J. Chemical Physics* **80**, 3726–3741.

12. Partridge, H., Bauschlicher, C.W., Stallcop, J.R., and Levin, E. (1993) *Ab initio* potential energy surface for H–H_2, *J. Chemical Physics* **99**, 5951–5960.

13. Stallcop, J.R. (1969) Semiclassical elastic scattering cross sections for a central field potential function, NASA SP-3052, Scientific and Technical Information Division.

14. Parker, G. and Pack, R.T. (1978) Rotationally and vibrationally inelastic scattering in the rotational IOS approximation. Ultrasimple calculation of total (differential, integral, and transport) cross sections for nonspherical molecules, *J. Chemical Physics* **68**, 1585–1601.

15. Stallcop, J.R. (1975) Inelastic scattering in atom–diatomic molecule collisions. I Rotational transitions in the sudden approximation, *J. Chemical Physics* **61**, 5085–5097.

16. Clark, G.B. and McCourt, F.R.W. (1995) Accurate calculation of diffusion and shear viscosity coefficients for H_2–H mixtures, *Chemical Physics Letters* **236**, 229–234.

17. Stallcop, J.R., Partridge, H., and Levin, E. (to be submitted).

18. Stallcop, J.R. (1971) N_2^+ Potential energy curves, J. Chemical Physics **54**, 2602–2605.

19. Stallcop, J.R., Partridge, H., and Levin, E. (in press) *Ab initio* potential energy surfaces and electron spin-exchange cross sections for H–O_2 interactions, *Physical Review A*.

20. Stallcop, J.R., Partridge, H., and Levin, E. (submitted) Analytical fits for the determination of the transport properties of air, *J. Thermophysics and Heat Transfer*.

21. Smith, F.T. (1972) Atomic distortion and the combining rule for repulsive potentials, *Physical Review A* **5**, 1708–1713.

MEASUREMENTS AND NATURE OF INTERMOLECULAR FORCES: THEIR ROLE IN GASEOUS PROPERTIES

F. PIRANI, D.CAPPELLETTI, AND V.AQUILANTI
Dipartimento di Chimica dell'Università
via Elce di Sotto n.8, 06123 Perugia, Italy

Abstract. The measurements of intermolecular forces by scattering experiments carried out with a molecular beam technique is illustrated. The case of aligned atomic and molecular beams is presented as an important tool to obtain the anisotropic components of the interactions. The paper reports also a simple and general approach to characterize the nature of the interactions and to estimate features of the potential for a variety of systems; applications to cases of interest for atmospheric phenomena are illustrated.

1. Introduction

The concept of intermolecular interaction potential is crucial for the understanding of many macroscopic and microscopic properties of the matter, such as the behaviour of real gases, gaseous transport properties, gas surface physical absorption, energy and structure of clusters, liquids and solids, and gas phase chemical kinetics [1]. An example is the Lennard-Jones collision frequency associated to energy transfer processes which controls unimolecular rate constants [2]. This frequency, as well as the transport properties, such as the binary diffusion coefficient and the interaction viscosity, are defined in terms of collision integrals which depend on the depth and location of the potential wells [1]. Another example is represented by several molecular systems which give laser action, e.g. the rare gas halides, which exhibit emission bands whose characteristic are affected by the interaction features both in the ground covalent and the excited ionic states [3]. As a third example of interest here, weak intermolecular forces are responsible of the selectivity in reactive and inelastic processes when dominated by in-

351

M. Capitelli (ed.), Molecular Physics and Hypersonic Flows, 351–360.
© *1996 Kluwer Academic Publishers. Printed in the Netherlands.*

termediate and long range effects [4].

The interactions involving molecules and open shell atoms, that is atoms with an internal angular momentum, are anisotropic and consequently their experimental and theoretical investigation is more demanding when compared with the case of isotropic forces, such as those operating in rare gas-rare gas systems. A manifold of potential energy surfaces describes the general behaviour of open shell atoms in gaseous baths [5]. Likewise, the course of reactive events has to be followed along a few specific potential energy surfaces of a properly defined symmetry [4].

In the case of open shell atoms exhibiting a sufficiently large spin-orbit splitting constants, their collisions with spherical symmetric particles can be shown to evolve on effective adiabatic curves which account for both electrostatic and spin-orbit terms [6, 7]. These curves are suitably given in terms of spherical and anisotropic components of the electrostatic interaction and of atomic spin-orbit constants [8]. Inelastic events, such as intramultiplet mixing, are also described as due to nonadiabatic effects between the effective curves [9].

In systems involving diatomic molecules, the interaction depends not only on the intermolecular distance, but also on the orientation of the molecule with respect to the intermolecular axis and on the molecular bond length. The dependence of the potential energy surface on these coordinates determines steric effects and energy transfer processes in collisions of the diatom with other particles. The interaction of a rotating diatomic molecule can also be given in terms of effective adiabatic curves if the molecule exhibits a strong quantization of the rotational motion and behaves like a rigid rotor [10]. These curves are obtained diagonalizing the total potential energy matrix given by the sum of electrostatic and rotational terms, and asymptotically correspond to different projections of the rotational angular momentum with respect to the intermolecular axis. Rotational transitions can again be described as due to nonadiabatic transitions between the adiabatic curves [10].

This paper describes the techniques employed in our laboratory in order to measure intermolecular forces and to discuss within a unified frame the nature and the main features of the interactions. Some interesting applications of this approach which allows to estimate potential parameters for a variety of systems, are also discussed.

2. Experimental Background

The characterization of the intermolecular potential exploits both spectroscopic and scattering methods [1]. Since each experiment probes particular features of the interactions, the simultaneous analysis of several experi-

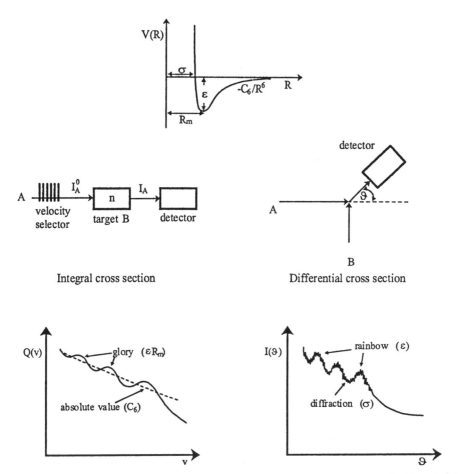

Figure 1. Sketch of the experimental techniques used to measure integral (left) and differential (right) cross sections. Specific features of the intermolecular potential obtainable from the analysis of the scattering results are also indicated.

mental data of different nature is an important key to obtain an accurate characterization of the interaction potentials over a wide intermolecular distance range [11].

Scattering experiments are performed using the molecular beam technique and measuring both differential and integral cross sections. The integral cross section experiment consists in the measurement of the attenuation of a velocity selected beam A (see Figure 1) by a target B contained, for example, in a scattering chamber. A two beams configuration is used for differential cross section measurements, where the intensity of the particles scattered at various angles is monitored by a rotating detector.

For the accurate characterization of the interaction potential features it

354

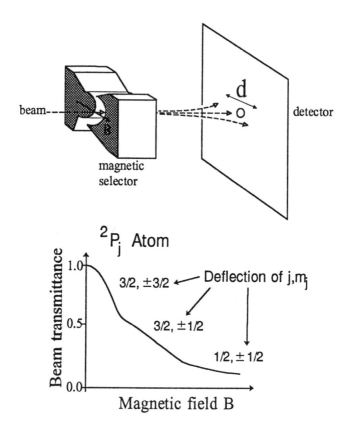

Figure 2. Schematic description of the magnetic analysis carried out in the transmission mode [9,12,13]. The particle intensity measured at a well defined velocity undergoes an attenuation when the beam travels through the inhomogeneous field of a Stern-Gerlach magnetic selector. The beam transmittance, defined as the ratio between beam intensities with and without magnetic field, is a decreasing function of the applied magnetic field strength B. The various slopes are related to the selective deflection d of the various sublevels, according to their different effective magnetic moments, and values depending on their population in the original beam. The lower part of the figure shows that for a ^2P_J atoms beam only species in the $^2P_{1/2}$ state remain in the transmitted beam if a sufficiently high magnetic field strength is applied.

is necessary the use of energy and/or angular high resolution conditions in order to measure interference effects, such as the glory structure in integral cross sections or rainbows and diffraction oscillations in differential cross sections [1].

In systems involving open shell atoms or molecules the anisotropy of the interaction can appreciably modify the measured cross sections, with quenching and shifting of the interference patterns. These effects are useful for obtaining information on the interaction through the analysis of scattering data but are often hard to unravel.

The use of polarized atomic and molecular beams in scattering experiments is therefore crucial for the characterization of both the collision dynamics and the interaction potentials.

In the case of beams of open shell atoms, the selective deflection analysis obtained by a Stern-Gerlach magnetic selector (see Figure 2) allowed us to measure and to control the magnetic sublevel populations in the atomic beams and to perform scattering experiments under such conditions [4, 9, 11, 12].

In the case of molecular beams, collisional alignment of the rotational angular momentum is naturally induced during the beam formation if supersonic expansions of mixtures containing molecules in excess of a lighter seeding gas are used (see Figure 3). Specifically, because of the large number of collisions with the seeding gases, molecules are cooled and aligned with the rotational angular momentum perpendicular to the flight direction. A recent application of the magnetic analysis technique on supersonic O_2 seeded beams shows that the molecular alignment is strongly dependent on the speed of molecules in the beam emerging out of the expansion [13].

This result allowed us to perform scattering experiments of rotationally aligned oxygen molecules, and therefore to obtain crucial information on the anisotropies of van der Waals forces [14].

3. Nature, Range and Strength of the Intermolecular Forces

Besides experimental determinations, another important complementary aspect of the study of intermolecular forces is to understand how the individual physical properties of each particle influences their mutual interactions: here the purpose is to understand the nature of intermolecular bonds and also to predict the main interaction features for systems hard to characterize experimentally and for which reliable theoretical information is lacking. The following illustrates this point from different viewpoints.

From a theoretical point of view the interaction potential can be considered as a sum of several terms. Some of these terms give a positive contribution and others give a negative contributions to the overall interaction [15]. So called *ab-initio* methods are generally unable to accurately represent the interaction potential, since an exact evaluation of all contributions is possible only in some simple cases, usually involving only a few electrons.

Considerable progress relies on the extensive use of semi-empirical methods [15, 16, 17] which attempt at relating the features of the wells from a knowledge of the asymptotic behaviours. In turn, accurate information on the asymptotic regions for complex systems is needed to put semi-

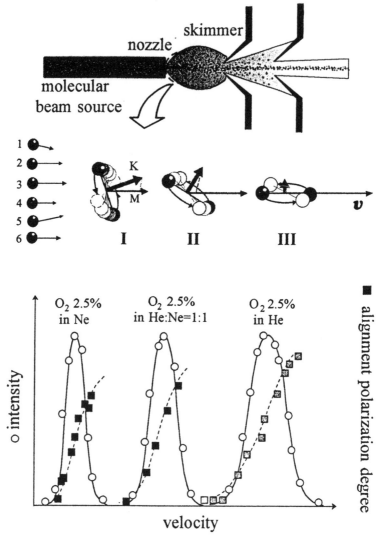

Figure 3. Blow up of a typical supersonic molecular beam expansion (upper part). Molecules emerging from gaseous mixtures, through the source nozzle, undergoes collisions with a faster seeding gas which produce the sequence of events (I, II, and III) illustrated in the middle part. Numbers refer to different values of impact parameters: collisions 3 and 4 give acceleration, while 2 and 5 originate molecular alignment and cooling; finally 1 and 6 focus the molecules in the forward direction. Results of magnetic analysis are shown (lower part) for seeded beams of O_2 molecules at total source pressure of 800 Torr. Open dots are relative beam intensities and filled squares are alignment polarization degree P, defined as $P = (w_0 - w_1)/(w_0 + w_1)$. w_0 and w_1 are are the relative populations of the molecular helicity states M equal to 0 and 1. Continuous curves are the best fit velocity distributions while dotted curves are a visual aid to describe the velocity dependence of P. The observed positive values of the latter are an indication that $w_0 > w_1$, *i.e.* molecules with their axes aligned along the flight direction prevail in front of the beam (adapted from Refs. 13, 14) .

empirically extracted intermolecular potentials on firmer grounds.

In absence of theoretical and experimental information, it is customary to use combination rules [1], based on semi-empirical or empirical considerations, which yield predictions for the interaction in asymmetric pairs based on the properties of the symmetric ones. These rules are valid when the interaction in symmetric and asymmetric pairs are of the same nature, for instance in rare gas-rare gas systems, and may fail in the generality of the cases.

Our approach takes into account that van der Waals well depth ϵ and its location R_m are determined by a critical balancing between the dispersion attraction and the repulsion due to the size of the external electronic clouds. Both these contributions can be given in terms of the polarizability of the interacting partners [18], which both describes the probability of induced dipole formation and is related to the atomic or molecular volume. On this grounds we established correlation formulas which are able to describe, in the intermediate and long distance range, the average interaction of the atom-atom systems experimentally investigated [19]. These correlations have been also extended, after appropriate modifications including the induction contribution, to ion-neutral systems obtaining again encouraging results [20].

A further extension of this study takes explicitly into account the anisotropic features of the interactions. In systems involving rare gas atoms and species with high electron affinity, such as oxygen and halogen atoms, the anisotropy comes from the open shell nature of the latter, which adds chemical contributions to the interaction. Specifically, this component is due to a configuration interaction (CI) between ground covalent and excited ionic states of the same symmetry: therefore, its scope is beyond the correlation formulas discussed so far, being their origin different from that of the typical van der Waals forces. The CI term is expected to be dependent on the energy separation among the interacting states and on the overlap integral between the orbitals which exchange the electron.

In the case of rare gas atom-homonuclear diatomic molecule, where no chemical contributions are present, it is important to consider that the molecular polarizability is strongly anisotropic since it is related to the electronic charge distribution around the molecular bond: such a distribution has to be represented by an appropriate ellipsoidal shape. In order to extend to this case the use of the polarizability correlations mentioned above, an effective angular dependent polarizability has to be introduced to the aim of representing the molecular size at any bond orientation, while the long-range attraction is described according to Buckingham [21].

The extension to atom-planar molecule systems is also possible by taking into account the additivity of the bond effects on the interaction. The

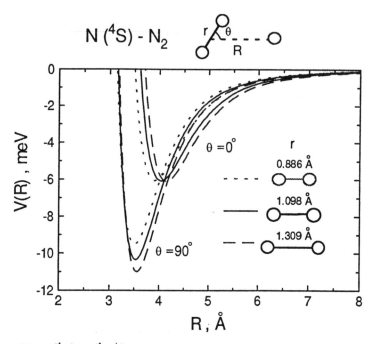

Figure 4. The $N(^4S)$-$N_2(^1\Sigma_g^+)$ interaction potential as a function of the intermolecular distance R, of the angle Θ between the approaching N atom and the N_2 molecule and the bond length r of N_2 molecule. V_0 is the spherically averaged interaction while $V_\perp$ and $V_\parallel$ are the perpendicular ($\Theta=90$) and parallel ($\Theta=0$) configurations respectively.

TABLE 1. Relevant potential features of the $N(^4S)$-N_2 interaction evaluated at the equilibrium distance of N_2 $r=1.098$ Å.

	R_m (Å)	ϵ (meV)	σ (Å)
V_0	3.78	8.8	3.33
$V_\perp$	3.54	10.4	3.15
$V_\parallel$	4.05	6.1	3.61

results obtained so far appear to be in a good agreement with the available accurate experimental determinations [22].

As indicated before this method can also be used to estimate the main interaction properties in systems not yet investigated experimentally: an example is provided by the atomic nitrogen-molecular nitrogen case. Here our correlation formulas give the features of the long range behaviour of the intermolecular potential, including its angular dependence (see Figure

4 and Table 1).

In addition, considering how the molecular nitrogen polarizability varies as a function of the bond length, it is possible to predict the dependence of the potential energy surface on this internal coordinate (see Figure 4 and Table 1). The same information on interaction properties can be obtained in the case of the oxygen atom-molecular nitrogen complex. Accounting for the open shell nature of the $O(^3P_J)$ atom it is possible to anticipate the main features of the manifold of the potential energy surfaces correlating, at large intermolecular distances, with the various sublevels $|J\Omega\rangle$ of the oxygen atom (where J is the total electronic angular momentum and Ω its projection along the intermolecular distance).

TABLE 2. Relevant potential features of the $O(^3P)$-N_2 interaction.

$|J\Omega\rangle$		R_m (Å)	ϵ (meV)	σ (Å)
	V_0	3.54	9.8	3.16
$|22\rangle=|10\rangle$	$V_\perp$	3.31	13.7	2.95
	$V_\parallel$	3.93	5.9	3.50
	V_0	3.67	7.2	3.26
$|21\rangle$	$V_\perp$	3.44	9.2	3.04
	$V_\parallel$	4.03	5.1	3.60
	V_0	3.75	6.4	3.33
$|20\rangle$	$V_\perp$	3.57	7.4	3.14
	$V_\parallel$	4.07	4.9	3.63
	V_0	3.75	6.7	3.37
$|11\rangle$	$V_\perp$	3.61	7.6	3.24
	$V_\parallel$	4.04	5.1	3.63
	V_0	3.69	7.5	3.31
$|00\rangle$	$V_\perp$	3.53	9.0	3.18
	$V_\parallel$	4.00	5.3	3.59

Table 2 summarizes results obtained for the interaction in two limiting cases corresponding to collinear and perpendicular configurations respectively and also for the interaction averaged over the N_2 orientations.

In conclusion this method provides the intermolecular interaction potential in a general, simple and realistic way. Some interesting applications of this approach have been already carried out in the study of energy transfer

360

processes [23, 24], ion-molecule reactions [25, 26] and ion-ion recombinations [27].

This work was supported by the Italian Consiglio Nazionale delle Ricerche and Ministero per l'Università, and by EU grants.

References

1. Maitland, G.C., Rugby, M., Smith, E.B., and Wakeham, W.A. (1987) *Intermolecular Forces*, Clarendon Press, Oxford.
2. See for instance Troe, J. (1977) *Journal of Chemical Physics* **66**, 4758.
3. Lo, G. and Setser, D.W. (1994) *Journal of Chemical Physics* **100**, 5432.
4. Aquilanti, V., Cappelletti, D., and Pirani, F. (1993) *Journal of the Chemical Society, Faraday Transaction* **89**, 1467.
5. Aquilanti, V. and Vecchiocattivi, F. (1989) *Chemical Physics Letters* **156**, 109.
6. Aquilanti, V. and Grossi, G. (1980) *Journal of Chemical Physics* **73**, 1165.
7. Aquilanti, V., Casavecchia, P., Grossi, G., and Laganá, A. (1980) *Journal of Chemical Physics* **73**, 1173.
8. Aquilanti, V., Liuti, G., Pirani, F., and Vecchiocattivi, F. (1989) *Journal of the Chemical Society, Faraday Transaction II* **85**, 925.
9. Aquilanti, V., Candori, R., and Pirani, F. (1988) *Journal of Chemical Physics* **89**, 6157.
10. Aquilanti, V., Beneventi, L., Grossi, G., and Vecchiocattivi, F. (1988) *Journal of Chemical Physics* **89**, 751.
11. Aquilanti, V., Cappelletti, D., Lorent, V., Luzzatti, E., and Pirani, F. (1992) *Chemical Physics Letters* **192**, 153.
12. Aquilanti, V., Cappelletti, D., Lorent, V., Luzzatti, E., and Pirani, F. (1993) *Journal of Physical Chemistry* **97**, 2063, and references therein.
13. Aquilanti, V., Ascenzi, D., Cappelletti, D., and Pirani, F. (1994) *Nature* **371**, 399.
14. Aquilanti, V., Ascenzi, D., Cappelletti, D., Franceschini, S., and Pirani, F. (1995) *Physical Review Letters* **74**, 2929.
15. Tang, K.T. and Toennies, J.P. (1984) *Journal of Chemical Physics* **80**, 3726.
16. Doukatis, C., Scoles, G., Marchetti, S., Zen, M., and Thakkard, A.J. (1982) *Journal of Chemical Physics* **76**, 3057.
17. Neyland, C. and Toennies, J.P. (1988) *Chemical Physics* **122**, 337.
18. Liuti, G. and Pirani, F. (1985) *Chemical Physics Letters* **122**, 245.
19. Cambi, R., Cappelletti, D., Liuti, G., and Pirani, F. (1991) *Journal of Chemical Physics* **95**, 1852.
20. Cappelletti, D., Liuti, G., and Pirani, F. (1991) *Chemical Physics Letters* **183**, 297.
21. Buckingham, A.D. (1967) *Advances in Chemical Physics* **12** 107.
22. Pirani, F. (1994) Discussion of the Faraday Society **97**.
23. Aquilanti, V., Candori, R., Pirani, F., Krümpelmann, T., and Ottinger, Ch. (1990) *Chemical Physics* **142**, 47.
24. Aquilanti, V., Candori, R., Pirani, F., and Ottinger, Ch. (1994) *Chemical Physics* **187**, 171.
25. Tosi, P., Eccher, F., Bassi, D., Pirani, F., Cappelletti, D., Aquilanti, V. (1991) *Physical Review Letters* **67**, 1254.
26. Tosi, P., Dmitrijev, O., Soldo, Y., Bassi, D., Cappelletti, D., Pirani, F., Aquilanti, V. (1993) *Journal of Chemical Physics* **99**, 985.
27. Aquilanti, V., Candori, R., Kumar, S.V.K., and Pirani, F. (1995) *Chemical Physics Letters* **237**, 456.

Transport Properties of Nonequilibrium Gas Flows

Raymond Brun
Laboratoire IUSTI - MHEQ
Université de Provence - Centre Saint Jérôme
13397 - Marseille - Cedex 20 - France

1 Introduction

In continuous reactive gaseous media, the macroscopic evaluation of the different quantities is classically obtained from the Navier Stokes equations coupled with kinetic equations (Clarke & Mc Chesney 1976),(Vincenti & Kruger 1965), (Lee 1984). However, until now, the expression of the transport terms related to the dissipative processes has not been clearly established in these reactive media but generally represents an extrapolation of results obtained in non-reactive media or in equilibrium situations (Dorrance 1962), (Hirschfelder et al. 1954), (Yos 1963).

It seems therefore essential to reexamine the foundations of the methods of determination of these terms in order to apply or to extend them to the case of nonequilibrium media. In non reactive cases, these methods are based on approximate solutions of the Boltzmann equation specific to each species of the medium obtained by expansion of the distribution function in series of a "small parameter" ε_I which, for continuous media, represents the ratio of a characteristic time between collisions τ_I and of a characteristic aerodynamic time - or flow time θ - (Chapman & Cowling 1970), (Ferziger & Kaper 1972). When ε_I is much smaller than one, the gaseous medium is "collisional" and non influenced - except geometrically - by the environment: It is said to be Maxwellian and in this simple case, the zeroth order distribution function in ε_I is a gaussian one, which corresponds to non dissipative Eulerian flows at the macroscopic level. The first order distribution function in ε_I takes into account the dynamic and thermal influence of the outer media and its propagation by collisions: this is the origin of dissipative phenomena such as viscosity, conduction and diffusion.

The corresponding macroscopic description is given by Navier-Stokes equations and transport terms may be calculated satisfyingly by well-established methods like Chapman-Enskog's and Gross-Jackson's (Chapman & Cowling 1970), (Ferziger &

M. Capitelli (ed.), Molecular Physics and Hypersonic Flows, 361–382.

Kaper 1972), (Mc Cormack 1973).

Now, when physico-chemical processes occur, like vibrational excitation, dissociation, ionization and chemical reactions, the ratio ε_{II} between the specific characteristic times ε_{II} (inelastic collisions) and the reference flow time θ can take any value. The problem is first to compare ε_I to ε_{II} and to insert the corresponding terms of physico-chemical production in the hierarchy given by an expansion in series of ε_I. Then, it is necessary to deduce the subsequent modifications brought to the zeroth and first order distribution functions and consequently to the transport terms themselves by application of Chapman-Enskog of Gross-Jackson methods (Kogan 1969), (Brun & Zappoli 1977), (Philippi & Brun 1981) .

It is obvious that the limiting cases for $\varepsilon_{II} \to 0$ and $\varepsilon_{II} \to \infty$ correspond respectively to the equilibrium and frozen cases in which the physico-chemical processes are either given by local conditions with a strong coupling with them (equilibrium chemistry) or completely uncoupled (frozen chemistry), this last case being of course practically equivalent to the previous non-reactive case. What is of interest here is, of course, the description of the intermediate nonequilibrium case at the zeroth and at the first order of the distribution function. This description must also assure the continuity with the equilibrium case, and this is a non trivial problem because of the used expansion methods. Futhermore, a large variety of nonequilibrium situations may exist due to numerous possible multi-scale physico-chemical processes.

It has been chosen here to consider one or two processes only so that a maximum of two parameters ε_I and ε_{II} have to be taken into account. Thus, considering a Chapman-Enskog expansion, two main cases are considered, a "weak nonequilibrium" (WNE) and a "strong nonequilibrium" (SNE) case depending on the absolute value of ε_{II} (Brun 1988)...

The example of vibrational nonequilibrium is taken as an application of the SNE case and corresponding to different collisional exchanges (TV, VV, resonant) having different probabilities, various vibrational nonequilibrium situations are examined at the zeroth order of the expansion.

Then, the application of Chapman-Enskog method enables one to express the first order distribution function and consequently the transport terms of the Navier-Stokes equations: By comparison with the usual case, modified and new terms appear such as cross heat flux terms and pressure terms. However it is easily seen that the matching of these nonequilibrium solutions with the equilibrium solution is not realized at the first order due to the classical expansion used here leading to expressions for transport coefficients μ, λ, unchanged as compared to the frozen case. Thus, a modification of this expansion is proposed which leads to uniformly valid expressions for these transport coefficients. At last, with the harmonic oscillator model and Mason and Monchick's type simplifications (Brun 1988), one can obtain relatively simple and tractable expressions for transport coefficients valid whatever the degree of nonequilibrium.

The cases of pure diatomic gases and binary mixtures are successively examined and examples of computation of transport coefficients more or less affected by the nonequilibrium are presented.

Finally the case of dissociative collisions is also examined but only from the point of view of the interaction with the vibrational nonequilibrium which may modify the dissociation rate constant usually employed: This modification may be determined rather simply by a WNE method: The example of nitrogen is proposed.

2   Fundamental considerations on nonequilibrium

Considering a reactive medium possessing only two collisional characteristic time scales ε_I and ε_{II}, all collisions are classified in two groups, for example elastic collisions I and inelastic ones II , TRV collisions I and reactive collisions II etc... If, by convenience, collisions I are assumed to be the most probable:$\varepsilon_I \ll \varepsilon_{II}$ and $\tau_I \ll \theta$ that is $\varepsilon_I = \varepsilon \ll 1$, which characteristizes a "collisional" or continuous medium. Now, three possible distinct cases may be distinguished:

$$\varepsilon_{II} \gg 1 : \quad \text{Frozen case} \rightarrow \varepsilon_{II} = \frac{1}{\varepsilon}$$

$$\varepsilon_{II} \ll 1 : \quad \text{Equilibrium case} \rightarrow \varepsilon_{II} = \varepsilon$$

$$\varepsilon_{II} \approx 1 : \quad \text{Nonequilibrium case} \rightarrow \varepsilon_{II} \approx 1$$

The Boltzmann equation for a species i characterizing also eventually one rotational -vibrational level, may be written under the following non-dimensional form:

$$\frac{d f_i^*}{dt} = \frac{1}{\varepsilon} J_I^* + \frac{1}{\varepsilon_{II}} J_{II}^* \tag{1}$$

f_i is classically expanded in series of ε up to the first order only. Then

$$f_i^* = f_i^{0*}(1 + \varepsilon \varphi_i^*) \tag{2}$$

where f_i^0 is the zeroth order distribution function and φ_i the perturbation with $f_i^1 = f_i^0 \varphi_i$ and $\varphi_i \ll 1$

The frozen case brings no new interest since the collisions II do not appear at the zeroth and first order. Now:

- For the equilibrium case:

$$\frac{d f_i^*}{d t^*} = \frac{1}{\varepsilon} (J_I^* + J_{II}^*) \tag{3}$$

The following system is deduced:

$$J_I^0 + J_{II}^0 = 0 \tag{4}$$

$$\frac{d f_i^0}{d t} = J_I^1 + J_{II}^1 \tag{5}$$

The integral equation (4) gives f_i^0 (Euler level) and the linear integro-differential equation (5) gives φ_i (Navier-Stokes level). The collisions I and II have the same role at each level and appear in the same manner in the transport terms. Furthermore, the species production equation is simply $\dot{w}_i = 0$.

- For the nonequilibrium case:

$$\frac{d f_i^*}{d t} = \frac{1}{\varepsilon} J_I^* + J_{II}^* \tag{6}$$

and,

$$J_I^0 = 0 \tag{7}$$

$$\frac{d f_i^0}{d t} = J_I^1 + J_I^0 \tag{8}$$

Thus a strong nonequilibrium appears at the zeroth and at the first order and Euler and Navier-Stokes equations are completed by kinetic equations giving the evolution of species. The collisions I and II will appear in a different way in the transport terms.

It is also to be noted that whatever f_i^0, in equilibrium or not, the first order solution f_i^1 always exhibits a nonequilibrium which remains small owing to the linearization. Therefore, the solution obtained for the transport terms with Eq.(4),(5) corresponds to a weak nonequilibrium case noted WNE case and the case corresponding to Eq.(7),(8) is denoted SNE case (strong nonequilibrium).

3 Vibrationally relaxing gases. SNE case

$i=i_r,i_v$ is a quantum number corresponding to the particular rotational and vibrational levels i_r and i_v, $f_i(\mathbf{v}, \mathbf{r}, t)$ is the distribution function of the i molecules depending on their velocity $\mathbf{v}$, the generalized coordinate $\mathbf{r}$ and the time t.

3.1 Zeroth-order solution; examples

If collisions I include only T-T (elastic) and T-R collisions, collisions II refer to collisions with vibration exchanges of all types:

$$M_{i_v} + M_{j_v} \rightarrow M_{k_v} + M_{l_v}$$

so that it is easily shown that the solution of Eq.(7) is:

$$f_i^0 = n_{i_v} \left(\frac{m}{2\pi kT}\right)^{3/2} \exp\left(-\frac{mc^2}{2kT}\right) \frac{g_{i_r} \exp\left(-\varepsilon_{i_r}/kT\right)}{Q_r(T)} \tag{9}$$

where the macroscopic quantities, vibrational population n_{i_v}, velocity $\mathbf{u}$ and translational -rotational temperature T are defined with f_i^0 and are given by Euler equations and general relaxation equations

$$\frac{\partial n_{i_v}}{\partial t} + \frac{\partial . n_{i_v}\mathbf{u}}{\partial \mathbf{r}} = \sum_{i_r jkl} \left(a_{kl}^{ij} n_{k_v} n_{l_v} - a_{ij}^{kl} n_{i_v} n_{j_v}\right) \tag{10}$$

where a is the rate of collisions II taking into account the summation on the rotational levels.

Now, if particular highly probable collisions with vibrational exchanges are included in collisions I, the vibrational populations n_{i_v} take specific forms.

Two important examples are the following:
- VV collisions are included in collisions I
- Resonant VV collisions only are included in collisions I.

The first case leads to a Treanor distribution encountered in gas dynamic lasers (Brun & Zappoli 1977).

$$n_{i_v} = \frac{n \exp\left(-\varepsilon_{i_v}/kT + Ki_v\right)}{\sum_{i_v} \exp\left(-\varepsilon_{i_v}/kT + Ki_v\right)} \tag{11}$$

The second case corresponds to a Boltzman, distribution at the vibrational temperature T_v

$$n_{i_v} = \frac{n \exp\left(-\varepsilon_{i_v}/kT_v\right)}{\sum_{i_v} \exp\left(-\varepsilon_{i_v}/kT_v\right)} \tag{12}$$

The two unknown macroscopic parameters K and T_v are respectively given by a unique relaxation equation replacing Eq.(10): These equations may be found in (Brun & Zappoli 1977).

As it is also well known, the problem is greatly simplified when, independently, the simple physical model of harmonic oscillator is used, so that Eq.(10) resolves also into a unique equation giving the vibrational energy E_v and, at the Euler level, we have simply:

$$\frac{d\,E_v}{d\,t} = \frac{\overline{E}_v - E_v}{\tau_v} \tag{13}$$

Here, the problem is not the choice of such or such model and details about this may be found in (Brun & Zappoli 1977).Below, we will use the most general model given by Eq.(12). This model is also very convenient because it uses the concept of vibrational temperature.

From a more general point of view, the macroscopic parameters appearing in f_i^0 are given by conservation equations including Euler and relaxation equations corresponding to the Fredholm alternative of Eq. (8), that is:

$$\sum_i \int \left(\frac{d\,f_i^0}{d\,t} - J_{II}^0 \right) C_{Ii} d_3 \mathbf{v} = 0 \tag{14}$$

where C_{Ii} are the collisional invariants of collisions I. A physical model is of course needed for collisions II.

3.2 First-order solutions. Transport terms

Starting from the zeroth order distribution function Eq. (9),(12) we have to solve Eq. (8). Three possible solving methods may be used, differing by the modelization of the collisional term J_I:

• A very simplified method, the B.G.K.M. method (Morse 1964), giving an analytic but rough expression for φ_i and therefore also for transport terms with one unique relaxation time for all dissipative effects, so that we obtain for Prandtl and Lewis numbers values equal to one (Brun & Zappoli 1977).

• A much more refined method -the Gross-Jackson method-consisting in expanding the collisional term J_I^1 in orthogonal functions assumed to be close to eigenfunctions of the attached operator, stopping the expansion at a given order N and replacing the higher order eigenvalues by a suitable constant. Two important advantages result from this method: More or less complex models of any order may be built and these models may be used beyond the collisional regime (Brun & Zappoli 1977), (Philippi & Brun 1981).

• The third method, used here, derives from the Chapman-Enskog method established for monatomic gases (Chapman & Cowling 1970) and later extended to the case of gases with internal energy in equilibrium (Wang-Chang & Uhlenbeck 1951) (WNE method).

Thus, taking into account the form of the known terms $\frac{d f_i^0}{dt}$ and J_{II}^0 written from the expression of f_i^0 Eq. (9),(12), φ_i may be written under the form:

$$\varphi_i = A_i \mathbf{c} \cdot \frac{\partial T}{\partial \mathbf{r}} + B_i \mathbf{cc} : \frac{\partial \mathbf{u}}{\partial \mathbf{r}} + D_i \frac{\partial . \mathbf{u}}{\partial \mathbf{r}} + F_i \, \mathbf{c} \cdot \frac{\partial T_v}{\partial \mathbf{r}} + G_i \tag{15}$$

where $X_i = A_i, B_i, D_i, F_i, G_i$ are unknown scalar functions of $\mathbf{r}, t, \mathbf{c}, , \varepsilon_{i_r}, \varepsilon_{i_v}$.

The last two terms in Eq. (15) come from the vibrational nonequilibrium already present at the zeroth order.

Each term X_i is expanded in Sonine-Wang-Chang-Uhlenbeck polynomials

$$X_i = \sum x_{mnp} S_t^m P_r^n P_v^p \tag{16}$$

with $x = a, b, d, f, g$. Zeroth and first order terms only are retained in the expansion Eq. (16), i.e. $x_{000}, x_{100}, x_{010}, x_{001}$, each index corresponding successively to the order of expansion for translation, rotation and vibration. These terms are determined from particular equations built with the corresponding terms of $\frac{d f_i^0}{dt} - J_{II}^0$

Finally, the coefficients x may be expressed as combinations of collisional integrals. These integrals depend only on collisions I for the coefficients a, b, c, d and f and on collisions II for g. The transport terms, stress tensor τ and heat flux $\mathbf{q}$ are then easily calculated and one obtains:

$$\tau = p\mathbf{I} - 2\mu \frac{\overline{\overline{\partial \mathbf{u}}}}{\partial \mathbf{r}} - \eta \frac{\partial . \mathbf{u}}{\partial \mathbf{r}} + P_R \mathbf{I} \tag{17}$$

Thus in the stress tensor, we have the following terms:
- A term of hydrostatic pressure: $p = n k T$
- A term of shear viscosity, with a viscosity coefficient $\mu = \rho b_{000} \left(\frac{kT}{m} \right)^2$
- A term of bulk viscosity, coming from the presence of the rotational mode assumed in equilibrium at the zeroth order, with a coefficient $\eta = n k T \, d_{100}$
- A term of "relaxation pressure" $P_R = n k T \, g_{100}$

This last term is the only one to contain collisional integrals of type II coming from the J_{II}^0 term.

For the heat flux terms, we have, per molecule:

$$\mathbf{q}_t = -\lambda_t \frac{\partial T}{\partial \mathbf{r}} - \lambda_{tv} \frac{\partial T_v}{\partial \mathbf{r}}$$

$$\mathbf{q}_r = -\lambda_r \frac{\partial T}{\partial \mathbf{r}} - \lambda_{rv} \frac{\partial T_v}{\partial \mathbf{r}} \tag{18}$$

$$\mathbf{q_v} = -\lambda_{\mathrm{rtr}}\frac{\partial T}{\partial \mathbf{r}} - \lambda_{\mathrm{v}}\frac{\partial T_{\mathrm{v}}}{\partial \mathbf{r}}$$

with

$$\lambda_{\mathrm{r}} = \frac{5}{2}n\,k\,T\frac{k}{m}a_{100} \qquad \lambda_{\mathrm{tv}} = \frac{5}{2}n\,k\,T\frac{k}{m}\frac{T}{T_{\mathrm{v}}}f_{100}$$

$$\lambda_{\mathrm{r}} = n\,k\,T\frac{C_{\mathrm{r}}}{m}a_{010} \qquad \lambda_{\mathrm{rv}} = n\,k\,T\frac{C_{\mathrm{r}}}{m}\frac{T}{T_{\mathrm{v}}}f_{010} \qquad (19)$$

$$\lambda_{\mathrm{vtr}} = n\,k\,T\frac{C_{\mathrm{v}}}{m}\frac{T}{T_{\mathrm{v}}}a_{001} \qquad \lambda_{\mathrm{v}} = n\,k\,T\frac{C_{\mathrm{v}}}{m}\frac{T}{T_{\mathrm{v}}}f_{001}$$

These transport terms appear in the Navier-Stokes equations which are completed by a relaxation equation giving the unknown T_{v} or E_{v}, that is:

$$n\frac{d\,E_{\mathrm{v}}}{d\,t} + \frac{\partial.\mathbf{q_v}}{\partial \mathbf{r}} = \sum_{i_{\mathrm{v}}}\varepsilon_{i_{\mathrm{v}}}\int J_{\mathrm{II}}d_3\mathbf{c} \qquad (20)$$

By comparison with the frozen or equilibrium cases, the following differences may be noted:
- The viscosity and conductivity coefficients depend only on collisions I.
- There are cross terms in heat flux expressions.
- The bulk viscosity is only due to the rotational mode.
- A relaxation equation does exist for determining T_{v}.
- A term of "relaxation pressure" appears.

Now, if we apply approximations of the Mason and Monchick type (Mason & Monchick 1962) which consist essentially in considering, in collisional integrals, the inelastic contribution as a perturbation, we can obtain for transport coefficients rather simple expressions which are given in Appendix I. Thus, it may be easily shown that no inelastic contribution appears in the terms not directly connected with internal energy exchanges: It is the case for μ, for example. Also, defining phenomenological rotational and vibrational times τ_{r} and τ_{v}, neglecting the coupling between the internal modes and assimilating the energy diffusion coefficients to the self diffusion coefficient D, the heat conductivities, the bulk viscosity and the relaxation pressure P_{R} may be written as functions of μ, τ_{r}, τ_{v} (Appendix I).

An example of computation of relaxation pressure is also presented in Fig.1, for the case of expansion of N_2 in a supersonic nozzle:P_{R}/P being proportional to the departure from equilibrium $\overline{E}_{\mathrm{v}} - E_{\mathrm{v}}$ and to the ratio $\tau_{\mathrm{I}}/\tau_{\mathrm{II}}$ i.e. here $\tau_{\mathrm{r}}/\tau_{\mathrm{v}}$, a maximum appears close to the throat but the absolute value remains very small. However for the case of reactive collisions included in the type II, this ratio could be much higher, but this remains an open point.

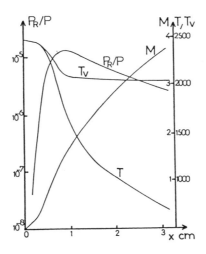

Figure 1: Relaxation pressure and temperatures in a supersonic nozzle flow: $T_0 = 2500$ K, $P_0 = 120$ Atm, N_2

It is to be noted also, that, since T is a translation-rotation temperature defined at the zeroth order, distinct translational and rotational temperatures, T_t and T_r appear at the first order, so that

$$E_t(T_t) + E_r(T_r) = E_t(T) + E_r(T) \tag{21}$$

This is also the cause of the existence of the bulk viscosity term (weak rotational nonequilibrium). T_t and T_r may be calculated like transport terms from their definition.(Brun 1986).

The very important last point which must be underlined is that the SNE case treated here cannot be valid in the asymptotic equilibrium case(WNE case, $T_v \to T$) due to the different hierarchy existing in Eq. (4), (5) and Eq. (7),(8). In particular, it is already obvious that the transport terms in WNE case will depend on collisions I and II and not only on collisions I as in SNE case. The same, no relaxation equation of Eq. (10) type is necessary in the WNE case.

Below, we examine briefly the results obtained in this last case.

4 Gases in vibrational equilibrium. WNE case

This is a more classical case (Wang-Chang & Uhlenbeck 1951). We start from the zeroth-order equilibrium solution of Eq. (4).

$$f_i^0 = n \left(\frac{m}{2\pi kT}\right)^{3/2} \exp\left(\frac{mc^2}{2kT}\right) \times \frac{g_{i_r} \exp\left(-\varepsilon_{i_r}/kT - \varepsilon_{i_v}/kT\right)}{Q_r(T)Q_v(T)} \tag{22}$$

Of course, no relaxation equation appears and the three unknown quantities n $\mathbf{u}$, T are given by classical Euler equations.

Then, from Eq. (5),φ_i is written:

$$\varphi_i = A_i' \mathbf{c}.\frac{\partial T}{\partial \mathbf{r}} + B_i' \mathbf{c}\,\mathbf{c} : \frac{\partial \mathbf{u}}{\partial \mathbf{r}} + D_i'\frac{\partial.\mathbf{u}}{\partial \mathbf{r}} \tag{23}$$

Expanding A_i', B_i' and D_i' as previously, we obtain for the transport terms:

$$\tau = \mathbf{pI} - 2\mu\overline{\frac{\overset{\circ}{\partial \mathbf{u}}}{\partial \mathbf{r}}} - \eta\frac{\partial.\mathbf{u}}{\partial \mathbf{r}}\mathbf{I} \tag{24}$$

$$\mathbf{q} = \mathbf{q_t} + \mathbf{q_r} + \mathbf{q_v} = -\lambda_{trv}\frac{\partial T}{\partial \mathbf{r}}$$

with:

$$\mu = \rho b_{000}' \left(\frac{kT}{m}\right)^2$$

$$\eta = nkTd_{000}' \tag{25}$$

$$\lambda_{trv} = \frac{nkT}{m} C_{trv} \left(a_{100}' + a_{010}' + a_{001}'\right)$$

Formally, these expressions are very similar to corresponding expressions of the SNE case with $T_v = T$, but, as previously noted, in a' b' and d', collisional integrals of type I and II are present and play the same role.

Mason and Monchick's original results (Brun 1988) have been obtained in this WNE case: they have been established for one internal mode. Transport coefficients taking into account the rotational and vibrational modes are presented in Appendix II: It is of course assumed that $\tau_v \gg \tau_r$.

Now, if it is possible to compute the macroscopic quantities n, $\mathbf{u}$, T, with Navier-Stokes equations written with one temperature T it is also possible to obtain the particular energies E_t ,E_r and E_v from their definition itself. For example, for E_v , we have (Brun et al. 1979) :

$$E_v = \overline{E}_v \left(1 + \frac{kC_{tr}}{C_{trv}^2}\tau_v\frac{\partial.\mathbf{u}}{\partial \mathbf{r}}\right) \tag{26}$$

where, as previously, τ_v is a phenomenological vibrational relaxation time.

Thus, at the first order we obtain a "weak" nonequilibrium given by Eq. (26). Similar expressions may be obtained for E_t and E_r. Then T_t, T_r and T_v are easily deduced (Wang-Chang & Uhlenbeck 1951). It is however to be noted that the difference between T_t and T_r is of the order of τ_r/θ that is generally very small.

A computational comparison has been made between the values of T_v obtained by SNE and WNE methods in the simple case of a one-dimensional steady expansion of N_2 in a nozzle (Brun et al. 1979): In Fig.2 the evolution of T_v/T_0 along the nozzle is represented for angles of the divergent part γ respectively equal to 4° and 8°: The following points may be observed:

• There is a difference between the frozen energies given by both methods, the freezing level being higher in the SNE case.

• The WNE method is applicable only for small departures from equilibrium, that is small divergent angles. For large angles (8°) τ_v/θ is no more small in the nozzle, where $\theta \approx (du/dx)^{-1}$ and oscillations appear.

No experimental result can guide a choice between the two methods but for N_2 it has been shown that SNE method may give too high a freezing level.

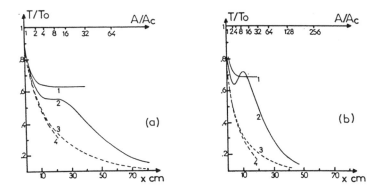

Figure 2: Temperature evolution along a nozzle: $T_0 = 2500$ K, $P_0 = 120$ Atm, N_2
1: T_v/T_0 (SNE method), 2: T_v/T_0 (WNE method),
3: T/T_0 (WNE method), 4: T/T_0 (SNE method),
(a): $\gamma = 4°$, (b): $\gamma = 8°$

It seems now obvious from theoretical and practical considerations that the results obtained from WNE and SNE methods cannot be the same in their common field of application. A more general method is needed and it has been indeed suggested. It is applied hereafter.

5 General method for vibrationally excited gases

A hierarchy different from this given by Eq. (4),(5) and Eq. (7),(8) must be written for the matching between WNE and SNE cases (Kogan et al. 1979), (Brun et al.1984).

Thus, simply, if the system giving successively the zeroth and first order solutions is written under the following form:

$$J_I^0 = 0 \tag{27}$$

$$\frac{d f_i^0}{d t} = J_I^1 + J_{II}^0 + J_{II}^1$$

where the term J_{II}^1 has been added in Eq. (5), the matching is realized because we have, for strong nonequilibrium:

$$J_{II}^{*0} \to 0 \qquad \text{and} \qquad J_{II}^{*1} \to 1$$

The zeroth-order solution of Eq (27) is identical to the SNE solution and the Euler system is completed by relaxation equations (10).

The perturbation φ_i is formally written under the same form as Eq. (15) but the coefficients $A_i \ldots G_i$ are now functions of integrals of type I and II. The transport terms are of course formally the same, Eq. (17) and Eq. (18).

After having used the previous assumptions the conductivities may be written:

$$\lambda_t = \frac{15}{4} \mu \frac{k}{m} \left\{ 1 - \frac{10}{3 \pi k} \left[\frac{C_r}{Z_r} + \frac{C_v}{Zv} \left(\frac{\overline{Q_v}}{Q_v} \right) F(T, T_v) \right] \right\} + \frac{5}{\pi} n D \frac{C_r}{Z_r}$$

$$\lambda_{tv} = \frac{5}{\pi} n D \frac{\overline{C_v}}{Z_v} \left(\frac{T}{T_v} \right)^2 \left(\frac{Q_v}{\overline{Q_v}} \right) F(T, T_v)$$

$$\lambda_r = n D C_r \left[1 - \frac{2 \rho D}{\pi} \frac{1}{\mu} \frac{1}{Z_r} \right] + \frac{5}{\pi} n D \frac{C_r}{Z_r}$$

$$\lambda_{rv} \approx 0 \tag{28}$$

$$\lambda_{vtr} = \frac{5}{\pi} n D \frac{\overline{C_v}}{Z_v} \left(\frac{Q_v}{\overline{Q_v}} \right) F(T, T_v)$$

$$\lambda_v = n D C_v \left[1 - \frac{2 \rho D}{\pi} \frac{1}{\mu} \frac{1}{Z_v} \left(\frac{T}{T_v} \right)^2 \right] \left[\left(\frac{Q_v}{\overline{Q_v}} \right) F(T, T_v) \right]$$

with

$$F\left(T, T_{v}\right) = \frac{1}{2}\left\{1 + \exp\left[\frac{h\nu}{kT}\left(1 - \frac{1}{T_{v}}\right)\right]\right\}$$

When $T_{v} \to T$, we find now the expressions of the WNE case (matching),

$$\lambda_{tr} = \lambda_{t} + \lambda_{r} = \frac{15}{4}\mu\frac{k}{m}\left[1 - \frac{10}{3\pi k}\left(\frac{C_{r}}{Z_{r}} + \frac{C_{v}}{Z_{v}}\right)\right] + nDC_{r}\left[1 - \frac{2}{\pi}\frac{\rho D}{\mu}\frac{1}{Z_{r}}\right] + \frac{10}{\pi}nD\frac{C_{r}}{Z_{r}}$$

$$\lambda_{v} = nDC_{v}\left[1 - \frac{2}{\pi}\frac{\rho D}{\mu}\frac{1}{Z_{v}}\right]$$

with $Z_{r} = \tau_{r}\backslash\tau$ $\qquad$ $Z_{v} = \tau_{v}/\tau$

Thus, λ_{tr} depends very weakly on the nonequilibrium, but λ_{v} depends strongly on it.

Numerical values of λ_{v} have been computed for CO_{2} and are represented in Fig.3 as a function of T and T_{v}. These results cover all practical nonequilibrium situations such as shocks and expansions. Thus, it is clear that not taking into account the nonequilibrium could lead to large errors in the determination of λ_{v} and consequently in the computed values of the heat flux.

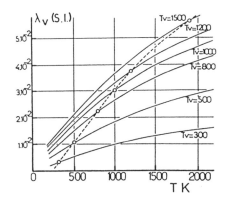

Figure 3: Vibrational conductivity λ_{v} for carbon dioxide. —o— Equilibrium curve

η (bulk viscosity) and P_{R} (relaxation pressure) remain unchanged when compared to the SNE case and $\mu \approx \mu_{el}$, $D_{r}, \approx D_{v} \approx D$

6   Nonequilibrium gas mixtures

The general Chapman-Enskog method (GCE) exposed in part 5 is also used for binary mixtures of diatomic gases in vibrational nonequilibrium. The main differences brought by the mixture lie in the coupling between both gases by VV exchanges and in the mass diffusion effects.(Pascal & Brun 1993) .

The flow of a mixture of two diatomic gases p and q, each possessing one rotational and one vibrational mode is considered. The collisions are assumed to be binary and the chemistry frozen. The quantum numbers $i = i_r$, i_v correspond to particular rotation i_r, and vibration i_v levels.

Physical considerations allow us to classify the collisions into the following two groups I and II. Collisions I (the most probable): $T - T$ collisions (elastic ones, translation-translation exchanges), $T - R$ collisions (translation-rotation exchanges), and $V - V$ collisions (resonant vibration-vibration collisions between two molecules of the same species: $p - p$ and $q - q$ and collisions II (the less probable): $T - V$ collisions (translation-vibration exchanges) and $V - V$ collisions (vibration-vibration collisions between molecules of different species: $p - q$ and $q - p$).

$$\frac{d f_{i_p}^*}{dt} = \frac{1}{\varepsilon_I} J_{I_p}^* + \frac{1}{\varepsilon_{II}} J_{II_p}^*$$

The GCE method gives:

$$\frac{d f_{i_p}^0}{dt} = J_{I_p}^1 + J_{II_p}^0 + J_{II_p}^1, \quad \text{first order,} \tag{29}$$

where

$$J_{II_p}^1 = J_{II_{pp}}^1 + J_{II_{pq}}^1$$

The self adjoint part of $J_{II_p}^1$ is only retained and the zeroth order solution $f_{i_p}^0$ and the first order perturbation φ_{i_p} are respectively written:

$$f_{i_p}^0 = n_p \left(\frac{m_p}{2\pi \, k \, T} \right)^{3/2} \exp\left(-\frac{m_p c_{ip}^2}{2 \, k \, T} \right)$$

$$\times \frac{g_{ir_p} \exp\left(-\varepsilon_{ir_p}/k\,T\right) \exp\left(-\varepsilon_{iv_p}/k\,T_{vp}\right)}{\displaystyle\sum_{ir_p} g_{ir_p} \exp\left(-\varepsilon_{ir_p}/k\,T\right) \sum_{iv_p} \exp\left(-\varepsilon_{iv_p}/k\,T_{vp}\right)}$$

with T_{vp} given by a relaxation equation including the TV and the VV exchanges with the component q.

$$\varphi_{i_p} = A_{i_p} \mathbf{c}_{i_p} \cdot \frac{\partial \ln T}{\partial \mathbf{r}} + B_{i_p} \mathbf{c}_{i_p} \mathbf{c}_{i_p} : \frac{\partial \mathbf{u}}{\partial \mathbf{r}} + D_{i_p} \frac{\partial \cdot \mathbf{u}}{\partial \mathbf{r}} + F_{i_p} \mathbf{c}_{i_p} \cdot \frac{\partial \ln T_{vp}}{\partial \mathbf{r}} + G_{i_p} + H_{i_p} + \mathbf{c}_{i_p} \cdot \mathbf{d_p}$$

with

$$\mathbf{d_p} = \frac{\partial}{\partial \mathbf{r}} \left[\frac{n_p}{n} \right] - \left[\frac{n_p}{n} - \frac{n_p m_p}{nm} \right] \frac{\partial \ln P}{\partial \mathbf{r}} \tag{30}$$

The transport coefficients deduced from Eq. (30) first written in terms of collision integrals are finally expressed as functions of macroscopic quantities, known or experimentally attainable.

For collision integrals of type I, Mason and Monchick type approximations are similar to the SNE method. For collision integrals of type II, which characterize the main vibrational energy exchanges, it may be shown that the use of the harmonic-oscillator model leads to relations between type II collision integrals and vibrational relaxation times τ_{TV} and τ_{VV}.

This may be obtained from a phenomenological equation of the Landau-Teller type for the vibrational energy, established by using the harmonic-oscillator model.

Applications are made for N_2/H_2 and N_2/O_2 mixtures. In the first case, the transport properties of both components are significantly different and the interest of the second case is obvious, this case having been up to now approximately treated.

The shear viscosity μ, practically insensitive to the internal structure, has been first computed in order to test the method and the results are summarized in Table 1 and compared to experimental results and popular correlations: Relative errors are only about 1 % for the present results and 8 % for the Wilke approximation. (Fig. 4).

Table 1: Shear viscosity in a N_2-H_2 mixture (108 kg/m s) at 273.16 K.

H_2 (mol %)	Present	Wilke	Expt.(5)
0	1666	1666	1688
15.9	1645	1660	1670
39.0	1587	1630	1600
65.2	1440	1520	1449
79.5	1284	1376	1285
80.3	1273	1364	1274
100	846	846	853

Then, other results are represented in (Fig. 5) to (Fig. 8) for various transport properties and nondimensional numbers which appear in the Navier-Stokes and relaxation equations. Further details and discussions may be found in (Pascal & Brun 1993).

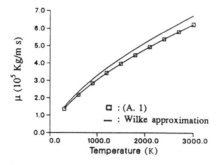

Figure 4: Shear viscosity in a N_2-H_2 mixture: $X_{H_2}=0.8$.

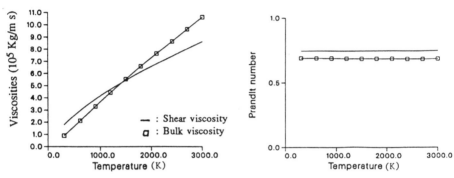

Figure 5: Viscosity for pure nitrogen

Figure 6: Prandlt number (air).
- : Wilke approximation
□ : Present results

7 Vibration-dissociation coupling

Another important problem involving reactive collisions and for which the GCE method may bring an important contribution concerns the vibration-dissociation coupling: The simultaneity of both phenomena, for example behind a shock contributes to delay the dissociation and the corresponding rate constant is strongly influenced

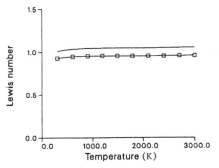

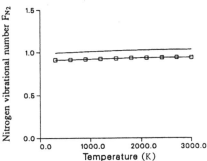

Figure 7: Lewis number (air)

Figure 8: Vibrational number for nitrogen in air mixture

by the vibrational nonequilibrium. The influence of the dissociation on vibration population is also non negligible.

The simplest case of a pure diatomic dissociating gas is considered here, in which elastic, inelastic and reactive collisions take place and possess different relaxation times. The recombination is not taken into account, so that the application of the method is restricted to the strong nonequilibrium part of shocked flows, excluding expansion flows for example, but the generalization to such flows is straightforward (Meolans et al, 1994).

In a wide range of temperature, the WNE method may be applied, due to the different order of magnitude of translation, rotation, vibration and dissociation characteristic times (Fig. 9): The corresponding nondimensional Boltzmann equation for f_{i_p} (molecules p on level i) is:

$$\frac{d f_{i_p}^*}{d t^*} = \frac{1}{\varepsilon} J_v^* + J_D^*$$

where J_V and J_D are respectively the collisional terms corresponding to vibrational transition and dissociation, J_v including also the translation-rotation exchanges, and with $\varepsilon = \tau_v / \theta$.

Applying the CE expansion, the following system is obtained:

$$J_v^0 = 0 \rightarrow f_{i_p}^0$$

$$\frac{d f_{i_p}^0}{d t} = J_v^1 + J_D^0 \rightarrow \varphi_{i_p}$$

$f_{i_p}^0$ is easily obtained and represents a Maxwell-Boltzmann distribution function for dissociating molecules and φ_{i_p} when expanded on the basis of Wang-Chang and Uhlenbeck polynomials for rotational and vibrational energies may be also obtained explicity.

When stopping at the first order, and assuming $\tau_v \gg \tau_r$ the expression for φ_{i_p} is the following

$$\varphi_{i_p} = g_{01} \left(\varepsilon_{i_p} - \overline{E}_{v_p} \right) / k T$$

where ε_{iv_p} is the vibrational energy of the level i_v, $\overline{E}_{v_p}$ the average equilibrium vibrational energy per molecule, k the Boltzmann constant, T the translation-rotation-vibration temperature defined at the zeroth order and g_{01} is a combination of collision integrals which may be easily connected to macroscopic parameters like τ_v and K_f^0, dissociation rate constant at the zeroth order i.e depending only on T.

Finally, taking also into account the zeroth order macroscopic conservation equations one obtains for g_{01}:

$$g_{01} = -\frac{n_p \tau_v K_f^{(0)}}{C_{v_p} T} \left[\left(\overline{E_{r_p}} + \overline{E_{v_p}} - \frac{3}{2} k T \right) \frac{C_{v_p} n_p}{C_{trv_p} n} + \overline{E_{v_{diss}}} - \overline{E_{v_p}} \right]$$

where n_p and n are respectively the density of p molecules and total density, C_{tp}, C_{rp} and C_{v_p} the translation, rotation and vibration specific heat per molecule p, with $C_{trvp} = C_{t_p} + C_{r_p} + C_{v_p}$ and $\overline{E_{vdiss}}$ being equal to the average vibration energy per molecule lost by dissociation, calculated at the zeroth order (equilibrium).

Knowing φ_{i_p}, the dissociation rate constant may be calculated at the first order of the expansion and

$$K_f^1 = K_f^0 (1 - \eta_v) \qquad \text{with} \qquad \eta_v = \frac{g_{01}}{k T} \left(\overline{E_{v_p}} - \overline{E_{v_{diss}}} \right)$$

It is also easily shown that g_{01} is proportional to the vibrational nonequilibrium appearing at the first order, so that:

$$E_{v_p} - \overline{E_{v_p}} = g_{01} C_{v_p} T$$

A complete calculation of the relaxing zone behind strong shocks in nitrogen is operated and the results for K_f^1 are represented in Fig.10. They are compared to the semi-empirical results of Park (Park 1989), (Park 1993) who replaces T by $T_a = (T T_v)^{0,5}$ or $T_b = T^{0,7} T_v^{0.3}$, in Arrhenius expression of K_f^0, .

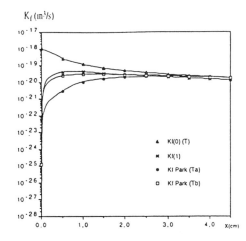

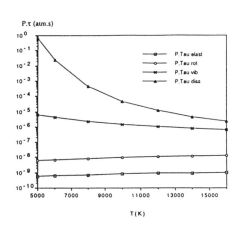

Figure 9: Nitrogen characteristic times

Figure 10: Nitrogen dissociation rate constant evolution behind a strong shock wave. Mach number: 19, Pressure: 10 Pa, Temperature: 235 K

8 Conclusion

A general Chapman-Enskog method has been proposed for nonequilibrium gas flows and applied to the case of vibrational nonequilibrium in pure gases and mixtures. Simplified expressions of the transport terms have also been proposed for the same cases.

Finally the influence of the vibrational nonequilibrium on the dissociation rate constant has been evaluated. The extension of the method of computation of the transport terms is now extended to chemical reactions with expected significant effects of nonequilibrium.

9 Appendix I. Simplified transport terms in SNE case

$$\mu = \mu_{el} \qquad D_r = D_v = D$$

$$\eta = n\,k\,T \frac{k\,C_r}{(C_{tr})^2} \tau_r$$

$$P_R = -\frac{n\,k\,C_r}{(C_{tr})^2} \tau_r \frac{\overline{E_v} - E_v}{\tau_v} = -\frac{\eta}{\tau_v} \frac{\overline{E_v} - E_v}{k\,T}$$

$$q_t = -\left\{ \frac{15}{4}\mu\frac{k}{m}\left[1 - \frac{10}{3\pi k}\frac{C_r}{Z_r}\right] + \frac{5}{\pi}nD\frac{C_r}{Z_r}\right\}\frac{\partial T}{\partial r}$$

$$q_r = -\left\{nDC_r\left[1 - \frac{2}{\pi}\frac{\rho D}{\mu}\frac{1}{Z_r}\right] + \frac{5}{\pi}nD\frac{C_r}{Z_r}\right\}\frac{\partial T}{\partial r}$$

$$q_v = -nDC_v\frac{\partial T_v}{\partial r}$$

The expression of q_v is simpler than those of q_t and q_r, because in collisions I, no net vibrational exchange occurs.

10 Appendix II. Simplified transport terms in WNE case

$$\mu = \mu_{el} \qquad\qquad D_r = D_v = D$$

$$\eta = nkT\frac{k}{(C_{trv})^2}(C_r\tau_r + C_v\tau_v)$$

$$\lambda_t = \frac{15}{4}\mu\frac{k}{m}\left[1 - \frac{10}{3\pi k}\left(\frac{C_r}{Z_r} + \frac{C_v}{Z_v}\right)\right] + \frac{5}{\pi}nD\left(\frac{C_r}{Z_r} + \frac{C_v}{Z_v}\right)$$

$$\lambda_r = -nDC_r\left[1 - \frac{2}{\pi}\frac{\rho D}{\mu}\frac{1}{Z_r}\right] + \frac{5}{\pi}nD\frac{C_r}{Z_r}$$

$$\lambda_v = -nDC_v\left[1 - \frac{2}{\pi}\frac{\rho D}{\mu}\frac{1}{Z_v}\right] + \frac{5}{\pi}nD\frac{C_v}{Z_v}$$

when $\tau_v \to \infty$, the conductivities are of course identical to the SNE case.

References

[Brun & Zappoli 1977] Brun, R. and Zappoli, B. (1977) Model equations for a vibrationally relaxing gas, The Physics of Fluids, 9, 1441 - 1448.

[Brun et al. 1979] Brun, R., Zappoli, B. and Zeitoun, D. (1979) Comparison of two computation methods for vibrational nonequilibrium flows, The Physics of Fluids, 4, 786 - 787.

[Brun et al.1984] Brun, R.Villa, M.P. and Meolans, J.G. (1984) Generalized transport terms in vibrationally relaxing flows, Rarefied Gas Dynamics, University of Tokyo Press.

[Brun 1986] Brun, R. (1986) Transport et Relaxation dans les Ecoulements Gazeux, Masson, Paris.

[Brun 1988] Brun, R. (1988) Transport properties in reactive gas flows, AIAA Paper, 88 - 2655.

[Chapman & Cowling 1970] Chapman, S. and Cowling, T. G. (1970) The Mathematical Theory of Non Uniform Gases, Cambridge University Press.

[Clarke & Mc. Chesney 1976] Clarke, J. F. and Mc. Chesney, M. (1976) Dynamics of Relaxing Gases, Butterworths, London.

[Dorrance 1962] Dorrance, W. H. (1962) Viscous Hypersonic Flows, McGraw-Hill, New- York.

[Ferziger & Kaper 1972] Ferziger, J. H. and Kaper, M. G. (1972) Mathematical Theory of Transport Processes in Gases, North Holland, Publishing Company, Amsterdam.

[Hirschfelder et al. 1954] Hirschfelder, J. O., Curtiss, C. F. and Bird, R. B. (1954) Molecular Theory of Gases and Liquids, J. Wiley and Sons, New-York.

[Kogan 1969] Kogan, M. N. (1969) Rarefied Gas Dynamics, Plenum Press, New-York.

[Kogan et al. 1979] Kogan, M. N., Galkin, V. S. and Makashev, N. K. (1979) Generalized Chapman-Enskog method, Rarefied Gas Dynamics, CEA Paris.

[Lee 1984] Lee, J. H. (1984) Basic governing equations for the flight regimes of aeroassisted orbital transfer vehicles, AIAA Paper, 84 - 1729.

[Mason & Monchick 1962] Mason, E. A. and Monchick, L. (1962) Heat conductivity of polyatomic and polar gases, J. Chemical Physics, 36, 1622 - 1640.

[Mc Cormack 1973] Mc Cormack, F. J. (1973) Construction of linearized kinetic models for gaseous mixtures and molecular gases, The Physics of Fluids, 16, 2095 - 2105.

[Meolans et al, 1994] Meolans, J. G., Brun, R., Mouti, M., Llorca, M. and Chauvin, A. (1994) Vibration-dissociation coupling in high temperature nonequilibrium flows, Aerothermodynamics for Space Vehicles, 2nd Symposium Proceedings ESTEC, Noordwijk ESA, 293 - 297.

[Morse 1964] Morse, T. F. (1964) Kinetic model for gases with internal degrees of freedom,The Physics of Fluids, 2, 159 - 169.

[Park C.1989] Park C. (1989) A Review of reaction rates in high temperature air, AIAA paper 89-1740.

[Park C. 1993] Park C. (1993) Review of chemical-kinetic problems of future NASA mission. Jour. Therm. and heat transf. 7,3, 385-398.

[Pascal & Brun 1993] Pascal, S. and Brun, R. (1993) Transport properties of nonequilibrium gas mixtures, Physical Review E, 5, 3251 - 3267

[Philippi & Brun 1981] Philippi, P. C. and Brun, R. (1981) Kinetic modeling of poly-atomic gas mixtures, Physica, 105A, 147-168.

[Vincenti & Kruger 1965] Vincenti, W. G. and Kruger, C. H. (1965) Introduction to Physical Gas Dynamics, J. Wiley and Sons, New-York.

[Wang-Chang & Uhlenbeck 1951] Wang-Chang, C. S. and Uhlenbeck, G. E. (1951) Transport phenomena in polyatomic gases. University of Michigan Report , CM 681.

[Yos 1963] Yos, J. M. (1963) Transport properties of nitrogen, hydrogen, oxygen and air to 30,000 K, Technical Memorandum, RAD TM 63 - 7 AVCO- RAD, Wilmington.

THE INFLUENCE OF NON-BOLTZMANN VIBRATIONAL DISTRIBUTION ON THERMAL CONDUCTIVITY AND VISCOSITY

E.V. KUSTOVA AND E.A. NAGNIBEDA
St.Petersburg University
198904, St-Petersburg, Petrodvoretz, Russia

The paper deals with the kinetic theory of transport processes in molecular gases with rotational and vibrational degrees of freedom excited. Considered are the conditions of strong vibrational nonequilibrium and weak rotational and translational one. In particular we investigate the case when vibrational energy store appears to be higher than its equilibrium value and anharmonic effects become to be important. In order to derive theoretical formulas for pressure tensor and heat flux a generalized Chapman-Enskog method is used. Non-Boltzmann distribution functions in zeroth approach are obtained with taking into account anharmonism of vibrations and different energy exchanges on various energy levels. Starting from these distributions, heat conductivity, bulk and shear viscosity coefficients and relaxation pressure are calculated in a wide range of vibrational and translational-rotational temperature and real gas effects are estimated. The comparison with other models and with experiment is given.

The model has been applied in study of boundary layer with nonequilibrium conditions on the wall and in free stream.

1. Introduction

Strong nonequilibrium effects must be taken into account when some of characteristic relaxation times become comparable with macroscopic flow time. Such conditions take place in expanding high enthalpy gas flows, in the hypersonic flows around spacecrafts and also in the case of some kind of vibrational energy pumping. If vibrational temperature in gas exceeds translational one then harmonic oscillator model is failed and anharmonic effects have to be considered. Moreover in the case of strong vibrational excitation different mechanism of energy exchanges on the various groups of vibrational energy levels is observed. It leads to non-Boltzmann distribution already in zeroth approach of generalized Chapman-Enskog method. In our previous papers [1,2] the kinetic model of distribution functions was constructed and transport processes in strong nonequilibrium gas were studied. In present paper on the basis of this model we estimate the influence

M. Capitelli (ed.), Molecular Physics and Hypersonic Flows, 383–392.
© *1996 Kluwer Academic Publishers. Printed in the Netherlands.*

of non-Boltzmann distribution, anharmonism and different energy transfers on all transport coefficients.

The possibility of using of simplified expressions for thermal conductivity coefficients is shown. The closed system of macroscopic equations is given. The flow in nonequilibrium boundary layer in nitrogen is investigated on the basis of present model. Heat flux to the surface is calculated under different conditions on the wall and in free stream.

2. Distribution functions in the zeroth approach

We consider the kinetic equations for distribution functions $f_{ij}(\mathbf{r},\mathbf{u},t)$ for every vibrational i and rotational j energy species in such a form

$$\frac{\partial f_{ij}}{\partial t} + \mathbf{u}\frac{\partial f_{ij}}{\partial \mathbf{r}} = \frac{1}{\varepsilon}J_{ij}^{(0)} + J_{ij}^{(1)}. \tag{1}$$

Here $J_{ij}^{(0)}$, $J_{ij}^{(1)}$ are collisional operators of rapid and slow processes, $\varepsilon = \tau_0/\tau_1$ is a small parameter, τ_0, τ_1 are average times between often and seldom collisions.

There exist different generalizations of Chapman-Enskog method for (1) [1,3]. It is known from the theory and experiment that in a system of anharmonic oscillators there are three groups of energy levels with different mechanism of energy exchanges [4]. On lower levels ($0 \leq i \leq i_*$) the most often collisions appear to be collisions with TT, RR, RT exchanges of translational and rotational energy and nonresonance VV' exchanges of vibrational quanta

$$k + l = (k \pm m) + (l \mp m),$$

k, l, are vibrational species, m is a number of exchanging quanta. On middle levels ($i_* < i \leq i_{**}$) the most often collisions are resonance VV'' exchanges between neighbour levels

$$k + (k \pm 1) = (k \pm 1) + k.$$

On upper levels all vibrational transfers have the same frequency as other collisions. Hence the operators $J_{ij}^{(0)}$, $J_{ij}^{(1)}$ appear to be different on the various groups of levels

$$J_{ij}^{(0)} = \begin{cases} J_{ij}^{TT} + J_{ij}^{TR} + J_{ij}^{RR} + J_{ij}^{VV'}, & 0 \leq i < i_* \\ J_{ij}^{TT} + J_{ij}^{TR} + J_{ij}^{RR} + J_{ij}^{VV}, & i_* \leq i < i_{**} \\ J_{ij}, & i \geq i_{**}, \end{cases} \tag{2}$$

$$J_{ij}^{(1)} = \begin{cases} J_{ij}^{VT} + J_{ij}^{VRT}, & 0 \leq i < i_{**} \\ 0, & i \geq i_{**}. \end{cases} \tag{3}$$

J_{ij} is a total collision operator.

In our previous papers [1,2] on the basis of analysis of additive invariants of operator $J_{ij}^{(0)}$ the distribution functions $f_{ij}^{(0)}$ in the zeroth approach of generalized Chapman-Enskog method are obtained as a combination of Treanor-Maxwell distribution on lower levels, plateau-Maxwell distribution on middle levels and Boltzmann-Maxwell on upper levels:

$$f_{ij}^{(0)} = \begin{cases} \alpha_1 s_{ij} \exp\left(-\dfrac{mc^2}{2kT} - \dfrac{\varepsilon_i + \varepsilon_j}{kT} - \gamma i\right), & 0 \le i < i_* \\[2ex] \dfrac{n_i}{Z_t Z_r} s_{ij} \exp\left(-\dfrac{mc^2}{2kT} - \dfrac{\varepsilon_j}{kT}\right), \quad n_i = \dfrac{\Gamma}{i+1} & i_* \le i < i_{**} \\[2ex] \alpha_2 s_{ij} \exp\left(-\dfrac{mc^2}{2kT} - \dfrac{\varepsilon_i + \varepsilon_j}{kT}\right), & i \ge i_{**} \end{cases} \tag{4}$$

$$\alpha_k = \frac{n_k}{Z_t Z_r Z_{v_k}}, \qquad k = 1,2$$

Here rotational ε_j and vibrational ε_i energy spectra are described by rigid rotator and anharmonic oscillator models

$$\varepsilon_j = (\hbar^2/2I_a)j(j+1), \qquad \varepsilon_i = \hbar\nu(1-\alpha)i - \alpha\hbar\nu i^2,$$

$\hbar$ is a Plank constant, ν is a frequency of molecular vibrations, I_a is an inertia momentum, α is an anharmonism parameter, k is a Boltzmann constant, $c = u - v$, $v(r,t)$ is macroscopic gas velocity, T is gas temperature, Z_t, Z_r, Z_{vk} are statistical sums, s_{ij} is a statistical weight, n_1, n_2, Γ are determined from continuty conditions for $f_{ij}^{(0)}$ at $i = i_*$ and $i = i_{**}$ and normalizing conditions.

Parameter γ is defined in terms of total number of vibrational quantum on lower levels W:

$$\rho W = \sum_{i=0}^{i_*} \sum_j i \int f_{ij}^{(0)} d\mathbf{u},$$

(ρ is a density) and may be presented traditionally as follows:

$$\gamma = \frac{\varepsilon_1}{k}\left(\frac{1}{T_1} - \frac{1}{T}\right).$$

Here T_1 denotes a temperature of the first vibrational level, $\varepsilon_1 = \hbar\nu(1-2\alpha)$. The distribution (4) may be written in unified form [2,1]:

$$f_{ij}^{(0)} = \frac{n s_{ij}}{Z} \exp\left(-\frac{mc^2}{2kT} - \frac{\varepsilon_j}{kT} - \frac{\varepsilon_\alpha + \varepsilon_\beta - \beta\varepsilon_1}{kT} - \frac{\beta\varepsilon_1}{kT_1} + \ln\frac{i_*+1}{\gamma+1}\right), \tag{5}$$

$$\alpha = \begin{cases} i_{**}, & 0 \le i < i_{**} \\ i, & i_{**} \le i < L, \end{cases} \quad \beta = \begin{cases} i, & 0 \le i < i_* \\ i_*, & i_* \le i < L, \end{cases} \quad \gamma = \begin{cases} i_*, & 0 \le i < i_* \\ i, & i_* \le i < i_{**} \\ i_{**}, & i_{**} \le i < L. \end{cases}$$

Here n is the total particles number, Z is total statistical sum.

In equilibrium state $(T = T_1)$ (5) transfers into Maxwell-Boltzmann distribution to the accuracy of normalizing factor. If anharmonism of vibrations is negligible ($\alpha = 0$, $\varepsilon_i = \hbar \nu i$) then we have two-temperature distribution. At decreasing of T_1/T the length of plateau part becomes to be small and may be neglected.

3. Macroscopic equations

Closed system of equations for macroparameters follows from (2),(3).

$$\frac{\partial \rho}{\partial t} + \frac{\partial}{\partial \mathbf{r}}(\rho \mathbf{v}) = 0$$

$$\frac{\partial (\rho \mathbf{v})}{\partial t} + \frac{\partial}{\partial \mathbf{r}}(\rho \mathbf{v}\mathbf{v}) + \frac{\partial}{\partial \mathbf{r}}\mathbf{P} = 0 \tag{6}$$

$$\frac{\partial}{\partial t}\left(\frac{3}{2}nkT + \rho E_r + \rho E_v\right) + \frac{\partial}{\partial \mathbf{r}}\left(\left(\frac{3}{2}nkT + \rho E_r + \rho E_v\right)\mathbf{v}\right) + \frac{\partial}{\partial \mathbf{r}}\mathbf{q} + \mathbf{P} : \frac{\partial}{\partial \mathbf{r}}\mathbf{v} = 0$$

$$\frac{\partial (\rho W)}{\partial t} + \frac{\partial}{\partial \mathbf{r}}(\rho W \mathbf{v}) + \frac{\partial}{\partial \mathbf{r}}\mathbf{q}_w = R_w$$

$$R_w = \sum_j \sum_{i=0}^{i_*} \int i J_{ij}^{(1)} \, d\mathbf{u} \tag{7}$$

Here $\mathbf{P}$ is pressure tensor, E_r and E_v are rotational and vibrational energy of unit mass, $\mathbf{q}$, $\mathbf{q}_w$ are total energy and quanta number fluxes.

The relaxation time τ_v in the flow of molecular gas consisted of anharmonic oscillators may be found from the equation:

$$\frac{1}{\tau_V} = \frac{R_w^{(0)}}{\rho W^{eq} - \rho W}, \tag{8}$$

$$W^{eq} = \frac{\sum_{i=0}^{i_*} i \exp(-\varepsilon_i/kT)}{\sum_i \exp(-\varepsilon_i/kT)}$$

4. Distribution functions in the first approach

In the first approach we have the follows distribution functions [1]:

$$f_{ij}^{(1)} = f_{ij}^{(0)}\left(-\frac{1}{n}\mathbf{A}_{ij}^{(1)}\nabla \ln T - \frac{1}{n}\mathbf{A}_{ij}^{(2)}\nabla \ln T_1 - \frac{1}{n}\mathbf{B}_{ij} : \nabla \mathbf{v} - \frac{1}{n}F_{ij}\nabla \mathbf{v} - \frac{1}{n}G_{ij}\right) \tag{9}$$

Functions $\mathbf{A}_{ij}^{(\gamma)}$ ($\gamma = 1, 2$), $\mathbf{B}_{ij}$, F_{ij}, G_{ij} may be found from the integral equations:

$$n I_{ij}(\mathbf{A}^{(1)}) = f_{ij}^{(0)}\mathbf{c}\left(\mathcal{C}^2 - \frac{5}{2} + \left[\frac{\varepsilon_j}{kT}\right]' + \left[\frac{\varepsilon_\alpha + \varepsilon_\beta - \beta\varepsilon_1}{kT}\right]'\right)$$

$$n I_{ij}(\mathbf{A}^{(2)}) = f_{ij}^{(0)} \mathbf{c} \left[\frac{\beta \varepsilon_1}{kT_1} \right]'$$

$$n I_{ij}(\mathbf{B}) = 2 f_{ij}^{(0)} \left(\mathbf{cc} - \frac{1}{3} \mathbf{c}^2 \mathbf{I} \right)$$

$$n I_{ij}(F) = f_{ij}^{(0)} \left\{ -\frac{nkT}{c_u^T c_w^{T_1} - c_w^T c_v^{T_1}} \left(\left(\mathbf{c}^2 - \frac{3}{2} + \left[\frac{\varepsilon_\alpha + \varepsilon_\beta - \beta \varepsilon_1}{kT} \right] \right)' + \left[\frac{\varepsilon_j}{kT} \right]' \right) \frac{c_w^{T_1}}{T} - \left[\frac{\beta \varepsilon_1}{kT_1} \right]' \frac{c_w^T}{T_1} \right) + \frac{2}{3} \mathbf{c}^2 - 1 \right\}$$

$$n I_{ij}(G) = -f_{ij}^{(0)} \left\{ \frac{R_w^{(0)}}{c_u^T c_w^{T_1} - c_w^T c_v^{T_1}} \left(\left(\mathbf{c}^2 - \frac{3}{2} + \left[\frac{\varepsilon_\alpha + \varepsilon_\beta - \beta \varepsilon_1}{kT} \right] \right)' + \left[\frac{\varepsilon_j}{kT} \right]' \right) \frac{c_v^{T_1}}{T} - \left[\frac{\beta \varepsilon_1}{kT_1} \right]' \frac{c_u^T}{T_1} \right) \right\} - J_{ij}^{(1)}(f^{(0)}, f^{(0)})$$

(10)

Here

$$\mathbf{c} = \sqrt{\frac{m}{2kT}} \mathbf{c},$$

$$[\alpha_i]' = \alpha_i - \langle \alpha \rangle_v, \qquad [\beta_j]' = \beta_j - \langle \beta \rangle_r,$$

$\langle \alpha \rangle_v$, $\langle \beta \rangle_r$ are values of α_i и β_j, averaged over vibrational and rotational spectrum respectively:

$$\langle \alpha \rangle_v = \frac{\sum_i \alpha_i \exp \left(-\frac{\varepsilon_\alpha + \varepsilon_\beta - \beta \varepsilon_1}{kT} - \frac{\beta \varepsilon_1}{kT_1} + \ln \frac{i_* + 1}{\gamma + 1} \right)}{\sum_i \exp \left(-\frac{\varepsilon_\alpha + \varepsilon_\beta - \beta \varepsilon_1}{kT} - \frac{\beta \varepsilon_1}{kT_1} + \ln \frac{i_* + 1}{\gamma + 1} \right)}$$

$$\langle \beta \rangle_r = \frac{\sum_j \beta_j \exp(-\varepsilon_j / kT)}{\sum_j \exp(-\varepsilon_j / kT)}$$

Modified specific heats are determined in such a form:

$$c_t^T = \frac{\partial E_t}{\partial T} = \frac{3}{2} \frac{k}{m}, \qquad c_r^T = \frac{\partial E_r}{\partial T}, \qquad c_v^T = \frac{\partial E_v}{\partial T},$$

$$c_v^{T_1} = \frac{\partial E_v}{\partial T_1}, \qquad c_w^T = \frac{\partial (\varepsilon_1 W)}{\partial T}, \qquad c_w^{T_1} = \frac{\partial (\varepsilon_1 W)}{\partial T_1},$$

$$c_u^T = \frac{\partial U}{\partial T} = c_t^T + c_r^T + c_v^T$$

$$\rho U = \frac{3}{2} nkT + \rho E_r + \rho E_v.$$

The equations (10) must be solved together with normalizing conditions.

5. Transport coefficients

In the first approach on the basis of (9) we obtain formulas for pressure tensor and heat flux:

$$\mathbf{P} = (p - p_{rel})\mathbf{I} - 2\eta\mathbf{S} - \zeta\nabla\mathbf{v}\cdot\mathbf{I}, \tag{11}$$

$$\mathbf{q} = -(\lambda_t + \lambda_r + \lambda_{vt})\nabla T - \lambda_v \nabla T_1. \tag{12}$$

Here $p = nkT$, $\mathbf{S}$, $\mathbf{I}$ are shear viscosities tensor and unit tensor. Shear and bulk viscosity η, ζ, relaxation pressure p_{rel} and all heat conductivity coefficients λ_s are expressed in terms of integral brackets.

$$\eta = \frac{1}{10}kT\,[B, B], \qquad \zeta = kT\,[F, F], \qquad p_{rel} = kT\,[G, F],$$

$$\lambda_t = \frac{k}{3}\left[\left(\frac{mc^2}{2kT} - \frac{5}{2}\right), A_{ij}^{(1)}\right], \qquad \lambda_r = \frac{k}{3}\left[\frac{\varepsilon_j}{kT} - \left\langle\frac{\varepsilon_j}{kT}\right\rangle_r, A_{ij}^{(1)}\right],$$

$$\lambda_{vt} = \frac{k}{3}\left[\frac{\varepsilon_i}{kT} - \left\langle\frac{\varepsilon_i}{kT}\right\rangle_v, A_{ij}^{(1)}\right], \qquad \lambda_v = \frac{k}{3}\left[\frac{\varepsilon_i}{kT} - \left\langle\frac{\varepsilon_i}{kT}\right\rangle_v, A_{ij}^{(2)}\right].$$

The integral bracket $[R, D]$ is defined as follows

$$[R, D] = \sum_{ij}\int I_{ij}(R)D_{ij}\,d\mathbf{u}. \tag{13}$$

The total heat flux may be written also as sum of the translational, rotational and vibrational energy fluxes.

$$\mathbf{q} = \mathbf{q}_t + \mathbf{q}_r + \mathbf{q}_v$$

$$\mathbf{q}_t = -\lambda_t\nabla T - \lambda_{tv}\nabla T_1 \quad \mathbf{q}_r = -\lambda_r\nabla T - \lambda_{rv}\nabla T_1 \quad \mathbf{q}_v = -\lambda_{vt}\nabla T - \lambda_v\nabla T_1 \tag{14}$$

Further solution shows that $\lambda_{tv} = \lambda_{rv} = 0$. One can see that $\mathbf{q}_v$ depend on both gas temperature T and vibrational temperature T_1 gradients. For harmonic oscillations $\mathbf{q}_v = \mathbf{q}_w$ and depends only on the vibrational temperature gradient ∇T_1.

Bulk viscosity and relaxation pressure appear due to rapid inelastic energy exchanges: rotational–translational (RT) and non-resonanse vibrational–vibrational (VV') for all energy spectrum and vibrational–translational (VT) on the upper vibrational levels ($i_* \leq i \leq L$). For harmonic oscillators without rotational degrees of freedom $p_{rel} = \zeta = 0$.

For solution of the equations (10) we use series of Sonine and Waldmann's polynomials with new basic functions:

$$A_{ij}^{(1)} = \frac{m}{2kT}\mathbf{c}\sum_{rpq}a_{rpq}^{(1)}S_{3/2}^{(r)}(\mathcal{C}^2)P_j^{(p)}\left(\frac{\varepsilon_j}{kT}\right)P_i^{(q)}\left(\frac{\varepsilon_\alpha + \varepsilon_\beta - \beta\varepsilon_1}{kT}\right)$$

$$A_{ij}^{(2)} = \frac{m}{2kT}\mathbf{c}\sum_r a_r^{(2)}P_i^{(r)}\left(\frac{\beta\varepsilon_1}{kT_1}\right)$$

$$B_{ij} = \sum_p b_p S_{5/2}^{(r)}(\mathcal{C}^2) \left(cc - \frac{1}{3} C^2 \mathbf{I} \right) \tag{15}$$

$$F_{ij} = \sum_{rpq} d_{rpq} S_{1/2}^{(r)}(\mathcal{C}^2) P_j^{(p)} \left(\frac{\varepsilon_j}{kT} \right) P_i^{(q)} \left(\frac{\varepsilon_\alpha + \varepsilon_\beta - \beta\varepsilon_1}{kT} - \frac{c_w^T}{c_w^{T_1}} \frac{T}{T_1} \frac{\beta\varepsilon_1}{kT_1} \right)$$

$$G_{ij} = \sum_{rpq} g_{rpq} S_{1/2}^{(r)}(\mathcal{C}^2) P_j^{(p)} \left(\frac{\varepsilon_j}{kT} \right) P_i^{(q)} \left(\frac{\varepsilon_\alpha + \varepsilon_\beta - \beta\varepsilon_1}{kT} - \frac{c_u^T}{c_v^{T_1}} \frac{T}{T_1} \frac{\beta\varepsilon_1}{kT_1} \right)$$

In so doing we express all transport coefficients in terms of standard integrals

$$\Omega^{(l,r)} = \left(\frac{kT}{\pi m} \right)^{1/2} \int_0^\infty \exp(-g_0^2) g_0^{2r+3} \mathcal{Q}^{(l)} \, dg_0, \tag{16}$$

$$\mathcal{Q}^{(l)} \equiv \mathcal{Q}^{(l)}(g) = 2\pi \int \{1 - \cos^l \chi(b,g)\} b \, db \tag{17}$$

and two integral brackets

$$\beta_r = \frac{2}{Z_r^2 Z_v^2} \left(\frac{kT}{\pi m} \right)^{1/2} \sum_{\substack{i\,j\,k\,l \\ i'\,j'\,k'\,l'}} s_{ij} s_{kl} \exp \left(-\frac{\varepsilon_{\alpha(i)} + \varepsilon_{\beta(i)} - \beta(i)\varepsilon_1}{kT} - \frac{\beta(i)\varepsilon_1}{kT_1} - \frac{\varepsilon_j}{kT} - \right.$$

$$\left. \frac{\varepsilon_{\alpha(k)} + \varepsilon_{\beta(k)} - \beta(k)\varepsilon_1}{kT} - \frac{\beta(k)\varepsilon_1}{kT_1} - \frac{\varepsilon_l}{kT} \right) \iint \exp(-g^2) g^3 (\Delta\varepsilon^r)^2 \sigma_{i\,j\,k\,l}^{i'\,j'\,k'\,l'} \, d^2\Omega \, dg, \tag{18}$$

$$\beta_v = \frac{2}{Z_r^2 Z_v^2} \left(\frac{kT}{\pi m} \right)^{1/2} \sum_{\substack{i\,j\,k\,l \\ i'\,j'\,k'\,l'}} s_{ij} s_{kl} \exp \left(-\frac{\varepsilon_{\alpha(i)} + \varepsilon_{\beta(i)} - \beta(i)\varepsilon_1}{kT} - \frac{\beta(i)\varepsilon_1}{kT_1} - \frac{\varepsilon_j}{kT} - \right.$$

$$\left. \frac{\varepsilon_{\alpha(k)} + \varepsilon_{\beta(k)} - \beta(k)\varepsilon_1}{kT} - \frac{\beta(k)\varepsilon_1}{kT_1} - \frac{\varepsilon_l}{kT} \right) \iint \exp(-g^2) g^3 (\Delta\varepsilon^v)^2 \sigma_{i\,j\,k\,l}^{i'\,j'\,k'\,l'} \, d^2\Omega \, dg. \tag{19}$$

Here g_0 is relative velocity of particles, $\sigma_{i\,j\,k\,l}^{i'\,j'\,k'\,l'} \, d^2\Omega$ is the differential scattering cross-section,

$$\Delta\varepsilon = \frac{1}{kT}(\varepsilon_{i'} + \varepsilon_{j'} + \varepsilon_{k'} + \varepsilon_{l'} - \varepsilon_i - \varepsilon_j - \varepsilon_k - \varepsilon_l) = \Delta\varepsilon^r + \Delta\varepsilon^v, \tag{20}$$

$\Delta\varepsilon^r$ and $\Delta\varepsilon^v$ are the resonance defects of rotational and vibrational energy in the inelastic collisions.

For the viscosity coefficients and relaxation pressure we get the following equations

$$\eta = \frac{5kT}{8\Omega^{(2,2)}} \tag{21}$$

$$\zeta = \zeta_r + \zeta_v \qquad p_{rel} = p_{rel}^r + p_{rel}^v \tag{22}$$

$$\zeta_r = \frac{kT}{\beta_r} \left(\frac{c_r^T c_w^{T_1}}{c_u^T c_w^{T_1} - c_w^T c_v^{T_1}} \right)^2 \qquad \zeta_v = \frac{kT}{\beta_v} \left(\frac{c_a^T c_w^{T_1} - c_w^T c_a^{T_1}}{c_u^T c_w^{T_1} - c_w^T c_v^{T_1}} \right)^2 \qquad (23)$$

$$c_a^T = c_v^T - c_w^T \qquad c_a^{T_1} = c_v^{T_1} - c_w^{T_1}$$

Integral bracket β_r may be expressed in terms of the rotational relaxation time τ_r with using Parker theory for τ_r [5]. Next, β_v is calculated with using SSH-theory modified for anharmonic oscillators [4].

The analysis of the relaxation pressure shows that $p_{rel} \sim 10^{-6} \div 10^{-3} p$ so its influence on the pressure tensor is small and may be neglected.

In order to estimate the role of β_r and β_v on the heat conductivity the limit case when $\Delta\varepsilon = \Delta\varepsilon^r + \Delta\varepsilon^v \to 0$ was treated. In this case $\beta_r = \beta_v = 0$ and formulas for heat conductivity coefficients have such a form:

$$\lambda_t = \frac{25kT}{16\Omega^{(2,2)}} c_t^T \qquad \lambda_r = \frac{3kT}{8\Omega^{(1,1)}} c_r^T$$

$$\qquad (24)$$

$$\lambda_{vt} = \frac{3kT}{8\Omega^{(1,1)}} c_v^T \qquad \lambda_v = \frac{3kT}{8\Omega^{(1,1)}} c_v^{T_1}$$

The calculation of heat conductivity coefficients with using both exact and approximate models shows that anharmonism of vibrations and nonresonance effects may be taken into account only in expressions for specific heats [6]. The contribution of integrals β_r and β_v in heat conductivity do not exceed 2%.

The dependence of viscosity and heat conductivity on the temperature T is given on Fig.1(a,b). It is seen that rotational bulk viscosity approaches to the shear viscosity and vibrational bulk viscosity may exceed them in strong nonequilibrium conditions. Heat conductivity coefficients λ_t and λ_r increase with the rising of the temperature T in such a way that λ_r comes to 40% of λ_t. Coefficients λ_v, λ_{vt} depend strongly from the deviations from the equilibrium and increase in strong nonequilibrium case. The comparison of $\lambda = \lambda_t + \lambda_r + \lambda_{vt}$ obtained on the basis

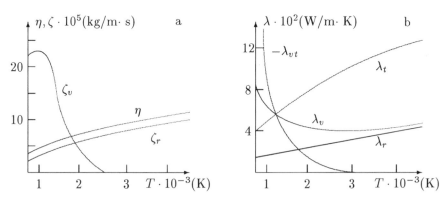

Figure 1: Shear and bulk viscosity (a) and heat conductivity (b) as a function of T at $T_1 = 5000K$

Molecule	$T(K)$	λ_{exp}	λ	λ_{VK}	λ_E	λ_H	λ_{MM}
N_2	77	7.4	7.48	7.5	7.7	8.2	7.4
	180	16.5	16.9	16.9	16.6	17.8	16.5
	260	23.0	22.9	23.0	22.4	23.9	22.6
	293	25.5	25.2	25.4	24.7	26.3	25.0
CO	77	6.7	6.78	7.2	8.0	8.5	—
	180	15.8	16.1	15.8	16.1	17.2	—
	260	22.2	22.2	21.7	21.9	23.3	—
	293	24.7	24.5	24.5	24.3	25.9	—
σ			2.24	3.0	6.3	12.1	3.2

Table 1: Heat conductivity coefficient $\lambda \cdot 10^3 W/m\ K$ calculated for different models. λ_{VK} – results of [8], λ_E computated with using Eucken's factor; λ_H – with modified one, λ_{MM} obtained on the basis of Mason-Monchick approximation. λ_{exp} presents an experimental data [7]. σ is root-mean square deviation.

of our model with experiment [7] and other models [8] is given in Table 1 for N_2 and CO.

The results are also compared with ones for harmonic oscillator and for cut anharmonic oscillator. In order to estimate the role of middle and upper levels three models of cut anharmonic oscillator were treated. The first one is based only on Treanor distribution (without taking into account middle and upper levels). This case is considered also in [9]. In the second model the middle levels are neglected ($i_* = i_{**}$) [6]. And the third model does not take into account upper levels ($i_{**} = L$). This examination shows that distribution on upper levels practically does not influence on transport coefficients. Harmonic oscillator model is valid when $T_1 < T$. Cut anharmonic oscillator model gives the good agreement with results based on (5) if $1 < T_1/T < 4$. Plateau part of distribution becomes important with increasing of T_1/T.

Finally the model is applied to study nonequilibrium boundary layer near flat plate. Heat fluxes q_t, q_r, q_v are given on Fig.2,3 in the case of different conditions in free stream and on the wall.

6. Conclusions

The existence of rapid inelastic RT, VV', VV'', VT exchanges lead to appearance of bulk viscosity and relaxation pressure as well as to additional heat conductivity coefficients. In the case of harmonic oscillator $\zeta_v = p_{rel}^v = \lambda_{vt} = 0$. Shear viscosity η and coefficients $\zeta_r, \lambda_t, \lambda_r$ do not depend on the degree of the vibrational excitation and have the same value as in the weak nonequilibrium case. Coefficients ζ_v, λ_v and λ_{vt} depend on the derivation from the equilibrium and increase with

T_1/T. At $T_1/T < 3 - 4$ λ_{vt} and ζ_v are negligible small but at $T_1/T \sim 5 - 7$ they may exceed coefficients λ_t and η at 5-6 times respectively. It means that the anharmonic effects are to be important in the case of strong vibrational nonequilibrium. At $T_1/T < 4$ cut anharmonic oscillator gives good results without taking into account plateau part of distribution. With the rising T_1/T the role of middle levels becomes essential.

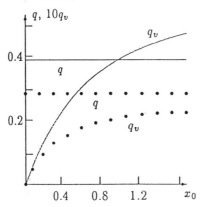

Figure 2: Heat flux as a function of x for catalytic wall at $T_\infty = T_{1\infty} = T_w = T_{1w} = 3336K$. Points show the results for harmonic oscillator

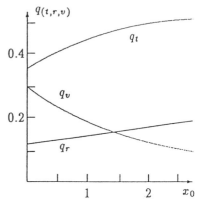

Figure 3: Heat flux as a function of x for catalytic wall at nonequilibrium conditions: $T_\infty = 500K$, $T_{1\infty} = 2000K$, $T_w = 500K$, $T_{1w} = 1000K$.

As to heat fluxes anharmonic effects are particularly important for the flux of vibrational energy and may enlarge it for 40–50%.

7. References

1. Kustova, E.V. and Nagnibeda, E.A. (1994) New kinetic model of transport processes in the strong nonequilibrium gas, in *19th Int. Symp. on RGD. Book of abstracts*, Oxford Univ. Press, England.
2. Kustova, E.V. (1995) Kinetic model of the molecular gas dynamics in the strong nonequilibrium conditions (in Russian), *Vestnik SPU. Math., mech., astr.* 1.
3. Pascal, S. and Brun, R. (1993) Transport properties of nonequilibrium gas mixtures, *Phys. Review E* 47, 5, 3251–3267.
4. Gordietz, B.F., Zhdanok, C.A. (1986) Vibrational kinetics of anharmonic oscillators, in M.Capitelli (ed.) *Nonequilibrium vibrational kinetics*, Springer-Verlag, Berlin, Heidelberg, New York, Tokyo, pp. 61–103.
5. Parker, J.G. (1959) Rotational and vibrational relaxation in diatomic gases, *Phys. Fluids* 2, 449–461.
6. Kustova, E.V. (1993) About the influence of vibrational and rotational nonequilibrium on the heat conductivity of molecular gas (in Russian), *Vestnik SPU. Math., mech., astr.* 4, 60–65.
7. Prangsma, G.J., Alberga, A.H., Beenaker, J.J. (1973) Ultrasonic determination of the volume viscosity of N_2, CO, CH, CO_2 between 77 and 300K, *Physics* 64, 278–288.
8. Van den Oord, R.J., Korving, J. (1988) The thermal conductivity of polyatomic molecules, *J. Chem. Phys*, 89, 7, 4333–4338.
9. Brun, R. (1990) Transport phenomena in relaxing gas mixtures: models and applications, in *Proceedings of the 17th International Symposium on RGD, Aachen*, VCH, Weinheim – New-York, pp.379–390.

THE ROLE OF INELASTIC ROTATIONAL AND VIBRATIONAL COLLISIONS ON TRANSPORT COEFFICIENTS

CARL NYELAND
Institute of Chemistry
University of Copenhagen
Denmark

Abstract

The Wang Chang-Uhlenbeck theory of rate coefficients for relaxation and transport (thermal conductivity, viscosity, self-diffusion) for molecular gases are considered. Firstly, the formal results of the theory are discussed for correctness in comparison with the proper theory of transport coefficients. Then some earlier published calculational results for a nitrogen gas using a Monte Carlo treatment of the bimolecular collisions following classical dynamics and also semiclassical dynamics are considered particularly for the meaning of the contribution from the vibrational degrees of freedom to the transport coefficients at high temperatures. These calculations were partly carried out for rotationally, non-vibrating molecules and partly for fully coupled rotational and vibrational degrees of freedom. From the collisional treatment one realised that the first-order treatment of the Wang Chang-Uhlenbeck theory is useful for calculations of transport coefficients for temperatures where the rigid rotor model is appropriate. The results at higher temperatures indicate that inclusion of a quantum mechanical treatment of vibration or eventually higher order terms in the Wang Chang-Uhlenbeck theory is necessary.

1. Introduction

The theory of polyatomic gases was brought into a very useful form by Wang Chang and Uhlenbeck [1] in a report written in 1951. Prior to that the most extended treatment of thermal conductivity in a polyatomic gas had been given by Hirschfelder [2] who obtained the expression

$$\lambda = \frac{5}{2}(\eta/m)c_{tr} + (\rho D/m)c_{int} , \tag{1}$$

where λ is the thermal conductivity, η is the viscosity, m and ρ are the molecular mass and mass density, respectively, and c_{tr} and c_{int} are the heat capacities for the translational and internal degrees of freedom, respectively. This result appeared from a model-type calculation, assuming that the contribution from the internal degrees of freedom can be added by using a local equilibrium approximation to the well-known result of Chapman and Enskog for monatomic gases. D in eq. 1 was considered to

M. Capitelli (ed.), Molecular Physics and Hypersonic Flows, 393–406.

be a diffusion constant for internal-state diffusion, neglecting inelastic collisions. Still earlier, Eucken [3] had considered the form

$$\lambda = (\eta/m)(\tfrac{5}{2}c_{tr} + c_{int}) , \qquad (2)$$

obtained by mean-free path arguments.

In the theory of Wang Chang, Uhlenbeck, and de Boer [1], one assumes a spherically symmetric distribution for the angular momentum of a molecule. The relaxation phenomena are thereby treated properly but the transport phenomena are not. Waldmann [4] and Snider [5] have set up a proper theory useful for outer-field effects but also for a proper treatment of zero-field transport coefficients. By focusing on the Wang Chang-Uhlenbeck description, we have neglected contributions from the spin polarizations in the inhomogeneous gases considered: This means that no results for external field problems can be obtained. In [6a] the errors expected in calculations following the Wang Chang-Uhlenbeck method for ordinary zero-field problems were briefly discussed. With a semiemprirical argument, the expected error was shown to be less than 1% for calculations of thermal conductivities and probably smaller for calculations of viscosities.

2. Wang Chang-Uhlenbeck Theory

The formal, bimolecular results for the transport coefficients obtained in the Wang Chang-Uhlenbeck theory have been discussed in [6a] in the classical limit and in [7] the semiclassical approach. Also the classical limit of the results for relaxation time and the pressure-broadening coefficient of the depolarized Rayleigh scattering (DPR) were given there. Here only the main points will be considered. The average symbol $\langle\!\langle\ \rangle\!\rangle$ for the collision integrals is defined by

$$\langle\!\langle\ \rangle\!\rangle = \left[\frac{kT}{\pi m}\right]^{1/2} Q_A^{-2} \sum_{ijkl} \int \int e^{-e_i-e_j-\gamma^2} \gamma_3\{\ \}\sigma_{ij}^{kl}\ d\Omega\ d\gamma , \qquad (3)$$

where the collision cross section σ_{ij}^{kl} for the processes

$$N_2(i) + N_2(j) \rightarrow N_2(k) + N_2(l) , \qquad (4)$$

of a nitrogen gas with i denoting the set of quantum numbers (v,j,m_j) is given by

$$\sigma_{ij}^{kl}\ d\Omega = 2\pi P_{ij}^{kl}(b)b\ db . \qquad (5)$$

$P_{ij}^{kl}(b)$ is the transition probability and is given as a function of the impact parameter b. Furthermore, we have

$$\epsilon_i = E_i/kT , \tag{6}$$

$$\gamma^2 = \frac{1}{4}mv^2/kT , \tag{7}$$

$$Q_A = \sum_i e^{-\epsilon_i} , \tag{8}$$

$$\Delta\epsilon = \epsilon_k + \epsilon_l - \epsilon_i - \epsilon_j , \tag{9}$$

where v is the relative velocity and Q_A the partition function for internal states. The following four collision integrals are considered.

$$\langle\!\langle (\Delta\epsilon)^2 \rangle\!\rangle , \tag{10}$$

$$\langle\!\langle \gamma^4 \sin^2 \chi - \frac{1}{2}(\Delta\epsilon)^2 \sin^2 \chi + \frac{1}{3}(\Delta\epsilon)^2 \rangle\!\rangle , \tag{11}$$

$$\langle\!\langle \gamma^2 - \gamma\gamma' \cos \chi \rangle\!\rangle , \tag{12}$$

$$\langle\!\langle (\epsilon_i - \bar{\epsilon})[(\epsilon_i - \epsilon_j)\gamma^2 - (\epsilon_k - \epsilon_j)\gamma\gamma' \cos \chi] \rangle\!\rangle , \tag{13}$$

where χ is the scattering angle in the relative system, γ' the reduced velocity after the collision, and $\bar{\epsilon}$ the average internal energy for a molecule.

The following formal first-order results from the Wang Chang-Uhlenbeck method can be written by use of the four collision integrals 10-13. The bulk viscosity η_v is given by

$$\eta_v = \tau n k^2 T c_{int}/c_v^2 , \tag{14}$$

where the relaxation time τ is given by

$$\tau^{-1} = \frac{2nk}{c_{int}} \langle\langle (\Delta\epsilon)^2 \rangle\rangle \,, \tag{15}$$

n is the density of molecules, c_v the heat capacity at constant volume and c_{int} the heat capacity of the internal degrees of freedom. The relaxation collision number ζ is defined as

$$\zeta = \frac{4}{\pi} \frac{p\tau}{\eta} \,, \tag{16}$$

where p is the pressure. The shear viscosity η is given by

$$\eta^{-1} = \frac{8}{5kT} \langle\langle \gamma^4 \sin^2 \chi - \frac{1}{2}(\Delta\epsilon)^2 \sin^2 \chi + \frac{1}{3}(\Delta\epsilon)^2 \rangle\rangle \,. \tag{17}$$

The coefficient of self-diffusion is given by

$$D^{-1} = \frac{8nm}{3kT} \langle\langle \gamma^2 - \gamma\gamma' \cos \chi \rangle\rangle \,. \tag{18}$$

The thermal conductivity is given by

$$\frac{\lambda m}{\eta k} = \frac{15}{4} + \beta c'_{int} - \frac{2c'_{int}(\frac{5}{2} - \beta)^2}{\pi\zeta + 2(\frac{5}{3}c'_{int} + \beta)} \,. \tag{19}$$

Here $\beta = \rho D_{int}/\eta$, where the coefficient of "internal diffusion" D_{int} is defined as

$$D_{int}^{-1} = \frac{8nm}{3kT} \Omega_{int}^{(1,1)} \,, \tag{20}$$

and the term $\Omega^{(1,1)}_{int}$ is given by

$$c'\Omega^{(1,1)}_{int} = \langle\langle(\epsilon_i - \bar{\epsilon})[(\epsilon_i - \epsilon_j)\gamma^2 - (\epsilon_k - \epsilon_l)\gamma\gamma'\cos\chi]\rangle\rangle , \qquad (21)$$

$c' = c_{int}/k.$

3. Intermolecular Potentials

For the potential between two nitrogen molecules two examples of short-range site-site potentials combined with a dispersion term and a quadrupole-quadrupole interaction term were considered in [6].

In potential model A we assumed the simple exponential site-site potential suggested by Berns and van der Avoird [8] from their ab initio calculations

$$V_{SR} = Ae^{-\alpha R} , \qquad (22)$$

where $A = 293.3$ au, $\alpha = 2.136$ au, and R is the distance between the atoms in different molecules. The C_6 dispersion constant was set to 24.4 au [8] and for the quadrupole moment Q the theoretical result [9] - 1.136 au was used. For vibrating molecules the quadrupole derivative also was taken into account. The theoretical value [9] $(\partial Q/\partial r)_{eq} = 0.632$ au was used.

In potential model B we assumed the extended site-site potential suggested by Ling and Rigby [10]

$$V_{SR} = A'e^{-\alpha' R - \beta R^2} , \qquad (23)$$

where $A' = 22.568$ au, $\alpha' = 1.186$ au, and $\beta = 0.09$ au. The same values as for model A were used for the dispersion constant C_6 and the quadrupole moment Q, but the quadrupole derivative was set to the experimental value, [11] $(\partial Q/\partial r)_{eq} = 0.933$ au. The two different potentials yield slightly different transport coefficients as will be discussed in the following sections.

4. Calculations and results

4.1 RIGID-ROTOR MODEL

From comparisons of results from quantum-mechanical close-coupling calculations of transport coefficients for He-N$_2$ mixtures [13] with results from classic mechanics calculations [14] one knows that results obtained for T > 100 K agree very well using the rigid-rotor model. For this model some results obtained [6b] for the viscosity and the thermal conductivity of pure nitrogen are shown on Figures 1 and 2 and compared with the best experimental results available.

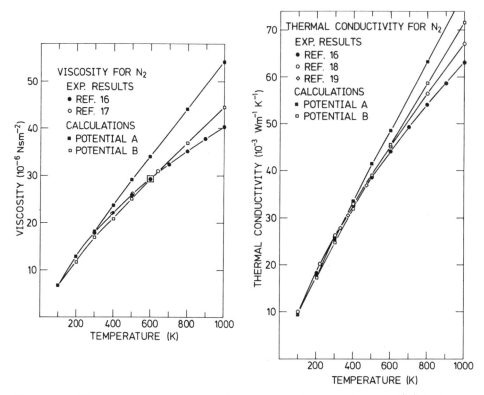

Figure 1,2 Viscosity and thermal conductivity for nitrogen. Results for rigid rotor molecular collision calculations compared with experimental results.

Below T ~600 K the experimental and the calculated results particularly for potential model B agree reasonably well. In Table 1 the results are presented as value for the Eucken number f. In comparison with the so-called Eucken approximation, f(Eucken) = 4.75, [3], and the Hirschfelder approximation, f(Hirschfelder) ≈ 5.2, [2], one gets the following relationships

$$f(Eucken) < f(Calc.) \sim f(Exp.) < f(Hirschfelder)$$

for the diatomic nitrogen gas.

TABLE 1. Results for the Eucken number ($f = \lambda m(\eta k)^{-1}$) for the rigid rotor model for nitrogen

T(K)	potential model A	potential model B	experimental
100	4-73	4.74	4.61
200	4.71	4.93	4.79
300	4.74	5.05	4.90
400	4.76	5.14	4.94
500	4.78	5.21	5.02
600	4.79	5.26	5.07

4.2 VIBRATING ROTOR MODEL

When T > 600 K the deviation on the Figures 1 and 2 are seen to be essential. The vibrational degree of freedom cannot be neglected. Van den Oord and Korving [12] have shown experimentally for a J_2 gas that the vibrational and the rotational degrees of freedom are that close coupled in a temperature gradient field that only the sum of internal energy has to be considered in the distribution functions of the results in chapter 3. A quantum mechanical or semiclassical treatment is needed for the collisions. Results from a coupled semiclassical treatment are seen on Figure 3 based on potential model B, [7]. This result shows a typical energy exchange effect: The addition of the vibrational degree of freedom reduces the thermal conductivity. Similar effects have been shown for electronic degrees of freedom, see appendix II.

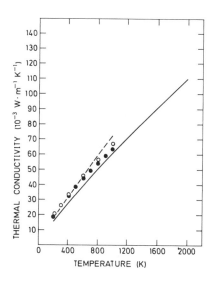

Figure 3. Full line is for coupled semiclassical vib-rot treatment. Broken line is for rigid-rotor classical mechanical treatment. Results from [7] by permission of the athors.

5. Discussion

From the discussion for nitrogen using the rigid rotor model, it can be argued that good agreement with experimental results for transport and relaxation coefficients for temperatures below 600 K can be obtained by minor adjustments of the intermolecular potential considered. Semiclassical treatment and eventually higher order Wang Chang-Uhlenbeck treatment are necessary in calculation of transport coefficients at temperature above 600 K. Estimates for technical use can easily be made in the low-temperature range based on any reasonable potential model eventually combined with an average value for the Eucken factor.

Appendix I

First-order results for transport coefficients of pure polyatomic gases in σ-symbols

Defining for convenience

$$\overline{C}_r = 4(\frac{kT}{\pi m})^{1/2} , \tag{I-1}$$

and

$$r = (\frac{2}{5} \frac{c_{int}}{k})^{1/2} , \tag{I-2}$$

one has [21]

$$\eta = \frac{kT}{\overline{C}_r} \sigma(2000)^{-1} , \tag{I-3}$$

$$\eta_v = \frac{k^2 c_{int} T}{c_v^2 \overline{C}_r} \sigma(0001)^{-1} , \tag{I-4}$$

$$D = \frac{kT}{nm \overline{C}_r} \sigma'(1000)^{-1} , \tag{I-5}$$

$$D_{\text{int}} = \frac{kT}{nm\,\overline{C_r}}\ [\sigma(1001) - \frac{1}{2}\,\sigma(0001)]^{-1}\ . \qquad \text{(I-6)}$$

These first-order results are identical to the results given in the text by use of wcu-integrals, as following relationships are exactly fulfilled

$$\sigma\binom{1010}{1001} = -\frac{5}{6}\,r\,\sigma(0001)\ , \qquad \text{(I-7)}$$

$$\sigma(1010) = \frac{2}{3}\,\sigma(2000) + \frac{25}{18}\,r^2\sigma(0001)\ . \qquad \text{(I-8)}$$

The wcu-integrals considered in the text can be written in σ-symbols as

$$\langle\!\langle \gamma^4\sin^2\chi - \tfrac{1}{2}(\Delta\epsilon)^2\ \sin^2\chi + \tfrac{1}{3}(\Delta\epsilon)^2\rangle\!\rangle = \tfrac{5}{8}\,\overline{C_r}\,\sigma(2000)\ , \qquad \text{(I-9)}$$

$$\langle\!\langle(\Delta\epsilon)^2\rangle\!\rangle = \frac{\overline{C_r}c_{\text{int}}}{2k}\,\sigma(0001)\ , \qquad \text{(I-10)}$$

$$\langle\!\langle(\epsilon_i - \bar{\epsilon})[(\epsilon_i - \epsilon_j)\gamma^2 - (\epsilon_k - \epsilon_l)\gamma\gamma'\cos\chi]\rangle\!\rangle = \\ \frac{3}{8}\frac{c_{\text{int}}}{k}\,\overline{C_r}[\sigma(1001) - \frac{1}{2}\,\sigma(0001)]\ ,$$

$$\text{(I-11)}$$

$$\langle\!\langle\gamma^2 - \gamma\gamma'\cos\chi\rangle\!\rangle = \frac{3}{8}\,\overline{C_r}\,\sigma'(1000)\ . \qquad \text{(I-12)}$$

Appendix II

Contributions to the thermal conductivity from electronic excitation transfer

In an early investigation [22] it was shown that the internal diffusion in an atomic gas with some thermal excitation is caused by an adiabatic excitation transfer process

$$A + A* \rightarrow A* + A . \tag{II-1}$$

The coefficient for internal diffusion was shown to be given by

$$nD_{int} = \frac{3}{16} \frac{(\pi m \ kT)^{\frac{1}{2}}}{\int_o^\infty Q_{ex} \ \gamma^5 \ e^{-\gamma^2} \ d\gamma} , \tag{II-2}$$

where Q_{ex} is the total cross section for the excitation transfer.
In the semiclassical approach Q_{ex} is given by [22]

$$Q_{ex} = 2\pi \int_o^\infty \sin^2\xi \ b \ db , \tag{II-3}$$

where

$$\xi = \frac{1}{\hbar v} \int_b^\infty \frac{\Delta E}{(r^2 - b^2)^{\frac{1}{2}}} \ r \ dr , \tag{II-4}$$

and $\Delta E(r)$ the r-dependent excitation energy, the energy between the 'gerade' and 'ungerade' states of the collision process (II-1). This energy difference can often be assumed to be exponential

$$\Delta E = A e^{-br} . \tag{II-5}$$

Numerical calculations following this method are relatively simple. For an atomic nitrogen gas for instance, where the first excited state $N(^2D)$ is located 2.38 eV above the ground state $N(^4S)$, the contribution to the thermal conductivity due to excitation transfer was estimated to be about 16%, [22] relative to a monoatomic gas at $T = 10000$ K, while the contribution due to the electron degree of freedom would be about 50% if excitation transfer was neglected as in the Eucken-Hirschfelder approximation.

Appendix III

The rotational relaxation time τ.

The Maxwell-Boltzmann distribution function f_2^o for an ideal polyatomic gas in equilibrium can be factorized as

$$f_i^o \ (T) = f_{i,tr}^{\,O}(T) \ f_{i,int}^{\,O}(T) \ , \tag{III-1}$$

For an isoterm, isobaric perturbation around equilibrium one often assumes that the distribution function is given by

$$f_i \ (T) = f_{i,tr}^{\,O}(T') \ f_{i,int}^{\,O}(T'') \ , \tag{III-2}$$

where

$$(T' - T) \ c_{tr} = (T - T'') \ c_{int} \ . \tag{III-3}$$

To first-order one has

$$f_i = f_i^o(T)(1 + \phi_i) \ , \tag{III-4}$$

where

$$\phi_i = [(\frac{3}{2} - \gamma^2) \ \frac{c_{int}}{c_v} + (\epsilon_i - \bar{\epsilon}) \ \frac{c_{tr}}{c_v}] \ \frac{T'' - T'}{T} \ , \tag{III-5}$$

for which one gets after some calculation that the internal energy follows a relaxation equation

$$\frac{dE_{int}}{dt} = -\frac{1}{\tau} \ [E_{int} \ (T'') - E_{int} \ (T')] \ , \tag{III-6}$$

where

$$\tau^{-1} = \frac{2kn}{c_{int}} <\{(\Delta\epsilon)^2\}> \ , \tag{III-7}$$

as mentioned in the text in the discussion of the bulk viscosity.

As the relaxation in a gas of vibrating rotors often needs to be described with more than one relaxation time the perturbation result of eq. III-7 should only be used for rotational relaxation of effectively rigid rotors.

Acknowledgment.

This research was supported in part by the Danish Natural Science Research Council.

References

1. Wang Chang, C,S., Uhlenbeck, G.E. and de Boer, J. (1964), The heat conductivity and viscosity of polyatomic gases, in J. de Boer and G.E. Uhlenbeck (eds.), *Studies in Statistical Mechanics*, vol. II, North-Holland Publishing Company, Amsterdam, pp. 242-268.

2. Hirschfelder, J.O. (1957) Heat conductivity in polyatomic or electronic excited gases, II, *J. Chem. Phys.* *26*, 282-285.

3. Eucken, A. (1913) Über das Wärmeleitvermöge, die specifische Wärme und die innere Reibung der Gase, *Phys. Z. 14*, 324-332.

4. Waldmann, L. (1957) Die Boltzmann-Gleichung für Gase mit rotierende Molekülen, *Z. Naturforschung A 12*, 660-662.

5. Snider, R.F. (1960) Quantum-Mechanical modified Boltzmann equation for degenerated internal states, *J. Chem. Phys.* *32*, 1051-1060.

6. a. Nyeland, C., Poulsen, L.L. and Billing, G.D. (1984) Rotational relaxation and transport coefficients for diatomic gases: Computations on nitrogen, *J. Phys. Chem.* *88*, 1216-1221, b. Nyeland, C. and Billing. G.D. (1988) Transport coefficients of diatomic gases: Internal-State analysis for rotational and vibrational degrees of freedom, *ibid 92*, 1752-1755.

7. Billing, G.D. and Wang, L. (1992) Semiclassical calculations of transport coefficients and rotational relaxation of nitrogen at high temperatures, *J. Phys. Chem.* *96*, 2572-2575.

8. Berns, R.M. and van der Avoird, A. (1980) N_2-N_2 interaction potential from *ab initio* calculations, with application to the structure of $(N_2)_2$, *J. Chem. Phys.* *72*, 6107-6116.

9. Billingsley, F.P. and Krauss, M. (1974) Quadrupole moment of CO, N_2, and NO^+, *J. Chem. Phys.* *60*, 2767-2772.

10. Ling, M.S.H. and Rigby, M. (1984) Towards an intermolecular potential in nitrogen, *Mol. Phys.* *51*, 855-882.

11. Reuter, D. and Jennings, D.E. (1986) The $v = 1 \leftarrow 0$ quadrupole spectrum of N_2, *J. Mol. Spectrosc.* *115*, 294-304.

12. van den Oord, R.J. and Korving, J (1988) The thermal conductivity of polyatomic molecules, *J. Chem. Phys.* *89*, 4333-4338.

13. Maitland, G.C., Mustafa, M., Wakeham, W.A. and McCourt, F.R.W. (1987) An essentially exact evaluation of transport cross sections for a model of He-N_2 interaction, *Mol. Phys.* *61*, 359-387.

14. Dickinson, A. and Lee, M.S. (1986) Classical trajectory calculations for anisotropy dependent cross-sections for He-N_2 mixtures, *J. Phys. B 19*, 3091-3107.

15. Billing, G.D. (1984) The semiclassical treatment of molecular roto-vibrational energy transfer *Comput. Phys. Rep. 1*, 237-296.

16. Touloukain, Y.S. (ed.) (1970, 1973) *Thermophysical Properties of Matter, Vol. 3, 11*, Plenum Press, New York.

17. Vogel, E. (1984) Präzisionsmessungen des Viskositätskoeffizienten von Stickstoff und den Edelgasen zwischen Raumtemperatur und 650 K, *Ber. Bunsen-Ges. Phys. Chem. 88*, 997-1002.

18. Shashkov, A.G. Abraminko, T.N. and Aleininikova, V.I. (1985) *J. Engineering Phys. (Engl. Transl.)* **49**, 83-93.

19. Haarman, J.W. (1973) Thermal conductivity measurements of He, Ne, Ar, Kr, N_2 and CO_2 with a transient hot-wire method, in *J. Kestin, (ed.) 'Transport-phenomena - 1973'*, Am. Inst. Phys. Conf. Proc. **11**, 193-198.

20. a. Monchick, L. and Mason, E.A. (1961) The transport properties of polar gases *J. Chem. Phys.* 35, 1676-1697. b. Monchick, L., Peireira A.N.G. and Mason, E.A. (1965) Heat conductivity of polyatomic and polar gases and gas mixtures, *J. Chem. Phys.* **42**, 3241-3256.

21. For a recent review, see McCourt, F.R.W., Beenakker, J.J.M., Köhler, W.E. and Kuščer, I. (1990) *Nonequilibrium Phenomena in Polyatomic Gases,* Clarendon Press, Oxford.

22. Nyeland, C. and Mason, E.A. (1967) Adiabatic excitation transfer in gases: Effects on transport, *Phys. Fluids.* **5**, 985-991.

ANALYSIS OF APPROXIMATION INTRODUCED USING SIMPLIFIED SUM RULES IN CALCULATION OF HIGH ENERGY AIR TRANSPORT COEFFICIENTS

F. DE FILIPPIS
C.I.R.A. - Centro Italiano Ricerche Aerospaziali
Via Maiorise - Capua (Italy)

1. Abstract

High energy air Transport Coefficients modelling is made either with semiempirical laws or following molecular theory of gases and liquids developments. Semiempirical models come from Sutherland's law for air viscosity. Instead, it is possible to develop different theoretical models using different approximation levels of the theory and different hypothesis on intermolecular potentials variations.

The calculation of these coefficients came from a particular weighed average of the contributions of the coefficients calculated for the single species present in the mixture with different concentrations. These sum rules are semiempirical simplification of the theoretical formulas, whose is impossible implementation in CFD hypersonic codes for the very expensive requested CPU times.

This work investigates the correctness to use some simplified expressions of the air transport coefficients at high energy, making comparison between results obtained using correct and approximate formulas.

2. Introduction

When space vehicles reenter the atmosphere at hypersonic speed, high energies are developed in air surrounding the body. Energy transfer in different flow zones are very important. For this reason a correct modelling of Transport Properties is necessary to provide the thermal and mechanical loads by CFD. We are speaking of temperatures higher than 1000 K.

M. Capitelli (ed.), Molecular Physics and Hypersonic Flows, 407–416.
© *1996 Kluwer Academic Publishers. Printed in the Netherlands.*

In the complete flow equations there are the following quantities: air viscosity (μ), air thermal conductivity (k) and air mass diffusivity (D). They are a measure of the transfer respectively of impulse, thermal energy and mass in the different flow regions.

The viscosity for a specie is defined by the relationship:

$$j_p = \mu_i \cdot \frac{dv}{dx} \tag{1}$$

where, v is the velocity flux, x the position in a direction perpendicular to the flux, j_p the impulse flux.

The thermal conductivity for a specie is defined by the Fourier law:

$$J_E = k_i \cdot \frac{dT}{dx} \tag{2}$$

where T is the temperature and J_E the Energy flux.

The binary diffusivity for two species by the Fick law:

$$J = \rho \cdot D_{ij} \frac{dc_i}{dx} \tag{3}$$

where ρ is the density, c_i the concentration of i-specie and J the mass flux.

They suppose linear proportionality between the impulse, energy and mass flow with velocity gradients, temperature and mass concentration respectively. The quantities μ_i, k_i and D_{ij} naturally depend from temperature that is the measure of the gas molecules agitation, responsible for the transport.

The air is a mixture of different species for which μ (air viscosity) and k (air conductivity) depend from μ_i and k_i respectively and from c_i (concentrations of different components in the gas mixture). The c_i depends from pressure and temperature and from the conditions in which the gas is considered : chemical equilibrium or chemical non-equilibrium.

For a mixture the multicomponent diffusion coefficients are defined:

$$J_i = \rho \cdot D_i \frac{dc_i}{dx} \tag{4}$$

for a specie in the mixture.

The quantities μ, k and D_i are present in the Navier-Stokes equations for which a correct modelling with temperature variation is required. At present, two kinds of approaches to model Transport Coefficient variation with temperature are possible: the semiempirical one and the theoretical one. Both gives us an expression of the variation with temperature of μ_i (viscosity of i-specie present in air mixture), of k_i (thermal conductivity of i-specie present in air mixture) and D_{ij} (mass diffusivity of i-specie in j-specie present in air mixture)

3. Sum Rules

Air is approximatively a mixture of five gases, of which compositions vary with pressure and temperature. The air Transport Coefficients are calculated opportunely summing the single specie or single pair ones.

Considering theoretical models the mixture viscosity come from a sequence of analytical relationships very laborious to calculate. They are reported in [1]. In particular it is necessary to calculate two determinants of 6x6 and 5x5 matrices and the ratio between these two quantities. The first matrix is A

$$
A = \begin{pmatrix}
H_{11} & H_{12} & H_{13} & H_{14} & H_{15} & x_1 \\
H_{12} & H_{22} & H_{23} & H_{24} & H_{25} & x_2 \\
H_{13} & H_{23} & H_{33} & H_{34} & H_{35} & x_2 \\
H_{14} & H_{24} & H_{34} & H_{44} & H_{45} & x_2 \\
H_{15} & H_{25} & H_{35} & H_{45} & H_{55} & x_2 \\
x1 & x_2 & x_3 & x_4 & x_5 & 0
\end{pmatrix}
\tag{5}
$$

and the second is B

$$
B = \begin{pmatrix}
H_{11} & H_{12} & H_{13} & H_{14} & H_{15} \\
H_{12} & H_{22} & H_{23} & H_{24} & H_{25} \\
H_{13} & H_{23} & H_{33} & H_{34} & H_{35} \\
H_{14} & H_{24} & H_{34} & H_{44} & H_{45} \\
H_{15} & H_{25} & H_{35} & H_{45} & H_{55}
\end{pmatrix}
\tag{6}
$$

The values of the H_{ii} terms and of the H_{ij} terms may be written as function of:

$$
H_{ii} = H_{ii}(\mu_i, \mu_j, x_i, x_j, M_i, M_j, T, R, A_{ij})
\tag{7}
$$

$$
H_{ij} = H_{ij}(\mu_i, \mu_j, x_i, x_j, M_i, M_j, T, R, A_{ij})
\tag{8}
$$

The explicit formulas for H_{ii} and H_{ij} are reported on [1], being x_i the mole fraction of the i-specie, M_i the molecular weight, T the temperature, R the gas constant and the quantities $A_{ij} = \frac{\Omega_{ii}(2,2)}{\Omega_{ij}(1,1)}$ are ratios between Collision Integral dependent by temperature and we remind on the [1] book for a detailed discussion about their significance.

The Wilke law is reported in [2]

$$\mu = \frac{\sum_i X_i \mu_i}{\sum_j X_j \cdot \left((8)^{\frac{-1}{2}} \cdot \left(1 + \frac{M_i}{M_j} \right)^{\frac{-1}{2}} \cdot \left(1 + \left(\frac{\mu_i}{\mu_j} \right)^{\frac{1}{2}} \cdot \left(\frac{M_j}{M_i} \right)^{\frac{1}{4}} \right)^2 \right)} \tag{9}$$

with M_i molecular weight of i-specie and represents an approximate semiempirical relationship to calculate mixture viscosity and it is at the moment the most used in literature. It can be implemented in CFD codes, without a catastrophic effort in terms of CPU time.

The air thermal conductivity is calculated from air viscosity using relationship (6).

The calculation of the five multicomponent diffusion coefficients for O, N, NO, N_2, O_2 in air is made using simple semiempirical laws. The most used rule is the following [3] :

$$D_{im} = \frac{1 - c_i}{\sum_{j \neq k} \frac{X_j}{D_{jk}}} \tag{10}$$

The knowledge of gas specie concentrations , gas specie molar fractions and binary diffusivities is required.

In general it is important even to correctly calculate the binary diffusivity of the fifteen pairs of species present in air. The theoretical formula to calculate this coefficient is the following:

$$D_{ij} = 2.6280 \cdot 10^{-3} \cdot \frac{\left(T^3 \cdot \left(\frac{1}{M_i} + \frac{1}{M_j} \right) \right)^{\frac{1}{2}}}{p \cdot \Omega_{ij}(1,1)} \tag{11}$$

for the binary diffusion between i-specie and j-specie. The quantities, whose above we spoke, $\Omega_{ii}(2,2)$ and $\Omega_{ij}(1,1)$ are function of temperature. They are Collision Integrals and depend from intermolecular potentials between equal species (ii) and between different species (ij). They require great calculation efforts and are tabulated only for fixed temperatures.

It is necessary to correct this last relationship considering the presence of various species in the mixture. The theory developed in [1] is able to give us correct formulas for D_{ij} of a multicomponent mixture and at page 541 there is a sequence of laborious relationships to make the calculation. Precisely required the knowledge of all the simple binary D_{ij} is required, of the molar fractions, of the molecular weights and the calculation of the minors for a particular 6x6 matrix.

4. Correct and approximate Air viscosity results

To make comparison between correct and approximate air viscosity calculation, we started from the variation of viscosity with temperature of the five species present in air mixture. They have been calculated using the more sophisticated theoretical model described in the works from [4] to [8].

At the first the air viscosity variation with temperature has been calculated using the Wilke law up to very high temperature values. Have been considered even different possible pressure values and the hypothesis of chemical equilibrium . The variation with temperature is shown in Figure 1. The pressure determines different concentrations values and different viscosity values and carries out the differencies in the curves, the chemical model that we used is reported in [9] and in [10]. The pressure effect is not catastrophic on the results. In the same figure there is also the air viscosity at different pressure calculated using the correct relationships reported in the previous section. The curves in this second graph don't differ substantially compared to the curve in the first graph. The Figure 2 is the ratio between correct and approximate air viscosity for different pressure, showing that the use of the Wilke law is a very good approximation to calculate air viscosity. In fact for all the pressure values and in the whole analyzed temperature range this ratio is very near to one.

5. Correct and Approximate air binary diffusivity results

Regarding the air diffusivity we point out that considering the air composed of five species it is consequently possible to determine fifteen binary diffusion coefficients. In a way similar to the viscosity calculation, we calculated these quantities using the theoretical formula (11) in a wide temperature range, considering the mixture in chemical equilibrium and fixing a pressure value of 1 atmosphere.

Successively we calculated all the coefficients following the elaborate algorithms reported on the [1] that practically permit to consider the effects due to the presence of various elements in the mixture. At this point we made the comparison.

The Figures from 3 to 7 is the set of all the obtained results. We considered ten of the fifteen binary diffusivity coefficients, because in the theory developed on [1] has not sense to calculate the self-diffusivity coefficients.

For all the coefficients we don't find very sensible differencies between the correct and approximate calculation. In general these are increasing as the temperature increases.

The binary diffusivity of Atomic Oxygen in Nitrogen Oxide, of Atomic Oxygen in Molecular Oxygen and of Atomic Oxygen in Molecular Nitrogen reach differencies up to the twenty percent after the 6000 Kelvin. The similar behavior of the last three coefficients is comprehensible because they regard the diffusion of atomic oxygen in similar air molecular species.

For the other seven calculated binary diffusivity coefficients the differencies doesn't superate the five percent.

Even for the air diffusion there are not substantial differences between the use of simple binary diffusivity or of 'correct' binary diffusivity.

6. Conclusions

This the work investigated the 'goodness' in the use of an approximate sum rule for air viscosity and air thermal conductivity calculations (the Wilke's law). A comparison with results obtained by complex, but correct sum rules has been made. We didn't find great differencies in these comparisons. It is correct the use Wilke's law for calculations regarding air mixture.

Moreover the air diffusion coefficients have been calculated using binary diffusivities, corrected for the presence of various species in the mixture. These coefficients don't differ with respect to the coefficients calculated with simplified binary diffusivity. This approximation can be made.

References

1. J. O. Hirschfelder, C. F. Curtiss, R. B. Bird "Molecular Theory of Gases and Liquids" - John Wiley & Sons (1963 - Second printing).
2. C.R. Wilke "A Viscosity Equation for Gas Mixtures", THE JOURNAL OF CHEMICAL PHYSICS VOL.18 NO. 4 (April, 1950).
3. R.J. Kee, J. Warnatz, J.A. Miller "A fortran computer code package for the evaluation of gas-phase viscosities, conductivities, and diffusion coefficients" - SAND83-8209 (1983).
4. F. De Filippis "Fenomeni di Trasporto - Manuale Teorico" - C.I.R.A. Internal document C.I.R.A. DILC-INT-TN-294 - November 1992.
5. F. De Filippis "Fenomeni di Trasporto - Modello di Yun e Mason .." - C.I.R.A. Internal document DILC-INT-TN-335 - May 1993.
6. K. S. Yun, E. A. Mason "Collision Integrals for the Transport Properties of Dissociating Air at High Temperatures" - The Physics of Fluids, vol. 5 - number 4 (April 1962).
7. F. De Filippis, S. Borrelli "Considerazioni su differenti modelli di trasporto per l'aria ad alte energie" - XII AIDAA (1993).
8. F. De Filippis, S. Borrelli "Analysis of different approximation levels introduced in the developments for transport coefficients models of high energy air" - XIX ICAS, Anaheim - CA (USA), September 1994.
9. J.V. Rakich., H.E. Bailey, C. Park "Computation of nonequilibrium,supersonic three-dimensional inviscid flow over blunt-nosed bodies",AIAA Journal,V0l.21,June 1983,pp.834-841.
10. A. Schettino, S. Borrelli, F. De Filippis "Influence of Transport and Thermokinetic Models in Free-flight and in Plasma Wind Tunnel Tests" - ISCFD August-September 1993 in Sendai-Japan).

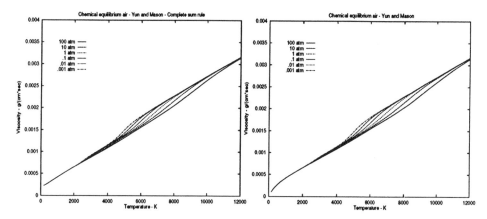

Figure 1. Air viscosity at different pressures calculated with approximate and correct sum rules respectively.

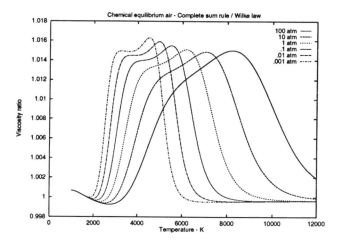

Figure 2. Air viscosity ratio between calculations made with approximate and correct sum rule at different pressures.

414

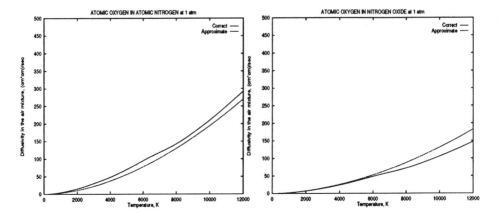

Figure 3. Binary diffusivity for atomic oxygen in atomic nitrogen and for atomic oxygen in nitrogen oxide. Comparison between correct and approximate calculations.

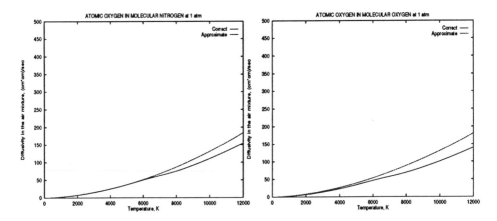

Figure 4. Binary diffusivity for atomic oxygen in molecular oxygen and for atomic oxygen in molecular nitrogen. Comparison between correct and approximate calculations.

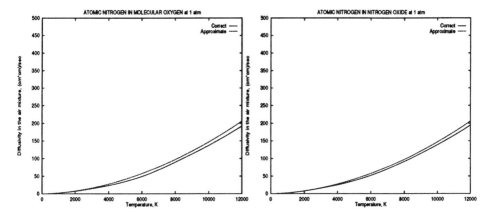

Figure 5. Binary diffusivity for atomic nitrogen in nitrogen oxide and for atomic nitrogen in molecular oxygen. Comparison between correct and approximate calculations.

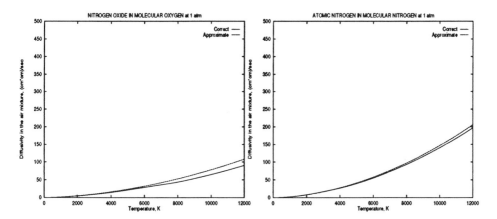

Figure 6. Binary diffusivity for atomic oxygen in molecular oxygen and for atomic oxygen in molecular nitrogen. Comparison between correct and approximate calculations.

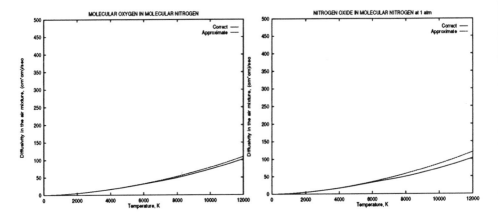

Figure 7. Binary diffusivity for atomic nitrogen in nitrogen oxide and for atomic nitrogen in molecular oxygen. Comparison between correct and approximate calculations.

THE ANALYTICAL APPROACHES IN THE NONEQUILIBRIUM VIBRATIONAL KINETICS

B.F.GORDIETS
Centro de Electrodinamica, Instituto Superior Tecnico, Lisboa, Portugal
(Permanent address: Lebedev Physical Institute of Russian Academy of Sciences, Leninsky Prospect 53, Moscow, Russia)

1. Introduction

A vibrational kinetics is a part of physical-chemical kinetics, which investigates the nonequilibrium populations of the quantum vibrational levels of molecules and their influence on the different parameters and evolution of systems. It must be used for study of the different phenomenons in the nonequilibrium molecular gases. The hypersonic flows of molecular gases is the example of the nonequilibrium system, where the nonequilibrium vibrational kinetics plays important role and must be taken into account in the modelling of the gas dynamic phenomenons.

If means that the modelling of gas dynamic phenomenons is usually the complex numerical problem, the use in such modelling relatively simple analytical approaches for vibrational kinetics is useful and desirable.

The main results will be presented below for the analytical solutions of the master equations for the vibrational level populations, rate of relaxation of the vibrational energy and for dissociation through boundary vibrational level. The some results in this area can be found in books and reviews [1-7].

2. The Vibrational Kinetics of Harmonic Oscillators

2.1. THE VIBRATIONAL RELAXATION OF DIATOMIC MOLECULES IN INERT GAS: VT- EXCHANGE

Let us designate the population of n-th vibrational level as N_n, and the probability of vibrational transition $n \to m$ per one collision of molecule with the atom of inert gas as P_{nm}. In this case, the system of kinetic

M. Capitelli (ed.), Molecular Physics and Hypersonic Flows, 417–435.
© *1996 Kluwer Academic Publishers. Printed in the Netherlands.*

equations for populations of vibrational levels in spatially homogeneous gas is following:

$$\frac{dN_n}{dn} = Z \sum_m (P_{mn} N_m - P_{nm} N_n), \tag{1}$$

where Z is the collision frequency. The first term on the right of eq. (1) describes the changing of number of molecules on n-th vibrational level due to the collisions, at which the transitions of molecules are carrying out on the level n from other vibrational levels. The second term describes the reverse processes, i.e., the departure of molecules from the level n due to the transitions on all other vibrational levels.

The binary atom-molecular collisions only at vibrational-translation energy exchange (VT-exchange) are taken into account in the eq. (1). The probabilities of direct and reverse transitions are connected by the relation of detailed balance:

$$P_{nm} exp \left(-\frac{E_n}{kT} \right) = P_{mn} exp \left(-\frac{E_m}{kT} \right) \tag{2}$$

for the conditions of Maxwell velocity distribution. Here E_n and E_m are the energies of vibrational levels and T is the temperature of translation degrees of freedom.

In the typical case, only one-quantum transitions with the probabilities

$$P_{n+1,n} = (n+1)P_{10} \tag{3}$$

are important for the harmonic oscillator.

Taking into account eqs. (2) and (3), it is possible to rewrite the system of kinetic equations (1) as

$$\frac{dN_n}{dn} = ZP_{10}\{(n+1)[N_{n+1} - N_n exp(-E_{10}/kT)] - n[N_n - N_{n-1} exp(-E_{10}/kT)]\}, \tag{4}$$

where $E_{10} = E_1 - E_0$ is energy of vibrational quanta. This system has two important peculiarities for the subsequent discussion.

1. The equations (4) have the property of canonical invariant i.e., keep the form of Boltzmann distribution function. Really, if in the initial moment of time we have Boltzmann distribution with vibrational temperature $T_v(0)$, the solution of eq. (4) will the Boltzmann function [8]

$$N_n(t) = N(1 - e^{-\Theta})e^{-n\Theta}, \tag{5}$$

where N is concentration of molecules, $\Theta = E_{10}/kT_v(t)$ and $T_v(t)$ is the current vibrational temperature of the oscillators. The parameter Θ will be changed in time as:

$$\Theta(t) = \ln \frac{e^{-t/\tau_{vt}}(1 - e^{E_{10}/kT - \Theta_0}) - e^{-E_{10}/kT}(1 - e^{-\Theta_0})}{e^{-t/\tau_{vt}}(1 - e^{E_{10}/kT - \Theta_0}) - (1 - e^{-\Theta_0})}, \tag{6}$$

where

$$\tau_{vt} = [ZP_{10}(1 - e^{-\Theta_0})]^{-1} \tag{7}$$

and $\Theta_0 = E_{10}/kT_v(0)$.

Note, that in general case, the solution of eq. (4) is determined by Hotlib polynomials [8].

2. It is possible to obtain the relaxation equation for the total vibrational energy from the system of equations (4). The vibrational energy of the unit of volume is

$$E_{vib} = \sum_n E_n N_n, \tag{8}$$

where $E_n = nE_{10}$ for harmonic oscillators.

If we multiply the equation (4) by nE_{10} and carry out the sum of it over all n, we will obtain the relaxation equation for the energy

$$\frac{E_{vib}}{dt} = -\frac{E_{vib} - E_{vib}^0}{\tau_{vt}}, \tag{9}$$

where E_{vib}^0 is equilibrium value of E_{vib} (for gas temperature T).

The solution of (9) has an exponential form (for $T = const.$) and depends upon the initial magnitude of vibrational energy only. Very often the equation (9) can be written for the average number of vibrational quanta α :

$$\frac{d\alpha}{dt} = -\frac{\alpha - \alpha^0}{\tau_{vt}}, \tag{10}$$

where

$$\alpha = \frac{1}{N} \sum_n n N_n. \tag{11}$$

2.2. ONE-COMPONENT SYSTEM: VV-EXCHANGE

The vibrational inelastic process - the energy exchange among the vibrational degrees of freedom of molecules (VV-exchange) - takes place as well as VT-exchange at the collisions among the diatomic molecules.

If the probabilities of VV- and VT- exchanges are small, it is possible to consider them as independent to a first approximation. In this case, the kinetic equations, defining the dependency of populations of vibrational levels upon the time, must additively take into account VV- and VT-exchanges.

Basing on the considerations of balance of number of particles on each vibrational level, one can write

$$\frac{dN_n}{dn} = Z \sum_m (P_{mn}N_m - P_{nm}N_n) + \frac{Z}{N} \sum_{m,s,l} (Q^{sl}_{mn}N_s N_m - Q^{ls}_{nm}N_l N_n). \quad (12)$$

Here $N = \sum_{m=0} N_m$; Q^{sl}_{mn} is the probability of VV-exchange per one collision, at which two molecules in the vibrational states n and s transform correspondingly into n and l states after collision.

The probabilities of VV-exchange are connected by relation of detailed balance:

$$Q^{sl}_{mn} exp \left(-\frac{E_s + E_m}{kT} \right) = Q^{ls}_{nm} exp \left(-\frac{E_l + E_n}{kT} \right) \quad (13)$$

and they have the following form for the harmonic oscillator:

$$Q^{sl}_{mn} = \begin{cases} (m+1)sQ_{10} & n = m+1, \ l = s-1, \\ m(s+1)Q_{10} & n = m-1, \ l = s+1, \\ 0 & n \neq m \pm 1, \ l \neq s \mp 1. \end{cases} \quad (14)$$

Taking into account (2),(3),(13) and (14), it is possible to rewrite the system of equations (12) as

$$\frac{dN_n}{dn} = ZP_{10}\{(n+1)[N_{n+1}-N_n exp(-E_{10}/kT)]-n[N_n-N_{n-1}exp(-E_{10}/kT)]\}+$$

$$ZQ_{10}\{(n+1)[(1+\alpha)N_{n+1} - \alpha N_n] - n[(1+\alpha)]N_n - \alpha N_{n-1}]\}.$$

The system (15) is the system of nonlinear equations, since α depends upon N_m. The vibrational relaxation of harmonic oscillators, described by the system (15), has two important peculiarities.

1. Two different relaxation time scales is appeared, because usually:

$$\tau_{vv} \approx \frac{1}{ZQ_{10}} \ll \frac{1}{ZP_{10}} \approx \tau_{vt} \quad (16)$$

The condition (16) allows to distinguish the fast and slow stages in the relaxation process. During the fast stage ($t \leq \tau_v$) VV-exchange plays the essential role, and VT-exchange can be neglected. On this stage the process of relaxation is described by the shorted system (without first sum in (15)).

2. The one-quantum VV-exchange has resonant character at the collision of harmonic oscillators (energy defect of resonance $(E_s + E_m) - (E_l + E_n) = 0$, therefore, the total number of vibrational quanta is keeping. It can be proved, if we multiply shorted equations (15) by n and carry out the sum over all n. Therefore, on the stage of VV-exchange ($t \leq \tau_{vv}$) the equations (15) are linear.

It is easy to obtain the stationary solution of shorted eq.(15) which is following Boltzmann distribution [9,10]:

$$N_n = N(1 - e^{-\Theta})e^{-n\Theta}, \tag{17}$$

where

$$e^{-\Theta} = \frac{\alpha}{1 + \alpha} \tag{18}$$

The establishment of Boltzmann distribution is the result of fast stage, i.e., the result of VV-exchange. Just this fact defines the fundamental role of VV-exchange in vibrational kinetics.

The second, slow stage of the vibrational relaxation is described by the system of total equations (15). The fast VV-exchange can not be taken into account at the consideration of this system in the time scale of $t \gg \tau_{vv}$. Only the results of this process - the formation of Boltzmann distribution with temperature, defined by the quantity of vibrational quanta - are essential for the subsequent discussion.

Therefore, the solution of (15) in the time scale can be found as the Boltzmann distribution function with vibrational temperature T_v, which time dependence is defined by (6) [11].

2. The relaxation of total vibrational energy ϵ_v (see (8)) in the one-component system of harmonic oscillators can be described by relaxation equation (9). This point can be proved, if to multiply (15) by $E_{10}n$ and to carry out the sum over all n. The one-quantum VV-exchange does not change the total vibrational energy at the equidistant arrangement of vibrational levels. The contribution of VT-processes to the relaxation of vibrational energy is essential only.

The model of harmonic oscillator very well describes the vibrational relaxation at the low levels of vibrational excitation, when it is possible to neglect the anharmonic effects. The influence of anharmonicity is growing at the increasing of average vibrational energy (energy per one molecule). The anharmonic effects are considered below.

2.3. BINARY MIXTURE OF DIATOMIC MOLECULES: VT-, VV- AND VV'-EXCHANGE

The processes of exchange of vibrational energies among the molecules of different components (VV'-exchange) take place in binary system of diatomic gases at the molecular collisions as well as VT- and VV-exchanges. If the probabilities of inelastic processes (VT, VV and VV') are small, they can be considered as the independent ones in the first approximation. In this case, the kinetic equations, defining the dependency of populations of vibrational levels upon the time, must additively take into account VT-, VV- and VV'-processes. The equations, describing VT- and VV-processes, were discussed before. The collision operator for VV'-process is analogous to the operator of VV-exchange and can be obtained by the replacement in VV-exchange (see the second term on the right of (12)) of one of the colliding molecules by the molecule of another sort. If N_n^A and N_n^B are the populations of n-th vibrational levels correspondingly of the molecules of sort A and B, the change of the population of A particles on the vibrational level n due to VV'-processes will be

$$\left(\frac{dN_n^A}{dt}\right)_{VV'} = \frac{Z_{AB}}{N_B} \sum_{m,s,l} (Q'_{mn}{}^{sl} N_s^B N_m^A - Q'_{nm}{}^{ls} N_l^B N_n^A), \qquad (19)$$

where $N_A = \sum_n N_n^A$, $N_B = \sum_n N_n^B$, $Z_{AB} = N_B Z_{AB}^0$; Z_{AB}^0 is the number of collisions of molecules A with molecules B in the unitary density of A and B particles; Qo_{mn}^{sl} is the probability of VVo-exchange per one collision, at which the molecule A transforms from vibrational state m in the state n, and the molecule B - from s in l.

The probabilities of VV'-exchange are connected by relation of detailed balance

$$Q'_{mn}{}^{sl} exp\left(-\frac{E_s^B + E_m^A}{kT}\right) = Q'_{nm}{}^{ls} exp\left(-\frac{E_l^B + E_n^A}{kT}\right), \qquad (20)$$

where E_i^M is the energy of i-th vibrational level of molecule M $(M = A; B)$.

The one-quantum transitions with probabilities similar (14) are typical for VV' exchange between two harmonic oscillators with energies of quanta

$$E_{10}^A > E_{10}^B; \quad E_{10}^A < 2E_{10}^B \qquad (21)$$

For such case, if means (20), the eq. (19) can be written as

$$\left(\frac{dN_n^A}{dt}\right)_{VV'} = Z_{AB}^0 N_B Q'_{10}\{(n+1)[(1+\alpha_B)e^{\Theta_A^0} N_{n+1}^A - \alpha_B e^{\Theta_B^0} N_n^A]$$

$$-n[(1+\alpha_B)e^{\Theta_A^0} N_n^A - \alpha_B e^{\Theta_B^0} N_{n-1}^A]\}, \qquad (22)$$

where $\Theta_A^0 = E_{10}^A/kT$, $\Theta_B^0 = E_{10}^B/kT$. The equations for N_n^B are obtained from eq. (22), if to interchange the positions of indexes A and B.

The process of vibrational relaxation, described by equations (22), has two peculiarities.

1. The equations (15) with additional VV' term (22) for A and for B molecules have three typical time scales:

$$\tau_{vv}^A \sim \frac{1}{Z_{AA}^0 N_A Q_{10}}, \quad \tau_{vv'}^A \sim \frac{1}{Z_{AB}^0 N_B Q'_{10}}, \quad \tau_{vt}^A \sim \frac{1}{Z_A P_{10}^A};$$

$$\tau_{vv}^B \sim \frac{1}{Z_{BB}^0 N_B Q_{10}}, \quad \tau_{vv'}^B \sim \frac{1}{Z_{BA}^0 N_A Q'_{10}}, \quad \tau_{vt}^B \sim \frac{1}{Z_B P_{10}^B}. \qquad (23)$$

It is possible to analyze the relaxation in three different time scales. In the time scale $t \leq \tau_{vv}^A$, τ_{vv}^B VV-exchange leads to the establishment of Boltzmann distribution in each of components with own vibrational temperature, defined by the value of vibrational energy in each component. The VV' exchange goes on in the time scale τ_{vv}^A, $\tau_{vv}^B \leq t \leq \tau_{vv'}^A$, $\tau_{vv'}^B$, and the synchronization of vibrational temperatures T_v^A and T_v^B is the result of this process. Really, the quasi-stationary distribution, formed under the action of VV'-exchange, is defined from the condition of detail equilibrium. This equilibrium, in the case of one-quantum transitions, gives relationship [12]

$$\frac{\alpha^A}{1 + \alpha^A} = \frac{\alpha^B}{1 + \alpha^B} exp \left(\frac{E_{10}^B - E_{10}^A}{kT} \right) \qquad (24)$$

where α^A and α^B are the average numbers of vibrational quanta in molecule A and B respectively. Taking into account the relation (18), one can rewrite eq.(24) as

$$\frac{E_{10}^A}{kT_v^A} - \frac{E_{10}^B}{kT_v^B} = \frac{E_{10}^A - E_{10}^B}{kT} \qquad (25)$$

which connects the temperatures T_v^A and T_v^B. Therefore, the processes of VV'-exchange lead to the synchronization of vibrational temperatures .

Finally, the establishment of total statistical equilibrium with the equalization of all temperatures goes on in the time scale of $t \sim \tau_{vt}$ order.

2.4. POLYATOMIC MOLECULES

The polyatomic molecules differ from diatomic ones by the presence of several vibrational degrees of freedom. The vibrational motion in molecule can be presented as superposition of normal vibrations at not very high level of vibrational excitation such that each normal vibration (normal mode) corresponds to an harmonic oscillator. Therefore, the vibrational relaxation in one-component system (or in the mixtures of polyatomic molecules) can be considered as vibrational relaxation in multi-component mixture of harmonic oscillators in this approximation. One can assume, that the intramode energy exchange (VV-exchange) goes on with considerably higher probabilities, than intermode VVo-exchange or VT-processes as well as in the case of binary mixtures of harmonic oscillators. The establishment of Boltzmann distribution with own vibrational temperature due to VV-exchange is the first stage of the process of vibrational relaxation in the mixtures of polyatomic molecules at these conditions. It is not necessary to consider VV-processes (as the finished ones) at the consideration of vibrational relaxation on the time interval, longer than the typical time of VV-exchange. In this situation, the large-scale behavior of vibrational relaxation, defined by VV'- and VT-exchange, consists of synchronization of vibrational temperatures and of establishment of total statistical equilibrium. It is convenient to describe this process by average number of vibrational quanta α_s for given vibration s. The equation for the change of α_s was deduced in the work [13] in detail (see also [1]). This equation is

$$\frac{d\alpha_s}{dt} = Z^0_{AB} N_B Q_{AB} \left\{ \begin{matrix} l_i \to 0 \\ 0 \to l_j \end{matrix} \right\} l_s \prod_{i=1}^{k}(r_i + \alpha_i^0)^{-l_i} \prod_{j=k+1}^{L} (\alpha_j^0)^{-l_j} \times$$

$$\times \{\prod_{i=1}^{k}[\alpha_i^0(r_i+\alpha_i)]^{l_i} \prod_{j=k+1}^{L} [\alpha_j(r_j+\alpha_j^0)]^{l_j} - \prod_{i=1}^{k}[\alpha_i(r_i+\alpha_i^0)]^{l_i} \prod_{j=k+1}^{L} [\alpha_j^0(r_j+\alpha_j)]^{l_j}\},$$

(26)

where $\alpha_s = r_s/[exp[\frac{E_{10}^s}{kT_v^s}] - 1]$, r_s is the degeneracy factor of the mode s, T_v^s is the vibrational temperature of s-th mode, α_s^0 is the equilibrium value of α_s, i.e., the value of α_s at $T_v^s = T$, where T is the temperature of translation degrees of freedom. Only the collisions of molecules A among molecules B were taken into account, and the transition of the system $A + B$ from the vibrational state $(v_1, v_2, ..., v_L)$ to the state $(v_1 \pm l_1, ..., v_k \pm l_k, v_{k+1} \mp l_{k+1}, ..., v_L \mp l_L)$ was assumed at the obtaining of (26).

In eq. (26) $Q_{AB} \left\{ \begin{matrix} l_i \to 0 \\ 0 \to l_j \end{matrix} \right\}$ is the probability of transition, calculated without consideration of degeneracy according to ordinary method by Shwarts, Slavskiy, Hertsfeld [3,7,14]. The equation (26) describes the relaxation of average number of vibrational quanta (vibrational energy) of mode s along the single channel, determined by the specification of numbers l_1, i_2, ..., l_L. It is necessary to carry out the sum of the right-hand side of eq. (26) over all possible l_1, i_2, ..., l_L at the presence of several relaxation channels. In a general case, the vibrational relaxation in the mixture of polyatomic molecules must be described by the system of equations (26), written for each mode.

The equation (26) is the most general expression of relaxation equation for vibrational energy (or for the average number of vibrational quanta), taking into account VV'- and VT-exchange at the collisions of different molecules. The equation (26) transforms into the relaxation Landau-Teller equation (9),(10) describing VT-exchange, at $L = 1$, $l_i = l_j = 0$ (for $i \neq s$), $l_s = 1$. The expression (26) coincides with relaxation equation (22), describing the one-quantum VV'-exchange in binary mixture at $L = 2$, $l_i =$

$1, l_j = 1, l_s = 1, r_s = 1, r_j = 1.$

3. The Vibrational Kinetics of Anharmonic Oscillators

In the present time, a special attention is given to the analysis of vibrational kinetics at low temperatures and large stock of vibrational energies. Under such conditions, the harmonic oscillator model was found to be unfit for description of the vibrational kinetics and the vibration anharmonicity must be taken into consideration. The anharmonicity dominates for high vibrational levels and influences on them populations, the vibrational energy relaxation rates, the rates of chemical reactions involving vibrationally excited molecules.

In a general case, the anharmonicity requires use numerical methods for analysis. However, let us consider the analytical solutions. Obviously, in the majority of cases, such solutions of nonlinear system of many equations like eq. (12) cannot be solved without simplifications. To do this, one should take into consideration a drastic difference in the process probabilities and use simple analytical dependencies on the number of a vibrational level of a molecule. For example, the fact that for a large group of vibrational levels of a molecule in a wide region of gas temperatures, the probabilities of one-quantum processes P_{mn}, Q_{mn}^{ij} are much greater than those multi-quantum processes, allows to simplify system (12)

$$\frac{1}{N}\frac{dN}{dt}f_n + \frac{df_n}{dt} = Z[(P_{n+1,n}f_{n+1} - P_{n,n-1}f_n) - (P_{n,n+1}f_n - P_{n-1,n}f_{n-1})] +$$

$$Z\sum_i[(Q_{n+1,n}^{i,i+1}f_i f_{n+1} - Q_{n,n-1}^{i,i+1}f_i f_n) - (Q_{n,n+1}^{i+1,i}f_{i+1}f_n - Q_{n-1,n}^{i+1,i}f_{i+1}f_{n-1})]$$

$$(27)$$

Expressions of (27) type are the equations for the normalized vibrational distribution function f_n related to the vibrational level populations N_n of a molecules by the connection $N_n = N f_n$, where N is the concentration of molecules. One simplification can be obtained using a simple model of the anharmonic Morse oscillator. In this case, the energy E_n of the n-th level in a diatomic molecule is given by:

$$E_n = nE_{10}\left[1 - \frac{\Delta E}{E_{10}}(n-1)\right],\qquad(28)$$

where ΔE is the molecular anharmonicity.

Using SSH-method [1,2,3,7,14] for calculations of collision transition probabilities, it is possible to obtain:

$$P_{n+1,n} \approx (n+1)P_{10}exp(\delta_{vt}n),\qquad(29)$$

$$Q_{n+1,n}^{i,i+1} \approx (n+1)(i+1)Q_{10}exp(-\delta_{vv}|n-i|)[3/2-1/2exp(-\delta_{vv}|n-i|)] \quad (30)$$

where $\delta_{vv} = \frac{0.427}{\alpha}(\frac{\mu}{T})^{\frac{1}{2}}\Delta E$ is the reduced mass of colliding particles (in *atomic units*), α is the constant in the exponential repulsive potential of the molecular interaction (in $\mathring{A}$); ΔE and T are measured in K. The similar expression is used for δ_{vt}, however, μ and α can be others if the V-T transitions are determined by collisions with an impurity gas.

3.1. V-T RELAXATION IN INERT GAS

Let us consider the evolution of the vibrational distribution function and the available vibrational energy of diatomic molecules being a small impurity to the inert gas. Then, the vibrational level populations are described by modified system (27) allowing only for the first brackets in the RHS. The analytical solution of this system for anharmonic oscillators can be obtained at low gas temperatures $kT < E_{10}$. If at $t = 0$ the initial level distribution of molecules has the Boltzmann profile at $T_0 \neq T$ $(kT_0 < E_{10})$, we obtain after time $t > \frac{1}{ZP_{10}}$ [15]:

$$f_n \simeq f_0 exp\left[-\frac{E_n}{kT} + \frac{E_1}{r_n}\left(\frac{1}{kT} - \frac{1}{kT_1(t)}\right)\right], \quad (31)$$

where

$$r_n = \prod_{i=2}^{n}\left(1 - \frac{P_{10}}{P_{i,i-1}}\right). \quad (32)$$

The time evolution of the distribution function f_n gives also an information on the relaxation rate of the vibrational energy $E_{vib} = N\sum_n E_n f_n$. Multiplying the shorted equations of type (27) by E_n with the utilization of Boltzmann distribution (it is well founded for low levels) we obtain the Losev formula [16] for the rate of a change of E_{vib}:

$$\frac{dE_{vib}}{dt} = -\frac{E_{vib} - E_{vib}^0}{\tau_{vt}^{anh}}, \quad (33)$$

$$\tau_{vt}^{anh} = \tau_{vt}^{h}\left[\frac{1 - exp(-E_{10}/kT_1 + \delta_{vt})}{1 - exp(-E_{10}/kT_1)}\right]^2, \quad (34)$$

where E_{vib}^0 is the equilibrium value of E_{vib}, τ_{vt}^{anh} and τ_{vt}^{h} are times of V-T relaxation for the models of anharmonic and harmonic oscillators, respectively. At high gas temperatures, the multi-quantum V-T transitions may dominate in the relaxation system. Then, the solution of system (27) for the one-quantum approximation may give substantial errors. In these cases, the vibrational relaxation can be easily governed by the Fokker-Plank diffusion equation. To the diffusion approximation, the vibrational kinetics of anharmonic oscillators in the inert gas is detailed in [17].

3.2. V-V EXCHANGE. THE TREANOR DISTRIBUTION

In a one-component system of anharmonic oscillators (or in that partially diluted with an inert gas), the V-V exchange dominates other processes for a group of low-lying vibrational levels. In this case, the term in the second square brackets in the RHS of system (27) must be reserved to find the vibrational distribution function. The nonequilibrium vibrational distribution function satisfying the system of equations for steady-state conditions was first found in [18]. Later, it was called the Treanor function. Its simplest derivation follows from the principle of detailed balance for any pair of direct and reverse V-V transitions. This means that the term in the second brackets of the RHS of eqs. (12) or (27) equals zero at any s, l, m, n As a result, one can easily get

$$f_n^{Tr} = f_0 exp \left(-\frac{nE_{10}}{kT_1} + \frac{nE_{10} - E_n}{kT} \right) = f_0 exp \left\{ -n \left[\frac{E_{10}}{kT_1} - (n-1)\frac{\Delta E}{kT} \right] \right\}. \tag{35}$$

Note that Treanor distribution (35) does not depend on transition probabilities and, therefore, has a general nature. For harmonic oscillators, $\Delta E = 0$ and distribution (35) is reduced to the Boltzmann one with a vibrational temperature T_1. At $T_1 > T$, the Treanor distribution has its minimum at a level n_0 and for $n > n_0$ gives the inverse population of vibrational states. The value of n_0 is derived from $\frac{df_n^{Tr}}{dn} = 0$ and is equal to:

$$n_0 = \frac{E_{10}}{2\Delta E} \frac{T}{T_1} + \frac{1}{2} \tag{36}$$

However, it is very difficult to obtain the absolute inverse population at $n > n_0$ in real systems since at the upper levels the dissipation of vibrational quanta is increased due to V-T and other processes. At equilibrium $(T_1 = T)$, the level n_0 coincides with the upper vibrational boundary level of a molecule.

3.3. V-V AND V-T EXCHANGE. MODERATE DEVIATION FROM EQUILIBRIUM

If V-V processes dominate at the lower levels of the anharmonic oscillator, and Treanor distribution (35) is valid, then for the upper levels the V-T probabilities are much greater than the exchange ones (formulas (29) and (30)). As a result, we obtain such a distribution that relative populations of the levels have the vibrational temperature equal to the temperature of a gas. System (27), containing in the RHS the first and second square brackets, must be solved to find the vibrational distribution function at all levels for a one-component system of anharmonic oscillators. The approximate analytical solution of such a system can be found only for individual regimes, characterized by the vibrational excitation degree and gas temperature.

First, let us consider the regime of a moderate deviation from equilibrium [19-21]. By this is understood the relaxation regime with so small excitation level populations that the V-V exchange involves only the lower states. On mathematical grounds, this means that for the terms in the second brackets of system (27) we have:

$$Q_{n+1,n}^{i,i+1} f_i f_{n+1} \gg Q_{n+1,n}^{n,n+1} f_n f_{n+1}, \qquad for \quad i \ll n \qquad (37)$$

If this inequality is satisfied for V-V processes which can affect the populations of high excited state ($n \gg 1$), then the non-resonance exchange will dominate. Then, the system (27) can be simplified using for summation the Boltzmann distribution for lower levels. Summation in the RHS of system (27) over i for the Boltzmann distribution f_i gives a linear system of equations for relative populations f_n. For the quasi-stationary regime the solution is following :

$$f_n = f_n^{Tr} \prod_{i=1}^{n-1} \varphi_{i+1} \qquad n \geq 2, \qquad (38)$$

$$\varphi_{i+1} = \frac{(3/2)\beta Q_{i+1,i}^{01} + P_{i+1,i} exp[(E_{10}/kT_1 - (E_{10}/kT)]}{(3/2)\beta Q_{i+1,i}^{01} + P_{i+1,i}}, \qquad (39)$$

$$\beta = \frac{1 - exp(-E_{10}/kT)}{[1 - exp(-E_{10}/kT + \delta_{vv})]^2} \qquad (40)$$

At lower levels the eq.(38) is the Treanor distribution f_n^{Tr} with temperatures T_1 and T. At upper levels the eq.(38) is the Boltzmann distribution with a gas temperature T and a certain effective number of particles.

3.4. RESONANCE V-V RELAXATION UNDER HIGH EXCITATION CONDITIONS

Besides the above-mentioned moderate excitation, the regime of a considerable deviation from equilibrium at low gas temperatures $kT < E_{10}$ and large available nonequilibrium vibrational energies can be realized.

It is convenient to use for description of such regime the diffusion approximation, assuming a smooth change in f_n along level number n to find the vibrational distribution function.

Let the gas temperature is such that

$$\frac{2\Delta E}{kT} \ll 1 \qquad (41)$$

If eq.(27) and probabilities (29),(30) to use for model of Morse oscillators and $exp(-\frac{2\Delta E(i-n)}{kT})$ and f_i to expand into series over $(i - n)$, then after replacing the sums by integrals and integration, we obtain the equation for f_n [22,23]:

$$\frac{3Q_{10}}{\delta_{vv}^3}\frac{d}{dn}\left[(n+1)^2 f_n^2\left(\frac{2\Delta E}{kT} - \frac{d^2\ln f_n}{dn^2}\right)\right] + P_{10}e^{\delta_{vt}n}(n+1)f_n = 0 \quad (42)$$

The equation (42) has an approximate analytical solution:

$$f_n \approx \begin{cases} f_n^{Tr}exp[-\frac{1}{2}(\frac{n}{n_0})^2] & \text{for} \quad n \le n_0 \quad (a) \\ \frac{\Gamma}{n+1} - \frac{P_{10}}{Q_{10}}\frac{kT}{12\Delta E}\frac{\delta_{vv}^3}{\delta_{vt}}\frac{exp(\delta_{vt}n)}{(n+1)} & \text{for} \quad n_0 < n < n_1 \quad (b) \end{cases}$$
$$(43)$$

where f_n^{Tr} and n_0 are given by formulas(35),(36) and the constant Γ is determined by coupling of eqs. (43 a) and (43 b) at the point n_0.

The level n_1 specifies the plateau length of the distribution function. The distribution at the levels $n > n_1$ should tend to the Boltzmann one with a gas temperature T. The value n_1 can be estimated from condition $\frac{1}{f_n}\frac{df_n}{dn}\big|_{n_1} = \frac{1}{f_n^B}\frac{df_n^B}{dn}\big|_{n_1}$, where f_n^B is the Boltzmann distribution function with a gas temperature T. It gives the following equation for n_1:

$$P[1+B(n_1)]exp(\delta_{vt}n_1) = (n_1+1)exp\left(-\frac{\Delta E}{kT}n_0^2 - \frac{1}{2}\right) + Pexp(\delta_{vt}n_0), \quad (44)$$

where

$$P = \frac{kT}{12\Delta E}\frac{\delta_{vv}^3}{\delta_{vt}}\frac{P_{10}}{Q_{10}}, \qquad (45)$$

$$B(n_1) = \delta_{vt}\left\{\frac{n_1[E_{10} - (n_1-1)\Delta E]}{kT} - \frac{1}{n_1+1}\right\}^{-1} \qquad (46)$$

and n_0 is given by (36). The equation (44) has simple solution for the typical case $B(n_1) \ll 1$:

$$n_1 \approx \frac{1}{\delta_{vt}}\ln\left[\frac{n_0+1}{P}exp\left(-\frac{\Delta E}{kT}n_0^2 - \frac{1}{2}\right) + exp(\delta_{vt}n_0)\right]. \qquad (47)$$

Note, that the regime of resonance V-V relaxation with distribution function (43) will be realized only for conditions

$$\frac{T_1}{T} > \frac{E_{10}}{2\Delta E(n_{min} - 0.5)},$$ (48)

where

$$n_{min} = min(n_2, n^*),$$ (49)

$$n^* = \frac{1}{\delta_{vv} + \delta_{vt}} \ln\left(\frac{3}{2}\beta\frac{Q_{10}}{P_{10}}\right)$$ (50)

and n_2 is the solution of following equation:

$$\frac{8}{\delta_{vv}^3}\frac{\Delta E}{kT}(n_2 + 1)\left[1 - exp\left(-\frac{2\Delta E}{kT}n_2 + \frac{\Delta E}{kT} + \delta_{vv}\right)\right]^2 \approx$$

$$\approx exp\left[\frac{\Delta E}{kT}n_2^2 - \left(\delta_{vv} + \frac{2\Delta E}{kT}\right)n_2 + \frac{\Delta E}{kT} + \frac{1}{2}\right].$$ (51)

For estimation n_2 it is possible to use the approximate solution of eq. (51):

$$n_2 \approx 1 + \frac{\delta_{vv}}{2}\frac{kT}{\Delta E} + \left\{\left(1 + \frac{\delta_{vv}}{2}\frac{kT}{\Delta E}\right)^2 + \left[\ln\left(\frac{8}{\delta_{vv}^2}\right) - \frac{1}{2}\right]\frac{kT}{\Delta E}\right\}^{1/2}.$$ (52)

Let us bring our attention to the determination of such important macroscopic parameters as average number of nonequilibrium vibrational quanta α and a "vibrational" temperature T_1. For strong nonequilibrium conditions the relationship between α and T_1 is obtained from the ordinary relation $\alpha = \sum nf_n$ by substituting into it the above distribution function(43). It gives:

$$\alpha \approx [exp(E_{10}/T_1) - 1]^{-1} + f_{n_0}^{Tr}e^{-1/2}(n_0 + 1)(n_1 - n_0)$$ (53)

The second term is responsible for the stored quanta on the plateau. The expression (53) can be practically employed also for the regime of the moderate excitation. In this case, the second term is always less than the first one and, the stock of vibrational quanta is close to those for the harmonic oscillator model.

Under steady-state conditions, the value of α can be found assuming that the sum of the excitation rates of vibrations due to the external source q and the dissipation rate of vibrational quanta $(d\alpha/dt)^{anh}$ is equal to zero

$$q + \left(\frac{d\alpha}{dt}\right)^{anh} = 0$$ (54)

For strong nonequilibrium conditions, the utilization of eqs. (29), (30), (43) and (53) yields [23]:

$$q = -\left(\frac{d\alpha}{dt}\right)^{anh} \approx \frac{\alpha - \alpha_0}{\tau_{vt}^{anh}} + \frac{2.2Q_{10}}{\delta_{vv}^3}\frac{\Delta E}{kT}(n_0 + 1)^2(f_{n_0}^{Tr})^2 \qquad (55)$$

where τ_{vt}^{anh} is determined from eq. (34) and α_0 is the equilibrium value of α. The first term in the RHS of eq. (55) describes relaxation at small deviations of α_0 from equilibrium. However, under highly nonequilibrium conditions, the second term responsible for the flux caused due to V-V processes at the level n_0 dominates, and the character of relaxation changes since the dissipation of quanta is now determined by the vibrational exchange probability, but not by the V-T one. On physical grounds, this means that the real V-T dissipation occurs at the upper levels $n > n_0$ and the flux of vibrational quanta towards these levels is determined by the resonance V-V process and is limited by a "narrow region", which corresponds to minimum level populations, i.e., by the Treanor distribution on level n_0. Though the second term in the RHS of eq. (55) describes strong nonequilibrium conditions it can be remained at the analysis of "weak" excitation regime since in this case it is always much less than the first term. This means that the relation (55) can be used in a wide range of parameters T_1 and T and extended to anharmonic oscillators the well-known Landau-Teller formula that describes the relaxation α for harmonic oscillator (see eq.(9),(10)).

Note, that the only steady-state conditions were analyzed in this section for relaxation of anharmonic Morse oscillators under the high excitation conditions. Non-steady-state regime have been investigated in [24] (see also [2]).

4. The Vibrational Kinetics and Dissociation involving Vibrationally Excited Molecules

The participation of vibrationally excited molecules in a chemical reaction means the presence of a negative source of such molecules. In the system (12) or (27) this presence is described by adding of the term ZNk_nf_n or Zk_nf_n, where k_n is the probability (or rate constant) of the reaction involving a molecule at the level n. The integral probability K of the reaction at the vibrational levels is

$$K = \sum_n k_nf_n \qquad (56)$$

If the terms k_nf_n are less than those standing for V-V or V-T processes, the chemical reaction does not affect the vibrational distribution function f_n, which can be used to calculate K by formula (56) with known probabilities k_n and distribution function f_n. However, in many cases, the probabilities k_n are so large that the chemical reaction distorts the distribution f_n. These cases are most interesting since they provide us with maximum nonequilibrium values of rate constants K. A typical example of the reaction is the molecule dissociation at the upper vibrational levels.

Below we will analyze the model of dissociation from the only upper boundary vibrational level n_b. The dissociation involves the molecular level-to-level one quantum transitions to the boundary level, at which the transition to a continuous vibrational energy spectrum occurs, i.e., the molecules dissociate.

The rate constant (or the probability) K_d of such a process under nonequilibrium conditions at the boundary level n_b can be calculated, in a general case, through the determined populations N_n of all vibrational states $n \leq n_b$. However, the analysis can be simplified by consideration of quasi-steady regime. In such a regime, the vibrational distribution f_n is dependent only on the formed average number of nonequilibrium quanta α and gas temperature T. The level populations N_n will vary with a time as:

$$N_n(t) = N(t)f_n; \qquad \sum_{n=0}^{n_b} f_n = 1 \qquad (57)$$

This means that the terms $\frac{1}{N}\frac{dN}{dt}f_n \equiv K_d f_n$ will appear in the LHS of eq. (27). The system can be solved analytically assuming that the regimes of moderate and strong deviations from equilibrium are studied independently. In this case, our problem is now to find the vibrational distribution function f_n with a correction χ_n for the dissociation effect $f_n = f_n^0(1+\chi_n)$. At a small deviation from equilibrium the function f_n^0 is determined by expressions (38)-(40). Using expressions (38)-(40) we can obtain [21]:

$$\chi_n = K_d \psi_n \qquad (58)$$

$$\psi_n = \sum_{m=1}^{n} \frac{1}{f_m^0(\beta Q_{m,m-1}^{01} + P_{m,m-1})}, \qquad (59)$$

where β is determined by (40). The eqs. (58),(59) can be used for calculation of the relative population of the boundary level n_b and rate constant (or probability) of dissociation K_d which occurs through this level with the rate constant (or probability) k_{n_b}:

$$K_d(T, T_1) = \frac{k_{n_b} f_{n_b}^0}{1 + k_{n_b} f_{n_b}^0 \psi_{n_d}} \qquad (60)$$

Note, that nonequilibrium rate constant K_d depends from two temperatures: T and T_1 through dependence from T and T_1 the functions $f_{n_b}^0$ and ψ_{n_d}. The value $K_d(T,T_1)$ can be presented as

$$K_d(T,T_1) = K^0(T)\frac{1 - exp(-E_{10}/kT_1)}{1 - exp(-E_{10}/kT)}\Phi(T,T_1),\qquad(61)$$

where K^0 is the rate constant of dissociation for vibrational equilibrium, when $T_1 = T$; $\Phi(T,T_1)$ is the nonequilibrium factor. At a moderate deviation from equilibrium it is possible to obtain from eqs. (38),(39),(58)-(61):

$$\Phi(T,T_1) \simeq exp\left[n^{**}E_{10}\left(\frac{1}{kT} - \frac{1}{kT_1}\right)\right],\qquad(62)$$

where

$$n^{**} = n^* + \frac{1}{(\delta_{vv} + \delta_{vt})}E_{10}\left(\frac{1}{kT} - \frac{1}{kT_1}\right) + \frac{1}{2}\qquad(63)$$

and n^* is determined by eq.(50).

It is possible also to estimate roughly the parameter $\Phi(T,T_1)$ for regime of resonance V-V relaxation with strong deviation from equilibrium. In this case

$$\Phi(T,T_1) \simeq \frac{B(n_{10})}{n_1 + 1}Pexp\left\{\frac{E_{n_1}}{kT} + \delta_{vt}n_1\right\},\qquad(64)$$

where $B(n_1)$, P and n_1 are determined by eq.(44)-(47).

In conclusion, it is necessary to note that the analytical approaches for solution of master equations for populations of vibrational levels and dissociation are given in present work for the single quantum vibrational transitions and for the dissociation from single upper boundary vibrational level. Such approximation is good for many cases but can be unfit for very high gas temperature regimes (for example, behind the front of very strong shock waves) and for non adiabatic vibrational transitions or vibrational transitions due to chemical exchange reactions. Polyquantum vibrational transitions and the dissociation from low levels must be taken into account for such cases. The some analytical approaches in vibrational kinetics for such regimes can be found in literature (see,for example, [17,25]) and in present book, but a number of interesting problems in this area must be solved in future.

5. References

1. Gordiets B.F., Osipov A.I. and Shelepin L.A. (1988) "Kinetic Processes in Gases and Molecular Lasers",Gordon and Breach sci. pub., N.Y.

434

2. Gordiets B.F. and Zhdanok S. (1986) Analytical Theory of Vibrational Kinetics of Anharmonic Oscillators, - in book *"Nonequilibrium Vibrational Kinetics"*, ed. by M.Capitelli , Springer-Verlag, Berlin, N.Y., p.47.

3. Herzfeld K.F. end Litovitz T.A. (1959) *"Absorption and Dispersion of Ultrasonic Waves"*, Academic Press, New York.

4. Cottrell T.L. and McCoubrey J.C. (1961) *"Molecular Energy Transfer in Gases"*, Butterworths, London.

5. Clark J.F. and McChesni M. (1964) *"The Dynamics of Real Gases"*, Butterworths, London.

6. Stupochenko E.V., Losev S.A. and Osipov A.I. (1965) *"Relaxation processes in Shock Waves"*, Nauka, Moscow.

7. Nikitin E.E. (1970) *Theory of Elementary Atomic and Molecular Processes in Gases"*, Chimiya, Moscow.

8. Montroll E.W. and Shuler K.E. (1957) Studies in Nonequilibrium Rate Processes. I. The Relaxation of a System of Harmonic Oscillators, *J. Chem. Phys.*, **26**, 454.

9. Osipov A.I. (1960) *Dokl. Akad. Nauk SSSR*, **130**, 523, (*in russian*).

10. Shuler K.E. (1960) *J. Chem. Phys.*, **32**, 1692.

11. Rankin C.C. and Light J.C. (1967) Relaxation of a Gas of Harmonic Oscillators, *J. Chem. Phys.*, **46**, 1305.

12. Osipov A.I. (1964) *J. Applied Mechanics and Technical Physics*, N1, 41 (*in russian*).

13. Biryukov A.S. and Gordiets B.F. (1972) The Kinetic Equations of Vibrational Relaxation in the Mixture of Polyatomic Gases, *J. Applied Mechanics and Technical Physics*, N6, 29 (*in russian*).

14. Schwartz R.N., Slawsky I. and Herzfeld K.F. (1952) *J. Chem. Phys.*, **20**, 1591.

15. Naidis G.N. (1976) *J. Applied Mechanics and Technical Physics*, **2**, 38 (*in russian*).

16. Losev S.A., Shatalov O.P. and Yalovik M.S. (1970) *Dokl. Akad. Nauk SSSR*, **195**, 585 (*in russian*).

17. Safaryan M.N. and Skrebkov O.V. (1975) *Fizika Goreniya Vzryva*, 4, 614 (*in russian*).

18. Treanor C.E., Rich I.W. and Rehm R.G. (1968) Vibrational Relaxation of Anharmonic Oscillators with Exchange-Dominated Collisions, *J. Chem. Phys.*, **48**, 1798.

19. Kuznetsov N.M. (1971) *Tekh. Eksp. Khim.*, **7**, 22 (*in russian*).

20. Gordiets B.F., Osipov A.I. and Shelepin L.A. (1971) Nonresonance Vibrational Exchange and Molecular Lasers, *J. Exp. Theor. Phys.*, **60**, 102 (*in russian*).

21. Gordiets B.F., Osipov A.I. and Shelepin L.A. (1971) The Processes of Nonequilibrium Dissociation and Molecular Lasers, *J. Exp. Theor. Phys.*,

61, 562 (*in russian*).

22. Gordiets B.F. and Mamedov Sh.S. (1974) Function Distribution and Vibrational Energy Relaxation of Anharmonic Oscillators, *J. Applied Mechanics and Technical Physics*, **3**, 13 (*in russian*).

23. Gordiets B.F., Mamedov Sh.S. and Shelepin L.A. (1974) Vibrational Relaxation of Anharmonic Oscillators for Strong Nonequilibrium Conditions, *J. Exp. Theor. Phys.*, **67**, 1287 (*in russian*).

24. Zhdanok S.A., Napartovich A.P. and Starostin A.N. (1979) Establishment of Vibrational Distribution of Two-atomic Molecules, *J. Exp. Theor. Phys.*, **76**, 130 (*in russian*).

25. Park C. (1990) *"Nonequilibrium Hypersonic Aerodynamics"*, Wiley, New York.

NON-EQUILIBRIUM VIBRATIONAL, ELECTRONIC AND DISSOCIATION KINETICS IN MOLECULAR PLASMAS AND THEIR COUPLING WITH THE ELECTRON ENERGY DISTRIBUTION FUNCTION

C. GORSE and M. CAPITELLI
Centro di Studio per la Chimica dei Plasmi del CNR
Department of Chemistry, Bari University, Bari, Italy 70126

Abstract

Self-consistent models for homogeneous discharges are illustrated considering N_2 as a case study. In particular the coupling between vibrational-dissociation kinetics, electronic kinetics and Boltzmann equation is illustrated. Extension of these models to hypersonic conditions is finally discussed.

1. Introduction

Vibrational kinetics under electric discharges has received much attention in these last years due to its importance in different fields of plasma technology.

Sophisticated models have been developed to this end either from the point of view of kinetics or from the point of view of input data (cross sections). A state to state vibrational kinetics is usually considered in these models, this kinetics being linked to an appropriate Boltzmann equation for the electron energy distribution function (eedf). Additional coupling of this kinetics to the dissociation and ionization ones as well as to the kinetics of electronically excited states has been considered.

The aim of this paper is to discuss the efforts made in the last 15 years to build up self consistent discharge models. The extension of these models to hypersonic conditions will be also considered to understand the main differences present for the different situations.

We will consider results only for N_2 even though the present approach can be extended to other systems as well.

M. Capitelli (ed.), Molecular Physics and Hypersonic Flows, 437–449.

2. Nitrogen: a case study

We limit our analysis to nitrogen that can be considered one of the most studied systems in molecular plasmas.

A serious kinetics for this system should include
- 1) a state to state vibrational kinetics of the ground electronic state $X^1\Sigma_g^+$,
- 2) a dissociation kinetics,
- 3) a collisional radiative model for the electronic states,
- 4) a description of the electron energy distribution function (eedf) by an adequate Boltzmann equation,
- 5) an ionization kinetics,
- 6) an estimation of the electric field necessary to sustain the discharge.

All these points have been separately analysed in the past by using uncoupled models. More recently however it appeared clear that only a self consistent treatment of all kinetics can realistically describe the properties of the system. In this context Cacciatore et al. [1] developed a model that couples the Boltzmann equation with the system of vibrational master equation, the coupling occurring through electron impact inelastic processes involving the vibrational levels as well as through second kind vibrational collisions (SVC). The vibrational kinetics included the following reactions

$$e + N_2(v) \rightarrow e + N_2(w) \qquad \text{(e-V)}$$

$$N_2(v) + N_2(w) \rightarrow N_2(v+1) + N_2(w-1) \qquad \text{(V-V)}$$

$$N_2(v) + N_2 \rightarrow N_2(v-1) + N_2 \qquad \text{(V-T)}$$

Dissociation was also allowed either in the so called pure vibrational mechanism (PVM) or by direct electron impact excitation (DEM).

According to PVM the vibrational quanta that cross a pseudo-level (v'+1) located just above the last bound level of the molecule (v') can dissociate i.e.

$$N_2(v') + N_2 \rightarrow N_2(v'+1) + N_2 \rightarrow 2N + N_2 \qquad \text{(VT)}$$

$$N_2(v') + N_2(v) \rightarrow N_2(v'+1) + N_2(v-1) \rightarrow 2N + N_2(v-1) \qquad \text{(VV)}$$

However dissociation can occur also through direct dissociation impact with free electrons (DEM)

$$e + N_2(v) \rightarrow e + 2N$$

The results obtained by this model showed a strong coupling between the eedf and the vibrational distribution function, this coupling coming mainly from the process

$$e + N_2(w) \rightarrow e + N_2(v) \qquad w > v$$

which is able to return energy to the electrons partially compensating that one they lost in the corresponding inelastic process. Practically the role of SVC is such to enlarge the bulk of eedf increasing the rate coefficients of electron-molecule processes with high energy thresholds.

A sample of results obtained operating this model, which can be considered as a mile stone in the study of vibrational kinetics, has been reported in figures 1-3. In particular figures 1-2 report the self consistent evolution of vibrational distribution and of electron energy distribution functions (eedf) for a specific electric condition (E/N is the electric field, T the gas temperature, p the filling gas pressure and n_e the electron number density).

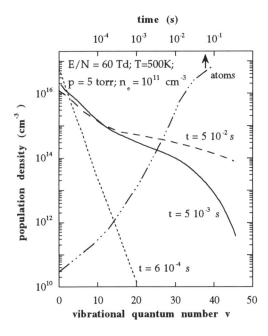

Fig. 1. Temporal vibrational populations and atom production

We start the evolution of both distributions in the so called "cold gas approximation", i.e. the integration of the system of kinetic equations begins

440

assuming that all the molecules are concentrated on the vibrational level v=0. Then e-V, V-V and V-T processes operate and shape the vibrational distributions according to the forms reported in figure 1. At the same time second kind (superelastic) vibrational collisions return energy to electrons so that the bulk of eedf strongly increases. Coming back to the vibrational distribution we can note that its first part satisfies a Treanor's form while the long plateau is induced by the quasi resonant V-V energy exchange processes unbalanced by V-T ones. It is worth noting that in these calculations V-T rates from atomic nitrogen were considered equal to the corresponding V-T ones involving molecules.

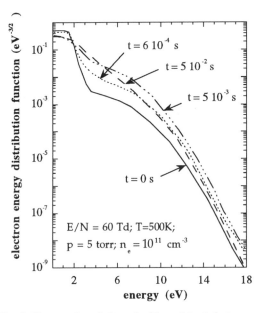

Fig. 2. Temporal evolution of self-consistent electron energy distribution function

Figure 3 reports the dissociation rates calculated by the electron impact dissociation process i) from v=0 ii) from all v and by the pure vibrational mechanism. We can see that the pure vibrational mechanism is dominant at long residence times. Note also the strong role of excited vibrational levels in enhancing the dissociation rate of the electron impact mechanism, this point being related to the strong dependence of reaction rates on the vibrational quantum number.

A similar situation was found for the ionization mechanism. In this case the electron impact mechanism, i.e. the process

$$e + N_2(v) \rightarrow e + N_2^+ + e$$

is not able to explain the ionization degree of a nitrogen plasma at low reduced electric field E/N (E/N < 60 Td ; 1 Td=10^{-17} Vcm^2) so that associative ionization processes involving vibrationally excited molecules and metastable electronic states were invoked to explain the ionization process.

It should be noted that the approach previously reported was able to basically reproduce the experimental CARS vibrational distributions of N$_2$ in flowing DC discharges [2] as well as the vibrational distributions of electronic states of N$_2$ [3-4]. Moreover a similar approach was also used to explain the highly non-equilibrium vibrational distributions of CO pumped by RF flowing He-CO discharges [5] as well as by nitrogen post discharges [6].

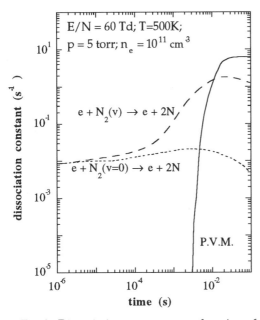

Fig. 3. Dissociation constants as a function of time

The model developed in ref. 1 contained many microscopic details able to reproduce several experimental observations. However Paniccia et al. [7] showed that not only second kind vibrational collisions but also second kind electronic collisions (SEC) can affect eedf and related quantities. The model developed in ref. 1 was therefore enriched by a collisional-radiative model involving the most important electronic states of N$_2$. In particular we developed a kinetics for A, B and C states of N$_2$ coupled to the vibrational kinetics as well as to the Boltzmann equation [8]. This new kinetics was

442

coupled to the other kinetics not only through second kind electronic collisions but also through the possibility of the inter-exchange of vibrational and electronic energy, a problem that is receiving increasing attention. A sample of eedf calculated with this model either in discharge or post discharge conditions is reported in figs. 4-5.

In particular figure 4 shows the temporal evolution of eedf calculated by coupling the Boltzmann equation to the vibrational master equations as well as to a kinetics of the most important electronic states of N_2. From this figure we can appreciate the effect of second kind vibrational collisions in enlarging eedf for energies less than 8 eV (essentially the same effect already encountered in figure 2). Additionaly second kind electronic collisions, i.e. processes

$$e + N_2^* \rightarrow e + N_2$$

enlarge the eedf for energies higher than 8 eV.

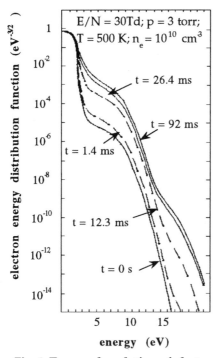

Fig. 4. Temporal evolution of electron energy distribution function

The effect of second kind electronic collisions in affecting eedf is well evident also in the post discharge conditions, i.e. when the reduced electric field E/N tends to zero. A strong peak at energy of approximately 6 eV due to the second kind collisions between cold electrons and $N_2(A^3\Sigma_u^+)$ metastable molecules is well evidenced on fig. 5. The results have been obtained switching off the electric field after a residence time of 0.13 ms in the discharge regime.

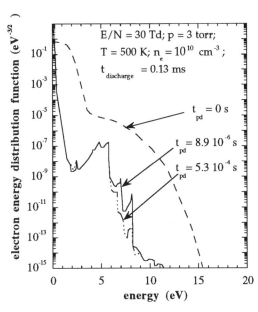

Fig. 5. Temporal evolution of the electron energy distribution function in the post-discharge regime

Apparently one could say that we know all about the elementary processes occurring in N_2 discharges so that we can now insert this kind of kinetics in more accurate models giving also the electric characteristics of the discharge. For instance one could add an ionization kinetics to derive the electron density n_e able to sustain a discharge at fixed values of E/N. This approach was followed by Gorse [9]. The results however, see figure 6, indicate a non stationary behaviour of eedf even at E/N = 80 Td, p=2 torr and a discharge radius tube R = 1 cm in contrast to the experimental results. Note however that the temporal evolution of eedf again contains all effects due to superelastic vibrational and electronic collisions previously discussed even though at longer times eedf converges toward the cold gas approximation since the abrupt decrease of electron concentration is not any more able to sustain considerable concentrations of vibrationally and electronically excited states.

This behaviour is probably due to the use of crude rate coefficients for some of the numerous processes inserted in the whole kinetics. This means that improvement in the modeling of N_2 discharges can be achieved by a corresponding improvement of the relevant cross sections. An example in this direction is to be found in the use of more reliable rate coefficients for vibrational deactivation of $N_2(v)$ by atomic nitrogen, i.e. the process

$$N_2(v) + N \rightarrow N_2(w) + N$$

Use of the recent trajectory study by Laganà et al [10] for the corresponding rates strongly cuts down the plateau in the vibrational distribution of N_2 under electric discharges [11], decreasing at the same time the importance of vibrational mechanisms in the dissociation and ionization processes. These points can be understood by looking at figures 7-8 that report the temporal evolutions of vibrational distributions and of dissociation rates resulting from insertion in the model of the rates calculated in reference 10.

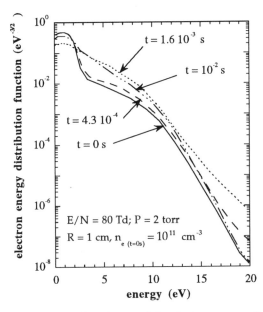

Fig. 6. Temporal evolution of the electron energy distribution function

Comparison between those last vibrational distributions and the ones reported on figure 1 evidences the role of V-T rates involving nitrogen atoms in deactivating part of the plateau of the vibrational distribution. As a consequence a tail in the vibrational distribution appears which will have strong consequences on the pure vibrational dissociation mechanism. This is indeed the case, as can be seen from figure 8, where we can appreciate the decline of pure vibrational mechanism as compared to electron impact dissociation ones. Note, however, that in any case the vibrational excitation of

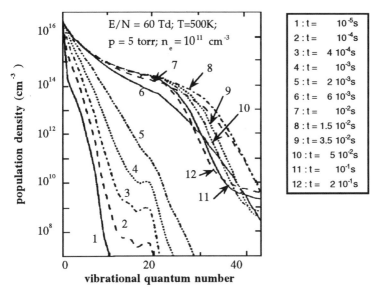

Fig. 7. Temporal evolution of vibrational populations

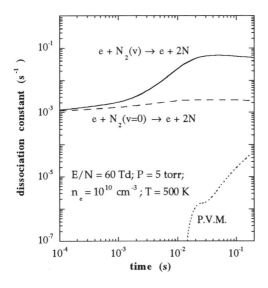

Fig. 8. Temporal evolution of the dissociation constants

the molecules enhances up to two orders of magnitude the DEM rate as compared with the v=0 contribution.

It should also be noted that improvement of the dissociation model used in our calculations could also raise the PVM rates in electric discharges. Our present approach is based on the so called "ladder climbing model", i.e. a model that considers dissociation as occurring only from the last bound level (v') of the molecule. Inclusion of heavy particle dissociation processes from all vibrational levels (specially from those belonging to the plateau of the distribution) should strongly increase the PVM rates.

Future studies of nitrogen system should be devoted to the ultimate fate of energy released during the recombination process of atomic nitrogen either in gas phase or on the surface.

In gas phase several mechanisms have been proposed for the recombination process, most of them describing the recombination process as occurring on electronically excited states. As a final state of this degradation one could obtain the metastable $A^3\Sigma_u^+$ state which could transfer its energy to the vibrationally excited states of $X^1\Sigma_g^+$ state through the following processes

$$N_2\,(A^3\Sigma_u^+) + N_2 \;\rightarrow\; N_2 + N_2(v{=}25)$$

$$N_2\,(A^3\Sigma_u^+) + N_2 \;\rightarrow\; N_2(v{=}12) + N_2(v{=}12)$$

In both cases the recombination energy should act as a pump for vibrational energy with strong consequences on the vibrational kinetics. This point has been recently analysed by Armenise et al. [12] in a completely different context. These authors studied the non equilibrium vibrational distributions of N_2 in the boundary layer of a body invested by a hypersonic flow of vibrationally excited molecules and atoms. Under these conditions highly non-equilibrium vibrational distributions develop as a result of selective pumping of vibrational levels during the recombination followed by a redistribution of the introduced quanta by V-T and V-V energy exchange processes.

At the same time Colonna and Capitelli [13] coupled the fluid dynamic and kinetic equations in the boundary layer approximation to a Boltzmann equation for the electron energy distribution function. The vibrational kinetics was limited to e-V, V-V, V-T and dissociation-recombination process and linked to the eedf through second kind vibrational collisions. The ionization degree was considered constant along the coordinate (η) perpendicular to the body while no electric fields were supposed to exist in the boundary layer.

Figures 9 and 10 report a sample of results which respectively show the electron energy distribution functions and the relative vibrational populations along the coordinate η normal to the body for the following conditions $T_{body} = T(\eta{=}0) = 1000$ K and $T_{ext.} = T(\eta{=}5) = 7000$ K, $p = 10^3$ Pa, ionization degree

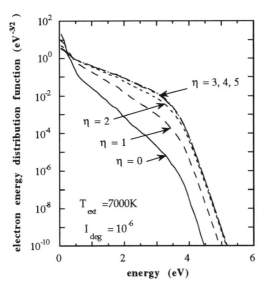

Fig. 9. Electron energy distribution functions along
the coordinate η perpendicular to the re-entering body

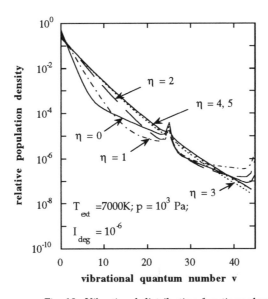

Fig. 10. Vibrational distribution functions along
the coordinate η perpendicular to the re-entering body

$I_{deg} = 10^{-6}$, β = 10 s^{-1} (see ref. 13 for a more complete definition of the parameters). We note that both distributions present a strong non-equilibrium character specially near the surface. Equilibrium distributions should be characterized by straight lines on both diagrams of figures 9 and 10.

Note also that the plateaux appearing in the vibrational distributions are the result of the recombination process rather than of V-V up pumping mechanism as in the previous cases.

3. Conclusions

In the present paper we have presented the efforts made in the last years to build up self consistent models for the different kinetics existing under non-equilibrium conditions.

To improve the models one should first improve the numerous input data. As an example an improvement of the potential energy surface for N-N$_2$ system urges to completely assess the dynamical calculations of Laganà et al. At the same time a better set of state to state electron impact cross sections for air components becomes more and more important specially when these models are to be extended to hypersonic conditions. In this case the electron average energy is less than 1 eV so that the behaviour of cross sections near the threshold energy contributes significantly to the relevant rate coefficients and this behaviour is quite difficult to calculate. Remind that the existing input data [1] for N$_2$ have been obtained by using the crude Gryzinskii approximation for electron impact (non resonant) cross sections involving the vibrational levels of N$_2$ while better quantum mechanical approaches have been used for calculating e-V cross sections as well as cross sections linking electronically excited states [14]. The situation is worse for the interaction of vibrationally excited molecules on surfaces as well as for the heterogeneous atomic recombination even though much work is being made in this direction [15].

Concerning the kinetics we can say that its insertion in complicated fluid dynamic codes is still at the infancy due to the resistance of fluid dynamic community to complicate their already complex equations. Preliminary efforts in this direction have been made by our group in the boundary equation approach and by Bellucci et al. [16] by using Euler equations. On the other hand Colonna and Capitelli extended the coupling of vibrational kinetics and Boltzmann equation for situations typical of re-entry problems. In this last case existence of an electric field due to space charge separation was completely ignored. This problem could be examined at the light of PIC(particle in cell)-Monte Carlo method presented by Longo [17].

These preliminary approaches should push the fluid dynamic community to develop dimensional codes including more details in the relevant kinetics.

Acknowledgments

This work has been partially supported by ASI (Agenzia Spaziale Italiana) and by MURST.

References

1. Cacciatore, M., Capitelli, M. and Gorse, C. (1982) Non equilibrium dissociation and ionization of nitrogen in electrical discharges: the role of electronic collisions from vibrationnally excited molecules, *Chem. Phys.* **66**, 141-151
2. Massabieaux, B., Gousset, G., Lefebvre, M. and Pealat, M. (1987) Determination of nitrogen $N_2(X)$ vibrational level populations and rotational temperatures using CARS in a d.c. low pressure discharge, *J. Phys. B* **48**, 1939-1949
3. Massabieaux, B., Plain, A., Ricard, A., Capitelli, M. and Gorse, C. (1983) Excitation of vibrational and electronic after-glow, *J. Phys. B: At. Mol. Phys.* **16**, 1863-1874
4. Plain, A., Gorse, C., Cacciatore, M., Capitelli, M., Massabieaux, B. and Ricard, A. (1985) On the coupling between $N_2(B)$ and $N_2(X)$ vibrational distributions in a glow discharge column, *J. Phys. B: At. Mol. Phys.* **18**, 843-849
5. De Benedictis, S., Capitelli, M., Cramarossa, F., d'Agostino, R., Gorse, C. and Brechignac, P. (1983) Vibrational kinetics in liquid nitrogen cooled 5% CO - He radio frequency discharges, *Optics Comm.* **47**, 107-110
6. De Benedictis, S., Capitelli, M., Cramarossa, F., Gorse, C. (1987) Non-equilibrium vibrational kinetics of CO pumped by vibrationnally excited nitrogen molecules: a comparison between theory and experiment, *Chem. Phys.* **111**, 361-70
7 Paniccia, F., Gorse, C., Bretagne, J. and Capitelli, M. (1986) Electron energy distribution functions in molecular nitrogen: the role of superelastic electronic collisions in discharge and post-discharge conditions, *J. Appl. Phys.* **59**, 4004-4006
8. Gorse, C., Cacciatore, M., Capitelli, M., De Benedictis S. and Dilecce, G. (1988) Electron energy distribution functions under N_2 discharges and post-discharge conditions: a self consistent approach, *Chem. Phys.* **119**, 63-70
9. Gorse, C. (1993) Non equilibrium plasma modeling, *Proceedings III - XXI International Conference on Phenomena in Ionized Gases*, Ecker, G., Arendt, U. and Boseler, J. eds., 141-148, Bochum
10. Laganà A., Garcia, E. and Ciccarelli, L. (1987) Deactivation of vibrationally excited nitrogen molecules by collision with nitrogen atoms, *J. Phys. Chem.* **91**, 312-314
11. Armenise, I., Capitelli, M., Garcia, E., Gorse, C., Laganà, A. and Longo, S. (1992) Deactivation dynamics of vibrationally excited molecules by nitrogen atoms. Effects on non-equilibrium vibrational distribution and dissociation rates of nitrogen under electrical discharges, *Chem. Phys. Lett.* **200**, 597-604
12. Armenise, I., Capitelli, M., Celiberto, R., Colonna, G., Gorse, C. and Laganà, A. (1994) The effect of $N + N_2$ collisions on the non-equilibrium vibrational distributions of nitrogen under re-entry conditions, *Chem. Phys. Lett.* **227**, 157-163
13 Colonna, G. and Capitelli, M. (1995) On the coupling of electron and vibrational kinetics in the boundary layer of hypersonic flow, *30th AIAA Thermophysics Conference* **AIAA 95-2071**, San Diego; (1996), *J. Thermophys. Heat Transfer* in press
14 Huo, W (1990) Electron collision cross sections involving excited states, *Non equilibrium Processes in Partially Ionized Gases*, Capitelli, M. and Bardsley J.N. eds, Plenum, New York
15 Billing, G.D. (1996) Semi-classical treatment of the dynamics of molecular surface interaction in Molecular Physics and Hypersonic flows, *Molecular Physics and Hypersonic flows*, Capitelli, M. ed., Kluver Academic Publishers, Dordrecht
16 Bellucci, V., Giordano, D., Colonna, G., Capitelli, M., Armenise, I. and Bruno, C. (1996) Vibrational kinetics for numerical simulation of thermal non equilibrium flows, *Molecular Physics and Hypersonic flows*, Capitelli, M. ed., Kluver Academic Publishers, Dordrecht
17 Longo, S. (1996) Self consistent description of the kinetics of free electrons, *Molecular Physics and Hypersonic flows*, Capitelli, M. ed., Kluver Academic Publishers, Dordrecht

SELF CONSISTENT MODELING OF FREE ELECTRON KINETICS

S.Longo

Dipartimento di Chimica dell' Università di
Bari and Centro di Studio per la Chimica dei
Plasmi del CNR, Via Orabona 4, 70126 Bari,
Italy

ABSTRACT

We show the fundamentals of non-equilibrium modeling of the electron
kinetics in rarefied molecular gases, under the action of both the self-
consistent electric field and the vibrational kinetics of molecules. In these
conditions the electron energy distribution function is far from the local
equilibrium one, and furthermore it cannot be calculated separately from the
vibrational manifold. As an example it is illustrated a self-consistent one
dimensional model for an electric discharge in nitrogen, realized by
coupling a Particle-in-Cell electron/ion dynamics including Monte Carlo
scattering on molecules and a simplified vibrational kinetics of nitrogen.

1. INTRODUCTION

The supersonic flow associated with reentry of a space vehicle in the Earth
atmosphere is characterized by a significant ionization of the gas, which is
associated with high vibrational excitation of molecules due to electron
impact and space charge effects: all these effects are generally recognized by
the supersonic fluid dynamics community, but the situation is different for
the non-equilibrium properties of the plasma in the flow.
What follows is a brief account on the fundamentals of non-equilibrium
electron kinetics in molecular gases [1,2] that the Plasma Chemistry group
operating in the University of Bari and in the *Centro di Studio per la Chimica
dei Plasmi* of the CNR has contributed to establish.
A non-equilibrium plasma is characterized by a strong difference between
the mean energies of its electron and neutral components. The possibility to
sustain a non-equilibrium plasma is due to the very large mass difference

451

M. Capitelli (ed.), Molecular Physics and Hypersonic Flows, 451–460.

between electron and molecules, which considerably slows down the translational energy transfer.

As soon as an even very dilute electron component appears in the gas, it determines almost exclusively the physical and chemical properties of the gas. In fact the electrons move very fast with respect to the neutral particles, and they can interact at long distances through the space-charge field. Furthermore, the cross sections for vibrational and electronic excitation of molecules by electron impact are very large. In many systems a small fraction of high temperature (10^4-10^5 K) electrons determines the chemical properties of the bulk gas at a by far lower temperature (less then 10^3 K): this non equilibrium, weakly ionized plasma is often called 'cold plasma'.

The cold plasma is a very complex system to study both experimentally and theoretically, its complexity being due to the numerous processes that occur in it and specially because of its far-from-equilibrium nature. In particular:

•   1) the distribution of population of electronic and vibrational states of molecules and electronic states of atoms is not described by a Boltzmann law,

•   2) the translational energy distribution of particles is not described by a Maxwell-Boltzmann law: this is specially true for charged particles, which are subjected to the effect of space-charge field (and, in the case of electric discharge plasmas, to the external electric field)

The statement (1) claims for a state-to-state description of the chemical kinetics in cold plasmas, but the stronger consequences for the description of the system follows from (2).

Let us in fact consider a generic two-body elementary process involving an electron e and another species A, of the kind $e + A \rightarrow P + \dots.$ The contribution to the rate of formation of the product species P, with dimensions $[l^{-3}t^{-1}]$ is given by

$$\frac{\partial n_P}{\partial t} = \dots + n_e n_A \sqrt{\frac{2}{m_e}} \int_{\varepsilon^*}^{\infty} f(\varepsilon)\sigma(\varepsilon)\varepsilon \, d\varepsilon$$

where n_A is the A-particle concentration, with dimensions $[l^{-3}]$. The production rate is a functional of the electron energy distribution function, or eedf, $f(\varepsilon)$. This function is defined in such a way that the fraction of electrons with kinetic energy between ε and $\varepsilon + d\varepsilon$ is given by $\varepsilon^{1/2} f(\varepsilon) d\varepsilon$.

It follows that the reaction rate does not simpy depend on the (electron) temperature through the Arrhenius law, as in the near-equilibrium case, because in the far-from-equilbrium regime we must take into account the shape of the eedf and the cross sections for several process as a function of the electron kinetic energy. The mathematical difficulty of studying the cold plasmas because of the far-from-equilibrium character of the electron component is now clearly understood: the reaction rates for electron-molecule processes are *functionals* of the eedf, and not simply *functions* of macroscopic parameters, as in the case of moderate non equilibrium.

The first step towards a rigorous description of the cold plasma is therefore the calculation of the eedf. This problem has been studied according to a *Boltzmann gas-like* description of the electron gas, which consists of the following two hypothesis:

- 1) The electron-neutral collision is an instantaneous process completely described in term of electron-molecule differential cross sections
- 2) The electron-neutral collision process is characterized by a succession of uncorrelated free-flight times t_1, t_2, etc. This means that the neutral particles are not explicitly described here, but introduced as a continuous field of implicit scattering particles.

These hypothesis can be introduced both in a transport kinetic equation and in particle dynamic model: accordingly, two approaches to the electron kinetics are obtained, which will be presented in the next two sections.

2. THE BOLTZMANN EQUATION FOR ELECTRON TRANSPORT

The first method we will examine to calculate the eedf consists in solving the Boltzmann equation for electron transport [3,4] under the effect of electric field and different kinds of electron collisions with neutral species and charged particles. Let us consider as starting point the Boltzmann equation for the configuration-space distribution of particles, that is the $f(r,v)$, where r and v are both 3-dimensional vectors in the electron position and velocity space respectively. The equation can be written:

$$\left(\frac{\partial}{\partial t} + \mathbf{v} \frac{\partial}{\partial \mathbf{r}} + \frac{\partial \mathbf{v}}{\partial t} \frac{\partial}{\partial \mathbf{v}} \right) f(\mathbf{r}, \mathbf{v}, t) = \left(\frac{\partial}{\partial t} \right)_{coll} f(\mathbf{r}, \mathbf{v}, t)$$

where the left side is total time derivative of $f(r,v)$, while on the right side we find a so-called 'collision' derivative representing the effect on the electron dynamics of the different kinds of collisions. In order to simplify the following considerations, we can use the 'two term approximation' which consists in approximating the $f(r,v)$ as:

$$f(\mathbf{r}, \mathbf{v}) = f_0(\mathbf{r}, v) + f_1(\mathbf{r}, v) \cos \theta$$

Where θ is the angle between **E** (the electric field) and **v**. Note that v on the right side is the modulus of the vector **v**.

After simplifying the eedf in this way, it is usually possible to perform an adiabatic elimination of the function f_1 with respect to the instantaneous f_0. It can be easily shown that in two term approximation the function $f_0((2\varepsilon/m_e)^{1/2})$, properly normalized, *is* the eedf: we are therefore lead to an equation for the eedf. Technical details can be found in ref. [3].

The resulting equation is a Fokker-Planck (or drift-diffusion) equation in the electron energy (ε) space and position. However, the equation assumes a more general form when including the inelastic collision terms [1] $(\partial f(\varepsilon)/\partial t)_{in}$, and the solution can be given as a rule only numerically. At this stage the transformed Boltzmann equation is nevertheless linear, and the numerical solution at least for the uniform case is straightforward.

The calculation procedure becomes more complex when one is interested in observing the effects on the eedf of the electron-electron interactions, because the equation becomes nonlinear. In the case of a uniform Boltzmann equation in e-space the Fokker-Planck term is explicitly introduced in the form given by Rosenbluth et al. [3,4]. In any case the transformed Boltzmann equation is formally a Master equation in ε-space.

3. MONTE CARLO METHOD

An alternative way of describing the electron dynamics in the framework of the Boltzmann electron gas description is using the Monte Carlo method [5]. The Monte Carlo method is a microscopic approach to solve the electron kinetics based on the numerical simulation of the motion of a large electron ensemble where the collision processes are introduced by generating appropriately distributed random numbers. At given times during the simulation the ensemble dynamic quantities $\{r,v\}_n$ are reported, and macroscopic quantities can later be calculated from this set of numbers. It is necessary to use many particles in the simulation, because the relative error on macroscopic quantites due to statistical fluctuations decreases slowly with the number n of elements in the set $\{r,v\}_n$, tipically as $n^{-1/2}$.

The time to the next electron-molecule collision must be calculated knowing the collision frequency as a function of the electron velocity: this problem is solved introducing a *null species* such that in an electron-null species collision (*null* process) the electron velocity does not change. The cross section for the electron-null species collision is chosen to obtain a constant total collisions frequency v_{tot}. The time to the next collision is then obtained in the form

$$t_c = -\nu_{tot}^{-1} \ln r$$

where r is an element of a set of random numbers uniformly distributed between the values 0 and 1. Such a set of random numbers can be generated by using a standard random-number-generator subroutine. The kind of collision process (including the *null* one) which occurs after the free flight is selected according to the related collision frequencies: let

$$\nu_k(\varepsilon) = \sqrt{2\varepsilon/m_e}\, \sigma_k(\varepsilon) n_A$$

be the contribution of the k-th electron collision process (involving the generic species A as collision partner) to the total collision frequency, and ν_{tot} the total collision frequency. After generating a random number r as before, the collision process selected will be the n-th, with n such that

$$\sum_{k=1}^{n-1} \frac{\nu_k}{\nu_{tot}} < r < \sum_{k=1}^{n} \frac{\nu_k}{\nu_{tot}}$$

After any collision, the new electron energy and the scattering angle is easily determined depending on the nature of the collision itself.

In the high pressure regime, that is when electron-neutral collisions are effective in driving the velocity distribution function towards almost spherical symmetry, the treatment of scattering can be simplified by different methods, as discussed in ref. [6]

4. PIC AND PIC/MCC METHODS

Most non-equilibrium plasmas of interest are characterized by a large number of electrons in the Debye sphere: in this conditions the interaction between charged particles can be approximated by a particle-space charge field interaction. This point of view is assumed in the so-called Particle-in-Cell, or PIC approach [7-9]. In the PIC approach the Newton equation for a large ensemble (10^4 - 10^5 particles) of electrons and positive ions are solved taking into account the local electric field as it results from linear interpolation within a cell of a mathematical mesh (from what follows the name 'Particle in Cell'). After any calculation step of the motion equations, the electric charge in any cell of the mesh is detemined from the number of electrons and positive ions found in the cell itself, according to their statistical weight w. Known the electric charge density, the electric potential and field are determined by solving the Poisson Equation on the same mesh. The PIC method is therefore fully equivalent to solve the Vlasov-Landau plasma problem assuming a *N-particle ensemble* solution for any particle species:

$$f(\mathbf{r}, \mathbf{v}, t) = \sum_{i=1}^{N} S(\mathbf{r} - \mathbf{r}_i) \delta(\mathbf{v} - \mathbf{v}_i)$$

Where r_i and v_i are the single-particle position and velocity, satisfying the Newton equations, δ is the Dirac kernel and S is a particle shape factor (a 'smoothed' δ). The Newton equations in PIC are usually solved by using a second order method with interlaced grids for velocity and position, the so-called *Leapfrog*, which is very simple and excellent in reproducing the electron plasma oscillations. The timestep of the PIC is determined by the necessity of correctly describing the electron plasma oscillations in order to avoid numerical instabilities. This last problem can be solved by using *implicit solvers*, which however have some difficulties in bounded systems.

As regards the solution of the Poisson equation many techniques are available, which are reviewed in ref.[8]

The simulation method becomes more powerful but also complex including the electron and ion collisions with neutral particles using the Monte Carlo Method described above. The resulting simulation technique is called PIC/MCC [7,8], where MCC stands for Monte Carlo Collisions. PIC/MCC simulation of plasmas produced by high-frequency electric discharge has given results in remarkable agreement with experiments for what regards the electron dynamic quantities, in particular the eedf.

Despite the impressive power and success of PIC/MCC approach, in order to apply it to the free electrons in hypersonix flows it is necessary to include the interaction between the gas chemistry and the electron dynamics in the non-equilibrium case, that is, with a state-to-state chemical kinetics and all the relevant electron-molecule collisions processes.

5. COUPLING PIC/MCC AND VIBRATIONAL KINETICS

The next step towards a self-consistent modeling of the free electron kinetics is the inclusion of vibrational or electronic excited molecules: these molecules are created by several elementary processes, many of which involving electron impact. On the other hand, the presence of excited species in the plasma can also have an effect on the electron motion. The problem is to find the concentration n_v of a given molecule in a well-defined vibrational state v.. The evolution of $\{n_v\}$ can be found by solving the master equation

$$\frac{\partial n_i}{\partial t} = \sum_j (p_{ji}n_j - p_{ij}n_i)$$

where p_{ij} are transition probabilities per unit time between vibrational levels i and j. The contribution of electron-molecule collision processes to the transition frequency p_{ij} is calculated as [1]

$$p_{ij}(x) = n_e \sqrt{\frac{2}{m_e}} \int_{\varepsilon_{ij}^*}^{\infty} f(\varepsilon, x)\sigma_{ij}(\varepsilon)\varepsilon \, d\varepsilon$$

where $\sigma_{ij}(\varepsilon)$ is the related cross section. It is useful to transform the master equation by introducing a transition matrix W given by

$$w_{ij} = p_{ij} - \delta_{ij} \sum_k p_{ik}$$

where δ_{ij} is the Kronecker symbol, in such a way that the master equation is written in the matrix form

$$\frac{\partial}{\partial t}\mathbf{n} = \mathbf{W}\mathbf{n}$$

The master equation is a linear equation for $\mathbf{n}$. The linear form holds when the VV processes, i.e. collisional vibrational energy exchange between molecules, are neglected, otherwise the matrix $\mathbf{W}$ will be a function of $\mathbf{n}$. The resulting non linear and 'stiff' system of differential rate equations (*rate* is a more appropriate name than *master* equation in this case) can be solved only using numerical methods. The situation is exactly analogous for the solution of the two-term Boltzmann equation (see above) either neglecting or taking into account e-e collisions.

The aim to which these efforts are devoted is the realization of a completely self-consistent model of a cold plasma. The model should include charged particle dynamics, collisions with neutrals, space charge effects, physical boundaries, fluidodynamics and state-to-state chemistry with all their mutual interactions.

As regards the solution of the master equation, one can take advantage of the very different relaxation times of chemical kinetics and electron dynamics by using a perturbative self-consistent approach. The procedure is the following: after simulating the electron dynamics by PIC/MCC for a sufficiently long time to get the space-dependent eedf $f(\varepsilon,r)$ with a low fluctuation level in the energy range of interest, the master equation is solved in any point of space (or globally solved if diffusion of species is taken into account). Afterwards, the PIC/MCC simulation is performed again taking into account the new gas composition, and so on. The whole cycle is repeated up to full convergence, that is when further changes are only due to Monte Carlo statistical fluctuations. Obviously this procedure gives physically meaningful results only for the steady state.

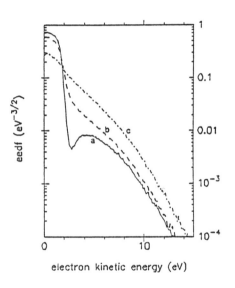

Figure 1. Eedf $f(\varepsilon)$ calculated for a radio frequency discharge in nitrogen, for different vibrational deactivation frequencies f_d: (a) 10^3 s^{-1}, (b) 10 s^{-1}, (c) 1 s^{-1}

Calculations of this kind has been realized by our group up to now for a 1D PIC/MCC model of a plasma produced by a high-frequency electric discharge in pure nitrogen, including a very simplified vibrational kinetics and neglecting diffusion of molecular species. This ideal discharge is made up between two plane particle-absorbing boundaries separated by a distance $d = 4$ cm. The space enclosed between the plates is filled with 0.1 torr N_2 gas at T = 300 K. Between the two plates is applied an oscillating voltage difference $V(t) = 200\ Volt\ cos\ (2\pi 13.56\ 10^6\ t)$.. The vibrational kinetics considered includes the processes:

$$N_2(0) + e <\text{---}> N_2 (v) + e$$

$$N_2(v) \text{---}> N_2(0)$$

with $v = 1,2,...,8$. The process in the second row is a phenomenological deactivation characterized by a frequency f_d, introduced to account for wall impact deactivation and Vibration-Translation energy transfer. The calculations have been performed assuming a time-step of 10^{-10} s and working with approximately 10^4 particles (electrons and ions). PIC/MCC steady state was attained after 10^{-3} s, the rate limiting-processes being the estabilishment of steady ambipolar diffusion to the plates.

In fig.1 are shown the time-averaged eedf at steady state in the centre of the systems, calculated assuming different values for f_d.

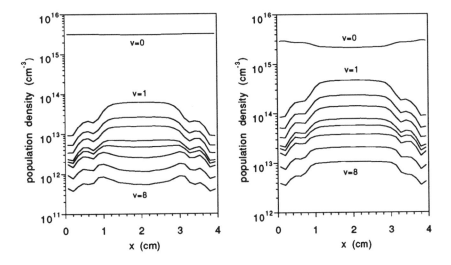

Figure 2. Vibrational distributions $\{n_i\}(x)$ calculated for a radio frequency discharge in nitrogen, for different deactivation frequencies f_d: (left) 10^2 s^{-1}, (right)10 s^{-1}

This figure should be compared with fig.2, which reports the steady-state concentrations of N_2 molecules in different vibrational states as a function of the position. Comparison with experiments suggests a situation intermediate between figs. 3a and 3b, nearer to fig.3b. The reader can notice the effect of the vibrational kinetics on the electron kinetics when the mean vibrational energy increses, due to the reduction of the quenching frequency. In all the cases shown, the eedf is far from a Maxwell distribution (a Maxwell distribution in the plot of fig.1 would be represented

460

by a straight line). This last observation confirms our idea that these plasmas are very far from thermodinamic equilibrium, and cannot be studied using macroscopic methods.

6. CONCLUSIONS

In this work we have analized some methods for mathematical modeling of non-equilibrium plasmas which goes beyond the assumption of local equilibrium and equilibrium distribution of population of molecular levels and electron kinetic energy. These methods have been applied to numerous real systems, and some of them demontrated to be really able to characterize the non-equilibrium kinetics of free electrons. This last can be properly described only taking into account the effect of the excited molecules in the gas mixture, as well as the space-charge interaction between free electrons and between electrons and positive ions. In particular, a PIC/MCC (or transport-equation) kinetics of electrons coupled to a Direct simulation Monte Carlo description of the state-to-state molecular kinetics can be indicated as a way to attain, in future, a satisfactory description of the non-equilibrium supersonic flow for reentry problems.

References

1. M.Capitelli and J.N.Bardsley (Editors): *Nonequilibrium Processes in Partially Ionized Gases*, NATO ASI Series, Plenum Press (1990)

2. M.J.Kushner and M.B.Graves (Editors): *Modeling Collisional Low Temperature Plasmas*, special issue of Trans.Plasma Science **19** (1991)

3. I.P.Shkarovsky, T.W.Johnson, M.P.Bachynski: *The Particle Kinetics of Plasmas*, Addison-Wesley (1966)

4. S.D.Rockwood, *Elastic and Inelastic Cross Sections for Electron-Hg Scattering from Hg Transport Data*, Phys.Rev.A **8**, 2348 (1973)

5. J.P.Boeuf: *Modelization de la Cinétique Electronique dans un Gas Faiblement Ionisé*, These d'Etat, Universitè de Paris Sud, Centre d'Orsay (1985)

6. S.Longo, M.Capitelli, *A Simple Approach to Treat Anisotropic Elastic Collisions in Monte Carlo Calculations of the Electron Energy Distribution Function in Cold Plasmas*, Plasma Chem. Plasma Proc. **14**, 1 (1994)

7. C.K.Birdsall and A.B.Langdon, *Plasma Physics via Computer Simulation*, New York: Mc. Graw- Hill (1985)

8. R.W.Hockney and J.W.Eastwood, *Computer Simulation Using Particles*, Adam Hilger, Bristol (1991)

9. S.Longo, D.Iasillo and M.Capitelli, work in preparation

ELECTRON-IMPACT IONIZATION OF AIR MOLECULES AND ITS APPLICATION TO THE ABATEMENT OF VOLATILE ORGANIC COMPOUNDS

B. M. Penetrante, M. C. Hsiao, J. N. Bardsley,
B. T. Merritt, G. E. Vogtlin and P. H. Wallman
Lawrence Livermore National Laboratory, Livermore, California 94550

A. Kuthi, C. P. Burkhart and J. R. Bayless
First Point Scientific, Inc., Agoura Hills, California 91301

INTRODUCTION

Volatile organic compounds (VOCs) are emitted from manufacturing the multitude of consumer products used every day. In most manufacturing processes, either for the raw materials, intermediates, or the finished product, VOC-containing materials are present as chemicals, solvents, release agents, coatings, and decomposition products that eventually must be disposed. In manufacturing, there is usually a gaseous effluent that contains low concentrations of organics and is vented into the atmosphere. Chlorinated VOCs are some of the most common solvents used, and are now found in hazardous concentrations at many industrial and government installations. Cost effective technologies for disposal of VOCs are therefore being sought by government and by industry, and there is a need for reliable data concerning the decomposition mechanisms associated with these compounds. Non-thermal plasma methods using electrical discharges and electron beams are some of the emerging technologies for the disposal of these toxic substances. The electron beam method has been applied to the removal of vinyl chloride [1], trichloroethylene [2-3], carbon tetrachloride [4-6] and other types of volatile hydrocarbons from industrial off-gases [7]. Some of the electrical discharge reactors that have been investigated for VOC abatement include the pulsed corona [8-10], ferroelectric packed bed [9-10] dielectric-barrier discharge [11-17], surface discharge [18-19], gliding arc [20-21] and microwave [22].

There are many types of non-thermal plasma techniques that are being investigated for VOC abatement applications. The basic principle that these techniques have in common is to produce a plasma in which a majority of the electrical energy goes into the production of energetic electrons, rather than into gas heating. Through electron-impact dissociation and ionization of the background gas molecules, the energetic electrons produce free radicals and electron-ion pairs that, in turn, decompose the VOC molecules. Whatever the type of reactor, the plasma can induce three basic types of reactions with the VOC molecules, as shown in Fig. 1. Electron-impact dissociation of oxygen molecules produces O (and OH radicals in the presence of water vapor) that could oxidize the VOC molecules. Electron-impact ionization of the background air molecules produces electron-ion pairs. The electrons could decompose the VOC molecules via dissociative electron attachment. The ions could decompose the VOC molecules via dissociative charge exchange. The decomposition pathway for a particular VOC depends on the reaction rate constants and the amount of radicals and electron-ion pairs produced in the plasma.

461

M. Capitelli (ed.), Molecular Physics and Hypersonic Flows, 461–475.

Radical-induced decomposition

$$e + O_2 \rightarrow e + O(^3P) + O(^1D)$$
$$O(^3P) + CCl_4 \rightarrow ClO + CCl_3$$
$$O(^1D) + H_2O \rightarrow OH + OH$$
$$OH + CCl_4 \rightarrow HOCl + CCl_3$$

Electron-induced decomposition

$$e + N_2 \rightarrow e + e + N_2^+$$
$$e + O_2 \rightarrow e + e + O_2^+$$
$$e + CCl_4 \rightarrow CCl_3 + Cl^-$$

Ion-induced decomposition

$$N_2^+ + CH_3OH => CH_3^+ + OH + N_2$$

Fig. 1. There are three basic types of chemical reactions responsible for the decomposition of volatile organic compounds: (a) decomposition via oxidation by O and OH free radicals or reduction by N atoms, (b) electron-induced decomposition via dissociative electron attachment, and (c) ion-induced decomposition via dissociative charge exchange.

The electron mean energy in a plasma reactor is very important because it determines the types of radicals produced in the plasma and the input electrical energy required to produce those radicals. Fig. 2 shows the dissipation of the input electrical power in a dry air discharge. Note that at low electron mean energies (< 5 eV) a large fraction of the input electrical energy is consumed in the vibrational excitation of N_2. Electron mean energies around 5 eV are optimum for the electron-impact dissociation of O_2, which is important for the production of O radicals. High electron mean energies are required to efficiently implement the dissociation of N_2. For VOCs that take advantage of electron-induced or ion-induced decomposition, high electron mean energies are required to efficiently implement the ionization of the background gas.

In terms of the electron energy distribution in the plasma, there are basically only two types of non-thermal atmospheric-pressure plasma reactors: electrical discharge reactors and electron beam reactors. Electrical discharge techniques can be implemented in many ways, depending on the electrode configuration and electrical power supply (pulsed, AC or DC). Two of the more extensively investigated types of electrical discharge reactors are the pulsed corona and the dielectric-barrier discharge, shown in Fig. 3. In the pulsed corona method, the reactor is driven by very short pulses of high voltage, thus creating short-lived discharge plasmas that consist of energetic electrons, which in turn produce the radicals responsible for the decomposition of the undesirable molecules. In a dielectric barrier discharge reactor, one or both of the electrodes are covered with a thin dielectric layer, such as glass or alumina. Dielectric-barrier discharge reactors, also referred to as silent discharge reactors, are now routinely used to produce commercial quantities of ozone. Whereas in the pulsed corona method the transient behavior of the plasma is controlled by the applied voltage pulse, the plasma that takes place in a dielectric-barrier discharge self-extinguishes when charge build-up on the dielectric layer reduces the local electric field.

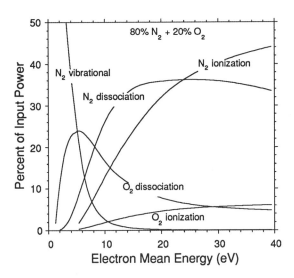

Fig. 2. Power dissipation in a dry air discharge, showing the percent of input power consumed in the electron-impact processes leading to vibrational excitation, dissociation and ionization of N_2 and O_2.

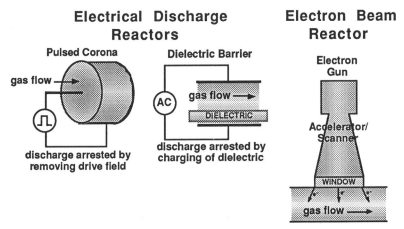

Fig. 3. There are basically two types of non-thermal atmospheric-pressure plasma reactors: electrical discharge reactors and electron beam reactors.

In this paper we present data on the non-thermal plasma processing of two representative VOCs: carbon tetrachloride and methanol. The investigation used a compact electron beam reactor, and two types of discharge reactors: a pulsed corona and a dielectric-barrier discharge. To the knowledge of the authors, this is the first comparison of the energy efficiency of electron beam, pulsed corona and dielectric-barrier discharge processing of these VOCs under identical gas conditions. For most electrical discharge reactors our analysis suggests that the attainable electron mean energy is rather limited and cannot be significantly enhanced by changing the electrode configuration or voltage waveform. Our experimental data confirms that there is no significant difference in the performance of our pulsed corona and dielectric-barrier discharge reactors. We observe that electron beam processing is remarkably more energy efficient than electrical discharge processing in decomposing either of these VOC molecules. During electron beam processing, the specific energy consumption is consistent with the energy required for the ionization of the background air molecules. For carbon tetrachloride, the dominant decomposition pathway is dissociative electron attachment. For methanol, the dominant decomposition pathway is dissociative charge exchange.

TEST FACILITY

All of our experiments were performed in a flow-through configuration. To characterize the energy consumption of the process for each VOC, the composition of the effluent gas was recorded as a function of the input energy density. The input energy density, Joules per standard liter, is the ratio of the power (deposited into the gas) to gas flow rate at standard conditions (25°C and 1 atm). The amount of VOC was quantified an FTIR analyzer and a gas chromatograph/mass spectrometer.

Our electron beam reactor used a cylindrical electron gun designed to deliver a cylindrically symmetric electron beam that is projected radially inward through a 5 cm wide annular window into a 17 cm diameter flow duct. An electron beam of 125 keV energy was introduced into the reaction chamber through a 0.7 mil thick titanium window. The electron beam current was produced from a low-pressure helium plasma in an annular vacuum chamber surrounding the flow duct.

Our pulsed corona reactor is a 1.5 mm diameter wire in a 60 mm diameter metal tube 300 mm long. The power supply is a magnetic pulse compression system capable of delivering up to 15-35 kV output into 100 ns FWHM pulses at repetition rates from 15 Hz to 1.5 kHz. The power input to the processor was varied by changing either the pulse energy or pulse repetition frequency. For the same energy density input, either method produced almost identical results. The gas mixtures were set with mass flow controllers. The gas and processor temperatures can be maintained at a temperature that can be controlled from 25°C to 300°C.

The dielectric-barrier discharge electrode structure has a similar electrode structure except that it has a dielectric material on the inside surface of the outer tube electrode. It consists of a 1.5 mm diameter wire in a 290 mm long alumina tube with inner and outer diameters of 53 mm and 58 mm, respectively. The middle 170 mm of the dielectric tube has aluminum foil coating the outside to form the other electrode.

ELECTRON AND CHEMICAL KINETICS

To calculate the ion and radical production yields by electrical discharge processing, we used the Boltzmann code ELENDIF [23] to calculate electron energy deposition. ELENDIF uses as input the specified gas composition and the electron-molecule collision cross sections. To calculate the ion and radical production yields by electron beam processing, we used the code DEGRAD [24]. DEGRAD also uses as input the specified gas composition and the electron-molecule collision cross sections. This code follows typical electrons as they perform successive collisions, and discrete energy bins are used to represent the energy degradation of an electron from a given beam energy. The procedure records the number of excitations, dissociations and ionizations, and the total number of all orders of secondary electrons. The chemical kinetics describing the subsequent interaction of the ions and radicals with the exhaust gas was studied using CHEMKIN-II [25].

RESULTS

In discharge processing, the rate coefficients for electron-impact dissociation and ionization reactions strongly depend on the electron mean energy in the discharge plasma. In pulsed corona and dielectric-barrier discharge reactors, the non-thermal plasma is produced through the formation of statistically distributed microdischarges. The electrons dissociate and ionize the background gas molecules within nanoseconds in the narrow channel formed by each microdischarge. The electron energy distribution in the plasma is complicated because the electric field is strongly non-uniform (e.g. because of strong space-charge field effects) and time dependent. However, most of the species responsible for the chemical processing are generated in the microdischarge channels already established during the main current flow. In each microdischarge column, the electrons acquire a drift velocity, v_d, and an average energy corresponding to an effective E/n, i.e., the value of the electric field E divided by the total gas density n. The efficiency for a particular electron-impact process can be expressed in terms of the G-value (number of dissociation or ionization reactions per 100 eV of input energy) defined as

$$G\text{-value} = 100\, k\, /\, (v_d\, E/n)$$

where k is the rate coefficient (cm³/molec-s). The rate coefficient k represents the number of reactions in a unit volume per unit time. The quantity $v_d\, E/n$ represents the amount of energy expended by the electrons in a unit volume per unit time. In Fig. 4 the calculated G-values for various electron-impact dissociation and ionization processes in dry air are shown as functions of the electron mean energy in the discharge plasma.

Under most conditions encountered in pulsed corona or dielectric-barrier discharge processing, the effective E/n is close to the value for breakdown (Paschen field) [26-27]. For dry air, the effective E/n is around 130 Td (1 Td = 10^{-17} V-cm²), which corresponds to an electron mean energy of about 4 eV. This analysis suggests that the attainable electron mean energy is rather limited and cannot be significantly enhanced by changing the electrode configuration or voltage pulse parameters. Fig. 5 shows the comparison between pulsed corona and dielectric-barrier discharge processing of methanol in dry air at 120°C.

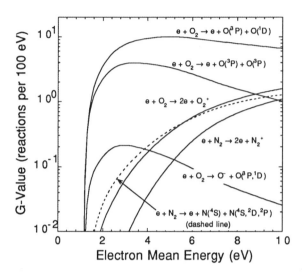

Fig. 4. Calculated G-values (number of reactions per 100 eV of input energy) for dissociation and ionization processes in dry air, shown as functions of the electron mean energy in a discharge plasma.

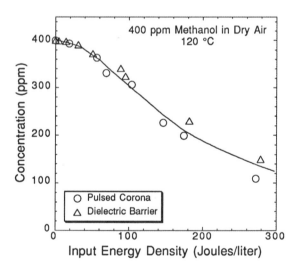

Fig. 5. Pulsed corona and dielectric-barrier discharge processing of 400 ppm methanol in dry air at 120°C. There is no significant difference in the performance of different types of electrical discharge reactors.

Our experimental data confirms that there is no significant difference in the performance of pulsed corona and dielectric-barrier discharge reactors.

In electron beam processing, the efficiency for a particular electron-impact process can be expressed in terms of the G-value, which is defined in the code DEGRAD as

$$\text{G-value} = 100\ N_j / E_p$$

where N_j is the number of dissociation or ionization events, and E_p is the primary electron energy.

Table 1 shows a comparison of the calculated G-values for dissociation processes in dry air using an electron beam and a discharge reactor. Discharge plasma conditions are optimum for the dissociation of O_2. The production of O radicals is higher in a pulsed corona reactor compared to that in an electron beam reactor. Table 2 shows a comparison of the calculated G-values for ionization processes in dry air using an electron beam and a discharge reactor. The efficiency for production of electron-ion pairs is much higher in an electron beam reactor compared to that in a pulsed corona reactor.

Table 1. Calculated G-values (number of reactions per 100 eV of input energy) for dissociation processes in dry air using an electron beam and an electrical discharge reactor.

REACTION	Electron Beam	Discharge
$e + N_2 \rightarrow e + N(^4S) + N(^4S, {}^2D, {}^2P)$	1.2	0.17
$e + O_2 \rightarrow e + O(^3P) + O(^3P)$	1.3	4.0
$e + O_2 \rightarrow e + O(^3P) + O(^1D)$	2.65	10.0
$e + O_2 \rightarrow O^- + O(^3P, {}^1D)$	0.11	0.19

Table 2. Calculated G-values (number of reactions per 100 eV of input energy) for ionization processes in dry air using an electron beam and an electrical discharge reactor.

REACTION	Electron Beam	Discharge
$e + N_2 \rightarrow 2e + N(^4S, {}^2D) + N^+$	0.69	$< 10^{-6}$
$e + N_2 \rightarrow 2e + N_2^+$	2.27	0.044
$e + O_2 \rightarrow 2e + O_2^+$	2.07	0.17
$e + O_2 \rightarrow 2e + O(^1D) + O^+$	1.23	0.0016

In non-thermal plasma processing of a mixture containing very dilute concentrations of VOC molecules, the input electrical energy is dissipated by the primary electrons mostly in interactions with the background gas molecules. The energetic primary electrons produce free radicals and electron-ion pairs through electron-impact dissociation and ionization. In a dry air mixture, electron-impact dissociation of molecular oxygen produces the ground state atomic oxygen $O(^3P)$ and excited atomic oxygen $O(^1D)$:

$$e + O_2 \rightarrow e + O(^3P) + O(^3P) \tag{1a}$$

$$e + O_2 \rightarrow e + O(^3P) + O(^1D) \tag{1b}$$

In addition, with energetic electrons, $O(^3P)$ and $O(^1D)$ can be produced via two-body dissociative attachment:

$$e + O_2 \rightarrow O^- + O(^3P, ^1D) \tag{2}$$

As seen in Fig. 4 the main contribution to O radical production comes from the dissociation reactions (1a) and (1b).

The O radicals can dissociate CCl_4 into ClO and CCl_3 [28-30]:

$$O(^3P) + CCl_4 \rightarrow ClO + CCl_3 \tag{3a}$$

$$O(^1D) + CCl_4 \rightarrow ClO + CCl_3 \tag{3b}$$

Another mechanism for the dissociation of CCl_4 is through the secondary electrons. Electron-ion pairs are produced through various electron-impact ionization processes [31-32]:

$$e + N_2 \rightarrow 2e + N_2^+ \tag{4}$$

$$e + N_2 \rightarrow 2e + N(^4S) + N^+ \tag{5}$$

$$e + N_2 \rightarrow 2e + N(^2D) + N^+ \tag{6}$$

$$e + O_2 \rightarrow 2e + O_2^+ \tag{7}$$

$$e + O_2 \rightarrow 2e + O(^1D) + O^+ \tag{8}$$

The secondary electrons can dissociate CCl_4 via dissociative electron attachment [17,18] to produce CCl_3 and a negative ion Cl^-:

$$e + CCl_4 \rightarrow CCl_3 + Cl^- \tag{9}$$

The rate coefficient for reaction (9) is on the order of 10^{-7} cm^3/(molec-s) [33-34]. The rate coefficient for reaction (3a) is less than 10^{-14} cm^3/(molec-s) [28], while that for reaction (3b) is around 10^{-10} cm^3/(molec-s) [29-30]. An examination of the G-values shown in Tables 1 and 2 indicates that dissociative electron attachment will dominate the initial decomposition of CCl_4 for both electron beam and electrical discharge reactor conditions.

The charge exchange reaction of positive ions, such as N_2^+, with the background O_2 is fast, resulting in mostly O_2^+ ions [35]:

$$N_2^+ + O_2 \rightarrow N_2 + O_2^+ \tag{10}$$

The positive ions react with Cl⁻ through the ion-ion neutralization reaction to produce Cl and O radicals:

$$Cl^- + O_2^+ \rightarrow Cl + 2O \tag{11}$$

In the absence of scavenging reactions for CCl_3, the input energy would be wasted because Cl and CCl_3 would simply recombine quickly to reform the original pollutant [36-37]:

$$Cl + CCl_3 + M \rightarrow CCl_4 + M \tag{12}$$

Fortunately, the presence of O_2 scavenges the CCl_3 through the fast reaction [38-39]:

$$CCl_3 + O_2 + M \rightarrow CCl_3O_2 + M \tag{13}$$

The CCl_3O_2 species undergoes a chain reaction involving the Cl radical and produces phosgene ($COCl_2$) as one of the main organic products [30,40]:

$$Cl + CCl_3O_2 \rightarrow CCl_3O + ClO \tag{14}$$

$$CCl_3O \rightarrow COCl_2 + Cl \tag{15}$$

The ClO species produces additional Cl radicals through a reaction with the O radicals [30,41]:

$$O + ClO \rightarrow Cl + O_2 \tag{16}$$

The other major product is Cl_2 which is formed by the reaction [41]:

$$Cl + Cl + M \rightarrow Cl_2 + M \tag{17}$$

The CCl_3 species from reaction (9) can also be scavenged by O [42]:

$$CCl_3 + O \rightarrow COCl_2 + Cl \tag{18}$$

and by N [43]:

$$CCl_3 + N \rightarrow ClCN + 2Cl \tag{19}$$

The apparent two-body rate constant for scavenging reaction (13) is $1.4 \times 10^{-9} \, T^{-1.1}$ (cm^3/molec-s). The rate constants are 4.2×10^{-11} and 1.7×10^{-11} (cm^3/molec-s) for scavenging reactions (18) and (19), respectively. Because of the much larger density of O_2 compared to O or N, the scavenging of CCl_3 by reactions (18) and (19) are therefore negligible compared to reaction (13) during processing in dry air.

Fig. 6 shows the results of experiments on electron beam and pulsed corona processing of 100 ppm of CCl_4 in dry air (20% O_2 80% N_2) at 25°C. The pulsed corona reactor requires 1277 Joules/liter for 90% decomposition of CCl_4, whereas the electron beam reactor requires only 20 Joules/liter to achieve the same level of decomposition.

An analysis of the rates of the reactions discussed above suggests that the rate limiting step in the decomposition of CCl_4 is determined by the dissociative attachment of CCl_4 to the thermalized electrons in the created plasma. The specific energy consumption for CCl_4 removal is therefore determined by the specific energy consumption (or G-value) for creating electron-ion pairs. Table 2 shows the calculated G-values for the ionization processes (4)-(8). For electron beam processing of dry air, the ionization G-value corresponds to a specific energy consumption of 33 eV per electron-ion pair produced. For pulsed corona processing, we calculate a specific energy consumption of around 1400 eV per electron-ion pair, assuming an effective electron mean energy of 4 eV in the discharge plasma. To first order, the calculated specific energy consumption for electron-ion pair production agrees very well with our experimentally observed specific energy consumption for CCl$_4$ decomposition. The results shown in Fig. 6 demonstrate that for VOCs requiring copious amounts of electrons for decomposition, electron beam processing is much more energy efficient than electrical discharge processing.

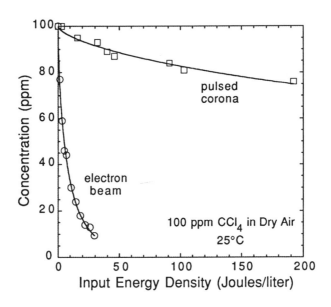

Fig. 6. Comparison between electron beam and pulsed corona processing of 100 ppm of carbon tetrachloride in dry air at 25°C.

After the concentration of CCl$_4$ has decreased to a few tens of ppm, the three-body attachment of thermal electrons to oxygen molecules [44]:

$$e + O_2 + O_2 \rightarrow O_2^- + O_2 \qquad (20)$$

$$e + O_2 + N_2 \rightarrow O_2^- + N_2 \qquad (21)$$

becomes a significant electron loss pathway compared to reaction (9). The rate constants for reactions (20) and (21) are $k_{(20)} = 2.5 \times 10^{-30}$ and $k_{(21)} = 0.16 \times 10^{-30}$ cm^6/s, respectively. The attachment frequency of thermal electrons to O$_2$ in dry air at atmospheric pressure is thus

$$\nu_{O2} = k_{(20)} [O_2]^2 + k_{(21)} [N_2] [O_2] \approx 0.8 \times 10^8 \text{ s}^{-1}.$$

The attachment rate coefficient for thermal electrons to CCl_4 is $k_{(9)} = 4 \times 10^{-7}$ cm^3/s. For 100 ppm CCl_4, the attachment frequency to CCl_4 is thus

$$\nu_{CCl_4} = k_{(9)} [CCl_4] \approx 10^9 \text{ s}^{-1}.$$

When the concentration of CCl_4 is down to around 10 ppm, the electrons will attach to oxygen molecules as frequently as to CCl_4 molecules.

Although the dominant pathway (dissociative electron attachment) for the initial decomposition of CCl_4 is the same in electron beam and pulsed corona processing, the composition of the final products are not the same. In pulsed corona processing, a larger amount of O radicals is produced relative to the amount of electrons. Even though these O radicals contribute only a small fraction to the initial decomposition of CCl_4, they do interact significantly with phosgene to change the composition of the final products [28,45]:

$$O + COCl_2 \rightarrow ClO + COCl \tag{22}$$

$$COCl + M \rightarrow CO + Cl + M \tag{23}$$

$$O + COCl \rightarrow CO_2 + Cl \tag{24}$$

Our model for the decomposition mechanism predicts a difference in product yields between electron beam and pulsed corona processing at the minimum energy required for near complete decomposition of CCl_4. For around 95% decomposition of 100 ppm CCl_4 in dry air by electron beam processing, the final products consist of around 100 ppm Cl_2 and 100 ppm $COCl_2$. For the same level decomposition of 100 ppm CCl_4 in dry air by pulsed corona processing, the final products consist of around 160 ppm Cl_2, 40 ppm $COCl_2$, 50 ppm CO and 10 ppm CO_2. Of course, with excessive energy deposition all the $COCl_2$ would eventually be converted into CO_x and Cl_2. However, as noted in References [46] and [47], the Cl_2 and $COCl_2$ products can be easily removed from the gas stream; e.g. they dissolve and/or dissociate in aqueous solutions and combine with $NaHCO_3$ in a scrubber solution to form $NaCl$ [47].

Fig. 7 shows the results of experiments on electron beam and pulsed corona processing of 100 ppm of methanol in dry air at 25°C. The pulsed corona reactor requires 450 Joules/liter for 90% decomposition of methanol, whereas the electron beam reactor requires only around 15 Joules/liter to achieve the same level of decomposition. In this case, the electron beam method is more efficient because the decomposition proceeds mainly via a dissociative charge exchange reaction

$$N_2^+ + CH_3OH \rightarrow CH_3^+ + OH + N_2 \tag{25}$$

The OH radicals resulting from the initial decomposition reaction (25) in turn may lead to additional decomposition of methanol via $OH + CH_3OH$.

To verify that the primary decomposition during electron beam processing does not proceed through an oxidation pathway using O radicals, we performed the experiment using N_2 as the background gas. As shown in Fig. 8, the specific energy consumption in dry air is almost identical to that in N_2.

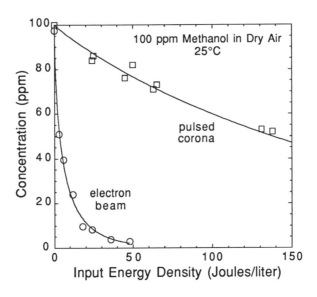

Fig. 7. Comparison between electron beam and pulsed corona processing of 100 ppm of methanol in dry air at 25°C.

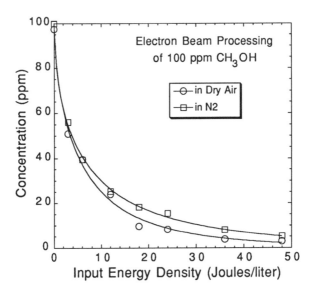

Fig. 8. Comparison between electron beam of 100 ppm of methanol in dry air and electron beam processing of 100 ppm of methanol in N_2. Gas temperature is 25°C.

ACKNOWLEDGMENTS

This work was performed in part at Lawrence Livermore National Laboratory under the auspices of the U.S. Department of Energy under Contract Number W-7405-ENG-48, with support from the Advanced Energy Projects Division of the Office of Energy Research. The electron beam processing equipment was developed under a National Science Foundation SBIR grant, Contract Number III-9122767.

REFERENCES

[1] R. C. Slater and D. H. Douglas-Hamilton, *J. Appl. Phys.* **52**, 5820 (1981).

[2] H. Scheytt, H. Esrom, L. Prager, R. Mehnert, and C. von Sonntag, in *Non-Thermal Plasma Techniques for Pollution Control: Part B - Electron Beam and Electrical Discharge Processing*, edited by B. M. Penetrante and S. E. Schultheis (Springer-Verlag, Heidelberg, 1993) pp. 91-102.

[3] S. M. Matthews, A. J. Boegel, J. A. Loftis, R. A. Caufield, B. J. Mincher, D. H. Meikrantz, and R. J. Murphy, *Radiat. Phys. Chem.* **42**, 689 (1993).

[4] L. Bromberg, D. R. Cohn, M. Kock, R. M. Patrick, and P. Thomas, *Phys. Lett. A* **173**, 293 (1993).

[5] M. Koch, D. R. Cohn, R. M. Patrick, M. P. Schuetze, L. Bromberg, D. Reilly and P. Thomas, *Phys. Lett. A* **184**, 109 (1993).

[6] B. M. Penetrante, M. C. Hsiao, B. T. Merritt, G. E. Vogtlin, P. H. Wallman, A. Kuthi, C. P. Burkhart and J. R. Bayless, *Phys. Lett. A* (in press, 1996).

[7] H.-R. Paur, in *Non-Thermal Plasma Techniques for Pollution Control: Part B - Electron Beam and Electrical Discharge Processing*, edited by B. M. Penetrante and S. E. Schultheis (Springer-Verlag, Heidelberg, 1993) pp. 77-90.

[8] M. C. Hsiao, B. T. Merritt, B. M. Penetrante, G. E. Vogtlin, and P. H. Wallman, *J. Appl. Phys.* **78**, 3451 (1995).

[9] T. Yamamoto, K. Ramanathan, P. A. Lawless, D. S. Ensor, J. R. Newsome, N. Plaks, G. H. Ramsey, C. A. Vogel, and L. Hamel, *IEEE Trans. on Ind. Appl.* **28**, 528 (1992).

[10] T. Yamamoto, P. A. Lawless, M. K. Owen, D. S. Ensor, and C. Boss, in *Non-Thermal Plasma Techniques for Pollution Control: Part B - Electron Beam and Electrical Discharge Processing*, edited by B. M. Penetrante and S. E. Schultheis (Springer-Verlag, Heidelberg, 1993) pp. 223-238.

[11] D. Evans, L. A. Rosocha, G. K. Anderson, J. J. Coogan, and M. J. Kushner, *J. Appl. Phys.* **74**, 5378 (1993).

[12] L. A. Rosocha, G. K. Anderson, L. A. Bechtold, J. J. Coogan, H. G. Heck, M. Kang, W. H. McCulla, R. A. Tennant, and P. J. Wantuck, in *Non-Thermal Plasma Techniques for Pollution Control: Part B - Electron Beam and Electrical Discharge Processing*, edited by B. M. Penetrante and S. E. Schultheis (Springer-Verlag,

474

Heidelberg, 1993) pp. 281-308.

[13] W. C. Neely, E. I. Newhouse, E. J. Clothiaux, and C. A. Gross, in *Non-Thermal Plasma Techniques for Pollution Control: Part B -Electron Beam and Electrical Discharge Processing*, edited by B. M. Penetrante and S. E. Schultheis (Springer-Verlag, Heidelberg, 1993) pp. 309-320.

[14] E. I. Newhouse, W. C. Neely, E. J. Clothiaux, and J. W. Rogers, in *ACS Symposium on Emerging Technologies in Hazardous Waste Management VI*, pp. 207-210.

[15] E. J. Clothiaux, J. A. Koropchak, and R. R. Moore, *Plasma Chem. Plasma Process.* **4**, 15 (1984).

[16] M. E. Fraser, D. A. Fee, and R. S. Sheinson, *Plasma Chem. Plasma Process.* **5**, 163 (1985).

[17] M. E. Fraser and R. S. Sheinson, *Plasma Chem. Plasma Process.* **6**, 27 (1986).

[18] S. Masuda, in *Non-Thermal Plasma Techniques for Pollution Control: Part B - Electron Beam and Electrical Discharge Processing*, edited by B. M. Penetrante and S. E. Schultheis (Springer-Verlag, Heidelberg, 1993) pp. 199-210.

[19] T. Oda, T. Takahashi, H. Nakano, and S. Masuda, in *Proceedings of the 1991 IEEE Industrial Application Society Meeting* (Dearborn, MI, September/October 1991) pp. 734-739.

[20] A. Czernichowski and H. Lesueur, in *Proceedings of the 10th International Symposium on Plasma Chemistry* (Bochum, Germany, 1991).

[21] A. Czernichowski and T. Czech, in *Proceedings of the 3rd International Symposium on High-Pressure Low-Temperature Plasma Chemistry* (Strasbourg, France, 1991).

[22] L. J. Bailin, M. E. Sibert, L. A. Jonas, and A. T. Bell, *Environ. Sci. Tech.* **9**, 254 (1975).

[23] W. L. Morgan and B. M. Penetrante, *Comp. Phys. Comm.* **58**, 127-152 (1990).

[24] B. M. Penetrante and J. N. Bardsley, *J. Appl. Phys.* **66**, 1871-1874 (1989).

[25] R. J. Kee, F. M. Rupley and J. A. Miller, "Chemkin-II: A FORTRAN Chemical Kinetics Package for the Analysis of Gas Phase Chemical Kinetics," Sandia National Laboratories Report No. SAND89-8009B UC-706 (April 1992).

[26] B. Eliasson and U. Kogelschatz, *J. Phys. B: At. Mol. Phys.* **19**, 1241 (1986).

[27] B. M. Penetrante, in *Non-Thermal Plasma Techniques for Pollution Control - Part A: Overview, Fundamentals and Supporting Technologies*, B. M. Penetrante and S. E. Schultheis, Eds. (Springer-Verlag, Berlin Heidelberg, 1993) pp. 65-90.

[28] J. T. Herron, *J. Phys. Chem. Ref. Data* **17**, 967 (1988).

[29] J. A. Davidson, H. I. Schiff, T. J. Brown, C. J. Howard, *J. Chem. Phys.* **69**, 4277 (1978).

[30] R. Atkinson, D. L. Baulch, R. A. Cox, R. F. Hampson, Jr., J. A. Kerr, J. Troe, *J. Phys. Chem. Ref. Data* **21**, 1125-1568 (1992).

[31] Y. Itikawa, M. Hayashi, A. Ichimura, K. Onda, D. Sakimoto, K. Takayanagi, M. Nakamura, H. Nishimura and T. Takayanagi, *J. Phys. Chem. Ref. Data* **15**, 985 (1986).

[32] Y. Itikawa, A. Ichimura, K. Ondaz, K. Sakimoto, K. Takayanagi, Y. Hatano, M. Hayashi, H. Nishimura and S. Tsurubuchi, , *J. Phys. Chem. Ref. Data* **18**, 23 (1989).

[33] A. A. Christodoulides and L. G. Christophorou, *J. Chem. Phys.* **54**, 4691 (1971).

[34] J. A. Ayala, W. E. Wentworth and E. C. M. Chen, *J. Phys. Chem.* **85**, 3989 (1981).

[35] D. L. Albritton, *Atom. Data Nucl. Data Tables* **22**, 1 (1978).

[36] F. Danis, F. Caralp, B. Veyret, H. Loirat, R. Lesclaux, *Int. J. Chem. Kinet.* **21**, 715 (1989).

[37] T. Ellermann, *Chem. Phys. Lett.* **189**, 175 (1992).

[38] F. Danis, F. Caralp, M. T. Rayez, R. Lesclaux, *J. Phys. Chem.* **95**, 7300-7307 (1991).

[39] F. F. Fenter, P. D. Lightfoot, J. T. Niiranen, D. Gutman, *J. Phys. Chem.* **97**, 313-5320 (1993).

[40] J. J. Russell, J. A. Seetula, D. Gutman, F. Danis, F. Caralp, P. D. Lightfoot, R. Lesclaux, C. F. Melius, S. M. Senkan, *J. Phys. Chem.* **94**, 3277 (1990).

[41] D. L. Baulch, J. Duxbury, S. J. Grant, D. C. Montague, *J. Phys. Chem. Ref. Data* **10**, Suppl. 1,1-1 (1981).

[42] J. T. Herron, *J. Phys. Chem. Ref. Data* **17**, 967 (1988).

[43] S. C. Jeoung, K. Y. Choo and S. W. Benson, *J. Phys. Chem.* **95**, 7282 (1991).

[44] L. M. Chanin, A. V. Phelps and M. A. Biondi, *Phys. Rev.* **128**, 219 (1962).

[45] W.-D. Chang and S. M. Senkan, *Environ. Sci. Tech.* **23**, 442 (1989).

[46] D. Evans, L. A. Rosocha, G. K. Anderson, J. J. Coogan and M. J. Kushner, *J. Appl. Phys.* **74**, 5378 (1993).

[47] R. M. Patrick and K. Hadidi, An Electron Beam Plasma System for Halogenated Hydrocarbon Vapor Destruction, Preprint No. 95-WP77B.04, 88th Annual Meeting & Exhibition of the Air & Waste Management Association, San Antonio, TX (18-23 June 1995).

FLUID DYNAMIC MODELING OF PLASMA REACTORS

P. VITELLO
Lawrence Livermore National Laboratory
7000 East Avenue, L-014
Livermore, CA 94550 USA

1. Introduction

The trend by the micro-electronic industry toward ultra large scale integrated circuits is driving the development of high plasma density sources with good plasma uniformity over a large area (up to 400 mm diameter in size). These new plasma reactor sources are used to deposit layers or to etch patterns. Characteristics of the plasma reactors under consideration are high plasma densities ($\geq 10^{10}$ cm^{-3}) to ensure large ion flux rates to surfaces, low plasma potential (10–30 eV) to minimize surface damage, and low neutral gas pressure ($\leq$ 30 mTorr) to minimize collisions by ions as they pass from the bulk of the plasma through the sheath so as to maintain high isotropy of the ion flux to surfaces. The high ion flux isotropy is essential for generating the sharp walled trenches used in integrated circuits. One of the most promising new sources is the inductively coupled plasma (ICP) source. One attractive feature of this source is its relative simplicity, e.g., no DC magnetic fields are required for their operation. Recent studies have reported experimental characterizations [1, 2] and computer modeling [3, 4] of these devices. Most of the studies have concentrated on smaller area ICPs. Here we present modeling of the LLNL 76 cm diameter ICP [5] and compare our results with recent measurements.

Interest in plasma modeling has grown in recent years due to the increase in sophistication in numerical simulation codes. Modeling provides an understanding of the fundamental physics of plasmas that can be used to aid in the design and optimization of plasma reactors and other devices. The new high density reactors require numerical models that are at least two-dimensional for accurate simulation of power coupling and particle transport. Models must treat electrons, and multiple ion and neutral species. Detailed volume and surface chemistry involving electron-ion,

477

M. Capitelli (ed.), Molecular Physics and Hypersonic Flows, 477–484.

electron-neutral, ion-ion, ion-neutral, and neutral-neutral reactions must be treated. To be applicable to industrial reactors, models are needed that can deal with the complex chemistry in gas mixtures such as Cl_2, HBr, CF_4, $TiCl_4$, and NH_3.

In a plasma reactor, such as an ICP source, the gas is only partially ionized, with the neutral density much larger than either the ion or electron density. Such plasmas are quasi-neutral, except for the sheath bordering surfaces, and have pressures and temperatures for the different species which are related as: $P_n >> P_e >> P_i$ and $T_e >> T_i > T_n$, where P_n is the neutral gas pressure, P_e is the electron pressure, P_i is the ion pressure, T_e is the electron temperature, T_i is the ion temperature, and T_n is the neutral temperature.

For low pressure plasma reactors, electrons and ions may have non-Maxwellian energy distributions. Because of the many in-elastic collision processes available for electrons, it is much more likely that the electron energy distribution function have non-linear features such as a high energy tail. Non-Maxwellian energy distributions can be accurately modeled using particle simulation schemes such as the Particle-In-Cell method. Particle simulation schemes however are extremely slow compared to fluid dynamic treatments, making knowledge of the regimes where fluid dynamic modeling is valid extremely important.

2. Discharge Model

We describe here the numerical plasma simulation model, INDUCT94 [4], which is based on a fluid dynamic treatment of a non-magnetized plasma. INDUCT94 solves a set of two-dimensional (cylindrically symmetric) time-dependent fluid equations for electrons and ions self-consistently with Poisson's equation for the electric potential. In addition, rf inductive heating is calculated from a time-averaged solution of Maxwell's equations. Ions are assumed to be isothermal and near the neutral species temperature.

Ion motion is governed by the equations of continuity and momentum conservation, which for ion species i are

$$\frac{\partial n_i}{\partial t} = -\vec{\nabla} \cdot n_i \vec{v}_i + \sum_{j=1}^{N_C} R_{ij}, \tag{1}$$

$$\frac{\partial n_i \vec{v}_i}{\partial t} = -\vec{\nabla} \cdot (n_i \vec{v}_i \vec{v}_i) + \frac{q_i n_i \vec{E}}{m_i} - \frac{1}{m_i} \vec{\nabla} (n_i k T_i) - \sum_{j=1}^{N_N} n_i \vec{v}_i \nu_{ij}. \tag{2}$$

Here n_i and $\vec{v}_i$ give the ion density and velocity, R_{ij} gives the chemical reaction rates leading to changes in the ion density, ν_{ij} is the ion neutral collision frequency, m_i is the ion mass, q_i is the ion charge, and $\vec{E}$ is the electric field. The sums run over the total number of chemical reactions, N_C, and the total number of neutral species, N_N. Chemical interactions treated including ionization, attachment, and recombination. The ion-neutral collision frequency is calculated from the corresponding cross-section using

$$\nu_{ij} = \frac{m_j}{m_i + m_j} \sigma_{ij} \bar{v}_i n_j, \tag{3}$$

where σ_{ij} is the ion-neutral cross-section between ion species i and neutral species j, n_j is the neutral density, and

$$\bar{v}_i = \left(\frac{8kT_i}{\pi m_i} + \vec{v}_i \cdot \vec{v}_i \right)^{1/2} \tag{4}$$

is the relative velocity between ions and neutrals.

The electron fluid model consists of the electron continuity and energy balance equations

$$\frac{\partial n_e}{\partial t} = -\vec{\nabla} \cdot \vec{\Gamma}_e + \sum_{j=1}^{N_C} R_{ej}, \tag{5}$$

$$\frac{\partial W_e}{\partial t} = -\vec{\nabla} \cdot \vec{Q} - e\vec{\Gamma}_e \cdot \vec{E} + P_{\text{ind}} - P_{coll}, \tag{6}$$

where

$$\vec{\Gamma}_e = -\frac{en_e}{m_e \nu_{eN}} \vec{E} - \frac{1}{m_e \nu_{eN}} \vec{\nabla}(n_e k T_e), \tag{7}$$

is the electron flux in the drift–diffusion approximation, and

$$\vec{Q} = \frac{5}{2} \left(1 - \frac{d \ln \nu_{eN}}{d \ln T_e} \right) \left[\vec{\Gamma}_e k T_e - \frac{n_e k T_e}{m_e \nu_{eN}} \vec{\nabla}(k T_e) \right] \tag{8}$$

is the electron energy flux [6]. The electron density is given by n_e, $W_e = 3n_e k T_e / 2$ is the electron thermal energy, m_e is the electron mass, ν_{eN} is the total electron-neutral collision frequency (summed over all neutral species), P_{ind} is the time-averaged power per unit volume absorbed by the electrons due to the inductive rf fields, and P_{coll} is the energy loss per unit volume due to electron-neutral collisions. Due to the slower ion response time, the drift-diffusion velocity approximation gives a poor representation of the ion velocity. We therefore solve the ion momentum equation directly.

The advective and chemistry terms in the ion and electron equations are solved separately (successively) by time splitting. The advective part of the electron continuity equation is solved implicitly to allow time steps greater than the CFL time limit and the Dielectric Relaxation time scale. The electron temperature equation is also solved implicitly to allow for large time step stability for thermal conduction and advection. We use time splitting to separately solve the electron continuity and electron temperature equations, with the temperature held fixed in the continuity equation and the density held fixed in the temperature equation. This allows for the most efficient solution of these equations. Each implicit equation is solved using a Alternating-Direction-Implicit scheme. Because of the high mass of the ions relative to the electrons, the ion velocities are orders of magnitude slower than the electron velocity. Due to the much longer ion dynamic time scales, an explicit temporal differencing scheme is used in the ion equations.

Neutral flow and chemistry is treated in INDUCT94 assuming constant total pressure and uniform spatial distribution. A density for each neutral species is maintained at each grid point, with volume changes due to chemistry calculated in the same manner as is done for electrons and ions. In determining changes in the mean neutral densities, the volume chemistry changes are summed and combined with changes due to flow input and surface chemistry changes. The surface chemistry makes use of total ion and neutral species wall currents.

Space-charge electric fields are determined self-consistently using Poisson's equation

$$\vec{\nabla} \cdot \epsilon \vec{\nabla} \phi = - \left(\sum_{i=1}^{N_I} q_i n_i - e n_e \right), \tag{9}$$

where ϵ is the local dielectric constant and N_I gives the number of ion species. The use of explicitly evaluated densities in the space-charge source term was found to be undesirable as this leads to a Dielectric Relaxation instability. The Dielectric Relaxation time scale is the electrostatic shielding time scale. For electron densities of the order of 10^{11}cm^{-3} the Dielectric Relaxations time scale is of the order of picoseconds. To avoid disastrous amplification of the electrostatic field we solve Poisson's equation at the future time level using the time advanced electron density

$$n_e^{n+1} = n_e^n + \delta t \frac{\partial n_e}{\partial t}, \tag{10}$$

$$= n_e^n - \delta t \vec{\nabla} \cdot \left(\frac{e n_e^n}{m_e \nu_{eN}^n} \vec{\nabla} \phi^{n+1} - \frac{1}{m_e \nu_N^n} \vec{\nabla} \left(n_e^n k T_e^n \right) \right) \tag{11}$$

where the superscript n implies the current time level variables, and superscript $n + 1$ signifies the future time level variables to be solved for. Due so their slower response to the electric field, the ion densities are treated explicitly and not time advanced. Only the potential ϕ (which comes from the drift velocity term) is actually evaluated at the future time level, allowing the solution for the potential to be evaluated separately from that of the other fluid variables. For large time steps this insures near ambipolar fields and for steady state reduces to the simple form of Poisson's equation.

The power absorption by the inductive fields, P_{ind}, is calculated using the electro-magnetic (EM) solver ORMAX [7]. ORMAX is a time harmonic code that solves a reduced set of Maxwell's equations in the ICP chamber and then calculates the power deposition in the plasma. ORMAX is fully integrated in the INDUCT94 model and is called as a subroutine. ORMAX requires plasma conductivity which can be easily found from the plasma temperature and density calculated in the plasma portion of the code. ORMAX also calculates the inductive voltage drop and capacitive current drop for each ICP coil turn. These quantities are used to advance a circuit model that couples the ICP coils to a current source. The current source can be scaled to achieve a fixed total power deposition in the plasma. The average voltage on the coils can also be coupled into Poisson's equation solved in the transport portion of the code.

3. Results

To explore the technology of large area plasmas for micro-electronic wafer fabrication or for flat panel display processing Lawrence Livermore National Laboratory has constructed a ICP device [5]. It uses a circular inductive coil with a 64 cm diameter top coil with four turns, couples 13.56 MHz rf electromagnetic energy from a source and matchbox to the plasma through a 5.5 cm top dielectric window. The plasma reactor chamber is 76 cm in diameter and 10 cm in height.

We modeled experiments at various pressures and powers for discharges in argon. Results are presented here for radial profiles of the electron density. These profiles correspond to the height in the chamber of 5.4 cm at which the experimental Langmuir probe measurements were taken. Figure 1 shows experimental and simulation results for for a range of inductive power at a fixed pressure of 10 mTorr. For fixed inductive power of 1200 W, we show in Figure 2 the experimental and simulation variation for the radial electron density as a function of pressure.

Comparing the density variation with inductive power shown in Figure 1 we observe the same general trends in profile variation with power. The agreement between the model and experimental density values is within the accuracy of the atomic data used in the INDUCT94 simulations. With increasing power at fixed pressure, the electron density increases roughly linearly with power. This scaling is consistent with power balance models [8]. The density profiles are very weakly dependent upon power, with a slight shifting of the density profile towards smaller radii as the power decreases. This profile effect may be explained qualitatively in terms of the

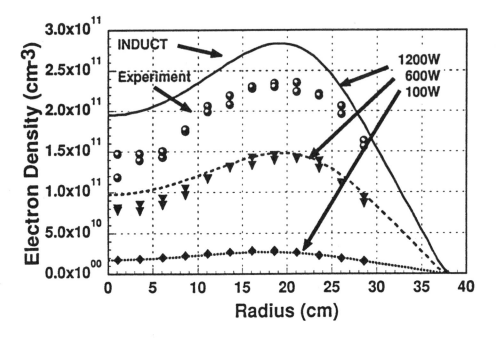

Figure 1. Radial profiles in Ar as a function of rf power at 10 mTorr pressure and z ~ 5.4 cm. Experimental data points are shown along with simulation results curves.

inductive skin depth, which is the scale length over which electro-magnetic energy is deposited. The skin depth is scales as the inverse of the plasma frequency and thus varies with electron density as $\delta \propto n_e^{-1/2}$. For low power, the lower electron density results in a longer power absorption scale length. Power is spread more evenly throughout the plasma at low power resulting in a shifting inwards of the density profile. At high power and short scale lengths, inductive power is deposited locally at a radius of roughly 20 cm.

Comparing the pressure variation of the density profiles shown in Figure 2, we again observe excellent agreement between experiment and model results. For higher pressure at fixed power, the density increases and the density profiles become more peaked.

The variation in the magnitude of the density can be explained qualitatively in terms of particle and power balances and the profile variations can explained in terms of thermal conductivity. For a fixed power, particle balance predicts decreasing electron temperature for increasing pressure in a simple gas such as Ar [8]. Furthermore, for a fixed power, the power balance equation can be solved for density to give $n_e \propto (u_B \epsilon_L)^{-1}$, where u_B is the Bohm velocity and ϵ_L is the energy loss per ionization. As pressure increases, temperature decreases which leads to a larger ϵ_L, but a smaller

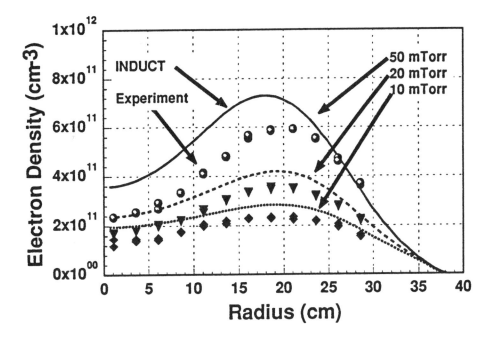

Figure 2. Radial profiles in Ar as a function of pressure for 1200 W inductive power and z ∼ 5.4 cm. Experimental data points are shown along with simulation curves.

u_B. For $T_e > 1$ eV in Ar, the product $u_B \epsilon_L$ goes down with decreasing temperature thus leading to an *increasing* density. Furthermore, for small skin depths, increasing pressure reduces the electron mean-free-path, decreasing the ability of thermal conductivity to spread the rf power throughout the plasma. This results in enhanced off axis peaking in the ionization rate and the density.

4. Conclusions

Our initial modeling of the LLNL Large Area ICP shows good agreement with the experiment in observed trends in density magnitude and profile over a large range of pressures and powers and has increased our general understanding of the source. Future works in modeling include adding more chemistry data to model electronegative gases and gas mixtures and extending the coil and circuit model to include the matching network. With these additions, we will gain an ability to more closely model practical devices and determine prescriptions for enhancing uniformity.

Acknowledgements

We wish to acknowledge helpful discussions with M. Surendra of the IBM Yorktown Heights Research Center. This work was performed under the auspices of the U. S. Department of Energy at the Lawrence Livermore National Laboratory under contract W-7405-ENG-48.

References

1. A. E. Wendt, L. J. Mahoney, and J. L. Shohet, 45th Gaseous Electronics Conference, paper LB-5, Boston, MA, 1992.
2. J. Hopwood, C. R. Guarnieri, S. J. Whitehair, and J. J. Cuomo, *J. Vac. Sci. Technol.* **A 11**, 147,152 (1993).
3. G. DiPeso, V. Vahedi. D. W. Hewett, and T.D. Rognlien (1994), *J. Vac. Sci. Technol. A* **Vol. no. A 12**, p. 1387.
4. R. A. Stewart, P. Vitello, D.B. Graves, E.F. Jaeger, and L. A. Berry (1995), *Plasma Sources Science and Technology*, Vol. no. 4, p. 36.
5. R. A. Richardson, P. O. Egan, and R. D. Benjamin, 12th Intl. Symposium on Plasma Chemistry, Minneapolis, MN, 1995.
6. V. E. Golant, A. P. Zhilinsky, & I. E. Sakharov, *Fundamentals of Plasma Physics*, John Wiley & Sons, New York, 1980, Chapter 6.
7. L. A. Berry, E. F. Jaegar, and J. S. Tolliver, preprint, 1995.
8. M. A. Lieberman and A. J. Lichtenberg, *Principals of Plasma Discharges and Material Proceseeing*, John Wiley & Sons, New York, 1994, Chapter 10.

MULTICOMPONENT REACTIVE GAS DYNAMIC MODEL FOR LOW-PRESSURE DISCHARGES IN FLOWING OXYGEN

M. J. PINHEIRO AND C. M. FERREIRA
Centro de Electrodinâmica, Instituto Superior Técnico, Lisboa, Portugal

AND

G. GOUSSET
Laboratoire de Physique des Gaz et des Plasmas, Université de Paris-Sud, Orsay, France

1. Introduction

Non-isothermal plasma sources under the presence of a gas flow are currently used in many surface processing applications. For this reason, the modelling of these discharges is of fundamental importance in order to understand the heavy-particle kinetics and the transport of species under gas flowing conditions. This paper is an extension of previous work carried out by our group [1-4] on the kinetic modeling of oxygen discharges to account for the effects of gas flow. Here, we present a basic model describing the main electron and heavy-particle kinetic and transport processes in a microwave discharge and we determine the axial concentration profiles, the steady-state discharge maintenance field and mean absorbed power per electron, and the axial distribution of the gas temperature as a function of the up-stream pressure, p (up to 3 Torr), the gas flow rate and the frequency of the sustaining field (390 to 2450 MHz).

In Section 2, we analyse the basic kinetic and transport processes and in Section 3 we derive a 1-D thermal balance equation for the gas that determines the axial distribution of the gas temperature. As in [1], the electron kinetics is described by the homogeneous electron Boltzmann equation which is coupled to the equations of change for the heavy particles. The above equations, together with the radial continuity and transport equations for the charged particles e, O_2^+ and O^- [3,4], allow the consistent cal-

M. Capitelli (ed.), Molecular Physics and Hypersonic Flows, 485–494.

culation of the steady-state maintenance electric field and mean absorbed power per electron. The inelastic and superelastic processes taken into account and the corresponding cross section data are the same as in [1,2].

The reactive fluid model for the heavy species described in Section 2.1 is based on the axial continuity and transport equations for $O_2(X^3\Sigma)$, $O(^3P)$, $O_2(a^1\Delta)$, $O_2(b^1\Sigma)$, O_3, O^- and $O(^1D)$. This model further accounts for the vibrational kinetics governing the populations of $O_2(X^3\Sigma, v)$ (Sect. 2.2) including $e - V$ and $V - V$ energy exchanges, and $V - T$ processes with molecules and dissociated atoms. The predicted characteristic vibrational temperature T_v vs. pressure and electron density is shown to be in agreement with measurements by CARS [5,6]. Finally, in Section 4 we report our calculations and discuss the relevant physical conclusions.

2. Kinetic Model with Gas Flow

2.1. KINETIC MODEL FOR HEAVY PARTICLES

The main motivation behind the development of the present hydrodynamical model relies on an apparent paradox present in the former model [1-4]. In fact, measurements by actinometry [6,7] have shown that the atomic concentration is nearly independent of the axial position in the surface wave discharge, although n_e increases from the discharge end ($\sim 2 \times 10^{10}$ cm^{-3}) to the gap ($\sim 2 \times 10^{11}$ cm^{-3}), where the sustaining surface wave is launched. The former static model [1-4] predicts an increase in dissociation with n_e. However, due to the long lifetime ($\sim 1.3\ ms$) of the oxygen atoms, the effect of the gas flow on the atomic concentration can be expected to be important. It is therefore necessary to account for the flow effects in order to reach realistic predictions.

As in [1-4], only the axial variations in the heavy species concentrations will be considered (radially homogeneous model). The pressure along the axis is calculated from the Navier-Stokes equation by radially averaging the gas velocity assuming a parabolic velocity profile and a laminar flow (the Reynold number, $\mathcal{R}$, is < 100).

The theoretical frame of this work is based on the axial equations of change for the concentrations of the above mentioned heavy species and for the total gas mass and momentum [8]. The equation for mass conservation can be written

$$\vec{\nabla}.\rho\vec{v_o} = 0 \tag{1}$$

where ρ designates the mass density ($\rho = \sum_{i=1}^{\nu=7} n_i m_i$) and $\vec{v_o}$ is the velocity of the center of mass. The equations of continuity for the chemical species $i = 1, ..., \nu$ are of the form

$$\vec{\nabla}.[n_i(\vec{v_o} + \vec{V_i})] = K_i = S_i - P_i n_i \tag{2}$$

where $\vec{V_i}$ denotes the diffusion velocity of the species in the center of mass frame, K_i is the net source of the ith chemical species per unit volume and per unit time, S_i and $P_i n_i$ denoting the creation and the destruction rate of the species, respectively.

The basic set of reactions is presented in Table 1 together with the corresponding rate coefficients and reaction enthalpies [9].

For a dilute gas, the diffusion velocities satisfy the equations [8]

$$\sum_{j=1}^{\nu} \frac{n_i n_j}{n^2 \mathcal{D}_{ij}} (\vec{V_j} - \vec{V_i}) = \vec{d_i} \tag{3}$$

where $\mathcal{D}_{ij}$ represents the binary diffusion coefficient for species i diffusing in species j and $n = \sum_{i=1}^{\nu=7} n_i$. Note that this expression is an approximation since we have neglected a term proportional to $\vec{\nabla} lnT$. The vectors $\vec{d_i}$ satisfy the approximate relations

$$\vec{d_i} = \vec{\nabla}\alpha_i - \frac{n_i m_i}{p\rho}\left(\frac{\rho}{m_i}\vec{X_i}\delta_{i6} - n_6\vec{X_6}\right) \tag{4}$$

where we have now neglected a term proportional to $\vec{\nabla} lnp$. Herein, $\alpha_i = n_i/n$ and δ_{i6} is the kronecker symbol ($n_{i=6} = [O^-]$). The axial force $\vec{X_6}$ acting on the negative ions is due to the axial space-charge field arising from the axial plasma inhomogeneity. The effects of this force are particularly important in axially confining the negative ions near the two extremities of the plasma column.

Finally, the Navier-Stokes equation for the present situation can be written as [10]

$$\frac{dp}{dz} = -\frac{4}{3}\rho\bar{v_o}\frac{d\bar{v_o}}{dz} - \frac{8\bar{\eta}}{R^2}\bar{v_o} \tag{5}$$

the bar denoting a radially averaged value..

2.2. VIBRATIONAL KINETIC MODEL FOR $O_2(X^3\Sigma, V)$

The continuity equation for the molecules $O_2(X^3\Sigma, v)$ in the vibrational state v is similar to Eqn.(2), the diffusion velocity V_v for each vibrational species satisfying a relation similar to Eqn.(3). For lack of data, it is assumed here that the vibrational species have the same diffusion coefficient, $\mathcal{D}_{vj} = \mathcal{D}_{Xj} = \mathcal{D}_{1j}$, $v = 0, ...l$ (16 vibrational states are taken into account, with $\sum_{v=0}^{15} n_v = n_X$). It can be shown [10] that the equation of continuity can be written in the form

$$\frac{d}{dz}\left[\alpha_v\frac{\mathcal{V}}{\sum_{j=1}^{\nu} m_j\alpha_j} - E\frac{d\alpha_v}{dz} + F\alpha_v\right] = S_v - P_v n_v \tag{6}$$

TABLE 1. List of reactions

Process	Rate coefficients	Enthalpy
$e + O_2(X^3\Sigma) \rightarrow O^- + O(^3P)$	$K_1 = f(E/N)$	–
$e + O_2(a^1\Delta) \rightarrow O^- + O(^3P)$	$K_2 = f(E/N)$	–
$O^- + O(^3P) \rightarrow O_2(X^3\Sigma) + e$	$K_3 = 1.4 \times 10^{-10} cm^3 s^{-1}$	$350.9 KJ/mol$
$O^- + O_2(a^1\Delta) \rightarrow O_3 + e$	$K_4 = 3 \times 10^{-10} cm^3 s^{-1}$	$53.4 KJ/mol$
$e + O_2(X^3\Sigma) \rightarrow 2O(^3P) + e$	$K_5 = f(E/N)$	$(6 - 5.1)eV$
$e + O_2(a^1\Delta) \rightarrow 2O(^3P) + e$	$K_6 = f(E/N)$	–
$O(^3P) + wall \rightarrow \frac{1}{2}O_2(X^3\Sigma)$	$K_7 = 800 \times \sqrt{\frac{T_g}{300}} s^{-1}$	$\frac{1}{2}5.1eV \times \beta_7$
$e + O_2(X^3\Sigma) \rightarrow O_2(a^1\Delta) + e$	$K_8 = f(E/N)$	$156.9 KJ/mol$
$e + O_2(a^1\Delta) \rightarrow O_2(X^3\Sigma) + e$	$k_9 = f(E/N)$	$94.3 KJ/mol$
$e + O_2(a^1\Delta) \rightarrow O_2(b^1\Sigma) + e$	$K_{10} = f(E/N)$	–
$e + O_2(b^1\Sigma) \rightarrow O_2(a^1\Delta) + e$	$k_{11} = f(E/N)$	–
$O_2(a^1\Delta) + wall \rightarrow O_2(X^3\Sigma)$	$K_{12} = 0.4 \times \sqrt{\frac{T_g}{300}} s^{-1}$	$0.98eV \times \beta_{12}$
$e + O_2(X^3\Sigma) \rightarrow O_2(b^1\Sigma) + e$	$K_{13} = f(E/N)$	$156.9 KJ/mol$
$e + O_2(b^1\Sigma) \rightarrow O_2(X^3\Sigma) + e$	$K_{14} = f(E/N)$	–
$O_2(b^1\Sigma) + wall \rightarrow O_2(X^3\Sigma)$	$K_{15} = 400 \times \sqrt{\frac{T_g}{300}} s^{-1}$	$1.63eV \times \beta_{15}$
$2O(^3P) + O_2(X^3\Sigma) \rightarrow O_3 + O(^3P)$	$K_{16} = 2.15 \times 10^{-34} exp(+345/T_g)$	$106.5 KJ/mol$
$O(^3P) + 2O_2(X^3\Sigma) \rightarrow O_3 + O_2(X^3\Sigma)$	$K_{17} = 6.4 \times 10^{-35} exp(663/T_g)$	$106.5 KJ/mol$
$O(^3P) + O_3 \rightarrow O_2(a^1\Delta) + O_2(X^3\Sigma)$	$K_{18} = 1.0 \times 10^{-11} exp(-2300/T_g)$	$297.6 KJ/mol$
$O(^3P) + O_3 \rightarrow 2O(X^3\Sigma)$	$K_{19} = 1.8 \times 10^{-11} exp(-2300/T_g)$	$391.9 KJ/mol$
$O_2(b^1\Sigma) + O_3 \rightarrow 2O_2(X^3\Sigma) + O(^3P)$	$K_{20} = 1.5 \times 10^{-11} cm^3 s^{-1}$	$50.4 KJ/mol$
$O_2(a^1\Delta) + O_2(X^3\Sigma) \rightarrow 2O_2(X^3\Sigma)$	$K_{21} = 6 \times 10^{-16} cm^3 s^{-1}$	$94.3 KJ/mol$
$O_2(a^1\Delta) + O_3 \rightarrow 2O_2(X^3\Sigma) + O(^3P)$	$K_{22} = 5.2 \times 10^{-11} exp(-2840/T_g)$	–
$e + O_2(X^3\Sigma) \rightarrow e + O(^3P) + O(^1D)$	$K_{23} = f(E/N)$	–
$e + O_2(a^1\Delta) \rightarrow e + O(^3P) + O(^1D)$	$K_{24} = f(E/N)$	$(8.4 - 7.06)eV$
$e + O(^3P) \rightarrow e + O(^1D)$	$K_{25} = f(E/N)$	–
$O(^1D) + O(^3P) \rightarrow 2O(^3P)$	$K_{26} = 8 \times 10^{-12} cm^3 s^{-1}$	$189.7 KJ/mol$
$O(^1D) + O_2(X^3\Sigma) \rightarrow O(^3P) + O_2(X^3\Sigma)$	$K_{27} = 7 \times 10^{-12} exp(67/T_g)$	$189.7 KJ/mol$
$O(^1D) + O_3 \rightarrow O(^3P) + O_3$	$K_{28} = 2.4 \times 10^{-10} cm^3 s^{-1}$	$189.7 KJ/mol$
$O(^1D) + O_2(X^3\Sigma) \rightarrow O(^3P) + O_2(b^1\Sigma)$	$K_{29} = 2.56 \times 10^{-11} exp(67/T_g)$	$32.8 KJ/mol$
$O(^1D) + O_2(X^3\Sigma) \rightarrow O(^3P) + O_2(a^1\Delta)$	$K_{30} = 10^{-12} cm^3 s^{-1}$	$95.4 KJ/mol$
$O(^1D) + O_3 \rightarrow 2O_2(X^3\Sigma)$	$K_{31} = 2.4 \times 10^{-10} cm^3 s^{-1}$	$189.7 KJ/mol$
$O(^1D) + O_3 \rightarrow O_2(X^3\Sigma) + 2O(^3P)$	$K_{32} = 1.2 \times 10^{-10} cm^3 s^{-1}$	$83.3 KJ/mol$
$O(^1D) + O_3 \rightarrow O_2(X^3\Sigma) + O_2(a^1\Delta)$	$K_{33} = 1.2 \times 10^{-10} cm^3 s^{-1}$	$487.3 KJ/mol$
$O(^1D) + wall \rightarrow O(^3P)$	$K_{34} = \frac{1.728 \times 10^3}{p} (\frac{T_g}{293})^{\frac{3}{2}} s^{-1}$	$1.97eV \times \beta_{34}$
$O(^3P) + O_2(b^1\Sigma) \rightarrow O(^3P) + O_2(X^3\Sigma)$	$K_{35} = 8 \times 10^{-14} cm^3 s^{-1}$	$156.9 KJ/mol$
$O^- + O_2(b^1\Sigma) \rightarrow O(^3P) + O_2(X^3\Sigma)$	$K_{36} = 6.9 \times 10^{-10} cm^3 s^{-1}$	$9.5 KJ/mol$
$e + O_2(X^3\Sigma) \rightarrow O_2^+ + 2e$	K_{37}	$12eV$
$O_2^+ + e \rightarrow O + O$	$K_{38} = 2 \times 10^{-7} (300/T_g) cm^3.s^{-1}$	$672.6 KJ/mol$
$O^- + O_2^+ \rightarrow O + O_2$	$K_{39} = 2 \times 10^{-7} (300/T_g)^{0.5} cm^3.s^{-1}$	$1023.5 KJ/mol$

where $\alpha_v = n_v/n_o$, and E and F are α_v-independents coefficients given by

$$E = 1/\left(\frac{\alpha_X}{\mathcal{D}_X^R} + \sum_{j=2}^{\nu} \frac{\alpha_j}{\mathcal{D}_{Xj}^R}\right) \tag{7}$$

and

$$F = \sum_{j=2}^{\nu} \Gamma_j \left[\frac{m_j}{m_X \mathcal{D}_X^R} - \frac{1}{\mathcal{D}_{Xj}^R}\right] E \tag{8}$$

Herein, the superscript R holds for the reduced diffusion coefficients $\mathcal{D}_X^R = \mathcal{D}_X n$.

3. Calculation of the Average Gas Temperature

It is important to take into account the discharge thermal balance in order to build a consistent description of the way the input power is dissipated in the medium. Therefore, we present here a unified description including the thermal energy balance equation for the gas. The most important gas heating source terms are taken into account [9] and presented in Table 1. The wall temperature, $T_w(z)$, and the electron density profiles are taken from experiments on surface wave discharge [7]. These are the only external data required in the model.

The 1-D thermal balance equation for the gas can be written as:

$$\rho c_p v \frac{\partial T}{\partial z} = v\frac{dP}{dz} + \frac{8\lambda}{R^2}(T - T_w) + \frac{4}{3}\frac{\mu v_o}{R} - \sum_{i=1}^{\nu} H_i R_i + Q_w + Q_e + Q_v \tag{9}$$

where c_p denotes the average specific heat of heavy particles at constant pressure ($c_p = \sum_{j=1}^{\nu} c_p^{(i)}$, $c_p^{(i)}$ being the specific heat for species i), v_o is the axial component of the gas velocity, λ is the $O_2(X^3\Sigma)$ thermal conductivity (taken from [11]), μ is the viscosity coefficient [8], Q_e is the net power transferred from the electrons to the gas translational mode in elastic collisions, rotational excitation, and O_2 dissociation (due to the production of hot atoms), Q_w is the net gas heating source due to wall reactions, and Q_v is the net gas heating source due to $V - T$ relaxation. $R_j = \sum_k^l R_{j,k}$ represents the net rate of creation of species j, where the sum extends over all the heavy particle reactions involving the species. $H_j = c_p^{(j)}(T - T^0) + H_j^0$ is the molecular enthalpy of species j (with a negative sign for exothermic reactions) and H_j^0 is the standard enthalpy of formation of the j-th species from its constituent elements at the enthalpy reference temperature $T^0 = 298\ K$ [8,9].

A complex aspect in the study of such a reacting system is the interplay of both volume and surface reactions (e.g., catalytic recombination of gaseous atoms on the wall). Surface reactions can partly be responsible for gas heating if and when there is a net transfer of heat from the wall to the gas phase. On the wall, energy is deposited by reassociation (mainly $O(^3P) + wall \rightarrow \frac{1}{2}O_2(X^3\Sigma)$ or by deactivation of excited states (e.g., $O_2(b^1\Sigma) + wall \rightarrow O_2(X^3\Sigma)$). A fraction $(1 - \beta)$ of this energy returns to gas phase, presumably through the production of vibrationally excited states $O_2(X^3\Sigma, v)$ that relax very fast and heat the gas [12]. It seems that the Rideal mechanism [13,14] (involving the collision of a gaseous atom with an adatom) is predominant for glass (perhaps due to a large activation energy for the Hinshelwood mechanism).

4. Results and Discussion

The hydrodynamical model previously described was applied for pressures between 0.39 and 2 Torr. The operating conditions are specified by the mean pressure at the tube entrance, the surface wave frequency, the flow rate and the type of flow, i.e., direct flow (FD) or inverse flow (FI) with respect to the wave propagation direction, and the tube diameter ($2R = 16$ mm). The symbols E, G and F appearing in the figures identify the positions of the entrance of fresh gas into the discharge region, of the exciter gap, and of the discharge end. Figs. 1(a)-(b) show experimental and calculated average gas temperature profiles as well as calculated vibrational temperatures for two pressures. T_w is the wall temperature taken from experiment. Note that at the lower pressure the experimental and the calculated profiles of the gas temperature are practically the same. We have found that the calculated $T_g(z)$ could be fitted to experiment only if a major contribution of wall atomic reassociation (in order, Reaction 7 in Table 1) to gas heating was assumed (source term: Q_w (eV.cm^{-3}.s^{-1}) $= \frac{1}{2}5.1\gamma[O]\frac{\bar{v}}{2R}\beta_7$, where $\bar{v}$ is the mean atomic velocity, γ is the probability of atomic reassociation at the wall and $(1 - \beta)$ is the wall accommodation coefficient for thermal energy). The agreement is obtained with $\beta_7 \simeq 9.68 \times 10^{-2}(T_w/300)^{1.5}$, for $400K < T_w < 900K$. Note that the vibrational temperature follows the axial dependence of the gas temperature except in the regions E and F: in the region E, we can observe a steap increase in T_v due to fast electron pumping of the $O_2(X^3\Sigma, v)$ states at the plasma entrance; in region F, T_v rapidly decays due to the strong decrease in electron density near the plasma end and to the fast vibrational deactivation by $V - T$ processes.

The heavy particle concentrations along the discharge tube are compared to available experimental data [6] in figs. 2(a)-(b). As it can be seen from this figure, due to the gas flow all these species are easily transported towards the afterglow region.

Figs. 3 and 4 show the O_3, $O(^1D)$, O^- and n_e (the electron density is from experiment) concentration profiles (FD) vs. z at 0.39 and 2 *Torr*, respectively. The ozone kinetics is strongly coupled to $O_2(b^1\Sigma)$. From Figs. 3 and 4 one observes the direct dependence of $[O(^1D)]$ on the electron density and the effect of the longitudinal space-charge field on the density profile of negative ions near the plasma edges (due to the strong electronic gradient at these edges).

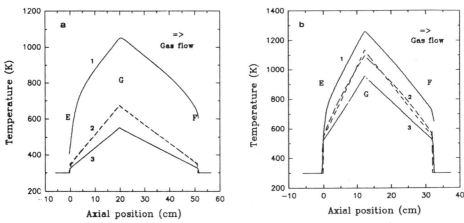

Figure 1. T_g and T_v vs. z at 390 MHz, for (a): $p = 0.39$ Torr (in this case the experimental curve is practically similar to the theoretical curve) and (b): $p = 2$ Torr. $Q = 75$ sccm (FD). Curve 1: T_v^{the}; curve 2: T_g^{exp}; curve 3: T_w^{exp}.

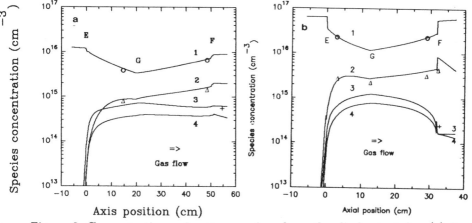

Figure 2. Concentrations of active species along the discharge tube. (a) for $p = 0.39$ Torr and (b) for $p = 2$ Torr, $f = 390$ MHz e $Q = 75$ sccm (FD). 1-$[O_2(X)]$; 2-$[O(^3P)]$; 3-$[O_2(a)]$; 4-$[O_2(b)]$. The experimental points are extracted from [6]: $o - O_2(X^3\Sigma)$; $\Delta - O_2(a^1\Delta)$; $+ - O(^3P)$.

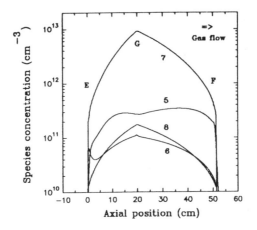

Figure 3. Concentrations of O_3 (5), O^- (6), $O(^1D)$(7) and n_e (8) vs. z at $p = 0.39$ Torr. $Q = 75$ *sccm* (FD) and $f = 390$ MHz.

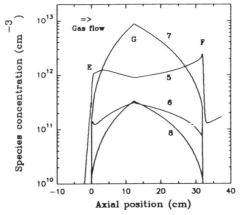

Figure 4. Concentrations of O_3 (5), O^-(6), $O(^1D)$(7) and n_e(8) vs. z at $p = 2$ Torr. $Q = 75$ *sccm* (FD) and $f = 390$ MHz.

Figures 5(a)-(b) show examples of calculated percentage contributions of various mechanisms to gas heating vs. axial position in the discharge column, for the same typical pressures. These calculations are based on a number of assumptions: we have used the cross sections for electron energy losses of 4.5, 6, and 8.4 eV proposed by Phelps but have assumed that only the later two correspond to dissociation, while the first one corresponds to the formation of bound vibrational levels of the states $c^1\Sigma_u^-$, $A^3\Sigma_u^+$, $C^3\Delta_u$. The kinetics of these bound levels is not yet well known (for example, they may be involved in the formation of $O_2(b)$ [15]), so we have made the working hypothesis that their relaxation ultimately results in gas heating (this is, of course, an extreme value for this contribution), maybe via the relaxation of $O_2(X, v)$ levels. Further, we have assumed energy accommodation coefficients of 50 % for wall deactivation of $O_2(a)$ and $O_2(b)$ and for electron-positive ion recombination at the wall.

In conclusion, wall processes seem to play an important role in the thermal balance of oxygen discharges, in addition to the role they are known to play in the particle balance.

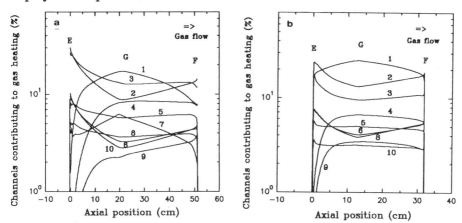

Figure 5. (a): $P = 0.39$ Torr, $f = 390$ MHz, $Q = 75$ sccm. Processes: 1- atomic reassociation at the wall; 2- relaxation of the 4.5 eV electron energy losses; 3 - 8.4 eV dissociation from $O_2(X)$; 4 - dissociation from $O_2(a)$; 5- $O(^1D) + O_2(X) \rightarrow O(^3P) + O_2(X)$; 6 - 6 eV dissociation from $O_2(X)$; 7 - electron-ion recombination at the wall; 8 - $O_2(X, v)$ vibrational relaxation; 9 - $O_2(a) \rightarrow O_2(X)$; 10 - $O_2(X) + O(^1D) \rightarrow O_2(b) + O(^3P)$. (b): As in (a) but for $p = 2$ Torr.

5. Acknowledgement

We are grateful to Dr. Boris F. Gordiets for very enlightening discussions.

6. References

1. Gousset G., Ferreira C.M., Pinheiro, M., Sa P.A., Touzeau M., Vialle M., and Loureiro J. (1991) Electron and heavy-particle kinetics in the low-pressure oxygen positive column, *J. Phys. D: Appl. Phys.* **24**, 290-300.
2. Ferreira C.M., Alves L.L, Pinheiro M. and Sá A.B. (1991) Modeling of low-pressure microwave discharges in Ar, He and O_2: similarity laws for the maintenance field and the mean power transfer, *IEEE Trans. on Plasma Sci.* **19**, 229-239.
3. Ferreira C.M., Gousset G., and Touzeau M. (1988) Quasi-neutral theory of positive columns in electronegative gases, *J.Phys.D:Appl.Phys.* **21**, 1403-1413.
4. Gousset G., Pinheiro M.J. and Ferreira C.M. (1991) A consistent kinetic

model of low pressure microwave discharges in oxygen, 10th International Symposium on Plasma Chemistry, Bochum, Germany.

5. Lefebvre M., Péalat, Gousset G., Touzeau M., and Vialle M. (1990) Vibration kinetics of oxygen discharges by CARS, *Proc.* 10th *ESCAMPIG*, Orléans, France, Europhys. Conf. Abstract, Vol 14E, 246-247.

6. Safari R., Boisse-Laporte C., Granier A., Lefebvre M. and Pealat M. (1990) Investigation of flowing microwave oxygen discharges by CARS, Proc. 10th Escampig, Europhys. Conf. Abstract, Vol.14-E, 258-259.

7. Granier A. (1992) Surface wave plasmas in $O_2 - N_2$ mixtures as active species sorces for surface treatments, NATO Workshop "Microwave discharges: Fundamentals and Applications", Vimeiro, Portugal.

8. Hirschfelder J.O., Curtiss C.F. and Bird R.B. (1954) *Molecular Theory of Gases and Liquids*, John Wiley and Sons, Inc., New York, USA.

9. Eliasson E. and Kogelschatz U. (1987) Brown Boveri Report, $N^o KLR83-40C$, Baden, Switzerland.

10. Pinheiro M.J. (1993) Modelização da cinética dos electrões e das espécies excitadas em descargas dc e hf de oxigénio a baixa pressão, PhD Thesis, Technical University of Lisbon, September, Portugal.

11. Engineering Sciences Data Unit **90014** (1990), publs: Institute of Chemical Engineering and Institute of Mechanical Engineers, UK.

12. Eenshuistra P. J., Bonnie J. H. M., Los J., and Hopman H. J. (1988) Observation of exceptionally high vibrational excitation of hydrogen molecules formed by wall recombination, *Phys. Rev. Lett.* **60**(4), 341-344

13. Greaves J.C. and Linnett J.W. (1959) Recombination of atoms and surfaces. Part 6.- recombination of oxygen atoms on silica from 20 ^0C to 600 ^0C, *Trans. Faraday Soc.* **55**, 1355-1361.

14. Gelb A. and Kim S.K. (1971) Theory of atomic recombination on surfaces, *J. Chem. Phys.* **55** (10), 4935-4939.

15. Lawton S.A. and Phelps A.V. (1978) Excitation of the $b^1\Sigma_g^+$ state of O_2 by low energy electrons, *J.Chem.Phys.* **69** (3), 1055- 1065.

MATHEMATICAL MODELS FOR PLASMA AND GAS FLOWS IN INDUCTION PLASMATRONS

S.A. VASIL'EVSKII, A.F. KOLESNIKOV, M.I. YAKUSHIN
Institute for Problems in Mechanics
Russian Academy of Sciences, Moscow, Russia

High frequency induction (electrodeless) plasmatrons provide the wide possibilities for modeling of nonequilibrium processes of dissociated gas flow thermochemical interaction with a surface at the conditions corresponding to hypersonic vehicle's flight in atmosphere at high altitude [1].

Plasmatrons IPG-4 and IPG-3 with power of 100 and 500 kW allow to produce sub- and supersonic flows of highenthalpy pure air, nitrogen, oxygen, carbon dioxide and other gases for the wide range of trajectory parameters - total enthalpy (5 - 40 MJ/kg) and stagnation pressure (0.01 - 1 atm). The main question in the problem of modeling of aerodynamic heating of vehicle surface is the question of correspondence between boundary layer local characteristics near a vehicle and a model.

Conditions for correspondence of heat transfer to stagnation points of a vehicle and a model have been formulated in [2] for the case of equilibrium flows outside boundary layers near a vehicle and a model.

In general, flows behind a shock wave and after plasmatron discharge channel exit are nonequilibrium, so the problem of thermochemical modeling may be solved only by combination of precise diagnostics of flow parameters with numerical simulation of gas and plasma flows for particular facility and flight conditions [3].

Numerical simulation of plasma and gas flows in induction plasmatrons is the integral part of modern experiments, in particular it is the

M. Capitelli (ed.), Molecular Physics and Hypersonic Flows, 495–504.

integral part of experiments on determination of materials catalytic properties with respect to atoms geterogeneous recombination at a surface [4].

Present work describes a method for numerical simulation of stationary laminar axisymmetric gas and plasma flows in regions from the inlet section of discharge channel to the model surface for geometry and physical parameters of our IPG-4 and IPG-3 plasmotrons at subsonic regimes.

It is natural to subdivide the whole flow field in plasmatron into three computing regions according to the geometry and physical and chemical processes in the flow: discharge channel (1); operating part of the jet after the discharge channel exit section (2); boundary layer at the model surface (3). Formulation of the problem of numerical simulation for gas and plasma flow can be simplified in each of these regions.

1. Plasma flow in the discharge channel

The previous measurements of translational, vibrational and rotational temperature values in air, nitrogen and oxygen subsonic flows revealed that these flows at plasmatron discharge channel exit section are nearly equilibrium at pressures $P \geq 0.1$ atm. On the basis of this experimental fact, we can formulate the problem of numerical simulation for the discharge channel flow in the frameworks of Navier - Stokes equations for equilibrium plasma flow. Then influence of the induced magnetic field within the inductor coils and Joule heat production in the discharge will be accounted by the source terms in the equations of impulse and energy conservation.

One temperature approximation is used for gas mixture. So, chemical composition at each point of the flow can be calculated as function of pressure and temperature according to chemical equilibrium equations for dissociated and partially ionized gas mixture.

Full Navier - Stokes equations are written in two-dimensional cylindrical coordinate system, but three velocity components are essential for the flow in discharge channel: along channel axes, radial and tangential velocity component due to flow spinning.

Maxwell equations are used to determine time-averaged complex amplitude of vortical electric field. The simple form of Ohm's law is used for averaged electric current and electric field strength.

Electric conductivity of equilibrium plasma mixture is the "interface" to couple the two subsystem of equations: Navier - Stokes and Maxwell equations.

At present, we don't know any work with strict formulation of electromagnetic part of the problem under consideration. Such a strict formulation should consider a plasma load as one of elements of equivalent electric circuit representing plasma generator on the whole. This means that electric current in inductor should be considered as one of unknown parameters in this problem because of the reverse influence of plasma load on electric parameters of high frequency generator.

In present work, simplified problem formulation is used [5]. We assume that $dE_\theta/dz \ll dE_\theta/dr$, so Maxwell equations can be simplified in local one-dimensional approximation:

$$\frac{d}{dr}\left(\frac{1}{r}\frac{d}{dr}(rE_\theta)\right) = -i\omega\,\mu_0\sigma\,E_\theta$$

$$i\,\omega\,\mu_0\,H_z = \frac{1}{r}\frac{d}{dr}(rE_\theta)$$

Here: r and z are radial and axial coordinates, E_θ - complex amplitude of the tangential component of vortical electric field, ω - angular frequency, μ_0 - magnetic constant, H_z is complex amplitude of the magnetic field axial component.

Terms with time-averaged electromagnetic force and Joule heat production in the equations of gasdynamic subsystem can be expressed by means of complex amplitude E_θ and it's radial derivative.

Unlike [5], boundary conditions for electric subsystem are symmetry conditions at the axis and the following condition at discharge channel wall:

$$r = r_w : \qquad \frac{1}{r}\frac{d}{dr}(rE_\theta) = i\,\omega\,\mu_0\,H_{zwo}$$

here H_{zwo} is complex amplitude of the magnetic field axial component at the channel wall produced by inductor coils without plasma flow within the channel, i.e with no account of the reverse influence of plasma flow in the channel on magnetic field at channel walls.

Finite difference analogies of Navier - Stokes equations for axisymmetric flow with spinning have been written for control volumes using staggered grid. Obtained equations have been solved by a method analogous to Patankar and Spalding SIMPLE method [6]. Final system of linear algebraic equations have been solved by the modified method of incomplete factorization (MIC).

Complicated vortical structure of equilibrium air plasma flow in IPG-4 discharge channel is shown in fig. 1: streamlines, isotherms and isolines of vortical electric field strength $|E_\theta|$ for pressure 0.1 atm, air flow rate 2.8 g/s, power input in plasma 30 KW.

Output parameters for solution in the first region are profiles of density, velocity, temperature and enthalpy at exit section of plasmatron discharge channel. These profiles are used then as inflow boundary conditions for solution in the second region.

2. Subsonic jet flow of viscous dissociated gas over a cylindrical model

For subsonic regimes of plasmatrons operation, jet flow over a model is characterized by low values of Mach and Reynolds numbers. So, full Navier - Stokes equations are used to simulate jet flow of multicomponent dissociated gas mixture in the second region. Ionization and tangential velocity component due to flow spinning are not accounted in this region because these processes are not essential for heat transfer to model surface.

Numerical simulation of the problem in the second region provides flow fields of enthalpy, temperature and species concentrations for nonequilibrium jet flow. These numerical data essentially complement results of measurements.

Calculations were made for noncatalytic, full catalytic surfaces and also for surfaces with catalycity discontinuity. Geterogeneous atoms recombination has been described as a first order reaction with two effective parameters - recombination rate constants for nitrogen and oxygen K_{wN} and K_{wO}.

Navier - Stokes equations in the second region have been solved also by Patankar and Spalding SIMPLE method [6].

Fig. 2 shows calculated temperature distribution on a plane perpendicular to jet flow of nonequilibrium dissociated nitrogen for the following three cases of heat exchange: full catalytic surface (1); low catalycity surface with $K_{wN} = 1$ m/s (2); low catalycity surface with high catalytic ring insert (3). Flow parameters at the jet symmetry axis correspond to experimental conditions in IPG-3: P=0,015 atm, H=23,5 MJ/kg, T=7400 K, V=650 m/s. For all three cases, including high catalytic ring with temperature jump due to the effect of "over-equilibrium" heating, we have a good agreement with experimental data on surface temperature measured by infrared scanning system AGA-780 [7].

Gasdynamic flow characteristics at outer edge of boundary layer near model can be obtained from numerical solution in the second region, because this region includes also boundary layer near the model. These characteristics are necessary for problem formulation in the third region.

3. Nonequilibrium boundary layer

Formally, solution obtained for the second region on the basis of Navier - Stokes equations provides all necessary information on flow parameters including heat transfer to model surface and parameters of nonequilibrium boundary layer near the surface for all operating regimes of IPG-4 and IPG-3. But in real practice, simplified method is necessary for quick and sufficiently precise calculation of heat transfer to model's stagnation point with arbitrary surface catalycity for the wide range of boundary layer nonequilibriumity from frozen to equilibrium.

The appropriate effective method has been developed [4] on the basis of one-dimensional equations of nonequilibrium boundary layer with finite thickness. These equations can properly simulate the flow, diffusion and heat transfer in the third region - within boundary layer along axis of symmetry near the stagnation point - for all operating regimes of IPG-4, IPG-3.

Mathematical formulation of the problem in the third region is based on the results of the previous Navier - Stokes solution for the second region, because the necessary boundary conditions for boundary layer finite thickness equations includes such dimensionless parameters as relative thickness of boundary layer and some velocity gradients at the outer edge of boundary layer. These dimensionless parameters are determined by the Navier - Stokes solution in the second region.

Finite difference analog of boundary layer finite thickness equations has been obtained with fourth order of approximation by means of Petukhov method [8]. This high order of approximation finite difference scheme provides a very high precision for heat flux value to the stagnation point even for the moderate number of grid points across the boundary layer. One - dimensional boundary value problem for the obtained finite difference equations can be solved by Thomas algorithm. So, it takes about 10 seconds to calculate heat flux to the stagnation point for the given wall catalycity using this effective method in the third region.

Fig 3 illustrates a method of determination of average effective geterogeneous recombination probability γ_{eff} for atoms recombination at surfaces with heat protection coatings at high temperatures by the results of surface temperature and heat flux measurements. The map of heat flux values in q_w and T_w variables has been calculated on the basis of previously described numerical solution technique for experimental conditions: pressure 0.1 atm, model radius 1.5 cm, jet flow enthalpy 29 MJ/kg and velocity 200 m/s. Solid curves on the map fig. 3 correspond to heat flux at constant values of γ_{eff}. For "Buran" heat protection tile coating we have $\gamma_{eff} \cong 0,004$, for carbon-carbon material with antioxidation coating $\gamma_{eff} \cong 0,006$.

Fig 4 shows a comparison between measured values of heat flux to the stagnation points of models with different metal cold surfaces and the results of numerical calculation of heat flux as function of the effective recombination probability rate $q_w(\gamma_{eff})$ for stagnation pressures 0.1 - 0.4 atm and surface temperature $T_w = 300$ K. Experiments and calculations were made for dissociated nitrogen in IPG-2 plasmatron for cilindrical model with 15 mm radius. It is clear that both experimental points and theoretical curves displace equally with pressure increase due to the appropriate increase of gas phase atoms recombination contribution to heat transfer.

Fig 5 illustrates effect of 20 % increase in heat flux to a titanium surface $T_w = 300$ K flowed by dissociated nitrogen at P=0.1 atm with molecular oxygen injection into nonequilibrium boundary layer [9]. This effect has been explained by interaction of gas-phase exchange reactions and oxygen atoms geterogeneous recombination at the titanium surface with low catalycity with respect to nitrogen atoms recombination and high catalycity with respect to oxygen atoms recombination. Weak injection of molecular oxygen initiates the exchange reactions with production of oxygen atoms near the surface. These atoms diffuse to the titanium surface and recombine with additional heat generation. A good correspondence of numerical results (solid curves) and experimental data (points) confirms that mathematical formulation of the problem is adequate to the conditions of the experiment.

REFERENCES

1. Gordeev, A.N.,Kolesnikov, A.F. and Yakushin, M.I. (1992) An induction plasma application to "BURAN's" heat protection tiles ground tests, *SAMPE Journal* **28** (3), 29-33.

2. Kolesnikov, A.F. (1993) Conditions of simulation of stagnation point heat transfer from a high - enthalpy flow, *Fluid Dynamics (English translation)* July, 131-137.

3. Kolesnikov, A.F. and Shchelin, V.S. (1990) Numerical analysis of simulation accuracy for hypersonic heat transfer in subsonic jets of dissociated nitrogen, *Fluid Dynamics (English translation)* September, 278-286.

4. Kolesnikov, A.F. and Yakushin, M.I. (1989) Determination of heterogeneous recombination effective probabilities of atoms from heat fluxes to the surface in dissociated air flow, *Matematicheskoye Modelirovanie (in Russian)* **1** (3), 44-60.

5. Boulos, M.I., Gagne, R. and Barnes, R.M. (1980) Effect of swirl and confinement on the flow and temperature fields in an inductively coupled r.f. plasma, *The Canadian Journal of Chemical Engineering* **58**, 367-381.

6. Patankar, S.V. and Spalding D.B. (1970) *Heat and Mass Transfer in Boundary Layers*, Intertext Books, London.

7. Baronetz, P.N., Kolesnikov, A.F., Kubarev, S.N., Pershin, I.S., Trukhanov, A.S. and Yakushin, M.I. (1991) Overequilibrium heating of the surface of a heat - shield tile in a subsonic jet of dissociated air, *Fluid Dynamics (English translation)* November, 437-442.

8. Petukhov, I.V. (1964) Numerical calculation of two-dimensional boundary layer flows, in *Numerical Methods of Solving Differential and Integral Equations and Quadrature Formulas (in Russian)*, Nauka, Moscow, pp.304-324.

9. Vasil'evskii, S.A, Kolesnikov, A.F. and Yakushin M.I. (1992) Increased heat transfer to a titanium surface with oxygen injection into the nonequilibrium boundary layer, *Fluid Dynamics (English translation)* Jan., 598-604.

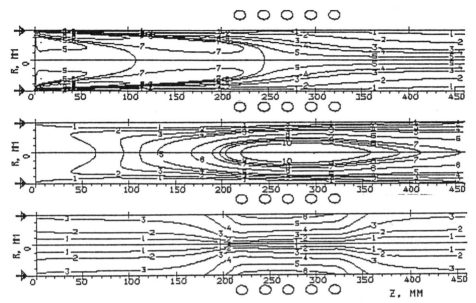

Fig. 1. Streamlines, isotherms and isolines of electric field strength in IPG-4 discharge channel. Dimensionless stream function values for curves: 0.7, 0.3, 0.1, 0.05, 0.01, 0.0, -0.01, -0.04. Temperature values for curves: from 1000 to 10 000 K with 1000 K increment.

$|E_\theta|$ values for curves:
50, 100, 200, 400, 800, 1200 V/m.

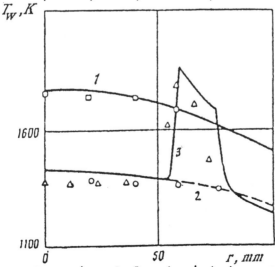

Fig. 2. Experimental (points) and numerical (curves) data on tile surface temperature

504

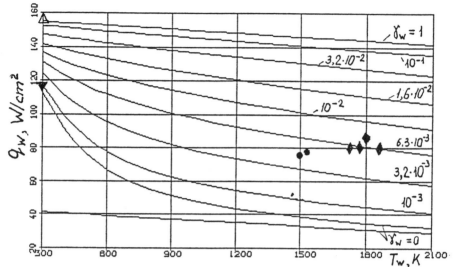

Fig. 3. Experimental (symbols) and numerical (curves) data on heat flux. Lower curve - frozen boundary layer. Symbols: - tile coating, - antioxidation silicate coating, - Mo, - Cu

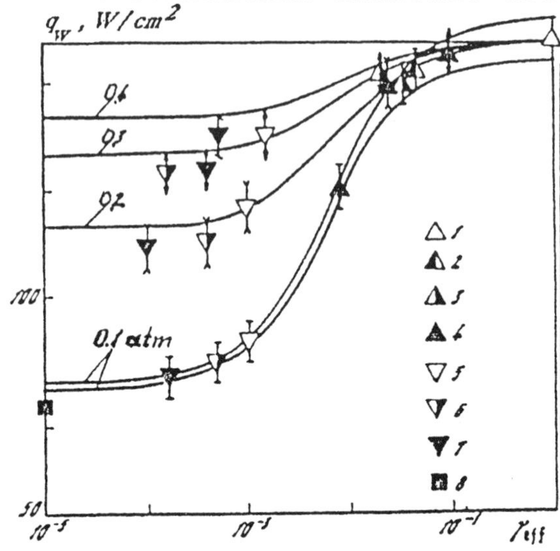

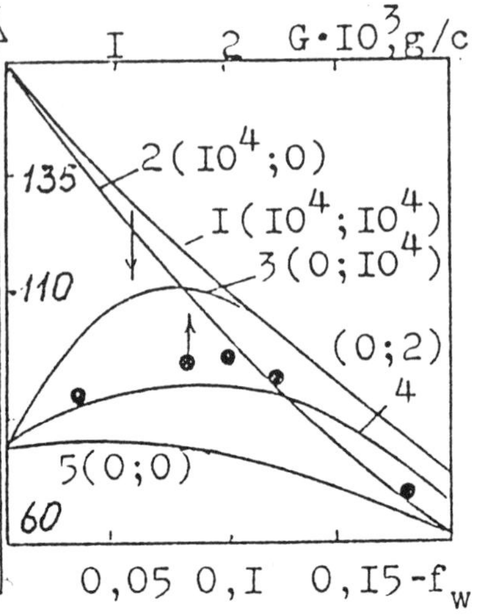

Fig. 4. Heat flux to metallic surfaces at model stagnation point. Experimental data (symbols) for pressures 0.1 - 0.4 atm:
1-Cu, 2-Ag, 3-Ni, 4-Be, 5- W, 6-Cr, 7-Mo, 8-Ti.

Fig. 5. Heat flux to model stagnation point via oxygen injection flow rate and injection parameter.

$N_2(A\ ^3\Sigma_u^+, v=0)$ DECAY IN N_2-O_2 PULSED POST-DISCHARGE

S. DE BENEDICTIS and G. DILECCE
Centro di Studio per la Chimica dei Plasmi C.N.R.

c/o Dipartimento di Chimica , Università di Bari
Via Orabona 4, 70126 Bari (Italy)

1. Introduction

Air like plasmas are currently investigated under a variety of experimental conditions because of the wide interest in the physical chemistry of the upper atmosphere [1], in the simulation of space shuttles re-entry conditions [2] and in the study of environmental chemistry [3]. Many elementary processes must be investigated for understanding this complex gas phase and gas-surface chemistry. In laboratory, glow and afterglow discharges and plasma jets in N_2-O_2 mixture are frequently used for these investigations.

Rate coefficients of many elementary reactions between species formed in N_2-O_2 system are actually available and the importance of many reactions can be estimated. However the modelling of these complex plasmas still requires a much larger data set, and particularly state to state rate coefficients for microscopic kinetics that are difficult to measure. In the case of N_2-O_2 plasma, species like O, N, NO, N_2O, NO_2, having a variety of internal excitation, as well as ions and electrons must be measured. Therefore, various diagnostic methods must be employed on the same system. Typical experiments would require mass spectrometry, radiative spectroscopy, Langmuir probes, ions analyser measurements. In general it is necessary to measure the density and the energy content of reactants and products for each reactive channel. As interesting examples we want to mention here the NO(X,v) formation and loss kinetics for which the role of N_2 vibrational excitation has been recently emphasised in N_2-O_2 discharge modelling [4], and the atomic recombination of fast N atoms in creating $N_2(X,v)$ and $N_2(A,v)$ distributions [5]. For evaluating the importance of these reactions it is necessary to perform more extensive direct measurements of the density and of the degree of internal excitation of species like $N_2(X\ ^1\Sigma_g^+,v)$ and NO, as well as an unambiguous measure of $N_2(A\ ^3\Sigma_u^+,v)$, N and O densities whose

M. Capitelli (ed.), Molecular Physics and Hypersonic Flows, 505–514.
© *1996 Kluwer Academic Publishers. Printed in the Netherlands.*

reaction can also produce NO. Unfortunately, such measurements by laser based diagnostics like CARS and LIF are not yet extensively carried out, so these problems are still open.

Our activity in this field aims to collect as much as possible experimental data on the effectiveness of kinetics involving metastable species formed in N_2-O_2 discharge. We believe that the pulsed discharge at variable duty cycle can be very useful to this concern because the analysis of the discharge and post discharge regimes allows a better discrimination from the electron kinetics. The present topic will be focused on the measure of the decay of the $N_2(A^3 \Sigma_u^+,v)$ metastable in N_2-O_2 pulsed rf post-discharge and the excitation of NO-γ band. Finally some insights on NO ground state vibrational excitation in electrical discharge will also be given. This fundamental subject is justified by the poor knowledge of nitric oxide V-V and V-T rate coefficients and the possibility of NO dissociation by V-V up-pumping, at least in comparison with that on CO [6], even though important efforts have been done in [7].

2. Decay of $N_2(A^3 \Sigma_u^+)$ metastable

The importance of $N_2(A)$ state in energy transfer with the electronic manifold of N_2 is still matter of study even though many elementary kinetics have been elucidated [8] [9]. Here we examine some experimental results on the quenching of $N_2(A)$ state in N_2 and N_2-O_2 discharges in the light of the current literature.

2.1 PURE NITROGEN DISCHARGE

The quenching of $N_2(A)$ metastable has been seen to take place by:

$$N_2(A,w) + N_2(X,v) \rightarrow N_2(X,w') + N_2(B,v') \tag{1}$$
$$N_2(A,w) + N_2(A,w') \rightarrow N_2(C \text{ and } B, w') + N_2(X,v') \tag{2}$$
$$N_2(A,w \geq 7) + N_2 \rightarrow N_2(B, w') + N_2 \tag{3}$$
$$N_2(A,v) + N \rightarrow N_2(X) + N^* \tag{4}$$

A study of $N_2(A)$ quenching carried out in pulsed rf discharge by LIF detection [10] (see the apparatus reported in fig. 1) revealed that process (1) is rather important while the contribution of (2) to the excitation of $(C^3\Pi_u)$ [11] and $B^3\Pi_g$ [12] with respect to the electron impact is important in post-discharge at low pressure. A typical decay of $N_2(A,v=0$ and $v=4)$ is shown in fig. 2. The relaxation of the higher level is faster than the lower one, being the former more involved in energy transfers with neighbouring electronic states.

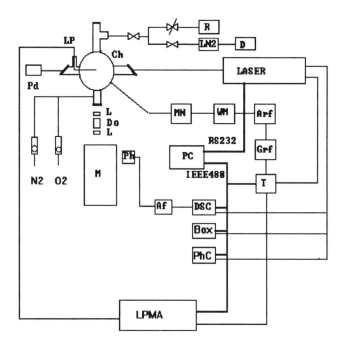

Figure 1. Experimental set-up currently employed in our laboratory for the measurement of the metastable states, radiative species and electrons. **Af**: fast amplifier; **Arf**: wide band rf power amplifier; **Box**: boxcar amplifier; **Ch**: vacuum chamber; **D**: diffusion pump; **DSC**: digitising O-scope; **Do**: Dove prism; **Grf**: low power rf generator; **L**: lens; **LN2**: liquid nitrogen trap; **Lp**: Langmuir probe; **LPMA**: Langmuir probe measurement apparatus; **M**: monochromator; **MN**: matching network; **PC**: personal computer; **Pd**: photodetector; **Ph**: photomultiplier; **PhC**: photon counter; **R**: rotative pump; **T**: timing cards; **WM**: wattmeter.

The little deviation from an exponential trend of $N_2(A,4)$ decay was explained in [10] as due to process (1), while the nitrogen dissociation was estimated to be small and (4) was unimportant.

2.2 N_2-O_2 DISCHARGE

The main quenching of $N_2(A)$ is by atomic and molecular oxygen [13]:

$$N_2(A,v) + O(^3P) \rightarrow NO(X,w) + N(^4S) \quad \Delta H= -2.9eV \qquad (5)$$
$$N_2(A,v) + O(^3P) \rightarrow NO(X,w) + N(^2D) \quad \Delta H= -0.5eV \qquad (6)$$
$$N_2(A,v) + O(^3P) \rightarrow N_2(X,w) + O \qquad (7)$$

508

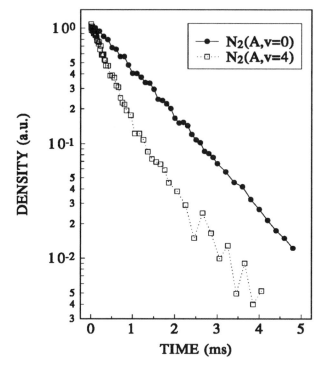

Figure 2. Decays of $N_2(A,v=0, 4)$ in 100 Hz pulsed 0.1 Torr N_2 discharge

$$N_2(A) + O_2 \rightarrow NO_2, NO, N_2(X,v), O \qquad (8)$$
$$N_2(A) + NO \rightarrow NO(A^2\Sigma) + N_2 \qquad (10)$$

The increase of deactivation rate of the metastable in the level v=0 in presence of oxygen in discharge is evident in the decays reported in fig. 3. The LIF signal intensity, measured as the integrated fluorescence pulse and properly corrected by the measured quenching rate of the laser produced emitting state, indicates that the density in the discharge of $N_2(A,v=0)$ decreases by more than one order of magnitude when O_2 varies from 0% to 20%.

3. Formation of NO and γ band excitation

Further important channels forming NO in N_2-O_2 system are the reactions of long lived nitrogen N atoms with atomic and molecular oxygen:

$$N(^4S) + O_2 \quad \rightarrow \quad NO(X,v \leq 7) + O(^3P) \qquad \Delta H= - 1.38 \text{ eV} \qquad (11)$$

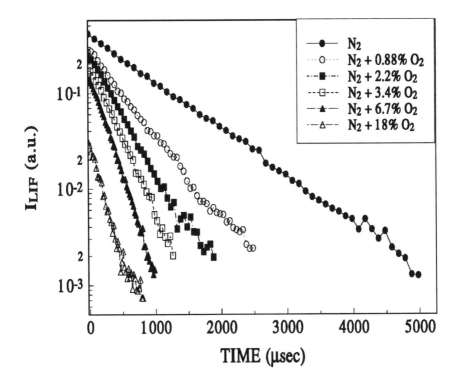

Figure 3. $N_2(A,v=0)$ density decays at various O_2 percentages in the mixture

$$N(^2D) + O_2 \quad \rightarrow \quad NO(X,v \leq 18) + O(^3P) \qquad \Delta H = -3.77 \text{ eV} \qquad (12)$$
$$\rightarrow \quad NO(X,v \leq 8) + O(^1D) \qquad \Delta H = -1.08 \text{ eV} \qquad (13)$$

$$N(^2P) + O_2 \quad \rightarrow \quad NO(X,v \leq 26) + O(^3P) \qquad \Delta H = -4.96 \text{ eV} \qquad (14)$$
$$\rightarrow \quad NO(X,v \leq 14) + O(^1D) \qquad \Delta H = -2.99 \text{ eV} \qquad (15)$$
$$\rightarrow \quad NO(X,v \leq 3) + O(^1S) \qquad \Delta H = -0.77 \text{ eV} \qquad (16)$$

A marker of many of these reactions is the degree of the NO ground state vibrational excitation that is expected to be different. But, as we will see later, getting such information requires a careful study of the nascent $NO(X,v)$ distribution.

We report also the endothermic channel :

$$N_2(X, v>12) + O \quad \rightarrow \quad NO(X) + N \qquad (17)$$

510

that, according to [5], is very effective in N_2-O_2 glow discharge when vibrational levels higher than 12 are pumped.

In our experiment, at present, we have only monitored the time resolved emission of NO-γ band for various N_2/O_2 compositions, and only preliminary NO(X,v=0) data have been obtained by LIF with 226 nm photons. The decay of NO-γ band, fig. 4, is predominantly single exponential. The presence of weak long time constant components is questionable. The decay in post-discharge of γ-band, anyway, agrees with a larger effectiveness of process (10) than the processes by electron impact (18) or direct atomic recombination (19), that would produce γ, β, and δ band emissions:

$$e + NO \rightarrow NO(\ A\ ^2\Sigma,\ B\ ^2\Pi,\ C\ ^2\Pi) + e \qquad (18)$$
$$N(^4S) + O(^3P) + M \rightarrow NO(\ A\ ^2\Sigma,\ B\ ^2\Pi,\ C\ ^2\Pi) + M \qquad (19)$$

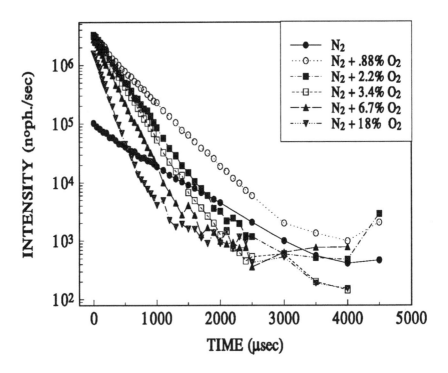

Figure 4. Decay in post-discharge of NO γ band emission at various O_2 percentages

The electron contribution would decay much faster than in the present experiment, as it is actually observed in the excitation and decay of NO γ band in the pulsed discharge of He-NO mixture. Moreover β band emissions have been found to be too much weaker than γ band .

The analysis of $N_2(A)$ and NO-γ band quenching rates reported in fig. 5, confirms to some extent the correlation of these species by process (10), at least in the composition range up to 8% O_2. Concerning the values of $N_2(A)$ quenching rate we have to distinguish two regimes: the first one at low O_2 percentage (up 8% O_2) in which quenching varies almost linearly by O atoms, processes (5) and (6), with a rate coefficient of about 2.2×10^{-11} cm^3 sec^{-1}; the second one, at higher O_2 content, where the quenching varies almost linearly by O_2 with a much lower rate coefficient, about 1.2×10^{-12} $cm^3 sec^{-1}$. This result indicates also that at low O_2 percentage O density increases with an estimated dissociation degree of about 20%, then reaching a saturation value at higher percentages. The rate coefficient values are in good agreements with literature data [13].

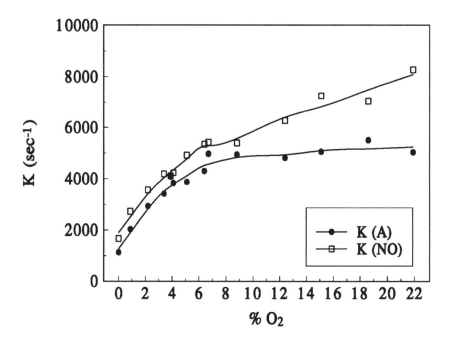

Figure 5. $N_2(A)$ and NO-γ quenching rates vs. O_2 percentage

512

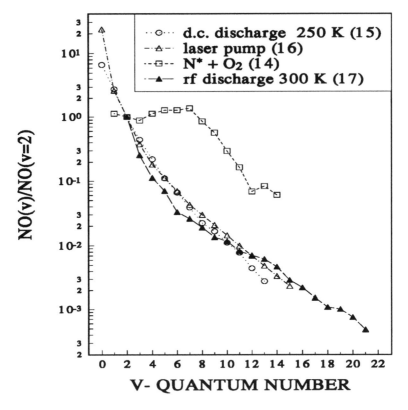

Figure 6. NO(X,v) distributions obtained in different experimental conditions

4. Vibrational excitation of NO(X)

The peculiarity of N_2-O_2 systems is that NO product is in most cases vibrationally excited. The detection of the infrared emission is anyway very difficult because of the small amount of the NO formed in this system, and also because the presence of very strong quenchers produces fast vibrational deactivation. In spite of this difficulty very accurate measurements have been reported in [11] by an experiment properly designed for the detection of very weak infrared signals from fundamental and overtone band systems. The measured nascent distributions show a very high vibrational non-equilibrium as shown in fig. 7. Much important is that the shape of the distribution is very different from those in which the vibrational distribution excitation is produced by V-V' vibrational up-pumping from the low vibrational levels. In the same figure in fact we have reported one NO(X,v) obtained by Rich's group [15] in an experiment in which NO(X,v=1) is pumped by a powerful resonant

CO infrared laser and two distributions obtained in d.c. [16] and rf [17] electrical discharges in He-NO mixture. A very high vibrational excitation is achieved in this latter experiment.

5. Conclusion

In this note we have emphasised the advantages of kinetic studies of the N_2-O_2 mixture carried out in a pulsed discharge by time resolved diagnostics. Even though this experimental results do not cover a fully kinetic analysis of N_2-O_2 mixture, which is still in progress, the correlation of the different decays by the time constant and the amplitudes of the components discriminates the various processes. Measurements via LIF of NO and O and N atoms as well as of $N_2(X,v)$ molecules by CARS will complete this study. This kind of work will contribute to the knowledge of the main features of N_2-O_2 discharges in order to approach their modelling.

6. Acknowledgements

The present work is partially supported by the Italian space agency (ASI)

7. References

1. Kossyi ,A. , Kostinsky, A. Yu, Matveryer, A.A., and Silakov, V.P. (1992) Kinetic scheme of the non-equilibrium discharge in nitrogen-oxygen mixture, *Plasma Source Sci. Technol.* **1**, 207-220.
2. Doroshenko, V.M., Kudryatvtsev, N.N., Smetanin, V.V.(1991) Equilibrium of Internal degree of freedom of molecules and atoms in hypersonic flight in the upper atmosphere, *High Temperature* **29**, 815-832
3. Atkinson, R., Baulch, D.L., Cox, R.A., Hampson, R.F. Jr., Kerr, J.A., Troe, J. (1992) Evaluated kinetic and photochemical data for atmospheric chemistry supplement IV, *J. Phys. Chem. Ref. Data* **21**, 1125-1147
4. Gordiets, B., Ferreira, C.M., Naorny, J., Pagnon, D., Touzeau, M., and Vialle, M. (1995) Surface and volume kinetics in a N_2/O_2 low pressure glow discharge, Heberlein J.V. Ernie D.W., Roberts J.T. (eds) *Proc.s ISPC-12*, Minneapolis, vol. **1**, pp. 445-456
   Guerra, V., Loureiro, J (1995) Self-consisten kinetic model for a N_2-O_2 dc glow discharge, Heberlein J.V. Ernie D.W., Roberts J.T. (eds) *Proc.s ISPC-12 conference*, Minneapolis, vol. **1**, pp. 451-456

514

5. Armenise, I, Capitelli, M., Celiberto, R., Colonna, G., Gorse, C., Lagana, A. (1994) The effect of $N+N_2$ collision on the non-equilibrium vibrational distributions of Nitrogen under re-entry conditions, *Chem: Phys. Letters* **227**, 157-1636

6. Cacciatore, M., Billing, G.D. (1981) Semiclassical calculation of V-V and V-T rate coefficients in CO, *Chem. Phys.* **58**, 395-407.

7. Yung, X., Kim, E.H., Wodtke, A.M. (1992) Vibrational energy transfer of very highly vibrationally excited NO, *J. Chem. Phys.* **96**, 5111-5128

8. Piper, G.L. (1992) Energy transfer studies on $N_2(X^1\Sigma_g^+)$ and $N_2(B^3\Pi_g)$. *J. Chem. Phys* **97**, 270-275

9. Bachmann, R., Li, X., Oettinger, Ch., Vilesov, A.F., (1992) Molecular-beam study of the collisional intramolecular coupling of $N_2(B^3\Pi_g)$ with $N_2(A^3\Sigma_u^+)$ and $N_2(W^3\Delta_u)$ states, *J. Chem. Phys* **96**, 1151-5164.

10. De Benedictis, S., Dilecce, G. and Simek, M., (1993) Time-resolved LIF spectroscopy on $N_2(A)$ metastable in a He/N_2 pulsed rf discharge, *Chem. Phys.* **178**, 547-560.

11. De Benedictis, S. and Dilecce, G., (1995) Vibrational relaxation of $N_2(C,v)$ state in N_2 pulsed rf discharge: electron impact and pooling reactions, *Chem. Phys.* **192**, 149-162

12. De Benedictis, S., Dilecce, G., Simek, M., and Cacciatore, M., (1995) Relaxation Kinetics of $N_2(B^3\Pi_g,v)$ in pulsed rf discharges: experimental results and mechanisms, Heberlein J.V. Ernie D.W., Roberts J.T. (eds) *Proc.s ISPC-12*, Minneapolis, vol. **1**, pp. 415-420

13. Piper, L.G., Caledonia, G.E., Kennealy, J.P. (1981), Rate constants for deactivation of $N_2(A^3\Sigma_u^+,v=0,1)$ by O, *J.Chem.Phys.* **75**, 2847-2853
    De Souza, A.R., Touzeau, M., Petitdider, M., (1985) Quenching reactions of metastable $N_2(A^3\Sigma_u^+,v=0,1,2)$ molecules by O_2, *Chem. Phys. Lett.* **121**, 423-428.

14. Rawlins, W.T., Fraser, M.E. and Miller, S.M., (1989) Rovibrational excitation of nitric oxide in the reaction of O_2 with metastable atomic nitrogen, *J. Phys. Chem.* **93**, 1097-1106

15. Bergman, R.C., Williams, M.J. and Rich, J.W., (1982) Measurement of vibration-vibration pumped population distributions in nitric oxide, *Chem. Phys.* **66** 357-364.

16. Soupe, S., Adamovich, I., Grassi, M.J., Rich,.J.W., Bergman, R.C. (1993) Vibrational and electronic excitation of nitric axide in optical pumping experiments, *Chem. Phys.* **174**, 219-228.

17. De Benedictis, S., Dilecce, G., and Cacciatore., M., (1995) NO(X,v) distribution in rf discharges: conditions for a very high vibrational pumping, Heberlein J.V. Ernie D.W., Roberts J.T. (eds) *Proc.s ISPC-12*, Minneapolis, vol. **1**, pp. 403-408.

INVESTIGATION OF SPECIES IN FLOWING NITROGEN AFTERGLOW BY A DISCHARGE PROBE

G. Dinescu, E. Aldea, D. Bivolaru*, G. Musa

Low Temperature Plasma Physics Department
Institute of Physics and Technology of Radiation Devices
*Space Science Laboratory
Institute of Gravity and Space Science
P.O.B. MG-7, Bucharest-Magurele,
RO-76900, ROMANIA

Abstract

A small oscillatory discharge is generated between the electrodes of a double probe that is introduced in a nitrogen flowing afterglow plasma. In the frame of a kinetic model for the afterglow the changes induced by the afterglow medium on the I-V characteristics of the double probe discharge were related to the actual concentration of species by considering that the produced effects are linearly dependent on the species number density. According to this hypothesis the reaction rate for de-excitation of vibrational excited nitrogen molecules by the wall have been inferred from experiment. Also the number densities of $N(^4S)$, $N(^2P)$, $N(^2D)$, $N_2(A^3\Sigma)$, along the flow axis have been computed.

1. Introduction

The diagnostic of rarefied gases by electrical discharges has been used for a long time. As an example Hirsch [1] has proposed a method (based on a alternative gas discharge) for static measurements of the vacuum.

515

M. Capitelli (ed.), Molecular Physics and Hypersonic Flows, 515–524.
© 1996 *Kluwer Academic Publishers. Printed in the Netherlands.*

Others [2] proposed a similar pulsed technique for steady state and transient measurements of mass density implying bodies immersed in subsonic and supersonic flows. The diagnostic of species in flowing postdischarges has been a subject of extensive studies too. Using the flow tube and the flowing afterglow techniques [3, 4] a basis for quantitative results in chemical gas phase kinetics has been created. The probing method for long life species in flowing postdischarges described in this paper is based on the effects induced by a flowing afterglow excited medium on the electrical properties of a small oscillatory discharge generated between the electrodes of a double probe.

In the following the changes in the I-V characteristics (breakdown and cutoff voltages, dynamic impedance) of a probing oscillatory discharge produced by species evolving from a N_2 flowing afterglow plasma are studied along the flow axis. On the basis of a kinetic model number densities of long life nitrogen species as $N(^4S)$, $N(^2P)$, $N(^2D)$, $N_2(A^3\Sigma)$ are calculated from measurements. By comparing the measurements with the model calculations it is possible to obtain information about the rate constants; for example the rate of de-excitation of vibrational excited nitrogen molecules by the wall has been obtained.

2. Experimental set-up

The experimental set-up consists in the discharge probe and associated circuitry and the flowing afterglow system (Fig.1).

The discharge probe looks like a double probe (two tungsten wires 12 mm long, 0.7 mm diameter and separated by 4 mm distance) and works with associated circuitry as relaxation oscillator [1,2]. In such a circuit a capacitor is repeatedly charged and discharged. During the discharge period the voltage over the probe electrodes changes between breakdown (V_b) and cutoff (V_c) values. Fig. 2 presents typical temporal waveforms for current intensity and for the voltage on the probe discharge during the cyclic operation of the probe in neutral flowing nitrogen at a pressure p=10 torr .The time constants for charging and discharging are roughly RC and $(r_d+R_m)C$ where r_d, R_m are respectively the values of the dynamic discharge resistance ($r_d=dU/dI$) and of measuring current resistance. The use of such circuitry is a meritorious possibility to obtain in short time many electrical parameters of the probe discharge.

The flowing afterglow has been obtained by expanding a subatmospheric plasma jet generated by a RF discharge [5] in a flow tube made of pyrex glass with 16 mm internal diameter. The discharge probe is mounted at the end flange of the tube and is movable along the flow.

The experimental parameter values ranged as follows: pressure p=1-20 torr, RF power P<100 w, flow rate F=50-200 sccm ,wall temperature T=300-320 K.

The measurements were performed with a 2432A Tektronix digital oscilloscope.

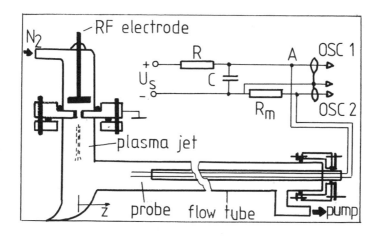

Figure 1 . Experimental set-up

3. Results and discussion

3.1 PROBE DISCHARGE MEASUREMENTS

One of the I-V characteristics of the discharge probe while running in unexcited neutral nitrogen (RF supply off) is shown in Fig.3, curve "*a*". Specific discharge regimes are observed on the characteristic. The region AB corresponds to the transition between the breakdown (V_b) and

518

burning (V_r) voltages (subnormal glow, negative internal resistance). The region BC corresponds to stable discharge working points (normal glow). The region CD corresponds to discharge extinction and is also a transition region (subnormal glow, negative internal resistance).

In Fig. 3, the curve "b" shows the IV characteristic obtained in the same conditions as curve "a" but the RF discharge is on. The modification of characteristic and the change of all electrical parameters of the discharge probe (V_b, V_r, V_c) are to be noticed, in comparison with curve "a". The influence of the RF electric field on the probe has been investigated by recording the RF signal induced in the probe and has been found reasonable small. Other processes influencing the experiment, for example the change of probe surface properties due nitridation have been avoided by doing the measurements after hours of operation, the state of surface being stabilized during this time.

One conclude that this modification of I-V characteristics is caused by the presence of afterglow species generated in the base RF discharge.

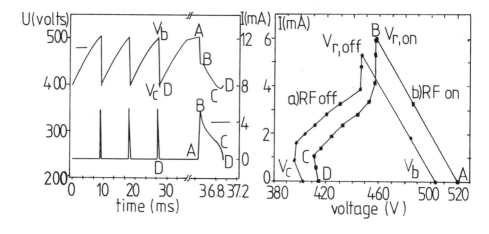

Figure2.
Time variation of current and
voltage of probe discharge

Figure 3.
Characteristics of probe discharge
in absence (curve a) and presence
(curve b) of base RF discharge

The key problem when one wants to use the discharge probe as a diagnostic instrument is to relate the changes of electrical parameters to the nature and density of species.

In order to study the afterglow in quite general situations the temporal area of current pulse A_{it} has been chosen to connect the experimental measurements to the kinetic model calculations. This parameter is the total charge passing through the discharge during the pulse and is consequently related to the V_b and V_c values, and can be expressed as :

$$A_{it} = C \cdot (V_b - V_c) \cdot (1 + \frac{r_d + R_m}{R}) \tag{1}$$

One finds it easy to measure and it shows measurable changes in presence of afterglow. The measurements have shown that in the range 1-15 torr the A_{it} parameter depends approximately linear on pressure. This is an indication that in a first approximation one can process the functional dependence between changes of A_{it} and densities of active species in the hypothesis of the linearity limit:

$$\Delta A_{it} = \sum_i \Delta n_i \cdot S_i \tag{2}$$

or in the second order approximation as :

$$\Delta A_{it} = \sum_i \Delta n_i \cdot S_i + \alpha \cdot (\sum_i \Delta n_i \cdot S_i)^2 \tag{3}$$

where ΔA_{it} is the variation of the electrical parameter due to the afterglow, Δn_i is the variation in the density of species, α is the nonlinearity coefficient and S_i are factors describing the sensitivity of probe to respective species.

During the experiments the temporal area of current pulse A_{it} has been measured for the same working conditions (p=10 torr, flow rate F=80 sccm, RF power P=50 w) but for different positions z along flow tube in two cases: RF discharge off and RF discharge on.

3.2. NITROGEN SPECIES IN AFTERGLOW

In the present description the species considered to have the most important number densities in the afterglow are taken into account (radicals and metastables): $N(^4S)$, $N(^2P)$, $N(^2D)$, $N_2(A^3\Sigma)$, $N_2(X^1\Sigma,v)$.

The following reactions, with the corresponding rate constants, contribute significantly to the afterglow chemistry:

(i) $N(^4S)+N(^4S)+N_2(X^1\Sigma,v)\text{---}>N_2(X^1\Sigma w)+N_2(X^1\Sigma,v')$

   $k_1=1.8\times10^{-33}$ cm^6 s^{-1} molec^{-2} [6,7,8]

(ii) $N(^4S)+N(^4S)+N_2(X^1\Sigma,v)$--->$N_2(A^3\Sigma)+N_2(X^1\Sigma,v')$
$k_2=5.4\times10^{-33}$ cm^6 s^{-1} molec^{-2} [6,7,8]

(iii) $N(^4S)+N_2(A^3\Sigma)$--->$N_2(X^1\Sigma,w)+N(^2P)$
$k_3=5.1\times10^{-11}$cm^{3}s^{-1}molec^{-1} [9]

(iv) $N(^2P)+N(^4S)$--->$N(^2D)+N(^4S)$
$k_4=1.8\times10^{-12}$ cm^{3}s^{-1}molec^{-1} [10]

(v) $N(^2D)+N_2(X^1\Sigma,v)$ --->$N(^4S)$ +$N_2(X^1\Sigma,w)$
$k_5=6\times10^{-15}$cm^{3}s-1molec^{-1} [10]

(vi) $N_2(X^1\Sigma,v)$+wall--->$N_2(X^1\Sigma,v')$ v' <v
k_w of order of 10^2 s^{-1} [11]

Taking into account the processes described above one can write the balance equations for number density as a system of coupled differential equations. In these reactions, except the last one, the radiative and the wall quenching have been neglected (because the radiative loss and the quenching at wall are small in comparison to quenching of these species by nitrogen atoms and molecules).

The analysis of the system of differential equations shows that initial $N(^4S)$ values of order of 10^{13}-10^{15} cm^{-3} cause the rapid disappearance (in a time of the order 1-10 ms) of the active species originating in discharge, excepting $N(^4S)$ and $N_2(X,w)$ for which longer decay times are requested.

After few milliseconds active species number densities are given by the following relationships:

$$N(^4S)]_t = \frac{[N(^4S)]_0}{(1 + At)} \tag{4}$$

$$N_2(A^3\Sigma)] = \frac{k_2}{k_3}\cdot[N(^4S)]\cdot\sum_w[N_2(X,w)] \tag{5}$$

$$N(^2P)] = \frac{k_2}{k_4}\cdot[N(^4S)]\cdot\sum_w[N_2(X,w)] \tag{6}$$

$$N(^2D)] = \frac{k_2}{k_5}\cdot[N(^4S)]^2 \tag{7}$$

$$N(X,w)]_t = [N(X,w)]_0\cdot\exp(-k_w t)+\frac{A\cdot P_w\cdot[N(^4S)]_0}{(1+At)^2\cdot k_w}\cdot(1-\exp(-k_w t)) \tag{8}$$

where
$$A = (k_1 + k_2) \cdot \sum_w [N_2(X,w]$$
(9)

and P_w is the probability of population of the w vibrational level by the reactions (i-v).

In the evolution of $[N(X,w)]$ two terms can be distinguished, a fast decay term (fd) due to wall quenching and a more slow term due to vibrational excitation by recombination (r) of nitrogen atoms. One can rewrite the solution for $[N(X,w)]$ by putting in evidence these terms as:

$$N(X,w)] = [N(X,w]_0^{fd} \cdot \exp(-k_w \cdot t) + \frac{[N(X,w]_0^r}{(1+At)^2}$$
(10)

After a time of about 100 ms the contribution of first term in (8) will become negligible and the concentration of active species will be determined by the local concentration of $N(^4S)$.

Due to small concentrations of active species in the afterglow $\sum_w [N_2(X,w)]$ is almost the same in afterglow and in unexcited gas

and it can be determined by gas pressure measurements as :

$$\sum_w [N_2(X,w)] = 3.53 \cdot 10^{16} \cdot p[torr]$$
(11)

3.3 THE FITTING PROCEDURE

From relations (4 - (8) which describe the kinetic of active species and from relations (1-3) the temporal dependence of A_{it} is given in the linear approximation by:

$$\Delta A_{it}^1(t) = \frac{A_1^1}{(1+At)} + \frac{A_2^1}{(1+At)^2} + A_3^1 \cdot \exp(-k_w \cdot t)$$
(12)

where :

$$A_1^1 = S_1 \cdot [N(^4S)]_0 + S_2 \cdot [N_2(A^3\Sigma)]_0 + S_3 \cdot [N(^2P)]_0$$
(13)

$$A_2^1 = S_3 \cdot [N(^2D)]_0 + \sum_w (S_w \cdot [N(X,w)]_0^r)$$
(14)

$$A_3^1 = \sum_w S_w \cdot [N(X,w)]_0^f$$
(15)

$$A_1^1 + A_2^1 + A_3^1 = \Delta A_{it}^1(t=0)$$
(16)

The afterglow time t is connected with the distance z from the plasma source to the point of measurements by $dt=dz/v_f$ where v_f is the local speed of the flowing afterglow (planar flow approximation)

Relation (12) was used as fitting function for the measured variation of A_{it} in the linear approximation. The fitting procedure is an iterative least square fit method with four fitting free variable: $[N(^4S)]_{t=0}$, k_w, A_1^1, A_1^2 in the case of linear approximation and with five fitting free variable ($[N(^4S)]_{t=0}$, k_w, A_1^2, A_2^2, α) in the case of second order approximation. The initial values for the fitting procedure, have been generated randomly, in the estimated range, for hundred times. The final results of the iterative procedure were always the same proving the unicity of solution. Also a small statistical dispersion was obtained. The final results of the fitting procedure in the case of linear approximation, are presented in Table 1

TABLE 1. Results of the fitting procedure in
the case of the first order   approximation

$[N(^4S)]_0$ 10^{15}[cm-3]	k_w [s^{-1}]	A_1^1 [mA·s]	A_2^1 [mA·s]	A_3^1 [mA·s]
8.5 +/- 0.5	45 +/- 3	56 +/-6	-219 +/-25	157 +- 12

By repeating the fitting procedure in the case of second order a low value of nonlinearity coefficient α which was obtained $(2[\mu As]^{-1})$ indicates that the approximation of linearity of ΔA_{it} with the density of active species is satisfactory. In Fig. 4 are presented both the experimental points and the fit curve obtained in the case of linear approximation. The number densities of $N(^4S)$, $N(^2P)$, $N(^2D)$, $N_2(A^3\Sigma)$ have been calculated using relations (1-5) and the initial number density of $N(^4S)$ have been obtained from the fitting procedure. Their dependence of position is shown in Fig.5.

One observes the relative weight of concentration of different species along the flow tube and the similar behavior of curves due to the control of kinetics by atomic nitrogen number density.

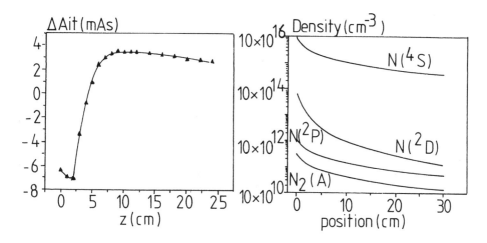

Figure 4.
Variation of ΔA_{it} with position
(points-experimental results)
(continuous line-- fit curve)

Figure 5.
Dependence of number densities
of active species on position

4. Conclusions

Investigation of flowing afterglow with a small discharge created between the electrodes of a double probe is an attractive way to take out supplementary information from gaseous excited medium. A meritorious possibility to obtain these information is to include the probe in a relaxation oscillator circuit and to operate it in a pulsed mode. In the case of flowing nitrogen afterglow important changes of electrical parameters of pulsed discharge probe (breakdown voltage, cutoff voltage, temporal waveform of current intensity) produced by afterglow species have been observed and measured along the flow axis. Afterwards due to superposition of effects of active species on the probe the main difficulty is to relate the variation of probe electrical parameters along the flow axis to the kinetics of each afterglow species. In the frame of a kinetic model for the afterglow has been possible to relate these changes to the actual concentration of species by considering that the produced effects are linearly dependent on the species number density. Acording to this hypothesis the reaction rate for de-excitation of vibrational excited nitrogen molecules by the wall has been inferred from experiment. Also the number densities of $N(^4S)$, $N(^2P)$, $N(^2D)$, $N_2(A^3\Sigma)$, species along the flow tube have been computed.

524

References

1. Hirsch, E.H. (1961) Vacuum measurements by means of alternating   gas discharges, *Rev.Sci.Instrum.* **32** 1373-1377
2. Rigney, D.S., Schwalb, A.J., and Levy, M. E. (1966) An electrical discharge probe for transient density measurements in rarefied flows in J.H. Leuw (eds) *Raref. Gas.Dynam.*, Academic Press, New York , pp 195-208
3. McDaniel, E.W., Cermak, W., Dalgarno, V., Ferguson, E., and Friedman, L., (1970) *Ion-Molecule Reactions* , Willey Interscience, New York
4. Ferguson, E., Fehsenfeld, E.C. and Schmeltekopf A.L (1969) in Adv.Atom. Mol.Phys , Academic Press, New York
5. Dinescu, G., Maris, O., and Musa, G., (1991) Spectral and electrical properties of low power , RF-generated postdischarge nitrogen plasma jet, *Contrib. Plasma. Phys.* **1**, 49-61
6. Ricard, A., Malvos, H., Bordeleau, S., and Hubert, J. (1994), Production of active species in common flowing postdischarge of an Ar-N_2 plasma and Ar-H_2-CH_4 plasma, *J. Phys. D: Appl. Phys.* **27**, 504-508
7. Yamashita, T., (1979), Rate of recombination of nitrogen atoms, *J.   Chem. Phys* **70**, 4248-4253
8. Dugan C. H., (1967), Metastable molecules in a nitrogen afterglow *J.Chem. Phys* **47** , 1512-1517
9. Young, R. A. and John, G .A. St, (1968), Experiments on $N_2(A^3\Sigma^+)$. Reaction with N*, *J.Chem.Phys.* 1968 **48** ,895-898
10. Kossyi, I. A., Kostinsky, A. Yu ., Matveyev, A. A, and Slakow, V. P., (1992) Kinetic scheme of the non-equilibrum discharge in nitrogen -oxygen mixtures, *Plasma Sources Sci.Technol.*, **1** 207-220
11. Gordiets, B.and Ricard, A.., (1993), Production of N,O and NO in N_2-O_2 flowing discharges, *Plasma Sources Sci.Tehnol.*, **2**, 156-163

THE RECOMBINATION OF IONIZED SPECIES
IN SUPERSONIC FLOWS

P. VERVISCH, A. BOURDON
CNRS URA 230 - CORIA
University of Rouen
76821 Mont Saint Aignan Cedex
FRANCE

1. Introduction

When charged particles are present the modeling of a supersonic or hypersonic flow remains a difficult issue. The evaluation of charged particles concentration requires a knowledge of numerous coupled elementary processes and a microscopic description more detailed than for neutral particles. The specificity of ionized flows is coming from the electric nature of charged particles and the weakness of the electron mass. The Poisson equation links strongly together electrons and ions and the Debye length is the scale length below which average charge neutrality cannot be guaranteed. Then if the collisional mean free path is greater than the Debye length, the simulation grid point and the fluid particle of the continuum description are within a neutral medium. This neutral assumption has consequences on the expression of conservation equations (ambipolar regime) and transport coefficients.

The slow rate of energy exchange between electrons and heavy particles by elastic collisions (huge difference between the electron mass and the heavy-particle mass) allows different electron and heavy-particle temperatures to exist. Then a multitemperature formalism is required to determine transport properties. Electrons are also very efficient to induce chemical reactions. For reactions with large energy thresholds, highly energetic electrons are involved. However as the electron-electron elastic collision cross section decreases as energy increases, the high energy tail of the electron distribution function may become easily non-Maxwellian. In these cases, the determination of the reaction rates remains a difficult issue.

M. Capitelli (ed.), Molecular Physics and Hypersonic Flows, 525–542.

An accurate description of the electron temperature appears to be difficult since in the electron energy conservation equation, different terms have the same order of magnitude and may strongly vary on short distances. So to calculate the electron temperature the modeling of these terms has to be accurate. Sometimes this required accuracy cannot be obtained easily. Finally, boundary conditions at the wall have to be modified in order to take into account the plasma sheath which insulates the electrons from the wall.

In the following, the existing models for transport coefficients, energy exchange terms, kinetic schemes and boundary conditions will be discussed and applied to a molecular nitrogen plasma. Then, to illustrate the influence of the different models, results of the comparison between computations and experiments of a nitrogen plasma jet and a boundary layer will be presented.

2. Physical model

2.1. TRANSPORT PROPERTIES

Multitemperature transport coefficients for a plasma medium depend on the local charge neutrality assumption in the fluid particle. The Poisson equation shows that when the electron density is high enough, the applied or induced electric field has to be very strong to create a substantial charge separation. In the case of zero current and no external electric field applied, the charged particles do not move independently. The electrons follow the mean velocity flow of heavy particles due to their strong electric coupling with ions and diffuse with them (ambipolar regime). Assuming a small Knudsen number in the resolution of Boltzmann's equation, macroscopic parameters as mass-averaged velocity and temperature are only defined by the zero order solution and transport coefficients are calculated with the first order solution, the zero order one being known. If electrons follow the heavy particles mean velocity with a different temperature, the zero order distribution function is defined and consequently transport coefficients may be expressed following the usual Chapman-Enskog method. For an ambipolar regime, the transport coefficients of heavy particles are the same as those that would result if the electrons were removed from the mixture. For electrons, usual one-temperature transport coefficients for the mixture have to be only slightly modified i.e. in the collision integrals where electrons are involved the electron temperature has to be used instead of the heavy particle temperature and in the Chapman bracket integrals $\frac{m_e}{m_i}$ must be set equal to zero.

To carry out numerical simulations, numerous simplified expressions of the transport coefficients have been proposed. In situations where the ionization degree is important, these models should be handle with care. A collision with a charged particle is characterized by a long range potential interaction. Then, the magnitude of collision integrals and their temperature dependence are very different from those involving only neutral particles. These differences may lead to serious discrepancies when usual mixing laws are used to calculate transport coefficients because these laws have been established for neutral mixtures.

Fig.1 shows the temperature evolution of the viscosity of a $N_2 - N^+$ mixture. The temperature and concentration values correspond to our experimental situation near the cooled flat plate. The pure viscosity of N_2 and N^+ are very different from each other by several orders of magnitude. In this example, the viscosity of the mixture (taking into account the real interaction potentials) remains very close to the viscosity of pure N_2. A viscosity calculation neglecting ionic species appears to be a good approximation. Crude mixture laws amplify the role of the weak viscosity component. Wilke's law [1] cannot be used in ionized medium. Armaly and Sutton's [2] law appears to be reliable only at high temperature. Fig.2 shows the same results for the thermal conductivity. The very popular Mason et Saxena [3] mixture law strongly underestimates the thermal conductivity and the same divergence is noted at low temperature with Armaly and Sutton's law. Near a cooled wall, the heavy particle temperature is low and crude estimates of transport coefficients may strongly modify the obtained results. Before computing an ionized flow it is therefore recommended to test the validity of mixing laws.

For heavy particle transport coefficients, the convergence of the Chapman expansions is very fast and first order approximations are usually sufficient. Conversely the convergence is slow for the electron transport coefficients and higher order approximations must be taken into account. When the ionization degree is very low, Fig.3 shows that a second order approximation for the electron thermal conductivity is sufficient. However only the third order approximation tends toward the Spitzer [4] value at a high ionization degree. Crude mixing laws derived by Jaffrin [5] and Yos [6] may be considered as second order approximations.

Generally, diffusion fluxes are described by a Fick's law and the diffusion coefficient of a species into the mixture is approximated by a molar average of the binary diffusion coefficients. To include ambipolar diffusion of ions the associated diffusion coefficient is multiplied by $\left(1 + \dfrac{T_e}{T}\right)$. This procedure is only valid when the ionization degree is very low. The main advantage of Fick's law is its simplicy which is very appealing for numerical simulations.

However it may be too crude for situations where the diffusion fluxes have the same magnitude as the convective ones. For a neutral medium Bird's [7] approximation is more accurate and gives results close to those obtained with the exact resolution of Stefan Boltzmann equation. Ramshaw and Chang [8] carried out recently a study on the multicomponent diffusion in a multitemperature plasma including pressure gradients and thermal diffusion terms. Their final rigorous expressions of the diffusion fluxes are unexpectedly easy to implement in codes.

2.2. ENERGY EXCHANGE MECHANISMS

In our molecular nitrogen plasma, three temperatures have been considered: a translational-rotational temperature T, a vibrational temperature for N_2 which represents the vibrational distribution and an electron temperature T_e. So there are three separate energy equations.

The rotational and vibrational thermal conductivities are derived from an Eucken relation. These relations are associated with a smooth collision hypothesis: energy exchange during a collision is a small part of the mean thermal energy. It is a good approximation for rotational collisions, but in an expanding flow where the temperature level is low, a vibrational smooth collision hypothesis is more discutable.

Equilibrium between rotational and translational modes induces an additional term in the viscous stress tensor namely the bulk viscosity. In an expansion or a compression the work of pressure affects directly the translational energy and modifies the rotational energy with a time lag. The bulk viscosity η has the same order of magnitude as the viscosity μ and should not be omitted as soon as the velocity divergence is significant.

For energy exchanges between vibration and translation V-T and between electrons and vibration e-V, Landau-Teller relaxation expressions are usually retained. The corresponding global relaxation times are representative of the inelastic collisions affecting the first vibrational levels. These models are associated to compression situations where only low energy levels are populated. In a recombining plasma the vibrational temperature is greater than the translational one. In such situations, high vibrational levels are populated significantly and their relaxation times are shorter than those of the low vibrational levels. Then, usual models for global relaxation times certainly overestimate the relaxation time in expanding flows.

We can also put forward that in molecular expanding flows, an important part of the total energy is stored in the vibrational mode. In this case, it is perhaps hazardous to represent the vibrational energy exchange with the crude Landau-Teller model. Presently more elaborated models are not available.

The global relaxation time is derived from the interspecies relaxation times with a simple molar averaged relation. For e-V, multilevel transitions have to be considered as in Lee [9] formulation. With the resonant nature of inelastic $N_2 - e$ collision cross sections, the e-V coupling is strong. In these conditions the accuracy of electron temperature calculations depends on our capability to determine correctly the vibrational energy evolution. Unfortunately, vibrational energy equation terms are the most roughly estimated. For V-T, the formulation proposed by Millikan and White [10] for the interspecies relaxation time is the most widely used. This formulation is based on experimental measurements of vibrational relaxation times. A semi empirical relation correlates the relaxation time to the reduced mass of colliding particle and to the characteristic vibrational temperature of the molecule. More recently, numerous approaches based on SSH formulation [11] have been developed to determine V-T and V-V rate constants. The accuracy of these approaches is well demonstrated on the $N_2 - N_2$ system [12]. The experimental data are well described and the $T^{-\frac{1}{3}}$ evolution of the relaxation time proposed by Millikan and White is respected down to 2000 K. For lower temperatures the long range attractive potential has to be included to reproduce experimental evolutions.

For other systems as $N_2 - N$, $N_2 - N^+$, $N_2 - N_2^+$ the situation is not so clear. Experimental data are very scarce for $N_2 - N$ and totally non-existent for $N_2 - N^+$ and $N_2 - N_2^+$. For $N_2 - N$ system a theoretical study carried out by Lagana [13] has shown the very strong temperature dependence of the relaxation time. Nitrogen atoms are very efficient to thermalize the vibrational energy. This is related to the very different shapes of the interaction potentials of $N_2 - N_2$ and $N_2 - N$ systems which prevent to extrapolate the results obtained for one system to the other one. Millikan and White correlations break down for these systems. Due to the lack of experimental or theoretical studies concerning $N_2 - ion$ colliding systems the relaxation time $\tau_v \left(N_2 - N^+ \right)$ and $\tau_v \left(N_2 - N_2^+ \right)$ have to be guessed. When the ionization degree is high, an efficient $N_2 - N^+$ coupling could impose a fast thermal equilibrium of the plasma i.e. $T_e = T_v$ by $e - N_2$ collisions and $T_v = T$ by $N^+ - N_2$ collisions.

Finally, the translation-electron energy exchange has the following expression:

$$Q_{T-e} = 3k \left(T - T_e \right) n_e \sum_{s \neq e} \frac{m_e}{m_s} \nu_{es}$$

where m_s is the mass of species s and ν_{es} is a collision frequency. The interactions between charged species are dependent on Coulomb potential which is well known. To avoid the divergence of the collision frequency all collisions with impact parameters greater than the Debye length are neglected. The elastic collision cross sections for $e - N_2$ [14] and $e - N$ [15] are well known and the details of resonant structures are well identified.

In a recombining situation, the recombination energy is divided between collision partners and radiation through collisional and radiative cascading elementary processes. Because of the radiation reabsorption of the plasma medium, a calculation of the radiative part depends on the experimental geometry. Then, it becomes necessary to solve the radiative-transport equation. In nitrogen, radiative effects are weak and a simple formulation with only transition probabilities is sufficient.

2.3. KINETIC SCHEMES

The set of reaction rate coefficients proposed by Park [16] is used as a initial reference for this work. Nevertheless this kinetic scheme has been developed and tested for flows undergoing compression which do not correspond to our physical situation. Moreover as ionization is important the associated excited states population density may modify the kinetic schemes. Here we focus on the reactions which mainly influence the charged species number density:

$$N_2 + e \rightleftharpoons 2N + e \tag{1}$$

$$N + e \rightleftharpoons N^+ + 2e \tag{2}$$

$$N^+ + N_2 \rightarrow N_2^+ + N \tag{3}$$

$$N_2^+ + e \rightarrow 2N \tag{4}$$

The dissociative recombinaison (4) is a very efficient channel to recombine charged species in a molecular plasma medium. Calculated recombination rate [17] for the level $v = 0$ of N_2^+ is in excellent agreement with experimental microwave results obtained by Zipf [19] for $100 < T_e < 1000\,K$. At higher temperatures the N_2^+ population density is distributed among the vibrational levels. So the global reaction rate depends on the efficiency of N_2^+ vibrational levels to dissociate.

Supposing a vibrational temperature equal to the electron one, Fig.4 shows the temperature dependence of the total reaction rate. Zipf has shown that the N_2^+ dissociative recombination coefficient increases very slowly for the first three vibrational levels. On Fig.4 is displayed the case where the three vibrational levels are supposed to have the same rate. Park's law is in this range of values. The accuracy of Dunn and Lordi [18] experimental results may be doubtful. These results have been obtained by an indirect way in an arc jet whose degree of impurity was questionable.

In conclusion, the dissociative recombination reaction rate may be considered as accuratly known, and the Park reaction rate will be used in our calculations.

The charge exchange reaction (3) is well known at high collisional energy, but at thermal energy results are scarce. Fig.5 reviews the existing reaction rates. Park's temperature evolution is only an estimate. Kossyi's value is deduced from a review of Kossyi et al [20] which estimates the backward reaction rate. Other reaction rates are deduced from collision experiments. Recent measurements carried out by Freysinger et al. [21] put forward an energy threshold higher than the thermodynamic threshold. First, they theoretically explained this particular point and then attributed the strong discrepancy of previous experimental studies, like the one carried out by Phelps [22] to the presence in these experiments of metastable nitrogen ions. In fact, a metastable ion, as a collision partner, lowers the reaction energy threshold. As the ionic metastable level may be easily populated by inelastic collisions with electrons, the charge exchange reaction becomes dependent on the electron temperature. Fig.5 points out that at 5000 K there is a three orders of magnitude difference between Park and Freysinger reaction rate values. At low pressure, the three-body recombination is relatively non-efficient and the charge exchange reaction followed by the fast dissociative recombination may play an important part in the recombination scheme.

The other recombination channel, namely the three-body recombination, is the result of elementary free-bound and bound-bound cascading processes. Experimental results are available only for a very limited range of plasma parameters [16]. In fact, this reaction has been mainly studied theoretically with the collisional-radiative model approach. As numerous elementary processes are involved, the accuracy of the reaction rates strongly depends on our knowledge of the atomic nitrogen structure and of elementary processes cross-sections. Park's reaction rate [16] has been determined in a collision-dominated plasma (radiation neglected) using Gryzinski [23] hydrogenic cross sections for inelastic collisions. This model, as the one proposed by Drawin [24] is adapted only for high energy levels of nitrogen. Therefore corrections have been taken into account for low energy levels. We wonder the reliability of Park's rate in our conditions.

The recent work of Kunc and Soon [25] on a collisional-radiative model for a stationary nitrogen plasma reviews the best possible cross sections and Einstein coefficients for nitrogen. When no experimental data or more accurate models are available, the inelastic cross sections are calculated following Sobelman formalism [26]. This approach, which uses a Born approximation, remains a rough method. In order to test the influence of the elementary processes models, we built up a new collisional-radiative model for nitrogen [27]. The calculations are carried out for a given electron temperature, but allow electron density and nitrogen levels density to evolve in time. We start the computations with a recombining situation. As expected, the general behaviour of the time-dependent solution is an early transient state followed by a quasi-stationary state for the excited levels of atomic nitrogen (except the fundamental and the two metastable levels). For an electron density equal to $10^{20} m^{-3}$ and an electron temperature equal to 5000 K the transient period lasts $10^{-8} s$. Only after this transient period i.e. when a quasi-steady state is obtained, the determination of reaction rates is meaningful. With a initial recombining condition, the steady-state gives the recombination reaction rate value. When time goes on, ionization becomes significant and the system evolves toward a stationary state. An ionization reaction rate may be deduced from the recombination reaction rate and the stationary densities. In a collision-dominated plasma, level populations in the final stationary state are in Saha equilibrium. In this case, the relation between the two rates i.e. ionization and recombination, is simple and depends only on the electron temperature.

Fig.6 shows the temperature dependence of the recombination rate for different sets of collision cross-sections. Only collisional processes have been considered. At low temperature, the recombination rate depends on the atomic levels near the ionization limit. Hydrogenic cross sections are sufficient, but the problem of the number of high energy levels to be considered, is more difficult. Conversely at high temperature, the levels near the ionization limit are in equilibrium with the continuum and the recombination rate depends on low energy levels. Then, the choice of a model for collision cross sections influences the recombination rate. For an electron temperature of about $10^4 K$ the recombination coefficient with our model is one order of magnitude less than Park's one (with Gryzinski cross sections). Differences are greater with Drawin cross sections. With such a sensitivity to collision cross section model, the determination of an accurate recombination or ionization rate appears difficult. Moreover radiative bound-bound and bound-free transitions have to be taken into account. Usually bound-free radiative transitions (dielectronic and radiative recombination) are treated separately. Fig.7 shows the bound-bound radiative effects on the recombination rate, supposing an optically thin medium. Bound-bound radiative transitions depopulate the excited nitrogen levels and then increase the recombination rate. This effect is significant only at high temperature and at low electron density. As U.V. lines are generally optically thick, these results overestimate the radiative effects on the recombination rate.

The dissociation of N_2 molecules by electron collisions (reaction 1) is also the result of numerous elementary reactions. It depends on the vibrational and electronic excitation of N_2 (triplets states). For flows undergoing compression, different models are available [28], but they do not correspond to our physical situation. In the modeling of the nitrogen jet, Park reaction rate with the one-temperature model and Winter [29] or Cosby [30] elementary reaction rates for $N_{2(v=0)} + e \rightarrow 2N + e$ will be tested.

2.4. BOUNDARY CONDITIONS

The properties of the plasma sheath surrounding an object in contact with a plasma have been extensively treated [31] and is still not fully understood. In its simplest form the interaction of a plasma with a wall can be characterized as follows: due to the high mobility of the electrons, the wall potential will adjust itself to be negative with respect to the surrounding plasma.

The repulsion of electrons results in the formation of a positive space-charge region shielding the neutral plasma from the negative wall. The typical extension of the sheath is given by the Debye length which is usually small compared to the mean free path. Ion distribution function is modified by wall losses rendering shielding impossible unless the Bohm criterion is fulfilled. This condition for sheath formation demands that the ions enter the sheath region with at least the sonic velocity of the electron-ion mixture.

Boundary conditions are obtained in writing at the last mean free path the continuity of ionic mass flux and electron energy flux between continuum and molecular description. To take into account the presence of the sheath, usually the Bohm velocity is used instead of ion thermal velocity. In this case sheath edge conditions have been transferred to the last mean free path location. This hypothesis neglects the influence of the presheath on the boundary condition.

As far as we know, there is no reliable work on the writing of the boundary condition at the last mean free path taking into account the presheath influence. Finally, it is important to point out that all these boundary conditions are obtained assuming a Maxwellian distribution function for electrons.

3. Test cases

3.1. NITROGEN PLASMA JET

Details on experiments and computations have been discussed elsewhere [32],[33],[34],[35]. Here we present only a few representative results which illustrate the sensitivity of the results to the models described hereabove. The plasma source is a d.c arc jet. The arc is initiated between a tungsten cathode and a nozzle-shaped copper anode. The plasma jet is underexpanded at the exit of the nozzle in a low pressure chamber. The jet undergoes first a quick relaxation followed by a recompression.

First, we have tested the influence of two sensitive reactions: the N_2 dissociation by electrons and the charge exchange reaction. Park's reaction rate values will be denoted set 1. In set 2 and 3 the N_2 dissociation rate is divided by 10^3 which corresponds to the reaction rate value associated to the elementary reaction $N_{2(v=0)} + e \rightarrow 2N + e$ and supposing a Maxwellian distribution function for electrons. Set 2 and 3 differ by the value of the reaction rate associated to the charge exchange reaction respectively Kossyi and Freysinger reaction rates.

Fig.8 and Fig.9 show the axial evolutions of nitrogen atom and electron concentration for the different sets. The three simulations are performed with Fick's law. In the first centimeter the important gap between nitrogen densities determined with set 1 or with set 2 or 3 is mainly due to the change in the N_2 dissociation reaction rate value which appears to be too high in set 1. Downstream nitrogen density is driven by the charge exchange reaction. As the value of the latter reaction rate is lowered, the atomic nitrogen concentration is depleted whereas the electron concentration increases. For nitrogen atom and electron density the set 3 gives the best agreement with the experimental results.

So the simple model which only retains a dissociation rate for N_2 in the order of the one from the v=0 level seems a good approximation. Park's formulation is representative of a compression situation and these results show that this reaction rate value cannot be used in a medium undergoing an expansion. For the charge exchange reaction rate value strong discrepancies exist between existing data. Our experiment confirm the recent value proposed by Freysinger. The charge exchange reaction appears to be frozen in our conditions

Fig.10 shows the influence of the diffusion flux model on the electron density evolution. The best agreement is obtained when Ramshaw's approach is implemented. The rough Fick's law neglects the influence of the temperature and pressure gradients and misfit our test case especially in the first ten centimeters where these gradients are significant.

In our experimental situation, the very efficient e-V coupling imposes permanently the thermal equilibrium $T_v = T_e$. Because of this strong coupling, the electron temperature depends on the magnitude of the V-T relaxation processes. In the plasma N_2 and N^+ are the major species. As the nitrogen atom is minor constituent its contribution to the V-T relaxation process is weak. Thus this experimental situation does not enable to test the modeling of its associated relaxation time: the electron temperature calculated with Lagana relaxation time is found to be very close to the one determined with Millikan and White relaxation time. If we retain Millikan and White formulation for the relaxation time of $N^+ - N_2$, Fig.11 shows that the comparison between the computed electron temperature and the experimental measurements is approximatively good. However the oscillations of the experimental temperature are not reproduced by the simulations. Translational temperature verifies the same oscillations due to expansion and compression. This suggests that the coupling between translational and vibrational mode is not well described. If we suppose that $\tau_{T-v}(N^+ - N_2) = \tau_{T-v}(N - N_2)$ with the Lagana expression, then $T_v = T_e = T$.
Information about the $\tau_{T-v}(N^+ - N_2)$ relaxation time would be of great interest to improve the modeling.

3.2. FLAT PLATE BOUNDARY LAYER

To study a boundary layer situation a stainless steel water cooled flat plate is set along the jet axis 200 mm from the nozzle exit. The plate is set 5 mm back from the axis in order to have an homogeneous region sufficiently wide over the plate where the boundary layer can be assumed to be two-dimensional. The flat plate is electrically insulated. A thorough comparison between computations and experiments has been carried out on electron temperature and density profiles for sections near the leading edge where the degree of ionization is significant.
The flow appears to be chemically frozen for charged species and electron density profiles depend only on the diffusion flux and wall conditions. In order to separate both contributions, first, the electron density is fixed at the wall to its experimental value and the influence of the modelling of the diffusion flux has been studied. In the experiment the electron number

density decreases strongly between x = 4 and 6 mm and decreases much slowlier downstream. On Fig.12 are shown the computed electron number density profiles with the classical ambipolar diffusion law and the more sophisticated expression derived by Ramshaw. With the classical Fick's law, the computed profiles strongly disagree with experiment. Conversely, with Ramshaw's expression a good agreement between computation and experiment is clearly shown, especially the strong electron density decrease between the 4 and 6 mm sections is well described.

For the downstream sections where the flow is only weakly ionized, electron density profiles mainly depend on the temperature dependence of the binary diffusion coefficient $D_{N^+ - N_2}$. Information about the interaction potential would be of great interest to improve the accuracy of the modeling.

In a second step, we tried to implement the classical wall conditions (Bohm's criterion). Fig.13 shows that this usual boundary condition does not describe experimental evolutions. Referring to the good agreement obtained when the electron density was fixed at the wall, we can conclude that classical wall conditions for charged species misfit our molecular nitrogen plasma.

For electron temperature profiles complexity is greater. The first problem comes from the lack of information on the vibrational mode. In the free stream, we may assume that the electron and vibrational temperatures are equal, but we have no experimental information about the wall accommodation. In Fig.14 the computed electron temperature is represented in two cases: total wall accommodation of the vibrational mode and no wall accommodation. In both cases the electron temperature is fixed at the wall to its experimental value.

Temperature evolutions show that the vibrational mode has a strong influence on the electron temperature through the electron-vibration energy exchange. In this case, a no wall accommodation hypothesis appears to be much more adapted than a total wall accommodation condition. Independently of boundary conditions, the electron temperature profile appears to be very sensitive to the e-V energy exchange term. We question the real accuracy of the Landau-Teller expression in our conditions.

If classical wall conditions on the electron temperature are implemented, the computed temperature is nearly constant in the boundary layer. As for electron number density, classical wall conditions disagree with experiment. The measured sheath potential is also greater than the one calculated with Bohm's criterion.

Classical wall conditions are written assuming that electrons are Maxwellian.

In our experimental situation, distortions of the distribution function may be caused by superelastic collisions with triplet states and vibrational levels of N_2 [36]. Information about the electron distribution function can be obtained by the second derivative of Langmuir probe characteristics. Fig.15 shows that the high energy tail of the distribution i.e. E > 4,5 eV is overpopulated.

This result is consistent with the high sheath potential value measured, since with an excess of energetic electrons, the sheath potential has to be increased to maintain a zero current at the wall. This fact could explain why classical wall conditions for the electrons fail in our case.

The ambipolar electric field can be deduced from Langmuir probe measurements i.e. the spatial derivative of plasma potential. In Fig.16 the spiky curve represents the experimental electric field. With usual approximations, the electric field is found to be proportional to $gradP_e$. In this case, the calculated electric field with experimental Te and Ne profiles strongly underestimates the experimental values. Following Ramshaw and Chang, in the expression for the ambipolar electric field an additional term due to thermal diffusion has to be taken into account. Using experimental Te and Ne profiles, the calculated electric field is in very good agreement with the experimental results. It is a additional confirmation of the validity of Ramshaw's approach.

4. Conclusion

This study shows the interest to have a quite exhaustive experimental study of a test case to carry out a thorough comparison with computations. When the amount of charged species is not negligible the modelling appears to be much more difficult than for neutral flows.

The electron temperature is very sensitive to strong source terms and much more accurate models than the ones available now are required to compute the electron temperature. This is particularly evident in boundary layer situations where significant gradients arise. Moreover, the problem of the writing of wall boundary conditions for electrons remains.

This nitrogen plasma analysis demonstrated the crucial need for accurate information on elementary processes. Without any effort in this direction, computations of thermo-chemical nonequilibrium partially ionized flows will remain uncertain.

538

Acknowledgements

The authors would like to thank Dr P. Domingo for providing figures and are grateful to Dr L. Robin for his kind contribution to this paper.

References

1. Wilke, C.R. (1950) A viscosity equation for gas mixtures, *J. Chem. Phys.*, **Vol. 18** no. 4, pp. 517–519
2. Armaly, B.F. and Sutton, K. (1980) Viscosity of muticomponent partially ionized gas mixtures, AIAA-80-1495
3. Mason, E.A. and Saxena, S.C. (1958) Approximate formula for the thermal conductivity of gas mixtures, *Phys. Fluids.*, **Vol. 1 no. 5**, pp. 361–369
4. Spitzer, L.J. (1962) *Physics of fully ionized gases*, 2nd. ed. Interscience, New York
5. Jaffrin, M.Y. (1963) Schock structure in a partially ionized gas, *Phys. Fluids*, **Vol. 8** no. 4, pp. 606
6. Yos, J.M. (1963) *Transport properties of nitrogen, hydrogen, oxygen and air to 30000 K*, Tech Memo RAD TM-63-7, AVCO.
7. Barlett, E.P., Kendall R.M. and Rindal R.A. (1985) *A unified approximation for the mixture transport properties for multicomponent boundary- layer applications*, NASA CR 1063.
8. Ramshaw, J. and Chang, C. (1993) Ambipolar diffusion in two-temperature multicomponent plasmas, *Plasma Chem. Plasma Process.*, **Vol. 13 no. 3**, pp. 489–498
9. Lee, J.H. (1980) *Electron-Impact vibrational excitation rates in the flowfield of aeroassisted orbital transfer vehicles*, J.N. Moss and C.D. Scott.
10. Millikan, R.C. and White, D.R. (1963) Systematics of vibrational relaxation, *J. Chem.Phys.*, **Vol. 39 no. 12**, pp. 3209–3213
11. Schwartz, R.N., Slawsky Z.I. and Herzfeld K.F. (1954) Calculation of vibrational relaxation times in gases, *J. Chem. Phys.*, **Vol. 22**, pp. 767–773
12. Dmitrieva, I., Pogrebnya, S. and Porshnev, P. (1990) V-T and V-V rate constants for energy transfer in diatomics. An accurate analytical approximation, *Chem. Phys.*, **Vol. 142**, pp. 25–33
13. Lagana, A., Garcia, E. and Ciccarelly, L. (1987) Deactivation of vibrationally excited molecules by collision with nitrogen atoms, *J. Phys. Chem.*, **Vol. 91**, pp. 312–314
14. Kennerly, R.E. (1980) Absolute total electron scattering cross sections for N_2 between 0.5 and 50 eV, *Physical Review A*, **Vol. 21 no. 6**, pp. 1876–1883
15. Ramsbottom, C.A. and Bell K.L. (1994) Low-energy electron scattering by atomic nitrogen, *Physica Scripta*, **Vol. 50**, pp. 666–671
16. Park, C. (1991) Chemical-kinetic problems of future NASA missions 1. Earth entries: A review, AIAA-91-0464
    Park, C. (1989) *Non equilibrium hypersonic aerothermodynamics*, Wiley, New York.
17. Guberman, S.L. (1991) Dissociative recombination of the ground state of N_2^+, *Geophysical Research Letters*, **Vol. 2 no. 6**, pp. 1051–1054
18. Dunn, M.G.and Lordi, J.A. (1970) Measurement of N_2^+ + e dissociative recombination in expanding nitrogen flows, *AIAA Journal*, **Vol. 8**, pp. 339–345
19. Zipf, E.C. (1980) The dissociative recombination of vibrational excited N_2^+ ions, *Geophysical Research Letters*, **Vol. 7 no. 9**, pp. 645–648
20. Kossyi, I.A., Kostinsky, A.Y., Matveyev, A.A. and Silakov, V.P. (1992) Kinetic scheme of the non-equilibrium discharge in nitrogen-oxygen mixtures, *Plasma Sources Sci. Technol.*, **Vol. 1**, pp. 207–220
21. Freysinger, W., Khan, F.A., Armentrout, P.B., Tosi P., Dmitriev O. and Bassi, D. (1994) Charge-transfer reaction of $N^+ + N_2$ from thermal to 100 eV. Crossed-beam

and scattering-cell guided-ion beam experiments, *J. Chem.Phys.*, **Vol. 101 no. 5**, pp. 3688–3695

22, Phelps, A.V (1991) Cross sections and swarm coefficients for nitrogen ions and neutrals in N_2 and argon ions and neutrals in Ar for energies from 0.1 eV to 10 keV, *J. Chem. Ref. Data.*, **Vol. 20 no. 3**, pp. 557–573

23. Gryzinski, M. (1965) Two-particle collisions.II. Coulomb collisions in the laboratory system of coordinates, *Physical Review A*, **Vol. 138**, pp. 322–335

24. Drawin, H.W. (1968) *Plasma Diagnostics*, W. Lochte-Hotgreven, Amsterdam.

25. Kunc, J.A. and Soon W.H. (1989) Collisional-radiative nonequilibrium in partially ionized atomic nitrogen, *Pysical Review A*, **Vol. 40 no. 10**, pp. 5822–5843

26. Sobelman, I.I. and Vainshtein L.A. and Yukov E.A. (1981) *Excitation of atoms and broadening of spectral lines*, Springer-verlag, New York.

27. Bourdon, A. and Vervisch P. (1995) Ionization and recombination rates of atomic nitrogen, work in preparation

28. Losev, S.A., Kovach, E.A., Makarov V.M., Pogobekjan M.Ju., Serglevskaya, A.L. and Shatalov, O.P. (1995) Chemistry models for air dissociation, this book.

29. Winters, H.F. (1966) Ionic adsorption and dissociation cross section for nitrogen, *J. Chem. Phys.*, **Vol. 44 no. 4**, pp. 1472–01476

30. Cosby, P.C. (1993) Electron-impact dissociation of nitrogen, *J. Chem. Phys.*, **Vol. 98 no. 12**, pp. 9544–9553

31. Riemann, K.U. (1991) The Bohm criterion and sheath formation, *J. Phys. D: Appl. Phys.*, pp. 493–518

32. Robin, L., Vervisch, P. and Cheron, B. (1994) Experimental study of a supersonic low pressure nitrogen plasma jet, *Phys. Plasmas*, **Vol. 1 no. 2**, pp. 444–459

33. Bultel, A. (1994) Ph.D. Thesis, University of Rouen, France.

34. Domingo, P., Bourdon, A. and Vervisch, P. (1995) Study of a low pressure nitrogen plasma jet, *Phys. Plasmas*, **Vol. 2 no. 7**

35. Robin, L., Cheron, B. and Vervisch, P. (1993) Measurements in a nitrogen plasma boundary layer: the boundary conditions of charged species, *Phys. of Fluid B: Plasma Physics*, **Vol. 5 no. 2**, pp. 610–620

36. Gorse, C., Cacciatore, M., Capitelli, M., De Benedictis, S., Dilecce, G. (1988) Electron energy distribution functions under N_2 discharge and post-discharge conditions: a self consistent approach, *Chem. Phys.*, **Vol. 119**, pp. 63–70

Fig 1: N$_2$-N$^+$ mixture viscosity, P=120 Pa,
10^5 μ [kg.m^{-1}.s^{-1}]=f(T[K]) , N$^+$/N$_2$=0,1

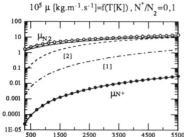

Fig 2: N$_2$-N$^+$ mixture thermal conductivity,
10^3 λ [W.m^{-1}.°K^{-1}]=f(T[K])

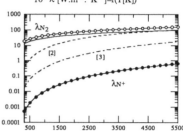

Fig 3: Thermal conductivity of electrons, 10^3 λe [W.m^{-1}.K^{-1}]=f(ξ=N$_{N2}$/Ne),
N$_2$, N$^+$, e$^-$ mixture, Te=5500 K, Ne=N$_{N^+}$=10^{19} part.m^{-3}

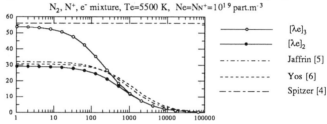

- ○ — [λe]$_3$
- ● — [λe]$_2$
- — · — · — Jaffrin [5]
- — — — — Yos [6]
- — · · — · · — Spitzer [4]

Fig 4: Dissociative recombination rate
kr [cm^3. mol^{-1}. s^{-1}]=f(Te[K])

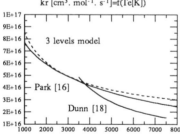

3 levels model

Park [16]

Dunn [18]

Fig 5: Charge-transfer reaction rate
kr[cm^3. mol^{-1}. s^{-1}]=f(T[K])

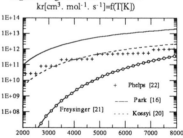

- + Phelps [22]
- — Park [16]
- · · · Kossyi [20]

Freysinger [21]

Fig 6: Collisions only
Three-body recombination rate [m^6.s^{-1}]=f(Te[K])
1-Drawin [24],
2-Gryzinski [23],
3-Park [16],
4-present model

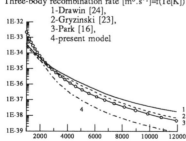

Fig 7: Collisions + bound-bound radiations
Three-body recombination rate[m^6.s^{-1}]=f(Te[K])
1-Ne=10^{17}m^{-3}, 2-Ne=10^{18}m^{-3},
3-Ne=10^{19}m^{-3} ,4-Ne=10^{20}m^{-3},
5-Ne=10^{22}m^{-3}, 6-Ne=10^{25}m^{-3},
-o- Park[16], •experiment [16]

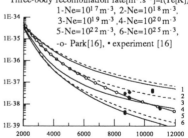

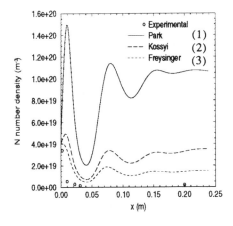

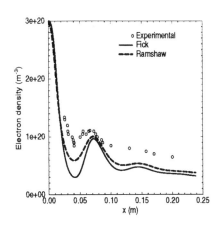

Figure 8: Axial evolution of the nitrogen density in the jet. Influence of the charge exchange reaction.

Figure 10: Influence of the ambipolar diffusion modelling on the electron density.

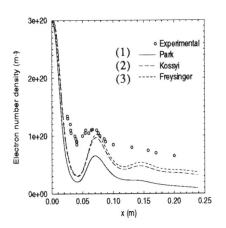

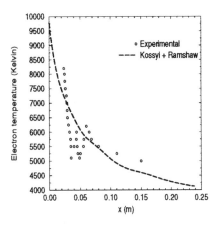

Figure 9: Axial evolution of the electron density in the jet. Influence of the charge exchange reaction.

Figure 11: Axial evolution of the electron temperature in the jet.

Fig 12: electron number density profiles Ne [m⁻³]=f(y[mm])

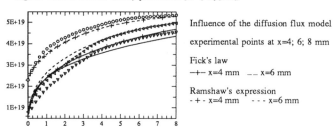

Influence of the diffusion flux model

experimental points at x=4; 6; 8 mm

Fick's law
—+— x=4 mm __ x=6 mm

Ramshaw's expression
- + - x=4 mm   - - - x=6 mm

Fig 13: electron number density profiles,
Ne [m⁻³]=f(y[mm])

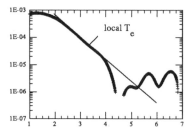

experimental points
o x=4 mm • x=6mm
Bohm's criterion
- - - x=4mm
- + - x=6 mm

Fig 14: Te and Ne fixed at the wall
Te[K]=f(y[mm]) at x=4mm

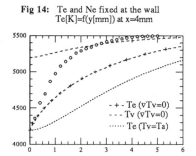

- + - Te (∇Tv=0)
- - - Tv (∇Tv=0)
· · · · · Te (Tv=Ta)

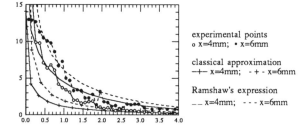

local T$_e$

Fig 15:

experimental second order derivate
of the probe current [A/V²],
function proportional to the
electron energy distribution
function f(ϵ[eV])
at x=4mm and y=0.2mm

Fig 16: ambipolar electric field E[V/cm]=f(y[mm])

experimental points
o x=4mm; • x=6mm

classical approximation
—+— x=4mm; - + - x=6mm

Ramshaw's expression
__ x=4mm; - - - x=6mm

MODERN COMPUTATIONAL TECHNIQUES
FOR 3D HYPERSONIC FLOWS

PASQUALE CINNELLA and CAREY F. COX
NSF Eng. Res. Center for Computational Field Simulation
P.O. Box 6176, Mississippi State, MS 39762 USA

1. INTRODUCTION

The present contribution to the NATO-sponsored Institute is focused on two main objectives: introducing the most commonly accepted form of the fluid dynamic governing equations for non-equilibrium hypersonic flows; and discussing some of more popular numerical algorithms that have been devised to solve these equations. The first task is significantly less trivial than what might be expected, as apparent from even a cursory examination of these Proceedings. The reason for this lies almost entirely with the thermodynamic non-equilibrium behavior of at least parts of the flow. In order to obtain engineering solutions of the hypersonic flow around or within even mildly complicated three-dimensional geometries, it becomes imperative to compromise on the sophistication of the non-equilibrium models employed in the numerical simulations. This is due to the limitations of current computational resources and to the necessity of obtaining solutions in a reasonable amount of time. In this context, two questions await a definite answer, and are the subject of on-going research: exactly what constitutes an acceptable simplification of the non-equilibrium models; and which models are more amenable to practical implementation in engineering codes.

In the following, a partial answer to the previous questions is attempted. A relatively general mathematical framework is introduced, which is able to accommodate most currently accepted non-equilibrium models for hypersonic flows, and several standard simplifications and approximations to the governing equations are discussed. An attempt is made throughout the exposition to underscore the portions of the fluid dynamic model that require input from molecular physicists and chemists. Then, the second task is tackled: introducing state-of-the-art numerical techniques for the solution of the (usually large) set of partial differential equations that are the net result of the previous task. The discussion is limited to finite-volume techniques, in conjunction with upwind algorithms, which seem to be the prevalent choice of the computational fluid dynamic community. Other approaches are possible, and the reader is referred to some of the literature cited for more details.

M. Capitelli (ed.), Molecular Physics and Hypersonic Flows, 543–560.
© 1996 *Kluwer Academic Publishers. Printed in the Netherlands.*

2. Mathematical and Physical Formulation

A gas mixture whose temperature is high enough to allow for the onset and/or sustenance of chemical reactions will typically deviate from the ideal gas behavior. Internal energy modes which behave non-linearly with temperature will be activated, and, in extreme cases encountered in hypersonic applications, local thermodynamic non-equilibrium will be established.

2.1. INTERNAL ENERGY

In recent years, a great deal of effort has been spent on physically accurate modeling of *real gas effects*, which is the somewhat misleading terminology used to indicate departures from the low temperature ideal gas laws. A useful review of the current literature is presented by Gnoffo *et al.* [1].

A general representation for the species internal energy per unit mass, e_s, suitable for the modeling of problems where thermodynamic non-equilibrium plays a key role, is the following

$$e_s = \tilde{e}_s(T) + e_{n_s}(T_{n_s}).$$ (1)

In the above, $\tilde{e}_s(T)$ is the contribution due to translation and internal modes that can be assumed to be in equilibrium at the translational temperature T. Moreover, $e_{n_s}(T_{n_s})$ is the contribution due to internal modes that are in thermodynamic non-equilibrium, meaning that they are not in equilibrium at the translational temperature T, but may be assumed to satisfy a Boltzmann distribution at a different temperature T_{n_s}.

Equation 1 is valid for a very broad range of physical conditions and is susceptible to simplification and reduction to simpler models when the flow regime is less extreme. This formulation is very general, and it allows the utilization of many practical thermodynamic models for the determination of the specific functional form of internal energy, specific heats, and related properties. Examples of specific models currently used can be found in the literature [2,3].

In the following, a gas mixture composed of N species will be considered, and two major representations of the thermodynamic state of the mixture will be studied. The first one is a *Non-Equilibrium Model*, whereby the first M species contain a non-equilibrium contribution e_{n_s}. A simplification of the previous thermodynamic model may be achieved for flows where there is enough time for the internal modes to reach equilibrium at the translational temperature T. This *Equilibrium Model* will feature only one contribution to the internal energy in Eq. 1, namely $\tilde{e}_s$. The usual ideal gas model is a special case of the Equilibrium Model, when vibrational and electronic modes are neglected, and translational and fully excited rotational contributions are included in $\tilde{e}_s$.

The mixture value for the internal energy per unit mass, e, may be written using a standard mass-fraction-averaged summation, as follows

$$e = \sum_{s=1}^{N} Y_s e_s = \sum_{s=1}^{N} Y_s \tilde{e}_s + \sum_{s=1}^{M} Y_s e_{n_s},$$ (2)

where the species mass fraction Y_s has been introduced, defined as the ratio of species density ρ_s to mixture density $\rho = \sum_s^N \rho_s$.

A few cautionary notes are in order at this point. The assumption that a Boltzmann distribution holds for the non-equilibrium portions of internal energy, albeit at a temperature different from the translational value, is questionable in many cases, as evident from other contributions to these Proceedings. Moreover, some models that have been proposed and utilized feature two separate non-equilibrium energy contributions to the species internal energy, one stemming from vibrational energy, and the other due to the electronic contributions [3,4]. The usual accompanying assumption in those cases is that the electronic contributions from different chemical species and any free electrons that might be present in the mixture will reach a common Boltzmann distribution at an electron/electronic temperature T_e. This results in a gas with two translational temperatures, T for heavy particles and T_e for electrons, and many ensuing complications [5]. Finally, models have been proposed whereby species s is not a chemical species in its entirety, but a subset of a chemical species at a given vibrational energy level, or in a range of energy levels. This results in a much larger set of governing equations for the fluid dynamic behavior of such a mixture, as will be discussed in the following, and makes the applicability of these models to three-dimensional problems all but impossible at the present time.

2.2. EQUATIONS OF STATE

Most hypersonic flows of practical interest will involve low to moderate pressures and densities. In these instances, *Dalton's Law* will yield the mixture pressure as the sum of species partial pressures

$$p = \sum_{s=1}^{N} \rho_s R_s T = \rho R T \,, \tag{3}$$

where R_s is the species gas constant, and the mixture gas constant R is defined as follows

$$R = \sum_{s=1}^{N} Y_s R_s \,. \tag{4}$$

Equation 3 is also known as the *Thermal Equation of State*.

Unlike the ideal gas case, the state relationship of the pressure to the specific internal energy cannot, in general, be written directly, but it occurs implicitly through the translational temperature. A *Caloric Equation of State* such as Eq. 2 will relate internal energy, or portions thereof, to temperature. Iterative procedures are necessary for the determination of the temperature, due to the inherently non-linear character of this equation [6]. Once T is found, the pressure is evaluated from Eq. 3.

2.3. FINITE-RATE CHEMISTRY

When chemically reacting flows are investigated, a classification of chemical phenomena according to the time available for the completion of reactions is usually employed for order-of-magnitude estimates of the flowfield properties. If reaction times are very large compared with the time scale at which the fluid dynamics is evolving, then reactions have virtually no time to occur, and a *frozen* flow can be assumed (perfect gas results are a particular class of frozen chemistry simulations). The other limiting case occurs when the reaction times are very small compared to the fluid dynamic times, which results in the reactions having enough time to reach their *local chemical equilibrium* values, given by the Laws of Mass Action [7].

The real situation, however, is the general case of *finite-rate chemistry* [8], whereby in at least portions of the flowfield and/or at some point in the time evolution of the flow, the reaction times are comparable to the fluid dynamic times. In this instance, it becomes necessary to simulate the actual kinetic behavior of the chemical system.

A general simulation of chemical effects can be achieved for a system containing N species where J reactions take place [9]

$$\nu'_{1,j}X_1+\nu'_{2,j}X_2+\cdots+\nu'_{N,j}X_N \rightleftharpoons \nu''_{1,j}X_1+\nu''_{2,j}X_2+\cdots+\nu''_{N,j}X_N , \qquad j=1,\ldots,J, \tag{5}$$

where the $\nu'_{s,j}$ and the $\nu''_{s,j}$ are stoichiometric coefficients of species X_s in the j-th reaction. Then, the rate of production of the s-th species, w_s, may be written as a summation over the reactions

$$w_s \equiv \frac{d\rho_s}{dt} = \mathcal{M}_s \sum_{j=1}^{J}(\nu''_{s,j} - \nu'_{s,j})\left[k_{f,j}\prod_{l=1}^{N}\left(\frac{\rho_l}{\mathcal{M}_l}\right)^{\nu'_{l,j}} - k_{b,j}\prod_{l=1}^{N}\left(\frac{\rho_l}{\mathcal{M}_l}\right)^{\nu''_{l,j}}\right],$$

$$s=1,\ldots,N, \tag{6}$$

where $\mathcal{M}_s$ is the species molecular mass. For reaction j, the forward and backward reaction rates, $k_{f,j}$ and $k_{b,j}$ respectively, are assumed to be known functions of temperature, and they are related by thermodynamics

$$k_{f,j}(T) = k_{b,j}(T)K_{e,j}(T) , \tag{7}$$

where $K_{e,j}$ is the equilibrium constant, which is a known function of the thermodynamic state. Incidentally, the reaction rates and their dependence upon temperature are pieces of information that the fluid dynamic community expects of the molecular physics and chemistry communities.

The finite-rate chemistry model described remains valid in conjunction with all of the thermodynamic models discussed. However, the presence of thermodynamic non-equilibrium in the flowfield is likely to exercise some effect upon the reaction rates, especially when diatomic or polyatomic molecules are involved, whose vibrational modes may be excited. Several attempts to model the interaction between chemical and thermodynamic non-equilibrium have been recorded [4,10]. In summary, the reaction rates become functions of the non-equilibrium temperatures T_{n_s}, as seen in several contributions to these Proceedings.

The previous framework was derived for homogeneous reactions in the gas phase. Surface reactions and catalytic effects are very important in many applications, although they are not fully understood [11]. More details on current modeling efforts in this area are given in other parts of these Proceedings.

2.4. PRACTICAL CHEMISTRY MODELS

It was previously mentioned that a detailed knowledge of the kinetic behavior of a chemically reacting gas mixture is necessary for finite-rate chemistry investigations. In recent years, much attention to the kinetics of air and of hydrogen/air mixtures has been registered in the scientific community, spurred by the drive towards hypersonic flight and the necessary propulsive tools to achieve it. A compilation of the most recent data for the kinetics of air may be found in Park [4], where attention is given to thermodynamic non-equilibrium and its influence upon the reaction rates. Detailed studies of hydrogen/air mixtures have been published recently [12]. Moreover, computerized databases are now available, involving several thousand reactions and hundreds of chemical species [13].

2.5. GOVERNING EQUATIONS

The governing equations for flows in thermo-chemical non-equilibrium represent the conservation and/or time variation of physical quantities such as mass of a species in the mixture, momentum, non-equilibrium energy contributions, and energy. They are established either for a control volume of arbitrary size (finite or infinitesimal) or for a generic point in the flowfield. The former approach results in an integro-differential form, the latter in a non-linear differential system. In the following, the integral form of the equations will be introduced for consistency with the finite-volume-based numerical techniques to be analyzed later. The differential form of the equations can be readily obtained from the integral form and has been extensively discussed in the literature (e.g., see Lee [14]).

The governing equations written in integral form are valid both in regions of smooth flows and across discontinuities, where they can be shown to reduce to the Rankine-Hugoniot *jump conditions* in the limit of infinitesimal volumes. The control volume employed for the derivation is allowed to move and deform with time, although many important applications take advantage of the simpler case of fixed control volumes. For an arbitrary volume $\mathcal{V}$, closed by a boundary $\mathcal{S}$, the governing equations in integral form read

$$\frac{\partial}{\partial t} \iiint_{\mathcal{V}} \mathbf{Q} \, d\mathcal{V} + \oint_{\mathcal{S}} (\mathbf{S} - \mathbf{S}_v) \cdot \mathbf{n} d\mathcal{S} = \iiint_{\mathcal{V}} \mathbf{W} d\mathcal{V}, \qquad (8)$$

where $\mathbf{Q}$ is the vector of conserved variables, $\mathbf{W}$ is the vector of source terms, and $\mathbf{S}$ and $\mathbf{S}_v$ are the inviscid and viscous flux vectors, respectively. The unit vector $\mathbf{n}$ is normal to the infinitesimal area $d\mathcal{S}$ and points outwards.

In conjunction with the Non-Equilibrium Model, and utilizing a Cartesian frame of reference (x, y, z) whose unit vectors are $\mathbf{i}$, $\mathbf{j}$, and $\mathbf{k}$, respectively, the

548

vectors Q and W in the previous equation read

$$
Q = \begin{pmatrix} \rho_1 \\ \rho_2 \\ \vdots \\ \rho_N \\ \rho u \\ \rho v \\ \rho w \\ \rho_1 e_{n_1} \\ \vdots \\ \rho_M e_{n_M} \\ \rho e_0 \end{pmatrix}, \quad
W = \begin{pmatrix} w_1 \\ w_2 \\ \vdots \\ w_N \\ \sum_s \rho_s g_{s_x} \\ \sum_s \rho_s g_{s_y} \\ \sum_s \rho_s g_{s_z} \\ Q_1 \\ \vdots \\ Q_M \\ \left(\sum_s \rho_s g_s\right) \cdot u \end{pmatrix}, \tag{9}
$$

and the flux vectors S, S_v read

$$
S = \begin{pmatrix} \rho_1 (u - u_v) \\ \rho_2 (u - u_v) \\ \vdots \\ \rho_N (u - u_v) \\ \rho u (u - u_v) + pi \\ \rho v (u - u_v) + pj \\ \rho w (u - u_v) + pk \\ \rho_1 e_{n_1} (u - u_v) \\ \vdots \\ \rho_M e_{n_M} (u - u_v) \\ \rho e_0 (u - u_v) + pu \end{pmatrix}, \quad
S_v = \begin{pmatrix} -\rho_1 v_1 \\ -\rho_2 v_2 \\ \vdots \\ -\rho_N v_N \\ \tau_{xx} i + \tau_{xy} j + \tau_{xz} k \\ \tau_{yx} i + \tau_{yy} j + \tau_{yz} k \\ \tau_{zx} i + \tau_{zy} j + \tau_{zz} k \\ -\rho_1 e_{n_1} v_1 - q_{n_1} \\ \vdots \\ -\rho_M e_{n_M} v_M - q_{n_M} \\ \Theta \end{pmatrix}, \tag{10}
$$

where

$$
\Theta = (\tau_{xx} u + \tau_{yx} v + \tau_{zx} w)i + (\tau_{xy} u + \tau_{yy} v + \tau_{zy} w)j +
$$
$$
(\tau_{xz} u + \tau_{yz} v + \tau_{zz} w)k - q - q^R - \sum_{s=1}^{M} q_{n_s} - \sum_{s=1}^{N} \rho_s h_s v_s. \tag{11}
$$

In the previous formulas, the *mass-averaged velocity* u for the mixture has been utilized, whose Cartesian components are (u, v, w) respectively. This velocity can be defined in terms of species velocities u_s as follows

$$
\rho u = \sum_{s=1}^{N} \rho_s u_s. \tag{12}
$$

The difference between species and mass-averaged velocities is a species *diffusion velocity*

$$
v_s = u_s - u, \qquad s = 1, \ldots, N. \tag{13}
$$

The velocity $u_\mathcal{V}$ that appears in the inviscid flux vector is the control surface velocity, which equals zero when the control volume is fixed with time. The first N equations are species continuity equations, relating the time rate of change of species densities ρ_s to convective and diffusive transport, and to the creation/destruction of the species due to chemical reactions. The species rates of production w_s have been defined by Eq. 6, when finite-rate chemistry models were discussed. Following species continuity are the three components of the momentum equation. The source term $\sum_s \rho_s g_s$ represents body forces, e.g. gravity and electric fields, and the viscous fluxes involve the components of the shear stress tensor. After momentum, M non-equilibrium energy equations are written, describing the creation and evolution in time and space of the non-equilibrium components of the internal energy. The viscous fluxes associated with these equations describe the heat transfer due to the transport of non-equilibrium energy by diffusion, $\rho_s e_{n_s} v_s$, as well as the conductive heat transfer, q_{n_s}. The source terms for the non-equilibrium energy equations, Q_s, will be described in more detail below. Finally, the global energy equation describes the time evolution of the total internal energy per unit volume $\rho e_0 = \rho e + \rho(u^2 + v^2 + w^2)/2$. The viscous flux in this case represents viscous dissipation, convective heat transfer due to both equilibrium and non-equilibrium portions of the internal energy, q and $\sum_s^M q_{n_s}$ respectively, radiative heat transfer, q^R, and diffusive heat transfer, $\sum_s^N \rho_s h_s v_s$, where h_s is the species enthalpy, $h_s = e_s + R_s T$.

For models where the electron/electronic energy modes are assumed to be in non-equilibrium at a common temperature T_e, a special form of the non-equilibrium energy equation is present [3,4]. The actual form of the equation, and possible simplified forms thereof, is the subject of some controversy, and on-going research [4,5].

In addition to the equations presented, thermal and caloric equations of state will relate pressure and temperatures to equilibrium and non-equilibrium internal energy components, as already discussed, and the viscous flux vector entries will be expressed in terms of the entries in the vector of conserved variables.

Simplifications in the physical model used for the analysis of non-equilibrium flows will bring about corresponding changes in the governing equations. For instance, the Equilibrium Model can be handled simply by dropping all of the non-equilibrium energy equations, along with the non-equilibrium contributions to the global energy equation. The whole process is equivalent to setting M equal to zero. Inviscid flows are modeled by dropping the viscous flux vector in conjunction with any of the thermodynamic models discussed.

2.6. THERMODYNAMIC NON-EQUILIBRIUM MODELS

Several non-equilibrium models have been utilized for the study of hypersonic flow and lasers. The most common ones are reviewed by Gnoffo et al. [1], and many more are discussed elsewhere in these Proceedings.

By far, the most common source of non-equilibrium behavior is vibrational energy. Two major physical phenomena will contribute to the source terms Q_s in Eq. 9, when it is assumed that vibrational non-equilibrium is present. The first one is the change in vibrational energy caused by inelastic collisions, that is, creation or destruction of vibrating particles due to chemical reactions; the second one is due to elastic collisions, and is tantamount to energy exchanges between the different internal energy modes. The most important contribution to the latter is the relaxation of non-equilibrium vibrational energy towards the equilibrium levels at the local translational temperature. Additional contributions include energy exchanges among the different vibrating species [14].

2.7. VISCOUS FLUXES AND TRANSPORT PROPERTIES

Once again, only the most basic models for the transport properties of a mixture of gases in thermo-chemical non-equilibrium are presented next. The search for better models that are not computationally overwhelming is in progress, and recent advances are detailed in these Proceedings. Physically reliable data on transport properties and how they are affected by thermo-chemical non-equilibrium represent another good example of the valuable contributions that the molecular physics and chemistry communities can provide to the fluid dynamics community.

The most common assumptions that are made when dealing with transport properties concern the treatment of the viscous stress tensor. Generally only *Newtonian* fluids are considered, where there is a linear relationship between stress and rate of deformation. Moreover, bulk viscosity effects are neglected. Under these assumptions, the stress tensor components are expressed as

$$\tau_{ij} = \mu\left(\frac{\partial u_i}{\partial x_j} + \frac{\partial u_j}{\partial x_i}\right) - \frac{2}{3}\mu\left(\frac{\partial u_1}{\partial x_1} + \frac{\partial u_2}{\partial x_2} + \frac{\partial u_3}{\partial x_3}\right)\delta_{ij}, \qquad i,j = 1,2,3, \qquad (14)$$

where μ is the viscosity coefficient and the symbol δ_{ij} stands for the Kronecker delta. The indices i and j refer to the x, y, and z space directions and velocity components.

For mixtures of gases in thermal equilibrium, the heat flux vector q is modeled by means of the product of thermal conductivity, λ, times the temperature gradient, a result known as *Fourier's Law*. The extension of this approach to flows in thermal non-equilibrium is usually done by considering similar contributions for the non-equilibrium temperatures. The resulting expressions read

$$q = -\lambda\nabla T, \qquad q_{n_s} = -\lambda_{n_s}\nabla T_{n_s}, \qquad s = 1,\ldots,M. \qquad (15)$$

Mass diffusion is a physical phenomenon that arises mostly because of the presence of gradients of mass or mole concentrations in the mixture. Assuming that the mixture behaves like a binary mixture yields the so-called *Fick's Law of Diffusion*

$$\rho_s v_s = -\rho D_s \nabla\left(\frac{\rho_s}{\rho}\right), \qquad s = 1,\ldots,N, \qquad (16)$$

where D_s is a diffusion coefficient.

Theoretical formulas for the viscosity coefficients of individual species and mixtures have been given as a result of the asymptotic analysis that leads to the Navier-Stokes equations, but they are very cumbersome to use. Curve-fit functional expressions have been proposed, such as the very popular one due to Blottner *et al.* [15]. Another very common approach is the extension of Sutherland's Law for a perfect gas to a generic component. Values of the viscosity coefficient for the mixture are usually recovered by means of *Wilke's rule* [16], which is an extension of a Sutherland-type equation to multi-component systems, obtained on the basis of kinetic theory and several simplifying assumptions.

Values of the thermal conductivity to be used in Eq. 15 are also obtained from an asymptotic analysis for small deviations from equilibrium. Curve-fit expressions have also been used for thermal conductivity, and a Sutherland-type formula proposed, similarly to what has been done for the viscosity coefficient [1]. Results for the mixture thermal conductivity can be obtained again by means of Wilke's rule.

The simplest modeling of the diffusion coefficients is obtained by assuming a constant Lewis number, where only one global diffusion coefficient is used. Multi-component effects can be partially taken into account by more complex choices of diffusion coefficients, relating those with binary diffusion coefficients [4].

Theoretical solutions of the Boltzmann equations for mixtures of gases out of thermal equilibrium have not been obtained at the present time. The numerical treatment of transport properties for these cases usually consists of fairly simple assumptions. Diffusion coefficients and viscosity are left unaltered, and thermal conductivity is modified in a straightforward manner: only the portions of the specific heats that are in thermal equilibrium are included in the equations for λ; the remaining portions contribute to the determination of λ_{n_s}.

The inclusion of turbulence in the viscous flux vectors is obtained in its simplest form by the addition of turbulent viscosity, thermal conductivity, and diffusion coefficients to their laminar counterparts. The turbulent contributions are then modeled by means of algebraic treatments, foremost among which is the model originated by Baldwin and Lomax [17], or by introducing and solving additional partial differential equations, for example for turbulent kinetic energy and dissipation. In general, the models for turbulence in the presence of thermo-chemical non-equilibrium have been straightforward extensions of the ones derived for an ideal gas. Recent departures from this tendency are the methods based on statistical techniques such as *Probability Distribution Function* [18].

3. Numerical Formulation

In order for a computer simulation to be possible, the governing equations presented in the previous section must be discretized, that is, reduced to a set of algebraic equations to be solved. A very convenient discretization approach is the finite-volume technique, whereby the integral form of the governing equations is solved for the unknown volume averages of conserved variables in some small, but finite, control volume. The advantage of this method is its use of the integral form of the equations, which allows a correct treatment of discontinuities and also consistently treats general grid topologies. Other discretizations are also possible: for example spectral techniques, and the more usual finite-difference approach [19].

3.1. THE FINITE-VOLUME TECHNIQUE

The finite-volume technique takes advantage of the integral form of the governing equations to achieve a discretized algebraic approximation that can be handled by a digital computer. The physical domain of interest is partitioned into a large set of sub-domains, volume-averaged values for the conservative variables Q are defined for each sub-domain (computational cell), and the governing equations described in Eq. 8 are written for each cell in terms of the volume averages. The surface integral that involves inviscid and viscous fluxes is partitioned into separate contributions from each of the faces that compose the boundary of the computational cell, and each contribution is written in terms of surface-averaged values of the fluxes. Then, the area-averaged values are related to the unknown volume-averaged values, typically by extrapolation. The resulting set of algebraic equations will be the discretized form of the original problem that is actually solved by means of a digital computer.

3.2. EXTRAPOLATION STRATEGIES

Relating the area-averaged values of inviscid and viscous fluxes to the volume-averaged values of the conserved variables is the key operation of any finite-volume strategy. A generalized extrapolation procedure to recover values at a face starting from quantities at the center of a volume can be formulated for a generic vector q, and this results in two typically different left and right extrapolated vectors, q^- and q^+ respectively. In order to minimize instabilities and non-monotonic behavior of the solution, high-order extrapolation formulas will involve limiters. A review of commonly used limiters and their properties is given by Sweby [20], and general extrapolation formulas are presented by Cinnella and Grossman [6].

Two different techniques have emerged in recent years for the treatment of the inviscid flux vector. The first one identifies q with the flux vector itself. Consequently, a vector of inviscid fluxes is created from volume-averaged values of the variables in a cell, and, subsequently, face values are created by extrapolation. The second method, called MUSCL extrapolation (Monotone Upstream-centered Schemes for Conservation Laws), identifies q with the vector of conserved variables, Q. The extrapolation produces face values for the conserved variables; then the fluxes are constructed from these values. It is useful to point out that for many applications, the extrapolation of *primitive* or even *characteristic* variables is preferred to the extrapolation of conserved variables.

3.3. FLUX-SPLIT ALGORITHMS

In recent years, flux-split techniques have obtained wide acceptance as an accurate means of discretizing the inviscid fluxes. Originally developed for a perfect gas, they have been extended to flows in chemical equilibrium, and to mixtures out of chemical and thermal equilibrium. They are found to be very accurate and robust when used for transonic, supersonic and hypersonic flows, and are fully compatible with conservative finite-volume, shock-capturing approaches.

Probably the two most popular schemes in the flux-vector splitting category are the ones due to Steger and Warming [21] and to Van Leer [22]. The basic premise is to split the inviscid flux vector in one space dimension into two parts, each containing the information that propagates downstream and upstream, respectively. The two parts are constructed using the extrapolation strategies already outlined, consistently with the direction of propagation. Another very popular scheme, less robust for hypersonic flows, but more accurate, is the flux-difference splitting technique due to Roe [23]. It consists of an approximate Riemann solver, where an arbitrary discontinuity is supposed to exist at the cell surface between the left and the right state, and an approximate solution for this situation is written in terms of waves propagating upstream and downstream. In this instance, the inviscid flux vector is not split, but reconstructed from the upstream and downstream contributions that constitute the left and the right states in the Riemann problem. A complete summary of the non-equilibrium version of these techniques can be found in Cinnella and Grossman [6], and derivations of similar algorithms have been published (e.g., see Liu and Vinokur [24]).

Extensions to two or three space directions are usually made by superimposing pseudo-one-dimensional problems, where the extrapolation formulas for left and right states keep track of the relevant direction only. In recent times, much effort has been spent towards the design of truly multi-dimensional algorithms, which would be at least approximately independent of the grid orientation, as will be discussed shortly.

3.4. RECENT ALGORITHMS

The algorithms presented do not exhaust the category of flux-split techniques. Osher-type formulations exist for perfect gases, and their extension to reacting flows has been attempted [25]. More recently, new kinds of flux splitting are emerging, whereby the treatment of the pressure gradient plays a central role, and they seem to show promise of improvement over the older schemes. Again, the original development is for perfect gases, although the extension to flows in thermo-chemical non-equilibrium is relatively straightforward [26].

All of the previously mentioned upwind algorithms originate from the Euler equations for gas dynamics. The Euler equations, however, can be obtained as moments of the Boltzmann equation, established in the kinetic theory of gases, provided that the velocity distribution function is Maxwellian. This connection between the Boltzmann equation and the Euler equations has lead to a new class of upwind *kinetic* schemes which are based on the principle that an upwind scheme at the Boltzmann level leads to an upwind scheme at the Euler level. Eppard and Grossman [27] have extended the kinetic schemes to flows with chemical and thermodynamic non-equilibrium by means of non-equilibrium kinetic theory.

Although the Roe scheme has been very successful, it does have well documented failings, such as expansion shocks and the carbuncle phenomenon [28]. In an attempt to address these failings, the positively conservative scheme of Harten-Lax-van Leer-Einfeldt (HLLE) has been recently introduced [29]. The HLLE scheme, although being very robust, is also highly dissipative at contact discontinuities and shear layers. Modifications to this algorithm in order to increase its resolution are in progress (e.g., see Obayashi and Wada [30]).

3.5. TRULY MULTI-DIMENSIONAL ALGORITHMS

To obtain accurate numerical simulations for complex flow problems, solvers must be capable of calculating strong and weak shock waves, shear and contact discontinuities, flame fronts, mixing layers, and boundary layers. While there has been considerable progress in these areas over the last decade, some improvements are still necessary. The current generation of Euler and Navier-Stokes solvers are limited by a noticeable degradation in the accuracy of resolution of these fluid-dynamic phenomena when they are oriented obliquely to the computational grid. This problem arises in central-difference codes because of the inability to properly tune the numerical dissipation, and in upwind codes because the current technology is essentially one-dimensional in nature and must be implemented in a dimensionally-split approach for calculations in two and three spatial dimensions. Consequently, excessive numbers of grid points and associated large amounts of computational time and memory are required to accurately resolve the critical flow features. This problem is magnified for calculations involving chemical and thermal non-equilibrium, since additional rate equations must be solved in a coupled fashion, and the species densities and non-equilibrium energies must be stored at each grid point.

One approach towards the development of multi-dimensional upwind solvers involves rotated or *multi-directional* Riemann solvers [31–34]. These schemes require the choice of a dominant upwinding direction and a local Riemann solution with left and right states that are functions of the upwinding angle. Solvers of this type have succeeded in calculating shocks oblique to the grid with nearly the same resolution as shocks which are aligned with the grid. However, for three-dimensional flows with complex fluid phenomena, the task of choosing only one pertinent upwinding direction is a formidable one. While these schemes represent an improvement over directionally-split schemes, they are not the final answer and still leave the CFD community in need of truly multi-dimensional ideas.

3.6. VISCOUS FLUXES

The viscous vectors are usually discretized using standard second-order accurate central differences. In most applications of engineering interest, the Thin-Layer version of the Navier-Stokes equations is actually discretized, whereby the viscous terms in the space directions that are not normal to a solid surface are neglected. On the other hand, when the full Navier-Stokes equations are analyzed, some mixed second-order derivative terms appear in the formulation, and they are usually treated in a diagonal-dominance-preserving fashion [35].

3.7. SOURCE TERMS

When using the finite-volume technique, the inclusion of source terms into the discretized equations is usually a trivial matter. Unless radiative transfer is included, source terms are local in nature, nor do they require the evaluation of space derivatives or gradients. The vector of source terms W, introduced in Eq. 8, is multiplied by the volume of the finite volume under consideration and algebraically summed to the unsteady and flux contributions.

Significant complications usually arise when radiative heat transfer is included in the source terms for the energy equations, because those contributions are *non-*

local: they depend on the thermo-chemical properties of the surrounding medium. Relatively drastic simplifications of the radiative source terms are common in these cases [36–38].

3.8. TIME INTEGRATION

After fluxes and source terms have been discretized and related to the vector of unknown volume-averaged variables, the remaining problem to be solved is how to advance the numerical solution in time, given a known initial state. The time integration schemes that are commonly used fall into two categories, *explicit* and *implicit*. Explicit schemes evaluate fluxes and source terms at the old time step prior to moving to a new time, whereas implicit techniques utilize linearized estimates of fluxes and source terms at the new time step when advancing the solution in time.

Two classes of problems usually arise when dealing with time integration. The first class contains the *unsteady* problems, where an accurate time integration is essential for the overall validity of the simulation. Moreover, some unsteady problems could involve moving boundaries and/or moving and deforming control volumes. The second class involves *steady* problems, where time accuracy is of no concern. A pseudo-transient problem is usually solved in this case, where the solution is advanced in a pseudo-time until convergence to a steady-state is reached. Most algorithms will be able to handle both steady and unsteady problems, provided that the discretized form of the *Geometric Conservation Law* [39] is enforced in the calculation of the time derivative of the control volumes for cases when the latter are deforming.

A popular and computationally inexpensive scheme that is second-order accurate in time is the m-stage, Jameson-style, explicit Runge-Kutta method. That and other algorithms used in CFD applications are discussed in Fletcher [19]. Unfortunately, all explicit schemes have the major drawback of becoming extremely inefficient for *stiff* problems, that is, for cases when there are order-of-magnitude differences among the characteristic times associated with the evolution of all the physical phenomena present [40,41]. In these conditions, the time step necessary for stability of the algorithm is limited by the *smallest* of the time scales present, and can become even orders of magnitude smaller than the value which would be necessary for an efficient resolution of the overall gas dynamic transient [42]. For this reason, time-implicit schemes are advocated for hypersonic flows, as they show an increase in efficiency that can be dramatic.

Hybrid schemes have also been investigated, whereby the source terms are treated implicitly and the fluxes remain explicit. These techniques show some promise, but they are still the object of current research [43].

Implicit time integration schemes are very popular especially for steady problems, due to their unconditional stability, the consequent increase in time step per iteration, and the overall efficiency that they make possible. Moreover, for reacting flow problems it has been shown that treating the source terms in an implicit fashion is tantamount to rescaling the governing equations so that all of the time scales in a flow problem become of the same order of magnitude [44]. Consequently, implicit algorithms are a viable solution to stiffness problems.

Considering only *structured* grids for simplicity, every finite volume in the computational field can be located by a set of three indices, i, j, and k. Writing the governing equations, Eq. 8, for the generic control volume (i, j, k) yields

$$\frac{\partial(\boldsymbol{Q}_{ijk}\mathcal{V}_{ijk})}{\partial t} + \boldsymbol{R}_{ijk} = 0 \,, \tag{17}$$

where a *steady state* residual, $\boldsymbol{R}$, has been introduced

$$\begin{aligned}
\boldsymbol{R}_{ijk} = [(\boldsymbol{S}-\boldsymbol{S}_v)_{i+1/2} \cdot \boldsymbol{n}_{i+1/2}\mathcal{S}_{i+1/2} + (\boldsymbol{S}-\boldsymbol{S}_v)_{i-1/2} \cdot \boldsymbol{n}_{i-1/2}\mathcal{S}_{i-1/2} + \\
(\boldsymbol{S}-\boldsymbol{S}_v)_{j+1/2} \cdot \boldsymbol{n}_{j+1/2}\mathcal{S}_{j+1/2} + (\boldsymbol{S}-\boldsymbol{S}_v)_{j-1/2} \cdot \boldsymbol{n}_{j-1/2}\mathcal{S}_{j-1/2} + \\
(\boldsymbol{S}-\boldsymbol{S}_v)_{k+1/2} \cdot \boldsymbol{n}_{k+1/2}\mathcal{S}_{k+1/2} + (\boldsymbol{S}-\boldsymbol{S}_v)_{k-1/2} \cdot \boldsymbol{n}_{k-1/2}\mathcal{S}_{k-1/2} - \\
\boldsymbol{W}_{ijk}\mathcal{V}_{ijk}.
\end{aligned} \tag{18}$$

and volume-averaged values for $\boldsymbol{Q}$ and $\boldsymbol{W}$ have been introduced. The surface integral has been broken into six contributions from the six boundary surfaces of control volume (i, j, k), whose unit normals have been indicated by $\boldsymbol{n}_{i\pm1/2}$, $\boldsymbol{n}_{j\pm1/2}$, and $\boldsymbol{n}_{k\pm1/2}$, respectively. Surface-averaged values of the fluxes have been utilized. In the above, the dot product between flux vectors and unit normals indicates that the operation is performed on each flux entry.

For cases where time accuracy is important, as in unsteady problems, second-order-accurate schemes in time have been developed, such as the implicit two-step Runge-Kutta scheme developed by Iannelli and Baker [45], or three-time-level schemes, such as the popular one proposed by Beam and Warming [46]. Examples abound of problems where even first-order algorithms such as Euler Implicit are deemed satisfactory. Incidentally, for steady problems, the Euler Implicit technique is almost universally accepted as the most efficient approach. Considering the three-time-level discretization of the governing equations as an example, Eq. 17 can be written as follows

$$\frac{(1+\psi)\Delta(\boldsymbol{QV})^n_{ijk} - \psi\Delta(\boldsymbol{QV})^{n-1}_{ijk}}{\Delta t^n} = (\theta - 1)\boldsymbol{R}^n_{ijk} - \theta\boldsymbol{R}^{n+1}_{ijk} \,, \tag{19}$$

where the superscript n refers to the time level, the *forward difference* operator $\Delta(\cdot)$ is defined as

$$\Delta(\cdot)^n = (\cdot)^{n+1} - (\cdot)^n \,, \tag{20}$$

and ψ, θ are two parameters that take different numerical values for different classical time integration schemes. Specifically, $\psi = 0$ and $\theta = 0$ yield the Euler Explicit scheme; $\psi = 0$ and $\theta = 1$ yield the Euler Implicit scheme; $\psi = 0$ and $\theta = 1/2$ the Trapezoidal scheme; $\psi = 1/2$ and $\theta = 1$ the 3-Point Backward scheme. It may be useful to point out that the first two schemes are first-order accurate in time, whereas the other two are second-order accurate.

For nonzero values of the parameter θ, implicit schemes are recovered. In those cases, the right hand side of the equation contains a contribution from the residual at the new time level, and this makes the equation *nonlinear* in the unknown Q^{n+1}. Nonlinear equations are usually solved by using a linearization algorithm, coupled with an iterative procedure applied to the linear equations until convergence to the nonlinear solution is achieved. Typically Newton's Method is applied to linearize the equations. Writing the previous equation as a function L of the unknown Q^{n+1}, yields

$$L(Q^{n+1}) = \frac{(1 + \psi)\Delta(QV)^n_{ijk} - \psi\Delta(QV)^{n-1}_{ijk}}{\Delta t^n} - \left[(\theta - 1)R^n_{ijk} - \theta R^{n+1}_{ijk}\right] = 0, \quad (21)$$

and the linear scheme is obtained

$$L'(Q^P)\Delta Q^P = -L(Q^P), \qquad P = 0, 1, \ldots, \quad (22)$$

where the iteration index P is such that $Q^{P=0} = Q^n$, and at convergence $Q^{P\to\infty} = Q^{n+1}$. The function $L(Q)$ is called the *unsteady* residual, and $L'(Q)$ is its first derivative. Their expressions are slightly complicated by the fact that the control volumes can be moving and deforming in time. After some algebra [47], it can be shown that the unsteady residual is given by

$$\begin{aligned} L(Q^P) =& \frac{V^{P+1}_{ijk}(Q^P_{ijk} - Q^n_{ijk})}{\Delta t^n} - \frac{\psi}{1 + \psi}\frac{V^{n-1}_{ijk}\Delta(Q)^{n-1}_{ijk}}{\Delta t^n} + \\ & \frac{Q^n_{ijk}}{1 + \psi}\left[(\theta - 1)\hat{R}^n_{ijk} - \theta\hat{R}^P_{ijk}\right] - \frac{1}{1 + \psi}\left[(\theta - 1)R^n_{ijk} - \theta R^P_{ijk}\right], \end{aligned} \quad (23)$$

where the *Geometric Conservation Law* established by Thomas and Lombard [39] has been utilized to describe the volume changes due to movement and deformation, yielding a *geometric* residual $\hat{R}$

$$\begin{aligned} \hat{R}_{ijk} = -[&u_V \cdot n_{i+1/2}S_{i+1/2} + u_V \cdot n_{i-1/2}S_{i-1/2} + \\ & u_V \cdot n_{j+1/2}S_{j+1/2} + u_V \cdot n_{j-1/2}S_{j-1/2} + \\ & u_V \cdot n_{k+1/2}S_{k+1/2} + u_V \cdot n_{k-1/2}S_{k-1/2}]. \end{aligned} \quad (24)$$

In summary, the unsteady residual includes fluxes and chemistry source terms (R), volume changes $(\hat{R})$, and unsteady contributions due to time rates of change of the vector Q.

The first derivative of the unsteady residual introduced in Eq. 22 can be evaluated when the dependence of geometric residual and cell volumes upon the vector Q^P is neglected. It will involve the Jacobians of the inviscid flux vectors, as well as contributions from viscous fluxes and source terms. Substitution of expressions for fluxes, source terms, and their Jacobians into the linearized equation, Eq. 22, results in a *large* set of coupled linear equations for the volume-averaged unknowns ΔQ^P. For three-dimensional problems, the storage requirements for an exact solution of this linear problem are too restrictive. Approximate solutions have been devised, namely solutions in a plane in conjunction with relaxation in the

558

third space direction, or approximate factorization schemes that reduce the original problem to a series of smaller problems [47,48].

Unsteady calculations are typically performed by iterating Eq. 22 until the unsteady residual is driven to zero, then moving to the next time step and repeating the procedure. In practice, a relatively small number of iterations has been shown to be sufficient for accurate calculations (less than 10, and as little as 2 in some cases [49]). Steady-state simulations can be performed within the same framework, however in this case only one iteration per pseudo-time step is typically utilized, and $\psi \equiv 0$ in Eq. 23. This is tantamount to reducing the unsteady residual to its steady-state counterpart.

4. Concluding Remarks

The present chapter has attempted to review the state of the art in the area of numerical methods for physically challenging flows such as those in thermo-chemical non-equilibrium. The exposition has been essentially limited to finite-volume techniques and upwind algorithms, although an effort has been made to mention other approaches.

Future work in this field is very promising due to the availability of larger and faster computers, and the advances in physical modeling and numerical algorithms. Parallel computers are likely to play a major role in making numerical simulations of reacting flows affordable. A better understanding of turbulent combustion and non-equilibrium aerothermodynamics is necessary, and some progress is being made in these fields. Several open questions of practical interest have been raised at the NATO Institute, concerning the interpretation of flight experiments, the behavior of electron temperature in shock tunnels, and the need for reliable three-body recombination reaction rates, to name a few examples. Solving some of these questions will result in tangible progress towards making better hypersonic non-equilibrium flow simulations.

References

1. Gnoffo, P.A., Gupta, R.N., and Shinn, J.L. (1989) Conservation Equations and Physical Models for Hypersonic Air Flows in Thermal and Chemical Nonequilibrium, NASA Technical Paper No. 2867.

2. Bertin, J.J., Glowinski, R., and Periaux, J. (Eds.) (1989) *Hypersonics*, Birkhäuser, Boston MA.

3. Candler, G.V. and MacCormack, R.W. (1988) The Computation of Hypersonic Ionized Flows in Chemical and Thermal Nonequilibrium, AIAA Paper No. 88-0511.

4. Park, C. (1991) *Nonequilibrium Hypersonic Aerothermodynamics*, John Wiley & Sons, New York NY.

5. Grossman, B., Cinnella, P., and Eppard, W.M. (1992) New Developments Pertaining to Algorithms for Non-Equilibrium Hypersonic Flows, *CFD Journal* 1(2):175–186.

6. Cinnella, P. and Grossman, B. (1991) Flux-Split Algorithms for Hypersonic Flows, in T.K.S. Murthy (ed.), *Computational Methods in Hypersonic Aerodynamics*, Computational Mechanics Publications, Southampton UK, pp. 153–202.

7. Cox, C.F. and Cinnella, P. (1994) General Solution procedure for Flows in Local Chemical Equilibrium, *AIAA Journal* 32(3):519–527.

8. Anderson, J.D.Jr. (1989) *Hypersonic and High Temperature Gas Dynamics*, McGraw-Hill, New York NY.

9. Vincenti, W.G. and Kruger C.H.Jr. (1986) *Introduction to Physical Gas Dynamics*, Robert E. Krieger, Malabar FL.

10. Marrone, P.V. and Treanor, C.E. (1963) Chemical Relaxation with Preferential Dissociation from Excited Vibrational Levels, *Physics of Fluids* **6**(9):1215–1221.

11. Bruno, C. (1989) Real Gas Effects, in J.J. Bertin, R. Glowinski, and J. Periaux (eds.), *Hypersonics*, Vol. 1, Birkhäuser, Boston MA, pp. 303–354.

12. Oldenborg, R., Chinitz, W., Friedman, M., Jaffe, R., Jachimowski, C., Rabinowitz, M., and Schott, G. (1990) Hypersonic Combustion Kinetics, NASA Technical Memorandum No. 1107.

13. Mallard, W.G., Westley, F., Herron, J.T., Hampson, R.F., and Frizzell, D.H. (1993) *NIST Chemical Kinetics Database: Version 5.0*, National Institute of Standards and Technology, Gaithersbug MD.

14. Lee, J.-H. (1985) Basic Governing Equations for the Flight Regimes of Aeroassisted Orbital Transfer Vehicles, *Progress in Astronautics and Aeronautics* **96**:3–53.

15. Blottner, F.G., Johnson, M., and Ellis, M. (1971) Chemically Reacting Viscous Flow Program for Multi-Component Gas Mixtures, Sandia Laboratories Report No. SC-RR-70-754.

16. Wilke, C.R. (1950) A Viscosity Equation for Gas Mixtures, *Journal of Chemical Physics* **18**(4):517–519.

17. Baldwin, B.S. and Lomax, H. (1978) Thin Layer Approximation and Algebraic Model for Separated Turbulent Flows, AIAA Paper No. 78-257.

18. Pope, S.B. (1985) PDF Methods for Turbulent Reactive Flows, *Progress in Energy and Combustion Sciences* **11**:119–192.

19. Fletcher, C.A.J. (1988) *Computational Techniques for Fluid Dynamics*, Springer-Verlag, New York NY.

20. Sweby, P.K. (1984) High Resolution Schemes Using Flux Limiters for Hyperbolic Conservation Laws, *SIAM Journal of Numerical Analysis* **21**(5):995–1011.

21. Steger, J.L. and Warming, R.F. (1981) Flux Vector Splitting of the Inviscid Gasdynamic Equations with Applications to Finite-Difference Methods, *Journal of Computational Physics* **40**:263–293.

22. Van Leer, B. (1982) Flux-Vector Splitting for the Euler Equations, in *Lecture Notes in Physics*, Vol. 170, Springer-Verlag, New York NY, pp. 507–512.

23. Roe, P.L. (1981) Approximate Riemann Solvers, Parameter Vectors, and Difference Schemes, *Journal of Computational Physics* **43**:357–372.

24. Liu, Y. and Vinokur, M. (1989) Upwind Algorithms for General Thermo-Chemical Nonequilibrium Flows, AIAA Paper No. 89-0201.

25. Abgrall, R. and Montagné, J.L. (1989) Generalization of Osher's Riemann Solver to Mixture of Perfect Gas and Real Gas, *La Recherche Aerospatiale* **1989-4**:1–13.

26. Bergamini, L., and Cinnella, P. (1994) Using the Liou-Steffen Algorithm for the Euler and Navier-Stokes Equations, *AIAA Journal* **32**(3):657–659.

27. Eppard, W.M. and Grossman, B. (1993) An Upwind Kinetic Flux-Vector Splitting Method for Flows in Chemical and Thermal Non-Equilibrium, AIAA Paper No. 93–0894.

28. Quirk, J.J. (1992) A Contribution to the Great Riemann Solver Debate, ICASE Report No. 92-64, NASA Langley Research Center, Langley VA.

29. Einfeldt, B., Munz, C.D., Roe, P.L., and Sjogreen, B. (1991) On Gudonov-Type Methods near Low Densities, *Journal of Computational Physics* **92**(2):273–295.

560

30. Obayashi, S. and Wada, Y. (1994) Practical Formulation of a Positively Conservative Scheme, *AIAA Journal* **32**(5):1093–1095.

31. Van Leer, B. (1993) Progress in Multidimensional Upwinding, in *Lecture Notes in Physics*, Vol. 414, Springer-Verlag, New York NY, pp. 1–26.

32. Dadone, A. and Grossman, B. (1992) A Rotated Upwind Scheme for the Euler Equations, *AIAA Journal* **30**(10):2219–2226.

33. Davis, S.F. (1984) A Rotationally-Biased Upwind Difference Scheme for the Euler Equations, *Journal of Computational Physics* **56**:65–92.

34. Struijs, R., Deconinck, H., De Palma, P., Roe, P.L., and Powell, K.G. (1991) Progress on Multidimensional Upwind Euler Solvers for Unstructured Grids, AIAA Paper No. 91-1550-CP.

35. Chakravarthy, S.R., Szema, K.-Y., Goldberg, U.C., Gorski, J.J., and Osher, S. (1985) Application of a New Class of High Accuracy TVD Schemes to the Navier-Stokes Equations, AIAA Paper No. 85-0165.

36. Anderson, J.D.Jr. (1969) An Engineering Survey of Radiating Shock Layers, *AIAA Journal* **7**(9):1665–1675.

37. Elbert, G.J. and Cinnella, P. (1995) Truly Two-Dimensional Algorithms for Radiative Heat-Transfer Calculations in Reactive Flows, *Computers and Fluids* **24**(5):523–552.

38. Hartung, L.C. and Hassan, H.A. (1992) Radiation Transport around Axisymmetric Blunt Body Vehicles Using a Modified Differential Approach, AIAA Paper No. 92-0119.

39. Thomas, P.D. and Lombard, C.K. (1979) Geometric Conservation Law and Its Application to Flow Computations on Moving Grids, *AIAA Journal* **17**(10):1030–1037.

40. Chang, S.-H. (1989) On the Application of Subcell Resolution to Conservation Laws with Stiff Source Terms, NASA Technical Memorandum No. 102384.

41. LeVeque, R.J. and Yee, H.C. (1988) A Study of Numerical Methods for Hyperbolic Conservation Laws with Stiff Source Terms, NASA Technical Memorandum No. 100075.

42. Oran, E.S. and Boris, J.P. (1987) *Numerical Simulation of Reactive Flow*, Elsevier, Amsterdam, The Netherlands.

43. Eklund, D.R., Drummond, J.P., and Hassan, H.A. (1986) The Efficient Calculation of Chemically Reacting Flow, AIAA Paper No. 86-0563.

44. Bussing, T.R.A. and Murman, E.M. (1985) A Finite-Volume Method for the Calculation of Compressible Chemically Reacting Flows, AIAA Paper No. 85-0331.

45. Iannelli, G.S. and Baker, A.J. (1988) A Stiffly-Stable Implicit Runge-Kutta Algorithm for CFD Applications, AIAA Paper No. 88-0416.

46. Beam, R.M. and Warming, R.F. (1978) An Implicit Factored Scheme for the Compressible Navier-Stokes Equations, *AIAA Journal* **16**(4):393–402.

47. Taylor, L.K. and Whitfield, D.L. (1991) Unsteady Three-Dimensional Incompressible Euler and Navier-Stokes Solver for Stationary and Dynamic Grids, AIAA Paper No. 91-1650.

48. Walters, R.W., Cinnella, P., Slack, D.C., and Halt, D. (1992) Characteristic Based Algorithms for Flows in Thermo-Chemical Nonequilibrium, *AIAA Journal* **30**(5):1304–1313.

49. Szego, S., Cinnella, P., and Cunningham, A.B. (1993) Numerical Simulation of Biofilm Processes in Closed Conduits, *Journal of Computational Physics* **108**(2):246–263.

CFD FOR SCIROCCO PROJECT

S. BORRELLI, F. DE FILIPPIS, M. MARINI, A. SCHETTINO
C.I.R.A. - Centro Italiano Ricerche Aerospaziali
Via Maiorise - 81043 Capua (CE) Italy

1. Abstract

The realization of the 70 MW Plasma Wind Tunnel SCIROCCO is going to be started. CIRA has recently started a systematic analysis of the aerothermodynamic process by means of advanced CFD codes. Nozzle flow has been studied for the 4 different exits of the designed conical nozzle and for 20 operating points of the facility. The flow around the 300 mm radius hemispherical model inside the test chamber has been investigated by assuming as asymptotic values those calculated at the nozzle exit in 6 different operating points. The results of both the studies have been compared with the estimations made during the design phase by means of engineering relationships. Good agreement has been obtained for Heat Fluxes and Pressures at the Stagnation point of the model, thus confirming the goodness of the design.

2. Introduction

During the activities carried out in the frame of the Hermes program, the European Space Agency (ESA) identified the need of a full scale arc-heated test facility for the development and qualification of thermal protection system (TPS). Since 1988 CIRA has been encharged to design the facility (called SCIROCCO) also because of the strong funding contribution allocated to the project by the Italian government. After different design phases [1], the reorientation of ESA programs toward capsule-like space transportation systems posed questions on the usefulness of SCIROCCO. In order to deal with this problem, ESA formed an international *Scirocco Advisory Committee,* with the aim to investigate the potentialities of the Hermes-tailored SCIROCCO with respect to other future ESA-considered programs. The Committee concluded that: i) SCIROCCO is fully useful as

561

M. Capitelli (ed.), Molecular Physics and Hypersonic Flows, 561–570.

material testing facility for almost all types of earth atmospheric reentry vehicle; ii) it is particularly useful for thermo-structure analysis because of its dimension and running time; iii) it can be also (but partially) used in the fields of Aerothermodynamics ([16] and [9]); iv) it is promising for applications in the Aeropropulsion. As a consequence, it was recommended to build the facility as it was designed.

The actual design is based on conservative engineering tools and semiempirical relationships for the description of the facility.

3. Modelling description

The SCIROCCO main nozzle has a 10 degrees semiangle conical shape and a throat diameter of 0.075 m. Being made in spools, it has 4 nominal exits defined by diameters of 0.900 m, 1.150 m, 1.350 m and 1.950 m [2]. The maximum Power of the facility is 70 MW. In the arc generator chamber it is possible to develop Pressures P_o in the range from 1 to 17.5 bar and Enthalpies H_o in the range from 3 to 40 MJ/kg.

The characterization of the SCIROCCO nozzle has been obtained by CFD simulations for thirty pairs of values H_o, P_o. The matrix of the runs executed is reported in Table 1 and shown in Figure 1 together with the SCIROCCO Arc Performance map as taken from reference [2]. Values of H_o and P_o out of this map have been used to investigate the SCIROCCO aerothermodynamic behaviour, if there will be future changes of the actual technical limitations.

For our applications we used an internal CIRA code called H2NS. It solves the full laminar Navier-Stokes equations, in 2D plane and axisymmetric cases, with thermo-chemical non-equilibrium. It is based on a finite volume technique, with an upwind ENO higher order formulation; the fluxes through the cell-faces are computed according to the Flux Difference Splitting formulation, with the Riemann solver proposed by Pandolfi and Borrelli [3]. A detailed description of this solver is given in [4]. The chemical model takes into account five species (O,N,NO,O_2,N_2); the ionized species are neglected because of their small amount. In the present application we use the Park-83 model [5] characterized by 18 reactions. The vibrational excitation of the particles has been taken into account using the model proposed by Millikan and White in 1963 [6]. The modelling of the Transport Properties has been made using a model developed at C.I.R.A. in 1993 [7, 8], based on the molecular theory of gases and liquids developed by Chapman & Cowling [12]: it takes into account the forces between molecules using Collision Integrals calculated by Yun and Mason in 1962. A large number of cross tests with other CFD codes and experimental data have been carried out to validate H2NS [9], [11], [13].

4. Nozzle analysis

For all cases we used the 100×20 grid represented in Figure 2. For each analyzed case, we used the exit nozzle values on symmetry axis. In this figure the core flow is quite conical and therefore the use of the centerline value is not a big limitation. In a next paragraph we will demonstrate this assumption. However, tests are in progress to evaluate the error induced by this assumption. Indications on the dynamic, thermal and chemical boundary layers at the exit of the largest nozzle configuration are reported in Figure 3, for the operating point $H_o = 40 \, MJ/kg$ and $P_o = 1 \, bar$. To completely describe the nozzle behaviour, the following aerothermodynamic parameters have been considered:

- Mach number
- Static Pressure (bar)
- Temperature (K)
- Density (kg/m^3)
- Flow velocity (m/s)
- Species concentrations
- Energy distribution

The obtained exit values of the Mach number as function of P_o and H_o are plotted in Figure 4. Mach number increases regularly as P_o decreases and H_o increases. In fact, for high energy and low pressure there is strong air dissociation that let the gas in the nozzle be almost completely monoatomic.

The entity of chemical air dissociation can be evaluated from Figure 5 that reports the exit frozen γ values as a function of P_o and H_o. We note higher γ values at high energies and low pressures in the Arc Chamber (i.e. completely dissociated air).

The static pressure at the exit of the $1.950 \, m$ diameter nozzle is reported on Figure 6. This quantity is principally modulated by the Arc Total Pressure. It increases with P_o maximum effects at medium Enthalpies (H_o). This behavior is to be ascribed directly to the Temperature influence. In fact, at nozzle exit the temperature shows a peculiar behaviour that can be understood by properly analyzing the Figure 7. T naturally increases with P_o , while it presents a maximum for a Total Enthalpy H_o of about $20 MJ/kg$. This means that for a fixed Total Pressure P_o in the Arc chamber, the H_o increase does not necessarily determine an exit temperature increase. This effect can be explained considering the chemical dissociation that acts on the energetic structure of the system with two opposite mechanisms: on one hand, the Temperature decreases as Mach increases and as the energy accumulated under thermochemical degrees increases (from Figures 4 and 5 we can see that M and γ increase with increasing Enthalpies). On the other hand, the Exit Temperature tends to increase when energy and thus the Total Enthalpy increases; the two combined effects determine the presence of a maximum at about $20MJ/kg$. It is interesting to note that the Temperature values are always very low, going down to 133 K. No assumption is provided to take into account the Vibration-Translation coupling. We expect that such an assumption brings to stronger expansion and consequently higher exit temperature.

Reference is made to [14] for the full analysis of the exit aerothermodynamic quantities obtained for the 0.900, 1.150, 1.350 m nozzle configurations. All the obtained results have been compared with the corresponding engineering ones obtained during the design phase [15]. A detailed critical review of the engineering results can be found in [14] where in general a sensible difference between engineering and CFD approaches can be noted.

The Exit Mach Number calculated by means of H2NS is always higher than the one calculated by engineering tools in the entire investigated domain: we have observed average differencies of about 50 per cent. Similar results have been obtained for the Exit Pressure. This time differencies of about 70 per cent have been estimated, being the H2NS values the lowest ones.

Also the temperature computed by H2NS is lower than the one estimated in [15]. The percentage disagreement between the two calculations is 65 per cent in average. In agreement with the facility specified dew point, the temperatures at the nozzle exit cannot be lower than $228K$. However, considering the very short time of permanence of eventual condensed water drops (very high velocities) and its very small quantity (low densities), the dew point does not affect the facility overall performances.

We did not find appreciable disagreement between H2NS and [15] calculations for what concerns exit Velocity, Density and Molecular Weight: the average differencies are of 15 per cent.

The described comparison is related to the maximum exit diameter (1.950 m). Comparison have been made also for the other nozzle configurations obtaining similar results, with percentage differencies slightly smaller.

These differencies are caused by the following reasons. In the engineering tools the expansion process along the nozzle is modelled as an equilibrium phenomena up to mach 3.6 and a successive frozen expansion up to the exit. This simplification is not always applicable, especially for high energy and low pressure, and it might cause a very sensible velocity reduction and pressure increase. As a consequence the boundary layer computed in the engineering case results generally thinner than H2NS calculations. This effect induced a less sensible difference because the effective area ratio used by H2NS is generally lower than the engineering tools. In particular these two opposite effects tend to balance themselves for the shorter nozzle while cause the above mentioned differencies at the nozzle exit.

5. Test model analysis

The ESA Requirements ask for 4 different test configurations in SCIROCCO test chamber [2]. We analyze here the *nose configuration* represented by a hemispheric model with a 300 *mm* nominal radius. Because of operational limits due to tunnel blockage, this radius is decreased to 240 *mm* when the 0.900 *m* exit diameter nozzle is used. We are principally interested to evaluate two aerothemodynamic quantities:

- the pressure at Stagnation Point (p_s)
- the heat flux at Stagnation Point ($\dot{q}_s$)

Also these parameters have been calculated by using the H2NS code. This effort is currently in progress and only 6 operating points are available. They are relative to 6 pairs H_o-P_o chosen opportunely at the boundary of the Scirocco Performance Map (see Table 2). We used a grid with 40 points along the flow direction and with 60 points along the sphere surface. The grid used and the geometry are represented in Figure 8.

In these computations, constant profiles corresponding to the centerline exit nozzle values have been taken as inlet conditions. Also in this case studies on the influence of this assumption on the flow field are in progress. The computed values of p_s and $\dot{q}_s$ for the largest nozzle configurations (1.950 *m*) are reported in Table 2 and graphically shown in Figure 9 in terms of *Performance Map*. In the same figure results obtained with engineering relationships [15] are also reported together with the *user requirements* relative to three Hermes reentry trajectories: the nominal, the extreme and the safety one. The engineering results have been obtained by using the classical equilibrium normal shock, and Fay-Riddel formulas. The motivation of these conclusion is the following. The values used to evaluate the performances of SCIROCCO depend mainly on total quantities (pressure and heat flux at the stagnation point) and not on their single contribution (chemical processes, kinetic energy contribution, etc).

On the other hand, the discrepancies found at the nozzle exit are related to single contributions. However, these discrepancies are "stored" in the bow shock causing a total pressure loss. Deeper considerations on this subject will be treated in forthcoming papers.

As it can be seen, there are some differencies between CFD calculations and the engineering estimations. In particular, we found SCIROCCO capable to cover higher Stagnation Heat Fluxes but lower Stagnation Pressures.

Similar results have been obtained for the exit diameter of 0.900 *m*. They show that the H2NS values are quite corresponding to those of the engineering tools in term of p_s but much higher in term of Stagnation Heat Flux. These disagreement derives from the assumption of Lewis number equal to 1.4 in engineering estimations, while a full modellization is provided by CFD codes. By superposing the two performance maps (Figure 9 and 10) it is easy to conclude that the *composed* CFD map guarantees the coverage of user requirements, in spite of what stated by means of engineering estimations.

6. Test chamber flow and nozzle exit profile

The effects of the plume on test chamber flow has been evaluated by means of CFD tools in [14], being the pressure inside the test chamber less than the nozzle exit one. Downstream of the nozzle a further expansion occurs along the stagnation line, resulting in a stronger bow shock ahead of the test article, causing a greater pressure loss. The pressure and heat flux at the stagnation point of the spherical model are affected by the plume. In particular, the SCIROCCO envelope in terms of p_s and $\dot{q}_s$ is reduced because of plume effects (about 10% in pressure and 15% in heat flux), as shown in Figure 11. Decreasing the distance between the nozzle exit section and the test article the values of p_s and $\dot{q}_s$ tend to the values predicted without plume simulation.

The presence of an irregular inlet profile in the nozzle (flow coming from arc heater) affects p_s and $\dot{q}_s$ at the stagnation point of the test article, because the diffusion phenomena occurring inside the nozzle do not make uniform the profiles of velocity and total enthalpy. As a consequence, being greater the energy level along the stagnation line, the values of p_s and $\dot{q}_s$ increase with respect to the uniform profile (see Figure 11). A global flowfield (nozzle and test chamber) is illustrated in Figure 12, where iso-Mach contours are reported.

7. Conclusions

This work is a first step toward a completely systematic characterization of the SCIROCCO Performance through CFD. The main aerothermodynamic quantities in the nozzle have been obtained for thirty possible couples P_o and H_o. The regularity of the Mach, Pressure, Temperature, Density etc. variations has been a confirmation of the goodness of the results. The comparison between our results and the calculations made in the design phase gave sensible differences in particular for Mach, Pressure and Temperature. The last one could produce relevant changes in the Performance Map because often the facility dew-point ($228K$) is expected to be reached and overcome.

For six values of H_o and P_o, the Static Pressure and the Heat Flux have been calculated at stagnation point of a hemi-sphere in Test Chamber for two exit diameters: the 0.900 m and the 1.950 m. The CFD results compare with the engineering ones better than for the nozzle because of the *"smoothing"* effects of the bow shock on p_s and $\dot{q}_s$ By superposing the two performance maps (Figure 9 and 10) it is easy to conclude that the *composed* CFD map guarantees 100 per cent coverage of user requirements, which was not the case for the engineering estimations.

The effects of plume in test chamber flow reduces the SCIROCCO envelope in terms of p_s and $\dot{q}_s$, whereas the presence of an irregular inlet profile in the nozzle is felt on the test article, causing an increase of p_s and $\dot{q}_s$.

References

1. T. J. Stahl, W. Winovich, G. Russo, S. Caristia - " Design and Performance Characteristics of the CIRA Plasma Wind Tunnel", AIAA-91-2272.
2. Aerotherm - "USER REQUIREMENTS TEST CONDITIONS ANALYSIS" - CIRA Doc number MC-3D-PSPZ-1-ST-T009 - 31/05/1993.
3. S.Borrelli, M. Pandolfi - "An Upwind Formulation for the numerical prediction of Non-Equilibrium Hypersonic Flows" - 12th International Conference on Numerical Methods in Fluid Dynamics, Oxford (UK), 1990.
4. S. Borrelli, De Filippis F., Schettino A. - "Viscous Non Equilibrium Hypersonic Flow Around Blunt Bodies .." Hermes R&D , Final Report, CIRA doc. num DILC-EST-TN-299 -15/12/1992.
5. Rakich J.V., Bailey H.E. and Park C. "Computation of Nonequilibrium, Supersonic Three Dimensional Inviscid Flow over Blunt Nosed Bodies", AIAA Journal N.6, 1983.
6. Millikan,R. C. , White, D. R. - " Systematics of vibrational relaxation", The Journal of Chemical Physics, vol. 39, No. 12, pp. 3209-3213, December 1963.
7. De Filippis F., S. Borrelli - "Considerazioni su differenti modelli di trasporto per l'aria ad alta energia " - 12° Congresso AIDAA Como (I)- 1993.
8. De Filippis F., S. Borrelli - " Analysis of Different Approximation Levels Introduced in the Development for Transport Coefficient Models of High Energy Air", 19th ICAS Congress, Anaheim, USA, 19-23 September 1994.
9. S.Borrelli "Numerical Analysis of Atmospheric Reentry Aerothermodynamics Supporting Experimental Tests in Plasma Wind Tunnels", in italian, Doctoral Thesis, 1993.
10. S.Borrelli, A.Schettino "Numerical Correlations between Flight and Plasma Wind Tunnel Tests", presented at the 25th AIAA Fluid Dynamics Conference, June 20-23 1994, AIAA 94-2355.
11. S.Borrelli, G.Russo, A.Schettino "A Contribution to the Analysis of High Enthalpy Nozzle Flows", published in "Hypersonic Flows for Reentry Problems", Springer-Verlag, 1992.
12. Chapman,S. & Cowling, T.,G., -" The Mathematical Theory of non-uniform gases" Cambridge University Press , 1970.
13. K. S. Yun, E. A. Mason - "Collision Integrals for the Transport Properties of Dissociating Air at High Temperatures" - The Physics of Fluids, vol.5, No. 4, April 1962.
14. S.Borrelli, F.De Filippis, G. Leone, M. Marini, C. Pascarella, A. Schettino -"Analisi dei Processi Termofluidodinamici e delle Prestazioni di PWT per la Realizzazione del Simulatore Dinamico" - CIRA Internal Document No. MC-3D-CIRA-A-TN-0031 - Febbraio 1995.
15. Aerotherm Corporation - "Arc Heater sizing Computer Program (arcsiz 94)" - Aerotherm User's Manual 0713-94-001, January 1994.
16. S.Borrelli, A.Schettino "Numerical Correlations between Flight and Plasma Wind Tunnel Tests", presented at the 25th AIAA Fluid Dynamics Conference, June 20-23

1994, AIAA 94-2355.

Table 1 - Matrix of used operating points for nozzle analysis

Nr.run	Po (bar)	Ho (MJ/kg)	Nr.run	Po (bar)	Ho (MJ/kg)
1	1	5	11	1	3
2	1	10	12	2.7	40
3	1	20	13	11	3
4	1	40	14	16.7	10
5	5	5	15	8.1	23
6	5	10	16	1	30
7	5	20	17	5	3
8	10	5	18	10	3
9	10	10	19	4.7	30
10	10	20	20	12.7	5

Table 2 - Matrix of operating points used for nose analysis at stagnation point for nozzle exit diameter of 1.950 m

number of point	Ho (Mj/kg)	Po (bar)	p_s (Pa)	$\dot{q}_s$ (kW/m^2)
1	40	1	490	860
2	40	2.67	1088	1294
3	23	8.1	2720	945
4	10	16.67	4900	438
5	3	10.94	3060	42.5
6	3	1	330	18.3

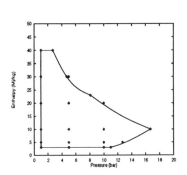

Figure 1. Arc Performance Map [1]

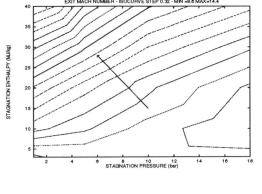

Figure 4.   Normalized radial distribution of the 1.950 m exit diameter for $H_o = 40 MJ/kg$ and $P_o = 1bar$

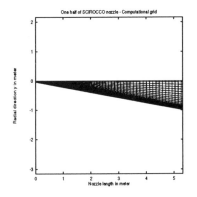

Figure 2.   Grid used nozzle computations

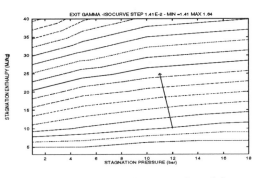

Figure 5. Centerline Gamma values (γ) as function of driving parameters P_o and H_o. Exit diameter of 1.950 m

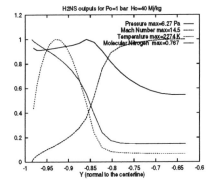

Figure 3.   Normalized quantities - nozzle exit - $H_o = 40 Mj/kg$ and $P_o = 1bar$

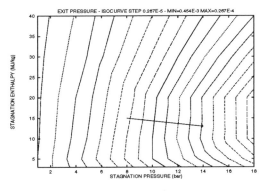

Figure 6. Centerline Static Pressure in *bar* as function of driving parameters P_o and H_o. Exit diameter of 1.950 m

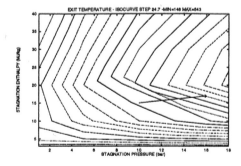

Figure 7. Centerline Exit Temperature in *Kelvin* as function of driving parameters P_o and H_o. Exit diameter of 1.950 m

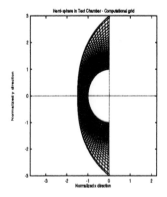

Figure 8. Computational grid used for the sphere geometry in Test Chamber.

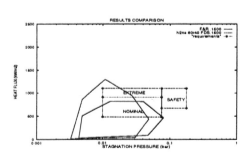

Figure 9. Performance Map for nozzle exit diameter of 1.950 m

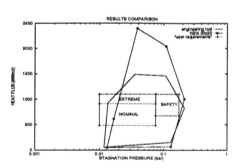

Figure 10. Performance Map for nozzle exit diameter of 0.900 m

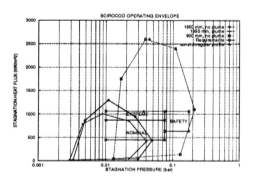

Figure 11. Final Performance Map of SCIROCCO

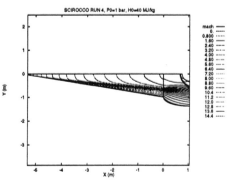

Figure 12. Iso-Mach contours of nozzle and test chamber

SOME EFFECTIVE METHODS FOR SOLVING 3D HYPERSONIC VISCOUS PERFECT GAS AND CHEMICALLY NONEQUILIBRIUM FLOWS

I.G.BRYKINA
Institute of Mechanics, Moscow State University
Michurinsky Prospect 1, Moscow 119899, Russia

1. Introduction

Two effective approximate methods for investigation 3D viscous hypersonic reactive flows over blunted bodies are proposed. The first method is developed for solving 3D flows by using axisymmetric solutions. It can be applied for arbitrary Reynolds numbers (when the continuum approach is valid), for different flow models, in consideration of physical and chemical processes and permits to use available axisymmetric methods and codes to solving 3D problems. The method is based on application of the obtained similarity relations, expressing heat flux, skin friction coefficient and species fractions on lateral surface of 3D blunted body by their values on surface of axisymmetric bodies, which depend only on the real body geometry and don't depend on values of gasdynamic flow parameters. Accuracy of the method is estimated by comparing with more accurate predictions, obtained from direct calculation of 3D governing equations. The method was tested for various flow models, including parabolized Navier-Stokes equations, viscous shock layer and boundary layer, for perfect gas and chemically reacting one, for different bodies and angles of attack. Freestream conditions have been corresponded to a reentry trajectory including frozen, nonequilibrium and closed to equilibrium flow regimes. Comparisons showed that the method allows to obtain an accurate predictions for space bodies regardless flow conditions, the model used, physical-chemical gas properties, wall catalycity and etc.

The present method is essentially different and simpler than approximate methods which use Cooke's axisymmetric analogue [1], since those methods require 3D inviscid solution, i.e., reduction of 3D problem to 2D one is not complete, and are applicable only for boundary layer model.

The very simple method is proposed for heat transfer prediction for moderate and high Reynolds numbers. The analytical solutions which depend only on body geometry were obtained for heat flux on lateral surface relative to it's value at the stagnation point. Comparing with numerical solutions showed that obtained formulas allow to predict with good accuracy heat flux distribution on fully catalytic surface of 3D bodies in reactive flows.

M. Capitelli (ed.), Molecular Physics and Hypersonic Flows, 571–580.
© 1996 Kluwer Academic Publishers. Printed in the Netherlands.

2. Similarity relations

Steady three-dimensional super- and hypersonic viscous flows without separation over blunted bodies are considered. At first 3D hypersonic viscous shock layer equations with modified Rankine-Hugoniot conditions at the shock and wall boundary conditions taking account of velocity slip and temperature jump were solved by the integral method of successive approximations [2] and analytical solution has been obtained [3]. Analysis of this solution showed that heat flux q depends on body geometry primarily by two parameters. One is an angle α between a normal to a surface and a freestream velocity vector V_∞; another is a combination Re/H, where H is the average of principal curvatures at a given point. These two parameters form the basis for derivation the similarity relations. For any line on the body surface starting at the stagnation point it is possible to find such equivalent axisymmetric body (EAB) for which the angle α^s between the normal to its generatrix and vector V_∞ would change along the generatrix as well as the angle α along the chosen line. Hence, the distribution of the heat flux along the line on the 3D body surface will be equal to one along the equivalent axisymmetric body ($\alpha^s = \alpha$), if the second parameters also coincide. It means that in calculating q on EAB $Re^* = ReH^s/H$ should be used as the Reynolds number. Here H^s is the average of principal curvatures of EAB at the point considered. It is convenient to construct EAB for meridional planes. Let a body surface has been given in the cylindrical coordinate system by the equation $r = r(z, \varphi)$.

Then heat flux q and shear stress τ along meridional plane of 3D body will be equal to heat flux q^s and shear stress τ^s on corresponding EAB:

$$q(Re) = q^s(Re^*), \quad \tau(Re) = \tau^s(Re^*) \tag{1}$$

$$Re^* = \frac{H^s}{H} Re$$

A shape $r^s(z)$ of EAB constructed for meridional plane $\varphi = \varphi^*$, is

$$r^s(z) = \int_0^z \frac{dr}{dz} \left[1 + \left(\frac{1}{r} \frac{dr}{d\varphi} \right)^2 \right]^{-1/2} dz \tag{2}$$

It's important that when we solve axisymmetric equations set for EAB to obtain q^s, τ^s, we must substitute instead the usual constant Re number the variable Re^* dependent on surface geometry at a given point.

Note that equivalent bodies depend only on body geometry and angle of attack and don't depend on values of gasdynamic flow parameters and therefore can be easily constructed.

So, if we have 3D body, for each meridional plane we can construct EAB, solve axisymmetric equations for this body with the variable, dependent on geometry Re number and thus obtain 3D solution.

For high Re numbers, when $q \sim Re^{-1/2}$ the asymptotic approximation for similarity relations (1) is:

$$q = \sqrt{H/H^s} \, q^s, \quad \tau = \sqrt{H/H^s} \, \tau^s \tag{3}$$

Here q^s, τ^s are determined for EAB at the same constant Re number as q, τ for a real body. This is the simplified form of similarity relations and in this form they can be applied, for example, for boundary layer model.

For a plane of symmetry of space body at zero or nonzero angle of attack EAB is a body formed by rotation of a corresponding branch of the centerline around the axis z.

At a stagnation point the similarity relation (1) for q is:

$$q^o(Re) = q^{so}(Re^*), \quad Re^* = \frac{2Re}{1+k} \tag{4}$$

Here k is the ratio of principal curvatures at a stagnation point, q^{so} is the heat flux to a stagnation point of an axisymmetric body, e.g., sphere. For high Re numbers the relation (4) is reduced to

$$q^o = \sqrt{\frac{1+k}{2}} \, q^{so}. \tag{5}$$

This relation coincides with analogous one obtained before in boundary layer theory [4].

3. Comparison of Approximate and Exact Solutions.

To estimate accuracy of the method a lot of comparisons were made between approximate and exact solutions. Heat flux and skin friction predictions, obtained from numerical solution of axisymmetric equations set with using similarity relations, compared with more accurate predictions, obtained from direct calculation of 3D governing equations set. Comparison was made for various flow models: parabolized Navier-Stokes equations, hypersonic viscous shock layer, boundary layer and for different bodies like ellipsoids, elliptical paraboloids and hyperboloids, and different angles of attack.

Results of comparing for perfect gas showed, that accuracy of the method is very good for all test cases. For example, for all flow models an error did not exceed 2% at a stagnation point and 4% on a elliptical paraboloid with $k = 0.4$ for wide variety of flow conditions: $Re_o = 1\text{-}10^5$, $T_w/T_o = 0.01\text{-}0.5$, $\gamma = 1.15\text{-}1.67$, $\omega = 0.5\text{-}1$ ($\mu \sim T^\omega$). Here $Re_o = \rho_\infty V_\infty R/\mu(T_o)$, T_o is freestream adiabatic stagnation temperature, R is one of the radii of principal curvatures at a stagnation point, T_w - wall temperature, γ - ratio of specific heats. Accuracy of the similarity relations is almost independent on flow parameters Re, γ, T_w, Pr. This allowed to

574

assume that similarity relations are valid not only for perfect gas, although they were obtained analytically in this case, but also for chemically reacting one.

To test the method for reactive flows, nonequilibrium chemical reactions and multicomponent diffusion for 5 species air were taken into account. The details of chemical kinetics, thermodynamic, and transport properties are given in [5]. Wall temperature is assumed to be at radiative equilibrium condition or constant. Slip boundary conditions for nonequilibrium gas [6] were used at a wall. Different models of surface catalycity were considered. Freestream conditions have been corresponded to altitudes h from 50 to 100 km over the reentry trajectory of Space Shuttle [7] including frozen, nonequilibrium and closed to equilibrium flow regimes. The numerical method of solving is given in [5]. Some results of comparing of approximate and exact solutions are given in Figures 1-4.

In Figure 1 heat flux distribution on a surface of ellipsoid is presented. Here and in the next figures solid lines correspond to exact 3D solution, light dots – to solution obtained by using similarity relations (1); x, y – Cartesian coordinates of a surface. In Figure 2 heat flux distributions are shown for different altitudes of reentry trajectory. It is seen that accuracy of the method is very good for all flow regimes. Dark dots correspond to simplified form of similarity relations (3) and disagreement with exact solution must be, it means that boundary layer theory is not applicable for such regimes. The range of applicability of relations (3) essentially depends on wall catalycity properties, so for noncatalytic wall they can be used at altitudes up to 65 km, for fully catalytic one – up to 90 km.

Comparisons made for different angles of attack showed that the method gives accurate heat prediction for angles of attack at least till 45°.

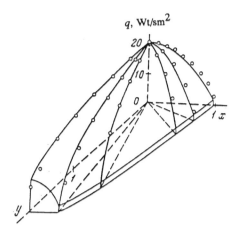

Figure 1. Heat flux distribution on an ellipsoid with axes ratio 1:1.5:1.2. Scott's catalycity model [8]; $h = 70$ km, $V_{\infty} = 7.25$ km/sec. $Re_{\infty} = 1.33 \cdot 10^4$.

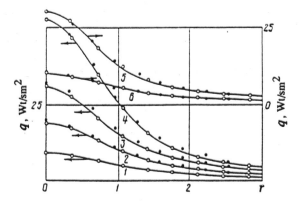

Figure 2. Comparison of heat flux distributions in the meridional plane $\varphi = 45°$ of paraboloid with $k = 0.5$ for different altitudes: lines 1-6 correspond $h = 100$, 90, 80, 70, 60, 50 km; Scott's catalycity model [8]; $r = \sqrt{x^2 + y^2}$.

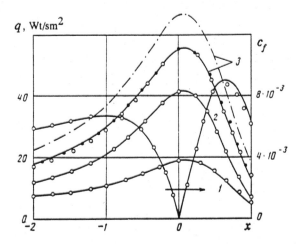

Figure 3. Heat flux and skin friction coefficient comparisons in the plane of symmetry of paraboloid with $k = 0.4$ at angle of attack 15°. Lines 1 — fully catalytic wall, 2 — Scott's model [8], 3 — noncatalytic; $h = 70$ km, $V_\infty = 7.25$ km/sec.

Figure 3 shows that for various catalycity models approximate and exact solutions are very closed and accuracy of the method does not depend on catalycity model used. c_f distribution is given only for fully catalytic wall as it depends only slightly on wall catalycity. Dashed-dotted line corresponds to usual axisymmetric solution and it will be discussed later.

Note that results of computations showed that the same similarity relations as those for heat flux also take place for species mass fractions on the wall.

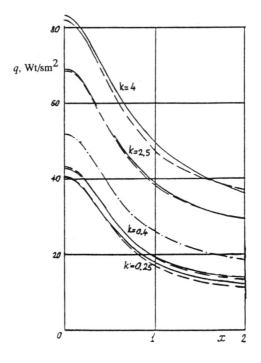

Figure 4. Heat flux distributions in planes of symmetry for elliptical hyperboloids with different *k*. Scott's catalycity model [8], *h* = 70 km, V_{∞} = 7.25 km/sec. Solid lines - exact solution, dashed - using (3), dashed-dotted - axisymmetric one.

Notice that for heat flux prediction in the plane of symmetry sometimes the usual axisymmetric solution is used for body formed by rotating a centerline, for instance in [9]. In Figure 4 comparison of three solutions − exact, obtained by the proposed method and axisymmetric is presented. Five considered bodies have the same centerline, but different transverse curvature. The axisymmetric solution for all these bodies is the same while the real heat flux values differ a few times. Thus the using of usual axisymmetric solution can lead to great errors because the influence of transverse curvature, or 3D effects, on heat transfer is very great. And this influence is well accounted by the similarity relations, introducing correction for the *Re* number dependent on geometry.

4. Analytical Solution for Relative Heat Flux.

The very simple method is proposed for heat transfer prediction for moderate and high Reynolds numbers ($Re > 10^2$). The analytical solution of 3D hypersonic viscous shock layer equations was obtained for heat flux on a lateral surface relative to it's value at a stagnation point. This solution depends only on body geometry, angle of attack and wall temperature; for cool wall it doesn't depend also on wall temperature. The formulas for relative heat flux have been obtained for wings at angles of slip and attack, for axisymmetric and 3D blunted bodies.

Distribution of relative heat flux along infinite wing ($v = 0$) at angles of slip (φ) and attack and along axisymmetric body ($v = 1$) can be predicted from:

$$\frac{q}{q_o} = \frac{\cos^2\alpha \sin\alpha \, r^v b}{\lambda [2(1+v)b_o \int\limits_0^s \lambda^{-1} b \cos^2\alpha \sin\alpha \, r^{2v} ds]^{1/2}} \tag{6}$$

$$\lambda = 1 + \frac{8 r \sin\alpha}{15 R \cos^2\alpha \, (1+v)}$$

$$v = 1: \quad b \equiv b_o \equiv 1;$$

$$v = 0: \quad b = 1 - (\sin^2\varphi + \sin^2\alpha \cos^2\varphi), \quad b_o = 1 - \sin^2\varphi$$

Here s is a length of arc from a stagnation point, α is an angle between a normal to surface and the freestream velocity ($v = 1$) or projection of the freestream velocity on plane perpendicular to generatrix ($v = 0$), R is radius of curvature of generatrix ($v=1$) or guide ($v=0$).

Relative heat flux distribution in a plane of symmetry of 3D body is:

$$\frac{q}{q_o} = \frac{\cos^2\alpha \, F^{1/2}}{2} \left[\frac{1+k}{2} \lambda \int\limits_0^s \frac{F \cos\alpha}{\lambda \, tg\alpha} ds \right]^{-1/2} \tag{7}$$

$$F = \exp \int\limits_0^s \frac{4H}{tg\alpha} ds, \quad \lambda = 1 + \frac{4 tg^2\alpha \, H^*}{15 H}$$

Here s is a length of arc from a stagnation point, α is an angle between a normal to surface and the freestream velocity, H is the average of principal curvatures at a given point, H^* is a curvature of centerline in the plane of symmetry, k is a ratio of principal curvatures at stagnation point.

Distribution of relative heat flux in each meridional plane $\varphi = \varphi^*$ of 3D blunted body is:

$$\frac{q}{q_o} = \frac{H^{1/2} \cos^2\alpha \sin\alpha \, r^o}{\lambda [2(1+k)H^o \int\limits_0^s \lambda^{-1} \cos^2\alpha \sin\alpha \, (r^o)^{2v} ds]^{1/2}} \tag{8}$$

$$r^0(z) = \int\limits_0^z \frac{dr}{dz} \left[1 + \left(\frac{1}{r}\frac{dr}{d\varphi}\right)^2\right]^{-1/2} dz, \quad \lambda = 1 + \frac{4tg^2\alpha}{15R^0H^0}$$

Equation $r = r(z, \varphi)$ governs body surface in cylindrical coordinate system. H^0 and R^0 are the average of principal curvatures and the radius of curvature of equivalent axisymmetric body $r^0(z)$.

All these formulas were obtained analytically and it is seen that relative heat flux depends only on body geometry and doesn't depend on gasdynamic flow parameters. Because the relative heat flux is very conservative. This is confirmed by numerical calculations of 3D equations. Figure 5 shows (strips contain all numerical solutions for $Re_o = 10^2$-10^5, $T_w/T_o = 0.01$-0.25, $\gamma = 1.15$-1.67) that really relative heat flux almost doesn't depend on Re, T_w/T_o, γ and is in good agreement with analytical solution (dots).

Moreover, numerical solutions of reactive gas flow in viscous shock layer showed, that relative heat distribution along fully catalytic surface is almost independent on chemical reactions taking place within the shock layer and differs very slightly from distribution obtained for perfect gas. It is seen from Figure 6 where strips contain all numerical solutions for altitudes from 50 to 90 km for two trajectories - reentry trajectory of Space Shuttle and with constant velocity 8 km/sec. All solutions are very closed to each other and to analytical solution (dots).

A lot of calculations were made, for instance in [10], to investigate how relative heat flux depends on flow regime (or altitude) for different wall catalycities. Value of relative heat flux on fully catalytic wall is almost independent on flow conditions and coincides with analytical solution. But if we consider another wall catalycities, the situation is differ: in some cases an error of using this solution can be enough large, about 30-40%.

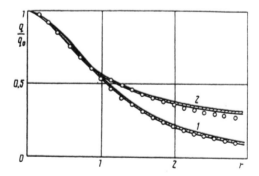

Figure 5. Relative heat flux distributions in meridional planes $\varphi = 45°$ of elliptical paraboloid with k = 0.4 (1) and hyperboloid with k = 0.5 and angle 80° in the plane y = 0 (2).

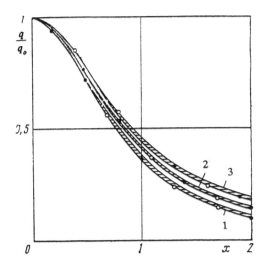

Figure 6. Relative heat flux distributions in planes of symmetry of different elliptical paraboloids with k = 0.4 (lines 1), k = 1(2) and k = 2.5(3).

Of course, absolute values of heat flux very differ for different flow regimes, but relative heat flux is very conservative value, therefore we can use obtained formulas for prediction heat fluxes also for frozen, equilibrium and non-
equilibrium flows in the case of fully catalytic wall. A lot of other comparisons
were made to estimate accuracy of formulas, which showed good agreement between numerical and analytical solutions for all flow conditions considered.

To predict absolute heat flux distribution on 3D body it's enough to know
accurate heat flux prediction only in a stagnation point of axisymmetric body, for
example, of sphere, and then to use the formulas.

Note, that more exact analytical solution, dependent not only on body
geometry, but also on pressure distribution, has also been obtained. And this
solution is in good agreement with numerical solution of Navier-Stokes equations.

5. Conclusion

The proposed method using similarity relations allows to obtain accurate heat
flux, skin friction coefficient and species fractions predictions for space bodies
regardless flow conditions, flow model used, physical-chemical properties of gases,
wall catalycity, etc. The method is simple, easy to apply, it represents significant
reduction in computational efforts over fully 3D methods and permit to use axisymmetric methods and codes to solving 3D hypersonic viscous flow problems.

The analytical solutions for relative heat flux allow to predict heat transfer on 3D body both for perfect gas and for chemically reacting one for fully catalytic surface if it's known heat flux at stagnation point of sphere.

6. Acknowledgements

This work is supported by the Russian Fundamental Investigations Foundation.

7. References

1. Cooke, J.C. (1959) An axially symmetric analogue for general three-dimensional boundary layers, *Aeronaut. Res. Council, Repts & Mem.* **3200**, 1-12.
2. Brykina, I.G. (1978) Solving of hypersonic viscous shock layer equations by the method of successive approximations, *J. Computational Mathematics and Mathematical Physics* **18**, No 1, 154-166.
3. Brykina, I.G., Rusakov, V.V., and Shcherbak, V.G. (1991) Analytical and numerical investigation of three-dimensional viscous shock layer on blunt bodies, *J. Applied Mechanics and Technical Physics* No 4, 81-88.
4. Brykina, I.G., Gershbein, E.A., and Peigin, S.V. (1980) Laminar three-dimensional boundary layer on the penetrated surface in the neighbourhood of the plane of symmetry, *J. Fluid Dynamics* **15**, No 5, 37-48.
5. Shcherbak, V.G., (1987) Numerical investigation of the structure of nonequilibrium three-dimensional hypersonic flow past blunt bodies, *J. Fluid Dynamics* **22**, No 5, 143-150.
6. Scott, C.D. (1975) Reacting shock layers with slip and catalytic boundary conditions, *AIAA Journal* **13**, 1271-1278.
7. Masek, R.V., Hender, D., and Forney, J.A. (1973) Evaluation of aerodynamic uncertainties for space shuttle, *AIAA Paper* **737**, 1-14.
8. Scott, C.D. (1980) Catalytic recombination of nitrogen and oxygen on high-temperature reusable surface insulation, *AIAA Paper* **1477**, 1-9.
9. Gupta, R.N., Moss, J.N., Simmonds, A.L., Shinn, J.L., Zoby, E.V. (1984) Space Shuttle heating analysis with variation in angle of attack and catalycity, *J. Spacecraft and Rockets* **21**, 217-219.
10. Brykina, I.G., Rusakov, V.V. and Shcherbak, V.G. (1989) Approximate formulas for heat fluxes to fully catalytic surface in the plane of symmetry, *J. Applied Mathematics and Mechanics* **53**, No 6, 956-962.

REVIEW OF FINITE-RATE CHEMISTRY MODELS FOR AIR

DISSOCIATION AND IONIZATION

C. PARK
Department of Aeronautics and Space Engineering
Tohoku University, Aoba-Ku, Sendai, 980 Japan

Nomenclature

A	An atomic species formed by dissociation
B	An atomic species formed by dissociation
D	Dissociation energy
e_v	Vibrational energy per unit mass
E	Energy level
k	Boltzmann constant
k_f	Quasi-steady state forward reaction rate coefficient
k_r	Quasi-steady state reverse reaction rate coefficient
k_f^*	One-way forward reaction rate coefficient
k_r^*	One-way reverse reaction rate coefficient
K_e	Equilibrium constant
K_{ij}	Collisional transition rate coefficient for i to j transition
m	Molecule
M	Second moment of transition rate coefficient, (20)
n	Number density
n	Preexponential power in rate coefficient
N	The number of internal states
p	Pressure, pascal
Q_m	Partitioin function of molecule m
Q_v	The contribution of the state v to partition function
s	The vibrational level bordering between the upper and lower states
t	Time, sec
T	Translational temperature, K
T_v	Vibrational temperature, K
v	Vibrational quantum number
δ	Average vibrational energy transferred in dissociation reaction
ϵ	Continuous vibrational energy level
ρ	Population density normalized by equilibrium value

Subscripts

1	First excited vibrational state
c	Continuum (free) state

M. Capitelli (ed.), Molecular Physics and Hypersonic Flows, 581–596.
© 1996 *Kluwer Academic Publishers. Printed in the Netherlands.*

e	Equilibrium
s	Settling chamber
x	Unspecified third body

1. Introduction: Finite-Rate Chemistry Problems in Hypersonic Flows

1.1. HISTORY OF FINITE-RATE CHEMISTRY PROBLEMS IN HYPERSONIC FLIGHT

The fact that air may undergo chemical changes such as vibrational excitation, dissociation, and ionization at high temperatures was made known to the aerospace community by physicists and chemists in the 1950s. The chemical phenomenon was studied seriously for the first time during the Apollo project in the United States. At first, thermodynamic and transport properties were calculated for equilibrium air [e.g., 1]. However, one soon learned the nonequilibrium nature of the problem when the heatshield for the Apollo entry vehicle was designed to withstand the heating by the radiation emitted by the hot air in the shock layer. The shock tube experiments conducted to determine the intensity of the radiation revealed that the shock layer flow is mostly in a finite-rate, nonequilibrium, chemistry regime. A concerted effort to understand the high temperature real-gas effects and the nonequilibrium chemistry problem was then begun to understand the chemical processes occurring behind shock waves and their influence on radiation phenomenon [e.g., 2].

The seriousness of the consequence of the real-gas effects was learned next during the first Earth-entry flight of the Apollo capsule. For a reason unknown at that time, the capsule attained a stable ungle of attack, named trim anagle, different from the planned angle, resulting in lift and drag different from the planned values. This resulted in a flight path different from the planned, and the capsule landed on the Pacifc ocean at many hundreds of kilometers away from the planned point. The astronauts, still caged in the floating capsule, had to wait several hours before they were picked up. Suggestions were made that this unexpected shift in the trim angle of attack was due to the high temperature real-gas effects [3,4]. However, the laboratory experiments conducted thereafter could not confirm it, probably because the facilities and the data reduction methods were inadequate [5,6].

A similar phenomenon occurred during the first few entry flights of the Space Shuttle vehicle. The vehicle started a nose-up motion despite that the body flaps were deflected downward to the maximum allowed value. The motion stopped short of reaching the stall angle for the vehicle. The motion was caused by the increase in the trim angle caused by the forward shift of the centere-of-pressure. Suggestions were made that this was due to the high temperature real-gas effects [7-9]. Some suspected that this was due to formation of separated flows over the compression corner formed by the deflected flaps [10]. In recent CFD calculations, it was shown that the phenomenon was indeed due to the chemical reactions, and that the flaps were perhaps more effective in the real-gas regime [11].

The fact that the convective heat transfer rates to the Shuttle vehicle could be made smaller than the values for the perfect-gas case by making the surface chemically noncatalytic to recombination of atomic oxygen was not quite believed in the beginning. There was initially a resistence to the proposal to build a high-enthalpy arc-heated wind tunnel at NASA Ames Research Center to test this problem. The flight of the Shuttle vehicle has proven the point, and the surface catalytic effect is now widely accepted. This is an example in which the high-temperature real-gas effect is taken advantage of.

Before the flight of the Shuttle vehicle, there was a heated debate on what Reynolds number the boundary layer flow over the vehicle will transition into turbulence. Most turbulence experts predicted the transition to occur at a small Reynolds number because of the rough nature of the vehicle surface due to the gaps between adjacent heat shield tiles. The flights demonstrated that the transition occurs at a Reynolds number even larger than for the smooth surface. Apparently, the chemical reactions occurring in the boundary layer suppress turbulent transition [12].

The radio blackout problem for a reentry body has been known from the days of Apollo. The electrons produced by the ionization of the species formed in the hot shock layer around a vehicle absorbs electromagnetic waves and thereby stops the waves from reaching the vehicle's antenna. In addition to numerous laboratory experiments [e.g., 13], a flight experiment, named RAM-C, was conducted to study this problem [14-17]. The results of the flight test were analyzed by many investigators [e.g., 18]. Though some aspects of the problem have been satisfactorily explained, there remain others that are yet to be explained.

Recently, two flight experiments, named Bow Shock Ultra-Violet 1 and 2, were conducted in which the radiation incident on the stagnation point of a hypersonic object was measured using a spectrometer of relatively high resolution at a flight speed of about 3.5 and 5.2 km/s, respectively [19,20]. The results obtained at 3.5 km/s could not be expplained; at high altitudes, the measured radiation was substantially stronger than calculated.

In order to study the real-gas problems, several different types of ground-based facilities have been built since the 1950s. Notable among these are the hot-shot tunnels, shock tunnels, arc-heated wind tunnels, and ballistic ranges. In the early years, the hot-shot tunnels and shock tunnels produced flows behaving so differently from those predicted that they were abandoned. Ballistic ranges were found to behave better, and the United States at one time seriously examined the option to build a very large such facility [21]. But the idea was abandoned partly because data-taking was difficult, and therefore not much use could be made of such facilities. Arc-heated wind tunnels were seen to produce flows so different from the flight case, but were used because there was no other choice. Recently, large hot shot tunnels and shock tunnels are being erected in Europe and Japan, as well as arc-heated wind tunnels. Understanding of the flows produced in these facilities is still far from perfect.

1.2. PROBLEMS OF MODELING OF POST-SHOCK FLOWS

Even though chemical reactions in the gas phase have been modeled in the chemical engineering community for decades preceding the space age, the methodologies developed by the community were found to be inadequate for use in the hypersonic flow environments. The reasons stem from the fact that the translational temperature immediately behind a shock wave is very high and that such a high temperature environment occurs mostly at high altitudes where density is very low. Entry vehicles are designed to decelerate at such high altitudes so that the heat transfer rates exerted on their surfaces are held at the values tolerable by the structural materials.

First, the temperatures encountered in hypersonic flight are too high for extrapolation of the experimentally-determined rate cofficients which are for a temperature of a few thousand degrees. The temperature immediately behind a normal shock wave at a flight speed of the Shuttle vehicle, which flies at about 7 km/s, is about 30,000 K, while it is nearly 60,000 K for Apollo vehicle, which flew at over 10 km/s. Experiments conducted at temperatures above 10,000 K [22,23] indicate that the reaction rates are smaller than those extrapolated from the low temperatures. When the rate of dissociation is expressed in an Arrhenius form

$$k_f = CT^n \exp(-\frac{E}{kT}) \qquad (1)$$

the pre-exponential temperature power n is usually between 0 and -1.5 at temperatures below about 5000 K. At temperatures above 5000 K, it seems to take a value smaller than -1.5.

Secondly, the dissociation and ionization rates are relatively so fast and air density is so low that the vibrational and possibly rotational modes seem to be still in the process of excitation while the reactions are in progress. That is, thermal nonequilibrium accompanies chemical nonequilibrium, and the two nonequilibrium phenomena are coupled.

Thirdly, as mentioned above in relation to the RAM-C fligh test, the behavior of electrons could not be explained satisfactorily. One presumes that this is because one does not understand fully the interactions between electrons and internal mode.

Fourthly, as mentioned above, at flight speeds of about 3.5 km/s, the radiation from the shock layer behaves strangely. One presumes that, at higher speeds, the interactions among the vibrational modes of O_2, N_2, and NO are fast, and therefore, the three molecules approximately attain a common vibrational temperature. At an intermediate hypersonic speed (between 2.5 to 4 km/s, say), the interactions among the three molecules are weak, and the three molecular species attain different vibrational temperatures. Interactions among the internal modes of the three molecules are not well known.

1.3. PROBLEMS OF MODELING OF COOLING FLOWS

Understanding the flows produced in the newly-erected hot shot tunnels, shock tunnels, and arc-heated wind tunnels is emerging as a new problem facing the aerospace community, especially in Europe and Japan. In the nozzle of such a facility, the gas mixture initially in equilibrium at a high temperature rapidly cools. The finite-rate chemical phenomena occurring in such a flow present many problems that were not seen in the flows behind a shock wave, as follows:

First, it is well established that the vibrational, electronic, and possibly rotational modes are in the process of nonequilibrium deexcitation while recombination processes are in progress in such a flow. The population distribution of the vibrational modes is known to depart far from the Boltzmann distribution, but its exact characterization is still incomplete. In addition, the rate of vibrational relaxation in such a flow was found to be much faster than that behind a shock wave. There is still a controversy as to the cause of this behavior.

Secondly, the existing experimental data on species mole fractions obtained in such a flow do not agree with calculations. The measured concentration of atomic oxygen is less, and those of atomic nitrogen and nitric oxide are larger than calculated. The measured flow velocities are correspondingly slower than calculated.

Thirdly, the spectral radiation emitted by the cooling flow seems to be complex and different from that predicted by the existing method.

Any theoretical model intent on characterizing finite-rate thermochemical reactions must explain all the phenomena listed above.

2. Quasi-Steady State (QSS) Rate Coefficient with Vibrational Equilibrium

In reviewing the existing theoretical models characterizing the coupling betweem chemical reactions and the internal mode distribution, one finds that the most important difference is in the definition of the reaction rate coefficients. It is well known that the reaction rate coefficients can be classified into two different groups: quasi-steady state (QSS) coefficients and intrinsic or one-way coefficients. Some models are for the QSS coefficients while others are for the one-way coefficients. The QSS coefficients have traditionally been defined for the one-temperature environment only. In order to define the QSS coefficients in the multi-temperature environment, and to demonstrate its legitimacy, the conventional QSS theory is reviewed here first.

Consider dissociation of molecule m to form two atoms A and B in the form

$$m \leftrightarrow A + B$$

The reaction rate is in general affected by the population distribution among the internal states. The population distribution is affected in turn by the state-to-state collisional transition rates. The transitional rates are affected further by the population distribution because the transition rates are affected by the state of the colliding particle. For simplicity, let us limit the consideration to the cases where the transition rates are functions of translational temperature only, unaffected by the internal population distribution. We shall consider only the case where the vibrational mode is involved. In addition, we stipulate that the first few vibrational states are in equilibrium with the translational temperature. By adopting the convention that the vibrational temperature is defined by the ratio of the populations between the states $v = 1$ and $v = 0$, this condition can be said to be a one-temperature environment, i.e. $T_v = T$. Such a case arises in the dissociation of O_2 in a large bath of argon behind a normal shock wave, at a temperature of about 2500 K, except immediately behind the shock wave. We question whether there are k_f and k_r which are functions of T only and are independent of time, $k_f = k_f(T)$ and $k_r = k_r(T)$, in the expression

$$\frac{dn_A}{dt} = \frac{dn_B}{dt} = -\frac{dn_m}{dt} = k_f n_x n_A n_B - k_r n_x n_A n_B \tag{2}$$

If such k_f and k_r exist, then they are automatically related by the equilibrium constant K_e in the form

$$k_r(T) = k_f(T)/K_e(T) \tag{3}$$

From three kinds of experiments, we know that such k_f and k_r exist. First, we monitor the temporal variations of number density of the species A, B, or m, and compare it with that described by equation (2). Second, we vary the initial pressure and repeat the experiment while maintaining T the same. If the resulting reaction rate coefficient values are the same, then $k_f(T)$ and $k_r(T)$ exist. Third, we can measure $k_f(T)$ and $k_r(T)$ independently and compare to see if equation (3) is satisfied.

On the other hand, we know that

$$\frac{dn_A}{dt} = \frac{dn_B}{dt} = -\frac{dn_m}{dt} = n_x \sum_{v=0}^{N} K_{vc} n_v - n_x \sum_{v=0}^{N'} K_{cv} n_A n_B$$

$$= n_x n_m \sum_{v=0}^{N} K_{vc} \frac{n_v}{n_m} - n_x n_A n_B \sum_{v=0}^{N} K_{cv} \tag{4}$$

is true. One can define a forward rate coefficient k_f^* as

$$k_f^* = \sum_{v=0}^{N} K_{vc} \frac{n_v}{n_m} = \sum_{v=0}^{N} K_{vc} \frac{Q_v}{Q_m} \rho_v \tag{5}$$

where ρ_v is the population of state v divided by its equilibrium value

$$\rho_v = \frac{n_v}{n_{ve}} \tag{6}$$

This is the intrinsic or one-way reaction rate coefficient. It is a function of t as well as T, because n_v or ρ_v is a function of t. Therefore k_f^* is not the $k_f(T)$ measured experimentally.

The time-independent k_f can exist only if n_v or ρ_v is a function only of n_m, n_A, n_B, and T. We know that

$$\frac{dn_v}{dt} = n_x \sum_{v'} K_{v'v} n_{v'} - n_x \sum_{v'} K_{vv'} n_v + n_x n_A n_B K_{cv} - n_x n_v K_{vc} \tag{7}$$

is true. Equation (7) is known as a master equation.

The QSS rate coefficients k_f and k_r can exist if the left-hand side (LHS) of (7) is zero for all v except $v = 0$, i.e.

$$n_x \sum_{v'} K_{v'v} n_{v'} - n_x \sum_{v'} K_{vv'} n_v + n_x n_A n_B K_{cv} - n_x n_v K_{vc} = 0, \quad v > 0 \qquad (8)$$

We know that

$$\sum n_v = n_m \sum \frac{Q_v}{Q_m} \rho_v = n_m \qquad (9)$$

The population distribution $n_v = n_m (Q_v/Q_m) \rho_v$ is now determined completely by n_m, n_A, n_B, and T, and therefore is the quasi-steady state (QSS) population. (6)

To solve equations (8) and (9) for n_v or ρ_v in a more symmetric fashion, one first expresses n_v in (8) by ρ_v using (6). Then one invokes the principle of detailed balance

$$n_{ve} K_{vv'} = n_{v'e} K_{v'v} \qquad (10)$$

$$n_{ve} K_{vc} = n_{Ae} n_{Be} K_{cv} \qquad (11)$$

One defines the normalized species densities ρ_A and ρ_B as

$$\rho_A = \frac{n_A}{n_{Ae}}, \qquad \rho_B = \frac{n_B}{n_{Be}} \qquad (12)$$

n_{Ae} and n_{Be} being the equilibrium number densities of species A and B for the given molecule density n_m at the given temperature T.

By substituting equations (10) to (12) into (8) and (9), one finds that (8) and (9) can be written in the form

$$C_{vv'} \rho_{v'} = D_v + \rho_A \rho_B E_v \qquad (13)$$

By inverting the matrix $C_{vv'}$ on the left-hand side and multiplying the resulting matrix to the two vectorial terms in the right-hand side, one obtains the solution vector ρ_v to equation (13) in the form

$$\rho_v = \rho_h(v, T) + \rho_A \rho_B \rho_p(v, T). \qquad (14)$$

The first term $\rho_h(v,T)$ is the QSS solution in the case where $\rho_A \rho_B$ is zero, that is, when the concentrations of the dissociated atoms are zero. One may call $\rho_h(v,T)$ a homogeneous solution because it is analogous to the case where the right-hand side is zero in an ordinary differential equation. It has a magnitude of the order of unity at the origin $v = 0$, because ρ_0 is nearly unity. At the dissociation limit $v = N$, its value is nearly zero, because it is nearly in equilibrium with the free (continuum, or dissociated) state which is zero in this case.

The function $\rho_p(v,T)$ in the second term of equation (14) represents the solution vector ρ_v when the degree of dissociation is much greater than its equilibrium value, such as in a cooling flow. Its numerical value is nearly unity at the dissociation limit, because the ρ_v is nearly equal to the product $\rho_A \rho_B$ at the dissociation limit $v = N$, because the uppermost vibrational level is nearly in equilibrium with the free state.

By substituting (14) into (4), and by using (6), one obtains

$$\frac{dn_A}{dt} = n_x n_m \sum_{v=0}^{N} K_{vc} \frac{Q_v}{Q_m} \rho_h - n_x n_A n_B \sum_{v=0}^{N} K_{vc} \frac{n_m}{n_{Ae} n_{Be}} \frac{Q_v}{Q_m} (1 - \rho_p). \qquad (15)$$

By comparing (15) with (2), one finds the QSS rate coefficients to be

$$k_f(T) = \sum_{v=0}^{N} K_{vc} \frac{Q_v}{Q_m} \rho_h \qquad (16)$$

$$k_r(T) = \sum_{v=0}^{N} K_{vc} \frac{n_m}{n_{Ae} n_{Be}} \frac{Q_v}{Q_m} (1 - \rho_p). \qquad (17)$$

By setting $\rho_h = 1$, one obtains the one-way rate coefficient for Boltzmann distribution. It is obvious that the QSS coefficients are smaller than such one-way rates, because ρ_h is always smaller than unity. The relative magnitudes of QSS coefficients k_f and k_r to those of the Boltzmann one-way coefficients k_f^* and k_r^* have been studied, among others, by Keck [24] and Haug and Truhlar [25] for the case of thermal dissociation reactions. The ratio between the two classes of rate coefficients are between 0.3 and 0.01 for typical hypersonic flight environments.

Identical procedures can be followed for all other types of reactions. For electron-impact ionization reactions [26,27], the ratio of the QSS coefficients to the one-way coefficients is between 10^{-2} and 10^{-6} for typical hypersonic flight environments.

The structure of the algebraic equation (13) resembles that of a linear integral equation. The quantity corresponding to the kernel of the integral equation is the state-to-state transition rate coefficients $K_{vv'}$ and K_{vc}. By the general nature of a linear integral equation, a change in the K-values affect the solution only to a square-root of the change. Thus, the QSS rate coefficients are relatively insensitive to the errors in the state-to-state transition rate coefficients. In comparison, the one-way rate coefficients are linearly affected by the errors in the state-to-state rate coefficients.

3. QSS Rate Coeficient with Vibrational Nonequilibrium

Now let us generalize the QSS formulation given above to the case where the first few vibrational states are not in equilibrium with the translational temperature, that is, $T_v \neq T$. Such an environment occurs immediately behind a shock wave. Various detailed calculations [e.g., 28] show that the upper vibrational levels, i.e. the levels above $v \approx 20$ for N_2, NO, or O_2, reach a quasi-steady state condition sooner than the low vibrational states. Therefore, except in the very vicinity of the shock wave, the QSS formulation given in the preceding section could be applied to the upper vibrational levels. But then how could we treat the low vibrational levels?

To answer this question, let us first express equation (7) using the normalized population ρ_v and invoking the principle of detailed balance. The master equation then becomes

$$\frac{1}{n_x}\frac{d\rho_v}{dt} = \sum_{v'=0}^{N} K_{vv'}(\rho_{v'} - \rho_v) + K_{vc}(\rho_A\rho_B - \rho_v) \qquad (18)$$

When the gas temperature is sufficiently high so that ΔE_1 is much less than T, energy levels can be considered to be continuously distributed. The continuous vibrational energy level can now expressed by ϵ, and the master equation (18) can be written as

$$\frac{1}{n_x}\frac{d\rho_\epsilon}{dt} = \int_{\epsilon'=0}^{N} K_{\epsilon\epsilon'}(\rho_{\epsilon'} - \rho_\epsilon)d\epsilon' + K_{\epsilon c}(\rho_A\rho_B - \rho_\epsilon) \qquad (19)$$

For vibrational quantum numbers below about 20, the collisional transitions occur preferentially between the neighboring states. For those states, the quantity

$$M(v) \approx M(\epsilon) = \frac{1}{2}\sum_{v'} K_{vv'}(v' - v)^2 \approx \frac{1}{2}\int_{\epsilon'} K_{\epsilon\epsilon'}(\epsilon' - \epsilon)^2 d\epsilon' \qquad (20)$$

attains a constant quantity at an ϵ' value only slightly different from ϵ, that is the integration converges quickly. M is the so-called the second moment of transition rate coefficient.

One expresses $K_{\epsilon\epsilon'}$ in equation (19) by a Taylor series in the powers of $(\epsilon' - \epsilon)$. For the low states, the higher powers are small because the transitions occur preferentially between the neighboring states. By performing the integration, one obtains $M(v) \approx M(\epsilon)$ also as a Taylor series. The resulting expression is a rapidly converging series in the powers of $(\epsilon' - \epsilon)$. One then expresses the $K_{\epsilon\epsilon'}$ under the integral in equation (18), and performs the integration. By comparing the resulting two expressions, and by neglecting the terms higher than $(\epsilon' - \epsilon)^3$, one concludes that, for the low vibrational states, the master equation can be approximated by

$$\frac{1}{n_x}\frac{d\rho}{dt} = \frac{\partial}{\partial\epsilon}[M(\epsilon)\frac{\partial\rho}{\partial\epsilon}] + K_{\epsilon c}(\rho_A\rho_B - \rho) \qquad (21)$$

For the low vibrational states, the second term in the right-hand side of equation (21) is much smaller than the first therm, and therefore can be neglected, resulting in

$$\frac{1}{n_x}\frac{d\rho_\epsilon}{dt} = \frac{\partial}{\partial\epsilon}[M(\epsilon)\frac{\partial\rho}{\partial\epsilon}] \qquad (22)$$

This is a diffusion equation.

The solution of the diffusion equation (22) defines a slope at origin, from which T_v can be defined. T_v varies with t as the solution evolves from $t = 0$. If M is assumed to be constant, then (22) can be written in the form

$$\frac{\partial\rho}{\partial t} = \kappa\frac{\partial^2\rho}{\partial\epsilon^2} \qquad (23)$$

Solution of (23) is

$$\rho = 1 - \text{erf}(\frac{\epsilon}{2\sqrt{\kappa t}}) \qquad (24)$$

Taking the value of M to be that at $v = 0$ and defining T_v by the ratio of the vibrational populations at $v = 0$ and $v = 1$, the corresponding solution of (21) becomes [29]

$$\frac{\partial T_v}{\partial t} = \frac{\pi}{4} n_x K_{10} \frac{E_1}{kT_v} \frac{E_1}{kT} \frac{(T - T_v)^2}{T^2} (T - T_v) \tag{25}$$

Noting that the Landau-Teller relaxation time τ_L is related to $n_x K_{10}$ by

$$\frac{1}{\tau_L} = n_x K_{10}[1 - \exp(-\frac{E_1}{kT})]$$

one can write (25) as

$$\frac{\partial T_v}{\partial t} = \frac{T - T_v}{\tau_D} \tag{26}$$

where

$$\frac{\tau_D}{\tau_L} = \frac{4}{\pi}[1 - \exp(-\frac{E_1}{kT})] \frac{kT_v}{E_1} \frac{kT}{E_1} \frac{T^2}{(T - T_v)^2} \tag{27}$$

The quantity τ_D is the vibrational relaxation time based on the diffusion model.

The diffusion model is valid when the post-shock translational temperature is much larger than E_1/k. To bridge the gap between the diffusion regime and the regime where the conventional Landau-Teller model is valid, a bridging formula was introduced in [29] as

$$\frac{\partial e_v}{\partial t} = \frac{e_{ve} - e_v}{\tau_L} |\frac{T_s - T_v}{T_s - T_{vs}}|^{s-1} \tag{28}$$

where s varies from unity at a low temperature and 3.5 at the high temperature limit.

We assume that the population distribution is dictated by the diffusion equation (22) below a vibrational level $v = s$. At levels above s, a QSS condition is assumed to prevail because the quantal effects are absent and transitions are allowed among all pairs of states. That is,

$$\int_{\epsilon'=0}^{s} K_{\epsilon\epsilon'}(\rho_{\epsilon'} - \rho_{\epsilon})d\epsilon' - K_{\epsilon c}\rho_{\epsilon} = -K_{\epsilon c}\rho_A\rho_B - \int_{\epsilon'=0}^{s} K_{\epsilon\epsilon'}\rho_{\epsilon'}d\epsilon' \tag{29}$$

The formal solution of (29) is

$$\rho = \rho_h(v, T, T_v) + \rho_A\rho_B\rho_p(v, T, T_v) \tag{30}$$

This is in the same form as the one-temperature case, equation (15), except that the summation must be taken from $v = s$ to $v = N$. The forward and reverse rate coefficients are also formally the same.

In reference 30, the forward and reverse rate coefficients were calculated using this procedure. These calculations, and similar calculations made, were used in obtaining the rate coefficient values over a range of typical T and T_v values. Over a range of T/T_v from 0.5 to 2, the forward rate coefficient was found to be expressible approximately as

$$k_f \approx k_f(\sqrt{TT_v}) \tag{31}$$

The model described above is commonly known as the two-temperature ther-mochemical model. Since the vibrational temperature is low immediately behind a shock wave even though the translational temperature is very high there, the model tends to predict a small reaction rate in the range where the translational temperature is very high. Thus the apparent smallness of the preexponential power n (see Section 1.2) is explained.

4. Average Vibrational Energy Feedback

The average vibrational energy transfered in dissociation reaction, δ, is deter-mined as

$$\delta = \frac{\sum K_{vc} n_v (D - E_v) - \sum K_{cv} n_A n_B (D - E_v)}{\sum n_v (D - E_v) - \sum n_A n_B (D - E_v)} \tag{32}$$

Using the normalized population ρ and the principle of detailed balance, this becomes

$$\delta = \frac{\sum K_{vc} Q_v (\rho - \rho_A \rho_B)(D - E_v)}{\sum K_{vc} Q_v (\rho - \rho_A \rho_B)(D - E_v)} \tag{33}$$

If $K_{vc} Q_v$ = constant, and the vibrational levels are equally spaced (harmonic), then (30) gives a value of $\delta = 0.5D$, as found in [31]. If $K_{vc} Q_v$ = constant and vibrational levels are denser at higher energies (anharmonic), then δ is larger than 0.5D.

When the QSS solution (14) or (30) is used, δ is less than 0.5D because the quantity $\rho - \rho_A \rho_B$ is very small near the dissociation limit. Referemce [30] considers that 0.3D is a good average value over the typical range of T and T_v.

In order to advance the above-described theoretical model of chemical reaction, one must know the state-to-state collisional transition rate coefficients $K_{vv'}$ and the bound-free transition rate coefficients K_{vc}, for all possible colliding partners, i.e., O, N, NO, N_2, and O_2. This must be done not only for the vibrational mode but also for rotational and electronic modes. When these are known, the QSS formuilation shown above must be repeated. Only then, not only the rate coefficients but also all energy transfer parameters can be determined.

5. Unsolved Problems Requiring New Modeling

5.1. BOW SHOCK UV FLIGHT EXPERIMENTS

As mentioned earlier, in order to determine the extent of radiation emission from the stagnation region of a shock layer over a spherical blunt body, two flight experiments, named Bow Shock UV Flight Experiments were conducted in the United States, the first at a flight velocity of 3.5 km/s and the second at 5.2 km/s [19,20]. The radiation incident on the stagnation point of the spherical nose was detected with photometers and spectrophotometers at the wavelengths between 200 and 250 nm. Most of the measured radiation consisted of that from the NO Gamma band. The intensity of the band was compared with the calculated values.

For the first flight, at relatively low altitudes, the measured radiation was nearly as expected. At high altitudes, the measured radiation was much stronger than calculated. It seemed that the NO Gamma band is excited instantly as the molecule formed.

In order to understand the observed phenomena, experiments were conducted both in a shock tube and a laser fluorescence chamber. The shock tube experiment was conducted between 3 and 4 km/s [32,33], and the IR radiation from NO emanating from behind the shock wave was measured. The results showed that vibrational levels of the NO molecules are excited immediately as they are formed through the Zeldovich reactions. The result could not be explained using a two-temperature model. A five temperature $(T, T_r, T_v(N2), T_v(O2), T_v(NO))$ was initiated [34]. The model most closely, but not completely, reproduced the IR data. It seems that, in order to satisfactorily explain the observed phenomena, one must correctly account for the transfer of energies in the translational, rotational, and vibrational modes in the Zeldovich reactions.

The laser fluorescence experiment was conducted in order to understand the coupling between the vibrational modes and the electronic modes [35]. The results showed that indeed there is a strong coupling between vibrational and the electonic modes. These newly acquired mechanisms were incorporated into the old model. These greatly reduced the discrepancy between the measurement and the calculation [19]. However, still considerable discrepancy remains unexplained.

In addition to the shock layer radiation, the radiation from the rocket exhaust was measured also in the second flight. The measured radiation was again much stronger than calculated [20].

5.2. BEHAVIOR OF ELECTRON TEMPERATURE IN COOLING FLOWS

In the late 1960s, electron temperature and electron density were measured in the test section of a shock tunnel at CALSPAN using N_2, O_2, and air as the test gases [36-38]. The conditions were: p_s = 17 - 35 atm, $T_s \approx 4000$ K for O_2, 7000 K for N_2 and air, nozzle length ≈ 1 m, nozzle area ratio ≈ 800, and the instruments used wereLangmuir probe and microwave transmitter and detector. In these flows, the dominant reactions removing electrons were the dissociative recombination reactions $N_2^+ + e \rightarrow N + N$ in the nitrogen flow, $O_2^+ + e \rightarrow O + O$ in the oxygen flow, and $NO^+ + e \rightarrow N + O$ in the air flow.

There were no surprises in electron density variation. The rate coefficients for the dissociative recombination reactions were determined from the experiments. The rate coefficients so determined were not much different from those determined from other experiments.

As for electron temperature, the test results were surprising: the three test gases showed considerably different behaviors. In nitrogen flow, electron temperature were not much different from the calculated vibrational temperature. Since the coupling between the vibrational mode of N_2 and free electrons is known to be very strong, the observed behavior had been expected.

However, the electron temperature values in the O_2 flow were much lower. The electron temperature values in the air flow were between those in the N_2 flow and those in the O_2 flow. The theoretical calculations based on the existing thermochemical model tended to predict the electron temperature values to be higher than the measured values [39]. Only by assuming that a strong radiative cooling occurred due to the optically-thin emission of the FeO molecular band, with an unlikely high transition probability value of $A = 2\times10^9$ sec^{-1}, was it possible to numerically reproduce the experimental data [39]. This is one of the unexplained problems of the cooling flow mentioned in Section 1.3.

An alternative explanation of the observed low electron temperatures in a flow containing O_2 is that there is a fairly strong coupling between the electron translational mode and the vibrational mode of O_2, that is, that the cross section for the e-v coupling for O_2 is about 1/10 of that for N_2. Whether this is so is not yet known. In addition, there is a question as to whether there are preferential electron energies for the dissociative recombination reactions. If the reaction occurs preferentially for high energy electrons, then the reaction tends to cool the remaining electrons. Whether this is so is not known either.

5.3. SPECIES CONCENTRATIONS IN COOLING FLOWS

In the late 1960s, the concentrations of O, N, N_2, O_2, and NO were measured in the test section of an arc-jet wind tunnel using a mass spectrometer. The operating conditions were: p_s = from 5 to 10 atm, T_s= from 3000 to 5000 K, nozzle length = about 1 m, and nozzle area ratio = 1000 to 5000. In 1994, a similar measurement was made in a shock tunnel which operated at much higher p_s.

In both these experiments, the measured O concentration was lower, and N and NO concentrations were higher than the values calcualted using the traditional method. As a result, the flow contained greater chemical energy than calculated, implying that the wind tunnel simulation was worse than calculated. This is another of the problems of the cooling flow mentioned in Section 1.3.

In reference 39, this phenomenon is attributed to the presence of the oxygen atoms electronically excited to its ^1D state. The excited O-atoms are likely to produce high reaction rates for the Zeldovich reactions. Assuming a certain dependence of the reaction rates to the population of the $O(^1D)$ atoms, the observed species concentrations were numerically reproduced fairly closely. To perfect the model, one must know the rate coefficients for the Zeldovich reactions for $O(^1D)$ in detail. We must know also how $O(^1D)$ is produced, and how the required concentration of $O(^1D)$ is maintained even when it is being removed through the Zeldovich reactions.

6. Conclusions

The differences among the existing reaction rate models are caused mainly by the differences in treating the existence of the quasi-steady state (QSS) distribution of internal states. When the QSS condition is recognized and its consequence is properly incorporated into the reaction rate calculation, the resulting rate coefficients agree approximately with the two-temperataure model of Park. This is because the QSS model forgives errors in the state-to-state transition rate coefficients.

There are three unsolved problems in hypersonic flight requiring new modeling of the rate processes. They are: 1) concentration of NO and its vibrational and electronic state populations in the speed range of 3 to 4 km/s, 2) the behavior of electrons in cooling flow, and 3) concentrations of O, N, and NO in a cooling flow.

In order to advance the knowledge on the rate process in hypersonic speed range, one must know the following: a) the traffic of mode energies in Zeldovich reactions, b) the rate of Zeldovich reactions with $O(^1D)$, c) the mechanism by which the $O(^1D)$ concentration is kept high in a cooling flow, d) The details of e-v energy transfer for O_2, e) the mechanism for excitation of NO $A^2\Sigma$, f) the state-to-state and the bound-free collisional transition rate coefficients for all internal states with O, N, NO, N_2, and O_2 as the colliding species.

7. References

1. Hansen, C. F. (1959) "Thermodynamic and Transport Properties of High Temperature Air," NASA TR R-50.

2. Allen, R. A., Rose, P. H., and Camm, J. C.(1962) "Nonequilibrium and Equilibrium Radiation at Super-Satellite Reentry Velocities," AVCO-Everett Research Laboratory RR 156.

3. Crowder, R. S., and Moote, J. D. (1969) "Apollo Entry Aerodynamics," *Journal of Spacecraft and Rockets,* **6,** 302-307.

4. Hillje, E. R., and Savage, R. (1968) "Status of Aerodynamic Characteristics of the Apollo Entry Configuration," AIAA Paper 68-1143.

5. Boylan, D. E., and Griffith, B. J. (1968) "Simulation of the Apollo Command Module Aerodynamics at Reentry Altitudes," *Proceedings of the 3rd National Conference on Aerospace Aerospace Meteorology,"* 370-378.

6. De Rose, C. E. (1969) "Trim Attitude, Lift, and Drag of the Apollo Command Module with Offset Center-of-Gravity Positions at Mach Numbers to 29," NASA TN D-5276.

7. Romere, P. O., and Whitnah, A. M. (1983) "Space Shuttle Entry Longitudinal Aerodynamic Comparison of Flights 1 - 4 with Preflight Predictions," *Shuttle Performance: lessons Learned,* NASA CP-2283, compiled byJ. P. Arrington and J. J. Jones, 283-307.

8. Woods, W. C., Arrington, J. P., and Hamilton, H. H. (1983) "A Review of Preflight Estimates of Rea-Gas Effects on Space Shuttle Aerodynamic Characteristics," ibid, 309-3456.

9. Griffith, B. J., Maus, J. R., and Best, J. T. (1983) "Explanation of the Hypersonic Longitudinal Stability Problem - Lessons Learned," ibid, 347-379.

10. Rault, D. (1994) "A Study of Shuuttle Body Flap Effectiveness at High Altitudes using DSMC," AIAA Paper 94-2021.

11. Weilmuenster, K. J., Gnoffo, P. A., and Greene, F. A. (1993) "Navier-Stokes Simulation of the Shuttle Orbiter Aerodynamic Characteristics with Emphasis on Pitch Trim and Body Flap," AIAA Paper 93-2814.

12. Goodrich, W. D., Derry, S. M., and Bertin, J. J. (1983) "Shuttle Orbiter Boundary-Layer Transition - A Comparison of Flight and Wind Tunnel Data," AIAA Paper 83-0485.

13. Eschenroeder, A., Hayami, R., Primich, R., and Chen, T.(1966) "Ionization in the Near Wakes of Spheres in Hypersonic Flight," AIAA Paper 66-0055.

14. Akey, N. D., and Cross, A. E. (1970) "Radio Blackout Alleviation and Plasma Diagnostic Results from a 25,000 Foot per Second Blunt-Body Reentry," NASA TN D 5615.

15. Grantham, W. L. (1970) "Flight Results of a 25,000 Foot per Second Reentry Experiment Using Microwave Reflectometers to Measure Plasma Electron Density and Standoff Distance," NASA TN D-6062.

16. Jones, W. L., and Cross, A. E. (1970) "Electrostatic Probe Measurements of Plasma for Two 25,000 Foot per Second Reentry Flight Experiments," NASA SP-252.

17. Jones, W. L., and Cross, A. E. (1972) "Electrostatic Probe Measurements of Plasma Surrounding Three 25,000 Foot per Second Reentry Flight Experiments," NASA TN D-6617.

18. Candler, G. V., and MacCormack, R. W. (1988) "The Computation of Hypersonic Ionized Flows in Chemical and Thermal Nonequilibrium," AIAA Paper 88-0511.

19. Levin, D. A., Candler, G. V., Collins, R. J., Erdman, P. W., Zipf, E. C., Epsy, P., and Howlett, C. (1993) "Comparison of Theory with Experiment for the Bow Shock Ultraviolet Rocket Flight," *Journal of Thermophysics and Heat Transfer,* **7**, 30-36.

20. Erdman, P. W., Zipf, E. C., Epsy, P. J., Howlett, C. E., Levin, D. A., Collins, R. J., and Candler, G. V. (1992) "Measurement of Ultraviolet Radiation from a 5 km/s Bow Shock," AIAA Paper 92-2870.

21. Witcofski, R. D. (1989) "Advanced Hypervelocity Aerophysics Facility Workshop: Proceedings of a Workshop Sponsored by NASA and held at Langley Research Center, May 10-11, 1988," NASA Conference Publication 10031.

22. Generalov, N. A., and Losev, S. A. (1966) "Vibrational Excitation and Molecular Dissociation of Gaseous Oxygen and Carbon Dioxide in a Shock Wave," *Journal of Quantitative Spectroscopy and Radiative Transfer,"* **6**, 101-125.

23. Appleton, J. P., Steinberg, M., and Liquornik, D. J. (1968) "Shock-Tube Study of the Vibrational Relaxation of Nitrogen Using Vacuum-Ultraviolet Light Absorption," *Journal of Chemical Physics,* **48**, 599-608.

24. Keck, J. C., and Carrier, G. (1965) "Diffusion Theory of Nonequilibrium Dissociation and Recombination," *Journal of Chemical Physics,* **43**, 2284-2298.

25. Haug, K., and Truhlar, G. (1987) "Mone Carlo Trajectory and Master Equation Simulation of the Nonequilibrium Dissociation Rate Coefficient for Ar + H_2 -¿ Ar + 2H at 4500 K," *Journal of Chemical Physics,* **86**, 2697-2716.

26. Bates, D. R., Kingston, A. E., and McWhirter, R. W. P. (1962) "Recombination Between Electron and Atomic Ions, 1. Optically Thin Plasmas," *Proceedings of the Royal Society (London), Series A,* **267**, 297-312.

27. Park, C. (1969) "Collisional Ionization and Recombination Rates of Atomic Nitrogen," *AIAA Journal,* **7**, 1653-1654.

28. Adamovich, I. V., Macheret, S. O., Macheret, S. O., Rich, J. W., and Treanor, C. E. (1995) "Vibrational Relaxation and Dissociatin Behind Shock Waves Part 2: Master Equation modeling," *AIAA Journal,* **33**, 1070-1075.

29. Park, C. (1990) *Nonequilibrium Hypersonic Aerothermodynamics,* John Wiley, New York, N.Y.

596

30. Sharma, S. P., Huo, W. M., and Park, C. (1992) "Rate Parameters for Coupled Vibration-Dissociation in a Generalized SSH Approximation," *Journal of Thermophysics and Heat Transfer,"* **6**, 9-21.

31. Marrone, P. V., and Treanor, C. E. (1963) "Chemical Relaxation with Preferential Dissociation from Excited Vibrational Levels," *Physics of Fluids,* **6**, 1215-1221.

32. Wurster, W. H., Treanor, C. E., and Williams, M. J. (1989) "Nonequilibrium UV Radiation and Kinetics Behind Shock Waves in Air," AIAA Paper 89-1918.

33. Wurster, W. H., Treanor, C. E., and Williams, M. J. (1990) "Kinetics of UV Production Behind Shock Waves in Air," AIAA Paper 90-1666.

34. Moreau, S. (1993) "Computation of High Altitude Hypersonic Flow-Field Radiation," Ph.D. Thesis, Department of Aeronautics and Astronautics, Stanford University.

35. Crosley, D. R., Eckstrom, D. J., Smith, G. P., Jusinski, L. E., Meier, U. E., and Raiche, G. A. (1991) "Energy Transfer and Kinetics in Radiating Shock-heated Air," SRI International Report No. MP91-099.

36. Dunn, M. J., and Lordi, J. A. (1970) "Measurement of N_2^+ + e Dissociative Recombination in Expanding Nitrogen Flows," *AIAA Journal,* **8**, 339-345.

37. Dunn, M. J., and Lordi, J. A. (1970) "Measurement of O_2^+ + e Dissociative Recombination in Expanding Oxygen Flows," *AIAA Journal,* **8**, 614-618.

38. Dunn, M. J., and Lordi, J. A. (1969) "Measurement of Electron Temperature and Number Density in Shock Tunnel Flows, Part 2, NO + e Dissociative Recombination in Expanding Nitrogen Flows," *AIAA Journal,* **7**, 2099-2104.

39. Park, C. (1993) "Validation of Multitemperature Nozzle Flow Coe NOZNT," AIAA Paper 93-2862.

CHEMISTRY MODELS FOR AIR DISSOCIATION

S.A. LOSEV, E.A. KOVACH, V.N. MAKAROV,
M.Ju. POGOSBEKJAN, A.L. SERGIEVSKAYA
Avogadro Center, Institute of Mechanics,
Moscow State University
1, Michurinsky pr., Moscow 117192, Russia

1. Introduction

The main problem of air dissociation description at high temperatures is the thermal non-equilibrium of reactant in strong shock waves. The concept of thermal non-equilibrium is peculiar to chemical reacting system without equilibrium between vibrational and translational degrees of freedom for reacting molecules. In order to elaborate and frame models of processes in high-temperature air, it is necessary to analyse the mechanism of vibrational relaxation, molecules dissociation, and chemical exchange reactions.

Microscopic, kinetic and macroscopic descriptions of gas are levels of dissociation modeling. The microscopic level of thermal non-equilibrium description is presented by the solutions of collision dynamic problem and using Direct Simulation Monte Carlo (DSMC) method. At the kinetic level of modeling the problem is solved using master equiations for vibrational level kinetics. Vibrational temperature T_v concept along with gas temperature T provide the possibility of more simple macroscopic description of thermal non-equilibrium gas.

At high temperatures when characteristic times of dissociation and vibrational relaxation become of the same order, the dissociation proceeds while vibrational relaxation is not complete so that vibrational temperature T_v is less than gas temperature T. Also, it is assumed that the Boltzmann distribution of vibrational excited molecules over lower vibrational levels of energy would not be significantly disturbed and, thus, there is the possibility of macroscopic description of the processes using the notion of vibrational temper-

M. Capitelli (ed.), Molecular Physics and Hypersonic Flows, 597–614.
© *1996 Kluwer Academic Publishers. Printed in the Netherlands.*

ature T_v for any molecular component. Therefore the models of processes occuring in high-temperature air is further considered in terms of two-temperature (T and T_v) chemical kinetics.

2. Microscopic Examination
of the Vibrational Temperature T_v Concept

Application of the translational temperature T concept should present no problems because the Maxwellian distribution is established very quickly. The vibrational temperature T_v concept for real conditions in high temperature gas may be justified using results of collision dynamics numerical solution. This problem is solved in terms of classical mechanics using $O_2 - Ar$ collisions as an axample [1]. For this purpose the real potentials of particle interaction were used : for O atoms in O_2 molecule the potential was obtained by the Ridberg-Klaine-Rees method with all spectroscopic data available [2], and for $O_2 - Ar$ interaction the Born-Meyer potential was taken in the dumb-bell model form together with the potential parameters measured by molecular beam scattering [3]. The results obtained by solving the dynamic impact problem at various values of collision energy , impact parameters, and O_2 initial rotation, and vibration before collisions were averaged over the Maxwellian distribution for translational motion and the Boltzmann distribution for molecular rotation at gas temperature T. Next the kinetic problem of vibrational excitation and dissociation of molecules behind shock wave front was solved.

The resulting population of the m-th vibrational level n_m is defined by conventional level temperature Θ_m as a Boltzmann population relation for this level to the zeroth one:

$$n_m = n_o exp(-E_m/\Theta_m),$$

where E_m is the vibrational energy of the m-th level (in degrees). The resulting vibrational level populations in terms of Θ_m values as a function of vibrational energy E_m/D (D is dissociation energy) are demonstrated in figure 1 for gas temperature behind shock front in $O_2 - Ar$ mixture. The figures in curves denote the number of molecular collisions in high temperature gas. The arrow indicates the area corresponding to the most intense loss of the vibrational energy in the dissociation process. 90% of molecules are to the left of the dashed line.

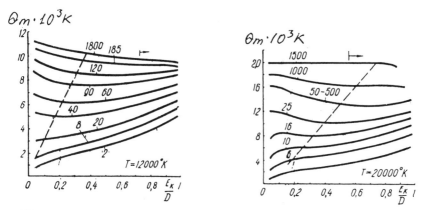

Figure 1. Level temperatures Θ_m of oxygen molecules versus relative vibrational energy in $O_2 \doteq Ar$ mixture behind shock front at T=12000 K and T=20000 K. Notations are indicated in the text.

The Boltzmann distribution of molecules over vibrational energy levels is fitting by the straight lines against on vibrational energy E_m/D. One can see that the Boltzmann distributions are not realized in real high temperature non-equilibrium gas behind strong shock front. Nevertheless in future we shall use the vibrational temperature T_v concept. T_v is considered to be a measure of vibrational energy volume as a level temperature for the first excited level. It is possible within certain limits because the large part of molecules are on the lower vibrational levels even at high temperature. It is significant that neglected here vibrational-vibrational (VV) energy exchange at collisions in molecular gas leads to more rapid establishment of the Boltzmann distribution. Hence the description problem of dissociation reacting system without thermal equilibrium between vibrational and translational degrees of freedom of reacting molecules leads to the concept of two-temperature chemical kinetics of gases in thermal non-equilibrium.

Another interesting solution result for $O_2 - Ar$ collisions is the difference between gas temperature T and vibrational temperature

T_v during the dissociation process in gas behind shock front (see figure 2). This result is appropriate to quasi-stationary state for vibrational energy balance.

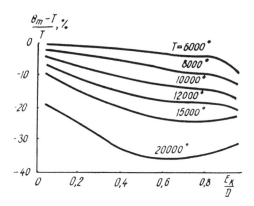

Figure 2. The difference between gas temperature T and level temperature Θ_m in quasi-stationary state behind shock for pointed temperature T values in $O_2 - Ar$ mixture.

The comparison of calculated vibrational relaxation times (as well as rate constant of dissociation) for $O_2 - Ar$ mixture and those measured using shock tubes [4,5] shows that they are in a reasonable agreement.

3. Experimental Modeling

The another line of modeling is experimental investigations. The experimental studies on air components (O_2, N_2) and some other molecules (Br_2, I_2) dissociation at maximum accessible temperatures and with necessary temporal resolution were performed in [6,7,8]. Such investigations were carried out on shock tubes of various power and diameters, in particular in Moscow State University. The shock tubes facility includes the systems for measurement of vibrational and translational gas temperatures, and gas molecules concentrations behind the shock wave front. These measurements are carried out by means of multichannel UV and visible absorption spectroscopy, and other methods. The example of experimentally recorded variations in vibrational and translational temperatures behind shock front in nitrogen is displayed in figure 3.

The main experimental results are the two-temperature rate constants $k(T, T_v)$ for dissociation of oxygen (up to 10500 K), nitrogen (up to 17000 K), and also bromine (up to 4000 K), iodine (up to 6000 K) molecules.

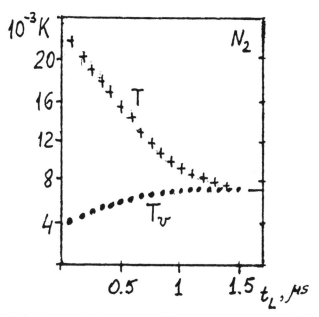

Figure 3. Measured vibrational T_v and translational T temperatures versus time t_L in laboratory system behind shock front in nitrogen for shock velocity V = 6.9 km/s and gas pressure before shock waves p = 1.13 torr [7].

Along with O_2 and N_2 these results for Br_2 and I_2 are used here for experimental validation of dissociation models because of relatively low dissociation energy. Notice that measured rate constants correlate with the quasi-stationary state of the dissociation and vibrational excitation balance in O_2, Br_2, and I_2, but it's not true in N_2 because of relatively small vibrational excitation rate for nitrogen.

The example for oxygen molecules dissociation rate constant is depicted in figure 4 in Arrhenius coordinates. At high temperatures the received values of two-temperature rate constant are much different from Arrhenius line because dissociation occurs in the absense of vibrational equilibrium. This difference is observed beyond temperatures $T^* = 4400K$ for O_2, $T^* = 6600K$ for N_2, $T^* = 1000K$ for Br_2 and I_2. Below this temperature T^* dissociation rate constant depends only on translational temperature T. In an effort to simulate thermal non-equilibrium dissociation for $T > T^*$ the two-temperature rate constant $k(T, T_v)$ is conveniently presented as two cofactors so that

$$k(T, T_v) = Z(T, T_v)k^o(T) ,$$

where $k^o(T)$ is the thermal equilibrium rate constant and $Z(T, T_v)$ is the non-equilibrium factor. The modeling of $Z(T, T_v)$ is the prime object of this work.

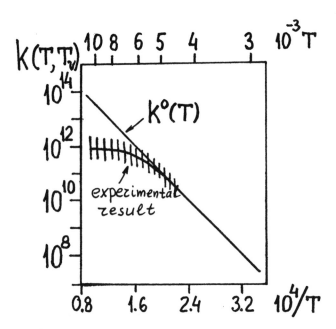

Figure 4. Measured values (vertical lines) of two-temperature rate constant $k(T, T_v)$ for oxygen dissociation in Arrhenius coordinates [6].

We can not measure the thermal equilibrium rate constant $k^o(T)$ at $T > T^*$ and use the thermal equilibrium theory for $k^o(T)$ evaluation and extrapolation to higher temperature using experimental results at $T < T^*$. This theory is the Nikitin ladder excitation model [9] within the frames of impulse approximation using the relations of classical mechanics for energy exchange at collision of a particle with each atom of a molecule. In the resulting dissociation rate constant value the effect of rotational and electronic states excitation is taken into account (see [10]). By this means the parameters of the Arrhenius generalized formula $k^o(T) = AT^n exp(-D/T)$ for the above mentioned molecules dissociation at collisions with similar molecules have been received, namely:

TABLE 1. Thermal equilibrium rate
constant $k^o(T)$ in dimension $[cm^3/mol \cdot s]$

Dissociating molecule	A (10^{22})	n	D
O_2	4.80	-1.85	59380
N_2	26.0	-1.81	113200
Br_2	0.280	-1.94	22910
I_2	0.093	-1.94	17740

These values have been applied in non-equilibrium factor $Z(T, T_v)$ calculation using experimental results for two-temperature rate constant $k(T, T_v)$.

Quantitative results of experimental modeling are presented below along with the results of analytical modeling.

4. Analytical Models for Two-temperature Dissociation

Avogadro Base of models of physical and chemical processes in gases and plasmas which was created in Moscow University includes 17 models for diatomic molecules dissociation in thermal non-equilibrium. The main concept of each model is the following:

HAMMERLING MODEL (1959) [11] - model for equiprobable dissociation of vibrational excited molecules: the cross sections for dissociation from individual vibrational states are independent of their vibrational quantum numbers.

LOSEV β - MODEL (1961) [12] - model for the truncated harmonic oscillator dissociation: the threshold of molecular dissociation is separated from the dissociation limit by a quantity βT as a mean thermal energy of gas (in Kelvin degrees).

MARRONE-TREANOR U-MODEL (1963) [13] - model for the distributed dissociation probability: the cross section for dissociation from various vibrational levels increases with the vibrational quantum number.

GORDIETS MODEL (1971) [14,15] - model for dissociation in terms of resonance and non-resonance vibrational energy exchange: the threshold of molecular dissociation is a result of VV- and VT-exchange balance deduced by SSH theory.

KUZNETSOV MODEL (1977) [16] - model for dissociation of highly rotational excited molecules from low vibrational levels.

SMEKHOV MODEL (1979) [17,18] - adiabatic model: the cross sections for dissociation are determined by the adiabatic Massey parameter with the energy defect between molecular vibrational state and dissociation limit.

KUZNETSOV MODEL (1981) [10] - model for anharmonic oscillator dissociation in terms of effective energy threshold of quick and slow vibrational exchange regions: only highly vibrational excited molecules are dissociating.

JAFFE MODEL-1 (1985) [19] - model for effective energy dissociation in terms of molecular rotation: the effective dissociation energy decreases because of centrifugal forces.

JAFFE MODEL-2 (1985) [19] - model providing for vibration and rotation overall contribution in molecules dissociation: the dissociation occurs if the sum of translational, vibrational and rotational energies exceeds dissociation energy.

PARK MODEL (1987) [20,21] - model for effective temperature: usual one-temperature rate constant is assumed to be dictated by the average temperature in terms of vibrational temperature.

FORD MODEL (1988) [22] - model for two-temperature analog of the generalized Arrhenius formula: the model parameters represent the contribution values of vibrational energy in activation energy and vibrational temperature in the Arrhenius preexponential factor.

HANSEN MODEL (1991) [23] - application of Landau-Zener theory: this theory defines the dependence of the cross section for dissociation on collision energy.

OLYNIC-HASSAN MODEL (1992) [24] - application of the Hinshelwood distribution over vibrational levels: minimum energy required for dissociation of a molecule with vibrational energy equal to e_v is $D_o - e_v$ (D_o is the dissociation energy).

KNAB MODEL (1992) [25] - generalized Marrone-Treanor model: the novelty is that the vibrational energy is considered as a part of dissociation energy.

LANDRUM-CANDLER MODEL (1992) [26] - bi-level model for dissociation of vibrational excited molecules: the threshold of dissociation is separated from dissociation limit by a mean thermal energy of gas.

MACHERET-RICH MODEL (1993) [27] - model for two dissociation mechanisms : one mechanism is the prevaling dissociation from low vibrational levels and the other one is equiprobable dissociation from all vibrational levels at $T_v \to T$.

MACHERET-FRIDMAN MODEL (1994) [28] - model for two dissociation mechanisms in terms of colliders configuration: one mechanism being significant at high temperature is the direct dissociation from low vibrational levels and the other one is the dissociation from high vibrational levels after vibrational exitation.

KUZNETSOV MODEL(1994) [29] - is the extension of the previous Kuznetsov model (1981) to the complex gas media contained not only dissociating molecules and inert particles, but also atoms, radicals, and ions.

Full description of each model includes a substantiation of the accepted assumptions and restrictions, appropriate mathematical formulas for rate constant $k(T, T_v)$ and non-equilibrium factor $Z(T, T_v)$, etc. This enables $k(T, T_v)$ and $Z(T, T_v)$ to be found at any temperatures T and T_v.

The analytical models mentioned above are of different theoretical justification. In the simplest Park model the vibrational temperature contribution is intuitively taken into account by means of substitution of the gas temperature T appearing in the one-temperature rate constant $k(T)$ expression for the effective temperature $T_{eff} = T^s T_v^{1-s}$ (s is the model parameter). In this model the non-equilibrium factor is equal to

$$Z(T, T_v) = (T_v/T)^{n(1-s)} exp[-(D/T)((T/T_v)^{1-s} - 1)] \ ,$$

where n is the exponent in the pre-exponential factor of the Arrhenius expression for $k^o(T)$.

A set of likewise simple models is based on application of general theoretical statements. The main models are those that include varying parameters: Marrone-Treanor, Losev, Knab, as well as Ford and Jaffe models. In Losev model the non-equilibrium factor is written as

$$Z(T, T_v) = [Q(T_v)/Q(T)] exp[-(D - \beta T)(1/T_v - 1/T)] \ ,$$

where $Q(T_v), Q(T)$ are partition functions, β is the model parameter. The Landrum-Candler model is reduced to this model for the fixed value $\beta = 1$. In Marrone-Treanor model we have:

$$Z(T, T_v) = \frac{Q(T)Q(T_F)}{Q(T_v)Q(-U)},$$

$$Q(T_m) = \frac{1 - exp(-D/T_m)}{1 - exp(-\Theta/T_m)},$$

$$T_m = T, T_v, T_F \ or \ -U, \ T_F = (1/T_v - 1/T - 1/U)^{-1}$$

Here Θ is the characteristic vibrational temperature, U is the model parameter. Hammerling model is reduced to this model for the fixed value $U = \infty$. In addition to the parameter U Knab model includes also the parameter α, which takes into account the vibrational energy contribution in activation energy of the considered reaction. The parameters s, β, U, α in Park, Losev, Marrone-Treanor and Knab models, respectively, are subject to vary. Fitting them to either the experimental results or data obtained using more physically proved models, one can attain the combination of simplicity with adequacy.

Finally, other models are based on using the fundamental physical principles and the results of solution of particular theoretical problems; these models do not include fitting parameters.

5. Comparison between Analytical Models and Experimental Results

Hence the analytical model results for the non-equilibrium factor $Z(T, T_v)$ are comparable with the experimental ones, see figures 5.

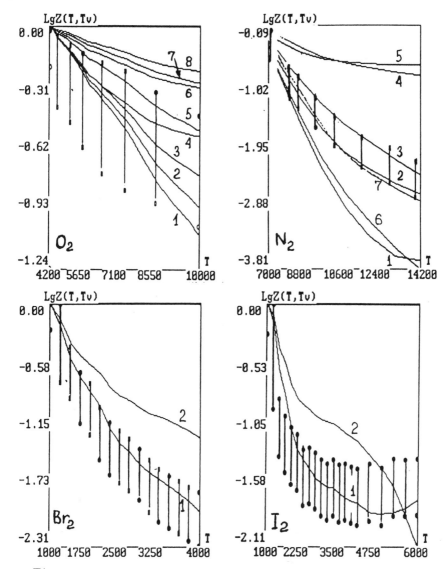

Figure 5. The comparison of the analytical model results with experimental ones. Notations are indicated in the text.

The experimental result are depicted by vertical lines for O_2 [6], N_2

[7], Br_2 and I_2 [8]. The figures on curves denote the models results, for O2 : 1 - Macheret-Fridman, 2 - Marrone-Treanor (with
U=T/1.7), 3 - Losev (with β=1.5), 4 - Smekhov, 5 - Kuznetsov
(1981), 6 - Hammerling, 7 - Olinik-Hassan, 8 - Park (with s=0.7);
for N_2 : 1 - Macheret-Fridman, 2 - Losev (with β =3), 3 - Kuznetsov
(1981), 4 - Park (with s=0.7), 5 - Hammerling, 6 - Marrone-
Treanor(with U=T/1.9), 7 - Gordiets; for Br_2 and I_2 : 1 - Macheret-
Fridman model, 2 - Kuznetsov (1981) model. The values of vibrational temperature T_v fit the experimental ones for O_2 and also for
other molecules. From figures 5 it is inferred that most of the analytical model results correlate to one another as well as to the experimental results. As indicated above, the results of some empirical
models (Losev, Park, Marrone-Treanor, etc.) are fitted to the experimental ones when the parameters β, s, U, etc. are varied.

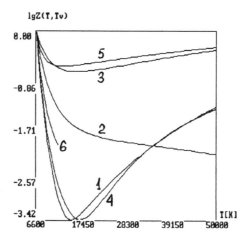

Figure 6. The non-equilibrium factor for N_2 dissociation at $T_v =$
6600 K as a result of models : 1 - Macheret-Fridman, 2 - Kuznetsov
(1981), 3 - Park (with s=0.7), 4 - Marrone-Treanor (with U=T/1.9), 5 -
Hammerling, 6 - Losev (with $\beta = 3$).

With the aim of high-temperature extrapolation procedure it is
interesting to consider the experimental results for I_2 dissociation
because in this case the temperature range extends up to 6000 K.
This high-temperature experimental limit is about 34% of I_2 dissociation energy that corresponds to about 20000 K for O_2 and about
38000 K for N_2. At high temperatures the results of Macheret-

Fridman model begin to increase because direct dissociation from low vibrational levels is taken into account. Kuznetsov (1981) model examined only highly vibrational excited molecules, and Kuznetsov results decrease at high temperature. More clearly this is evident from the next figure 6 for the extrapolation the models results to very high temperature. In the figure 6 the U-parameter in Marrone-Treanor model is consistent with Macheret-Fridman model results for $U = T/1.9$.

6. Recommendation for Two-temperature Description of Dissociation

The main result of the drawn examination is that for solution of applied gasdynamic problems we can recommend to use more simple (and old!) Marrone-Treanor model on thermal non-equilibrium dissociation with U value about $U = T/2.2$ for O_2 and $U = T/1.9$ for N_2. Notice that from the results of the dynamic problem solution for $O_2 - Ar$ collisions [1] it follows that $U = T/1.7$ which is close to the above mentioned recommendation.

7. Air Dissociation in Shock Waves

The mechanism of air dissociation in strong shock waves includes also chemical exchange reaction (Zel'dovich mechanism). Bimolecular endothermic chemical exchange reaction between nitrogen molecules and oxygen atoms starts with the oxygen molecule dissociation, but nitrogen molecules may be in thermal non-equilibrium because vibrational excitation of N_2 occurs simultaneously with O_2 dissociation. This raises the question as to thermal non-equilibrium impact on the rate of the reaction $N_2(v) + O$.

There are many analytical models for chemical exchange reactions: model of vibrational energy efficiency and Macheret formulas [30], Knab model [25], etc, but we considere the results of numerical solution of collision dynamics problem for this reaction .

The simulation of exchange reaction $N_2(v) + O \rightarrow NO + N$ was carried out by quasi-classical trajectory method within the framework of adiabatic approach. The potential energy surface constructed in [31] have been used in this calculations in form of extended LEPS model. The resulting two-temperature rate constant assuming the Boltzmann distribution of N_2 molecules over vibrational levels is depicted in figure 7. As is easy to see, the values $k(T, T_v)$ differs little from single-temperature values $k(T, T_v = T)$.

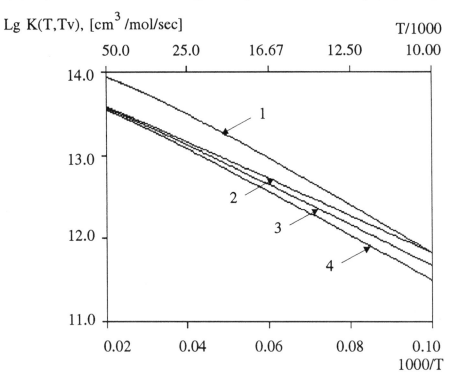

Figure 7. Two-temperature rate constant of reaction $N_2(v) + O \rightarrow NO + N$ as a result of quasi-classical trajectory method. The figures denote : 1 - $T_v = T$, 2 - $T_v = 10000$, 3 - $T_v = 8000$, 4 - $T_v = 4000$ K.

The expert training of the above mentioned analytical and numerical models and experimental results makes possible to extrapolate air chemical rate constans to very high temperature. One example of such expert training is the implementation of Marrone-Treanor model for strong shock waves in the Earth's atmosphere at altitude H =70 km and for shock velosity V=9 km/s , see figure 8 . For this purpose U-parameter was being varied over a wide range from $U = \infty$ (as for Hammerling model dot-and-dash lines) to $U = T/5$ (enture line); in this figure dashed lines correspond to $U = T/1.9$. The effect of this varing is evident near the shock front.

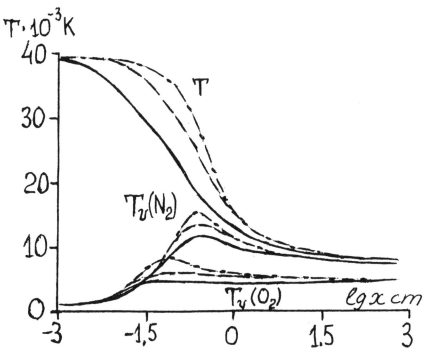

Figure 8. Translational temperature T and vibrational temperatures $T_v(N_2), T_v(O_2)$ versus distance x from the shock front.

8.Acknowledgements

This work was supported by Russian Foundation for Fundamental Research (Grant N 94-01-00633).

The authors would like to acknowledge helpful discussions with A.A. Fridman, N.M. Kuznetsov, B.F. Gordiets, G.D. Smekhov, O.P. Shatalov, J.W. Rich, C.E. Treanor, C. Park, S.O. Macheret, S.V. Zhluctov, I. Adamovich , S.Ja. Umanskiy,. The authors are grateful to N.A.Khrapak and I.N.Maisuradze for helping.

9.References

1. Kuksenko B.V. and Losev S.A. (1969) Vibrational excitation and dissociation on diatomic molecules on atom-molecule collisions at high- temperature gas, *Dokl.Akad.Nauk SSSR* **185**, 69-72.

2. Vanderslice J.T., Mason E.A. and Maich W.O. (1960) *J. Chem. Phys.* **32**, 515.

3. Beljaev Ju.N. and Leonas V.B. (1966) The short-range intermolecular interaction of oxygen and nitrogen molecules, *Dokl.Akad.Nauk SSSR* **170**, 1039-1043.

4. Losev S.A. and Shatalov O.P. (1970) The vibrational relaxation of oxygen molecules in mixture with argon up to temperature about 10,000 degrees, *Chimia vysokich energiy* **4**, 263-267

5. Wray K.L. (1962) Shock tube study of the coupling of the $O_2 - Ar$ rates of dissociation and vibrational relaxation, *J. Chem.Phys.* **37**, 1254-1262.

6. Shatalov O.P. (1973) Oxygen molecules dissociation in the absence of vibrational equilibrium, *Fisika gorenia i vzryva* **9**, 699-704

7. Yalovik M.S. and Losev S.A. (1972) Vibrational kinetics and dissociation of nitrogen molecules at high temperature, *Scientific Proceedings, Institute of Mechanics, Mosc.State Univ.* **18**, 3-34 .

8. Generalov N.A. (1981) The investigation of non-equilibrium state of gas medium stimulated by shock waves and glow dischardge, Thesis, Inst. for Problem of Mech. Acad. Sci. USSR, Moscow.

9. Nikitin E.E. (1974) *Theory of Elementary Atomic and Molecular Processes in Gases,* Clarendon Press, Oxford.

10. Kuznetsov N.M. (1981) *Kinetics of Unimolecular Reactions,* Nauka, Moscow .

11. Hammerling P., Teare J.D. and Kivel B. (1959) Theory of radition from luminous shock waves in nitrogen, *Phys.of Fluids,* **2**, 422-426.

12. Losev S.A. and Generalov N.A. (1961) On phenomena of molecular oxygen vibrational excitation and dissociation at high tem-

peratures, *Dokl. Akad. Nauk SSSR.* **141,** 1072-1076.

13. Marrone P.V. and Treanor C.E. (1963) Chemical relaxation with preferential dissociation from excited vibrational levels, *Phys.of Fluids* **6,** 1215.

14. Gordiets B.F., Osipov A.I. and Shelepin L.A. (1972) Non-equilibrium dissociation and molecular lasers, *Sov. Phys.-JETP* **34,**299-306.

15. Gordiets B.F. and Zhdanok S.A. (1986) , Analytical description on vibrational kinetics of anharmonic oscillators, in M.Capitelli (ed.), *Nonequilibrium Vibrational Kinetics,* Springer - Verlag, Berlin, pp.60-103

16. Kuznetsov N.M. (1977) Dissociation kinetics with high energy transitions, *Dokl.Akad.Nauk SSSR* **233,** 413-416.

17. Smekhov G.D. and Losev S.A. (1979) The role of vibrational-rotational excitation on dissociation of diatomic molecules, *Teoret. Experiment. Chimija.* **15,** 492-497.

18. Smekhov G.D. and Zhluktov S.V. (1993) Dissociation of diatomic molecules at very high temperatures, *Khim. Fizika* **12,**337-339.

19. Jaffe R.L. (1986) Rate constant for chemical reactions in high temperature nonequilibrium air, *Progress in Aeronautics and Astronautics* **103,** 123-151.

20. Park C. (1987) Assessment of two-temperature kinetic model for ionizing air, *AIAA-Paper* N87-1574.

21. Park C. (1990) *Nonequilibrium Hypersonic Aerothermodynamics,* Wiley-Intern. Publ., N.Y.

22. Ford D.I., Johnson R.E. (1988) Dependance of rate constants on vibrational temperature : an Arrehnius description, *AIAA-Paper* N 88-0461.

23. Hansen C.F. (1991) Dissociation of diatomic gases, *J.Chem. Phys.* **95,**7226-7233.

24. Olynic D.R., Hassan H.A. (1992) A new two-temperature dissociation model for reacting flows, *AIAA-Paper* N92-2943.

25. Knab O., Fruhauf H.-H. and Jonas S. (1992) Multiple temperature descriptions of reaction rate constants with regard to consistent chemical-vibrational coupling, *AIAA-Paper* N 92-2947.

26. Landrum D.V. and Candler G.V. (1992) Development of a new model for vibrational-dissociation coupling in nitrogen, *AIAA-Paper* N92-2853.

27. Macheret S.O. and Rich J.W. (1993) Theory of nonequilibrium dissociation rates behind strong shock waves, *AIAA-Paper* N93-2860.

28. Macheret S.O., Fridman A.A., Adamovich I.V., Rich J.W., Treanor C.E.(1994) Mechanisms of nonequilibrium dissociation of diatomic molecules, *AIAA-Paper* N94-1984.

29. Kuznetsov N.M. and Sergievskaya A.L. (1994) Threshold of quick vibrational region and rate constant of dissociation of diatomic molecules in complex gas media, *Khim. Fizika.* **13,**15-23.

30. Rusanov V.D. and Fridman A.A. (1984) *Physics of Chemically Active Plasma,* Nauka, Moscow.

31. Levitsky A.A. (1986) Mathematical simulation of plasmachemical processes. Thesis, Inst. Oil-Chemistry Synthesis, Acad. Sci. USSR, Moscow.

PRIMARY-PRINCIPLES BASED SIMULATION OF CHEMICAL AND RELAXATION KINETICS IN HYPERSONIC FLOWS. SCALE OF AMBIGUITY DUE TO THE SOURCE TERMS.

S.V. ZHLUKTOV* and G.D. SMEKHOV**
*Institute for Computer-Aided Design (Russian Academy of Sciences),
2-d Brestskaya 19/18, Moscow 123056, Russia,
Phone: +7(095)939-38-66, Fax: +7(095)939-01-65
**Institute of Mechanics at Moscow State University,
Michurinskiy pr. 1, Moscow 119899, Russia,
Phone: +7(095)939-27-13, Fax: +7(095)939-01-65

1. Dissociation Reactions

Adiabatic Model for dissociation of diatomic molecules under thermal nonequilibrium has been developed in [1,2]. Outline here the main ideas of the approach used. Consider reactions

$$1. \quad O_2 + M \Leftrightarrow O + O + M$$
$$2. \quad N_2 + M \Leftrightarrow N + N + M \tag{1}$$
$$3. \quad NO + M \Leftrightarrow N + O + M$$

Here M is a third particle. We assume that: 1) all the molecules are in the ground electronic state; 2) rotational modes of the molecules are in equilibrium with translational ones at translational temperature T; 3) vibrational quantum states are populated according to Boltzmann distribution at a vibrational temperature T_v common for all the air molecules; 4) vibrations and rotations do not depend on each other, i. e.

$$E_{IJ} = E_I + E_J \tag{2}$$
$$E_I = Ik\Theta_v(1 - Ik\Theta_v / 4D) \tag{3}$$
$$E_J = k\Theta_r J(J+1) \approx k\Theta_r J^2 \tag{4}$$

where E_I, E_J, and E_{IJ} are potential energies respectively of the I-th vibrational level, J-th rotational level, and (I,J)-th vibrational-rotational level of unharmonic oscillator; Θ_v is characteristic vibrational temperature; Θ_r is characteristic rotational temperature; k is the Boltzmann constant; $D=kT_D$ is dissociation energy; T_D is characteristic dissociation temperature. Due to very small values of rotational quanta of the air molecules (about 2K for O_2, N_2, and NO), we shall assume that $J \gg 1$. Under the assumptions, dissociation rate constant of a diatomic molecule can be represented as follows:

$$K_f = \sum_{I=0}^{I_{max}} \sum_{J=0}^{J_{max}} k_{IJ} f_{IJ} \tag{5}$$

M. Capitelli (ed.), Molecular Physics and Hypersonic Flows, 615–624.
© 1996 Kluwer Academic Publishers. Printed in the Netherlands.

In this expression I_{max} is the maximum vibrational quantum number corresponding to dissociation limit; J_{max} is the maximum rotational quantum number corresponding to given I and to dissociation limit; subscripts f and b are used for the forward and backward reactions respectively; k_{IJ} is the constant of dissociation from the (I,J)-th vibrational-rotational level; f_{IJ} is the distribution function:

$$f_{IJ} = \frac{1}{Z_v(T_v)Z_r(T)} g_I g_J \exp(-\frac{E_I}{kT_v} - \frac{E_J}{kT}) \tag{6}$$

$$Z_v(T_v) \approx \frac{1}{1 - \exp(-\Theta_v / T_v)}, \quad Z_r(T) \approx T / \Theta_r \tag{7}$$

$$g_I = 1, \quad g_J = 2J + 1 \approx 2J \tag{8}$$

where Z_v and Z_r are vibrational and rotational partition functions; g_I and g_J are degeneracies of the I-th vibrational and J-th rotational levels respectively.

By the definition

$$k_{IJ} = \int_0^\infty \sigma_{IJ}(u) u f(u) du \tag{9}$$

Here σ_{IJ} is the cross-section of dissociation from the (I,J)-th vibrational-rotational level; u is the speed of relative motion of colliding particles; $f(u)$ is Maxwellian distribution function:

$$f(u) = 4\pi (\frac{m}{2\pi kT})^{3/2} \exp(-\frac{mu^2}{2kT}) u^2 \tag{10}$$

where m is the reduced mass of colliding particles. For σ_{IJ} we assume that

$$\sigma_{IJ} = \sigma_0 P_{IJ}, \quad P_{IJ} = \exp(-\chi), \quad \chi = \frac{\Delta E_{IJ}}{\alpha_0 h u}, \quad \Delta E_{IJ} = D - E_I - E_J \tag{11}$$

Here σ_0 is an asymptotic value of the cross-section at $u \to \infty$; P_{IJ} is the probability of dissociation from the (I,J)-th vibrational-rotational level; χ is the Massey adiabatic parameter; α_0 is the constant in the repulsive interaction potential: $V = V_0 \exp(-\alpha_0 r)$, r is the distance between colliding particles, h is the Plank constant.

Let us assume that: $P_{IJ} = 0$ at $u < u^*$, and $P_{IJ} = P_{IJ}(u^*) = const$ at $u \geq u^*$, where

$$u^* = \sqrt{2\Delta E_{IJ} / m} \tag{12}$$

In this case Eq. (9) can be easily integrated:

$$k_{IJ} = \sigma_0 \sqrt{\frac{8kT}{\pi m}} \exp(-\chi - \frac{\Delta E_{IJ}}{kT}) \left[\frac{\Delta E_{IJ}}{kT} + 1 \right] \tag{13}$$

$$\chi = \frac{\Delta E_{IJ}}{\alpha_0 h u^*} = \frac{1}{\alpha_0 h} \sqrt{\frac{m \Delta E_{IJ}}{2}} \approx \frac{1}{\alpha_0 h} \sqrt{\frac{m(D - E_I)}{2}} \tag{14}$$

In order to simplify the expression, we shall assume that the reaction cross-section (11) do not depend on the rotational quantum number J (see Eq. (14)). Then we have:

$$K_f = \sigma_0 \sqrt{\frac{8kT}{\pi m}} \frac{1}{Z_v(T_v)Z_r(T)} \exp(-D / kT) \quad \times$$

$$\times \sum_{I=0}^{I_{max}} \exp\left[-\frac{1}{\alpha_0 h}\sqrt{\frac{m(D-E_I)}{2}}\frac{E_I}{k}(\frac{1}{T_v}-\frac{1}{T})\right]\sum_{J=0}^{J_{max}}2J\left[\frac{D-E_I-E_J}{kT}+1\right] \quad (15)$$

Due to the quasi-classic character of rotational motion (i.e. the small values of rotational quanta), we can replace summing by integration with respect to J in Eq. (15). This yields:

$$K_f = \sigma_0\sqrt{\frac{8kT}{\pi m n}}\frac{1}{Z_v(T_v)(kT)^2}\exp(-D/kT)\,\Phi \quad (16)$$

$$\Phi = \frac{1}{2}\sum_{I=0}^{I_{max}}\exp\left[-\frac{1}{\alpha_0 h}\sqrt{\frac{m(D-E_I)}{2}}\frac{E_I}{k}(\frac{1}{T_v}-\frac{1}{T})\right]\times(D-E_I)(D-E_I+2kT) \quad (17)$$

σ_0 is close to gas-kinetic cross-section. Therefore, we have the model totally determined by physical quantities: σ_0, α_0, m, D, Θ_v, T, T_v. Note that the model do not contain any empirical parameter. The following values of characteristic temperatures are used in the present study: T_D=59364 K, Θ_v=2256 K for O_2; T_D=113250 K, Θ_v=3372 K for N_2; T_D=75392 K, Θ_v=2720 K for NO. Calculations with real σ_0 and α_0 have shown that in the case of thermal equilibrium, Adiabatic Model yields dissociation rate constants within the scatter in the published experimental data on these constants [2]. Below are the used values of gas-kinetic cross-sections and parameters of repulsive potentials: $\sigma_0(O_2-O_2)$=3.519 10^{-15} cm^2, $\alpha_0(O_2-O_2)$ =2.85 A^{-1}; $\sigma_0(O_2-N_2)$=3.285 10^{-15} cm^2, $\alpha_0(O_2-N_2)$ =3.02 A^{-1}; $\sigma_0(O_2-NO)$=3.866 10^{-15} cm^2, $\alpha_0(O_2-NO)$ =3.78 A^{-1}; $\sigma_0(O_2-O)$=2.807 10^{-15} cm^2, $\alpha_0(O_2-O)$ =4.85 A^{-1}; $\sigma_0(O_2-N)$=2.680 10^{-15} cm^2, $\alpha_0(O_2-N)$ =4.13 A^{-1}; $\sigma_0(N_2-N_2)$=3.058 10^{-15} cm^2, $\alpha_0(N_2-N_2)$ =3.16 A^{-1}; $\sigma_0(N_2-NO)$=3.620 10^{-15} cm^2, $\alpha_0(N_2-NO)$ =3.64 A^{-1}; $\sigma_0(N_2-O)$=2.598 10^{-15} cm^2, $\alpha_0(N_2-O)$ =5.12 A^{-1}; $\sigma_0(N_2-N)$=2.476 10^{-15} cm^2, $\alpha_0(N_2-N)$ =3.310 A^{-1}; $\sigma_0(NO-NO)$=4.228 10^{-15} cm^2, $\alpha_0(NO-NO)$ =3.26 A^{-1}; $\sigma_0(NO-O)$=3.117 10^{-15} cm^2, $\alpha_0(NO-O)$ =3.95 A^{-1}; $\sigma_0(NO-N)$=2.984 10^{-15} cm^2, $\alpha_0(NO-N)$ =3.72 A^{-1}. It is necessary to emphasize here that the good agreement with published rate constants has been obtained with immediate use of these values of σ_0, α_0, without any fitting.

2. Exchange Reactions

In the present study the Adiabatic Model is extended to the exchange reactions:
$$N_2 + O \Leftrightarrow NO + N$$
$$NO + O \Leftrightarrow O_2 + N \quad (18)$$

In order to describe the forward (endothermic) process, let us break down the spectrum of the corresponding molecule into two parts. The lower part is bounded by an effective border $D^* = kT_D^*$, and the upper part lies between the effective border D^* and dissociation limit D. In other words

$$K_f = K_{f1} + K_{f2} = \sum_{I=0}^{I^*}\sum_{J=0}^{J^*}k_{IJ}f_{IJ} + \sum_{I=I^*+1}^{I_{max}}\sum_{J=0}^{J_{max}}k_{IJ}f_{IJ} \quad (19)$$

Here I^* and J^* correspond to the effective border $D^* < D$ of the given exchange reaction in the endothermic direction. The analysis similar to that for dissociation yields the following expression for the first term in Eq. (19):

$$K_{f1} = \sigma_0^* \sqrt{\frac{8kT}{\pi m}} \frac{1}{Z_v(T_v)(kT)^2} \exp(-D^*/kT)\, \Phi^* \tag{20}$$

$$\Phi^* = \frac{1}{2}\sum_{I=0}^{I^*} \exp\left[-\frac{1}{\alpha_0 h}\sqrt{\frac{m(D^*-E_I)}{2}} - \frac{E_I}{k}(\frac{1}{T_v}-\frac{1}{T})\right](D^*-E_I)(D^*-E_I+2kT) \tag{21}$$

Here σ_0^* is the reaction cross-section. Consider the second term in Eq. (19). It describes the reactions proceeding from vibrational-rotational levels higher then D^*. Let us use Eq. (9) where we assume that $\sigma_{IJ}(u)=\sigma_0^*=\text{const}$ for $E_{IJ} > D^*$. Then integrating Eq. (9), we obtain:

$$k_{IJ} = \sigma_0^* \sqrt{\frac{8kT}{\pi m}}$$

$$K_{f2} = \sigma_0^* \sqrt{\frac{8kT}{\pi m}} \frac{1}{Z(T_v)} \frac{\Theta_r}{T} \sum_{I=I^*+1}^{I_{max}} \exp(-\frac{E_I}{kT_v}) \sum_{J=0}^{J_{max}} 2J \exp(-\frac{E_J}{kT}) =$$

$$= \sigma_0^* \sqrt{\frac{8kT}{\pi m}} \frac{1}{Z(T_v)}(1-\exp(\frac{T_D}{T})) \sum_{I=I^*+1}^{I_{max}} \exp(-\frac{E_I}{kT_v}) \tag{22}$$

It turns out impossible to use σ_0 values from section 1 as σ_0^*. Apparently, cross-sections of exchange reactions differ from gas-kinetic ones. A good agreement between the reaction rate constants at thermal equilibrium and published experimental data has been obtained when the cross-sections were taken to be $\sigma_0^*(N_2,O)= \sigma_0(N_2,O)/10$, $T_D^*(N_2,O)=44933$ K; $\sigma_0^*(NO,O)= \sigma_0(NO,O)/17.5$, $T_D^*(NO,O)=31214$ K

3. Approximation of Adiabatic Model

Expressions for thermal nonequilibrium endothermic reactions imply summing over vibrational levels of unharmonic oscillator. In calculating a real flow it should be performed several times at every node of computational mesh for every molecular species. This requires substantial computational time. Therefore it is reasonable to derive an approximate formulae for the suggested Adiabatic Model. The following expressions for the terms containing summing over unharmonic vibrational levels are recommended:

$$\Phi = 1/2\Big\{\varphi_1\big[T_D + T_D Z_1(T_m)\exp(-\Theta_1/T_m)\big] +$$

$$+ \varphi_2 0.15(T_D - T_D^*)Z_2(T_m)\exp(-(T_D + \Theta_2)/T_m)\Big\} \tag{23}$$

$$\Phi^* = 1/2\varphi_3\Big\{T_D^* + Z_1(T_m)\big[T_D^* \exp(-\Theta_1/T_m) - \varepsilon_0(T_m)\big]\Big\} \tag{24}$$

$$K_{f2} = \sigma_0^* \sqrt{\frac{8kT}{\pi m}} \frac{Z_2(T_v)}{Z_v(T_v)}(1-\exp(-\frac{T_D}{T}))\exp(-\frac{T_D^*+\Theta_2}{T_v}) \tag{25}$$

where

$$\varphi_1 = k^2(T_D + 2T)\exp(-\beta_{ij}\sqrt{T_D}), \quad \beta_{ij} = 0.016158\sqrt{m_{ij}(g/mol)}/\alpha_{ij}(A^{-1}) \quad (26)$$

$$\varphi_2 = k^2(0.15(T_D - T_D^*) + 2T)\exp(-\beta_{ij}\sqrt{0.15(T_D - T_D^*)}), \quad (27)$$

$$\varphi_3 = k^2(T_D^* + 2T)\exp(-\beta_{ij}\sqrt{T_D^*}), \quad (28)$$

$$Z_1(T_m) = \frac{1-\exp(-T_D^*/T_m)}{1-\exp(-\Theta_1/T_m)}, \quad Z_2(T_m) = \frac{1-\exp(-(T_D - T_D^*)/T_m)}{1-\exp(-\Theta_2/T_m)}, \quad (29)$$

$$\varepsilon_0(T_m) = \frac{\Theta_1}{\exp(\Theta_1/T_m)-1} - \frac{T_D^*\exp(-\Theta_1/T_m)}{\exp(T_D^*/T_m)-1}, \quad \frac{1}{T_m} = \frac{1}{T_v} - \frac{1}{T} \quad (30)$$

$Z_2(T_v)$ in Eq. (25) is calculated with use of the first formula in (29). The idea of the approximation is to break down the vibrational spectrum of the molecule into three parts: ground state, the lower part of the spectrum bounded by an energy $D^* < D$ with an effective vibrational temperature Θ_1, and the upper part of the spectrum with its own vibrational characteristic temperature Θ_2. In the case of dissociation reactions it is recommended to take $T_D^* = 46364$ K, $\Theta_1 = 1656$ K, and $\Theta_2 = 542$ K for O_2; $T_D^* = 90402$ K, $\Theta_1 = 2443$ K, and $\Theta_2 = 762$ K for N_2; $T_D^* = 59520$ K, $\Theta_1 = 1984$ K, and $\Theta_2 = 635$ K for NO. In the case of exchange reactions it is suggested to take $T_D^* = 44933$ K, $\Theta_1 = 2995$ K, and $\Theta_2 = 1971$ K for N_2; $T_D^* = 31214$ K, $\Theta_1 = 2401$ K, and $\Theta_2 = 1578$ K for NO.

Eqs. (23), (24), and (25) with given effective characteristic temperatures approximate Eqs. (17), (21), and (22) with the accuracy defined by the multiple 1.5. The comparisons have been carried out in independent variation of T and T_v within 2000 K $\leq T$, $T_v \leq 50000$ K.

It is assumed that the backward (exothermic) reactions are dictated by the translational temperature T only, and the corresponding reaction rate constants are calculated as follows:

$$K_{bi} = K_{fi}(T,T)/K_{ci}(T) \quad (31)$$

where K_c is the equilibrium constant.

Note also that the real dissociation rate constants in the five-species air are:

$$K_{fi}(T,T_v) = \sum_{j=1}^{5} K_{f,ij}(T,T_v)c_j/m_j, \quad i = O_2, N_2, NO, \quad j = O_2, N_2, NO, O, N \quad (32)$$

4. Governing Equations and Boundary Conditions

Multicomponent Nonequilibrium Viscous Shock Layer (MNVSL) equation set [2,3] is used in the present study for simulating flows past axisymmetric blunt bodies. Consider the equation of vibrational relaxation. It has the following form in the orthogonal coordinate system normally connected to the body surface:

$$\rho[\frac{v_1}{H_1}\frac{\partial}{\partial x}(\varepsilon_i c_i) + v_2\frac{\partial}{\partial y}(\varepsilon_i c_i)] + \frac{1}{H_1 r^v}\frac{\partial}{\partial y}[H_1 r^v(\varepsilon_i J_{iy} - \mu\frac{c_i}{S_i}\frac{\partial \varepsilon_i}{\partial y})] = W_i^{(v)} \quad (33)$$

where x is the distance along the body surface, y is the distance along the normal measured from the body; ρ is the mass density; v_1 and v_2 are the tangential and normal components of the velocity; $H_1 = 1 + y/R$ and $r = r_w + y \cos\alpha$ are the metric coefficients, $R(x)$ is the radius of the surface curvature, $r (x,y)$ is the distance from given point to the axis of symmetry, $r_w (x)$ is the distance from the body surface to the axis of symmetry, α is the angle between the tangent to the surface and the axis of symmetry, $v = 0$ for a planar flow and $v = 1$ for an axisymmetric flow; J_{iy} is the normal component of the diffusion flux of species i; c_i is the mass fraction of this species; N is the total number of species; $\varepsilon_i (T_v)$ is the average vibrational energy of the i-th species. In order to calculate this quantity, the model of the harmonic oscillator cut-off at the dissociation limit is used:

$$\varepsilon_i = R_A [\frac{\Theta_{vi}}{\exp(\Theta_{vi} / T_v) - 1} - \frac{T_{Di}}{\exp(T_{Di} / T_v) - 1}]; \qquad S_i = \sum_{j=1}^{N} S_{ij} x_j , \qquad (34)$$

S_{ij} is binary Schmidt's number for the i-j-th colliding pair, x_j is the molar fraction of species j; $W^{(v)}$ in (33) is the vibrational source term, which is calculated as follows [2]:

$$W_i^{(v)} = c_i (\varepsilon_i(T) - \varepsilon_i(T_v)) / \tau_i - W_{Di}, \qquad (35)$$

$$\tau_i = \{P \sum_r x_r \exp[18.42 - 1.16 \times 10^{-3} m_{ir}^{1/2} \Theta_i^{4/3} (T^{-1/3} - 0.015 m_{ir}^{1/4})]\}^{-1} + \{\sigma_v \langle V_i \rangle n\}^{-1} \qquad (36)$$

$$m_{ir} = m_i m_r / (m_i + m_r), \qquad \sigma_v = 10^{-17} (50000 / T)^2, cm^2 \qquad (37)$$

$$W_{D,O_2} = \rho^2 m_{O_2} \{[K_{f1} E_1(T_m) c_{O_2} / m_{O_2} - \rho K_{b1} E_1(\infty)(c_o / m_o)^2] -$$
$$- [K_{f5} \varepsilon_{O_2}(T_v) c_{NO} c_O / m_{NO} m_O - K_{b5} \varepsilon_{O_2}(T_v) c_N c_{O_2} / m_N m_{O_2}]\} \qquad (38)$$

$$W_{D,N_2} = \rho^2 m_{N_2} \{[K_{f2} E_2(T_m) c_{N_2} / m_{N_2} - \rho K_{b2} E_2(\infty)(c_N / m_N)^2] +$$
$$+ [K_{f4} E_4(T_m) c_{N_2} c_O / m_{N_2} m_O - K_{b4} E_4(\infty) c_N c_{NO} / m_N m_{NO}]\} \qquad (39)$$

$$W_{D,NO} = \rho^2 m_{NO} \{[K_{f3} E_3(T_m) c_{NO} / m_{NO} - \rho K_{b3} E_3(\infty) c_N c_O / m_N m_O] -$$
$$- [K_{f4} \varepsilon_{NO}(T_v) c_O c_{N_2} / m_O m_{N_2} - K_{b4} \varepsilon_{NO}(T_v) c_N c_{NO} / m_N m_{NO}] +$$
$$+ [K_{f5} E_5(T_m) c_O c_{NO} / m_O m_{NO} - K_{b5} E_5(\infty) c_N c_{O_2} / m_N m_{O_2}]\} \qquad (40)$$

An advantage of Adiabatic Model model is that it "automatically" yields values of the energy lost / obtained in endothermic / exothermic reactions. By the definition, the average energy lost by the molecule in a single endothermic reaction can be calculated as follows:

$$E_{lost} = (K_f(T, T_v))^{-1} \sum_{I=0}^{I_{max}} k_{fI}(T, T_v) E_I , \text{ where } K_f(T, T_v) = \sum_{I=0}^{I_{max}} k_{fI}(T, T_v) \qquad (41)$$

Assuming that the rate constants of exothermic reactions are determined only by translational temperature, we have the following expression for the average energy obtained by the molecule in a single exothermic reaction:

$$E_{obt} = (K_f(T, T))^{-1} \sum_{I=0}^{I_{max}} k_{fI}(T, T) E_I \qquad (42)$$

The following approximations are recommended for E_{lost} in dissociation:

$$E_{lost} = \frac{k}{2\Phi}\left[T_D\varphi_1 Z_1(T_m)\varepsilon_1(T_m) + 0.15(T_D - T_D^{\cdot})\varphi_2 Z_2(T_m)\exp(-\frac{T_D^{\cdot} + \Theta_2}{T_m})\varepsilon_2(T_m)\right] \quad (43)$$

$$\varepsilon_1(T_m) = \frac{\Theta_1}{\exp(\Theta_1/T_m)-1} - \frac{T_D^{\cdot}}{\exp(T_D^{\cdot}/T_m)-1}, \quad (44)$$

$$\varepsilon_2(T_m) = \frac{\Theta_2}{\exp(\Theta_2/T_m)-1} - \frac{T_D - T_D^{\cdot}}{\exp((T_D - T_D^{\cdot})/T_m)-1} \quad (45)$$

and in exchange reactions in endothermic direction:

$$E_{lost} = k(A+B)/C \quad (46)$$

$$A = T^{-1.5}\exp(-T_D^{\cdot}/T)\varphi_3 Z_1(T_m)\varepsilon_1(T_m)(T_D^{\cdot} - \varepsilon_1(T_m))/2$$

$$B = T^{0.5}(1-\exp(-T_D/T))\exp(-T_D^{\cdot}/T_v)Z_2(T_v)(T_D^{\cdot}\exp(-\Theta_2/T_v) + \varepsilon_2(T_v))$$

$$C = T^{-1.5}\exp(-T_D^{\cdot}/T)\Phi^{\cdot} + T^{0.5}(1-\exp(-T_D/T))\exp(-(T_D^{\cdot}+\Theta_2)/T_v)Z_2(T_v)$$

In Eqs. (38)-(40) $E_i(T_m) = E_{i,lost}(T,T_v)$, $E_i(\infty) = E_{i,obt} = E_{i,lost}(T,T)$.

The equation for T_v is obtained by summing Eqs. (33) over molecular species and linearization with respect to T_v.

Submit the expression for the heat flux:

$$J_{qv} = -\lambda^{tr,rot}/C_p^{tr,rot}\left(\partial H/\partial y - v_1\partial v_1/\partial y - v_2\partial v_2/\partial y - \sum_{i=1}^{N}h_i\partial c_i/\partial y\right) + \sum_{i=1}^{N}h_i J_{iy} -$$
$$- \lambda^{tr,rot}/C_p^{tr,rot}\sum_{j=Molecules}\partial\varepsilon_j/\partial y\left(Pr^{tr,rot}/S_j - 1\right)c_j/m_j \quad (47)$$

Here H is the total enthalpy of the mixture; h_i is specific enthalpy of the i-th species; $\lambda^{tr,rot}$, $C_p^{tr,rot}$, $Pr^{tr,rot}$ are respectively heat conductivity, specific heat, and Prandtl number of the mixture due to translational modes of particles. As it seen from this expression, the term due to vibrational energy flux is taken explicitly. It is assumed that viscosity, thermal conductivity and binary diffusion coefficients do not depend on nonequilibrium excitation of vibrations.

Transport properties necessary for calculation of viscous flow (the viscosity and thermal conductivity of partially dissociated air mixture due to translational modes of particles) are determined using the approximate formulae suggested in [4]. The formulae yield these coefficients with quite a good accuracy in a wide range of conditions. The contribution of rotational modes (and vibrational ones in the case of thermal equilibrium) to the heat conductivity of molecular species is accounted for by the Eucken correction in the linear approximation [5].

The Rankine-Hugoniot shock-slip relationships are used on the shock wave [6]. On the body surface no-slip and no-flow conditions are used for the momentum equation, condition of radiation equilibrium of the wall is implemented for the energy conservation equation, conditions of zeroth catalytic activity of the surface are used for the mass conservation equations and extreme conditions of thermal equilibrium wall and of zeroth flux of vibrational energy to the wall are used for the equation of vibrational relaxation.

Method of global iterations [7] is implemented to integrate the governing equations. The used differential scheme has the second order of approximation in the streamwise direction and the fourth order across the shock layer.

5. Results and Discussion

The calculations of the flow over 1-m sphere with non-catalytic surface have been carried out at two freestream conditions: ρ_∞=4.30 10^{-5} kg/m^3, P_∞=2.54 Pa, T_∞=205.3K, (regime 1) and ρ_∞=1.63 10^{-4} kg/m^3, P_∞=10.9 Pa, T_∞=233.3K (regime 2). Mach number was M_∞=25 in both cases. The regimes correspond to the maximum heat loads on a reentry vehicle. First of all, the calculations with use of exact expressions (16-17,19-22,41,42) have been compared with those performed with use of approximations (23-25,43,46). The difference in the heat flux values turns out to be no more that 3%, and that in the shock stand-off distance no more than 1%. But CPU time necessary to solve MNVSL equations with the exact formulae is 4 times greater than in the case of using the approximations.

Than calculations have been carried out in order to appreciate the effect of thermal nonequilibrium and boundary condition used for the equation of vibrational relaxation on heat flux and shock stand-off distance. The results of these calculations are presented in Tables 1-4:

Table 1. Heat flux to non-catalytic wall, W/m^2 (regime 1)

	x=0	x=0.4510 m	x=1.047 m	x=1.571 m
T_v=T	1.823 10^5	1.543 10^5	6.422 10^4	1.404 10^4
$T_v|_w$=T_w	1.899 10^5	1.610 10^5	6.645 10^4	1.390 10^4
$J_{vib}|_w$=0	1.818 10^5	1.544 10^5	6.381 10^4	1.329 10^4

Table 2. Shock stand-off distance, m (regime 1)

	x=0	x=0.4510 m	x=1.047 m	x=1.571 m
T_v=T	6.392 10^{-2}	7.363 10^{-2}	1.469 10^{-1}	3.826 10^{-1}
$T_v|_w$=T_w	6.932 10^{-2}	7.922 10^{-2}	1.552 10^{-1}	3.982 10^{-1}
$J_{vib}|_w$=0	6.935 10^{-2}	7.928 10^{-2}	1.553 10^{-1}	3.985 10^{-1}

Table 3. Heat flux to non-catalytic wall, W/m^2 (regime 2)

	x=0	x=0.4510 m	x=1.047 m	x=1.571 m
T_v=T	5.288 10^5	4.192 10^5	1.325 10^5	2.112 10^4
$T_v|_w$=T_w	5.419 10^5	4.301 10^5	1.341 10^5	2.014 10^4
$J_{vib}|_w$=0	5.121 10^5	4.072 10^5	1.281 10^5	1.946 10^4

Table 4. Shock stand-off distance, m (regime 2)

	$x=0$	$x=0.4510$ m	$x=1.047$ m	$x=1.571$ m
$T_v=T$	$5.467 \ 10^{-2}$	$6.285 \ 10^{-2}$	$1.277 \ 10^{-1}$	$3.463 \ 10^{-1}$
$T_v|_w=T_w$	$5.681 \ 10^{-2}$	$6.489 \ 10^{-2}$	$1.307 \ 10^{-1}$	$3.526 \ 10^{-1}$
$J_{vib}|_w=0$	$5.682 \ 10^{-2}$	$6.493 \ 10^{-2}$	$1.308 \ 10^{-1}$	$3.529 \ 10^{-1}$

From the tables one can see that the difference in the heating rates is within 4.5% and that in the shock stand-off distance is within 8.5% at these conditions. Thus, the effect of thermal nonequilibrium on the flow characteristics is rather small under the conditions discussed. In the case of thermal nonequilibrium, the difference in the heat flux values between the calculations with the aforementioned boundary conditions for the equation of vibrational relaxation is within 6%.

The other comparison has been done to appreciate sensitivity of the MNVSL model with respect to the values of vibrational energy lost/obtained in endothermic/exothermic reactions. The exact values of E_{lost} and E_{obt} have been multiplied by 1.5 and divided by 1.5. The difference between the corresponding calculations is within 2% both for the heating rate and for the shock stand-off distance.

The last comparison has been done to appreciate the influence of possible uncertainty in the gas-kinetic cross-sections and interaction potentials on the flow characteristics. The values of α_0 and σ_0 have been multiplied by 1.25 and divided by 1.25. Corresponding dissociation rate constants differ from each other by the order of magnitude. Calculations show that this variation of reaction rate constants leads to variation of the hating rate within 32% and variation of the shock stand-off distance within 14% at freestream conditions discussed.

6. Conclusions

The Adiabatic Model is suggested to calculate rate constants of dissociation-recombination and exchange reactions under thermal nonequilibrium together with the values of the average energy lost/obtained in endothermic/exothermic reactions. In order to save CPU time, approximations of these quantities are recommended.

Adiabatic Model has three main peculiarities: 1) it is based on the experimentally measured quantities α_0, σ_0 and do not contain any empirical parameter; 2) the model uses adiabatic principle (11), therefore, it permits the molecule to dissociate from any vibrational level, including the ground state, so, even at very small T_v reaction rate constants remain finite; 3) the model allows centrifugal dissociation.

Adiabatic Model has been implemented in calculations of the flowfield over a blunt body within the scope of Viscous Shock Layer equation system. Five-species air has been considered. The calculations have shown that the effect of vibrational nonequilibrium on the flow characteristics is relatively small if accurate balancing of vibrational energy is done within CVDV approach. The effect of the surface activity with respect to accommodation of vibrational energy is also small. The flow in the viscous shock layer over a blunt body is weakly sensitive to variations of the energy lost and obtained in reactions. In contrast, the effect of possible uncertainties in gas-kinetic cross-sections and repulsive intermolecular potentials may be considerable.

Because of small influence of vibrational nonequilibrium on the flow characteristics, it is recommended to carry out engineering calculations under the assumption of thermal equilibrium, especially in 3D problems. Nevertheless, the Adiabatic Model holds its value in hypersonic calculations even under thermal equilibrium, since the model yields a physical extrapolation of experimental data to high temperatures (up to 50000 K) occurring behind strong shock waves. Due to a relative simplicity of the model, it may find its implementation in calculations of other high-enthalpy multi-species gas flows.

7. Acknowledgments

The work is supported by Russian Foundation for Fundamental Researches.
Grant No: 93-01-17646, 95-01-01261.

8. References

1. Smekhov, G.D. and Zhluktov, S.V. (1992) The Rate Constants of Adiabatic Dissociation of Diatomic Molecules, *Khimicheskaya Fizika* **11**, No. 5, 1171-1179.
2. Zhluktov, S.V., Smekhov, G.D., and Tirskiy, G.A. (1994) Rotation-Vibration-Dissociation Coupling in Multicomponent Nonequilibrium Viscous Shock Layer, *Izvestija Rossiyskoy Akademii Nauk. Mekhanika Zhidkosti i Gaza* **5**, 166-180.
3. Zhluktov, S.V. and Tirskiy, G.A. (1990) An Effect of Vibration-Dissociation Coupling on the Heat Transfer and Air Drag in Hypersonic Flows past Blunt Bodies, *Izvestija Akademii Nauk USSR. Mekhanika Zhidkosti i Gaza* **3**, 141-151.
4. Andriatis, A.V., Zhluktov, S.V., and Sokolova, I.A. (1992) Transport Coefficients of Nonequilibrium Air Mixture, *Matematicheskoe Modelirovanie* **4**, No. 1, 44-65.
5. Girshfelder, G.U. (1957) Heat Conductivity in Polyatomic and Electronically Excited Gases, *Journal of Chemical Physics* **26**, No. 2, 282-285.
6. Tirsky, G.A. (1993) Up-to-Date Gasdynamic Models of Hypersonic Aerodynamics and Heat Transfer with Real Gas Properties, *Annual Review Fluid Mechanics* **25**, 151-181.
7. Tirskiy, G.A., Utyuzhnikov, S.V., and Yamaleev, N.K. (1994) Efficient Numerical Method for Simulation of Supersonic Flow past Blunted Body at Small Angle of Attack, *Computers & Fluids* **23**, No. 1, 103-114.

INTERFACING NONEQUILIBRIUM MODELS WITH COMPUTATIONAL FLUID DYNAMICS METHODS

G.V. CANDLER, D. BOSE AND J. OLEJNICZAK
Department of Aerospace Engineering & Mechanics and Army HPC Research Center, University of Minnesota Minneapolis, MN 55455

1. Introduction

In this paper we discuss how thermo-chemical nonequilibrium models must be chosen to suit specific hypersonic flows. Current models can be very complicated, involving a large chemical kinetics model and many internal energy modes. Also, realistic hypersonic flows have a very large range of length scales, making them difficult to simulate with computational methods. Therefore, it is important to use thermo-chemical models that are appropriate; the models should have sufficient complexity to capture the relevant properties of the flow, with a minimum of computational cost. Thus, there is no single thermo-chemical model that is appropriate for all re-entry flows, and more complicated models may not be useful because they make the problem computationally intractable.

This article first discusses some examples of hypersonic flows to illustrate the complexity of the flow field and the interaction between the gas motion and the thermo-chemical state of the flow. Then, the conservation equations are outlined briefly and the vibrational energy conservation equation is discussed in detail to show how the vibration affects the chemical reactions, and *vice versa*. Then some recent work on the modeling of nitric oxide formation in air is discussed to show that in some cases very complicated reaction models may be required. Finally, a new approach for computing flows with trace species is presented.

2. Examples of Nonequilibrium Flows

In this section, we discuss several hypersonic flows to illustrate some of the processes that must be modeled when the flow field is chemically reacting.

M. Capitelli (ed.), Molecular Physics and Hypersonic Flows, 625–644.
© 1996 Kluwer Academic Publishers. Printed in the Netherlands.

2.1. APOLLO COMMAND MODULE AERODYNAMICS

Recently, Hassan *et al.*[10] used a five-species, thermo-chemical nonequilibrium air model to compute the flow over the Apollo Command Module geometry at one trajectory point and for one wind-tunnel flow condition. Table 1 shows that the computational vehicle trims at the flight and wind tunnel conditions, and the computed lift and drag coefficients are within 10% of the data for the reacting calculation, and within 5% for the perfect-gas model. The source of the differences is unknown, however insufficient grid resolution and small geometric differences between the flight vehicle and the computation are two possible reasons. These calculations show that the trim angle of attack is changed by the chemical reactions that occur in the flow field. The reactions increase the density rise across the bow shock, changing its shape and altering the surface pressure distribution and the trim angle of attack.

TABLE 1. Comparison of Apollo Command Module aerodynamic coefficients

	5-species Air	Flight Data	Perfect Gas	Tunnel Data
α (deg)	17.5	17.5	21.0	21.0
C_L	-0.354	-0.33	-0.387	-0.406
C_D	1.397	1.26	1.272	1.27
L/D	-0.253	-0.265	-0.304	-0.32
$C_{M_{cg}}$	-0.001	0.0	-0.012	0.0

This calculation pushed the limits of vector supercomputing even though the Apollo Command Module is a relatively simple vehicle. If a more complicated, and presumably, more accurate, thermo-chemical model had been used, the calculation would have been impossible.

2.2. REACTING N_2 FLOW OVER A CYLINDER

Many hypersonic flows are difficult to model because the chemical reactions in the flow field occur at rates which are similar to the rate of the gas motion. In this situation, a reaction may be initiated at one point in the flow field, but the gas moves to a new location before the reaction is complete. This chemical nonequilibrium is illustrated in Fig. 1, which is a plot of hypersonic flow of nitrogen over a cylinder. In the upper half of the

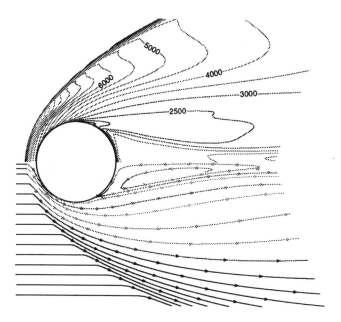

Figure 1. Reacting N$_2$ flow over a 2 inch cylinder. Top: temperature contours (K). Bottom: streamlines shaded to indicate degree of chemical reaction. (u_∞ = 5600 m/s, T_∞ = 1960 K)

figure, the temperature in the flow field is plotted. The peak temperature is 13,300 K, and the wake is much cooler, at about 3000 K. The lower half of the figure plots the streamlines which have been shaded to show the degree of chemical reaction, with black indicating no reaction, and light gray the most. The gas is highly reacted in the stagnation region; this gas flows around the cylinder, slowly recombining as the temperature falls. However, the recombination rate is slower than the fluid motion rate, resulting in a super-equilibrium concentration of reacted gas in a portion of the wake.

A gas is in thermal nonequilibrium if its internal energy cannot be characterized by a single temperature, and it is in chemical nonequilibrium if its chemical state does not satisfy local chemical equilibrium conditions. Portions of many external hypersonic flows are in thermal and chemical nonequilibrium because as the gas passes through the bow shock wave, much of its kinetic energy is converted to random translational motion. Then, collisions transfer translational energy to rotational, vibrational, electronic, and chemical energy. This energy transfer takes some number of collisions, during which time the gas moves to a new location where the temperature and density may be different. Thus, the internal energy modes and chemical composition of the gas lag the changes in the translational temperature. We can determine if a flow will be in thermal or chemical

628

nonequilibrium by constructing the Dahmköhler number, Da, which is the ratio of the fluid motion time scale to the internal energy relaxation or chemical reaction time scale.

Consider the steady-state mass conservation equation for species s

$$\frac{\partial}{\partial x_j}(\rho_s u_{sj}) = w_s, \tag{1}$$

where ρ_s is the mass density of species s, u_{sj} is the species s velocity in the x_j direction, and w_s is the rate of production of species s per unit volume due to chemical reactions. We can non-dimensionalize this equation using the total density, ρ, the speed, V, and the nose radius, r_n

$$\frac{\partial}{\partial \bar{x}_j}(\bar{\rho}_s \bar{u}_{sj}) = \frac{r_n w_s}{\rho V} = \frac{\tau_f}{\tau_c} = Da. \tag{2}$$

Thus, Da represents the ratio of the fluid motion time scale, $\tau_f = r_n/V$, to the chemical reaction time scale, τ_c; or it is the ratio of the chemical reaction rate to the fluid motion rate. A similar expression can be derived to describe the relative rate of internal energy relaxation.

When $Da \to \infty$, the internal energy relaxation or chemical reaction time scale approaches zero (becomes infinitely fast), and the gas is in equilibrium. That is, its thermal or chemical state adjusts instantaneously to changes in the flow. When $Da \to 0$, the reaction time scale approaches infinity, the gas is frozen and does not adjust to changes in the flow.

If the chemical source term, w_s, is proportional to the density squared (as it is for dissociation reactions), the binary scaling law can be derived from the above expression. Let us write $w_s = C\rho^2 k_f$, where k_f is a temperature-dependent reaction rate, and C is a constant. Then we have

$$Da = \rho r_n \frac{Ck_f}{V} \tag{3}$$

Thus, the reaction rate is proportional to the density-length scale product. k_f depends exponentially on temperature, which for hypersonic flows depends on the free-stream kinetic energy $\frac{1}{2}V_\infty^2$. Therefore, for a dissociation-dominated flow, the Dahmköhler number depends on the binary scaling parameter, ρr_n, and the free-stream kinetic energy.

The use of an appropriate Dahmköhler number to assess the relative rates of internal relaxation and chemical reaction processes can be very effective for determining the appropriate thermo-chemical model. In many cases, some finite-rate processes are relatively fast so that they may be treated as being in equilibrium; this reduces the complexity of the model.

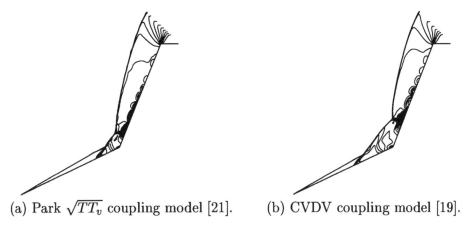

(a) Park $\sqrt{TT_v}$ coupling model [21].   (b) CVDV coupling model [19].

Figure 2.   Pressure contours for double-cone geometry at typical re-entry conditions: $\rho_\infty = 0.01$ kg/m^3, $T_\infty = 2260$ K, $u_\infty = 6690$ m/s, base height is 2 inches (5.08 cm) [20].

2.3. EFFECT OF VIBRATION-DISSOCIATION COUPLING

One of the primary uncertainties in modeling hypersonic flows is how the vibrational state of the gas affects its dissociation rate. There are many models for this vibration-dissociation coupling, but none have been adequately validated using experiments.

As an example of vibration-dissociation coupling effects, consider Fig. 2, which plots the pressure contours in the flow field of a double-cone geometry at typical re-entry flow conditions. The shock wave-boundary layer interaction produces a region of separated flow at the intersection of the cones. Also, the shock-shock interaction produces a dramatic pressure rise, followed by an oscillating pressure distribution on the surface of the second cone. The only difference between these two calculations is the vibration-dissociation coupling model. The Park coupling model [21] produces a flow field with a smaller separation zone than the CVDV model [19]. The difference in the separation zone causes a large change in the surface pressure distribution, as seen in Fig. 3; there is a similar change in the convective heat transfer rate. This example shows that there is still uncertainty in the modeling of the vibration-dissociation coupling process which may lead to important differences in the surface pressure and heat transfer rate.

From a computational standpoint, the flows discussed above are difficult and relatively expensive because there is a wide range of length scales present, and a large computational mesh must be used. (In grid refinement studies, a 1024 × 1024 mesh was used.) Had a more detailed state-to-state vibrational relaxation model been used, even these two-dimensional calculations would have been intractable. Therefore, for practical problems, we must rely on simple vibration-dissociation coupling models.

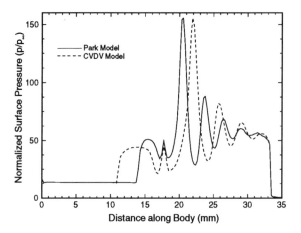

Figure 3. Surface pressure distribution on double-cone geometry shown in Fig. 2 [20].

2.4. NONEQUILIBRIUM RADIATION

The recent Bow Shock UltraViolet (BSUV) flight experiments [15] demonstrate that it is difficult to predict the radiation emitted from air in thermochemical nonequilibrium. These two experiments measured the radiation emitted from the shock layer of a 10 cm radius sphere flying at 3.5km/s and 5.1 km/s. Levin *et al.* discuss the experiment in detail; Fig. 3 is a summary of the results and plots the 230 ± 25 nm radiation emitted from the stagnation region. Under these flow conditions this emission is produced by the NO γ and β systems. Two radiation models are plotted in this figure; the 'baseline' curve was the state-of-the-art before the first experiment was flown, and the 'QSS-T' is a modified model. Clearly, the QSS-T model represents the data (solid symbols) much better, especially at high altitude. The primary modification that was required involved the assumptions about what is the characteristic temperature that governs NO excited state formation. The QSS-T model uses a quasi-steady-state approach with the translational temperature as the governing temperature.

The QSS-T model works reasonably well at low altitudes. But at high altitudes, the model does not compare very well with the data. It is about a factor of 20 in error at 88 km, and the variation of the radiance with altitude is wrong. (Recent work [3] indicates that this difference is primarily caused by errors in the vibration-dissociation coupling model used to calculate the oxygen dissociation rate.) This example shows that there is a lot of uncertainty in the modeling of nonequilibrium NO radiation, and that the models must be validated by comparison with experiments.

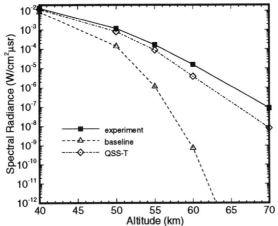

Figure 4. Comparison of radiation models with the first Bow Shock UltraViolet flight experiment (3.5 km/s). 230 nm, 0°-viewing photometer.

3. Conservation Equations

In this section, we discuss several elements of the conservation equations for a thermo-chemical nonequilibrium flow. The derivation of the entire set of conservation equations is not given here; see other parts of this volume and [14] for further details.

We briefly discuss the assumptions required to derive the continuum conservation equations. Then, we derive the vibrational energy equation because it illustrates the strong interaction between the vibrational state and the chemical reactions in a nonequilibrium gas. Then, we discuss a new model for nitric oxide formation in low-density, hypersonic flows. This work shows that in some cases very detailed thermo-chemical models may be required to adequately represent the physics of a nonequilibrium flow.

3.1. ASSUMPTIONS

We assume that the gas is described by the Navier-Stokes equations extended to account for the presence of chemical reactions and internal energy relaxation. For these equations to be valid, the flow must satisfy:

1. The gas must be a continuum. If we relate the mean-free-path, λ, to a local length scale that is determined by the normalized density gradient, we can form the gradient-length-local Knudsen number as [2]:

$$(\text{Kn})_{\text{GLL}} = \frac{\lambda}{\rho} \left| \frac{d\rho}{d\ell} \right|, \tag{4}$$

where ℓ is in the direction of the steepest density gradient. Boyd *et al.* [2] show that when $(\text{Kn})_{\text{GLL}} > 0.05$ the Navier-Stokes formulation fails. Failure typically occurs in shock waves and near the body surface in low-density flows. Also, the wake region of blunt bodies may have regions of continuum formulation failure. In these regions, either a higher-order continuum formulation must be used [17], [26] or a particle-based method such as the direct simulation Monte Carlo (DSMC) method is required.

2. The mass diffusion fluxes, shear stresses, and heat fluxes must be proportional to the first derivatives of the flow quantities. If not, a higher-order continuum approach, such as the Burnett equations [4], must be employed.

3. The internal energy modes must be separable. That is, we can describe each energy mode by its own temperature.

4. Finally, the flow is only weakly ionized. In this case, the Coulomb cross-section is small relative to the electron-neutral collision cross-section.

These limitations of the conservation equations are easily quantified and understood; it is much more difficult to assess the validity of the thermo-chemical model. The next section shows that the details of thermo-chemical models must be considered in the context of the entire flow model.

3.2. VIBRATIONAL ENERGY CONSERVATION

It is useful to consider the derivation of the vibrational energy conservation equation in some detail. This will show how the chemical reactions affect the relaxation of the vibrational energy, and *vice versa*. We assume that there is a single diatomic species in a multi-species gas mixture.

As in [5], we let $f_{s_\alpha}(x_i, v_{s_\alpha}, t)\, dx\, dv_{s_\alpha}$ represent the number of particles in vibrational level α of species s in a volume $dx \equiv dx_1\, dx_2\, dx_3$ and $dv_{s_\alpha} \equiv dv_{s_{\alpha_1}}\, dv_{2_{\alpha_2}}\, dv_{s_{\alpha_3}}$. The velocity distribution function f_{s_α} is defined such that

$$\int_{-\infty}^{+\infty} f_{s_\alpha}\, dv_{s_\alpha} = n_{s_\alpha}, \tag{5}$$

where n_{s_α} is the number density of molecules in level α of species s.

Under the assumption that the translational energy is a classical energy mode and that all velocities in the range $-\infty$ to $+\infty$ are allowed, we can show that f_{s_α} obeys the Boltzmann equation:

$$\frac{\partial f_{s_\alpha}}{\partial t} + v_{s_{\alpha_i}} \frac{\partial f_{s_\alpha}}{\partial x_i} + F_{s_{\alpha_i}} \frac{\partial f_{s_\alpha}}{\partial v_{s_{\alpha_i}}} = C_{s_\alpha}^+ - C_{s_\alpha}^-, \tag{6}$$

where $F_{s_{\alpha_i}}$ is the external force per unit mass acting in the i direction on the particles, and $C_{s_\alpha}^+$ and $C_{s_\alpha}^-$ represent the number of particles created and destroyed due to collisions of species s in level α per unit time and per unit phase space volume. These terms are the collision integrals.

The conservation of vibrational energy equation for species s is found by taking the moment of the Boltzmann equation with respect to ϵ_{s_α} and summing over all vibrational energy levels α. This yields

$$\sum_\alpha \int_{-\infty}^{\infty} \epsilon_{s_\alpha} \frac{\partial f_{s_\alpha}}{\partial t} \, dv_{s_\alpha} + \sum_\alpha \int_{-\infty}^{\infty} \epsilon_{s_\alpha} v_{s_{\alpha_i}} \frac{\partial f_{s_\alpha}}{\partial x_i} \, dv_{s_\alpha} \tag{7}$$

$$+ \sum_\alpha \int_{-\infty}^{\infty} \epsilon_{s_\alpha} F_{s_{\alpha_i}} \frac{\partial f_{s_\alpha}}{\partial v_{s_{\alpha_i}}} \, dv_{s_\alpha} = \sum_\alpha \int_{-\infty}^{\infty} \epsilon_{s_\alpha} (C_{s_\alpha}^+ - C_{s_\alpha}^-) \, dv_{s_\alpha}.$$

where ϵ_{s_α} is the vibrational energy per molecule of species s in level α. The convection term in (7) can be manipulated to yield

$$\sum_\alpha \int_{-\infty}^{\infty} \epsilon_{s_\alpha} v_{s_{\alpha_i}} \frac{\partial f_{s_\alpha}}{\partial x_i} \, dv_{s_\alpha} = \frac{\partial q_{vsi}}{\partial x_i} + \frac{\partial}{\partial x_i} \Big(E_{vs}(v_{si} + u_i) \Big), \tag{8}$$

where $E_{vs} = \sum_\alpha n_{s_\alpha} \epsilon_{s_\alpha}$ is the vibrational energy per unit volume, and $q_{vsi} = \sum_\alpha n_{s_\alpha} \epsilon_{s_\alpha} u_{s_{\alpha_i}}$ is the vibrational heat flux in the x_i direction. If the external forces $F_{s_{\alpha_i}}$ are independent of the velocity, then the force term in (7) is identically zero as it is assumed that the distribution function f_{s_α} vanishes at infinity. The source term on the right hand side of (7) represents the rate of change of the number of molecules in level α in species s. This can be written as

$$\sum_\alpha \int_{-\infty}^{\infty} \epsilon_{s_\alpha} (C_{s_\alpha}^+ - C_{s_\alpha}^-) \, dv_{s_\alpha} = \sum_\alpha \epsilon_{s_\alpha} \left(\frac{\partial n_{s_\alpha}}{\partial t} \right)_{coll}. \tag{9}$$

Combining the above expressions we have

$$\frac{\partial E_{vs}}{\partial t} + \frac{\partial}{\partial x_i} (E_{vs} u_i + E_{vs} v_{si} + q_{vsi}) = \sum_\alpha \epsilon_{s_\alpha} \left(\frac{\partial n_{s_\alpha}}{\partial t} \right)_{coll} = Q_{Vib}. \tag{10}$$

To find an expression for Q_{Vib} we must make the following assumptions:
1. The system of interest is a dilute mixture of vibrating and dissociating molecules and atoms weakly interacting with an infinite heat bath.

2. The internal energy modes are separable.

3. The interaction Hamiltonian which causes transition between vibrational levels can be treated as a perturbation on the energy of the vibrating molecules. Thus, quantum mechanical perturbation theory can be used to derive the master relaxation equation.

Under these assumptions Heims [11] showed

$$\frac{\partial n_{s_\alpha}}{\partial t} = \nu_{s_\alpha} (N_s - 2 n_s)^2 - \mu_{s_\alpha} n_{s_\alpha} + \sum_i (a_{s_{i\alpha}} n_{si} - a_{s_{\alpha i}} n_{s_\alpha}), \tag{11}$$

where N_s is the number of s atoms, n_s is the number molecules of species s, $a_{s_{lk}}$ is the transition probability per unit time from vibrational level l to k, ν_{s_α} is the recombination rate of atoms to molecules in level α, and μ_{s_α} is the dissociation rate from level α.

Evaluating the source term Q_{Vib} in (10) yields

$$Q_{Vib} = \sum_\alpha \epsilon_{s_\alpha} \left[\nu_{s_\alpha} (N_s - 2n_s)^2 - \mu_{s_\alpha} n_{s_\alpha} + \sum_i (a_{s_{i\alpha}} n_{s_i} - a_{s_{\alpha i}} n_{s_\alpha}) \right], \quad (12)$$

The first two terms of (12) are respectively the gain and loss of vibrational energy due to chemical reactions. The last two terms of (12) account for the exchange of vibrational and translational energy due to collisions. At the microscopic level, these two processes are not linked and can be treated independently. Thus we define

$$Q_{Chem} = \sum_\alpha \epsilon_{s_\alpha} \left(\nu_{s_\alpha} (N_s - 2n_s)^2 - \mu_{s_\alpha} n_{s_\alpha} \right), \quad (13)$$

$$Q_{V-T} = \sum_\alpha \epsilon_{s_\alpha} \sum_i (a_{s_{i\alpha}} n_{s_i} - a_{s_{\alpha i}} n_{s_\alpha}). \quad (14)$$

To find an expression for Q_{Chem} we sum (11) over all α levels and define

$$\gamma_s = \sum_\alpha \nu_{s_\alpha}, \qquad \mu_s = \sum_\alpha \frac{\mu_{s_\alpha} n_{s_\alpha}}{n_s}, \quad (15)$$

where γ_s is the recombination rate and μ_s is the dissociation rate

$$\left(\frac{\partial n_s}{\partial t} \right)_{coll} = \gamma_s (N_s - 2n_s)^2 - \mu_s n_s, \quad (16)$$

where $\gamma_s (N_s - 2n_s)^2$ and $n_s \mu_s$ represent the rate of change of the number of molecules of species s per unit time and unit volume due to the forward and backward chemical reactions, respectively. Let the source term due to chemical reactions of species s, w_s, be written as $w_s = w_{f_s} + w_{b_s}$ so that w_{f_s} and w_{b_s} are the rates of change of the mass of species s per unit time and unit volume due to the forward and backward chemical reactions, respectively. This implies

$$\gamma_s (N_s - 2n_s)^2 = \frac{w_{b_s}}{m_s}, \qquad n_s \mu_s = -\frac{w_{f_s}}{m_s}. \quad (17)$$

The average vibrational energy gained during recombination and the average vibrational energy lost during dissociation are the weighted averages of the level-specific dissociation rates multiplied by the vibrational energy of each level. They are respectively:

$$\overline{G} = \frac{\sum_\alpha \epsilon_{s_\alpha} \nu_{s_\alpha}}{\sum_\alpha \nu_{s_\alpha}} = \frac{\sum_\alpha \epsilon_{s_\alpha} \nu_{s_\alpha}}{\gamma_s} \tag{18}$$

$$\overline{E} = \frac{\sum_\alpha \mu_{s_\alpha} n_{s_\alpha} \epsilon_{s_\alpha}}{\sum_\alpha \mu_{s_\alpha} n_{s_\alpha}} = \frac{\sum_\alpha \mu_{s_\alpha} n_{s_\alpha} \epsilon_{s_\alpha}}{n_s \mu_s}.$$

In general, the n_{s_α} in the above equations are functions of T and T_v. However at equilibrium $T = T_v$, and we can define $n_{s_\alpha}^*$ as the number density at equilibrium. Thus we have

$$\nu_{s_\alpha} (N_s - 2 n_s^*)^2 - \mu_{s_\alpha} n_{s_\alpha}^* + \sum_i (a_{s_{i\alpha}} n_{s_i}^* - a_{s_{\alpha i}} n_{s_\alpha}^*) = 0. \tag{19}$$

Now we make the assumption that at equilibrium each process must be in equilibrium independent of the other process.

$$\nu_{s_\alpha} (N_s - 2 n_s^*)^2 - \mu_{s_\alpha} n_{s_\alpha}^* = 0. \tag{20}$$

If the transition probabilities are independent of time, (19) must hold for all time. If we substitute the expressions for ν_{s_α} from the above equation into (13), after some manipulation we find

$$\overline{E} = \frac{\sum_\alpha \mu_{s_\alpha} n_{s_\alpha}^* \epsilon_{s_\alpha}}{\sum_\alpha \mu_{s_\alpha} n_{s_\alpha}^*}. \tag{21}$$

If we write $\overline{E} = \overline{E}(T, T_v)$, we immediately see that $\overline{G} = \overline{E}(T, T)$.

Therefore, all physically consistent models for the vibrational energy source term due to chemical reactions must be of the form

$$Q_{Chem} = \frac{1}{m_s} \left(\overline{E}(T, T_v) \, w_{f_s} + \overline{E}(T, T) \, w_{b_s} \right). \tag{22}$$

An expression for Q_{V-T} was originally derived by Landau and Teller for simple harmonic oscillators not undergoing dissociation. They found

$$Q_{V-T} = \frac{E_{v_s}(T) - E_{v_s}(T_v)}{\tau_{vib}}, \tag{23}$$

where τ_{vib} is the vibrational relaxation time, and is given by a theoretically determined expression as a function of the local thermodynamic state of the gas. Under more general conditions, Q_{V-T} will have the above form, but τ_{vib} will be different and will also depend on the oscillator model used.

The equation for the conservation of vibrational energy of species s is

$$\frac{\partial E_{vs}}{\partial t} + \frac{\partial}{\partial x_i}(E_{vs} u_i + E_{vs} v_{si} + q_{vsi}) = Q_{V-T} + Q_{Chem}. \tag{24}$$

Many models for Q_{Chem} exist; see other portions of this volume for details of these models. This derivation shows that the thermo-chemical state of the gas must be included at detailed level in order to properly derive the conservation equations. When the collision integral in the Boltzmann equation is non-trivial, as in these flows, it becomes very difficult to write down appropriate conservation equations; these equations inevitably involve models. These models generally represent detailed state-to-state kinetics in a very simplified form. As such, they have limitations and a range of applicability which must be understood by the user.

3.3. KINETICS OF THE $N_2 + O \rightarrow NO + N$ REACTION

In this section, we discuss elements of a new model for the formation of nitric oxide under strong thermal nonequilibrium conditions. This work shows that in some cases it is necessary to use a very complicated model to represent the relevant physics. As illustrated in Fig. 4 above, the Bow-Shock UltraViolet (BSUV) experiments show the limitations of current models for NO radiation prediction.

The theoretical estimates using state-of-the-art flow simulation codes were found to be 200 times lower than the high-altitude experimental measurements. A part of the difference was due to the breakdown of the continuum flow equations at high altitudes. On using the DSMC technique, Boyd *et al.* [3] found that the theoretical prediction was still 60 times lower than the measurements. In the BSUV flight regime, NO is primarily formed by the Zeldovich reactions

$$N_2 + O \rightarrow NO + N, \qquad (25)$$
$$O_2 + N \rightarrow NO + O,$$

and the first Zeldovich reaction is dominant. This reaction is driven by the availability of O atoms, which are obtained from the O_2 dissociation process. Boyd *et al.* [3] showed that using the Macheret and Rich [18] vibration-dissociation coupling model instead of the Park model [21] significantly improved the ultraviolet radiation predictions. However, they found that the theoretical predictions were still a factor of 2 to 4 in error. The remaining error is probably due to an improper treatment of the first Zeldovich reaction under strong nonequilibrium conditions. Analysis of the BSUV spectrally-resolved radiation data has shown that the NO molecules have higher vibrational and rotational temperatures than predicted by the present thermo-chemical models.

In this section we present results from a study of the first Zeldovich reaction under thermal equilibrium and nonequilibrium conditions. Emphasis is given to investigating the effect of low T_v and T_r on the rate of the first Zeldovich reaction. In addition, we analyze the energy disposal among the products of this reaction. The results show that despite the NO being formed from an endothermic reaction, its vibrational and rotational temperatures are much higher than the average gas vibrational and rotational temperatures, in accordance with the experimental predictions. This can lead to a significant increase in electronically excited NO and radiative emission in the flow field.

A quasiclassical trajectory method (QCT) is chosen to study this reaction [13], [23], [24]. This method is attractive because it yields the reaction rates at any initial state of the reactants and it provides useful insight into the details of the reaction process. We use two previously-developed *ab initio* potential energy surfaces for this system [25]. A quasiclassical trajectory study has been done for the second Zeldovich reaction [7] and the first reaction in the backward direction by Gilibert *et al.* [8], [9].

3.3.1. *Classical Trajectory Calculations*
To study the dynamics of reaction (25), we solve the equations of motion for the NNO triatomic system. The appropriate form of the equations in the center of mass coordinate system is given elsewhere [12]. Six equations for positions and six for momenta are solved at a every time step over the potential energy surface using a fourth-order Runge-Kutta scheme with a constant time step. A trajectory code is specifically developed for execution in data-parallel mode on the Thinking Machines CM-5. As many as one million trajectories are run for each case.

3.3.2. *Reaction Attributes*
In order to calculate the reaction cross-section, one has to estimate a reasonable value of the maximum impact parameter, b_{max}, such that collisions with impact parameter, $b > b_{max}$, cannot possibly react. The reaction cross-section is then defined as the product of the area, πb_{max}^2, and an average probability of reaction over all possible collisions. For atom-diatom collisions with relative translational energy, E_t, and diatom rovibrational state (v, j), the reaction cross-section is given by

$$\sigma_r(E_t, v, j) = \int_{\xi=0}^{2\pi} \int_{\eta=0}^{2\pi} \int_{\phi=0}^{2\pi} \int_{\theta=0}^{\pi} \int_{b=0}^{b_{max}} \tag{26}$$

$$P_r(b, \theta, \phi, \eta, \xi, E_t, v, j) 2\pi b \, db \, \frac{\sin\theta}{2} \, d\theta \, \frac{d\phi}{2\pi} \, \frac{d\eta}{2\pi} \, \frac{d\xi}{2\pi},$$

where θ and ϕ define the orientation of the diatomic molecule, ξ represents the initial phase of vibration, and η describes the orientation of the angular momentum of the rotating diatom. The quantity P_r represents the probability of reaction under the given collision conditions. We use the *Monte*

Carlo method to evaluate this multi-dimensional integral. In this method the above integral is approximated by a summation

$$\sigma_r(E_t, v, j) = \frac{\sum_N f(\bar{\beta}_i)}{N} \tag{27}$$

where $f(\bar{\beta}_i)$ is a function of $\bar{\beta}_i$, the set of all collision parameters. This summation is evaluated using an appropriately large set of N trajectories. The probability function, P_r, is set equal to 1 if reaction occurs in a trajectory, otherwise it is set equal to 0. An *importance-sampling* function [6] is used to choose the impact parameter for each trajectory. A state-dependent thermal rate constant of reaction (25) can be obtained by averaging the reaction cross-section over the Maxwellian translational energy distribution function at a given temperature.

The temperature in a typical hypersonic flow may be as high as 30,000 K. In order to make reasonable predictions of the concentration of NO in the flow field, experimental or theoretical rates are required at very high temperatures. Most of the experimental rates available are below $T = 5000$ K, a very few are between 5000-8000 K, and none are above 10,000 K. In practice the computational fluid dynamics flow simulations are made using empirical rate constants. These rates are obtained by extrapolating the experimental data using an Arrhenius expression and the validity of the rates above 8000 K is unknown.

The net rate constant due to the reaction on the two potential energy surfaces is calculated using the following expression

$$k = \frac{1}{3}k(^3A'') + \frac{1}{3}k(^3A'). \tag{28}$$

In order to keep the statistical uncertainty below a reasonable limit, a large number of trajectories are sampled. The standard deviations for all of the evaluated points are below 4%, except for $T = 4000$ and 3000 K, where the standard deviations are around 10%.

Figure 5 shows a comparison of the computed rate constant data with the available experimental and empirical data. The QCT rate constants are calculated down to a temperature of 3000 K. Since there is a sharp drop in the rate constant at lower temperatures, it is extremely difficult to get a reasonable number of reactive trajectories at temperatures below 3000 K. The agreement of the QCT data with the available data is very good. Also, the QCT data points lie within the experimental zone of uncertainty throughout the temperature range of $3000 - 8000$ K. This demonstrates the validity of the analytical surface fits of $^3A''$ and $^3A'$ surfaces based on the CCI *ab initio* data. The QCT calculations disagree with the empirically obtained rates of Park *et al.* [22] at temperatures above 7000 K. The difference is up

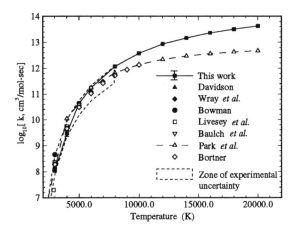

Figure 5. Comparison of the rate constant of reaction (25) obtained in this work with the available data.

to an order of magnitude at $20,000\,\mathrm{K}$. Thus, the new rate constants can significantly alter the prediction of the NO concentration in re-entry flow computations.

A modified Arrhenius expression is used to find an appropriate fit to the computed rate constant data over the entire temperature range. The expression, obtained using a least-squares algorithm in the log scale, is

$$k = (5.69 \pm 0.19) \times 10^{12}\, T^{0.42} \exp\left(-\frac{42938 \pm 147}{T}\right) \quad \mathrm{cm^3\,mol^{-1}\,s^{-1}} \quad (29)$$

The pre-exponential temperature factor is due to a small curvature of the rate constant curve when represented versus $1/T$. Also, the exponential power in the Arrhenius expression is larger than the ones obtained by other investigators. The coefficients in (29) are results of a least square fit of the QCT data and should be viewed merely as curve-fit parameters.

These calculations show that the ratio of the contributions of the $^3A'$ and $^3A''$ surfaces to the rate constant, $k(^3A')/k(^3A'')$, is $1/15$ at a temperature of $5000\,\mathrm{K}$ and increases asymptotically to $1/2$ above $12,000\,\mathrm{K}$. Thus, it is important to include both surfaces in order to adequately predict the dynamics of reaction (25) at high temperatures.

In most shock-heated experiments and shock dominated flows, there is often a substantial degree of thermodynamic nonequilibrium due to a finite-rate internal energy relaxation. Under such circumstances, the energy distribution functions and species concentrations can only be obtained by solving a complete master equation for all rovibrational states. However, these calculations are often impracticable due to the non-availability of

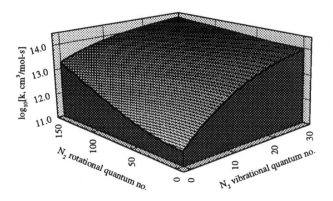

Figure 6. Surface describing the variation of the rate constant of reaction (25) with rovibrational levels of N_2 molecules. The translational energy is distributed according to a Boltzmann distribution at $T = 10,000$ K.

the vast amount of rate data required and a huge computational effort to do the calculations. The thermal rate constant under such conditions is evaluated by averaging the state-dependent rate constant, $k(v, j, E_t)$, over the relevant distribution functions. In order to get a qualitative insight into the rate constant dependence on the rovibrational state of N_2, Fig. 6 plots the rate constant at $10,000$ K versus the vibrational and rotational quantum levels. The vibrational energy is more efficient in increasing the rate constant than the rotational energy; this shows that the vibrational energy is more effective in overcoming the late barriers of the $^3A''$ and $^3A'$ surfaces. Note that from Fig. 6, we see that the reaction rate may be significantly reduced when the gas is not in thermal equilibrium.

We determine the effect of the N_2 vibrational excitation on the product NO vibrational energy distribution in a gas where the reactant translational and rotational levels are thermally populated at $10,000$ K. The reaction is

$$N_2(v) + O \rightarrow NO[P(v')] + N, \qquad (30)$$

where $P(v')$ is the normalized vibrational distribution of NO. In Fig. 7 we plot $\log_{10}[P(v')]$ versus the NO vibrational quantum level, v', for various values of the N_2 vibrational level, v. Initially, we set the N_2 vibrational energy to its ground state, $v = 0$. The product NO molecules show a nearly Boltzmann distribution. The slope of this distribution is primarily dependent on the translational temperature of the reactants and remains almost invariant with an increase in N_2 vibrational energy up to $N_2(v = 10)$.

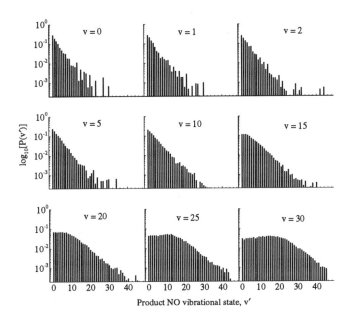

Figure 7. Product NO vibrational energy distribution at different reagent N_2 vibrational states.

However, as v is raised above 10, it produces a uniform NO vibrational distribution in low energy levels, while the high levels still show a linear behavior (see Fig. 10). Above $v = 10$, the vibrational energy of N_2 becomes greater than the reaction energy barrier. Consequently when N_2 has a high vibrational energy, those NO vibrational levels with energy less than the N_2 vibrational energy relative to the dissociated atoms are equally populated.

The results of this study will be used to develop a model for high-temperature air that explicitly allows the rate of reaction (25) to depend on the vibrational and rotational temperature of the reactants [1]. Also, product NO will have the appropriate amounts of energy placed in its internal energy modes. This example shows that when experimental or theoretical evidence dictates, it may be necessary to develop a very complicated model for a specific process.

4. Overlay Method for Trace Species Formation

In certain hypersonic external flows, some species may be present in very small concentrations. For example, it may be necessary to compute the radiative emission from the OH that is formed by the dissociation of the water vapor molecules that are present in the free-stream air. In this case, the trace species do not alter the overall flow field.

In this situation, we can analyze the flow more efficiently by first computing the flow using the major chemical species. Then, we "overlay" the formation of the trace species on this flow field. That is we hold fixed the density, velocity, temperatures, and mass fractions of the major species, and compute the formation of the trace species. We solve the species mass conservation equation,

$$\frac{\partial \rho_s}{\partial t} + \frac{\partial}{\partial x_j}(\rho_s u_j + \rho_s v_{sj}) = w_s, \tag{31}$$

for each of the trace species, where u_j comes from the major species solution. More details may be found in Levin *et al.* [16].

There are several advantages to the overlay approach. First, we must solve the underlying flow field only once. Then we can overlay very complicated chemical kinetics models for the trace species. The solution of the overlay equations is much simpler than that of the entire equation set because the convective term is linear since the velocity is held fixed by the previous solution. The implicit methods discussed above may be used, and they will converge very well.

Boyd *et al.* [3] use the overlay method in conjunction with a underlying DSMC solution to compute the formation of NO in a low density hypersonic flow. This approach is necessary because there are so few NO molecules produced in the flow field that they could not be resolved using the relatively small ($< 10^6$) computational particles.

This approach could be used to compute the nonequilibrium population of the excited electronic states in a high temperature flow field. This would amount to a direct calculation of the radiation process, where the population of all of the excited states is computed. This is opposed to the conventional approach where it is assumed that the excited state population and de-population occurs much more quickly than the fluid moves.

The overlay method shows promise for use in decoupling some complicated reaction or relaxation processes from the full flow field calculation. This would reduce the cost of doing the computation, without sacrificing accuracy of the thermo-chemical model. Of course, this method can only be used where the decoupling is valid.

5. Conclusions

This chapter shows that there is no unique formula for interfacing the thermo-chemical model with computational methods. Instead, it is necessary to understand the important features of a flow, and tailor the model so that these processes are captured accurately. It is important to realize that overly complex models are not beneficial because they dramatically increase the cost of the computations.

The examples discussed show that it is important to develop the thermo-chemical models in the context of the flow field conservation equations. Also, if experimental or theoretical evidence dictates, it may be necessary to develop a highly complicated model for a specific aspect of the thermo-chemistry. And finally, in some conditions the computational cost of the flow field simulations may be dramatically reduced by decoupling part of the nonequilibrium processes from the flow field solution.

Computational methods can improve our understanding of nonequilibrium flows. However, it is important not to treat the thermo-chemical modeling in a superficial fashion. Instead, it is necessary to understand the limitations of the model so that an appropriate model can be designed for the flow field of interest. Only in this way can accurate simulations of realistic flows be obtained.

Acknowledgment

The authors would like to thank Prof. Iain Boyd of Cornell University and Dr. Deborah Levin of the Institute for Defense Analyses for their contributions to the work discussed herein. We would also like to acknowledge support from the Air Force Office of Scientific Grant No. F49620-93-1-0338 and the Army Research Office Grant No. DAAH04-93-G-0089. Computer time was provided by the University of Minnesota Supercomputer Institute. This work was also sponsored by the Army High Performance Computing Research Center under the auspices of the Department of the Army, Army Research Laboratory cooperative agreement number DAAH04-95-2-0003 / contract number DAAH04-95-C-0008, the content of which does not necessarily reflect the position or the policy of the government, and no official endorsement should be inferred.

References

1. Bose, D. and Candler, G.V. (1996) Kinetics of the $N_2 + O \rightarrow NO + N$ Reaction in Nonequilibrium Flows, *AIAA Paper No. 96-0104*.
2. Boyd, I.D., Chen, G. and Candler, G.V. (1995) Predicting Failure of the Continuum Fluid Equations in Transitional Hypersonic Flows, *Physics of Fluids* 7 pp. 210–219.
3. Boyd, I.D., Candler, G.V. and Levin, D.A. (1995) Dissociation Modeling in Low Density Hypersonic Flows of Air, *Physics of Fluids* 7 pp. 1757–1763.
4. Burnett, D. (1936) The Distribution of Molecular Velocities and the Mean Motion in a Non-Uniform Gas, *Proc. London Mathematical Soc.* 40 pp. 382.
5. Clarke, J.F. and McChesney, M. (1975) *Dynamics of Relaxing Gases* Butterworths, London.
6. Faist, M.B., Muckerman, J.T. and Schubert, F.E. (1978) Importance Sampling and Histogrammic Representations of Reactivity Functions and Product Distributions in Monte Carlo Quasiclassical Trajectory Calculations, *J. Chem. Phys.* 69 pp. 4087–4096.
7. Gilibert, M., Aguilar, A., González, M. and Sayós, R. (1993) Quasiclassical Trajectory Study of the $N(^4S_u) + O_2(X^3\Sigma_g^-) \rightarrow NO(X^2\Pi) + O(^3P_g)$ Atmospheric Reaction on

the $^2A'$ Ground Potential Energy Surface Employing an Analytical Sorbie-Murrell Potential, *Chem. Phys.* **172** pp. 99–115.

8. Gilibert, M., Aguilar, A., González, M. and Sayós, R. (1993) Dynamics of the $N(^4S_u)+NO(X^2\Pi) \to N_2(X^1\Sigma_g^+)+O(^3P_g)$ Atmospheric Reaction on the $^3A''$ Ground Potential Energy Surface II. Analytical Potential Energy Surface and Preliminary Quasiclassical Trajectory Calculations, *J. Chem. Phys.* **99** pp. 1719–1733.

9. Gilibert, M., Aguilar, A., González, M., Mota, F. and Sayós, R. (1993) Dynamics of the $N(^4S_u) + NO(X^2\Pi) \to N_2(X^1\Sigma_g^+) + O(^3P_g)$ Atmospheric Reaction on the $^3A''$ Ground Potential Energy Surface I. Analytical Potential Energy Surface and Preliminary Quasiclassical Trajectory Calculations, *J. Chem. Phys.* **97** pp. 5542–5553.

10. Hassan, B., Candler, G.V. and Olynick, D.R. (1993) Thermo-Chemical Nonequilibrium Effects on the Aerothermodynamics of Aerobraking Vehicles, *J. Spacecraft and Rockets* **30** pp. 647–655.

11. Heims, S.P. (1963) Moment Equations for Vibrational Relaxation Coupled with Dissociation, *J. Chem. Phys.* **38** pp. 603.

12. Hirst, D.M. (1985) *Potential Energy Surfaces: Molecular Structure and Reaction Dynamics* Taylor and Francis, London.

13. Karplus, M., Porter, R.N. and Sharma, R.D. (1965) Exchange Reactions with Activation Energy, *J. Chem. Phys.* **43** pp. 3259–3287.

14. Lee, J.H. (1985) Basic Governing Equations for the Flight Regimes of Aeroassisted Orbital Transfer Vehicles, *Thermal Design of Aeroassisted Orbital Transfer Vehicles,* ed. H.F. Nelson, *Progress in Aeronautics and Astronautics* **96** pp. 3–53.

15. Levin, D.A., Candler, G.V. Collins, R.J., Erdman, P.W., Zipf, E.C. and Howlett, L.C. (1994) Examination of Theory for Bow Shock Ultraviolet Rocket Experiments–I, *Journal of Thermophysics and Heat Transfer* **8** pp. 447–452.

16. Levin, D.A., Candler, G.C. and Collins, R.J. (1995) An Overlay Method for Calculating Excited State Species Properties in Hypersonic Flows, *AIAA Paper No. 95-2073.*

17. Lumpkin, F. E., III and Chapman, D.R. (1992) Accuracy of the Burnett Equations for Hypersonic Real Gas Flows, *J. Thermophysics and Heat Transfer* **6** pp. 419–425.

18. Macheret, S.O. and Rich, J.W. (1993) Nonequilibrium Dissociation Rates Behind Strong Shock Waves: Classical Model, *J. Chem. Phys.* **174** pp. 25.

19. Marrone, P.V. and Treanor, C.E. (1963) Chemical Relaxation With Preferential Dissociation from Excited Vibrational Levels, *Physics of Fluids* **6** pp. 1215–1221.

20. Olejniczak, J., Candler, G.V., Hornung, H.G. and Wen, C. (1994) Experimental Evaluation of Vibration-Dissociation Coupling Models, *AIAA Paper No. 94-1983.*

21. Park, C. (1986) Assessment of Two-Temperature Kinetic Model for Dissociating and Weakly Ionizing Nitrogen, *AIAA Paper No. 86-1347.*

22. Park, C., Howe, J.T., Jaffe, R.L. and Candler, G.V. Review of Chemical-Kinetic Problems of Future NASA Missions, II: Mars Entries, *J. Thermophysics and Heat Transfer* **8** pp. 9–23.

23. Porter, R.N. and Raff, L.M. (1976) Classical Trajectory Methods in Molecular Collisions, *Dynamics of Molecular Collisions Part B*, ed. W.H. Miller, Plenum, New York, pp. 1–50.

24. Truhlar, D.G. and Muckerman, J.T. (1979) Reactive Scattering Cross Sections III: Quasiclassical and Semiclassical Methods, *Atom Molecule Collision Theory*, ed. R.B. Bernstein, Plenum, New York, pp. 505–561.

25. Walch, S.P. and Jaffe, R.L. (1986) Calculated Potential Surfaces for the Reactions: $O + N_2 \to NO + N$ and $N + O_2 \to NO + O$," *J. Chem. Phys* **86** pp. 6946–6956.

26. Zhong, X., MacCormack, R.W. and Chapman, D.R. (1993) Stabilization of the Burnett Equations and Applications to Hypersonic Flows, *AIAA J.* **31** pp. 1036.

COMPUTATION OF HIGH-TEMPERATURE NONEQUILIBRIUM FLOWS

H.-H. FRÜHAUF
Institut für Raumfahrtsysteme, Universität Stuttgart
Pfaffenwaldring 31, 70550 Stuttgart, Germany

1. Introduction

The development of a robust, efficient and accurate nonequilibrium flow solver for reentry flows in a wide altitude and velocity range is a challenging task. A detailed and consistent thermochemical modeling is needed in the gas-phase. Moreover, a detailed gas–wall interaction model has to be developed which models at least the most important physico-chemical processes at the wall in the continuum and in the slip flow regime. Finally a robust, efficient and accurate solver has to be set up in order to handle strong shocks and the extremely different time scales which may occur in nonequilibrium flows.

In the last years the URANUS (Upwind Relaxation Algorithm for Nonequilibrium Flows of the University Stuttgart) code has been developed. It includes an efficient and accurate fully coupled, fully implicit upwind solver, a new physically consistent multiple temperature model for the gas-phase and a new multiple temperature gas-wall slip condition model, containing an advanced finite rate chemistry model.

Thermochemical relaxation processes in the gas-phase are accounted for by the multiple temperature Coupled Vibration-Chemistry- Vibration (CVCV) model [15]. Its basic modeling concepts are adopted from Treanor and Marrone's CVDV-model [17]. The vibrational excitation is taken into account not only in dissociation but also in exchange and associative ionization reactions. The influence of vibrational excitation on chemistry and the influence of chemical reactions on vibration are modeled consistently.

In the newly developed gas–wall model [6, 7] thermal nonequilibrium effects at surfaces of finite catalycity are taken into account consistent with

M. Capitelli (ed.), Molecular Physics and Hypersonic Flows, 645–663.
© *1996 Kluwer Academic Publishers. Printed in the Netherlands.*

the gas-phase modeling [15]. Physically reasonable boundary conditions for vibrational and electron energy equations in the continuum regime are provided by the gas kinetic slip condition model for nonequilibrium flows, which is an extension of the model of Gupta and Scott [9]. An advanced finite rate chemistry model for the description of surface catalysis is implemented. The model is based on a detailed reaction scheme for $N_2, O_2, NO,$ N and O distinguishing adsorption-desorption processes, recombination reactions due to the Eley-Rideal and Langmuir-Hinshelwood mechanisms as well as dissociation and dissociative adsorption reactions.

In the paper, the numerical and thermochemical modelings will be briefly described. Details of the gas-phase and gas-wall interaction models are presented in the contributions of O. Knab [16] and A. Daiss [7] at this conference. The major part of the paper will summarize important results of an elaborate validation of the new models. This includes significant results of sensitivity studies which demonstrate the influence of the new modelings in comparison with existing ones.

2. Navier-Stokes Solver

The new two-dimensional/axisymmetric Navier-Stokes solver URANUS has been developed for the simulation of nonequilibrium flows around reentry vehicles in a wide altitude and velocity range [10, 5]. The unsteady, compressible full Navier-Stokes equations for the 11-species 5-temperature gas-phase model

$$\frac{\partial \vec{Q}}{\partial t} + \frac{\partial (\vec{E} - \vec{E}_v)}{\partial x} + \frac{\partial (\vec{F} - \vec{F}_v)}{\partial y} = \vec{S} \tag{1}$$

Conservation vector

$$\vec{Q} = [\rho_s, \rho u, \rho v, \rho e_t, \rho_k e_{v,k}, \rho_e e_e]^T \tag{2}$$

Convective fluxes

$$\vec{E} = \begin{bmatrix} \rho_s u \\ \rho u^2 + p \\ \rho uv \\ \rho u h_t \\ \rho_k u e_{v,k} \\ (\rho_e e_e + p_e) u \end{bmatrix} \qquad \vec{F} = \begin{bmatrix} \rho_s v \\ \rho uv \\ \rho v^2 + p \\ \rho v h_t \\ \rho_k v e_{v,k} \\ (\rho_e e_e + p_e) v \end{bmatrix} \tag{3}$$

Viscous fluxes

$$
\vec{E}_v =
\begin{bmatrix}
\rho D_s \frac{\partial \psi_s}{\partial x} \\
\tau_{xx} \\
\tau_{xy} \\
q_x + \sum_{i=1}^{11} \rho h_i D_i \frac{\partial \psi_i}{\partial x} + u\tau_{xx} + v\tau_{xy} \\
\lambda_{v,k} \frac{\partial T_{v,k}}{\partial x} + \rho h_{v,k} D_k \frac{\partial \psi_k}{\partial x} \\
\lambda_e \frac{\partial T_e}{\partial x} + \rho h_e D_e \frac{\partial \psi_e}{\partial x}
\end{bmatrix}
\tag{4}
$$

$$
\vec{F}_v =
\begin{bmatrix}
\rho D_s \frac{\partial \psi_s}{\partial y} \\
\tau_{yx} \\
\tau_{yy} \\
q_y + \sum_{i=1}^{11} \rho h_i D_i \frac{\partial \psi_i}{\partial y} + v\tau_{yy} + u\tau_{xy} \\
\lambda_{v,k} \frac{\partial T_{v,k}}{\partial y} + \rho h_{v,k} D_k \frac{\partial \psi_k}{\partial y} \\
\lambda_e \frac{\partial T_e}{\partial y} + \rho h_e D_e \frac{\partial \psi_e}{\partial y}
\end{bmatrix}
\tag{5}
$$

Source term vector

$$
\vec{S} =
\begin{bmatrix}
\dot{\omega}_s \\
0 \\
0 \\
0 \\
Q_{t-v,k} + Q_{v-v,k} + Q_{c-v,k} + Q_{e-v,k} \\
Q_{e-t} - \sum_{k=mol} Q_{e-v,k} + Q_{c-e}
\end{bmatrix}
\tag{6}
$$

are discretized in space using the cell-centered Finite-Volume approach. The inviscid fluxes normal to the faces of the quadrilateral cells (second term in eq. 7) are formulated in the physical coordinate system and calculated with Roe/Abgrall's approximate Riemann solver [23, 1]. Second order accuracy is achieved by a linear extrapolation of the variables from the cell center to the cell faces. TVD limiter functions applied on forward, backward and central differences in characteristic variables formulated for non-equidistant meshes are used to determine the corresponding slopes inside the cells thus characterizing the direction of information propagation and preventing oscillations at discontinuities. The viscous fluxes (third term in eq. 7) are discretized in the computational domain using classical central and one-sided difference formulas of second order accuracy.

Time integration is accomplished by the Euler backward scheme and the resulting implicit system of equations

$$
V_{ij} \frac{Q_{ij}^{n+1} - Q_{ij}^n}{\Delta t} + \left[\sum_{k=1}^{4} \vec{\Phi}_k \, \Delta s_k \right]_{ij}^{n+1} + \vec{\mathcal{F}}_{v,ij}^{n+1} - V_{ij} \vec{S}_{ij}^{n+1} = \vec{R}_{ij}^{n+1} = 0
\tag{7}
$$

is solved iteratively by Newton's method with underrelaxation

$$\sum_{l,m} \frac{\partial \vec{R}_{ij}^{\alpha}}{\partial \vec{Q}_{lm}^{\alpha}} \Delta \vec{Q}_{lm}^{\alpha} = -\vec{R}_{ij}^{\alpha} \quad \text{with} \quad \vec{Q}_{lm}^{\alpha+1} = \vec{Q}_{lm}^{\alpha} + \omega \Delta \vec{Q}_{lm}^{\alpha} \quad (8)$$

which theoretically provides the possibility of quadratic convergence for initial guesses close to the final solution. Since presently only steady state solutions are of interest, the time index n is used as the Newton iteration index α. The time step is computed locally in each cell from a given CFL number. To gain full advantage of the convergence behaviour of Newton's method, the exact Jacobians of the flux terms and the source term have to be determined. While the source term Jacobians are implemented exactly, the Jacobians of the flux terms are only computed approximately. The Jacobians of the Roe-fluxes are determined approximately following Barth [2] by assuming a locally constant Roe-matrix. The viscous Jacobians are obtained from the Thin-Layer approximation of the viscous fluxes. Furthermore in deriving the viscous Jacobians, the transport coefficients are considered to be independent of the conservation variables. The resulting block-pentadiagonal linear system of equations

$$A_{i,j}^{-1,0} \Delta \vec{Q}_{i-1,j}^{n} + A_{i,j}^{0,-1} \Delta \vec{Q}_{i,j-1}^{n} + A_{i,j}^{1,0} \Delta \vec{Q}_{i+1,j}^{n}$$
$$+ A_{i,j}^{0,1} \Delta \vec{Q}_{i,j+1}^{n} + A_{i,j}^{0,0} \Delta \vec{Q}_{i,j}^{n} = -\vec{R}_{ij}^{n} \quad (9)$$

where the coefficients $A_{i,j}^{\beta,\gamma}$ are defined by

$$A_{i,j}^{0,0} = \frac{\partial \vec{R}_{ij}^{n,*}}{\partial \vec{Q}_{ij}^{n}} + I \frac{V_{ij}}{\Delta t_{ij}} \quad \text{for} \quad (\beta,\gamma) = (0,0) \quad (10)$$

$$A_{i,j}^{\beta,\gamma} = \frac{\partial \vec{R}_{ij}^{n,*}}{\partial \vec{Q}_{i+\beta,j+\gamma}^{n}} \quad \text{for} \quad (\beta,\gamma) = (-1,0),(1,0),(0,-1) \text{ and } (0,1) \quad (11)$$

is iteratively solved by the Jacobi line relaxation method and an underlying subiteration scheme is used to minimize the inversion error. A simple preconditioning technique (multiplying each equation with the inverse matrix of the main diagonal) is implemented which improves the condition of the linear system and simplifies the LU-decomposition of the block-tridiagonal matrices to be solved in every line relaxation step. Thus the block-pentadiagonal system 9 can be solved efficiently with arbitrary accuracy.

All boundary conditions are formulated in a fully implicit manner to preserve the convergence behaviour of Newton's method. Different wall boundary conditions and relations have been implemented. No-slip condition and thermal equilibrium are usually prescribed at the wall. In order to predict the influence of wall catalytic effects the two limiting cases of a fully and a noncatalytic wall are implemented. A radiation equilibrium wall boundary condition can be prescribed as well. Finally, a slip condition model is implemented which allows for thermal nonequilibrium at the wall in the continuum regime and for slip effects in the slip flow regime. The coupled boundary equations are of the form

$$\overline{\overline{\vec{J}}} \cdot \vec{n}^T = \vec{S}^w. \tag{12}$$

The wall source term vector $\vec{S}^w$ is computed as difference of the fluxes to and from the wall, see section 5. The boundary equation for the species s is

$$\vec{j}_s \cdot \vec{n}^T = \dot{w}_s^w \tag{13}$$

with the net mass flux to the wall

$$\vec{j}_s = \rho D_s \nabla \Psi_s \tag{14}$$

and the wall source term

$$\dot{w}_s^w = \overline{Z}_s^- - \overline{Z}_s^+, \tag{15}$$

$\overline{Z}_s^-$ and $\overline{Z}_s^+$ being mass fluxes of particles to and from the wall, respectively.

3. Gas-Phase Multiple Temperature Model

Relaxation processes of internal degrees of freedom and chemical relaxation processes do occur simultanously. For high-temperature flows a thermochemical relaxation model is needed, which models the influence of excited internal degrees of freedom on chemical relaxation processes and vice versa consistently. The thermochemical relaxations are described by the source term vector (see eq. 6) in the Navier-Stokes equations (the components of the source term vector are coupled).

Thermochemical relaxation processes in the gas-phase are accounted for by the multiple temperature Coupled Vibration-Chemistry- Vibration (CVCV) model [15, 16]. Its basic modeling concepts are adopted from Treanor and Marrone's CVDV model [17]. The vibrational excitation is taken into account not only in dissociation but also in exchange and associative ionization reactions. The influence of vibrational excitation on chemistry and the influence of chemical reactions on vibration are modeled consistently in the source terms of the conservation equations. Molecules are approximated as truncated harmonic oscillators. Due to a cross section enlargement at higher excitation, chemical reactions predominately occur from the highly vibrationally excited states (preferential reaction). It is assumed that the vibrational modes of different molecular species relax through series of Boltzmann distributions characterized by separate vibrational temperatures $T_{v,k}$. The vibrational energy contribution for overcoming a chemical activation threshold is limited (i.e. a minimal energy contribution is required from the translational motion of the particles). Both assumptions of $T_{v,k}$-Boltzmann populated vibrational energy states and truncated harmonic oscillators are persistently maintained in modeling the energy exchange mechanisms between translational-vibrational, vibrational-vibrational and chemical-vibrational modes, respectively.

Vibrational energy equations are modelled for the neutral molecules of nitrogen, oxygen and nitric oxide.

The electron energy equation takes into account all important energy exchange mechanisms, see Ref. [10].

3.1. CONSISTENT VIBRATION–CHEMISTRY–VIBRATION COUPLING

The **vibration–chemistry coupling** is modeled through the nonequilibrium rate coefficient

$$k_f(T, T_{v,k}) = \Psi_f(T, T_{v,k}) \cdot k_f^{eq}(T) \tag{16}$$

with the equilibrium rate coefficient

$$k_f^{eq}(T) = C\, T^s e^{-\frac{A}{\Re T}} \tag{17}$$

and the nonequilibrium factor

$$\Psi_f = \frac{k_f(T, T_{v,k})}{k_f^{eq}(T)} = \frac{Q_{v,k}^{D_k}(T)}{Q_{v,k}^{D_k}(T_{v,k})} \tag{18}$$

$$* \quad \frac{e^{-\frac{\alpha A}{\Re T}} Q_{v,k}^{\alpha A}(\Gamma_k) + Q_{v,k}^{D_k}(T_k^0) - Q_{v,k}^{\alpha A}(T_k^0)}{e^{-\frac{\alpha A}{\Re T}} Q_{v,k}^{\alpha A}(-U_k) + Q_{v,k}^{D_k}(T_k^*) - Q_{v,k}^{\alpha A}(T_k^*)}$$

with the vibrational partition function of a truncated harmonic oscillator $Q_{v,k}^{\alpha A}$ and the Pseudo temperatures Γ_k, T_k^0, T_k^*

$$\frac{1}{\Gamma_k} = \frac{1}{T_{v,k}} - \frac{1}{T} - \frac{1}{U_k} \tag{19}$$

$$\frac{1}{T_k^0} = \frac{1}{T_{v,k}} - \frac{1}{U_k} \tag{20}$$

$$\frac{1}{T_k^*} = \frac{1}{T} - \frac{1}{U_k}. \tag{21}$$

α limits the maximum vibrational energy contribution for overcoming the activation threshold A to αA, $\alpha = 0.8 = const$.
U_k is a measure of the extend to which the upper vibrational levels are more reactive, $U_k = \frac{D_k}{5\Re} = const$.
The **chemistry–vibration coupling** is modeled through the energy exchange term

$$Q_{c-v,k} = \sum_{r=1}^{47} \{\dot{\omega}_{app,kr}\, G_{app,kr} - \dot{\omega}_{va,kr}\, G_{va,kr}\} \tag{22}$$

with $\dot{\omega}_{app,kr}$ and $\dot{\omega}_{va,kr}$ representing the molar production and decomposition rates of the molecules in the reaction r. $G_{va,kr}$ and $G_{app,kr}$ denote vanishing and appearing vibrational energies which are on average lost or gained by a molecule X_k beeing destroyed or created in a chemical reaction with

$$
\begin{aligned}
G_{va,k} =\ & \frac{e^{-\frac{\alpha A}{\Re T}}\, Q_{v,k}^{\alpha A}(\Gamma_k)\, L_k^{\alpha A}(\Gamma)}{e^{-\frac{\alpha A}{\Re T}} Q_{v,k}^{\alpha A}(\Gamma_k) + Q_{v,k}^{D_k}(T_k^0) - Q_{v,k}^{\alpha A}(T_k^0)} \\
& + \frac{Q_{v,k}^{D_k}(T_k^0)\, L_k^{D_k}(T_k^0) - Q_{v,k}^{\alpha A}(T_k^0)\, L_k^{\alpha A}(T_k^0)}{e^{-\frac{\alpha A}{\Re T}}\, Q_{v,k}^{\alpha A}(\Gamma_k) + Q_{v,k}^{D_k}(T_k^0) - Q_{v,k}^{\alpha A}(T_k^0)}
\end{aligned}
\tag{23}
$$

and the molar vibrational energy content of a harmonic oscillator truncated one quantum state below the energy Υ

$$L_k^{\Upsilon}(T) = \frac{\Re\Theta_{v,k}}{\exp\frac{\Theta_{v,k}}{T} - 1} - \frac{\Upsilon}{\exp\frac{\Upsilon}{\Re T} - 1}. \tag{24}$$

In the CVCV model $G_{va,k}(T,T_{v,k})$ is computed physically consistent with the nonequilibrium factor $\Psi_f(T,T_{v,k})$ according to the relation

$$-\frac{\partial \ln k_f(T,T_{v,k})}{\partial(1/T_{v,k})} = \frac{G_{va,k}}{\Re} - \frac{\varepsilon_{v,k}}{\Re}. \tag{25}$$

Hence the chemistry–vibration coupling is modeled consistently with the vibration–chemistry coupling. This is important for an accurate computation of thermochemical relaxations.

4. Transport coefficients

The transport coefficients are computed for the 11-species air mixture following Gupta and Yos however with several enhancements [8]. The collision integrals have been updated, primarily for ion–neutral atom interactions

652

and ion–ion interactions. An accurate mixing law is implemented in order to compute the transport coefficients of partly ionized air mixtures more accurately. Viscosity and translational heat conductivity are computed from the collision integrals and the mixing rule. The heat conductivities due to internal degrees of freedom are computed via specific heat capacities, neglecting the coupling between different internal degrees of freedom. The effective diffusion coefficient in the heat flux vector is approximately computed from binary diffusion coefficients.

5. Gaskinetic Gas–Wall Interaction Model

Up to now thermal equilibrium or adiabatic surface assumptions have served as boundary conditions for the vibrational and electron energy equations. These assumptions are based rather on the lack of information about surface processes than on physical considerations. Therefore, in the newly developed gas–wall model [6, 7] thermal nonequilibrium effects at surfaces of finite catalycity are taken into account consistently with the gas-phase modeling [15]. Physically reasonable boundary conditions for vibrational and electron energy equations in the continuum regime are provided by the gas kinetic slip condition model for nonequilibrium flows. The slip condition model, which is an extension of the model of Gupta and Scott [9] is based on the calculation of particles, momentum and energy fluxes of particles moving to or coming from the surface see fig. 1. The motion of

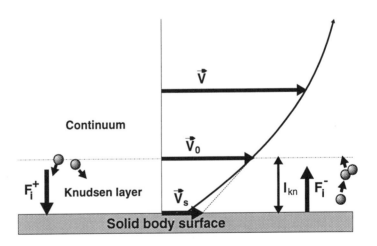

Figure 1. Gaskinetic gas–wall interaction model

particles colliding with the surface can be described by velocity distribution functions provided by the Enskog approximation. Mass-, momentum- and energy fluxes due to particles moving to a surface are

$$F_s^+ = - \int\limits_{(\vec{v}_s \cdot \vec{n}^T) < 0} \mathcal{M}_s (\vec{v}_s \cdot \vec{n}^T) f_s \, dv_{s,z} \, dv_{s,y} \, dv_{s,x} \qquad (26)$$

where $\mathcal{M}_s$ stands for
particle mass
particle momentum.
particle energy

For $\mathcal{M}_s = m_s \rightarrow$ mass flux of species s: $F_s^+ = \overline{Z}_s^+$.

For the total energy flux the contribution of internal degrees of freedom has to be added.

Perturbation Ansatz for the distribution function f_s (Enskog-approximation)

$$f_s = f_s^{[0]}(1 + \phi_s) \qquad (27)$$

$$f_s^{[0]} = n_s \left(\frac{m_s}{2\pi kT} \right)^{3/2} \exp\left[-\frac{m_s}{2kT}(c_{s,x}^2 + c_{s,y}^2 + c_{s,z}^2) \right] \qquad (28)$$

with $f_s^{[0]}$ local Maxwellian distribution of the species s

ϕ_s perturbation term

k Boltzmann constant

c_s peculiar velocity.

Fluxes from the wall result

i) from the surface reaction model with a detailed reaction scheme for N_2, O_2, NO, N and O and rate coefficient of species s dependent on surface coverage with adsorbed atoms, particle flux to the wall $\overline{Z}_s^+$ and reaction rates,

ii) from the momentum/energy exchange model for nonreacting particles containing accommodation coefficients.

Surface dissociation reactions which are enabled by thermal nonequilibrium and temperature slip effects are included [7]. The slip condition model can be applied in the continuum regime and in the slip flow regime. The results obtained in this paper were mostly computed with the conventional gas-wall interaction models. First results which were obtained with the slip condition model are presented as well. More details including results on finite catalysis are given in [7].

6. Code Validation

The validation concept is to test the physically advanced gas-phase and gas-wall interaction models with constant model parameters for a variety of well known test cases like Sharma's nozzle flow, Vetter's sphere flow, shuttle equivalent hyperboloid flow and RAM C-II and FIRE-II flows. The extensive validation is performed with the eleven-species five-temperature gas-phase model throughout using Park's rate constants and equilibrium constants of 1985. The parameters of the CVCV multiple temperature model are calibrated against a solution of the state selective model of Warnatz and Riedel and are kept constant throughout the validation process. The transport coefficients are computed according to Gupta and Yos with some enhancements. In recomputing the well known test cases the gas-phase model and the gas-wall interaction model are validated for nonequilibrium flows around geometrically simple blunt bodies in a wide density and velocity regime. Weakly up to strongly dissociated and ionized flows in the continuum regime are considered, as well as flows in the slip flow regime. Up to now there are unfortunately only a few detailed measurements available which will be used for a detailed validation of thermochemical relaxation processes. Otherwise the validation will be performed by comparing measured and predicted shock stand-off distances, wall heat fluxes and concentrations.

6.1. CONVERGENCE OF THE FULLY COUPLED FULLY IMPLICIT URANUS CODE FOR NONEQUILIBRIUM FOREBODY FLOWS IN THE LOW/MEDIUM REYNOLDS NUMBER REGIME

The physically consistent computation of the source term vector within the CVCV model and the exact computation of the source term Jacobian allow CFL numbers as great as 10^3. Thus 'Newton-type' convergence rates are obtained, once the shock location is steady. The computation of complex nonequilibrium forebody flows is started from uniform flow with the first order accurate scheme. The TVD limiter [10] is used to obtain second order accurate solutions. In the case of complex nonequilibrium flows the limiter parameter is increased continuously from the first to the second order accurate scheme. The favorable convergence properties of the URANUS solver are demonstrated in figs. 2 and 3 and in table 1.

6.2. VALIDATION OF THE BOLTZMANN DISTRIBUTION

The state-selective model of Warnatz and Riedel [22] has been used to calibrate important constants of the CVCV multiple temperature model and to check the applicability of the assumption of Boltzmann populated vibrational energy levels [12]. An isochore adiabatic relaxation process of air is considered. This is a transient relaxation process in a homogeneous system without any spatial gradients.

The initial conditions are: Pressure of 0.0256 *bar*, translational temperature

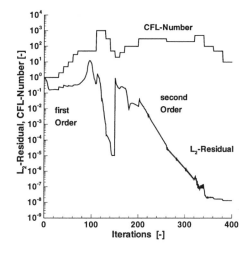

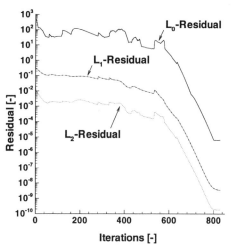

Figure 2. Vetter's sphere flow $D = 0.1\ m$, $\rho_\infty = 1.6 \cdot 10^{-3}\ \frac{kg}{m^3}$, $M_\infty = 12.7$, $Re_{D,\infty} = 4.3 \cdot 10^4$, Fully catalytic wall, $T_w = 300\ K$, $71 \times 35 = 2485$ mesh

Figure 3. Double ellipse flow: Antibes-Workshop test case VI.3, $M_\infty = 25$, $Re_\infty = 2.2 \cdot 10^4$, $66 \times 74 = 12284$ mesh

of $22000\ K$ and vibrational temperatures of $300\ K$. These correspond to the post-shock conditions of a reentry flight at $71\ km$ height.

As can be seen from fig. 4, good agreement is obtained for the relaxation of the heavy particle temperatures T. Small deviations are due to the use of different equilibrium constants. Moreover, the vibrational temperatures of the molecular species N_2 and O_2 calculated with the CVCV model illustrate the strong thermal nonequilibrium character of the process. Now, inserting these vibrational temperatures into the Boltzmann distribution function one gets the population of selected vibrational levels. Fig. 5 presents the mole fraction profiles of different vibrational energy levels of N_2. This plot as well as the O_2-population profiles (not included here) justify the assumption of Boltzmann like populated vibrational energy states for the post-shock relaxation. Small deviation from a Boltzmann distribution have only a small influence on the relaxation process. The inelastic cross section in Park's high temperature correction formula [21] for vibrational relaxation times influence the relaxation significantly. They were calibrated against the state-selective solution as well.

6.3. NITROGEN NOZZLE FLOW

The URANUS code is now applied to a recombining expansion flow in order to demonstrate that the CVCV multiple temperature model is capable to predict this type of flow as well [13]. A pure nitrogen flow is computed

	M_∞	Re_∞	Number of mesh points	max. CFL number	$\|L_2\|$, Its/O	Remarks
Double ellipse[*]	8.15	$1.67 \cdot 10^7$	29000	100	75	High Reynolds number, fine mesh
Vetter's Sphere	12.7	$4.3 \cdot 10^4$	1525	> 1000	20	CVCV model allows for CFL > 1000
Double ellipse	25	$2.2 \cdot 10^4$	12284	70	30	Comparable convergence rates of perfect gas and non-equilibrium flow computations
Space Shuttle equivalent hyperboloid	~ 25	$1.7 \cdot 10^3$	4968	70	50	Complex gaskinetic boundary equation does not lead to significant drop in the convergence rate
FIRE-II	37	$1.7 \cdot 10^5$	2485	300	35	Convergence rate does not deteriorate for significantly ionized flows

TABLE 1. Convergence properties of the URANUS code for forebody nonequilibrium flows in a wide velocity and density regime ([*] perfect gas flow)

through a plane nozzle for which vibrational temperature measurements [24] are available. The nozzle length is $0.125\,m$, the throat height is $0.64\,cm$ and the expansion ratio 12.5. Reservoir conditions of $5600\,K$ and $100\,atm$ which imply a gas composition in mole fractions of $\psi_{N_2} = 0.9864$ and $\psi_N = 0.0136$ were considered. A 74×58 grid with $9.5 \cdot 10^{-8}\,m$ cell width at the wall was used to compute the half nozzle flow. Fig. 6 illustrates that a strong mesh refinement is necessary in order to resolve the very thin boundary layer in the vicinity of the throat. As can be seen from fig. 7 the vibrational temperature predictions along the axis are in good agreement with measurements of Sharma et al. [24]. In this only weakly recombining flow, the vibrational-translational energy exchange mechanism is the dominant energy exchange mechanism. Thus these results confirm the applicability of the Landau-Teller formula in predicting fast expanding flows, a subject, which was recently under discussion.

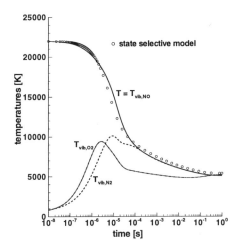

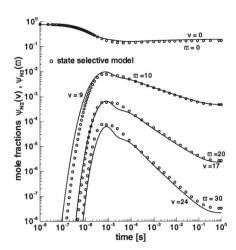

Figure 4. Heavy particle temperatures T and vibrational temperatures T_{vib,N_2}, T_{vib,O_2} profiles for an transient isochore, adiabatic relaxation of air with the initial conditions $T_0 = 22000 \ K$, $p_0 = 0.0256 \ bar$ and $T_{vib,i_0} = 300 \ K$.

Figure 5. Population of selected N_2 vibrational energy levels calculated by the state-selective model proposed by Warnatz and Riedel [13] and by the multiple temperature CVCV model.

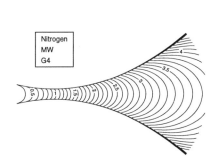

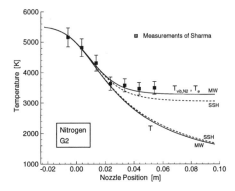

Figure 6. Sharma nozzle: Mach number contours

Figure 7. Temperature distribution along the Sharma nozzle centerline

6.4. RAMC-II FLIGHT EXPERIMENT

The RAMC-II free-flight experiment [11] is recomputed in order to investigate whether the measured electron density distributions can be predicted with the CVCV multiple temperature model [18]. RAMC-II is a sphere-cone configuration with a nose radius of $0.1524 \ m$ and a half-cone angle of $9°$. The total length is $1.3 \ m$. The flowfield was computed for three trajectory points $H = 81, 71$ and $61 \ km$. The upstream velocity of $7.65 \ km/s$

is constant for these trajectory points. A fully catalytic wall with a constant wall temperature $T_w = 1000\ K$ was assumed in the computations. The no-slip condition is assumed at the wall. The post-shock temperatures range from 25000 K to 20500 K for this weakly ionized flow with maximal electron mole fractions ranging from $3 \cdot 10^{-4}$ to $2 \cdot 10^{-3}$. Park's 1985 reaction rate coefficients [20] which are used throughout in the CVCV model lead to a good agreement with the electron density measurement. This is not true for Park's 1991 reaction rate coefficients which lead to an underprediction of the electron density by a factor about 5 to 7. Even though thermal nonequilibrium is significant in all flow cases different multi-temperature models like the CVCV model or Park's two-temperature model, do affect the electron density distribution only very weakly.

The computed electron density distributions along the body and normal to it close to the base are shown in figs. 8 and 9 together with the measured values. The agreement is reasonable. The deviation between the predicted and measured electron density profiles close to the base at higher altitudes will be reinvestigated.

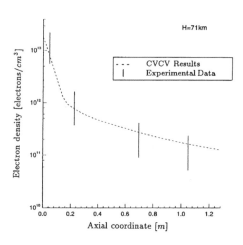 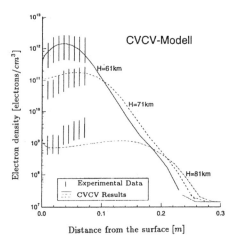

Figure 8. RAMC-II flight experiment: Wall electron density distribution over the body axis.

Figure 9. RAMC-II flight experiment: Electron density distribution normal to the body close to the base.

6.5. FIRE-II FLIGHT EXPERIMENT

The FIRE-II flight experiment [4] is recomputed [14], in order to validate the CVCV model in the case of a significantly ionized high-temperature flow at an altitude of 67.05 km. The free-stream conditions are given by $v_\infty = 11.25\ km/s$, $\varrho_\infty = 1.47 \cdot 10^{-4}\ kg/m^3$ and $T_\infty = 228\ K$. The URANUS calculations assume a fully catalytic wall boundary condition. Moreover, the wall temperatures are prescribed to be in equilibrium at 1030 K. The FIRE-II forebody is approximated by a sphere with an effective radius of 0.747 m. The grid employed for the axisymmetric computations consists of 35 cells along the body and 71 cells normal to the surface. In order to resolve the steep temperature gradients expected near the wall the mesh is strongly refined with a $1 \cdot 10^{-6}$ m spacing at the wall. Furthermore, a refinement has also been performed in the shock region. Good convergence properties were obtained with the URANUS code for this significantly ionized flow as well. The L_2-residual dropped six orders of magnitude within 300 iterations. The stagnation line temperature distributions are given in fig. 10. A nonequilibrium/equilibrium post shock flowfield can be detected from

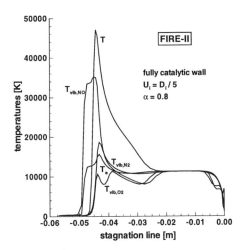

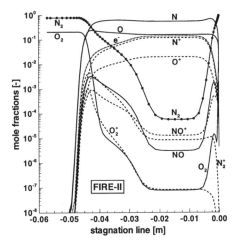

Figure 10. FIRE-II flight experiment: Stagnation line temperature distributions

Figure 11. FIRE-II flight experiment: Stagnation line mole fractions

this figure. The maximum translational post shock temperature is about 50000 K and the maximum electron mole fraction is about 0.14. Detailed sensitivity studies [14] revealed that the consistent vibration-dissociation-vibration coupling and the consideration of nonequilibrium rate constants for exchange and associative ionization reactions are important for an accurate prediction of the shock stand-off distance and the vibrational exitation and deexitation of the molecules in the shock vicinity. In the boundary layer next to the cold and fully catalytic wall the translational temperature decreases very steeply. Thus recombination processes occur in this layer. The consistent vibration-dissociation-vibration coupling leads to a supression of the vibrational temperatures of molecular oxygen and nitric oxide at the boundary layer edge. Nonequilibrium rate constants for exchange and associative ionization reactions significantly increase the vibrational excitation of the molecules in the immediate vicinity of the cold and fully catalytic wall, leading to a 14 % higher stagnation point convective heat flux as compared with standard multiple temperature models which do not account for the above effects.

6.6. SHUTTLE FLIGHT EXPERIMENT

In order to validate the multiple temperature gas-wall slip condition model [6] computations have been performed for flows around Shuttle-equivalent hyperboloids for different flow conditions ranging from a near continuum flow with a free-stream Knudsen number of about 0.03 to transitional flows with Knudsen numbers of about 0.54. The free-stream Knudsen number Kn_∞ is calculated as the ratio of the mean free path λ_∞ to the hyperboloid nose radius R_N which varies slightly from trajectory point to trajectory point as the angle of attack of the Shuttle changes during reentry. At an altitude of 92.35 km the nose radius is 1.295 m. In fig. 12 the heat

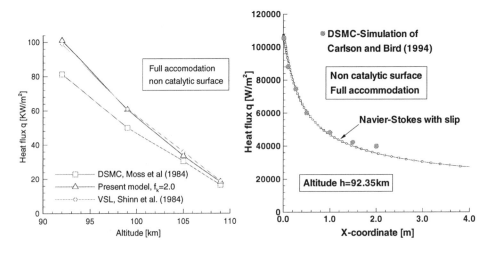

Figure 12. Shuttle equivalent hyperboloid flow: heat fluxes

fluxes calculated with the present model are compared to Viscous-Shock-Layer (VSL) calculations with slip condition performed by Shinn et al. in 1984 [25] and DSMC results obtained by Moss and Bird [19] (1984) for the considered trajectory points. All methods assume full accommodation of the scattered particles and a noncatalytic surface behavior.

The URANUS-slip and VSL-slip (1984) stagnation point heat fluxes are in good agreement in the altitude regime $92.35 \leq H \leq 109.35\ km$ while the DSMC method of Moss and Bird (1984) predicts a lower stagnation point heat flux.

A good agreement is obtained however between the heat fluxes, computed by the URANUS-slip and the Carlson-Bird [3] (1994) methods for an altitude of 92.35 km. In contrast to the old method the new method of Bird accounts for the vibration–dissociation coupling. Therefore at this altitude, where thermochemical relaxations are still important, DSMC and NS-slip solutions do agree, if an equivalent thermochemical modeling is used in the gas-phase.

The measured stagnation point heat flux is smaller than the heat flux computed under the assumption of a non catalylic wall with full accommodation. The gaskinetic gas–wall interaction model with the new surface reaction scheme [7] is expected to lead to a better agreement between predicted and measured heat fluxes.

7. Conclusions

– The efficient fully coupled, fully implicit Navier-Stokes code URANUS has been recently developed for nonequilibrium flows.
– Due to the detailed eleven-species, five-temperature formulation in the gas-phase model and in the gas-wall slip condition model the URANUS code is applicable to predict flowfields around reentry vehicles in a wide altitude and velocity range.
– The newly developed coupled vibration–chemistry–vibration (CVCV) multiple temperature model predicts thermochemical relaxations, shock stand-off distances and wall heat fluxes more accurately.
– The extension of thermal nonequilibrium in the CVCV model to exchange and associative ionization reactions tends to increase the vibrational excitation of molecules in the immediate wall vicinity and also tends to increase the wall heat flux.
– The physically consistent formulation of the source term vector in the CVCV model admits CFL numbers greater than 1000 and thus ensures a Newton-type convergence rate in predicting a wide class of nonequilibrium forebody flows. These include the significantly ionized FIRE II flow which is close to thermochemical equilibrium.
– The new slip model is able to predict chemical and thermal nonequilibrium effects and provides a basis for a more accurate modeling of catalytic surface reactions and quenching effects at solid surfaces.
– The wall heat flux in the slip flow regime is computed in accordance with Carlson-Bird's new DSMC solution in the case of a non catalytic wall with full accommodation.
– The spatial resolution requirements at the wall are significantly reduced by the new slip model and residuals drop sufficient orders of magnitude, thus ensuring an accurate heat flux computation.

662

8. Acknowledgements

This work is supported by the Deutsche Forschungsgemeinschaft.

References

1. Abgrall, R. (1991) Hypersonic Calculations by Riemann Solver Techniques, *Computer Physics Communications* **65**, pp. 1–7
2. Barth, T.J. (1987) Analysis of Implicit Local Linearization Techniques for Upwind and TVD Algorithms, *AIAA-Paper 87-0595, AIAA 25th Aerospace Sciences Meeting, Reno, Nevada*
3. Carlson, A.B. and Bird, G.A. (1994) Implementation of a Vibrationally Linked Chemical Reaction Model for DSMC, *NASA TM 109109*
4. Cauchon, D.L. (1967) Radiative Heating Results from the FIRE-II Flight Experiment at a Reentry Velocity of 11.4 Kilometer per Second, *NASA TM X-1402*
5. Daiß, A., Schöll, E., Frühauf, H.-H. and Knab, O. (1993) Validation of the URANUS Navier-Stokes Code for High-Temperature Nonequilibrium Flows, *AIAA-Paper 93-5070, München*
6. Daiß, A., Frühauf, H.-H. and Messerschmid, E. W. (1994) A New Gas/Wall Interaction Model for Air Flows in Chemical and Thermal Nonequilibrium, *Proccedings of the Second European Symposium on Aerothermodynamics for Space Vehicles and Fourth European High-Velocity Database Workshop, ESTEC, The Netherlands*
7. Daiß, A., Frühauf, H.-H. and Messerschmid, E. W. (1995) Chemical Reactions and Thermal Nonequilibrium on Silica Surfaces, *NATO ASI, Molecular Physics and Hypersonic Flows*
8. Fertig, M. (1994) Analyse von Transportkoeffizientenmodellen für Luft in einem Temperaturbereich von 250 - 30000K, *Internal Report IRS 94-S-01, Institut für Raumfahrtsysteme, Universität Stuttgart*
9. Gupta, R.N., Scott, C.D. and Moss, J.N. (1985) Surface-Slip Equations for Multicomponent Nonequilibrium Air Flow, *NASA Technical Memorandum 85820*
10. Jonas, S. (1993) Implizites Godunov-Typ-Verfahren zur voll gekoppelten Berechnung reibungsfreier Hyperschallströmungen im thermo-chemischen Nichtgleichgewicht, *Dissertation, Fakultät für Luft- und Raumfahrttechnik, Universität Stuttgart*
11. Jones, S. and Cross, A.E. (1972) Electrostatic Probe Measurements of Plasma Parameters for Two Reentry Flight Experiments at 2500 Feet per Second, *NASA TN D-6617*
12. Knab, O., Frühauf, H.-H. and Messerschmid, E. W. (1993) Validation of a Physically Consistent Coupled Vibration-Chemistry-Vibration Model for Ionized Air in Thermo-Chemical Nonequilibrium, *AIAA-Paper 93-2866*
13. Knab, O., Frühauf, H.-H. and Messerschmid, E. W. (1994) URANUS/CVCV Code Validation by Means of Thermochemical Nonequilibrium Nozzle Airflow Calculations *Proceedings of the Second European Symposium on Aerothermodynamics for Space Vehicles and Fourth European High-Velocity Database Workshop, ESTEC, The Netherlands*
14. Knab, O., Gogel, T., Frühauf, H.-H. and Messerschmid, E. W. (1995) CVCV Model Validation by Means of Radiative Heating Calculations, *AIAA-Paper 95-0623, AIAA 33rd Aerospace Sciences Meeting and Exhibit, Reno, Nevada*
15. Knab, O., Frühauf, H.-H. and Messerschmid, E. W. (1995-2) Theory and Validation of the Physically Consistent Coupled Vibration-Chemistry-Vibration Model *Journal of Thermophysics and Heat Transfer* **Vol. 9, No. 2**, pp. 219-226.
16. Knab, O. (1995-3) A Physically Consistent Vibration–Chemistry–Vibration Model, *NATO ASI, Molecular Physics and Hypersonic Flows, Maratea May 21 - June 3*
17. Marrone, P.V. and Treanor, C.E. (1963) Chemical Relaxation with Preferential Dis-

sociation from Excited Vibrational Levels, *Physics of Fluids* **Vol. 6, No. 9**, pp. 1215-1221

18.   Mewes B. (1994) Implementierung des Luftreaktionsmodells nach Park in das URANUS-Nichtgleichgewichtsrechenverfahren, *Institut für Raumfahrtsysteme, Internal Report IRS 94-S-12*

19.   Moss, J.N. and Bird, G.A. (1984) Direct Simulation of Transitional Flow for Hypersonic Reentry Conditions, *AIAA-84-0223, 22nd Aerospace Sciences Meeting, Reno, Nevada*

20.   Park, C. (1986) Convergence of Computation of Chemical Reacting Flows, *Thermophysical Aspects of Reentry Flows, Progress in Astronautics and Aeronautics* **Vol. 103**, pp. 478-513

21.   Park, C. (1987) Assessment of Two-Temperature Kinetic Model for Ionizing Air, *AIAA-Paper 87-1574*

22.   Riedel, U. (1992) Numerische Simulation reaktiver Hyperschallströmungen mit detaillierten Reaktionsmechanismen, *Ph.D. Thesis, University of Heidelberg*

23.   Roe, P.L. (1981) Approximate Riemann Solvers, Parameter Vectors and Difference Schemes, *J. Comp. Phys.* **43**, pp. 357-372

24.   Sharma, S.P., Ruffin, S.M., Gillespie, W.D. and Meyer, S.A. (1992) Nonequilibrium Vibrational Population Measurements in an Expanding Flow Using Spontaneous Raman Spectroscopy, *AIAA-Paper 92-2855*

25.   Shinn, J.L. and Simmonds, A.L. (1984) Comparison of Viscous Shock-Layer Heating Analysis with Shuttle Flight Data in Slip Flow Regime, *AIAA-84-0226, AIAA 22nd Aerospace Sciences Meeting, Reno, Nevada*

NONEQUILIBRIUM VIBRATION-DISSOCIATION COUPLING

IN HYPERSONIC FLOWS

C. PARK
Department of Aeronautics and Space Engineering
Tohoku University, Aoba-Ku, Sendai, 980 Japan

Nomenclature

A	An atomic species formed by dissociation
B	An atomic species formed by dissociation
D	Dissociation energy
e_v	Vibrational energy per unit mass
E	Energy level
E_x	Hidden energy per unit mass
H	Enthalpy
k	Boltzmann constant
K_{ij}	Collisional transition rate coefficient for i to j transition
m	Molecule
M	Second moment of transition rate coefficient
n	Number density
N	The number of internal states
p	Pressure, pascal
t	Time, sec
T	Translational temperature, K
T_v	Vibrational temperature, K
v	Vibrational quantum number
V	Flow velocity
ϕ	Vibrational relaxation rate enhancement factor
ρ	Population density normalized by equilibrium value

Subscripts

1	First excited vibrational state
c	Continuum (free) state
e	Equilibrium
r	Rotation
s	Settling chamber
v	Vibration

665

M. Capitelli (ed.), Molecular Physics and Hypersonic Flows, 665–678.
© 1996 *Kluwer Academic Publishers. Printed in the Netherlands.*

1. Introduction

This lecture is a sequel to an earlier lecture entitled "Review of Finite-Rate Chemistry Models for Air Dissociation and Ionization," which appears in this volume and is identified as reference 1 in the present lecture. In contrast to that lecture, which focused on the problem of modeling of vibration-dissociation coupling in a post-shock flow, this lecture examines mostly the modeling problems associated with cooling flows.

2. Review of Diffusion Model for Vibrational Relaxation

As was shown in reference 1, the mathematical nature of the temporal evolution of vibrational population in a post-shock, i.e. behind a shock wave, nonequilibrium flow is best understood by expressing the vibrational populations n_v in terms of the normalized population ρ defined as the ratio of n_v to that under equilibrium:

$$\rho_v = \frac{n_v}{n_{ve}}$$

For dissociation of molecule m to form two atoms A and B in the form

$$m \leftrightarrow A + B$$

the nonequilibrium number densities of the molecules A and B are first normalized by

$$\rho_A = n_A/n_{Ae}, \qquad \rho_B = n_B/n_{Be}$$

For vibrational states below about $v = 20$, the second moment of transition M defined as [1]

$$M(v) \approx M(\epsilon) = \frac{1}{2}\sum_{v'} K_{vv'}(v' - v)^2 \approx \frac{1}{2}\int_{\epsilon'} K_{\epsilon\epsilon'}(\epsilon' - \epsilon)^2 d\epsilon'$$

is finite, because transitions occur preferentially between the neighboring states. Reference 1 shows that, for those low vibrational states, the population function ρ satisfies a diffusion equation of the form

$$\frac{1}{n_x}\frac{d\rho_\epsilon}{dt} = \frac{\partial}{\partial\epsilon}[M(\epsilon)\frac{\partial\rho}{\partial\epsilon}]$$

Assuming $M = $ constant, the vibrational temperature defined by the slope of ρ at origin, $v = 0$, satisfies the equation

$$\frac{\partial T_v}{\partial t} = \frac{\pi}{4}n_x K_{10}\frac{E_1}{kT_v}\frac{E_1}{kT}\frac{(T - T_v)^2}{T^2}(T - T_v)$$

This is expressed in reference 1 as

$$\frac{\partial T_v}{\partial t} = \frac{T - T_v}{\tau_D}$$

where

$$\frac{\tau_D}{\tau_L} = \frac{4}{\pi}[1 - \exp(-\frac{E_1}{kT})]\frac{kT_v}{E_1}\frac{kT}{E_1}\frac{T^2}{(T - T_v)^2}$$

τ_L being the Landau-Teller relaxation time. This equation is valid at high temperatures where T is much greater than E_1/k, and replaces the conventional Landau-Teller equation there. A formula bridging between the high temperature regime and the conventional low temperature (T of the order of E_1/k) regime was proposed in reference 1 in the form

$$\frac{\partial e_v}{\partial t} = \frac{e_{ve} - e_v}{\tau_L}|\frac{T_s - T_v}{T_s - T_{vs}}|^{s-1}$$

where s varies from unity at a low temperature and 3.5 at the high temperature limit. This constitutes the so-called diffusion model for vibrational relaxation.

The Landau-Teller relaxation time τ_L is determined empirically through experiments conducted behind a shock wave. In reference 2, it is expressed using two parameters a and b in the form

$$\tau_L = \frac{1}{p}[a(T^{-1/3} - b) - 18.42] \quad \text{sec}$$

where p is the partial pressure of the colliding species in atm. Reference 2 shows that the parameters a and b are simple functions of molecular constants for most collision combinations. Exceptions to this rule is shown in reference 3 and several other recent works of the present author.

3. Treanor Distribution

A chemically reacting cooling flow is produced by rapidly cooling a hot flow in which significant vibrational excitation, dissociation, or ionization is present. The cooling is achieved by a rapid adiabatic expansion. Such flows are found in the leeward side of a hypersonic vehicle or in the nozzle of a hypersonic wind tunnel. There, the translational temperature is not much more, or sometimes much less, than the characteristic vibrational temperature E_1/k. In such an environment, the diffusion model described above is irrelevant, because the condition that the translational temperature be much greater than E_1/k is violated.

Instead, the vibrational temperature is equal to or higher than the local translational temperature in a cooling flow. The presence of high concentration of vibrationally excited molecules promotes resonance or near-resonance vibration-vibration energy transfers. This necessitates modeling of the thermochemical behavior in a manner different from the diffusion model given above. Consider, for instance, a collision between two N_2 molecules, both at the same vibrational level v. Assume that the molecules are harmonic, so that vibrational transitions are allowed only to the neighboring levels v-1 or v+1. The collision can produce $N_2(v)$ + $N_2(v)$ or $N_2(v-1)$ + $N_2(v+1)$. The former is an elastic collision, and the latter is a resonant vibrational transfer. In either case, the sum of the vibrational energies is unchanged during collision.

However, N_2 is in reality slightly anharmonic, so that the vibrational energy gaps become smaller as v increases. As a result, the collision leading to $N_2(v-1)$ + $N_2(v+1)$ is no longer strictly resonant, but is only nearly resonant, and the sum of vibrational energies is reduced slightly through the collision. The energy decrement is discharged into the translational mode. Conversely, if an $N_2(v-1)$ molecule collides with an $N_2(v+1)$ molecule, a near-resonant vibrational transfer can produce two $N_2(v)$ molecules, accompanied by a slight increase in the sum of the vibrational energies. The vibrational energy increase is complemented by a decrease in translational energy of the same magnitude.

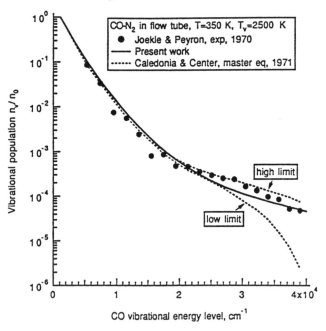

Figure 1. Measured and calculated population distribution in a typical cooling CO flow showing Treanor distribution, taken from reference 5.

Such behavior is of no consequence in a post-shock flow because most molecules are in the ground vibrational state and there can be no v-1 state for the ground state v = 0. However, in a cooling flow, vibrational levels above the ground state are normally significantly populated, and hence the transitions involving v-1 state can take place in significant frequencies.

Even though the real molecules are anharmonic, the transitions among the low states occur still mostly between the neighboring states, and so most vibrational transitions tend to occur in the near-resonant fashion. The quasi-steady state distribution caused by these near-resonant transitions is one known as Treanor distribution [4] which is curved concave upward in a semi-logarithmic plot. Such distribution has been confirmed by many experiments to exist to a vibrational level typically of 15 to 20 as shown in figure 1, which was taken from reference 5.

4. State-to-State Transition Probabilities

For a harmonic oscillator collided by a monatomic atom, the state-to-state transition probability or rate coefficient varies linearly with the quantum number. Even in a real molecule which is anharmonic, this trend exists up to a quantum number of about 4 or 5. However, since the number density of vibrationally excited molecules diminish with increasing vibrational quantum number, the absolute transition rates decrease with vibrational quantum number beyond v = 4 or 5. At a certain intermediate quantum number, the near-resonance is almost totally lost, and the selection rule favoring the neiboring transitions begins to weaken, that is, the energy exchange processes become classical. Now all colliding molecules contribute to a state-to-state transition, and therefore the resulting rate increases with increasing vibrational quantum numbere.

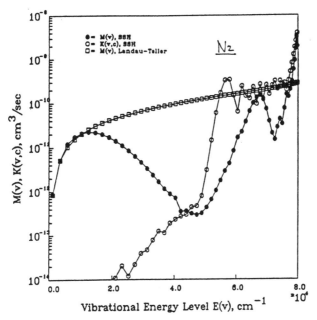

Figure 2. M(v) and K_{vc} for N_2 at T = 4000 K, taken from reference 3.

This behavior is best illustrated in a plot showing the second moment of transition M(v) as a function of vibrational quantum number v, shown in figure 2 for N_2, which was taken from reference 3. As seen in the figure, M(v) has two peaks, one at about v = 5 and the other at the dissociation limit v = N. This has been verified experimentally in reference 6, in which the rate of relaxation of vibrational states are measured for O_2 using laser fluorescence (see, figure 3). This point is corroborated in several contributed articles appearing in this volume.

The absolute magnitudes of the quantity M(v), or more fundamentally the state-to-state transition rate coefficients $K_{vv'}$, have hitherto been uncertain. There are several lectures and contributed articles on this subject in this volume.

5. Vibrational Relaxation in Hypersonic Nozzle

It is well known that the distribution of populations of any internal states becomes a straight line on a semi-logarithmic plot shown in figure 1 if the states are populated according to a Boltzmann distribution. As shown in figure 1, the distribution of vibrational populations in a nonequilibrium cooling flow takes a nonlinear form in such a plot. It is necessary to characterize such a distribution, assess consequences, and incorporate the knowledge in the design of hypersonic vehicles.

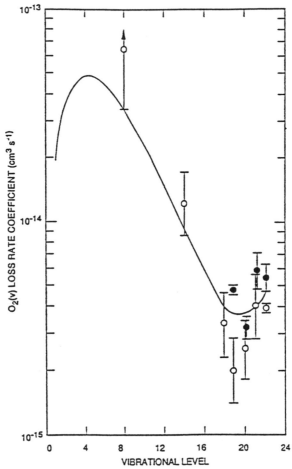

Figure 3. The observed rate of relaxation of vibrational levels of O_2, taken from reference 6.

The first characterstics of the nonequilibrium vibrational distribution needed is the slope of the distribution at origin, $v = 0$, that is, the ratio of number densities of the $v = 1$ state to that of the $v = 0$ state. In many equilibrium flow environment, vibrational temperature is deduced from this ratio. For this reason, we shall define vibrational temperature T_v in a nonequilibrium environment also by this ratio. This definition of vibrational temperature is the same as that for the post-shock flow given in Section 2.

Vibrational temperature in the hypersonic nozzle of a shock tunnel has been measured since the 1960s. In the early studies, sodium or chromium was seeded into the flow, and the line-reversal temperature was measured [see, e.g., 7]. The measured vibrational temperature values were compared with the calculated values. The measured values were generally lower than the values calculated using the relaxation time values determined in the post-shock flows. The calculations were repeated with relaxation rate values increased by an arbitrary factor ϕ. The ϕ value that reproduced the measured vibrational temperature values was found to be between 5 and 1000 in shock tunnels [8]. Presence of the seed species and other impurities such as water vapor was suspected of being the cause of this large discrepancy. This was partly confirmed in separate experiments [9].

In order to determine the intrinsic vibrational relaxation rate in a cooling flow in the absence of such impurities, carefully controlled experiments were carried out in references 10 and 11 with carbon monoxide in a clean environment. The vibrational temperature was determined from the radiation emitted by the vibrational excited states of the CO molecule, thereby eliminating the need for seed species. The ϕ value was found to be no more than 3 in these experiments.

The vibrational relaxation in N_2 was measured in reference 12 in a cooling flow of pure nitrogen using Raman spectroscopy, without introducing seed species. The parameter ϕ was found to be nearly unity in that experiment (see figure 4). The vibrational temperatures in an arc-jet wind tunnel flow [13] are found also to be consistent with $\phi \approx 1$ [14] (see figure 5).

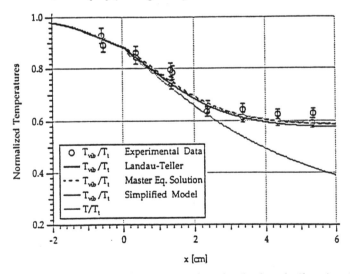

Figure 4. Comparison between the measured and calculated vibrational temperatures in N_2 shock tunnel flow, taken from reference 12. The figure shows that the vibratinal relaxation in a clean cooling flow occurs with the same Landau-Teller relaxation rate as behind a shock wave.

The theoretical treatment of this problem varies among investigators. Some view that the large ϕ values occur even for clean flows, and are a natural consequence of the fast v-v transfers [e.g., 15]. Such theories discredit the accuracy of those experiments that found $\phi \approx 1$. Others rationalize $\phi \approx 1$ [16], on account that, even though the v-v transfers are fast, the net vibrational energy change per near-resonant v-v collision is small.

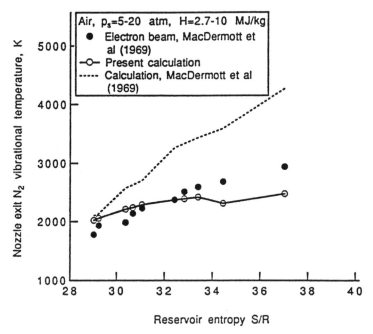

Figure 5. Comparison between the measured and calculated vibrational temperatures in air arc-jet wind tunnel flow [13], taken from reference 14. The figure shows that the vibratinal relaxation in an arc-jet flow occurs with the same Landau-Teller relaxation rate as behind a shock wave.

In practical operation of shock tunnels, perhaps the issue is irrelevant because the flows produced therein always contain impurities. If the impurities indeed enhance the vibrational relaxation rates by a large factor, then perhaps it is advantageous to have small amount of such impurities in a shock tunnel, because it would bring the flow in the test section closer to vibrational equilibrium. The necessary premise to this view is that the amount of such imurities be so small that it would not affect other aspects of the flow. In order to exploit this point, it is necessary to know more in detail how each impurity species enhances vibrational relaxation rate. Further future work is needed along this line.

6. Energy Contained in High Vibrational States

If the state-to-state transition rate coefficients $K_{vv'}$ and K_{vc} are given, then one can compute the temporal variation of the vibrational distribution in a cooling flow by integrating the master equation. Many such calculations have been performed in the past [e.g., 16,17]. Such calculations always predict that the distribution is S-shaped in the semi-logarithmic plot as shown in figure 1, and that significant amount of highly vibrationally excited molecules exists as a result. The most important consequnce of this phenomenon is the fact that the vibrational energy contained by molecules is larger than that inferred from the calculated vibrational temperature.

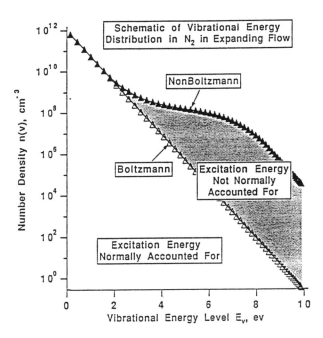

Figure 6. Definition of hidden vibrational energy, taken from reference 5.

When only the vibrational temperature defined by the populations in the $v = 1$ and $v = 0$ states is given, as would be in most flow property calculations, one can only calculate the equilibrium vibrational energy assuming a Boltzmann distribution corresponding to the vibrational temperature e_{ve}. The true vibrational energy in a cooling flow is significantly larger than e_{ve} because of the presence of the large number of highly vibrationally excited molelcules. The difference between the true vibrational energy and e_{ve} is termed hidden vibrational energy in reference 5 (see figure 6). In a real flow, the energy available for translational mode is lower by this hidden vibrational energy compared with that calculated using the traditional method in which such energy is not accounted for. This means that, by ignoring the hidden energy, one overestimates temperature, and hence pressure, and likely the flow velocity. A small decrease in flow velocity is highly undesirable in certain aerospace applications such as scramjet engine nozzle. Therefore it is worthwhile to quantify this hidden energy as accurately as possible.

Even though the hidden vibrational energy can be calculated through integration of master equation, such calculation is prohibitively expensive when it is

674

to be carried out for a multidimensinal flow. An approximate method is devised in reference 5 to replace the master equation integration with an algebraic equation. In that method, not only the vibrational but rotational nonequilibrium was allowed. The vibration-rotation space was drawn in two dimensions as shown in figure 7 and characterized by the normalized radius r = R/D and the polar angle angle θ = atan (E_r/E_v). The normalized population $\rho(r,\theta)$ was expressed in the form

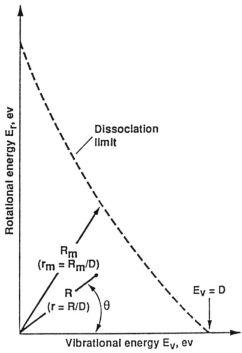

Figure 7. Vibrational-rotational space over which population distribution is calculated, taken from reference 5.

$$\ln(\rho) = a_0(\theta) + a_1(\theta)r + a_2(\theta)r^2 + a_3(\theta)r^m + a_4(\theta)r^n$$

To determine the seven parameters a_i, i = 0 to 4, and m and n, six known physical constraints were imposed. They were:

1) the fact that the sum of all vibrational-rotational states must equal the known number density of the molecule,

2) the slope at the origin, r = 0, must give the known vibrational and rotational temperatures,

3) the curvature at the origin must correspond to that given by the Treanor distribution,

4) the states in the dissociation limit are in equilibrium with the free state,

5) the slope of the population distribution at the dissociation limit correspond to that of the translational temperature, and

6) the curvature at the dissociation limit equals -0.5 D/kT.

The condition 6) was obtained by solving the master equation written in an integral equation form [5]. The procedure leaves one degree of freedom to choose. The power m was varied over a range, and was chosen to be between 4 and 12 by comparing with the master equation solution.

The method enables determination of the hidden energy from the local thermodynamic variables, T, T_v, n_m, n_A, and n_B. The calculation can be coupled with the flow calculation. The result of a typical such calculation is shown in figure 8, taken from reference 5, in which the magnitude of the hidden energy normalized by the total enthalpy of the flow and the decrement in flow velocity ΔV normalized by the local velocity V, are shown. As seen here, the hidden energy can be a few percent of the total enthalpy, and the velocity can decrease by a several percent because of it. A velocity decrement of this magnitude is very serious in aerospace applications, and so this result underscores the need for further refinement of this line of investigation.

7. Interaction of Highly Vibrationally Excited Molecules on Chemical Reaction Rates

In reference 1, it was pointed out that the species concentrations observed at the exit of a shock tunnel or arc-jet wind tunnel are quite different from those calculated using the known method. It was speculated therein that perhaps there is a relatively large concentration of oxygen atoms electronically excited to its ^1D state in a cooling flow, despite that they are constantly being removed through the Zeldovich reactions. In order for this hypothesis to be valid, the $O(^1D)$ atoms must be continuously produced while the Zeldovich reactions are in progress.

One can speculate that $O(^1D)$ atoms are created by an eneregy-exchange process of the type

$$O(^3P) + N_2(\text{high v}) \rightarrow O(^1D) + N_2(\text{low v})$$

Highly vibrationally excited O_2 molecules can also function in place of N_2. Presence of large concentration of highly vibrationally excited N_2 and O_2 then could lead to constant supply of $O(^1D)$.

The v-e exchange similar to that shown above could happen between an electronic states and a vibrational state of the same molecule. N_2 has two prominent electronic states $a^3\Sigma_u^+$ and $b^3\Pi_g$ below the dissociation limit. There can be energy exchanges of the following kind:

$$N_2(a^3\Sigma_u^+, \text{low v}) + N_2(X^1\Sigma_g^+, \text{low v}) \rightarrow N_2(X^1\Sigma_g^+, \text{low v}) + N_2(X^1\Sigma_g^+, \text{high v})$$

$$N_2(b^3\Pi_g, \text{low v}) + N_2(X^1\Sigma_g^+, \text{low v}) \rightarrow N_2(X^1\Sigma_g^+, \text{low v}) + N_2(X^1\Sigma_g^+, \text{high v})$$

$$N_2(b^3\Pi_g, \text{low v}) + N_2(X^1\Sigma_g^+, \text{low v}) \rightarrow N_2(a^3\Sigma_u^+, \text{low v}) + N_2(X^1\Sigma_g^+, \text{high v})$$

The $a^3\Sigma_u^+$ and $b^3\Pi_g$ states have statistical weights of 3 and 6, respectively, compared with 1 for the $X^1\Sigma_g^+$ state. Hence, it is most likely that most three-body recombination $N + N + X \rightarrow N_2 + X$ occurs into the a and b states instead of the X state. The above v-e exchanges may then contribute to increasing the population of the highly vibrationally excited $N_2(X)$ molecules, leading ultimately to the overpopulation of the $O(^1D)$ atoms.

The O_2 molecule also has several electronic states slightly below the dissociation limit. These states will be populated through the three-body recombination reaction. Through the vibration-to-electronic energy exchange of the kind shown above, the highly vibrationally-excited $O_2(X)$ molecules will be created. In the process, $O_2(X)$ state may be excited to $O_2(b^1\Sigma_g^+)$ state. The $O_2(b^1\Sigma_g^+)$ can then produce $O(^1D)$ atoms through

$$O_2(b^1\Sigma_g^+) + O(^3P) \rightarrow O_2(X^3\Sigma_u^-) + O(^1D)$$

Finally, similar exchanges between the vibrational or electronic energies of N_2, NO, and O_2 are also possible.

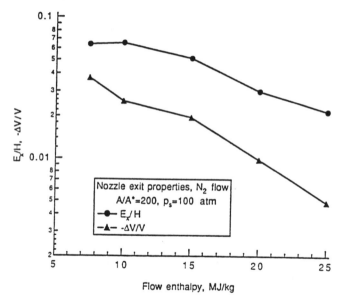

Figure 8. The hidden energy E_x and velocity decrement ΔV for a typical shock tunnel flow of nitrogen, taken from reference 5.

All these possibilities lead to the speculation that, in a cooling flow, most of the three-body recombination takes place to form N_2, NO, or O_2 in their high vibrational or electronic states wich form reservoirs of internal energy. The energy exchanges among these states occur in a manner of cascade, and can ultimately cause the upward electronic transition of $O(^3P)$ into $O(^1D)$. Through these processes, $O(^1D)$ will be continuously produced as long as the energy reservoirs are populated. In order to quantify the rate of production of $O(^1D)$ through such energy cascade processes, one must know a great deal more about the energy exchanges among those states.

8. Conclusions

In a cooling flow produced in shock tunnels, which is believed to contain imurities, experimental evidences indicate that the vibrational temperature defined by the ratio of populations of the v = 1 and v = 0 states relaxes with a rate much faster than those observed behind a shock wave. However, in a clean flow, the evidences show that the relaxation rate is nearly the same as those behind a shock wave. Theory describing this behavior is seemingly incomplete. For practical application, it would be desirable to characterize accurately how the impurities affect vibrational relaxation.

Because of the anharmonicity of molecular vibration among the low vibrational states and classical nature of vibrational motion among the high vibrational states, the state-to-state vibrational transition rates form a bimodal distribution when plotted against vibrational quantum number, with a minimum at an intermediate vibrational quantum number. This fact and the fact that three-body recombination takes place preferentially into high vibrational states produce an S-shaped distribution of vibrational population in a cooling flow, with a result that a large number of the molecules formed by recombination is in highly vibrationally excited states.

The energy contained in the high vibrational states, referred to as hidden energy, has been calculated using an approximate method, and was shown to be a significant fraction of the total enthalpy of the flow. The flow velocity is significantly reduced as a result. This shows a need for further refinement of the technique for calculating the concentration of the highly excited molecules.

It is likely that three-body recombinations to form N_2, NO, and O_2 mostly occurs to form those molecules in highly excited vibrational and electronic states. Through the vibration-to-vibration (v-v) and vibration-to-electronic (v-e) energy exchanges, these internal energies cascade down in the energy level. In the process, $O(^1D)$ is likely to be produced in a significant concentration. The $O(^1D)$ atoms then undergo Zeldovich reactions with rates much different from those of the $O(^3P)$ atoms. They may be the cause of the anomaly in the observed species concentrations in the flows produced in shock tunnels and arc-jet wind tunnels mentioned in reference 1.

9. References

1. Park, C. (1995) "Review of Finite-Rate Chemistry Models for Air Dissociation and Ionization," *Molecular Physics and Hypersonic Flow*, edited by M. Capitelli, xxxx-xxxx.

2. Millikan, R. C., and White, D. R. (1963) "Systematics of Vibrational Relaxation," *Journal of Chemical Physics,* **139**, 3209-3213.

3. Park, C. (1990) *Nonequilibrium Hypersonic Aerothermodynamics*, John Wiley, New York, N.Y.

4. Treanor, C. E., Rich, J. W., and Rehm, R. G. (1968) "Vibrational Relaxation of Anharmonic Oscillators with Exchange-Dominated Collisions," *Journal*

678

of Chemical Physics, **48,** 1798-1807.

5. Park, C. (1995) "Estimation of Excitation Energy of Diatomic Molecules in Expanding Nonequilibrium Flows," *Journal of Thermophysics and Heat Transfer,* **9,** 17-25.

6. Park, H., and Slanger, T. G. (1994) "O_2(X,v=8-22) Quenching Rate Coefficients for O_2 and N_2, and O_2(X) Vibrational Distribution from 248 nm O_3 Photodissociation," *Journal of Chemical Physics,* **100,** 287-302.

7. Hulre, I. R., Russo, A. L., and Hall, J. G. (1964) "Spectroscopic Studies of Vibrational Nonequilibrium in Supersonic Nozzle Flows," *Journal of Chemical Physics,* 40, 2076-2089.

8. Hurle, I. R. (1971) "Nonequilibrium Flows with Special Reference to The Nozzle-Flow Problem," *Proceedings of the 8th International Shock Tube Symposium,* Chapman and Hall, London, 3/1-3/37.

9. von Rosenberg, C. W., Jr., Bray, K. N. C., and Pratt, N. H. (1971) "The Effect of Water Vapor on the Vibrational Relaxation of CO," *13th Symposium (International) on Combustion,* Combustion Institute, Pittsburgh, Pennsylvania, 89-98.

10. Blom, A. P., Bray, K. N. C., and Pratt, N. H. (1970) "Rapid Vibrational De-Excitation Influenced by Gasdynamic Coupling," *Astronautica Acta,* **5,** 487-494.

11. McLaren, T. I., and Appleton, J. P. (1970) "Vibrational Relaxation Measurement of Carbon Monoxide in a Shock-Tube Expansion Wave," *Journal of Chemical Physics,* **53,** 2850-2857.

12. Gillespie, W. D., Bershader, D., Sharma, S. P., and Ruffin, S. (1993) "Raman Scattering Measurements of Vibrational and Rotational Distributions in Expanding Nitrogen," AIAA Paper 93-0274.

13. MacDermott, W. N., and Marshall, J. G. (1969) "Nonequilibrium Nozzle Expansions of Partially Dissociated Air: A Comparison of Theory and Electron-Beam Experiments," AEDC-TR-69-66.

14. Park, C., and Lee, S. H. (1995) "Validation of Multitemperature Nozzle Flow Code," *Journal of Thermophysics and Heat Transfere,* **9,** 9-16.

15. Godiets, B. F., Osipov, V. A., and Shelepin, L. A. (1988) *Kinetic Processes in Gases and Molecular Lasers,* Gordon and Breach, London.

16. Ruffin, S. M. (1993) "Vibrational Energy Transfer of Diatomic Gases in Hypersonic Expanding Flows," Ph.D. Thesis, Dept. of Aeronautics and Astronautics, Stanford University.

17. de Gavelle de Roany, A. C., Flament, C., Rich, J. W., Subramaniam, V. V., and Warren, W. R. (1993) "Strong Vibrational Nonequilibrium in Supersonic Nozzle Flows," *AIAA Journal,* **31,** 119-128.

VIBRATIONAL KINETICS FOR NUMERICAL SIMULATION
OF THERMAL NON EQUILIBRIUM FLOWS

V. BELLUCCI
CRS4 Research Center
Cagliari, Italy

D. GIORDANO
ESA-ESTEC
Noordwijk, The Netherlands

G. COLONNA, M. CAPITELLI AND I. ARMENISE
University of Bari
Bari, Italy

AND

C. BRUNO
University of Roma I
Roma, Italy

1. INTRODUCTION

In numerical simulation of subsonic and supersonic flows thermodynamic models built on the Boltzmann distribution are widely used. In these models the population distributions of the quantum states associated with each independent molecular degree of freedom follow a Boltzmann distribution and the major benefit deriving from this hypothesis is that the thermodynamics of a gas mixture can be described by a set of algebraic functions. However, for sufficiently high Mach numbers, the energy density of the flow field becomes considerably large and leads to more marked non-equilibrium effects; therefore, one should expect the Boltzmann distribution assumption to become less tenable.

A more fundamental and rigorous approach to the problem is to calculate the population distributions from the master equations [1, 4] governing the kinetics of the particle exchanges among quantum states. In principle, this idea applies indifferently to any molecular degree of freedom. How-

679

M. Capitelli (ed.), Molecular Physics and Hypersonic Flows, 679–689.

ever, the attention in the literature has been prevalently focused on the vibrational degree of freedom of diatomic molecules; thus, the term *vibrational kinetics* has come into use to refer to the subject. In this approach, thermodynamics and gas dynamics are inseparably coupled because, in a gasdynamical context, the master equations assume the form of balance equations just like those of mass, momentum and energy. Unfortunately, such a rigorous approach is affected by the high computational cost required to integrate the huge number of associated differential equations.

In this work we define a relatively simple but meaningful test case, for which the computational load imposed by the master equations can still be within the reach of modern computers, and we study it by using both the *traditional* harmonic oscillator model and the rigorous approach based on the master equations.

2. DESCRIPTION OF THE TEST CASE

The test case we have studied numerically is the steady, two-dimensional flow of N_2 at $M_\infty = 6.5$ past an infinite cylinder of radius $r = 1\ m$ in an uniform stream in thermal equilibrium. The main reasons behind our choice are: (a) the flow pattern in the stagnation region behind the bow shock is representative of blunt body flows and, therefore, is of specific relevance to reentry aerothermodynamics; (b) the possibility to analyse in a single test case both compression and expansion regions.

As a first step in the analysis of the problem, viscosity and thermal conduction effects have been disregarded; moreover, the stream temperature T_∞ was set to 300 K in order to keep, for the assumed $M_\infty = 6.5$, the translational temperature peak below 3000 K in the flow region behind the shock wave: this bound ensures N_2 is chemically inert. Calculations have been performed for an asymptotic stream pressure $p_\infty = 50\ Pa$.

Consistent with our main objective, vibrational relaxation has been calculated both from the standard vibrational energy rate equation, assuming an harmonic oscillator behaviour of N_2, and from a set of vibrational master equations which account for the first ten vibrational energy levels of the N_2 molecule, assumed as an anharmonic oscillator. The relaxation times proposed by Millikan and White [5] and by Blackman [6] have been used for the traditional harmonic oscillator model.

3. GOVERNING EQUATIONS

The thermodynamic energy of N_2 is made up by the translational, the rotational and the internal contributions. The N_2 molecule is assumed to behave as a rigid rotator and the population distribution over its rotational quantum states is assumed to follow a Boltzmann distribution. Moreover, the corresponding rotational temperature is taken equal to the translational one [7]. The thermodynamic pressure is not affected by the internal degrees of freedom and is given in terms of the translational temperature, according to the perfect gas expression.

When the N_2 molecule is assumed to behave as an harmonic oscillator, the population distribution over its vibrational quantum states follows a Boltzmann distribution, with a vibrational temperature which, in general, differs from the translational one when thermal nonequilibrium prevails [7]. The non viscous balance equations of mass, momentum, total energy and vibrational energy [7, 8] are integrated. The vibrational energy production term is given in accordance with the harmonic oscillator vibrational relaxation theory. The relaxation time is determined from an expression which is an approximation of that in the Landau and Teller's theory. The sets of constants proposed by Millikan and White [5] and by Blackman [6] have been used in this study to evaluate the relaxation time.

In the vibrational kinetics model, the populations of the vibrational quantum states are unknowns of the problem and have to be determined via the master equations governing the kinetics of the particle exchanges among those states. In a fluid dynamics context, the master equations replace the balance equations for the mass and the vibrational energy [9, 10]. The quantum level energies are determined from a third order approximating formula that considers anharmonic effects [11, 12]. The spectrum is truncated when the value of the energy ε_ℓ becomes greater than the N_2 dissociation energy ($9.62 \ eV = 1.541 \cdot 10^{-18} \ J$); this occurs when $\ell = 45$ but the computational load becomes very heavy if all 45 levels are taken into account. However, for the particular flow conditions assumed in this study, i.e. absence of dissipative effects and of chemical activity of N_2, preliminary calculations have shown that there are no substantial differences in the results if a reduced number of levels is considered; according to these preliminary results, the choice $\ell = 10$ appears a reasonable compromise between the accuracy deriving from considering the whole vibrational spectrum, i.e. $\ell = 45$, and the need of affordable computations. The master equations production terms are associated with the kinetic mechanisms according to which the particle exchanges among the quantum states take place. There are two kinds of processes relevant to the case considered in this study: V-T processes, in which a molecule loses or gains a vibrational quantum as a result of a collision with another molecule, and V-V processes, in which a vibrational quantum is exchanged between the colliding molecules [9, 10].

The numerical discretization of the system of governing equations is obtained by a a cell-centered, finite volume method, and the system is integrated over an elemental volume of the computational domain by a central difference scheme. The adaptive dissipation model of [13] is included; this model uses a blend of second and fourth differences: the former prevent oscillations near the location of shock waves, the latter are important for stability and convergence to steady state. The scheme is second-order accurate in space on cartesian grids. The time integration is performed by a five-stage, point-implicit Runge-Kutta scheme. The initial conditions ($t = 0$) assumed for the calculation of the flow field correspond to those of the asymptotic stream. In the case of the vibrational kinetics approach, the partial densities are initialised to the values corresponding to a Boltzmann distribution at the asymptotic temperature, on the first ℓ levels of the spectrum. The boundary conditions at infinity coincide with the initial conditions. Consistently with the assumption of no viscosity, the cylinder contour is a streamline. Finally, outflow variables (outflow boundary conditions) are obtained from first order extrapolation.

4. DISCUSSION OF THE RESULTS

4.1. STAGNATION LINE

The stagnation line is a zone of strong compression. The nondimensional pressure along the stagnation line is shown in Fig. 1. The agreement among the different models is quite satisfactory, although the harmonic oscillator model with the Millikan and White's relaxation time produces a slightly larger stand-off distance.

The population distributions over the vibrational energy levels at various stations along the stagnation line are shown in Figs. 2; these figures illustrate the vibrational relaxation of the generic fluid particle moving toward the stagnation point. The plain solid and dashed lines refer to the harmonic oscillator model with, respectively, Millikan and White's and Blackman's relaxation time; their lack of curvature is typical of harmonic oscillators. The solid lines with circles are relative to the vibrational kinetics model. Frame (a) shows the situation in the asymptotic stream, where thermal equilibrium prevails, i.e. $T = T_v = T_\infty$. The harmonic oscillators lines are obviously superposed. The slight curvature shown by the solid circled line is due to the anharmonicity of the energy levels. The curvature disappears if the populations are plotted versus the energy of the vibrational levels rather than the vibrational quantum number. Frame (b) illustrates the situation across the shock wave. The lower three lines describe the vibrational relaxation. The lower plain solid and dashed lines (marked with T_v) are Boltzmann distributions calculated at the vibrational temperatures associated with the two harmonic oscillator models at that particular station. The lower solid circled line (marked m.e.) is the population distribution produced by the master equations. The upper three lines (marked

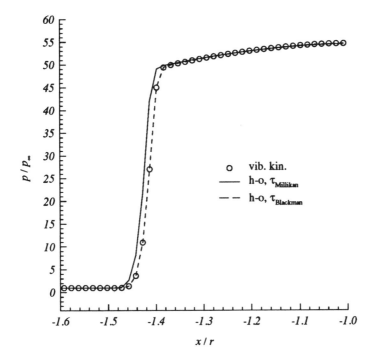

Figure 1. Nondimensional pressure profile along the stagnation line.

with T) are Boltzmann distributions based on the translational temperatures associated with each model at that particular station. In vibrational
kinetics approach, they are Boltzmann distributions on the first ten levels
of the spectrum.

The difference in the slopes of the distributions is indicative of how the
corresponding model predicts the thermal nonequilibrium. Harmonic oscillators translational and vibrational temperatures differ, as shown by the
different slopes of the corresponding plain solid and dashed lines. Looking
at the vibrational kinetics model, the population distribution produced by
the master equations (line marked m.e.) differs from the Boltzmann distribution at the corresponding translational temperature (upper solid circled
line) and becomes non-Boltzmann from the third vibrational level on.

Frames (c) to (f) show relaxation towards thermal equilibrium: the lower
three lines tend to superpose to the upper three lines but the stand-off distance is not sufficient to achieve thermal equilibrium at the stagnation
point. Furthermore, the population distributions predicted by the master
equations coincide satisfactorily with a Boltzmann distribution downstream

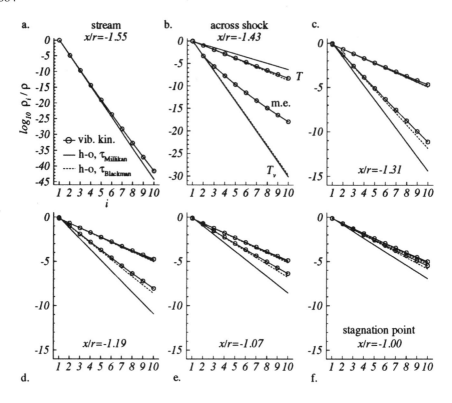

Figure 2. Population distributions over the vibrational levels. Stagnation line.

of the shock wave, at least for the ten levels accounted for in this study, and they are found to be in substantial agreement with the Boltzmann distributions deriving from the harmonic oscillator model with the Blackman's relaxation time. The harmonic oscillator model with the Millikan and White's relaxation time, instead, over estimates the thermal nonequilibrium with respect to the other models.

4.2. CYLINDER WALL

The cylinder wall is a zone of strong expansion. The nondimensional pressure profile is shown in Fig. 3 and the agreement among the different models is quite satisfactory.

The population distributions over the vibrational energy levels at the stagnation point and at the cylinder shoulder are shown in Fig. 4; in these figures, the line style convention (see paragraph Sec. 4.1) adopted for Figs. 2 applies again. Frame (a) is a copy of frame (f) from Fig. 2; it has been plotted in a different scale for convenience of comparison. Frame (b) shows

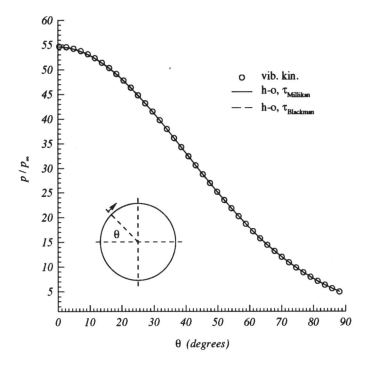

Figure 3. Nondimensional pressure profile along the cylinder wall.

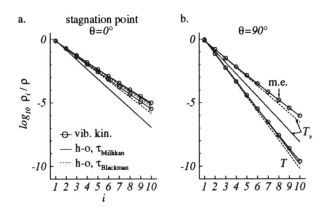

Figure 4. Population distributions over the vibrational levels. Cylinder wall.

the situation at the cylinder shoulder: the expansion along the cylinder wall drives vibrational over relaxation overshooting the lines based on the translational temperatures. Here again the population distributions produced by the master equations coincide satisfactorily with a Boltzmann distribution.

Fig. 5 yields a clearer understanding of vibrational relaxation along the cylinder wall. The profiles of the nondimensional translational, vibrational

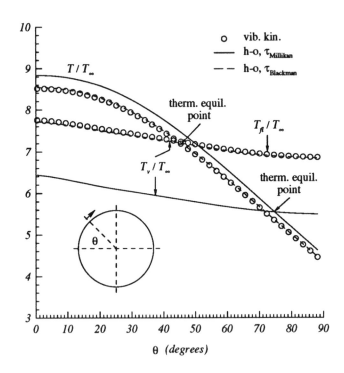

Figure 5. Nondimensional translational, vibrational and first-level temperature profiles along the cylinder wall.

and first-level temperatures are shown. The vibrational kinetics model does not require the introduction of a "vibrational" temperature. However, a "first-level" temperature T_{fl} can be defined to compare results in terms of temperatures. The first-level temperature is defined as the temperature corresponding to the Boltzmann distribution which can be accommodated on the population ρ_1/ρ of the first vibrational level.

No spatial region of thermal equilibrium exists and the vibrational and first-level temperatures overshoot the corresponding translational temperatures in a single point, i.e. the point of thermal equilibrium, which is predicted at an angular position $\theta \cong 45°$ by the vibrational kinetics model and the harmonic oscillator model with Blackman's relaxation time, and at an angular position $\theta \cong 74°$ by the harmonic oscillator model with Millikan and White's relaxation time. Once again, this latter model over estimates the thermal nonequilibrium with respect to the other models.

4.3. FLOW PATTERN PAST THE CYLINDER

The translational isotherms $T/T_\infty = const$ are illustrated in Figs. 6. They

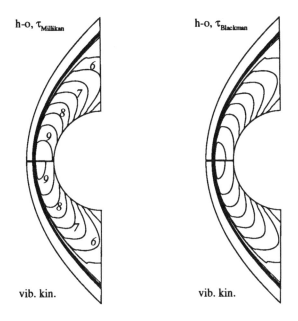

Figure 6. Translational (T/T_∞) isotherms. $p_\infty = 50\ Pa$.

reveal how the harmonic oscillator model with Millikan and White's relaxation time over estimates the translational temperature field with respect to the other two models which are found in remarkable agreement.

The opposite situation exists relatively to the vibrational isotherms $T_v/T_\infty = const$ and the first-level isotherms $T_{fl}/T_\infty = const$; these are shown in Fig. 7. The harmonic oscillator model with Millikan and White's relaxation time under estimates, in this case, the vibrational temperature field with respect to the other two models.

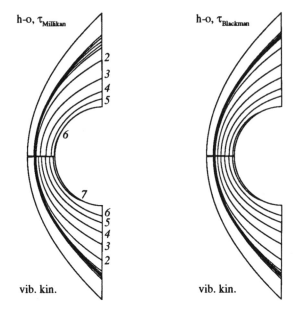

h-o, τ_{Millikan}

2
3
4
5

6

7

6
5
4
3
2

vib. kin.

h-o, τ_{Blackman}

vib. kin.

Figure 7. Vibrational (T_v/T_∞) and first-level (T_{f1}/T_∞) isotherms. $p_\infty = 50\ Pa$.

5. CONCLUSIONS

The results presented in this work show a basic Boltzmannian behaviour of the vibrational population distributions, although the harmonic oscillator model with Millikan and White's relaxation time shows differences with respect to the one with the Blackman's relaxation time and the vibrational kinetics model; these latter models, in turn, are found in remarkable agreement.

However, caution should be exercised to generalise this conclusion to flow situations in which the differences between translational and first-level temperatures, in the vibrational kinetics model, are much larger than those found in the case studied here. In fact, it should be kept in mind that larger differences [2] are responsible for the strong non-Boltzmannian character of the vibrational population distributions, as, for example, in flows expanding through a nozzle [14, 15].

ACKNOWLEDGMENTS

The authors wish to acknowledge the courtesy of the ESTEC Aerothermodynamics section (YPA) for providing the computing facilities. This work has been partially supported by ASI (Agenzia Spaziale Italiana) and by the Regional Government of Sardinia.

References

1. W. E. Meador, G. A Miner and J. H. Heinbockel (1993) Vibrational relaxation in hypersonic flow fields, NASA TP-3367.
2. M. Cacciatore, M. Capitelli, S. De Benedictis, M. Dilonardo, C. Gorse (1986) Vibrational kinetics, dissociation and ionization of diatomic molecules under nonequilibrium conditions, in M. Capitelli, *Topics in Current Physics, Vol. 39: Nonequilibrium Vibrational Kinetics*, Springer Verlag, Berlin, pp. 5–46.
3. C. F. Hansen (1983) Rate processes in gas phase, NASA RP-1090.
4. M. Capitelli and E. Molinari (1980) Kinetics of dissociation processes in plasmas in the low and intermediate pressure range, in S. Veprek and M. Venugopalan, *Topics in Current Chemistry, Vol. 94: Plasma Chemistry III*, Springer Verlag, Berlin, pp. 59–109.
5. R. C. Millikan and D. R. White (1963) Systematics of vibrational relaxation, *J. Chem. Phys.* **39**, 3209.
6. V. Blackman (1956) Vibrational relaxation in oxygen and nitrogen, *J. Fluid Mech.* **1**, 61.
7. W. G. Vincenti and C. H. Kruger, Jr. (1965) *Introduction to physical gas dynamics*, Krieger Publishing Company, Florida.
8. J. D. Anderson, Jr. (1989) *Hypersonic and high temperature gas dynamics*, McGraw Hill Book Co., New York.
9. I. Armenise, M. Capitelli, R. Celiberto, G. Colonna, C. Gorse, A. Laganà (1994) The effect of $N + N_2$ collisions on the non-equilibrium vibrational distributions of nitrogen under reentry conditions, *Chem. Phys. Lett.* **227**, 157.
10. I. Armenise, M. Capitelli (1995) On the coupling of non-equilibrium vibrational kinetics and dissociation-recombination processes in the boundary layer surrounding an hypersonic reentry vehicle, 2nd European Symposium on Aerothermodynamics for Space Vehicles, ESA-SP-367.
11. G. Herzberg (1963) *Molecular spectra and molecular structure, I. Spectra of diatomic molecules*, D. Van Nostrand, Inc., New York.
12. K. P. Huber and G. Herzberg (1979) *Constants of diatomic molecules*, Van Nostrand Rheinold, New York.
13. R. C. Swanson and E. Turkel (1992) On central-difference and upwind schemes, *J. Comput. Phys.* **101**, 292.
14. S. P. Sharma, S. M. Ruffin, W. D. Gillespie and S. A. Meyer (1992) Nonequilibrium vibrational population measurements in an expanding flow using spontaneous Raman spectroscopy, 27th Thermophysics Conference, AIAA Paper 92-2855.
15. R. A. Jones, L. N. Myrabo and H. T. Nagamatsu (1995) A numerical investigation of the effect of vibrational nonequilibrium in expanding flows, 30th Thermophysics Conference, AIAA Paper 95-2076.

A PHYSICALLY CONSISTENT

VIBRATION-CHEMISTRY-VIBRATION COUPLING MODEL

O. KNAB
Institut für Raumfahrtsysteme, Universität Stuttgart
Pfaffenwaldring 31, 70550 Stuttgart, Germany

1. Introduction

It is well–known that the relaxation processes occuring in reentry flows are crucially affected by the interdependence of vibrational excitation and chemical reactions. Since the energy stored in the vibrational mode of a molecule contributes to overcome the activation threshold of a chemical reaction, vibrational excitation determines rate constants. On the other hand, chemical reactions remove or add a certain amount of vibrational energy and thus affect the average vibrational state of a molecular species. While vibration–dissociation–vibration coupling effects are accounted for by many models used in CFD, the influence of vibrational excitation on exchange reactions $X_i + M_1 \rightleftharpoons X_r + M_2$ and the influence of these reactions on the vibrational state of the participating molecules X_i and X_r is mostly neglected. This is unsatisfactory for two reasons: low–energetic exchange reactions as e.g. the Zeldovich reactions

$$NO + O \rightleftharpoons O_2 + N \qquad \text{and} \qquad N_2 + O \rightleftharpoons NO + N \tag{1}$$

firstly, strongly contribute to the processes occuring in the low–temperature part of the boundary layer, and secondly, govern the thermochemical relaxation process of nitric oxide. Consequently, when one neglects the vibration–chemistry–vibration coupling effects of these reactions, important nonequilibrium phenomena may not be recognized by numerical flow simulations.

Motivated by these deficiencies, the Coupled Vibration–Chemistry–Vibration (CVCV) model has been derived on the basis of $T_{vib,i}$–Boltzmann populated vibrational energy levels and truncated harmonic oscillators [5]. It specifies multiple temperature rate constants and vibrational energies transferred due to chemical reactions consistently for both dissociation and

M. Capitelli (ed.), Molecular Physics and Hypersonic Flows, 691–702.
© *1996 Kluwer Academic Publishers. Printed in the Netherlands.*

exchange reactions. Simple analytic expressions for rate constants as well as transferred vibrational energies recommend the model for CFD applications.

In this paper the importance of a consistent vibration–chemistry–vibration coupling model for accurate reentry flow simulations will be pointed out. In particular the effect of the nonequilibrium modeling of the exchange reactions on thermochemical relaxation processes will be discussed.

2. Vibration–Chemistry–Vibration Coupling Concepts

2.1. THE CVCV MODEL

The basis of the CVCV model is the formulation of the state–specific rate constants [3, 5]

$$k_{f_j}^{(E)}(v) = \Lambda\, k_{f_j}^{eq}(T) \exp\left[\frac{\varepsilon_{vib,i}(v)}{\Re}\left(\frac{1}{T} + \frac{1}{U_i}\right)\right] \qquad \varepsilon_{vib,i}(v) \le \alpha\, A_j \quad (2)$$

$$k_{f_j}^{(E)}(v) = \Lambda\, k_{f_j}^{eq}(T) \exp\left[\frac{\varepsilon_{vib,i}(v)}{\Re U_i} + \frac{\alpha A_j}{\Re T}\right] \qquad \varepsilon_{vib,i}(v) > \alpha\, A_j \quad (3)$$

of an endothermic exchange reaction (superscript (E))

$$X_i(v) + M_1 \rightarrow X_r(w) + M_2 \qquad (4)$$

which describes the decomposition of a molecule X_i excited in the particular vth vibrational quantum state and the production of a molecule X_r excited in the wth vibrational state. Note that the activation energy A_j (here in $J/mole$) of this type of reaction is less than the dissociation energy D_i of the molecule X_i. Following the modeling of Marrone and Treanor [7] the variable U_i (with dimension of a temperature) is taken to be a measure of the extent to which the upper vibrational levels are more reactive due to a cross section enlargement at higher excitation. $U_i = \infty$ corresponds to an equal probability assumption or so-called nonpreferential reaction modeling. The parameter α limits the maximum vibrational energy contribution for overcoming the activation threshold A_j of the reaction j to αA_j $(0 \le \alpha \le 1)$ and consequently requires a minimum energy contribution, namely $(1-\alpha)A_j$, from the translational motion of the particles. With other words, α is a measure of the effectiveness vibrational energy is contributing to activation: $\alpha = 0$ means that vibrational excitation has no effect on chemistry, $\alpha = 1$ assumes a rather strong dependence of chemical rates on vibration.

Summing up the state–specific rate constants (2-3) over the finite number of energy levels of a truncated harmonic oscillator by means of a $T_{vib,i}$-

Boltzmann distribution leads to the nonequilibrium rate constant

$$k_{f_j}^{(E)}(T, T_{vib,i}) = \Psi_{f_j}^{(E)}(T, T_{vib,i}) \, k_{f_j}^{eq}(T) = \Psi_{f_j}^{(E)}(T, T_{vib,i}) \, C \, T^s \, e^{-\frac{A_j}{\mathscr{R}T}} \quad (5)$$

with the nonequilibrium factor [4, 5]

$$\Psi_{f_j}^{(E)}(T, T_{vib,i}) = \frac{Q_{vib,i}^{D_i}(T)}{Q_{vib,i}^{D_i}(T_{vib,i})} \frac{Q_{vib,i}^{\alpha A_j}(\Gamma_i) + e^{-\frac{\alpha A_j}{\mathscr{R}\Gamma_i}} Q_{vib,i}^{D_i - \alpha A_j}(T_i^\circ)}{Q_{vib,i}^{\alpha A_j}(-U_i) + e^{-\frac{\alpha A_j}{\mathscr{R}(-U_i)}} Q_{vib,i}^{D_i - \alpha A_j}(T_i^\star)} \quad (6)$$

being a measure of the deviation from thermal equilibrium. This expression which allows for the **vibration–chemistry** coupling phenomena is determined by vibrational partition functions

$$Q_{vib,i}^\Upsilon(\vartheta) = \frac{1 - e^{-\frac{\Upsilon}{\mathscr{R}\vartheta}}}{1 - e^{-\frac{\Theta_{vib,i}}{\vartheta}}} \quad (7)$$

of a harmonic oscillator cut off one quantum state below the energies Υ and calculated at various temperatures ϑ. The pseudotemperatures Γ_i, T_i° and $T_i^\star$ are defined according to

$$\frac{1}{\Gamma_i} = \frac{1}{T_{vib,i}} - \frac{1}{T} - \frac{1}{U_i}, \qquad \frac{1}{T_i^\circ} = \frac{1}{T_{vib,i}} - \frac{1}{U_i}, \qquad \frac{1}{T_i^\star} = \frac{1}{T} - \frac{1}{U_i}. \quad (8)$$

Usually multiple temperature models determine the vibrational temperatures of the molecules from the solution of their global vibrational energy conservation equation [1, 5]

$$\frac{\partial E_{vib,i}}{\partial t} + \nabla(E_{vib,i} \vec{v}) = -\nabla \vec{q}_{vib,i} - \nabla(E_{vib,i} \vec{V_i})$$
$$+ Q_{T-V,i} + Q_{e-V,i} + Q_{V-V,i} + Q_{C-V,i}. \quad (9)$$

Among others, one has to formulate the vibrational energy exchange rate $Q_{C-V,i}$ resulting from chemical reactions. In previous papers it has already been outlined that this **chemistry–vibration** coupling term may not be modeled independently of the state–specific rate constants (2-3). Moreover, chemistry–vibration coupling depends also on the assumptions of $T_{vib,i}$–Boltzmann populated vibrational energy levels and truncated harmonic oscillators which have been introduced in order to describe the vibration–chemistry coupling phenomena (6). The dependence of both vibration–chemistry and chemistry–vibration coupling terms on the same modeling assumptions characterizes a physically consistent vibration–chemistry–vibration coupling model. In the formulation [5, 7]

$$Q_{C-V,i} = \sum_{j=1}^{m} \{\dot{\omega}_{app,ij} \, G_{app,ij} - \dot{\omega}_{va,ij} \, G_{va,ij}\}. \quad (10)$$

the quantities $\dot{\omega}_{app,ij}$ and $\dot{\omega}_{va,ij}$ denote the molar production and decomposition rates of the molecule X_i in the reaction j while $G_{app,ij}$ and $G_{va,ij}$ represent the molar vibrational energies which are on average gained or lost by a molecule X_i being created or destroyed in the reaction j. Maintaining the assumptions of $T_{vib,i}$–Boltzmann populated vibrational energy levels and truncated harmonic oscillators the physically consistent CVCV model expression for $G_{va,ij}$ in exchange reactions reads [4, 5]

$$G_{va,ij}^{(E)}(T, T_{vib,i}) = \frac{Q_{vib,i}^{\alpha A_j}(\Gamma_i) L_i^{\alpha A_j}(\Gamma_i) + e^{-\frac{\alpha A_j}{\Re \Gamma_i}} Q_{vib,i}^{D_i - \alpha A_j}(T_i^\circ) \left(L_i^{D_i - \alpha A_j}(T_i^\circ) + \alpha A_j \right)}{Q_{vib,i}^{\alpha A_j}(\Gamma_i) + e^{-\frac{\alpha A_j}{\Re \Gamma_i}} Q_{vib,i}^{D_i - \alpha A_j}(T_i^\circ)}$$

(11)

with

$$L_i^{\Upsilon}(\vartheta) = \frac{\Re \Theta_{vib,i}}{e^{\frac{\Theta_{vib,i}}{\vartheta}} - 1} - \frac{\Upsilon}{e^{\frac{\Upsilon}{\Re \vartheta}} - 1}$$

(12)

being the average molar vibrational energy content of a harmonic oscillator truncated one quantum state below the energy Υ at the temperature ϑ. The quantity $L_i^{D_i}(T_{vib,i}) = \varepsilon_{vib,i}$ determines the average vibrational energy of the molecule X_i. Applying the principle of detailed balance the vibrational energy on average gained in a formation process is obtained by the limit consideration [5, 7]

$$G_{app,ij}^{(E)} = \lim_{T_{vib,i} \to T} \left\{ G_{va,ij}^{(E)}(T, T_{vib,i}) \right\} = G_{va,ij}^{(E),eq}(T)$$

(13)

to

$$G_{app,ij}^{(E)}(T) = \frac{Q_{vib,i}^{\alpha A_j}(-U_i) L_i^{\alpha A_j}(-U_i) + e^{\frac{\alpha A_j}{\Re U_i}} Q_{vib,i}^{D_i - \alpha A_j}(T_i^\star) \left(L_i^{D_i - \alpha A_j}(T_i^\star) + \alpha A_j \right)}{Q_{vib,i}^{\alpha A_j}(-U_i) + e^{\frac{\alpha A_j}{\Re U_i}} Q_{vib,i}^{D_i - \alpha A_j}(T_i^\star)}$$

(14)

It is emphasized that the expressions (6), (11) and (14) are able to describe the complete nonequilibrium modeling of the CVCV model. Since exothermic backward reactions are characterized by an approximately disappearing activation barrier ($A_j \approx 0$), their nonequilibrium rate constants

$$k_{b_j}^{(E)}(T, T_{vib,r}) = \Psi_{b_j}^{(E)}(T, T_{vib,r}) k_{b_j}^{eq}(T) = \Psi_{b_j}^{(E)}(T, T_{vib,r}) \frac{k_{f_j}^{eq}(T)}{K_{e_j}(T)}$$

(15)

can be calculated by means of the limit consideration

$$\Psi_{b_j}^{(E)}(T, T_{vib,r}) = \lim_{\substack{T_{vib,i} \to T_{vib,r} \\ D_i \to D_r}} \left[\lim_{A_j \to 0} \left\{ \Psi_{f_j}^{(E)}(T, T_{vib,i}) \right\} \right]$$

(16)

with the index r being appointed to the molecular reactant X_r. The same procedure yields the transferred vibrational energies of the molecule X_r to

$$G_{va,rj}^{(E)}(T_{vib,r}) = \lim_{\substack{T_{vib,i} \to T_{vib,r} \\ D_i \to D_r}} \left[\lim_{A_j \to 0} \left\{ G_{va,ij}^{(E)}(T, T_{vib,i}) \right\} \right] = L_r^{D_r}(T_r^\circ) \quad (17)$$

$$G_{app,rj}^{(E)}(T) = \lim_{\substack{T_{vib,i} \to T_{vib,r} \\ D_i \to D_r}} \left[\lim_{A_j \to 0} \left\{ G_{app,ij}^{(E)}(T) \right\} \right] = L_r^{D_r}(T_r^\star) \; . \quad (18)$$

Dissociation reactions (superscript (D)) of the type $X_i + M_1 \rightleftharpoons M_1 + M_2 + M_3$ are characterized by activation energies $A_j = D_i$. Consequently, the nonequilibrium factors of the forward reactions are given by

$$\Psi_{f_j}^{(D)}(T, T_{vib,i}) = \lim_{A_j \to D_i} \left\{ \Psi_{f_j}^{(E)}(T, T_{vib,i}) \right\} \; . \quad (19)$$

Since recombination reactions do not depend on excited vibrational states it follows that $\Psi_{b_j}^{(D)} = 1$ [5, 6]. The energies on average lost in dissociation processes are determined by

$$G_{va,ij}^{(D)}(T, T_{vib,i}) = \lim_{A_j \to D_i} \left\{ G_{va,ij}^{(E)}(T, T_{vib,i}) \right\} \quad (20)$$

whereas the energies on average gained in recombination processes are given by

$$G_{app,ij}^{(D)}(T) = \lim_{A_j \to D_i} \left\{ G_{app,ij}^{(E)}(T) \right\} \; . \quad (21)$$

Electron impact dissociation reactions $X_i + e^- \rightleftharpoons e^- + M_2 + M_3$ and associative ionization reactions $M_1 + M_2 \rightleftharpoons X_r^+ + e^-$ depend on the translational motion of the free electrons [8]. In order to estimate the influence of vibrational nonequilibrium on these types of chemical reactions one has to replace the heavy particle temperature T by the electron temperature T_e in the equations (19-21) and (16-18). For a more detailed description of the CVCV model the reader should consult the original literature [3, 5, 6].

All CVCV–model validation calculations [2, 4, 5, 6] carried out with the nonequilibrium Navier–Stokes code URANUS (**U**pwind **R**elaxation **A**lgorithm for **N**onequilibrium Flows of the **U**niversity of **S**tuttgart) [1] so far, sustained a preferential reaction modeling expressed by the parameters $U_i = D_i/(5\Re)$ and $\alpha = 0.8$. These values are maintained throughout all CVCV–model calculations presented in this work.

2.2. THE CVDV AND CVD MODELS

A special case of the CVCV model is the so–called **C**oupled **V**ibration–**D**issociation–**V**ibration (CVDV) model. This modeling is characterized by

model	dissociation reaktions		exchange reactions	
	k_{f_j}	$Q_{C-V,i}$	k_{f_j}, k_{b_j}	$Q_{C-V,i}$
CVD	$U_i = D_i/(5\Re)$, $\alpha=0.8$	incon	$U_i=\infty$, $\alpha=0$	incon
CVDV	$U_i = D_i/(5\Re)$, $\alpha=0.8$	con	$U_i=\infty$, $\alpha=0$	incon
CVCV	$U_i = D_i/(5\Re)$, $\alpha=0.8$	con	$U_i = D_i/(5\Re)$, $\alpha=0.8$	con

TABLE 1. Summary of CVD-, CVDV- und CVCV–model characteristics; con $\equiv \dot{\omega}_{app,ij}\, G_{app,ij} - \dot{\omega}_{va,ij}\, G_{va,ij}$, incon $\equiv \dot{\omega}_{ij}\, \varepsilon_{vib,i}$

the use of equilibrium rate constants for exchange and associative ionization reactions regardless the vibrational states of the molecules X_i and X_r. Therefore, the nonequilibrium factors must be

$$\Psi_{f_j}^{(E)} = \Psi_{b_j}^{(E)} = 1 \ . \tag{22}$$

This corresponds to the assumption that vibrational excitation has no effect on these types of chemical reactions. Selecting the CVCV–model parameters to $U_i = \infty$ and $\alpha = 0$ the equations (6) and (16) simulate the CVDV coupling concept specified in (22). In addition, the contribution of exchange and associative ionization reactions to the $Q_{C-V,i}$–term is calculated by the expression $\dot{\omega}_{ij}\, \varepsilon_{vib,i} = (\dot{\omega}_{app,ij} - \dot{\omega}_{va,ij})\, \varepsilon_{vib,i}$. As can be shown, this formulation does not account for an effective chemistry–vibration coupling and is physically inconsistent with the formulations (22) [5, 6]. Nevertheless, it is widely used in the literature. From Table 1, however, it becomes apparent that the CVDV model retains the consistent vibration–dissociation–vibration coupling concept of the CVCV modeling.

The **Coupled Vibration–Dissociation** (CVD) model furthermore neglects the dissociation–vibration coupling phenomena employing the inconsistent $Q_{C-V,i}$–formulation $\dot{\omega}_{ij}\, \varepsilon_{vib,i}$ also for dissociation reactions. In this case the coupling term is determined by $Q_{C-V,i} = \sum_j \dot{\omega}_{ij}\, \varepsilon_{vib,i} = \dot{\omega}_i\, \varepsilon_{vib,i}$. Table 1 summarizes the CVD-, CVDV- and CVCV–model characteristics.

3. Discussion

In the following discussion the extended vibration–chemistry–vibration coupling concept of the CVCV model will be investigated simulating two reentry flows. Thereby, emphasize is laid on the nonequilibrium processes occuring in the vicinity of the wall. Both computations have been performed by the URANUS code [1].

3.1. FIRE II FLIGHT EXPERIMENT

The FIRE II flight experiment has been recomputed in order to evaluate the nonequilibrium modeling of the exchange reactions [4]. The considered trajectory conditions at an altitude of $67.05\,km$ are given by $v_\infty = 11.25\,km/s$, $\varrho_\infty = 1.47 \cdot 10^{-4}\,kg/m^3$ and $T_\infty = 228\,K$. Due to the beryllium heat shield of the FIRE II vehicle the surface was assumed to be fully catalytic. Moreover, the wall temperatures were prescribed to be in equilibrium at $1030\,K$. The differences in predicted shock stand–off distances as well as convective and radiative heat fluxes resulting from CVD-, CVDV- and CVCV–coupling concepts have already been pointed out in Ref. [4]. Figure 1 shows the vibrational temperature profiles of nitric oxide along the stagnation line calculated by the three coupling methods. Great discrepancies are obvious both in the shock region and in the vicinity of the wall. Figure 2 compares the profiles close to the wall. In this region the translational temperature profiles (not included here) of all three model calculations approximately coincide with the $T_{vib,NO}$–profile labeled 1. Thus, only the CVDV and CVCV model predict thermal nonequilibrium in the near wall layer.

Since the CVD model accounts for no consistent **chemistry–vibration** coupling neither for dissociation nor for exchange and associative ionization reactions these coupling phenomena are found to be responsible for the nonequilibrium state. This statement is confirmed by Fig. 3 which shows the coupling term $Q_{C-V,NO}$ of nitric oxide. Vibrational energy removal due to dissociation reactions leads to a negative source term $Q_{C-V,NO}$ and hence to a $T_{vib,NO}$–profile being suppressed under the translational temperature profile. Figure 4 specifies the main contributions to the $Q_{C-V,NO}$–term of the CVCV–model calculation. Obviously, the reactions 1, 2 and 5 remove vibrational energy whereas only the Zeldovich mechanism 4 adds vibrational energy. The superposition, however, leads to a negative coupling term and consequently to an overall energy removal. This effect is of course more distinct for the CVDV modeling, because the vibrational energy gain due to the exchange reaction 4 is underestimated by the inconsistent formulation $\dot{\omega}_{ij}\,\varepsilon_{vib,i}$, providing the lowest $T_{vib,NO}$–profile in Fig. 2.

As can be seen from Fig. 3, a strong chemistry–vibration coupling phenomenon immediately at the wall $x < 0.5\,mm$ is predicted by the CVCV–model calculation only. Thus, it must be an effect of the nonequilibrium modeling of the exchange reactions. From Fig. 4 it becomes evident that nitric oxide is formed in the exchange reaction $NO + O \rightleftharpoons O_2 + N$ but also destroyed in the second Zeldovich mechanism $N_2 + O \rightleftharpoons NO + N$. So both reactions take place in the reverse direction. Since the CVCV model predicts the average vibrational energy gain $G^{(E)}_{app,ij}(T)$ in an endothermic exchange reaction to be greater than the average vibrational energy loss

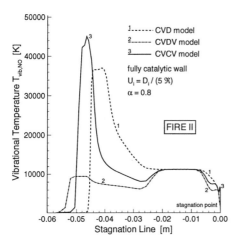

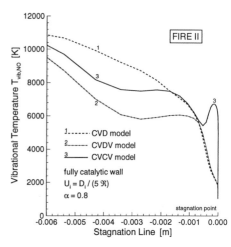

Figure 1. $T_{vib,NO}$-profiles along the stagnation line

Figure 2. $T_{vib,NO}$-profiles close to the stagnation point

$G^{(E)}_{va,rj}(T, T_{vib,r})$ in an exothermic exchange reaction [6] one gets a positive coupling term $Q_{C-V,NO}$, leading to an increase in the vibrational temperature of nitric oxide (see Fig. 2).

In conclusion, the extended consideration of vibrational nonequilibrium effects in exchange reactions by the recommended CVCV model significantly affects the prediction of thermochemical relaxation processes especially in the near wall region. As has been pointed out in [4] this extension further leads to an increase of the convective heat flux in the stagnation

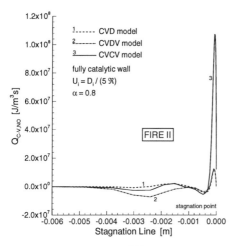

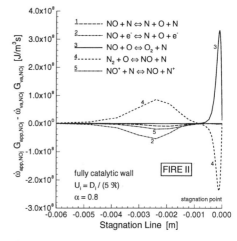

Figure 3. Chemistry–Vibration coupling term $Q_{C-V,NO}$ close to the wall

Figure 4. Contributions of various reactions to $Q_{C-V,NO}$ (CVCV model)

region of 10%.

3.2. MIRKA CAPSULE FLOW

MIRKA is a $1\,m$ diameter ballistic capsule which is planned to fly diagnostic experiments in order to develope freeflight measurement techniques and to validate thermochemical modelings. The simulated trajectory point corresponds to the flight phase with expected maximal heat flux [6]: $h = 58\,km$, $v_\infty = 6.32\,km/s$, $\varrho_\infty = 4.0 \cdot 10^{-4}\,kg/m^3$ and $p_\infty = 0.3\,mbar$. The URANUS/CVCV–model computation was performed with the assumption of a radiation equilibrium wall characterized by an emission coefficient of $\varepsilon = 0.8$. Figure 5 displays the temperature distributions along the stagnation line calculated with a non catalytic wall assumption. Remarkable is the steep vibrational temperature increase of oxygen and nitric oxide close to the wall. Figure 6 compares the temperature profiles with those obtained by a fully catalytic wall calculation. Apparently, the wall catalycity is crucial for the thermal state of the gas phase near the surface. This is further confirmed by the concentration distributions shown in Fig. 7. While a fully catalytic surface provides a monotonous increase of oxygen molecules, a non catalytic wall yields molecular oxygen production just in front of the wall. The Figures 9 und 10 specify the most important contributions to the chemistry–vibration coupling term Q_{C-V,O_2} of oxygen. Due to the different orders of magnitude of oxygen molecules predicted by the two wall conditions the source terms are devided by the partial density ϱ_{O_2}. The superpositions of all contributions are shown in Fig. 8.

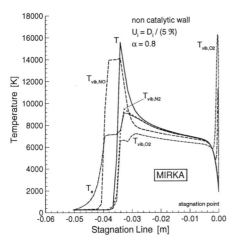

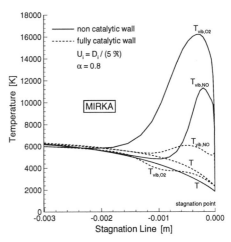

Figure 5. Temperature distributions along the stagnation line (non catalytic wall)

Figure 6. Comparison of temperature profiles resulting from a non and fully catalytic wall assumption

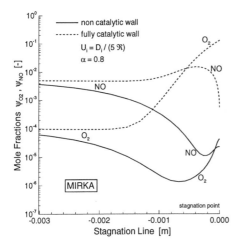

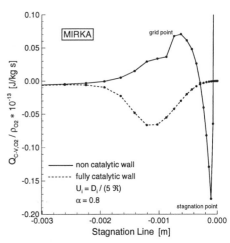

Figure 7. Mole fractions of O_2 and NO in the vicinity of the wall

Figure 8. Comparison of the coupling terms Q_{C-V,O_2} predicted by a non and fully catalytic wall assumption

In both cases the exchange reaction $NO + O \rightleftharpoons O_2 + N$ leads to a reduction of O_2 and thus to a vibrational energy loss. In the case of a non catalytic surface, however, recombination reactions provide for a vibrational energy gain and hence a positive overall chemistry–vibration coupling term Q_{C-V,O_2}. Supported by low translation–vibration and vibration–vibration energy exchange rates in this area there appears a vibrational temperature increase. The reason is that the average vibrational energy $G^{(D)}_{app,O_2}$ of a

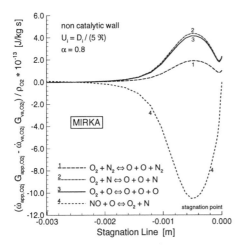

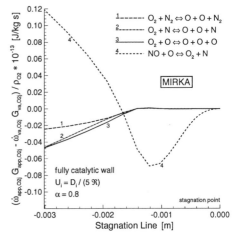

Figure 9. Contributions to the coupling term Q_{C-V,O_2} (non catalytic wall)

Figure 10. Contributions to the coupling term Q_{C-V,O_2} (fully catalytic wall)

recombined O_2-molecule exceeds the vibrational energy $G_{va,O_2}^{(E)}$ on average lost by a O_2-molecule in the Zeldovich mechanism labeled 4. In the case of a fully catalytic surface there are much more oxygen molecules in the gas phase (see Fig. 7). This means that the potential for recombination processes is lower and the gas mixture is closer to chemical equilibrium. Consequently, dissociation/recombination reactions do not yield an important contribution to the chemistry–vibration coupling term Q_{C-V,O_2}, as can be seen in Fig. 10. The vibrational energy loss in the exchange reaction 4 dominates the source term und gives the temperature distribution $T_{vib,O_2} < T$ (see Fig. 6).

At a first glance the vibrational temperature maxima near the wall look somehow questionable. However, the detailed analysis of the processes explained their cause. Moreover, they could be interpreted as an indication for non Boltzmann populated energy levels. The recombination or generally formation processes predominantly populate the upper vibrational levels. This is represented by the CVCV modeling through $G_{app,ij}(T) \gg \varepsilon_{vib,i}$. If translation–vibration and vibration–vibration energy exchange mechanisms do not redistribute the levels, the upper levels remain overpopulated. Approximating the population of these upper levels by a $T_{vib,i}$–Boltzmann distribution yields high vibrational temperatures which describe the upper levels quite well but underestimate the population of the lower vibrational levels [6]. State–specific nonequilibrium flow calculations as presented at this conference should contribute to confirm these multiple temperature CVCV–model results.

4. Conclusions

- Vibration–chemistry–vibration coupling concepts significantly affect the prediction of thermochemical relaxation processes of molecules. Great differences are obvious both in the shock region and in the vicinity of the wall.
- The consideration of vibrational nonequilibrium effects in exchange reactions as suggested by the CVCV model influences the prediction of thermochemical relaxation processes especially in the near wall region.
- High vibrational temperatures near the wall could be interpreted as an indication for non Boltzmann populated vibrational energy levels. These can be generated either by formation processes in exchange reactions, as shown by the FIRE II calculation, or by recombination reactions, as demonstrated by the MIRKA simulation.
- Comparisons with state–specific nonequilibrium flow simulations as presented at this conference will help to validate the multiple temperature CVCV model further on.

702

5. Acknowledgements

This work is supported by the Deutsche Forschungsgemeinschaft DFG. The author wishs to express his sincere thanks to M. Fertig for his assistance calculating the MIRKA capsule flow.

References

1. Daiß, A., Schöll, E., Frühauf, H.-H. and Knab, O. (1993) Validation of the URANUS Navier-Stokes Code for High–Temperature Nonequilibrium Flows, *AIAA-93-5070*.
2. Frühauf, H.-H., Daiß, A., Gerlinger, U., Knab, O. und Schöll, E. (1994) Computation of Reentry Nonequilibrium Flows in a Wide Altitude and Velocity Regime, *AIAA-94-1961*.
3. Knab, O., Frühauf, H.-H. and Jonas, S. (1992) Multiple Temperature Descriptions of Reaction Rate Constants with Regard to Consistent Chemical–Vibrational Coupling, *AIAA-92-2947*.
4. Knab, O., Gogel, T.H., Frühauf, H.-H. und Messerschmid, E.W. (1995) CVCV–Model Validation by Means of Radiative Heating Calculations, *AIAA-95-0623*.
5. Knab, O., Frühauf, H.-H. and Messerschmid, E.W. (1995) Theory and Validation of the Physically Consistent Coupled Vibration-Chemistry-Vibration Model, *Journal of Thermophysics and Heat Transfer* 9, No. 2, pp. 219-226.
6. Knab, O. (1995) Konsistente Mehrtemperatur–Modellierung von thermochemischen Relaxationsprozessen in Hyperschallströmungen, Ph.D. thesis, Universität Stuttgart.
7. Marrone, P.V. and Treanor, C.E. (1963) Chemical Relaxation with Preferential Dissociation from Excited Vibrational Levels, *Physics of Fluids* 6, No. 9, pp. 1215-1221.
8. Park, C. (1986) Convergence of Computation of Chemical Reacting Flows, in J.N. Moss and C.D. Scott (eds.), *Thermophysical Aspects of Reentry Flows*, Progress in Astronautics and Aeronautics, AIAA, New York, Vol. 103, pp. 478-513.

FUNDAMENTAL ASPECTS OF THE COUPLING OF NON-EQUILIBRIUM VIBRATIONAL KINETICS AND DISSOCIATION-RECOMBINATION PROCESSES WITH THE BOUNDARY LAYER FLUIDYNAMICS IN N_2 AND AIR HYPERSONIC FLOWS

I.ARMENISE, M.CAPITELLI and C.GORSE

Centro di Studio per la Chimica dei Plasmi del CNR and Department of Chemistry - University of Bari (Italy)

Abstract

A ladder climbing model including V-V (vibration-vibration), V-T (vibration-translation) energy exchange processes linked to a dissociation recombination kinetics has been developed and inserted in the fluid dynamics equations describing the boundary layer surrounding a non-catalitic surface hitted by an hypersonic flow of atomic and vibrationally excited molecular nitrogen.
The results show a strong overpopulation of vibrational levels with respect to Boltzmann distributions along the coordinate perpendicular to the surface and a corresponding non-Arrhenius behaviour of dissociation constants.
The extension of the model to the air system confirm the strong non-equilibrium character of the relevant kinetics even though the reaction between vibrationally excited nitrogen and atomic oxygen controls both N_2 dissociation and NO formation

1. Introduction

We have recently shown the importance of the recombination-dissociation process in creating non-equilibrium vibrational distributions in the boundary layer surrounding a body flying at hypersonic velocity /1-5/.
These results were obtained by inserting in a simplified fluid dynamic model appropriate kinetic equations describing the relaxation of vibrational levels

M. Capitelli (ed.), Molecular Physics and Hypersonic Flows, 703–716.

of N_2 as well as their coupling to the recombination-dissociation processes. Recombination considers selective pumping of vibrational levels v=45 and v=25 as a result of the introduction of vibrational quanta over the last bound level of the molecule and the formation and subsequent relaxation of the electronically excited state $A^3\Sigma_u^+$ of nitrogen on level v=25.

This kind of approach determines not only strong non-equilibrium vibrational distributions near the surface but also a population inversion at v=25.

To understand the sensitivity of our results to different recombination-dissociation models we present in this paper a new model of dissociation-recombination based on the so called ladder climbing model/6/. According to this model dissociation occurs when the vibrational quanta overcome a pseudo level located above the last vibrational level of the molecule, while the recombination is calculated through the detailed balance principle. The kinetic equations are then inserted in the fluid dynamic model described in refs. 1-5. Particular emphasis will be given to the possibility of a non-Arrhenius behaviour of dissociation constants along the coordinate perpendicular to the surface.

Extension of the present model to air kinetics will be also discussed even though the bulk of the present results refer to N_2.

2. The ladder climbing model

We assume that the dissociation of N_2 occurs through the excitation of the vibrational ladder of the molecule ending in the dissociation continuum represented by a pseudo level (v=46 in our equations) located over the last bound vibrational level (v'=45) of the molecule. The following processes are then responsible for the dissociation

VT molecule-molecule energy exchange processes

$$N_2(v'=45) + N_2 \ \text{-->} \ N_2(46) + N_2 \ \text{-->} \ N + N + N_2 \tag{1}$$

VT atom-molecule energy exchange processes

$$N_2(v) + N \ \text{-->} \ N_2(46) + N \ \text{-->} \ N + N + N \tag{2}$$

VV energy exchange processes

$$N_2(v) + N_2(v'=45) \text{-->} \ N_2(v-1) + N_2(46) \text{-->} \ N_2(v-1) + N + N \tag{3}$$

The dissociation rate is then written as

$$\frac{dN(46)}{dt} = \sum_w P_{w,w-1}^{45,N} N_2(45) N_2(w) + P2_{45,N} N_2(45) N_2 +$$

$$\sum_w P1_{w,N} N_2(w) N = k_d N_{tot} \tag{4}$$

where the first therm is the contribution of V-V processes while the other two terms are the contributions coming from V-T processes from molecules and atoms. k_d is a phenomenological first order dissociation constant.

In the previous formula N_{tot} is the total number density of the gas, $N_2(w)$ is the number density of molecules in the wth vibrational level, N is the number density of atoms and P, P2 and P1 are respectively the rate coefficients of V-V, V-T molecule-molecule and V-T atom-molecule energy exchange processes linking the wth level with the continuum (v=46=N).

We can separate the contributions coming from molecules and atoms i.e. we can study separately the two reactions

$$N_2 + N_2 \longrightarrow 2N + N_2 \tag{5}$$

$$N_2 + N \longrightarrow 3N \tag{6}$$

In the first case we solve the vibrational master equation of nitrogen only in the presence of nitrogen molecules, while in the second case we solve the vibrational master equation in the presence of a large quantity of atomic nitrogen. So we can define a pseudo-first order dissociation constant from molecules as

$$k_d^{N_2} = \frac{\sum_v P_{v,v-1}^{45,N} N_2(45) N_2(v)}{N_{tot}} + \frac{P2_{45,N} N_2(45) N_2}{N_{tot}} \tag{7}$$

and from atoms as

$$k_d^N = \frac{\sum_v P1_{v,N} N_2(v) N}{N_{tot}} \tag{8}$$

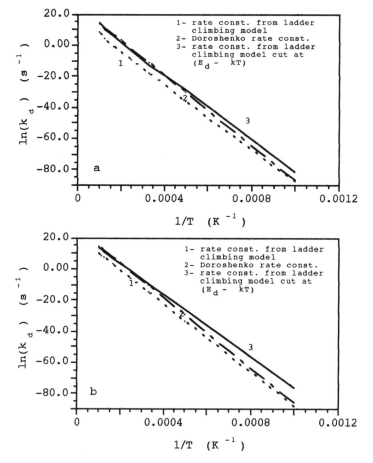

fig.1a-b - Logarithms of pseudo-first order dissociation constants versus $1/T$
($N_{tot} = 7.24*10^{18}$ part/cm^3) for the reaction:
a) $N_2 + N_2 \rightarrow N_2 + 2N$; b) $N + N_2 \rightarrow 3N$.

To obtain the two constants we must know the concentration of all vibrational levels including the last bound one ($v'=45$) together with the relevant rate coefficients connecting bound levels with the pseudolevel. These rates have been obtained by extrapolating bound-bound rates to the continuum. This procedure probably underestimates the rates.

To calculate the population densities we have coupled the equation for the pseudolevel with the system of vibrational master equations giving the temporal evolution of each vibrational level under the presence of V-V and V-T energy exchange processes.

We start the integration of the system of first order differential equations with all molecules concentrated in $v=0$. Then V-V and V-T energy exchange processes propagate the vibrational quanta over the whole vibrational manifold ending in dissociation. A time dependent dissociation constant can be therefore calculated: at the same time a quasistationary value of dissociation constant can be defined when the dissociation constant reaches a plateau (see ref.7). The pseudo first order quasistationary dissociation constants have been reported against 1/T in figures 1a-b obtaining an Arrhenius behavior for the dissociation process induced by molecules (fig.1a) and by atoms (fig.1b).This equilibrium behaviour is due to the fact that the corresponding quasistationary vibrational distributions (those corresponding to the plateau regime) satisfy Boltzmann laws at the corresponding translational temperature.

In the same figures we have also reported the corresponding "experimental" results recommended in ref./5/. We can see that our results which have been fitted by the following equations

$$K_D{}^N = \exp(-22.18) * \exp(-1.0928*10^5/T) \ cm^3 sec^{-1}/particle \qquad (9)$$

$$K_D{}^{N2} = \exp(-24.745) * \exp(-1.1124*10^5/T) \ cm^3 sec^{-1}/particle \qquad (10)$$

underestimate up to two orders of magnitude the dissociation constants compared with the "experimental" results. This point however deserves some comments. The accuracy of experimental data for nitrogen dissociation is still an open problem . From the theoretical point of view the ladder climbing model is an oversimplification. The corresponding results strongly depend on the used V-V and V-T rates. A better agreement with the experimental results was found many years ago by using an other set of V-V and V-T rates /7/.

The weakness of ladder climbing model lies in the fact that only transitions involving the last bound level of molecule result in dissociation. This point could be eliminated by imposing as last bound vibrational level the dissociation limit decreased by kT. Results from this new model which have been fitted by the following equations

$$K_D{}^N = \exp(-19.642) * \exp(-1.0*10^5/T) \ cm^3 sec^{-1}/particle \qquad (11)$$

$$K_D{}^{N2} = \exp(-19.845) * \exp(-1.048*10^5/T) \ cm^3 sec^{-1}/particle \qquad (12)$$

have been also reported in figures 1a-b resulting in a better agreement with the experimental values.

To summarize we have obtained two independent sets of theoretical dissociation constants to be joined to the experimental values of Doroshenko et al./5/.

The three sets of values can be used in our previous model (that one which selectively pumps levels v=25 and v=45) /1-5/ for checking the sensitivity of vibrational distributions to different k_d (and therefore recombination rates) values.

The corresponding results present a qualitative similar behaviour even though strong quantitative differences are observed.

3. The insertion of ladder climbing model in the fluid dynamics

We want now to study the vibrational distributions which develop when the ladder climbing model is inserted in the equations describing the boundary layer surrounding a body flying at hypersonic velocity. To this end we have slightly modified the set of second order differential equations described in our previous papers: in particular the relevant equation for atom formation is replaced by a continuity equation for the pseudo level while the recombination is spread over all the vibrational manifold of the molecule (see also ref./6/ for the zerodimensional model). The relevant sets of second order differential equations are written as

$$C_v'' + f * Sc * C_v' = -\sum_{i=0}^{46} B(v,i) * C_i \qquad\qquad v=0\div46$$

$$\theta'' + f * Pr * \theta' = \sum_v \frac{Le * E_v}{c_p * T_e}\left(\sum_i B(v,i)*C_i\right) \qquad\qquad (13)$$

where the first 47 ones are the continuity equations of the bounded vibrational levels ($v=0\div45$) and of atoms ($v=46$) whereas the last equation is the energy one. The derivatives have been done with respect to the coordinate normal to the surface (η), C_v ($=\rho_v/\rho$) is the mass concentration of molecules on the vth vibrational level, C_N ($=\rho_N/\rho$) the atom mass concentration, $\theta = T/T_e$ (T is the temperature and subscript "e" is referred to quantities at the edge of the boundary layer), Pr, Le and Sc are Prandtl, Lewis and Schmidt numbers, f is a stream function, c_p is the gas translational and rotational heat capacity (see refs.[1-5] for details)

The coefficients appearing on the right hand side part of the equations read as

$$B(v,k) = (\sum_w C_w P_{w,w-1}^{k,v} + P2_{k,v}(1-C_N) + 2C_N P1_{k,v}$$

$$+ 4C_N^2 P_{N,45}^{v-1,v} \frac{p_e}{R*T*(1+C_N)} \quad)*A \qquad\qquad \text{if } k<v$$

$$B(v,k) = (\sum_w C_w P_{w,w+1}^{k,v} + P2_{k,v}(1-C_N) + 2C_N P1_{k,v} + C_{45}P_{k,v}^{45,N})*A \quad \text{if } k>v$$

$$B(v,v) = -(\sum_w C_w P_{w,w-1}^{v,v+1} + \sum_w C_w P_{v,v-1}^{w,w+1} + (P2_{v,v+1} + P2_{v,v-1})*(1-C_N)$$

$$+ 2C_N \sum_{k \neq v} P1_{v,k} + 4C_N^2 P_{N,45}^{v,v+1} \frac{p_e}{R*T*(1+C_N)})*A$$

$$B(v,46) = (2C_N P2_{N,v}(1-C_N) + 4C_N^2 P1_{N,v}$$

$$+ 2C_N \sum_{w=1}^{46} C_{w-1} P_{N,v}^{w-1,w} \frac{p_e}{R*T*(1+C_N)})*A$$

$$B(46,46) = -(\sum_w C_w P_{46,45}^{w,w+1} + P2_{46,45}(1-C_N) + 2C_N \sum_{k \neq 46} P1_{46,k}$$

$$- 2C_N \sum_{w=0}^{45} C_w P_{N,45}^{w,w+1} \frac{p_e}{R*T*(1+C_N)})*A$$

where

$$A = \frac{Sc*p_e}{R*T*(1+C_N)\beta}$$

and p_e represents the pressure both at the edge of the boundary layer and along the normal to the surface, R the gas constant, β the velocity gradient along the surface (du_e/dx). As we can notice, the symbols adopted are the same used in our previous works. Here the subscript N is equal to 46.

710

Let us now examine the corresponding vibrational distributions for fixed parameters.

Figures 2 and 3 report the vibrational distributions along η for low and high pressure conditions showing a strong non-equilibrium character near to the surface (remind that η=0 represents the surface), this behaviour being due to the recombination process. This point can be better understood by looking at figure 4 where the fluid dynamic problem has been solved by completely neglecting the dissociation-recombination processes. We can see that the vibrational distributions closely follow Boltzmann distributions at the local gas temperature.

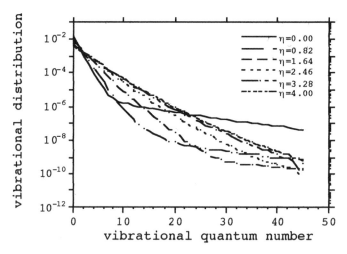

fig.2 - Normalized vibrational distributions C_v versus vibrational quantum number v at different η values using ladder climbing model with T_w = 1000 K, T_e = 7000 K, p_e = 1000 N/m^2, β = 5000 s^{-1}

We must be aware, however, that non-equilibrium vibrational distributions can occurr also in the boundary layer as a result of the so called V-V up pumping mechanism. The conditions suitable for this mechanism occurr when the surface is kept at low temperature and when the concentration of atomic nitrogen is not important. The two conditions can be obtained by decreasing both the temperature at the edge of the boundary layer and that one on the surface.

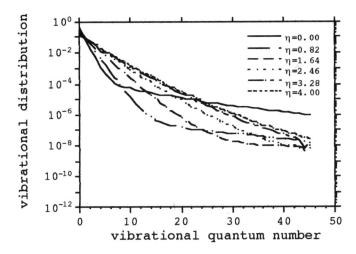

fig.3 - Normalized vibrational distributions C_v versus vibrational quantum number v at different η values using ladder climbing model with $T_w = 1000$ K, $T_e = 7000$ K, $p_e = 10^5$ N/m^2, $\beta = 10^5$ s^{-1}

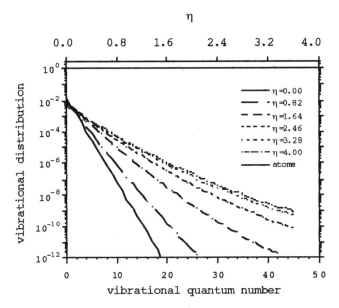

fig.4- Normalized atomic density C_N versus η (upper axis) and normalized vibrational distributions C_v versus vibrational quantum number v at different η values neglecting dissociation recombination processes with $T_w = 1000$ K, $T_e = 7000$ K, $p_e = 1000$ N/m^2, $\beta = 5000$ s^{-1}

4. Dissociation constants in the boundary layer

In the previous section we have reported the vibrational distributions of N_2 along the coordinate η of the boundary layer, emphasizing their strong non-equilibrium character. This means that also the dissociation constants should show a non-equilibrium behaviour along the same coordinate or equivalently along the temperature corresponding to each η value.

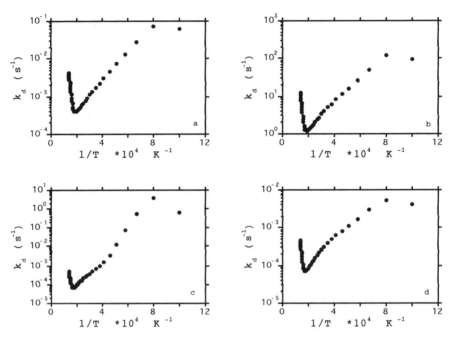

fig.5a-d - Total pseudo-first order dissociation rate constants for the following conditions : 5a) T_e=7,000K, T_w=1,000K, p_e = 1000 N/m^2, β = 5000 s^{-1}; 5b) T_e=7,000K, T_w=1,000K, p_e = 10^5 N/m^2, β = 10^5 s^{-1}; 5c) T_e=5,000K, T_w=300K, p_e = 1000 N/m^2, β = 5000 s^{-1}; 5d) T_e=5,000K, T_w=1000K, p_e = 1000 N/m^2, β = 5000 s^{-1}

This behaviour can be appreciated in figure 5a-d where we have reported the total pseudo-first order dissociation rate constant (i.e. from eq.4) versus 1/T in the boundary layer for different conditions . We can see that a non-Arrhenius behaviour exists: in particular dissociation constants can also increase with decreasing gas temperature as a result of the non-equilibrium character of vibrational distributions along η. This behaviour is due to the interplay of recombination process and of V-V up pumping mechanism as well as to the fact that under the present conditions the residence time $(1/\beta)$ is not sufficient to thermalize the molecules.

To better understand the last point we have repeated the calculations of figure 5c by eliminating both recombination and V-V energy transfer i.e. we are considering only V-T processes including dissociation. In this case we should expect Boltzmann distributions for the vibrational distributions and an Arrhenius behaviour for the dissociation rates. This trend is partially verified. In particular we observe vibrational distributions which satisfy Boltzmann laws and pseudofirst order dissociation rates which follow an Arrhenius behaviour with apparently two slopes. In any case dissociation constants do not increase with decreasing gas temperature.

5. Air kinetics

Extension to air flow of the previous ideas is not so obvious due to the numerous elementary kinetic processes occurring in the N_2/O_2 reacting mixture.

Preliminary results in this direction have been obtained by considering the following elementary processes

$$N_2(v) + N_2(w) \longleftrightarrow N_2(v-1) + N_2(w+1) \tag{14}$$

$$N_2(v) + N_2 \longleftrightarrow N_2(v-1) + N_2 \tag{15}$$

$$N_2(v) + N \longleftrightarrow N_2(v-1) + N \tag{16}$$

$$O_2(v) + O_2(w) \longleftrightarrow O_2(v-1) + O_2(w+1) \tag{17}$$

$$O_2(v) + O_2 \longleftrightarrow O_2(v-1) + O_2 \tag{18}$$

$$O_2(v) + O \longleftrightarrow O_2(v-1) + O \tag{19}$$

$$O_2(v) + N_2 \longleftrightarrow O_2(v-1) + N_2 \tag{20}$$

$$N_2(v) + O \longleftrightarrow N_2(v-1) + O \tag{21}$$

$$N_2(v) + O \longleftrightarrow NO + N \tag{22}$$

The model in particular includes a ladder climbing model for nitrogen (47 levels), a similar one for oxygen (34) and some coupling reactions between the two systems. In particular the formation of NO through the reaction of vibrationally excited molecules and oxygen atoms is of particular interest. Details of these calculations will be presented elsewere.

714

A sample of results has been reported in figure 6a-c where the pseudo-first order dissociation constants for N_2 and O_2 as well as the NO formation rate have ben plotted as a function of instantaneous $1/T$ value.

We can see that still a non-Arrhenius behaviour is present in the different rates. A better insight of the results, however, shows that N_2 dissociation is due to process (22) rather than to the ladder climbing model. This point is confirmed by the fact that the NO formation rate is the same as the dissociation rate of N_2. It should be noted that the small dissociation constants of O_2 compared to N_2 and NO are simply due to the fact that the concentration of oxygen molecules in the boundary layer is very small.

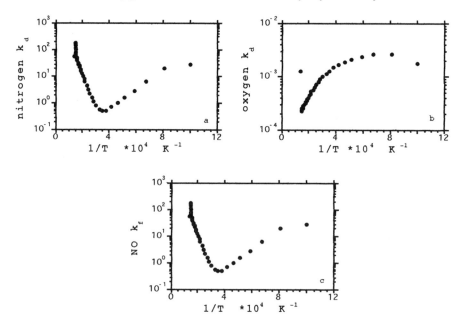

fig.6a-c - Total pseudo-first order dissociation rate constants of nitrogen (6a) and oxygen (6b) and total formation rate of NO (6c) for the following conditions: $T_w = 1000$ K, $T_e = 7000$ K, $p_e=10^3$ N/m^2, $\beta=5*10^3 s^{-1}$

6.Conclusions

We have shown in the previous pages that strong non equilibrium vibrational distributions can arise in the boundary layer surrounding a body flying at hypersonic flow either as the result of recombination process or of V-V up pumping mechanism.

In the first case these distributions depend on the dissociation-recombination rates as well as on the adopted model. Qualitatively, however, the distributions follow the same trend presenting a large non-equilibrium character near to the surface.

Another interesting point is the strong non-Arrhenius behaviour of dissociation rates versus the instantaneous $1/T$ values met in the boundary layer. This trend confirms similar results obtained by our group during the development of pure vibrational mechanisms under electrical conditions /6-7/. Extension of these results to air flow shows a similar behaviour in the dissociation constants of N_2 and O_2 and on the formation rate of NO even though in this case the reaction between vibrationally excited molecules and oxygen atoms controls the situation.

Acknowledgments

This work has been partially supported by ASI (Agenzia Spaziale Italiana). The authors also thank Dr. G.Colonna for usefull discussions.

References

[1] - Armenise, I., Capitelli,M., Celiberto,R., Colonna,G.
   Gorse, C., and Laganà , A. (1994) The effect of $N + N_2$ collisions on the non-equilibrium vibrational distributions of nitrogen under reentry conditions Chem. Phys. Lett. 227, pp. 157-163
[2] - Armenise, I., Capitelli,M., Celiberto,R., Colonna,G.
   Gorse, C.(1994) Non-equilibrium vibrational distributions of N2 under reentry conditions: the role of atomic nitrogen AIAA/ASME 6th Joint Thermophysics and Heat Transfer Conference - June 20-23, 1994/Colorado Springs, Colorado, paper n.94-1987
[3] - Armenise, I., Capitelli,M., and Gorse, C. (1995) On the coupling of non-equilibrium vibrational kinetics and dissociation-recombination processes in the boundary layer surrounding an hypersonic reentry vehicule ESA SP-367, pp.287-292
[4] - Armenise,I., Capitelli,M., Colonna,G., Koudriavtsev, N. and Smetanin,V. (1995) Non-equilibrium vibrational kinetics during hypersonic flow of a solid body in nitrogen ant its influence on the surface heat transfer Plasma Chem. Plasma Proc. Vol.15 No.3, pp. 501-528

716

[5] - Armenise, I., Capitelli, M., Colonna, G., Gorse, C. (1996) Non-equilibrium vibrational kinetics in the boundary layer of reentering bodies J.Thermophysics and Heat Transfer, in press

[6] -Capitelli , M. and Molinari , E. (1980) Kinetics of dissociation processes in plasmas in the low and intermediate pressure range Topics in Curr. Chem.90, pp. 59 -109

[7] -Capitelli, M. and Dilonardo , M. (1978) Non-equilibrium dissociation of nitrogen Rev. Phys. Appl. (Paris) 13, pp. 115-123

STATISTICAL SIMULATION OF HIGHLY NONEQUILIBRIUM HYPERSONIC RAREFIED FLOWS

M.S. IVANOV, S.G. ANTONOV, S.F. GIMELSHEIN,
A.V. KASHKOVSKY AND G.N. MARKELOV
Institute of Theoretical and Applied Mechanics,
Novosibirsk 630090, Russia

1. Introduction

The study of space vehicle aerodynamics at the altitudes 80-120 km becomes more and more important. This is primarily caused by the development of new generation of space transportation systems, for which the mentioned altitudes are determining. This requires more comprehensive and detailed understanding of the problems of hypersonic aerodynamics at high flight altitudes.

Modern experimental facilities (hypersonic wind tunnels of various types, hot-shot wind tunnels, etc.) cannot duplicate the conditions of hypersonic flight in rarefied atmosphere with due account of physico- chemical processes. Therefore, the computational methods of rarefied gas aerodynamics are practically the only means to obtain data on the complex hypersonic rarefied flows of chemically reacting gas.

Presently the Direct Simulation Monte Carlo (DSMC) method is the main tools for investigation of multidimensional rarefied gas flows. This is related to the fact that the methods of continual computational aerodynamics based on Navier-Stokes equations cannot be applied to study the flow around a space vehicle at the altitudes higher than 85-90 km because of considerable atmosphere rarefaction.

Many authors (see, e.g., [1-3]) are now deeply involved in the numerical study of real gas effects on the rarefied hypersonic flow about blunted bodies. Real gas effects change considerably the flow structure and the heat flux distributions but stipulate no noticeable changes in aerodynamic characteristics.

When a hypersonic rarefied gas flows about a concave body at incidence, the excitation of internal degrees of freedom and chemical reactions may

717

M. Capitelli (ed.), Molecular Physics and Hypersonic Flows, 717–736.
© *1996 Kluwer Academic Publishers. Printed in the Netherlands.*

change considerably not only the flow structure but also the distributed and total aerodynamic characteristics, in particular, the pitching moment. The main objective of the present paper is the investigation of the influence of real gas effects on the flow pattern over concave bodies.

The results of numerical simulation of 2D, axisymmetric and 3D flows of this kind about concave bodies are presented in the paper. The computations are performed with the rarefied computational tools [4,5]. Various models describing the internal degrees of freedom of molecules and chemical reactions are presented and used for the computations.

2. Conditions for Calculations and Numerical Method

A special set of concave bodies was used in calculations. A flat plate with a flap and 10% elliptic profile with a deflected rear part were used for 2-D calculations. Calculations of axisymmetrical flows were carried out for a cone with a flare. A flat delta-wing (span 70%) with a flap was used for 3-D calculations. Such a set of aerodynamic models allowed us to reveal the influence of flow dimensionality on the distributed aerodynamic characteristics and flowfields structure.

The length of deflected surfaces was $0.2L$ where L was the total body length and the angle of deflection was $\delta = 0°, 15°$ and $30°$. The flap was deflected towards the free stream. The free stream velocity and temperature amounted to $U_\infty = 7600$ m/s and $T_\infty = 189$ K, respectively. The wall temperature was assumed to be constant and equal to $T_w = 1000$ K. The Knudsen number determined from the characteristic length L was $Kn = 0.01$. Such conditions are much the same as during the flight of space vehicles at the altitudes 85-100 km, depending on the characteristic size of a vehicle.

Numerical schemes [6] of the DSMC method based on the majorant principle were used for the calculations in this work. A Variable Hard Sphere Model with an exponent of 0.75 in viscosity-temperature dependence was chosen as a potential of intermolecular collisions. The particle reflection on the body surface was modeled according to the diffuse law with complete accommodation of translational and internal energies.

As the first step the phenomenological Larsen-Borgnakke model was used for account of the internal degrees of freedom of molecules (Sections 3.1–3.3). Rotation and vibration relaxation collision numbers were supposed to be 5 and 50, respectively. The simple model of chemical reactions [7] was taken for these calculations.

To study the effect of internal degrees of freedom of molecules and chemical reactions on the flow, calculations were carried out for the following gas models: monoatomic gas (T-T gas); gas with translation-rotation exchanges (T-R gas); gas with translation-rotation-vibration exchanges (T-R-V gas); reacting oxygen (two dissociation reactions). The calculations were also carried out for reacting nitrogen and air, but only results for oxygen are presented below since it is the most active air component from the viewpoint of chemical reactions.

The quantum character of vibrational energy mode is rather important for high-temperature flows, and as the second step two models for DSMC method which make use of a level-to-level description of vibrational mode are introduced (Section 3.4). Model of chemical reactions with the vibration-dissociation coupling is also described in Section 3.4.

3. Results and Discussion

3.1. THE INFLUENCE OF ATTACK FOR VARIOUS GAS MODELS

The results of calculations the flow about a flat plate and a plate with a flap (the deflection angle $\delta = 30°$) for the angles of attack $0°, 20°$ and $40°$ and various gas models are presented in this Section.

When T-T gas flows about a plate and a plate with a flap (Fig. 1a) at the angle of attack $40°$, a detached shock wave is formed. As a result, a vast subsonic region is formed behind the bow shock wave. In case T-R gas flows about a flat plate, the shock wave is attached to the body. However, the flap deflection (Fig. 1b) increases considerably the pressure in the upstream direction, which, in turn, causes a marked increase in the shock wave inclination. A vast subsonic region is formed behind the bow shock wave similarly to T-T gas. The excitation of vibrational degrees of freedom of molecules changes the qualitative flow structure near a plate with a flap (Fig. 1c), as compared with the previous cases. The shock wave is practically attached to the body. The subsonic region is divided into two subregions. The inclusion of chemical reactions (Fig.1d) decreases the shock wave slope preserving the flow structure, as compared with T-R-V gas.

Thus, when a plate with a flap is flown about at the angle of attack $40°$, the excitation of internal degrees of freedom alters the entire flow pattern. However, this conclusion cannot be applied when the angle of attack is $0°$ or $20°$. For the flow about a plate and a plate with a flap at the angle of attack $0°$ or $20°$ the bow shock wave is attached to the body for all gas models. The gas model complication slightly decreases the shock wave slope for $20°$ and hardly changes the flow structure for $0°$. The subsonic region in this case is insignificant and adjacent to the body.

Figures 2, 3 show the impact of the angle of attack changes on the distributed aerodynamic characteristics. Most essentially these changes are reflected in the pressure coefficient C_p distribution along the body (Fig. 2).

A drastic decrease in pressure along the entire body is observed for all gas models with the angle of attack decreasing. Figure 2 also shows that the gas model influence decreases with the angle of attack decreasing. Thus, the monotonic change of the pressure coefficient depending on the angle of attack should be noted for all gas models. The behavior of the friction coefficient C_f (Fig. 3) is far more complicated. The detached shock wave formation during T-T gas flowing about a plate and a plate with a flap at the angle of attack $40°$ results in smaller values of C_f up to the flap influence zone than those for the angle of attack $20°$. For $20°$, a modest-sized separated flow region is observed near the corner point. For $0°$, the coefficient C_f near the corner point is slightly smaller for a plate with a flap than that for a flat plate. When T-R, T-R-V gases or oxygen flows about a flat plate, C_f values are fairly close for the angles of attack $20°$ and $40°$, whereas C_f behavior for $0°$ is qualitatively different.

The total aerodynamic characteristics for the flow about a flat plate and a plate with a flap at the angle of attack $40°$, $20°$ and $0°$ are compared in Fig. 4. For the flow about a flat plate at the angle of attack $40°$ the gas model complication stipulates the monotonic decrease of the drag coefficient C_D and lift coefficient C_L. For a plate with a flap C_L continues decrease while C_D increases when the internal degrees of freedom are excited and then decreases when chemical reactions are taken into account.

The pitching moment coefficient $C_m(X_{c.g.})$ with respect to the body center of gravity $X_{c.g.}$ is one of the most important aerodynamic characteristics, as it determines the trimming angle of a flying vehicle. Both for a flat plate and for a plate with a flap we observed the decrease in $C_m(X_{c.g.})$ as the gas model is complicated. The largest reduction of $C_m(X_{c.g.})$ is observed when the flow structure is qualitatively changed. The center of pressure $X_{c.p.}$ (figures show $X'_{c.p.} = X_{c.p.} - X_{c.g.}$) shifts backwards for the both configurations.

The flap efficiency may be simply estimated on the basis of the analysis of the change in the pitching moment (ΔC_m) and $X_{c.p.}$ location due to the flap deflection. Figure 4 shows that for the angle of attack $40°$ the flap efficiency increases with the gas model complication. The main reason for such small flap efficiency in T-T gas is the strong upstream influence of flap disturbances and the corresponding pressure increase on the front part of the plate.

The angle of attack decreasing to $20°$, the gas model influence on aerodynamic coefficients is reduced both for a flat plate and a plate with a flap (Fig. 4b). Note that for the cases of T-R-V gas and reacting oxygen $|\Delta C_m|$ decreases, as compared with the angle of attack $40°$. Nevertheless, since the reduction of aerodynamic forces is even more essential, the center of pressure location shifts further to the trailing edge of the body than for

40°. For 0° (Fig. 4c) the center of pressure location shifts even further than for 40° and 20° when the flap is deflected. Thus, we may speak about the increase in the flap efficiency when the angle of attack is reduced.

3.2. THE INFLUENCE OF THE FLAP DEFLECTION ANGLE AND THE BODY BLUNTNESS

The results of calculations of the flow about a flat plate and an elliptic profile for various flap deflection angles are presented in this Section. The calculations were carried out for T-R gas and for the angle of attack 40° because in this case, as was mentioned above, there appears a qualitative change in the flow structure for a plate when the flap is deflected by 30°.

Figure 5 shows the Mach number flowfield for the flap deflection 15°. As is seen from the figure, the flow structure corresponds to the case of the attached shock wave.

The flow structure near an elliptic profile for various deflection angles of the rear part of the profile is shown in Fig. 6. The flow about such a body is characterized by the detached shock wave and a local subsonic region increase near the tip for all flap deflection angles. The profile curvature leads to a certain downstream decrease of the subsonic region and, hence, to weak influence of the deflected flap on the flow structure near the front half of the body.

As is shown in Fig. 7, the flap deflection impact is smaller for an elliptic profile than for a plate, this especially refers to the case $\delta = 30°$.

The integral aerodynamic characteristics for a plate and an elliptic profile are shown in Fig. 8. The successive increase in the flap deflection angle stipulates the growth of the drag and lift coefficients for the two configurations. The flap efficiency for a plate, as was noted above, drops sharply after the shock wave detachment and formation of a vast subsonic region. This explains $|\Delta C_m|$ decrease during the transition from $\delta = 15°$ to $\delta = 30°$ and the forward shift of the center of pressure. For an elliptic profile the pressure maximum is observed in the body tip region (see Fig. 7b). This accounts for the fact that the center of pressure shifts forward (about 5%), as compared with the flat plate case. The flap deflection increases $|\Delta C_m|$ and monotonically shifts the center of pressure backwards when the flap deflection angle increases.

3.3. FLOW RAREFACTION AND DIMENSIONALITY EFFECTS

The flow about an elliptic profile ($\delta = 0°$, $\delta = 30°$) at the angle of attack 40° for the Knudsen numbers $Kn =0.03, 0.01, 0.0033$ is considered in Figs. 9, 10. As is seen from Fig. 9a, the pressure coefficient C_p distribution remains practically the same for $\delta = 0°$ and for $\delta = 30°$ when the Knudsen number decreases. The friction coefficient C_f (Fig. 9b) is considerably reduced along the entire body surface.

The smaller flow rarefaction stipulates C_D reduction and C_L increase for both configurations considered (Fig. 10). Note that for $\delta = 0°$ the center of pressure shifts forward, while for $\delta = 30°$ it shifts backward and, correspondingly, $|\Delta C_m|$ increases. Thus, we may speak of the increase in the flap efficiency when the Knudsen number decreases.

In what follows there are the results for plane, axisymmetrical and spatial flow of T-R-V gas about bodies with flap for $\delta = 30°$.

Figure 11 shows the Mach number fields for the flow about a cone with a flare and a flat delta wing with a flap. The slope of the bow shock wave for the axisymmetrical case is considerably smaller than for a plate with a flap (see Fig. 1c) due to spatial spread. The spread is especially essential for 3D case, which causes the further reduction of the shock wave slope (Fig. 11b) and the shift of the wave inflection point downstream.

The distributed characteristics for the studied set of bodies are shown in Fig. 12 (no results are given for a flat delta wing since they are practically coincident with those for a cone). Due to the reduction of the shock wave slope C_p decreases and C_f increases for spatial flows ($\delta = 0°$), as compared with a plane case.

The upstream influence zone of the deflected surface is the largest for 2D case and the smallest for 3D case. The pressure maximum on the flap is the largest for 3D case while the minimum C_f value is nearly equal for the models under consideration.

The total aerodynamic characteristics of a delta wing for various gas models are shown in Fig. 13. The gas model effect is much smaller here than for the plane flow. Unlike the flat plate case, C_D increase is observed here when the internal degrees of freedom are excited.

3.4. MODELS OF VT ENERGY TRANSFER AND CHEMICAL REACTIONS

A great number of models were suggested to describe the excitation of internal degrees of freedom and chemical reactions by DSMC method, varying in their labour intensity and physical validity. The variety of models and algorithms draws aside one of the most interesting problems: to what extent the choice of this or that new model influence on the aerodynamics of various bodies. In the present paper the attempt of such analysis is made. New models for probabilities of the above mentioned processes available for DSMC methods are described and their effect on the flow about a 2D concave body (a plate with a flap) is investigated. The 2D model was selected due to its simplicity and more illustrative possibilities in comparison with 3D case.

To describe the VT energy exchange we have used the following models:

V1. Larsen-Borgnakke model with constant $Z_v = 50$;

V2. Larsen-Borgnakke model with variable $Z_v(T)$ [8];

V3. Model taking into account the quantum nature of vibrations [9,10]; molecules are considered as anharmonic oscillators (AHO) and only one-step vibrational transitions are involved. The energy dependence of VT transition probabilities required in DSMC method is found from the temperature dependence law of the VT rate constants [11] with account of high-temperature correction [12];

V4. A new model for VT exchanges with multi-step transitions. It employs the energy-dependent form for VT transition cross-sections found using quasiclassical theory [13] for the Morse intermolecular potential.

Depending on the adiabatic parameter λ the collisions are classified into three groups: slow ($\lambda_i > \lambda_2 > 1$ and $\lambda_f > \lambda_2 > 1$), fast ($\lambda_i < \lambda_1 < 1$ and/or $\lambda_f < \lambda_1 < 1$) and intermediate. Here the adiabaticity parameter is

$$\lambda_k = \omega(n_k)\frac{d}{v_k}$$

where $\omega(n_k) = \omega_e(1 - 2x_e(n_k + 0.5))$; $v_k = \sqrt{2E_k/\mu}$; n is the vibrational quantum number; subindexes i, f refer to the initial or final channels (pre- and post- collision states) respectively; v_k and E_k are the velocity and energy of translational motion in the channel k; ω_e is the oscillator frequency; x_e is the anharmonicity parameter; μ is the reduced mass of colliding particles; d is the parameter in the Morse potential

$$V(R) = D\{\exp[-2(R - R_0)/d] - 2\exp[-(R - R_0)/d]\}$$

For slow collisions the VT cross-section may be written as [13]:

$$\sigma_{n_i \to n_f} = \sigma_0 \frac{v_f}{v_i} A_{VT} \exp\left[-2\Delta_n \lambda_k \arctan\left(\sqrt{\frac{E_k}{D}}\right)\right] \Bigg/$$

$$\Bigg/ \left[\Delta_n \lambda_k \left(\arctan\left(\sqrt{\frac{E_k}{D}}\right)\sqrt{\frac{E_k}{D}}\Big/\left(\frac{E_k}{D} + 1\right)\right)\right]$$

where

$$A_{VT} = \frac{\epsilon_k^{\Delta_n}}{(\Delta_n!)^2}\left[\frac{\pi d}{4\hbar}\left[a_1 \mu d\omega(n_k) - (2a_2 - a_1)\sqrt{2\mu D}\right]\right]^{2\Delta_n},$$

$\Delta_n = |n_f - n_i|$, $\epsilon_k = (k + 0.5)(1 - x_e(k + 0.5))$, a_1, a_2 are the potential anisotropy parameters; $k = \min(i, f)$, σ_0 is the total elastic cross-section.

For fast collisions

$$\sigma_{n_i \to n_f} = 2\pi d^2 \frac{v_f}{v_i} \int_{x_0}^{\infty} x \left[1 - \frac{D}{E_i} e^{-(x-x_{00})} \left(e^{-(x-x_{00})}(1-x) - 2 + x \right) \right] \times$$

$$\times (J_{\Delta_n}(|F(k,x)|))^2 dx,$$

where $J_m(x)$ are Bessel functions, $x_0 = x_{00} - \ln\{1 + \sqrt{1 + E_i/D}\}, x_{00} = R_0/d$,

$$F(k,x) = \frac{DSd}{2\hbar} \sqrt{\pi x \epsilon_k} e^{-(x-x_{00})} (a_1 e^{-(x-x_{00})} - 2\sqrt{2} a_2),$$

$$S = \begin{cases} \frac{1}{v_k} & \text{for } \lambda_k < \lambda_1, \ \lambda'_k > \lambda_1, \ k' \neq k; \\ \sqrt{\frac{1}{v_i^2} + \frac{1}{v_f^2}} & \text{for } \lambda_i < \lambda_1, \ \lambda_f < \lambda_1. \end{cases}$$

In the intermediate region $\lambda_2 > \lambda_k > \lambda_1$ the VT cross-section is found by interpolation of cross-sections for slow and fast collisions:

$$\sigma_{n_i \to n_f, \lambda_k} = \exp\left[\frac{\lambda_2 - \lambda_k}{\lambda_2 - \lambda_1} \ln(\sigma_{n_i \to n_f, \lambda_1}) + \frac{\lambda_k - \lambda_1}{\lambda_2 - \lambda_1} \ln(\sigma_{n_i \to n_f, \lambda_2}) \right].$$

Since the expressions for VT cross-sections are rather cumbersome, before starting the computations we calculated a special array for VT transition probabilities for the further usage.

We chose the following values of constants for the case of O_2–O_2 interaction: $D = 1.65 \cdot 10^{-21} J, d = 6.09 \cdot 10^{-11} m, R_0 = 3.86 \cdot 10^{-10} m, a_1 = 0.33, a_2 = 0$. For O_2–O interaction the anisotropy parameter a_1 increases: $a_1 = 0.66$. For both interaction types $\lambda_1 = 1.0, \lambda_2 = 3.5$.

To describe the gas-phase chemical reactions we have used the following models:

C1. Model with the probabilities that depends only on the total energy of colliding pair [7]. The collision dynamics and algorithm for this case was described in details in [14]. This model was used together with models V1 and V2 of vibrational excitations (in the previous Sections of the paper V1-C1 models were used).

C2. Model of chemical reactions taking into account the vibration-dissociation coupling (VDC) [9, 10]. When constructing this model, an attempt is made to extend the VDC approach for continual flows [15] to the case of rarefied gas. To this end, the expansion of chemical reactions rate constants in terms of vibrational levels is used like in [15]:

$$k(T) = \sum_{v=0}^{v_m} f_B(v, T) k(v, T),$$

where

$$k(T) = a T^B \exp\left[\frac{-E_D}{kT}\right],$$

$$k(v,T) = A_v T^\beta \exp\left[\frac{-E_a(v)}{kT}\right],$$

$$A_v = A_0(v+1),$$

v_m is the highest vibrational level with the energy less than the dissociation energy threshold.

For a dissociation reaction

$$E_a(v) = E_d - (E_v - E_0),$$

$$A_0 = \frac{a T_R^{B-\beta} Q(T_R) \exp\left[\frac{E_0}{kT}\right]}{v_m(v_m+3)/2 + 1},$$

where E_v, E_0 are the energies of the v-th and 0-th vibrational levels respectively; Q is the vibrational partition function, T_R and β are the fit constants.

For an exchange reaction

$$E_a(v) = \begin{cases} E_e - (E_v - E_0), & E_v - E_0 < E_e \\ 0, & E_v - E_0 > E_e \end{cases}$$

$$A_0 = a T_R^{B-\beta} Q(T_R) \, exp\left[\frac{E_0}{kT}\right] \times$$

$$\times \left[v_1(v_1+3)/2 + 1 + \exp\left[\frac{E_e}{kT}\right] \sum_{v=v_1+1}^{v_m} (v+1) \exp\left[-\frac{(E_v - E_0)}{kT}\right] \right]^{-1}$$

where v_1 is the highest vibrational level with the energy less than the exchange reaction energy threshold E_e.

The reaction probabilities may be obtained on the basis of collision theory. For example, for the dissociation reaction $AB(v) + C \rightarrow A + B + C$

$$P_d(v) \propto (v+1)\left(1 - \frac{E_a}{E_c - E_{vib}(AB)}\right)^{\xi+1-\alpha} (E_c - E_a - E_{vib}(AB))^{\beta-0.5+\alpha}$$

where E_c is the total energy of colliding pair, $\xi = (\xi_{rot}(AB) + \xi_{int}(C))/2$

Thus, the molecule dissociation probability increases with its vibrational level increase (even when the total energy of colliding pair is constant).

This model was used together with models V3 and V4 of vibrational excitation.

726

3.5. EFFECT OF VT TRANSFER AND CHEMICAL REACTION MODELS

At the first step the calculations of the state-to-state rate constants were made. The comparison of those results with the corespondent experimental and theoretical data shows that for O_2–O_2 interaction in the temperature range 300–10000 K the rate constants K_{10} of the transition from the first vibrational level to the zeroth one for models V3 and V4 are very close to each other and practically coincide with well known semi-empirical estimates given in [16] (see Fig. 14). For the temperatures $T > 10000$ K K_{10} of V3 model is nearly the same as K_{10} calculated by the analytical expression [17] obtained for the high-temperature region, while K_{10} for V4 model lies slightly lower. Experimental data on O_2–O interaction are rather limited and the estimates from [16] seem inapplicable to this case. The K_{10} for V3 and V4 models together with the expression [17] are compared in Fig. 14. Thus, the state-to-state transition constants for V3 and V4 are in reasonable agreement with the currently available data.

The results of computations of the thermal relaxation of heated chemically non-reacting molecular oxygen (at the initial moment $T_{transl} = T_{rotat} = 15,000$ K, $T_{vibrat} = 0$) are presented in Fig. 15 for various vibrational models. The fastest relaxation is observed for Larsen-Borgnakke model, the slowest one - for the model with one-level VT transitions. After the equilibrium was reached for all models the translational and internal temperatures were equal to each other, that means that the detailed balance law was obeyed.

In what follows there are the results of calculation of the flow about a flat plate with a flap by non-reacting O_2 for various models of VT energy exchange. The influence of the model choice is most strongly revealed in the temperature fields, in particular, those of vibrational temperature (Fig. 16). The temperature relaxation process is much slower when the discrete nature of vibrations is taken into account, and vibrational temperature in the shock front and in the boundary layer for V3 model is considerably lower than the corresponding values for V1 model (cf. Figs. 16a and 16b). Besides, note the increase in the upstream influence of the deflected flap for V3 model, which stipulates the shift of the shock wave inflection point to the body tip. When multi-step VT transitions are taken into account (model V4, Fig. 16c), the gas relaxation is faster, as compared with V3 model.

Figure 17 shows the distributed aerodynamic coefficients for a plate with a flap and five various models of non-reacting O_2: T-R gas (gas without vibrational degrees of freedom) and V1, V2, V3, V4 gases. The comparison of T-R gas model with others shows the reduction of the influence zone of the flap when vibrations are considered. This zone is the smallest for V1 and V2 models whose the pressure coefficient distributions are rather close to each other (Fig. 17a) and the distributions of the skin friction C_f and heat transfer C_h coefficients practically coincide. Note also the formation of a separation region near the corner point of the body for these models (Fig. 17b). The results for V3 model differ essentially from the results for both T-R model and V1, V2 models (Figs. 17a-17c). The results for V4 model occupy the intermediate position between those for V1-V2 and V3. Considering V4 to be the most realistic model, it should be noted that V3 model gives the much closer to V4 results than V1 and V2 ones.

The total aerodynamic characteristics (drag coefficient C_D, lift coefficient C_L, lift-to-drag ratio C_L/C_D, negative values of the pitching moment with respect to the tip $-C_m(0)$ and location of the center of pressure $X_{c.p.}$) for T-R, V1, V3 and V4 models are shown on Fig. 18. The aerodynamic characteristics for V2 model are not shown because they are mostly the same as for V1 model. The aerodynamic characteristics for V3 and V4 models are slightly more different, but the difference is not very essential (the maximum difference is observed for $C_m(0)$ and amounts to 2%). At the same time the C_L difference for V1-V2 and V3-V4 models is about 10%. Note also that the values of C_L, C_L/C_D and $X_{c.p.}$ for V3 and V4 models lye between the corresponding values for T-R and V1 models, whereas C_D and $-C_m(0)$ are larger for V3 and V4 models.

Figure 18 also shows the results of the calculation of the flow about a plate with a flap by chemically reacting O_2 for various models of chemical reactions (V1+C1, V3+C2, V4+C2 cases were considered). The inclusion of chemical reactions for all three models (V1, V3 and V4) independently of the choice of the chemical reactions model (with or without VDC) decreases $C_D, C_L, C_L/C_D$ and $C_m(0)$ and slightly increases $X_{c.p.}$. The choice of the chemical reactions model has the greatest effect on $C_m(0)$: the contribution of chemical reactions to $C_m(0)$ is the largest for V4 and the smallest for V1 models respectively. At the same time the changes in C_L and C_L/C_D values for chemically reacting gases, obtained in the framework of V1+C1, V3+C2, and V4+C2 models are approximately the same for all gas models as compared with the corresponding chemically non-reacting cases V1, V3, and V4. The influence of the choice of the chemical reactions model is most pronounced for the flowfields structures, especially for the fields of species concentrations (see Fig. 19). Only large total energy of colliding pair is sufficient for reactions to take place within the frame of V1+C1 model, therefore, we observe a considerable increase in the O mole fraction near the plate tip (Fig. 19a). The VDC model implies chemical reactions to take place only under the condition of sufficient excitation of vibrational molecular mode. This stipulates an essential delay of reactions as compared with V1+C1 model (cf. Figs. 19a and 19b, c). The same effect is seen when the distributed aerodynamic characteristics are considered (Fig. 20). The inclusion of C1 model leads to a considerable decrease in C_p and C_h at the first half of the plate whereas the inclusion of C2 model weakly changes the aerodynamic characteristics in this region. On the whole, the influence of the model of chemical reactions on the distributed aerodynamic characteristics is less essential than on the flowfields.

4. Conclusion

The flow structure and aerodynamic characteristics of 2D, axisymmetrical and 3D models of bodies with flap are calculated in the present paper. A notable difference between the plane and spatial flows is shown. The spatial spread stipulates the reduction of the influence zone of the deflected surface. The gas model has a smaller effect on the flap efficiency in 3D case than in 2D case.

The smaller flow rarefaction is shown to cause the increase in the control surface efficiency. The influence of the body surface shape and the flap deflection angle are illustrated by the flow about a plate and a 10% elliptic profile. A large impact of the flow structure (detached - attached shock waves) on the control surface efficiency is shown.

The influence of the angle of attack on the flow structure near a flat plate and a plate with a flap is investigated for various gas models. The increase in the angle of attack is shown to cause the decrease in the flap efficiency.

New models describing the vibrational degrees of freedom and chemical reactions for DSMC method are presented. They are compared with conventional models by the example of the flow about a plate with a flap by oxygen. The oxygen flow about a 2D concave body is the case where the influence of the models of internal degrees of freedom and chemical reactions on the flow structure is most pronounced.

The performed study allows the estimation of the effect of vibration and chemistry models on the aerodynamic characteristics and flowfields near a plate with a flap. On this basis we can draw some conclusions:

- the inclusion of the discrete nature of molecule vibrational levels changes considerably both the flowfields and the aerodynamic characteristics of a plate with a flap;

- the influence of multi-step VT transitions on the overall aerodynamic characteristics is not essential - about 1-2%, while they have greater effect on the distributed aerodynamic characteristics (especially on C_p) and flowfields;

- the model of chemical reactions (namely, the account of VDC) has much weaker influence on the aerodynamic characteristics than the model of vibrational excitation (i.e., the aerodynamic characteristics are more dependent on the vibration model than on the chemistry model).

Models V4 and C2 should be considered to be most realistic among examined ones. New experimental evidence on VT and chemical relaxation processes will provide a possibility of refining some parameters used in the models but this will hardly have much effect on the results of calculations.

5. Acknowledgment

This work was supported in part by the Russian Foundation of Fundamental Investigations and International Science Foundation (Grant No. RBW000). This support is gratefully acknowledged.

1. Marriot, P.M. and Harvey, J.K. (1991) New Approach for Modelling Energy Exchange and Chemical Reactions in the Direct Simulation Monte Carlo Method, *Proc. XVII Int. Symp. on Rarefied Gas Dynamics*, Aachen, Germany, 784-791.
2. Dogra, V.K., Wilmoth, R.G., and Moss, J.N. (1991) Aerothermodynamics of a 1.6-m-Diameter Sphere an Hypersonic Rarefied Flow, *AIAA Paper 91-0773*.
3. Haas, B.L. and McDonald J.D. (1991) Validation of chemistry models employed in a particle simulation method, *AIAA Paper 91-1367*.
4. Ivanov, M.S., Antonov, S.G., Gimelshein, S.F., and Kashkovsky A.V. (1992) Rarefied numerical aerodynamic tools for reentry problems, *Proc. I Europ. Comp. Fluid Dyn. Conf.*, Brussels, Belgium, 1121-1128.
5. Ivanov, M.S., Antonov, S.G., Gimelshein S.F., and Kashkovsky A.V. (1994) Computational Tools for Rarefied Aerodynamics, Rarefied Gas Dynamics: Theory and Simulations, edited by B.D. Shizgal and D.P. Weaver, Vol. 159, Progress in Astronautics and Aeronautics, AIAA, Washington, DC.
6. Ivanov, M.S. and Rogasinsky, S.V. (1991) Theoretical analysis of traditional and modern schemes of the DSMC method, *Proc. XVII Intern. Symp. on Rarefied Gas Dynamics*, edited by A.E. Beylich, Aachen, 629-642.
7. Bird, G.A. (1979) Simulation of multi-dimensional and chemically reacting flows, *Proc. XI Intern. symp. on Rarefied Gas Dynamics*, Paris, 365-388.
8. Parker, J.G. (1959) Rotational and vibrational relaxation in diatomic gases, *Phys. Fluids* **2**, No.4, 449-462.
9. Ivanov, M.S. (1993) Transitional Regime Aerodynamics and Real Gas Effects, *Proc. Int. Symp. on Aerospace and Fluid Science*, Sendai, Japan, 77-101.

10. Ivanov, M.S., Gimelshein, S.F., Markelov, G.N., Antonov, S.G., and Titov, E.V. (1994) Real Gas Effects on Rarefied Flow over Simple Concave Bodies, *Thermophysics and Aeromechanics* bf 1, No. 1, 19-34.
11. Doroshenko, V.M., Kudryavtsev, N.N., and Smetanin V.V. (1991) Equilibrium of internal degrees of freedom of molecules and atoms at high altitude hypersonic flights, Teplofisika Visokih Temperatur **29**, No.5, 1013-1026.
12. Park., C. (1985) Problems of Rate Chemistry in the Flight Regimes of Aeroassisted Orbital Transfer Vehicles, *Progr. in Astronautics and Aeronautics* **96**, AIAA, New York, 511-537.
13. Bogdanov, A.V., Dubrovskii, G.V., Gorbachev, Yu.E., and Strelchenya, V.M. (1989) Theory of vibrational and rotational excitation of polyatomic molecules, *Physical Reports* **181**, No.3, 121-206.
14. Gimelshein, S.F., and Ivanov, M.S. (1994) Simulation of Chemically Reacting Gas Flow Using Majorant Frequency Scheme of DSMC, *Rarefied Gas Dynamics: Theory and Simulations*, edited by B.D. Shizgal and D.P. Weaver, Vol.159, Progress in Astronautics and Aeronautics, AIAA, Washington, DC, 218-233.
15. Warnatz, J., Riedel, U., and Schmidt, R. (1991) Different Levels of Air Dissociation Chemistry and Its Coupling with Flow Models, *Proceedings of the 2nd Joint Europe/U.S. Short Course in Hypersonics*, Preprint No. 4.
16. Millikan, R.C. and White, D.R. (1966) Systematics of Vibrational Relaxation, *J. of Chem. Phys.* **39**, No.12, pp. 3209-3213.
17. Losev, S.A., Makarov, V.N., Pogosbekyan M.Ju., Shatalov, O.P., Nikolsky, V.S. (1990) Thermochemical Nonequilibrium Kinetic Models in Strong Shock Waves on Air, *AIAA Paper 94-1990*.

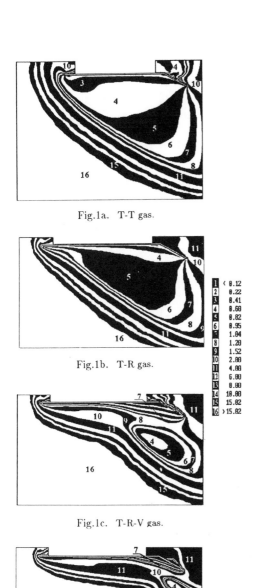

Fig.1a. T-T gas.

Fig.1b. T-R gas.

1	< 0.12
2	0.22
3	0.41
4	0.68
5	0.82
6	0.95
7	1.04
8	1.20
9	1.52
10	2.00
11	4.00
12	6.00
13	8.00
14	10.00
15	15.02
16	> 15.02

Fig.1c. T-R-V gas.

Fig.1d. Oxygen.

Fig.1. Mach number flow structure
for plate with a flap.

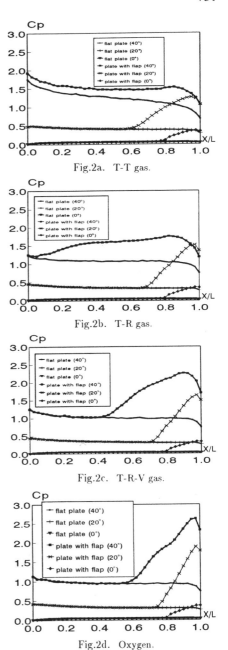

Fig.2a. T-T gas.

Fig.2b. T-R gas.

Fig.2c. T-R-V gas.

Fig.2d. Oxygen.

Fig.2. Pressure coefficient.

732

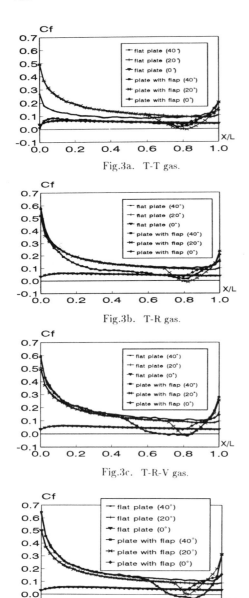

Fig.3a. T-T gas.

Fig.3b. T-R gas.

Fig.3c. T-R-V gas.

Fig.3d. Oxygen.

Fig.3. Skin friction coefficient.

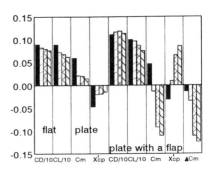

Fig.4a. $\alpha = 40^o$.

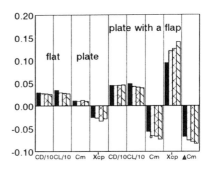

Fig.4b. $\alpha = 20^o$.

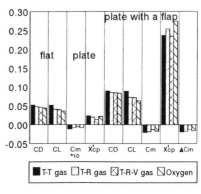

Fig.4c. $\alpha = 0^o$.

Fig.4. Aerodynamic coefficients
for various angles of attack.

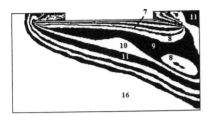

Fig.5. Mach number flow structure
for plate wuth a flap 15 deg.
T-R gas.

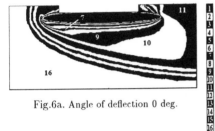

Fig.6a. Angle of deflection 0 deg.

1	⟨ 0.12
2	0.22
3	0.41
4	0.60
5	0.82
6	0.95
7	1.04
8	1.20
9	1.52
10	2.00
11	4.00
12	6.00
13	8.00
14	10.00
15	15.02
16	⟩ 15.02

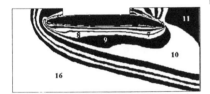

Fig.6b. Angle of deflection 15 deg.

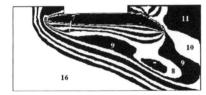

Fig.6c. Angle of deflection 30 deg.

Fig.6. Mach number flow structure
for elliptic profile with
deflection rear part T-R gas.

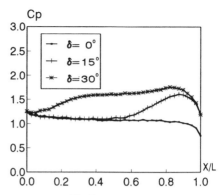

Fig.7a. Plate.

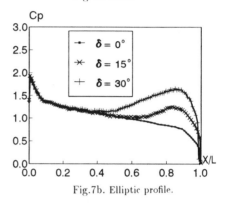

Fig.7b. Elliptic profile.

Fig.7. Pressure coefficient.

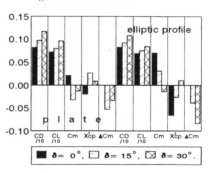

Fig.8. Aerodynamic coefficients for
various angles of flap deflection.

734

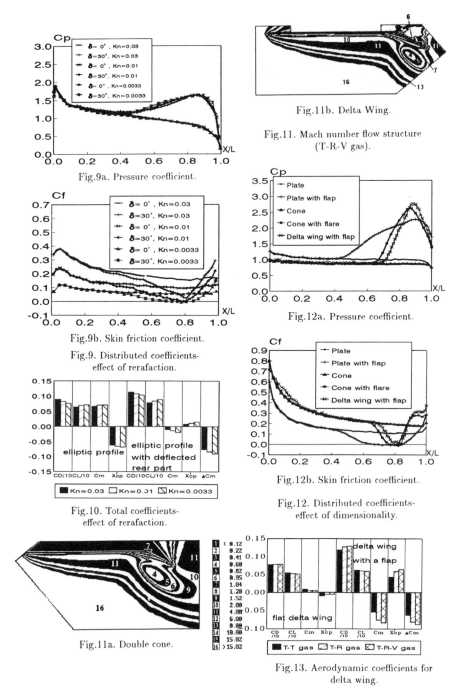

Fig.9a. Pressure coefficient.

Fig.9b. Skin friction coefficient.

Fig.9. Distributed coefficients-
effect of rerafaction.

Fig.10. Total coefficients-
effect of rerafaction.

Fig.11a. Double cone.

Fig.11b. Delta Wing.

Fig.11. Mach number flow structure
(T-R-V gas).

Fig.12a. Pressure coefficient.

Fig.12b. Skin friction coefficient.

Fig.12. Distributed coefficients-
effect of dimensionality.

Fig.13. Aerodynamic coefficients for
delta wing.

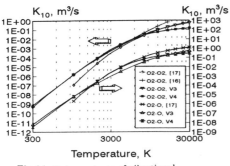

Fig.14. Rate constants of vibrational transition $1 \rightarrow 0$.

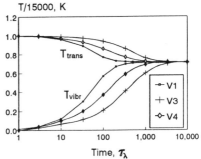

Fig.15. Thermal relaxation of O_2.

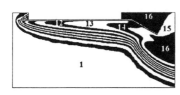

Fig.16a. **V1** model.

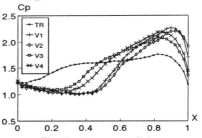

Fig.17a. Pressure coefficient.

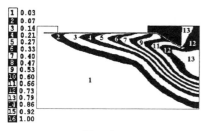

Fig.16b. **V3** model.

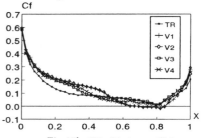

Fig.17b. Skin friction coefficient.

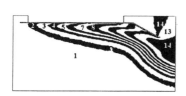

Fig.16c. **V4** model.

Fig.16. Vibrational temperature fields.

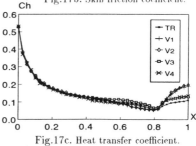

Fig.17c. Heat transfer coefficient.

Fig.17. Aerodynamic coefficients for various vibration models.

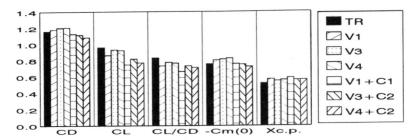

Fig.18. Total aerodynamic coefficients for various models.

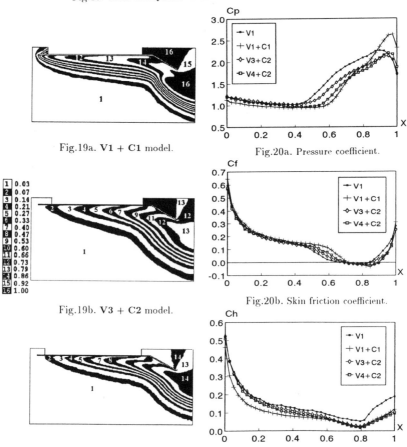

Fig.19a. **V1 + C1** model.

Fig.20a. Pressure coefficient.

Fig.19b. **V3 + C2** model.

Fig.20b. Skin friction coefficient.

Fig.19c. **V4 + C2** model.

Fig.20c. Heat transfer coefficient.

Fig.19. Atomic oxygen mole fraction.

Fig.20. Aerodynamic coefficients for various chemistry models.

EXTENSION OF THE Δ–ε METHOD TO DIATOMIC GASES

Marc Monroe
Philip L. Varghese
Department of Aerospace Engineering & Engineering Mechanics
The University of Texas, Austin, Texas

Abstract

The delta-epsilon method for solving the Boltzmann equation is extended to allow for internal energy modes. The delta-epsilon method utilizes a discrete approximation of the infinite velocity phase space. The velocity distribution function is only defined over regions where its values are deemed large enough to be of importance to the physics of the problem. This allows distribution points to exist anywhere in the phase space. A fourth order finite-difference scheme is used to model the convection terms and a Monte Carlo-like method is applied to the discrete velocity space to model the elastic collision integral. Internal energy mode effects on the distribution function are modeled by an inelastic contribution to the collision integral. The exchange of energy between translational and internal modes is modeled empirically by a relaxation rate equation that contains a characteristic relaxation time and the mean internal mode energies. Using this approach the internal mode energy distributions do not have to be computed as only mean values are used. A relation governing the variation of the velocity distribution function with temperature is developed from the Maxwellian distribution. Numerical examples of homogeneous relaxation are given and compared to solutions obtained via Direct Simulation Monte Carlo (DSMC).

1. Introduction

The Boltzmann equation represents the basis for the solution of problems involving rarefied flow fields. Applications can be found in areas as varied as the study of rocket plumes and semi-conductor processing. Due to the non-linearity of the equation and a large number of degrees-of-freedom, analytical solutions of the full Boltzmann equation are hard to obtain. Solutions methods to practical problems, with rare exceptions, must be numerical in nature. Descriptions of various methods are given, for example, in [1-9].

The exchange of internal and translational energy frequently plays an important role in nonequilibrium flows such as nozzle expansions and hypersonic flows

M. Capitelli (ed.), Molecular Physics and Hypersonic Flows, 737–747.

in the rarefied regime. The direct simulation Monte Carlo (DSMC) method is widely used for the simulation of rarefied nonequilibrium flows. The Larsen-Borgnakke phemonological model is commonly used to model the energy exchange [10]. Various modifications to this model have been proposed for both rotational and vibrational modes [11-18]. In these models a fraction of collisions are considered inelastic and the probability of energy exchange between internal and translational modes is a function of the rotational collision number Z_R. The calculation of an inelastic collision is then performed by statistically sampling values using the equilibrium distribution functions and the total collision energy. Each of these models introduces a translational energy dependent Z_R. These approaches have been shown to perform well in modelling internal/translational energy exchange. However, Monte Carlo simulation methods suffer from statistical fluctuations and low convergence rates. Furthermore, they are very expensive to implement for transient problems and in domains where Knudsen number $Kn \ll 1$. Although Navier-Stokes codes are more suitable in these domains, they are not easily interfaced with DSMC codes because of the particle nature of the latter.

The delta-epsilon $(\Delta-\varepsilon)$ method represents an alternative to the DSMC method for simulating rarefied flow fields by directly solving the Boltzmann equation. The application of this method to monatomic gases has been described by Tan and Varghese [7]. The $\Delta-\varepsilon$ method discretizes the velocity phase space as in other discrete velocity approximations [8,9]. The difference is that the distribution function is only defined over regions where its values are deemed large enough to be of importance to the solution calculation. This allows distribution points to exist anywhere in the phase space, yet reduces the dimension of the distribution function vector to a manageable size for computational purposes. The time integration is split into collisionless convection step followed by a homogeneous collision process. A fourth order explicit finite-difference scheme is used to model the convection terms. An implicit scheme can be applied, but conservation laws are no longer strictly satisfied. A conservation correction such as that of Aristov and Cheremissin [19] would be needed to remedy this shortcoming. A Monte Carlo-like method described in [7] is applied to the discrete velocity space to model the collision integral, though other collision integral models could be used.

This paper describes an extension of the $\Delta-\varepsilon$ method to include inelastic collision processes. The inelastic model is strictly phenomological in nature and is based on a local characteristic relaxation time constant. The model is implemented so as to ensure conservation of mass, momentum, and energy. This approach involves a substantial reduction in the computational effort necessary to calculate internal mode interactions. Example calculations for homogeneous relaxation including rotational and vibrational internal modes are given in Section 3.

2. Formulation

The dimensionless Boltzmann equation for a single species diatomic gas is

$$\frac{D f}{D t} \equiv \frac{\partial f}{\partial t} + \nabla_r \cdot (vf) = I \tag{1}$$

where $I = I_{elastic} + I_{inelastic}$ is the contribution of elastic and inelastic collision processes to the velocity distribution function. $I_{elastic}$ is defined as

$$I_{elastic} \equiv \frac{1}{Kn} \int_{S^2} \frac{d\Omega}{4\pi} \int_{\Re^3} [f(r,v',t)f(r,w',t) \\ -f(r,v,t)f(r,w,t)]g\sigma(g,\theta)dw . \tag{2}$$

where $f = f(r, v, t)$ is the velocity distribution function normalized to the number density, r is the position vector in physical space, v is the velocity vector, and t is the time. S^2 denotes the surface of the unit sphere. The relative speed g is given by $\|v - w\|$. The quantities θ, v', w', and σ are binary collision parameters; θ is the collision deflection angle, v' and w' are the post-collisional velocities of v and w, respectively, and $\sigma = \sigma(g, \theta)$ is the differential cross-section for binary collisions. Kn is the Knudsen number and is defined as $Kn = \lambda/L$ with $\lambda = 1 / (\sqrt{2}n_0\sigma_0)$. λ is the mean free path, L is a characteristic length, and σ_0 is a reference cross-section. For simplicity the variable hard sphere (VHS) model of Bird [6] is used for the differential cross-section. $I_{inelastic}$ is described in the following discussion of the implementation of internal energy in the $\Delta-\varepsilon$ model.

The coupling between the translational and molecular internal energy must be accounted for when modeling diatomics and this corresponds to the inelastic collisions described by $I_{inelastic}$. Both the internal rotational and vibrational energy are modeled. The energy distribution functions are not modeled themselves -- each mode is described entirely by a single parameter, i.e. the mean rotational and vibrational energy respectively.

Energy conservation is modeled using a split convection and relaxation process that parallels the treatment of the Boltzmann equation. For example the rotational energy convection step is modeled by the following equation

$$\frac{\partial e_{rot}}{\partial t} + \bar{c} \cdot \frac{\partial e_{rot}}{\partial r} = 0 \tag{3}$$

where the internal rotational energy e_{rot} is related to the rotational temperature by $e_{rot} = RT_{rot}$. A convection process is not needed for the homogeneous relaxation test cases of this paper. In this work, vibrational energy is modeled using a quantized simple harmonic oscillator model (SHO) that relates e_{vib} to vibrational temperature by

$$e_{vib} = \frac{R\theta_v}{\exp\left(\dfrac{\theta_v}{T_{vib}}\right) - 1}.$$ (4)

However, the extension to anharmonic oscillators is trivial by an appropriate modification of equation (4). The relaxation represents the exchange of internal energy and translational energy via inelastic collistions. This process is modeled by an exponential decay of the internal energy towards the local equilibrium energy value, with a rate governed by the local internal relaxation time (i.e. $\tau_{int} = Z_{int}\tau_{el}$) where Z_{int} is the internal mode collision number and τ_{el} is the elastic collision time:

$$\frac{de_{rot}}{dt} = \frac{e_{rot}^{eq} - e_{rot}}{\tau_{rot}} \; ; \; \frac{de_{vib}}{dt} = \frac{e_{vib}^{eq} - e_{vib}}{\tau_{vib}}.$$ (5)

The local changes in rotational and vibrational energy during any time interval are computed from these rate models. The corresponding change in translational energy follows from energy conservation. This model is the continuum analog of the relaxation models used in the particle simulation methods. The internal mode collision number (Z_{int}) may be a constant or defined as a function of the translational energy. More complex models with $Z_{int} = Z_{int}(T_{tr}, T_{int})$ can also be implemented with little increase in computational effort.

The only additional modeling requirement for the translational-internal energy interchange is the determination of its effect on the velocity distribution. This is the model used in computing $I_{inelastic}$. A model for this process is obtained by examining the effect of temperature variations on the equilibrium velocity distribution. This change in the distribution function due to a change in translational temperature is governed by

$$\frac{\delta f}{\delta T_{tr}} = f_{max}(T_{tr}) \cdot \frac{1}{T_{tr}}\left[\frac{3\xi^2}{4\,e_{tr}} - \frac{3}{2}\right]$$ (6)

where f_{max} is a Maxwellian distribution at the translational temperature T_{tr}, e_{tr} is the translational internal energy, and ξ is the thermal speed. This model guarantees that a gas with an initial Maxwellian velocity distribution relaxes via a series of Maxwellian distributions changes and while it is truly valid only for equilibrium distributions it offers a good starting point for calculations away from equilibrium.

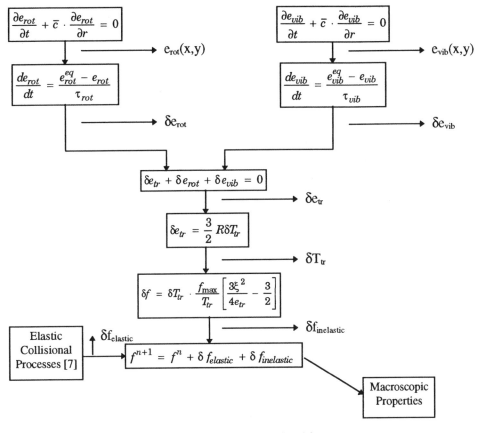

Figure 1. Relaxation algorithm

The entire procedure for updating the distribution function at each time step consists of computing the change in the distribution function by accounting for the elastic and the inelastic effects separately. The exchange process is indicated schematically in figure 1 which illustrates the simplicity of the algorithm.

3. Example of the Δ–ε Method

This methodology has been successfully applied to modeling the homogeneous rotational relaxation and combined rotational/vibrational relaxation of a simple gas with a Maxwellian initial velocity distribution. These problems have been solved by several authors using DSMC with various energy exchange mechanisms [6,11,12,14-18]. Results of the Δ–ε calculations will be discussed.

HOMOGENEOUS ROTATIONAL RELAXATION TO EQUILIBRIUM

The test case for homogeneous relaxation consisted of a simple diatomic gas with a constant rotational relaxation collision number $Z_R = 5$ and the collision cross-section of a Maxwell molecule. Initially, the translational temperature of the gas was set to 455 K and the rotational temperature to 0 K. This leads to an equilibrium temperature of 273 K. The non-dimensionalized time step for the Δ–ε procedure was set to $t' = t/t_0$ where $t_0 = \lambda_0/v_0$, $v_0 = \sqrt{2RT_0}$ and λ_0 is the hard sphere mean free-path at the equilibrium temperature. This works out to approximately one mean collision time. The time non-dimensionalization actually leads to a time step slightly (~12%) greater than one mean collision time, but it is of no significance for the results reported here since all calculations use consistent time scales.

Figure 2 compares the rotational temperature relaxation rates obtained by the Δ–ε method, the DSMC code of Bird [6] using the Larsen-Borgnakke (L-B) model, and a theoretical exponential relaxation. The theoretical curve is the analytical solution to equation (5) given by:

$$T_{rot} = T_{eq} \cdot (1 - \exp(- \frac{2}{\sqrt{\pi} \cdot Kn \cdot \beta} \cdot \frac{t}{Z_R})) \tag{7}$$

where the constant $\dfrac{2}{\sqrt{\pi} \cdot Kn}$ is due to the time non-dimensionalization and Z_R is the rotational collision number. β accounts for the influence of the molecular collision model on the collision time constant and is equal to unity for this case of Maxwell molecules. The Δ–ε method relaxation follows the exponential relaxation closely. This is because the Δ–ε energy exchange model is the numerical integration of the first-order rate equation (4) while the theoretical solution is an analytical solution to this same equation. The Δ–ε model also is in reasonable agreement with the DSMC solution for the temperature profile. The differences during the relaxation are of the order of 5%. As noted by Bird [6], the difference arises because the relaxation collision number in the DSMC calculation is only approximated by the reciprocal of the fraction of inelastic collisions in the L-B model. This fraction was set to equal Z_R in these calculations.

For this problem, the Δ–ε code, is a factor of 20 times faster than the DSMC code when running on the same workstation. This is due to the large amount of computer time necessary to reduce the statistical fluctuations in the DSMC calculations. The collision integral in the Δ–ε method suffers from smaller fluctuations than the DSMC methods [7]. The Δ–ε method is very efficient for unsteady problems and has demonstrated a higher efficiency than direct simulation methods in these cases. It is however less efficient for steady state problems [7] though efficiency could be further improved using various acceleration schemes such as multi-grid iterations.

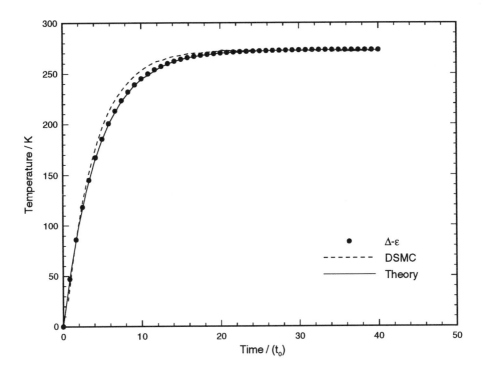

Figure 2. Homogeneous relaxation of rotational temperature in a diatomic gas with $Z_R = 5$.

ROTATIONAL/VIBRATIONAL RELAXATION TO EQUILIBRIUM

The rovibrational test case consists of a simple diatomic gas with a constant rotational relaxation collision number $Z_R = 5$, a vibrational relaxation collision number $Z_V = 50$, and the collision cross-section of a Maxwell molecule. The initial translational temperature of the gas was set to 1500 K and the rotational/vibrational temperatures to 0 K. This leads to an equilibrium temperature of 868 K. The time step for the Δ–ε procedure was set as in the rotational relaxation test case.

Figure 3 compares the rotation and vibration temperature relaxation rates obtained by the Δ–ε method and the DSMC code of Bird [6] using a discrete energy level quantum approach in the Larsen-Borgnakke (L-B) model. The vibrational temperature is seen to increase rapidly at very early times. The rotational temperature then quickly overtakes the vibrational temperature and equilibrates with the translational temperature. The two then relax together to the final equilibrium with the vibrational temperature at a rate determined by the vibrational relaxation time.

744

The Δ-ε method is again in reasonable agreement with the DSMC solution for both the rotational and vibrational temperature profiles. There are differences in the vibrational temperatures over early times in the relaxation process but these disappear quickly. The differences in the relaxation behavior arise because the relaxation collision number is only approximated by the reciprocal of the fraction of inelastic collisions in the L-B model as discussed in Bird [6]. Similar departures of DSMC from a similar (but not identical) relaxation equation have been described by Lumpkin, et al. [20].

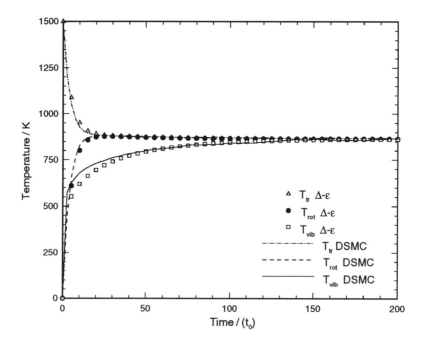

Figure 3. Homogeneous relaxation of rotational/vibrational temperature in a diatomic gas. $T_{vib} = 0$ K, $T_{rot} = 0$ K, $T_{tr} = 1500$ K, $Z_R = 5$, $Z_V = 50$.

The very large initial increase in vibrational temperature seen in figure 3 is due to the sudden increase in vibrational energy from the initial zero vibrational energy. For a quantum oscillator the vibrational specific heat c_{vib} is zero in the limit $T_{vib} \rightarrow 0$. Hence at low temperatures, small changes in e_{vib} result in large changes in T_{vib}. Figure 4 shows a similar calculation when the initial rotational and vibrational temperature are 273 K. In this case the initial increase in T_{vib} is much diminished.

The addition of the vibrational model increases the computational burden of the method negligibly and the execution time per time step is comparable to the pure rotational computation.

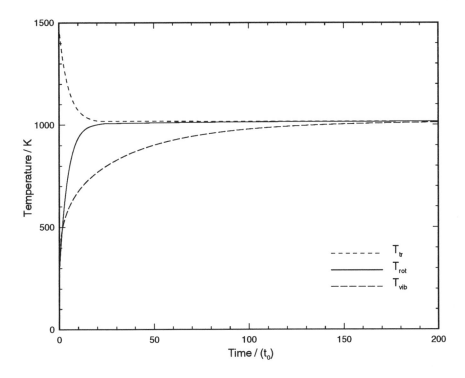

Figure 4. Homogeneous relaxation of rotational/vibrational temperature in a diatomic gas. $T_{vib} = 273$ K, $T_{rot} = 273$ K, $T_{tr} = 1500$ K, $Z_R = 5$, $Z_V = 50$.

4. Conclusions

We have extended the Δ–ε method of direct solution of the full non-linear Boltzmann equation by accounting approximately for internal energy modes. Internal energy is modeled as a macroscopic quantity at each point in the flowfield. This extension provides an efficient procedure for the effect of internal energy on the calculation. The test case of rotational homogeneous relaxation is shown to correspond well to DSMC solutions of the same problem. Combined vibrational-rotational relaxation has been modeled and the Δ–ε method has been shown to handle vibrational energy very efficiently.

The successful implementation of internal energy modes permits the Δ–ε method to be used for more practical problems. Work is in progress to apply the method to 1d shocks and axi-symmetric flows and we plan to extend it to handle multiple species. In addition, a modified relaxation model has been developed that will use the actual distribution values rather than the equilibrium values to account for a non-equilibrium situation with greater accuracy.

746

5. References

1. Roger, F. and Schneider, J. (1994) Deterministic Method for Solving the Boltzmann Equation, in B.D. Shizgal and D.P. Weaver (eds.), *Rarefied Gas Dynamics: Theory and Simulations* **159**, AIAA, pp. 335 - 343.

2. Ohwada, T. (1994) Numerical Analysis of Normal Shock Waves on the Basis of the Boltzmann Equation for Hard-Sphere Molecules, in B.D. Shizgal and D.P. Weaver (eds.), *Rarefied Gas Dynamics: Theory and Simulations* **159**, AIAA, pp. 482 - 488.

3. Aristov, V.V. and Tcheremissine, F.G. (1994) Solution of the Boltzmann Equation for Study of Inclined Shock-Wave Reflection, in B.D. Shizgal and D.P. Weaver (eds.), *Rarefied Gas Dynamics: Theory and Simulations* **158**, AIAA, pp. 448 - 460.

4. Perthame, B. (1995) Introduction to the Theory of Random Particle Methods for Boltzmann Equation.

5. Buet, C. (1995) A Discrete-Velocity Scheme for the Boltzmann Operator of Rarefied Gas Dynamics.

6. Bird, G.A. (1994) *Molecular Gas Dynamics and the Direct Simulation of Gas Flows*, Clarendon Press, Oxford.

7. Tan, Z. and Varghese, P. (1994) The $\Delta-\varepsilon$ Method for the Boltzmann Equation, *Journal of Computational Physics* **110**, 327 - 340.

8. Yen, S.M. and Lee, K.D. (1989) Direct Numerical Solution of the Boltzmann Equation for Complex Gas Flow Problems, in E.P. Muntz, D.P. Weaver, and D.H. Campbell (eds.), *Rarefied Gas Dynamics: Theoretical and Computational Techniques* **118**, AIAA, pp. 337 - 342.

9. Tcheremissine, F.G. (1989) Advancement of the Method of Direct Numerical Solving of the Boltzmann Equation, in E.P. Muntz, D.P. Weaver, and D.H. Campbell (eds.), *Rarefied Gas Dynamics: Theoretical and Computational Techniques* **118**, AIAA, pp. 343 - 358.

10. Borgnakke, C. And Larsen, P.S. (1975) Statistical Collision Model for Monte Carlo Simulation of Polyatomic Gas Mixture, *Journal of Computational Physics* **18**, 405 - 420.

11. Abe, T. (1993) Direct Simulation Monte Carlo Method for Internal-Translational Energy Exchange in Nonequilibrium Flow, in B.D. Shizgal and D.P. Weaver (eds.), *Rarefied Gas Dynamics: Theory and Simulations* **159**, AIAA, pp. 103 - 113.

12. Boyd, I.D. (1990) Rotational-translational energy transfer in rarefied nonequilibrium flows, *Physics of Fluids A: Fluid Dynamics* **2**, 447 - 452.

13. Koura, K. (1993) Statistical Inelastic Cross Section Model for Molecules with Discrete Rotational Energy, in B.D. Shizgal and D.P. Weaver (eds.), *Rarefied Gas Dynamics: Experimental Techniques and Physical Systems* **158**, AIAA, pp. 164 - 173.

14. Boyd, I.D. (1993) Relaxation of Discrete Rotational Energy Distributions Using a Monte Carlo Method, *Physics of Fluids A: Fluid Dynamics* **5**, 2278 - 2286.

15. Choquet, I. (1994) Thermal nonequilibrium modeling using the direct simulation Monte Carlo method: Application to rotational energy, *Physics of Fluids A: Fluid Dynamics* **6**, 4042 - 4053.

16. Boyd, I.D. and Bergemann, F. (1994) New Discrete Vibrational Energy Model for the Direct Simulation Monte Carlo Method, in B.D. Shizgal and D.P. Weaver (eds.), *Rarefied Gas Dynamics: Experimental Techniques and Physical Systems* **158**, AIAA, pp. 174 - 183.

17. Boyd, I.D. (1994) Particle Simulation of Vibrational Relaxation, in B.D. Shizgal and D.P. Weaver (eds.), *Rarefied Gas Dynamics: Theory and Simulations* **159**, AIAA, pp. 78 - 86.

18. Choquet, I. and Marmignon, C. (1994) Discrete Modeling of Vibrational Relaxation Processes in Rarefied Nonequilibrium Flows, *Rarefied Gas Dynamics: Theory and Simulations* **159**, AIAA, pp. 87 - 96.

19. Aristov, V.V. and Cheremisin, F.G. (1980) *USSR Compu. Meths. Math. Phys.* **20**, 208.

20. Lumpkin, F.E., Haas, B.L. and Boyd, I.D. (1991) Resolution of differences between collision number definitions in particle and continuum simulations, *Physics of Fluids A: Fluid Dynamics* **3**, 2282 - 2284.

DIRECT NUMERICAL SIMULATION OF LOW-DENSITY ATMOSPHERIC FLOW ON IO

J. V. AUSTIN AND D. B. GOLDSTEIN
Center for Aeromechanics Research
Dept. of Aerospace Engineering and Engineering Mechanics
The University of Texas at Austin
Austin, TX 78712

Abstract. We explore the effects of a non-condensible gas on the SO_2 atmospheric flow near volcanic plumes and sublimating frost patches on Io. The flows involve rarefied/continuum transition, compressibility, gas mixtures, and radiation. Preliminary results are presented which show that a non-condensible gas can either limit or enhance the tranport of SO_2 away from a local source. Such analysis is of importance for determining the distribution of gases near the surface and, hence, the flux of atmosphere into space. The Direct Simulation Monte Carlo (DSMC) method is used.

1. Background

The atmosphere of the Jovian moon Io was first detected by the radio-occultation experiments of Pioneer 10. Since then, SO_2 has been clearly found as both a surface frost [1, 2] and a gas [3]. Many models of the SO_2 atmosphere on Io have been subsequently proposed. These models generally fall into three categories: "sputtered," "sublimated," or "volcanic" [4]. In sputtered models, a highly rarefied, collisionally thin atmosphere is produced by the impact of energetic particles on Io's surface. In sublimated models, such as those of Ingersoll *et al.* [5, 6], the SO_2 atmosphere is in vapor pressure equilibrium with the SO_2 surface frost. In volcanic models, such as those of Moreno *et al.* [7], volcanos provide the primary source of SO_2 for the atmosphere. Both the sublimation and volcanic models are capable of producing a collisionally thick atmosphere. Near-ultraviolet spectra [8] and infrared observations [9] of Io both suggest that Io's atmosphere is col-

749

M. Capitelli (ed.), Molecular Physics and Hypersonic Flows, 749–758.

lisionally thick and thus discount sputtering as the primary source of Io's atmosphere.

Ingersoll *et al.* [5] modeled the circumplanetary flow around Io as due to the sublimation of SO_2 frost on the day side and its condensation on the night side. At the sub-solar point the surface temperature reaches ~130 K and the vapor pressure of SO_2 exceeds 10^{-7} bar, while on the dark side the surface is cold and virtually all of the gas condenses. Solving the vertically integrated conservation equations in one dimension, Ingersoll *et al.* found that in the process of expanding away from the sub-solar point, the flow pressure and temperature drop and the Mach number may reach four. In a second paper [6], Ingersoll modified his earlier work to account for nonuniform surface properties found in observational studies—local volcanic plumes and patchy frost cover are handled.

Moreno *et al.* [7] used computer simulations to model SO_2 frost sublimation atmospheres as well as day and night-side volcanic atmospheres. These simulations incorporated a crude radiative transport model within a time-explicit, finite-volume formulation of the equations of inviscid compressible gas dynamics. In common with Ingersoll [6], Moreno *et al.* found that non-condensing flow over a hot dry surface travels much further than does condensing flow over a cold surface. Moreno *et al.* also found that an interesting oblique "re-entry" shock forms circumferentially around a volcanic vent as the cooled plume falls supersonically back toward the surface and used this to explain ring deposits observed around the vents.

The problem with most SO_2 atmospheric models is they do not produce sufficiently large densities near the evening terminator to reproduce the ionospheric profile measured by Pioneer 10. In order to resolve this apparent contradiction, it has been proposed that a second, high vapor-pressure/non-condensible gas may be present on Io. Matson *et al.* [10] proposed that H_2S may be the second gas for the following reasons: the sulfur to oxygen ratio in the plasma torus around Jupiter suggests another source of sulfur, H_2S is a common volcanic gas, and the dark polar caps may be H_2S frost darkened by UV exposure. Kumar and Hunten [11] and Ingersoll *et al.* [5] suggested O_2, a photochemical by-product of SO_2, as a non-condensible atmospheric constituent. Lellouch *et al.* [9] found that CO and SO may also be present. Moreno *et al.* [7] were able to match the Pioneer 10 data using an atmosphere of all H_2S, however they did not look at the effects of an H_2S/SO_2 mixture. Although its presence has often been suggested, there have been no intensive attempts to model the effect of a non-condensible gas on the flows of the presumably substantial SO_2 atmosphere.

2. Effects of Non-condensible Gas on Atmospheric Flow

A non-condensible gas could exert a discernible hydrodynamic influence on SO_2 flows by impeding the flow and substantially limiting its extent. The non-condensible gas could completely obstruct the SO_2 behind an atmospheric discontinuity—a hydraulic jump if the flow were supercritical or perhaps a combined hydraulic jump/shock wave if the SO_2 flow were supersonic as well at that point. The effect may occur locally wherein a non-condensible gas confines the flow of SO_2 from frost patches or volcanic plumes. Radial outflow of SO_2 along the surface may thereby be confined to a smaller region than that predicted by Ingersoll's single species model [6]. If the SO_2 were so confined, the efflux of S and O from Io would be reduced because the cross-section of SO_2 exposed to sputtering would be smaller.

Another possibility also exists. Instead of blocking the flow of SO_2 with an atmospheric discontinuity, a small amount of non-condensible gas (perhaps less than a few hundred mean free paths thick) may partially blanket the surface and permit the SO_2 to temporarily accumulate above. A "puddle" of SO_2 would thus be held away from the surface by the non-condensible gas. The SO_2 would penetrate the blanket layer to the surface by only molecular or perhaps turbulent diffusion and could thus travel *greater* distances from a source than would be otherwise expected. The temperature of the blanket layer of non-condensible gas would be determined by a balance of conduction and radiation up to the overlying SO_2 and down to the surface. The overlying SO_2 may be warmer due to the heat imparted at the source and/or bulk heating due to ion penetration into the upper atmosphere. Such a thermal inversion layer (warm SO_2 overlying cool non-condensible gas) would stabilize the atmosphere and retard vertical mixing. Lellouch has, in fact, found that portions of the upper atmosphere are actually quite warm (500-600 K at 40 km) [9].

The possibility that a non-condensible gas could either appreciably confine or, conversely, significantly aid in the distribution of SO_2 can best be investigated via computer simulation. It is possible that a sublimating frost (wherein gas is cold and enters the atmosphere gradually from below) and a volcanic plume (for which gas, after passing through the re-entry shock, is appreciably warmed) would exhibit opposite effects in the presence of a non-condensible gas.

3. Computational Techniques

Various atmospheric flows discussed above span the transition regime from continuum to collisionless. In such low density flows the Navier-Stokes equations break down because the scale of the gradients of the flow variables becomes comparable to the mean free path. An efficient rarefied gas dynamics technique is the method of choice for such flows. The Direct Simulation Monte Carlo (DSMC) technique pioneered by Bird [12] models low density gas flow by simulating the movement and collisions of thousands or millions of particles. Hence, a slightly modified Boltzmann equation is solved not by a finite-difference scheme but by directly tracking the motions of gas molecules. This approach is now a standard technique for solving complex low density flows which may involve multiple species, polyatomic molecules with internal energy, radiation, and a wide variety of chemical phenomena. The DSMC technique is well suited for simulating various portions of the atmospheric flow on Io principally because it can handle mixtures of gases [13] (e.g. SO_2, H_2S, and O_2) and can incorporate gravity fields, variable molecule/surface adhesion coefficients, and radiative heat transfer. The technique is designed for modestly rarefied flows, however continuum flows may also be computed with a corresponding loss of efficiency. We also have the novel ability to run a merged DSMC/Euler finite volume solver to model flows which exhibit both fully continuum and transitional character (e.g. a fully continuum volcanic plume expanding into a vacuum).

4. Results

We present preliminary results for simulated atmospheric flows on Io to determine the effects of a non-condensible gas on volcanic plumes and frost patches. We examine how a non-condensible gas may *confine* or *enhance* the distribution of SO_2 and the effect of such a gas on the SO_2 density and temperature profiles.

The result from a DSMC simulation of a strong volcanic plume on the night side of Io is shown in Figure 1. The surface is fully diffuse with full thermal accommodation and a 20% sticking coefficient (a 20% chance of an incident particle condensing onto the surface). For this preliminary calculation the molecules were modeled as simple monatomic hard spheres. The flow is axisymmetric; only half of the plume is seen. Several features are apparent: immediately above the orifice the plume expands out to about ten times its initial radius of 8 km. As the rising gas decelerates due to gravitational forces, it is abruptly slowed by a nearly normal canopy shock at about 250 km altitude [14]. In the subsonic region above the canopy shock the gas expands out radially and begins to accelerate downwards. Just above the surface the high speed descending gas passes through a curving oblique re-entry shock [7] to again flow radially outwards. The shocked gas has sufficient thermal energy to rebound vertically and a minor "bounce" occurs at around 500 km radius. The character of the re-entry shock appears sensitive to the local surface temperature: at low temperatures there is no shock and the gas pours into the surface unimpeded, while at higher

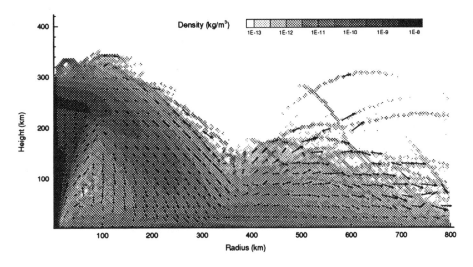

Figure 1. A low density, supersonic, axisymmetric, volcanic plume exhausting vertically and condensing onto the surface. The contours show density and the local velocity vectors are overlaid. The model was run at low density in order to keep the computation time short (less than an hour)—much higher densities can be simulated with longer computation time. At the exit the plume is 8 kilometers in radius and has a velocity of 1000 m/s (~Mach 7) and a temperature of 100 K. The lower boundary is condensing with the temperature held at 90 K, representing the night-side of Io.

temperatures the gas does not condense as readily and the outflowing gas from the surface blocks the downfalling volcanic plume thus forming the re-entry shock. The remainder of the expansion (beyond 500 km radius) is uneventful as the gas slowly condenses out upon the surface.

The canopy and re-entry shocks can be better seen in Figure 2. This figure shows temperature contours for a volcanic plume erupting into an atmosphere which contains a non-condensible gas. For simplicity, both gases are modeled as monatomic hard spherical molecules of the same mass; one species is allowed to condense and the other is not (a sticking factor of 0%).

Figures 3 and 4 show the results from six simulations of a plume exhausting into a non-condensible atmosphere. The plume velocity is different in each case, but the mass outflux is held the same. For a very weak plume (lower left image) most of the domain (> 150 km radius) is a simple isothermal exponential atmosphere of the non-condensible species. This situation is analogous to the flow near a sublimating frost patch. In the corresponding image of Figure 4 we see that the plume gas is confined to a cylindrical region of roughly 80 km radius behind a vertical species discontinuity. The vertical expansion of the sublimating material provides cooling, which leads to a decrease in scale height over the frost patch. Stronger plumes can pen-

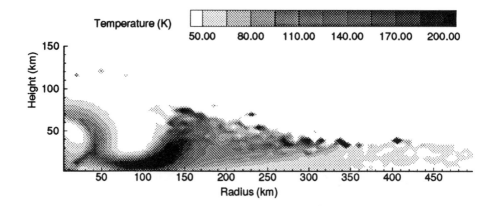

Figure 2. Temperature contours for a plume of condensible gas (SO_2) exhausting into a gravitationally settled atmosphere of non-condensible gas (H_2S). At the exit the plume is 8 kilometers in radius and has a velocity of 500 m/s ($\sim$Mach 3.5) and a temperature of 100 K. The SO_2 is allowed to condense on the surface with a sticking factor of 20% but the H_2S is considered noncondensible and is diffusely reflected with a temperature of 50 K. The hot re-entry shock is plainly seen, as is the dramatic spread of hot plume gas over the blanketing layer of cold non-condensible.

etrate the non-condensible background gas to varying degrees before they collapse back. The flow pattern appears to depend on the plume strength relative to the scale height of the atmosphere, and therefore it depends on the Reynolds number. As the plume strength is increased the plume rises to a substantial height and finally takes on the mushroom shape that characterized the single species plume. Stronger plumes are seen to spread to a larger radius. For the stronger plumes we find that the species discontinuity becomes thicker and tilts over as the plume gas rides up over the top of the non-condensible gas. The over-riding plume gas is appreciably warmer than the gas below and forms a local inversion layer, as seen in Figure 2.

Our simulations of volcanos and frost patches are being expanded and thoroughly compared to the results of Ingersoll [6] and Moreno *et al.* [7]. Preliminary comparisons show reasonable agreement.

5. Summary

The SO_2 atmosphere is likely the source of high altitude sulfur and oxygen atoms. Neutral atoms near Io may be ionized by one or more processes (e.g. charge exchange reactions with torus ions or UV sunlight) and then scoured away by Jupiter's rotating magnetic field. Hence, to answer questions about the exact mechanism for the supply of ions to the plasma torus, a detailed model of Io's atmosphere is needed. A detailed understanding of the flow

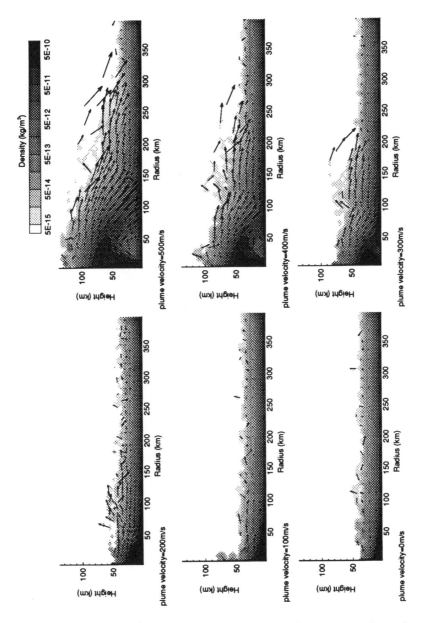

Figure 3. A series of simulations of a condensible gas (SO₂) erupting with varying exit velocities into a non-condensible gas (H₂S). At the exit the plume is 8 kilometers in radius and has a temperature of 100 K. The SO₂ is allowed to condense on the surface with a sticking factor of 20% but the H₂S is considered noncondensible and is diffusely reflected with a temperature of 50 K. The contours represent density and the local velocity vectors are overlaid. As in Figure 1, the density was lowered to decrease computation time.

756

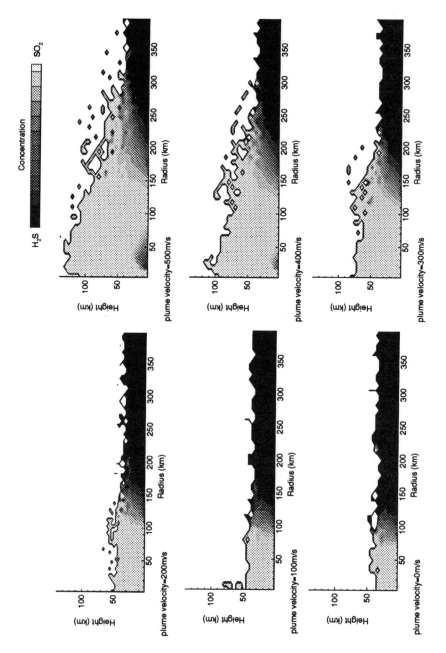

Figure 4. Relative concentration of the non-condensible gas (H_2S) and the condensible gas (SO_2) for the same conditions as Figure 3. The dark areas represent 100% H_2S and the light areas are 100% SO_2. A species discontinuity is clearly visible. The H_2S trapped below the canopies of the strongest plumes ($\sim$ 30 km radius) diffuses away slowly ($\sim$1 hour after plume startup).

associated with the volcanic plumes—ultimately the major source of atmospheric volatiles—is also essential to understanding Io and its distribution of surface frost. We suggest that these problems are coupled with the distribution of a non-condensible gas about Io and we have run a preliminary set of computer simulations to investigate this. We find that strong plumes can rise above and be partially supported by the non-condensible gas atmosphere, enhancing the transport of SO_2 away from the volcano. We also find that the outflow from frost patches can be confined by the non-condensible gas, thereby reducing the transport of SO_2. By varying the concentration of non-condensible gas we can alter the structure of the atmospheric discontinuities. Our preliminary studies have demonstrated that such flows are far from simple.

6. Acknowledgments

We would like to thank Wyatt Phillips for technical assistance in running these simulations. The simulations presented were run on DEC Alpha workstations.

References

1. Fanale, F.P., Brown, R.H., Cruikshank, D.P. and Clark, R.N. (1979) Significance of absorption features in Io's IR reflectance spectrum, *Nature* **280**, 760–761.
2. Smythe, W.D., Nelson, R.M. and Nash, D.B. (1979) Spectral evidence for SO_2 frost or absorbate on Io's surface, *Nature* **280**, 766–767.
3. Pearl, J.C., Hanle, R., Kunde, V., Maguire, W., Fox, K., Gupta, S., Ponnamperuma C. and Raulin, F. (1979) Identification of gaseous SO_2 and new limits for other gases on Io, *Nature* **288**, 757–758.
4. Schneider, N.M., Smyth, W.H. and McGrath, M. A. (1989) Io's atmosphere and neutral clouds, In M.J.S. Belton, R.A. West, and J. Rahe (eds.), *Time-Variable Phenomena in the Jovian System* NASA, Washington, D.C, pp. 75–99.
5. Ingersoll, A.P., Summers, M.E. and Schlipf, S.G. (1985) Supersonic meteorology of Io: Sublimation driven flow of SO_2, *Icarus* **64**, 375–390.
6. Ingersoll, A.P. (1989) Io meteorology: How atmospheric pressure is controlled locally by volcanos and surface frosts, *Icarus* **81**, 298–313.
7. Moreno, M.A., Schubert, G., Baumgardner, J., Kivelson, M.G. and Paige, D.A. (1991) Io's volcanic and sublimation atmospheres, *Icarus* **93**, 63–81.
8. Ballester, G.E., Strobel, D.F., Moos, H.W. and Feldman, P. D. (1990) The atmospheric abundance of SO_2 on Io, *Icarus* **88**, 1–23.
9. Lellouch, E., Belton, M., de Pater, I., Paubert, G., Gulkis, S. and Encrenaz, T. (1992) The structure, stability, and global distribution of Io's atmosphere, *Icarus* **98**, 271–295.
10. Matson, D.L., Johnson, T.V., McEwen, A.S. and Soderblom, L.A. (1988) Io: The case for an H_2S atmosphere, *Bull. Am. Astron. Soc.* **20**, 1125.
11. Kumar, S., and Hunten, D.M. (1982) The atmospheres of Io and other satellites., in D. Morrison (ed.), *Satellites of Jupiter*, Univ. of Arizona Press, Tucson, pp. 782–806.
12. Bird, G.A. (1976) *Molecular Gas Dynamics*, Oxford University Press, London.
13. Goldstein, D.B. (1992) Discrete-velocity collision dynamics for polyatomic molecules, *Physics of Fluids* A, 4 (8) August.

14. Cook, A.F., Shoemaker, E. and Smith, B. (1979) Dynamics of volcanic plumes on Io, *Nature*, **280**, 743–746.

SINGLE PARTICLE CELLULAR AUTOMATA MODELS FOR SIMULATION OF THE MASTER EQUATION

Non-Reacting Single Specie Systems

HIMANSHU AGRAWAL AND E. RATHAKRISHNAN
Department of Aerospace Engineering
Indian Institute of Technology, Kanpur—208016, India

Abstract. We present a hierarchy of cellular automata models for simulation of non-reacting single specie particle dynamical systems governed by the master equation. In these models lattice site values are interpreted to be representing single particles moving over an arbitrary regular spatial lattice with any one of the finite number of velocities belonging to a discrete velocity set at any time step. The particle interaction rules of these models lead to construction of *state transition probability* (STP) tables similar to those encountered in discrete Markov processes. Consequently, these models mimic the dynamics represented by the master equation when the STP tables allow the process to be homogeneous in time. The lattice site values in these models are *globally coupled*; making them inherently sequential simulation tools. Within the framework of *single particle representation* (SPR) interpretation of lattice site values, a generalized model development methodology has been formalized and demonstrated by developing a synchronous two-dimensional single speed SPR-LGA model over square spatial lattice for simulation of single specie non-reacting dynamical systems. In this model particles are permitted to have any one of the four velocities belonging to the discrete velocity set $\{(\pm 1, 0), (0, \pm 1)\}$ at any time step. Using this model simulations have been carried out on system of particles enclosed in a container at equilibrium. Simulation results are found to be in excellent agreement with the available theoretical results.

1. Introduction

Cellular automata (CA) are abstract systems introduced by Ulam and von Neumann in connection with evolutionary biological systems [1]. Since their

M. Capitelli (ed.), Molecular Physics and Hypersonic Flows, 759–770.

inception CA have evolved rapidly and found numerous applications in both biological and non-biological systems. CA have been recognized as a formalism presenting a framework for developing extremely fast, simple, and massively parallel computer models for simulation of complex systems [2]. Various aspects of CA modeling have been studied extensively and number of models for simulating various phenomena occurring in systems of varied complexity have been developed [3].

CA along with many other rule based modeling systems that work in discrete time over discrete spatial lattices with discrete lattice sites values, when applied to fluid dynamical systems are known as *lattice gas automata* (LGA) in the literature. The first LGA model was developed by Hardy, de Pazzis, and Pomeau [4] over square spatial lattice (henceforth, the HPP model). This model lacks isotropy, Galilean invariance, and time reversibility and the impact parameter during particle collisions is always zero. Consequently, its actual dynamics is one-dimensional despite the underlying two-dimensional spatial lattice. This has been confirmed in computer simulations by Margolus *et al.* [5]. Through simulations Margolus *et al.* found that another LGA model, the TM gas model [6], that also exists over square spatial lattice like the HPP model but has different collision rules and nonzero impact parameter shows truly two-dimensional behavior. It will be beneficial to note that, within our knowledge, the micro-dynamical equations for TM gas model have not been formulated yet.

Following the HPP model, Frisch, Hasslacher, and Pomeau developed another LGA model [7] (henceforth, the FHP model). They observed that square spatial lattice lacks required symmetries to be isotropic in two-dimensions and hence used the more appropriate triangular spatial lattice for their model. Consequently, Navier-Stokes equations in two-dimensions have been recovered in the FHP model. Despite this, computer simulations by Margolus *et al.* [5] show that the FHP model has an abnormally high dynamical exponent to be truly two-dimensional model.

Following these early attempts large number of one, two, three, and four-dimensional LGA models have been developed and studied [8, 9, 10]. At present perhaps the most notable model among these is that by Appert and Zaleski [9] (henceforth, the AZ model). This model incorporates finite range attractive interactions and shows liquid-gas phase transitions. Further comments on these models are beyond the scope of this paper.

In this paper our attention is focused on a specific aspect of LGA modeling, namely, the interpretation of lattice site values employed during model development. Analysis of the fundamental elements of CA in view of the available literature reveals that the existing LGA models can be classified in two categories depending upon the interpretation of lattice site values employed in them. These categories are: (i) the models in which lattice site values are interpreted to be representing multiple particles—*multiple particle representation* (MPR) LGA models, and (ii) the models in which lattice site values are interpreted to be representing single particle—*single particle representation* (SPR) LGA models.

Extensive investigations have been carried out on MPR-LGA models in the literature; some notable examples being the HPP, TM, FHP, and AZ models and their extensions. These models can be projected in alternate forms in which only one particle occupies a spatial lattice site by imposing appropriate partitioning on a group of lattice sites as has been done for the HPP, TM, and FHP models in [6]. As a result, the MPR-LGA models can be viewed as efforts towards developing SPR-LGA models; though, this view has never been adopted before.

Except for the above view and few studies that qualify to be on SPR-LGA models [11], these models have largely remained neglected. This is because of their complexity which arises from *globally coupled* particle dynamics in these automata, *i.e.*, the new state of a particle depends upon the current state of all the particles occupying the spatial lattice, in general. This makes the SPR-LGA models important in modeling physics and physical systems because properties (*e.g.*, momentum) can be transported over arbitrarily long distances in one time step independent of mass transport. Another notable fact is that for the potential energy of a system to be defined by an Ising-type Hamiltonian each lattice site should necessarily be occupied by exactly one particle or state [12]. Furthermore, because of externally imposed constraints MPR-LGA models can never give complete dynamical information. Hence, to study all possible aspects of particle dynamics over given spatial lattices development of SPR-LGA models is necessary. In view of these observations, a hierarchy of SPR-LGA models has been introduced and studied in this paper.

2. Hierarchy of SPR-LGA Models

The hierarchy of SPR-LGA models is a set of models in which only one particle is allowed to occupy a lattice site of an underlying spatial lattice at a time step. The particles occupying the spatial lattice move along the lattice links with any one of the finitely many velocity vectors belonging to a discrete velocity set at any time step. Velocity of any particle is not correlated to or restricted by the velocity of other particles at any time step by externally imposed constraints. The motion of particles along the lattice links spontaneously leads to collisions which are resolved using a set of collision rules. Particle dynamics over the spatial lattice during one evolution, *i.e.*, from the beginning of one time step till its end, is decomposed into two steps, namely: (i) collision resolution, and (ii) particle translation.

3. Types of Particle Collisions

Collisions occurring among particles in SPR-LGA models can be subdivided in two fundamentally distinct categories depending upon their cause of origin and phenomenological representations. These are: (i) *annihilating* or *hard-core* collisions, and (ii) *grazing* or *soft-core* collisions.

Annihilating collisions are those collisions which, if left unresolved, spontaneously lead to annihilation (loss) of at least one or more of the colliding particles. These collisions occur in following two situations: (i) when two or more particles try to reach the same lattice site simultaneously, and (ii) when two or more particles moving along the same axis, and at least one of which is moving opposite to others, try to pass through the same lattice link simultaneously. Since only these collisions occur among particles interacting via hard-core[1] interaction potential $\phi_H(r)$, they collisions are also referred to as *hard-core collisions*. Collision rules must necessarily incorporate elements that are capable of resolving these collisions.

Grazing collisions occur among particles separated by arbitrary distances over the spatial lattice and moving with arbitrary velocity vectors. These collisions occur because of the soft-core[2] interaction part $\phi_S(r)$ of interaction potential $\phi(r)$ among particles. Incorrect resolution of these collisions leads to incorrect modeling of soft-core interaction potential and not loss of particles. Hence, elements to resolve these collisions need to be incorporated into collision rules only if interaction potential among particles differs from hard-core interaction potential.

Aforementioned discussion shows that by incorporating appropriate elements in collision rules for both the hard-core and soft-core collisions it is (in principle) possible to develop SPR-LGA models for simulating dynamical systems with arbitrary interaction potential among particles. We have not be able to develop such SPR-LGA models. Only soft-core collisions have been considered in the models developed in this paper. Henceforth, the prefix *soft-core* will be omitted.

4. States of LGA-Modeled Dynamical Systems

Henceforth, given state of all the particles over spatial lattice which includes their species, positions, and velocity vectors, will be referred to as *spatial lattice configuration* or merely as *configuration*. The breakup of dynamics

---

[1]Here, definition of hard-core interaction potential ϕ_H between two particles separated by a distance r has been extended to: $\phi_H(r) = \begin{cases} \phi_a & \text{for } r \leq a \\ 0 & \text{for } r > a \end{cases}$, where $0 \leq \phi_a \leq \infty$, and a is critical distance.

[2]The interaction potential $\phi(r)$ can be decomposed into hard-core interaction potential $\phi_H(r)$ and soft-core interaction potential $\phi_S(r)$ as: $\phi(r) = \phi_H(r) + \phi_S(r)$.

of particles over the spatial lattice during one evolution into two steps (*c.f.*, Sec. 2) implies that depending upon the instant of time at which spatial lattice configuration is frozen, three different configurations are possible in any simulation during one evolution. These are: (i) *pre-collision* or *collisional configuration*, (ii) *post-collision* or *collision-less configuration*, and (iii) *post-translation configuration*.

The *collisional configuration* is spatial lattice configuration at the beginning of an evolution. It is starting configuration for collision resolution because particles collisions occur in it. The *collision-less configuration* is obtained from collisional configuration after collision resolution is over. This configuration marks the beginning of particle translation. The *post-translation configuration* is obtained by repositioning particles in collision-less configuration to new lattice sites pointed out by their velocity vectors. It marks the beginning of next evolution.

5. Collision Resolution

Collision resolution can be addressed under three different starting assumptions, namely: (i) the assumption that the laws of conservation of mass, momentum, and energy hold during individual collisions (*classical conservation approach* or *CCA*), (ii) the assumption that the law of conservation of mass holds during individual collisions and laws of conservation of momentum and energy hold statistically over spatial lattice (*weak statistical conservation approach* or *WSCA*), or (iii) the assumption that the laws of conservation of mass, momentum and energy hold statistically over spatial lattice (*strong statistical conservation approach* or *SSCA*).

Under each of these three assumptions, both synchronous and asynchronous methods are possible for collision resolution. In asynchronous methods global coupling among lattice site values is bypassed by randomly selecting few particles over spatial lattice and replacing their states with new states in accordance with appropriate pre-selected criteria to complete an evolution. On the other hand in synchronous methods all the collisions are resolved simultaneously to complete an evolution. SPR-LGA models with asynchronous collision resolution algorithms that satisfy the assumptions of WSCA/SSCA exist in the literature [11] (*c.f.*, Sec. 1). In this paper we only consider SPR-LGA models with synchronous collision resolution algorithms that satisfy the assumption of CCA.

5.1. SYNCHRONOUS COLLISION RESOLUTION USING CCA

In view of the ergodic hypothesis all the collision-less configurations having same mass, momentum, and energy as pre-collision configuration qualify as valid solutions of the pre-collision configuration, if not *forbidden*.[3] This implies that, collision resolution is essentially transformation of arbitrary pre-collision configuration to one plausible collision-less configuration. All collision-less configurations which are solutions of arbitrary pre-collision configuration can be obtained by changing the states of particles (lattice site values) in it. The ergodic hypothesis does not put restriction on how and under what assumptions the transformation has to be carried out. However, restrictions may be imposed by the approach being followed for carrying out the transformation, *e.g.*, as with the assumptions of CCA.

Within the framework of CCA two types of methods (algorithms) are possible for synchronously resolving inter-particle collisions occurring in SPR-LGA models—(i) *single step methods*, and (ii) *multi-step methods*.

Single step methods claim that all the collisions can be resolved in one step (*i.e.*, in one tick of system clock of an arbitrary processor) while keeping the assumptions of CCA intact. However, it has been shown in [13] that it is practically impossible to develop such algorithms. Consequently, further discussion on these methods will be completely skipped from this paper.

5.2. MULTI-STEP METHODS

Multi-step methods attempt on-line construction of exactly one plausible collision-less configuration for arbitrary pre-collision configurations. These methods work under the assumption that all the collisions occurring in arbitrary pre-collision configurations can be resolved to desired accuracy by iteratively searching for and resolving an appropriate set of configurations called *fundamental collision configurations* (FCCs). This assumption is well justified under ergodic hypothesis at the microscopic level. However, at macroscopic level, justification for this assumption comes from the observation that micro-dynamic rules for multi-particle collision systems, though supposedly existing, are unknown except for two-particle systems. Hence, any systematic attempt at iterative transformation of arbitrary pre-collision configurations to collision-less configurations must start by decomposing the pre-collision configurations in terms of smaller configurations, *e.g.*, the *fundamental collision configurations* here.

The FCCs are defined to be spatial lattice configurations having just one collision among a group of particles occupying the spatial lattice. Particles undergoing direct collision in FCCs are called *primary* or I^{st}-*order collision*

---

[3]Spatial lattice configurations that cannot be obtained in one evolution through dynamical rules are termed *forbidden* configurations, *e.g.*, in non-reacting systems, spatial lattice configurations in which particles have moved to lattice sites that cannot be reached in one evolution by any of the velocity vectors belonging to the selected discrete velocity set are *forbidden* configurations.

partners. Nearest neighbors[4] of primary collision partners are called *secondary* or IInd*-order collision partners*, and so on. Based on maximum order of collision partners present, the FCCs are subdivided into Ist-order FCCs, IInd-order FCCs, and so on. Spatial lattice of Ist-order FCCs is smallest and contains particles undergoing collision and the site of collision if collision occurs at a lattice site. Besides members of the underlying Ist-order FCC, spatial lattice of IInd-order FCCs contains nearest neighbors of colliding particles as well; and so on for others.

Identification of a set of FCCs for resolving collisions fixes a SPR-LGA model. Following this, all possible solutions are tabulated for each of these FCCs. Within the framework of CCA, all collision-less configurations having same mass, momentum, and energy as that of the FCC are its valid plausible solutions. These can be obtained by arbitrary changes in velocities of particles belonging to the FCC. While computing solutions of a FCC, it is treated as an isolated system, or alternatively, lying in free space. Following the tabulation, *state transition probabilities* are assigned to various FCC ↦ SOLUTION state transition pairs. Finally, relative priorities are assigned to the FCCs to fix the order of search during simulations.

6. Particle Translation and Boundary Conditions

Particle translation necessarily follows collision resolution to ensures that the assumptions of CCA are not violated. In this step, particles are repositioned at new lattice sites pointed out by their velocity vectors.

Free boundaries are treated probabilistically. Solid boundaries are taken care of by introducing additional particle species and velocity vectors corresponding to the boundary particles and appropriately extending the collision rules to incorporate these. Moving boundaries can be implemented through appropriate state transition probabilities as well by assigning non-zero velocity vectors to boundary particles.

7. Dynamics of SPR-LGA Models

The foregoing discussion shows that when the collision rules are *time invariant*,[5] the sequence of states computed by the multi-step methods comprise a discrete stationary Markov process which is homogeneous in time, *i.e.*, it depends only on the time interval and not on initial and final times. This

---

[4]*Nearest neighbors* (NNs) of a lattice site (i, j) (or, particle occupying this lattice site) are defined to be all those lattice sites (or, particles) whose states might be affected by an arbitrary change in the direction of motion of the particle occupying the lattice site (i, j).

[5]The collision rules are *time invariant* iff the same set of FCCs and the same state transition probability table are valid at all time steps during a simulation.

dynamics is represented by the *discrete master equation* [14]

$$\Pi(z|z_0; t + \Delta t) - \Pi(z|z_0; t) = \Delta t \sum_{z'} \left[w_{z'z} \Pi(z'|z_0; t) - w_{zz'} \Pi(z|z_0; t) \right] \quad (1)$$

where, z, z', and z_0 are phase vectors representing the state of the system at times t, $t < t' < 0$, and 0, respectively; $\Pi(z|z_0; t)dz$ represents the probability of finding the system in the phase volume dz around the point z at time t, granted that it was in the state z_0 at $t = 0$; and the coefficient $w_{z'z}$ represents the probable rate of transitions from the state z' to the state z.

In the above equation, $\Pi(z|z'; t)$ is *conditional N-particle joint probability distribution function*. Clearly, further reductions of this equation for specific SPR-LGA models and state transition probability tables is a non-trivial task; and, so is specification of weights $w_{zz'}$ and $w_{z'z}$.

8. Single Speed SPR-LGA Model

In this model identical neutral particles move over square spatial lattice with velocity vectors belonging to the discrete velocity set $\{(\pm 1, 0), (0, \pm 1)\}$ at any time step. Collisions among fluid particles are resolved using rules outlined in Fig. 1. It can be readily verified that all these rules satisfy the assumptions of CCA. In this Fig. only one out of many possible equivalent FCCs has been shown in the column (0) entitled *FCC*. All the valid solutions of the FCCs are shown in corresponding rows in columns (1)–(8). Similarly, collisions among fluid particles and stationary boundary particles are resolved using rules outlined in Fig. 2. For systems having stationary solid walls, these Figs. are merged together by appropriately reassigning relative priorities of all the FCCs belonging to them (*e.g.*, as in Fig. 2).

During simulation, collisions are resolved iteratively. In an iteration, the spatial lattice is searched for all the occurrences of FCCs shown in the Figs. in the order of their relative priorities. Any node of the FCCs can be searched. However, for FCCs shown in Fig. 2, searching for the fluid particle node reduces computation. If a FCC is found, its solution is identified using a then generated pseudo random number and the pre-selected state transition probabilities. Following this, states of all the primary collision partners in the FCC are replaced with the states of corresponding particles in its solution. Each lattice site is touched exactly once. The iterations are terminated when no FCCs that can lead to violation of assumptions of classical conservation are found over the spatial lattice.

9. Simulation Results and Discussion

Simulations were carried out for an athermal diffusive dynamical system comprising of particles randomly distributed at equilibrium density of n

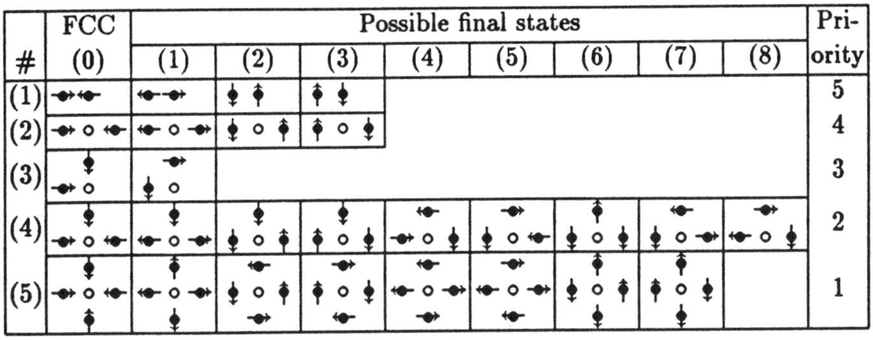

Figure 1. Rules for resolving collisions among simple fluid particles in SPR-LGA model for simulation of isothermal dynamics over square spatial lattice. Symbols stand for: o space or fluid particle with velocity such that it does not collide with other particles in the FCC, ● fluid particle, ←↑↓→ direction of motion of particles. In the priority column lower numbers indicate higher priority.

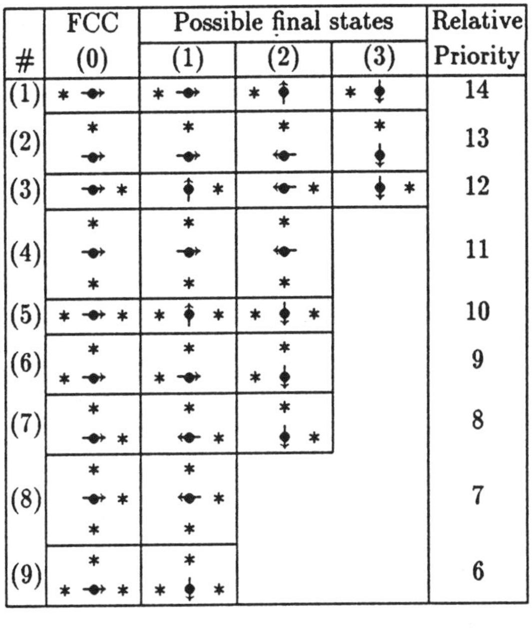

Figure 2. Rules for resolving collisions among simple fluid particles and fixed solid boundary particles. Symbols stand for: * fixed boundary particle, ● fluid particle, ←↑↓→ direction of motion of particles. Format is identical with Fig. 1.

particles per lattice site in a rectangular enclosure of dimensions $(N_x, N_y) = (254, 254)$ with 2 lattice site thick boundary walls. Particle collisions were resolved in accordance with the rules outlined in Figs. 1 and 2. The FCCs of Fig. 1 were searched before those of Fig. 2, in the order of their relative priorities. In the state transition probability table all the solutions of a FCC

768

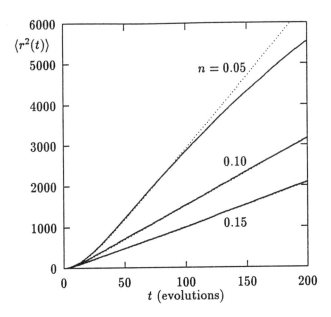

Figure 3. Time variation of the mean squared displacement. Sampled data (solid), and the Langevin solution (dotted).

were assigned equal probability. In the following discussion all quantities are measured in natural CA units.

Behavior of the system was studied for $n = 0.05, 0.10$, and 0.15. In each simulation 253 particles lying nearest to the center of the box $(127, 127)$ were tagged and traced for 1000 time steps. Mean squared displacement of particles $\langle r^2(t) \rangle \equiv \langle [\mathbf{r}(t) - \mathbf{r}(0)]^2 \rangle$ was sampled. Since the system is Newtonian, the velocity autocorrelation function $\langle \mathbf{v}(0) \cdot \mathbf{v}(t) \rangle$ identifies with $\partial^2 \langle r^2(t) \rangle / \partial t^2$ [12]. In this modeled system two, three, and four particle collisions occur simultaneously and particle velocities change after collisions. Consequently, the mean collision time τ for tagged particle dynamics is best calculated as $\tau = -[\ln(1 - \lambda)]^{-1}$, where λ is given by Eqn. (13) of [15]. This gives $\tau = 17.84, 8.31$, and 5.34, for $n = 0.05, 0.10$, and 0.15, respectively.

Fig. 3 shows time variation of $\langle r^2(t) \rangle$ averaged over 25 ensembles, *i.e.*, 6325 particles. The dotted lines show the Langevin variation [16]

$$\langle r^2(t) \rangle = \frac{2kT\tau}{m}[t - \tau(1 - e^{-t/\tau})]$$

where $k = 1$ is the Boltzmann constant, $T = 1$ is temperature, $m = 1$ is mass of particles, and τ is calculated as above. This gives Eqn. (2) of [15].

The agreement of sampled data and Langevin results is nearly exact. Lines representing sampled data start curving down following first collision of any of the tagged particles with system boundaries (seen for $n = 0.05$

in Fig. 3). These results show that: (i) Particles perform random walk over spatial lattice, hence, diffusion equation is recovered in the macroscopic limit for state transition probability tables used. (ii) Dynamical exponent is 1, hence, system is truly two-dimensional.

The average number of iterations required per time step for these simulations were 4.14, 10.88, and 58.57 for $n = 0.05, 0.10$, and 0.15, respectively. It can be readily shown that in open systems, and closed systems with at least 2×2 spaces, the iterations will always converge. Upper bounds on number of iterations cannot be derived because random numbers are used. For n above ≈ 0.15 simulations could not be carried out because too many iterations were required.

Because of global coupling among lattice site values, termination of iterations requires global communication over spatial lattice. This makes SPR-LGA models inherently sequential. This, however, is not a major setback as far as modeling of physics and physical systems is concerned. Methods of parallelization of these automata can be found, perhaps by incorporating appropriate modifications [17].

The models are nonlinear and capable of simulating highly compressible systems, *e.g.*, shock waves. Simulations with different state transition probability tables show that both the Gibbs and non-Gibbs equilibrium states are reachable for same initial conditions. It is readily seen that besides the master equation, SPR-LGA models can simulate non-stationary discrete Markov processes also when collision rules and state transition probability tables are made time dependent.

H.A. wishes to acknowledge G. D. Doolen for having freely provided his reference database on cellular automata, lattice gas automata and related topics. CSIR and DST provided partial financial support for travel.

References

1. J. von Neumann. *Theory of Self-Reproducing Automata*. Univ. of Illinois Press, 1966.
2. S. Wolfram, (ed). *Theory and Applications of Cellular Automata*. World Sci., 1986.
3. S. Wolfram. *Rev. Mod. Phys.*, 55(3):601–644, 1983; D. Farmer, T. Toffoli, and S. Wolfram, (eds). *Cellular Automata*, vol. 10 of *Physica D*, 1984; H. Gutowitz, (ed). *Cellular Automata: Theory and Experiment*, vol. 45 of *Physica D*, 1990; G. Doolen, (ed). *Lattice Gas Methods for PDE's: Theory, Application, and Hardware*, vol. 47 of *Physica D*, 1991; T. M. M. Verheggen, (ed). *Numerical Methods for the Simulation of Multi-Phase and Complex Flow*, vol. 398 of *Lecture Notes in Physics*, Springer-Verlag, 1992; M. A. van der Hoef, M. Dijkstra, and D. Frenkel. *Europhys. Lett.*, 17(1):39–43, 1992; T. Naitoh *et al. Phys. Rev. E*, 47(6):4098–4103, 1993; S. P. Das, H. J. Bussemaker, and M. H. Ernst. *Phys. Rev. E*, 48(1):245–255, 1993; P. Grosfils *et al. Phys. Rev. E*, 48(4):2655–2668, 1993.
4. J. Hardy and Y. Pomeau. *J. Math. Phys.*, 13(7):1042–1051, 1972; J. Hardy, Y. Pomeau, and O. de Pazzis. *J. Math. Phys.*, 14(12):1746–1759, 1973; J. Hardy, O. de Pazzis, and Y. Pomeau. *Phys. Rev. A*, 13(5):1949–1961, 1976.
5. N. Margolus, T. Toffoli, and G. Vichniac. *Phys. Rev. Lett.*, 56(16):1694–1696, 1986.

770

6. T. Toffoli and N. Margolus. *Cellular Automata Machines: A New Environment for Modeling.* MIT Press, 1987.

7. U. Frisch, B. Hasslacher, and Y. Pomeau. *Phys. Rev. Lett.*, 56(14):1505–1508, 1986.

8. D. d'Humières, P. Lallemand, and U. Frisch. *Europhys. Lett.*, 2(4):291–297, 1986; D. d'Humières and P. Lallemand. *Helv. Phys. Act.*, 59(6/7):1231–1234, 1986; G. Y. Vichniac, P. Tamayo, and H. Hartman. *J. Stat. Phys.*, 45(5/6):875–883, 1986; B. Chopard and M. Droz. *Phys. Lett. A*, 126(8/9):476–480, 1988.

9. C. Appert and S. Zaleski. *Phys. Rev. Lett.*, 64:1–4, 1990.

10. M. Gerits, M. H. Ernst, and D. Frenkel. *Phys. Rev. E*, 48(2):988–999, 1993; B. Chopard, P. Luthi, and M. Droz. *Phys. Rev. Lett.*, 72(9):1384–1387, 1994.

11. F. J. Alexander *et al. Phys. Rev. E*, 47(1):403–410, 1993; J. J. Alonso, J. Marro, and J. M. González-Miranda. *Phys. Rev. E*, 47(2):885–898, 1993; B. Chopard *et al. Phys. Rev. E*, 47(1):R40–R43, 1993; W. Kob and H. C. Andersen. *Phys. Rev. E*, 47(5):3281–3302, 1993; W. Kob and H. C. Andersen. *Phys. Rev. E*, 48(6):4364–4377, 1993; M. Nieswand, W. Dieterich, and A. Majhofer. *Phys. Rev. E*, 47(1):718–720, 1993.

12. J. R. Gunn, C. M. McCallum, and K. A. Dawson. *Phys. Rev. E*, 47(5):3069, 1993.

13. H. Agrawal and E. Rathakrishnan. Submitted to *Phys. Rev. Lett.*.

14. R. L. Liboff. *Kinetic Theory—Classical, Quantum, and Relativistic Descriptions.* Prentice Hall, 1990.

15. M. H. Ernst and T. Naitoh. *J. Phys. A: Math. Gen.*, 24(11):2555–2564, 1991.

16. H. L. Friedman. *A Course in Statistical Mechanics.* Prentice-Hall, 1985.

17. H. Agrawal and E. Rathakrishnan. Under preparation.

APPLICATION OF MONTE CARLO SENSITIVITY ANALYSIS TO ELECTRON-MOLECULE CROSS SECTIONS IN NITROGEN DISCHARGES

F.Esposito, M.Capitelli
Centro Studi Chimica dei Plasmi del C.N.R. and Dipartimento di Chimica - Università di Bari
via Orabona, 4 - 70126 Bari - Italy

1 - Introduction

It is quite frequent in chemical kinetics, in combustion studies or in meteorology, to use large computational models that depend - generally in a *non-linear* fashion - on a certain number of "parameters" , that is quantities, fixed during the model calculation, that allow to modify the model behaviour in order to fit it to the specific problem studied. Often these parameters are not well known (typically, cross sections or rate constants in chemical kinetics). When the number of parameters to be fitted is low (less than three or four, for example), generally it is relatively easy to search for the better values by simply varying the parameters in some ordered way and comparing the obtained results with experimental knowledge of the investigated phenomenon. But that is impossible to do if the parameters to be varied are tenths, because of the huge computational effort and memory storage capacity requested, which grow as a *power* of the number of varied parameters. In this case, a workable strategy is to reduce as much as possible the number of parameters to be adjusted, leaving fixed the "less important" ones, that is those parameters not largely affecting the output of the model, when restricted to specified intervals. Sensitivity analysis is a tool to know the "importance" of each parameter, once its variation interval has been assigned (this last could be the parameter incertitude around the experimental value, or the allowed range of variation of a controllable quantity).
We have performed sensitivity analysis on a numerical model, based on the Boltzmann transport equation applied to calculate the electron energy distribution function (eedf) and some rate coefficients in a nitrogen discharge, in order to search for the most important parameters, chosen among the vibrational temperature T_V of molecules and cross sections of 20 electron-molecule collision processes, the former linearly variable between 0 K and 5000 K, the latter varing of ±50% of their respective original values, taken from Phelps [1].
To perform sensitivity analysis, after the experience of some works on this topic [2,3,4, 5], we have elaborated a *global* method (that is a method that allows *simultaneous* variation of all the considered parameters) that works efficiently and has allowed to obtain good results with reasonable calculational time. It is worth noting that with 21 varied parameters, a sistematic scan of the 21 dimensions of the parameter space needs, *at least*, $2^{21} \approx 2*10^6$ calculations on the numerical model, the F.A.S.T. method would require, with *not negligible error* (difficult to be evaluated) on the output (interference order M=6, see [6,7,8,9]), *not less than* 40,000 calculations. We have used

771

M. Capitelli (ed.), Molecular Physics and Hypersonic Flows, 771–781.

only 1000 points, obtaining results with numerical errors of small entity and easy to be calculated and improved. Our method is based on Monte Carlo evaluation of some integrals, with a particular sensitivity coefficient and an original method for *largely* decreasing numerical errors without increasing the number of model calculations required.

2 - Monte Carlo sensitivity method

The mathematical model used (Boltzmann equation in this work) can be considered as a function of a chosen group of parameters:

$$f = f(k_1, k_2, ... k_n) = f(\underline{K}) \qquad (1)$$

(here parameters eventually not considered are left fixed and not explicitly indicated). We have considered the value of the output averaged on all but the j-th parameter, indicated as $\langle f \rangle(k_j)$:

$$\langle f \rangle(k_r) = \int ... \int f(\underline{K}) \prod_{j \neq r} P_j(k_j) dk_j \qquad (2)$$

Here all the parameters are supposed to be indipendent, therefore their global distribution $P(\underline{K})$ is simply the product of the distribution $P_j(k_j)$ of each of them. As sensitivity coefficient (SC), in order to get an unambiguos sensitivity measure (see [5,6,9] for related problems) of the model to j-th parameter variation, we take the maximum value of the absolute difference between $\langle f \rangle(k_j)$ and $\langle f \rangle$, when the j-th parameter is varied in the preassigned range L_j:

$$SC(k_j) = \max | \langle f \rangle(k_j) - \langle f \rangle |, \ k_j \in L_j \qquad (3)$$

This is a measure of the displacement from the mean value of the output, due to the j-th parameter. It is easy to compare this measure with a prefixed maximum error ε acceptable for the f output of the model: if $SC(k_j) < \varepsilon$, then the j-th parameter is *not* a "sensible parameter", and its precise value can be neglected (that is, it will be assigned to it any value in the preassigned interval of variation). On the other hand, if $SC(k_j)$ is high, it is interesting to know the trend of $\langle f \rangle(k_j)$, which can give very useful information. In fact, while the sensitivity coefficient gives only a measure of how much the output is perturbed by j-th parameter, $\langle f \rangle(k_j)$ can make easy to search for appropriate *restricted* intervals into which the j-th parameter can vary without meaningfully affecting the model output. Another feature of $\langle f \rangle(k_j)$ when k_j is a sensible parameter is that it already represents a monodimensional approximation of model behaviour as a function of all the considered parameters, provided all the remaining SCs are negligible (therefore, the trend $\langle f \rangle(k_j)$ and the related SC have an immediate physical interpretation in the context of the studied model, not being simply mathematical indicators of the relative importance of one parameter with respect to the remaining ones, as in other sensitivity methods).

The problem is to calculate the integrals in (2), minimizing the number of calculations required on the model. We have found convenient a Monte Carlo method for this purpose: we vary randomly all the preselected parameters simultaneously and indipendently (by means of *indipendent* random generators) in their respective preassigned intervals, then for the j-th parameter we "ordinate" the obtained outputs by distributing them into a convenient number of "bins", each bin corresponding to a different small subinterval of the range L_j of allowed variation. Into each bin I_j^i of the j-th parameter, k_j is restricted between $(k_j^i - \Delta_j/2, k_j^i + \Delta_j/2) = I_j^i$, being Δ_j the width of the bin for the j-th parameter, k_j^i the mean value of k_j in I_j^i. It is easy now to obtain an approximation of $\langle f \rangle (k_j)$ as the collection of values $\langle f \rangle (k_j^i)$:

$$\langle f \rangle (k_j^i) \equiv \frac{1}{N_i} \sum_{k_j \in I_j^i} f(\underline{K}) \tag{4}$$

(summation is over the model output values having $k_j \in I_j^i$; N_i is the number of these values). Maximum numerical error on $\langle f \rangle (k_j^i)$ is easily calculated as three times the square root of the variance of data in the i-th bin of the j-th parameter, divided by N_i:

$$\sigma_j^i = 3 * \sqrt{\frac{\left\langle f(\underline{K}) \Big|_{k_j \in I_j^i}^2 \right\rangle - \left\langle f(\underline{K}) \Big|_{k_j \in I_j^i} \right\rangle^2}{N_i}} \tag{5}$$

because of the gaussian character of repeated sums involved in (4). Relation (5) insures that numerical error decreases as $1/\sqrt{N_i}$: this allows "continuous" and controlled improving of results by *adding* new model calculations in (4) and checking errors with (5). This is a very important feature, lacking in other sensitivity methods like FAST [6,7,8,9], because it allows to exploit all the calculations already performed on the model, and to preventivate computational time required for a prefixed error to be reached. On the other side, in sensitivity calculations with large computational models, this rate of result improving is critically slow. Therefore it is very useful to search for conditions allowing an optimization of any part of sensitivity evaluation method.

An interesting possibility is offered by improving the rate of error decreasing with an appropriate sequence of points in parameter space, different by a randomly generated one. The so called "LP sequences" [10] are precalculated as a function only of the number n of space dimensions (number of considered parameters in this case), and are obtained following a criterion of maximum uniform spread into the n-dimensional space. According to Sobol [10], these sequences allow an error decreasing of the order of nearly 1/N. This method will be applied in future works on sensitivity requiring very large computational effort.

The method that we propose in this work for lessening the noise in Monte Carlo sensitivity results is addressed to the decreasing of the numerator appearing in (5), that is the variance of the i-th bin data:

$$\left\langle f(\underline{K}) \Big|_{k_j \in I_j^i}^2 \right\rangle - \left\langle f(\underline{K}) \Big|_{k_j \in I_j^i} \right\rangle^2 \tag{6}$$

This quantity, theoretically indipendent on the number N_i of samplings, is determined only by the amount of output variation due to all but the j-th parameter. Considering the definition (2), it is easy to realize that reducing the variance of $f(\underline{K})$ due to all the parameters not equal to k_r will produce results much less noisy on $\langle f \rangle(k_r)$. Most of noise comes from sensible parameters, because they clearly spread the distribution of $f(\underline{K})$: but really it is easy to remove their contribution. In fact, by subtracting $\langle f \rangle(k_q)$ from $f(\underline{K})$, being k_q the most sensible parameter, it is straightforward to demonstrate that:

$$\int \cdots \int [\, f(\underline{K}) - \langle f \rangle(k_q)\,] \prod_{j \neq r} P_j(k_j) dk_j = \langle f \rangle(k_r) - \langle f \rangle \qquad (7)$$

Of course, in order to obtain $\langle f \rangle(k_r)$ two quantities are to be known with *small* error: $\langle f \rangle$ and $\langle f \rangle(k_q)$. About the former there is no problem, because that average is computed on the whole set of calculated points in parameter space, not only on the points in a bin. Concerning the latter, it is not difficult to obtain, because it is very likely that only a little group of parameters be the "leading" ones, that is the most sensible. They produce the most relevant variations in (6), provided their SC is high: therefore they cause the largest amount of noise on the remaining parameter SC evaluations. On the contrary, less important parameters produce small variations in (6), hence for a given number of calculations on the model, relevant parameters have the smallest numerical errors on averaged outputs $\langle f \rangle(k_j)$. This feature is of large usefulness, because it allows to apply the preceding considerations on error decreasing to less important parameters, SCs of which are affected by large errors and cause misunderstanding about *the whole set* of the calculated SCs. Instead of enormously increasing the number of model calculations, we first obtain the trend $\langle f \rangle(k_q)$, that is the trend with the lowest error smaller than a prefixed value, generally with little number of calculations, then we remove large errors on the remaining trends by applying the explained method. An important feature is that the procedure is repeatable for parameters in order of increasing numerical errors. In fact, once the large noise of the most sensible parameter k_q has been removed, all the trends $\langle f \rangle(k_j)$ (but the q-th one) will be recalculated with much smaller errors, so it is now possible to apply the method for the second low noise parameter, and so on. The results of this technique are really of large utility in order to significantly decreasing the number of required calculations on the model with a prefixed error in sensitivity results.

3. The model

The mathematical model to which sensitivity analysis has been applied is the Boltzmann Transport Equation for electrons in an electrical discharge in nitrogen, in the so-called two terms approximation [11]:

$$\partial n(\varepsilon,t)/\partial t = -\partial J_E/\partial \varepsilon - \partial J_e/\partial \varepsilon - S_{in} - S_{sup} \qquad (8)$$

Here $n(\varepsilon,t)$ is the eedf, which has been calculated at stationarity according to the method given in [12]. The two first terms in the second member of (8) are flows of electrons in the energy space due respectively to electric field and elastic collisions. S_{in} and S_{sup} are respectively the inelastic and superelastic collision terms. Collision processes of electrons are considered only with gas molecules (e-M collision) in one of the states listed in tab. 1, with molecular transition from or towards ground state (respectively in case of inelastic or superelastic electron collision). Of course, neglecting all the state-to-state transitions is an approximation, that we have used in this work because we are principally interested in relative variations rather than absolute values of model outputs; besides, we suppose that the model behaviour is not deeply modified, with respect to cross section variations of the considered processes, by the approximation used here. The molecular population is determined, for the vibrational part, by the temperature T_v, which is a parameter of our analysis varying between 0K and 5000K, while the electronic states are left empty constantly. The electron-electron collisions, that would require a time expensive iterative method, are totally neglected by taking a sufficiently low electron density (ionization degree: 10^{-6}). The electric field is fixed at $E/N = 50$ Td. Sensitivity analysis has been performed not only on 100-step discretized eedf, but also on inelastic rate coefficients of all the e-M collision processes considered, calculated as:

$$R = \int_{\varepsilon^*}^{\infty} \frac{n(\varepsilon)}{n_0} \, v(\varepsilon) \, \sigma(\varepsilon) \, d\varepsilon \qquad (9)$$

(ε^* is the process threshold, n_0 is the electron density, $v(\varepsilon)$ the classical electron speed). All the cross sections shown in fig.1 are multiplied, regardless of electron energy, by indipendent factors ranging from 0.5 to 1.5, uniformly distributed. It is straightforward to realize that there is a linear sensitivity of the rate of a certain process to the cross section of that process: the interest of this analysis is in rate sensitivity to all the remaining cross sections, through eedf sensitivity.

4. Results

In this paragraph all the presented sensitivity results are obtained by means of noise reduction technique applied as many times as the number of varied parameters, that has allowed a decreasing of numerical error of about six times on the most noisy trends $\langle f \rangle (k_j)$, corresponding to a decreasing in model required calculations of not less than 36 times.

symbol	initial or final state of molecules	threshold ev	symbol	initial or final state of molecules	threshold ev
1 v	1° vibr.	0.29	1 e	$A^3\Sigma$	6.17
2 v	2° vibr.	0.59	2 e	$B^3\Pi$	7.35
3 v	3° vibr.	0.88	3 e	$W^3\Delta$	7.36
4 v	4° vibr.	1.17	4 e	$B^3\Sigma$	8.16
5 v	5° vibr.	1.47	5 e	$a^1\Sigma$	8.40
6 v	6° vibr.	1.76	6 e	$a^1\Pi$	8.55
7 v	7° vibr.	2.06	7 e	$w^1\Delta$	8.89
8 v	8° vibr.	2.35	8 e	$c^3\Pi$	11.03
			9 e	$E^3\Sigma$	11.88
			10 e	$a"^1\Sigma$	12.25
			11 e	sum of singlets	13.00
			Ion	N_2^+	15.60

Table 1.

776

In fig. 2a it is shown extrema of eedf spread as a consequence of simultaneous variations of all the selected parameters, together with mean values and the eedf obtained with all the parameters exactly at the center of the respective variation range (for cross section data, this means they are not varied at all from their original values). In fig. 2b a total measure of eedf sensitivity is shown, calculated as maximum absolute displacement from mean value for each discretized energy value. For high energy values total sensitivity reaches several orders of magnitude. In fig. 3a-b ((b) is the flat projection of (a)) it is shown the eedf sensitivity to each parameter. On vertical axis there is the sensitivity coefficient of decimal logarithm of eedf for the energy value indicated on the "eV" axis, referred to the variation of the parameter indicated on the remaining axis. Taking into account that the mean error on eedf sensitivity results has a mean value of ±0.05, with maximum value of ±0.1 for the highest energy values, it is straightforward to note that:

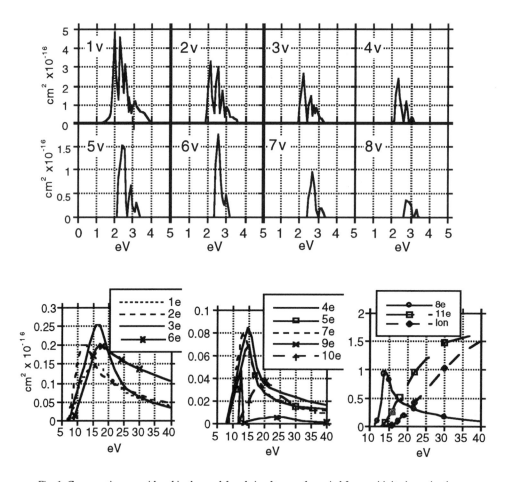

Fig. 1. Cross sections considered in the model and simultaneously varied for sensitivity investigation.

1) there is eedf sensitivity of nearly 0.1 (that is $10^{\pm 0.1} \approx \pm 20\%$ because of the use of decimal logarithm) caused by variations of the first three vibrational cross sections of tab.1 (1v,2v,3v), constant through energy axis starting beyond 2 eV. These results are not very sure, because of the presence of noise of the same magnitude of the obtained value. Anyway, sensitivity analysis rarely is applied to the study of so small variations. For example, in F.A.S.T. method only sensitivities differing by *orders* of magnitude are considered sure [6] ;

2) there is eedf sensitivity to 1e (up to 0.2) and 2e,3e,6e (up to 0.3), to 8e (up to 0.65), to 11e (up to 1.7) and to ionization cross section variation (up to 1.1);

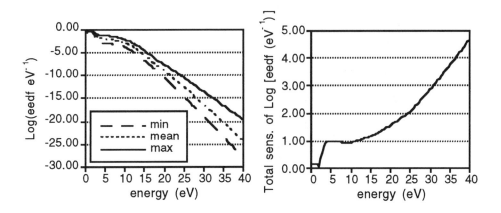

Fig. 2a-b. Extrema of eedf variation (a) and corresponding eedf total sensitivity (b).

3) sensitivity to vibrational temperature T_V has a constant value of 0.6 beyond about 2 eV;

4) cross section sensitivities as a function of energy have an initial increasing trend (starting at energy values increasing by order of the process list in tab.1), followed in all cases, but 11e and ionization, by a constant trend.

Concerning last point, the variation of any given cross section has not negligible effect starting from the energy value at which the cross section is appreciable, as it is clear by comparing fig. 3a-b with fig. 1 (for example, in the case of all vibrational cross sections this energy value is about 2 eV). It is possible to note an increasing trend as a function of energy in the neighboring of this energy value for each cross section sensitivity. It is less easy to explain the following constant section of each sensitivity trend, once the cross section has become negligible (11e and ionization cross sections have not negligible values at the end of energy axis, therefore it is not surprising that their sensitivities do not reach a flat trend as all the other ones). One possible explanation could be that higher energy part of eedf is scarcely populated, so it is largely sensible to changes in lower energy eedf parts, which are "directly" modified by cross section variations. But this could be also a second (or higher) order sensitivity effect, that is the interaction of two (or more) simultaneously varying parameters: to study this possibility a higher order sensitivity analysis is required.

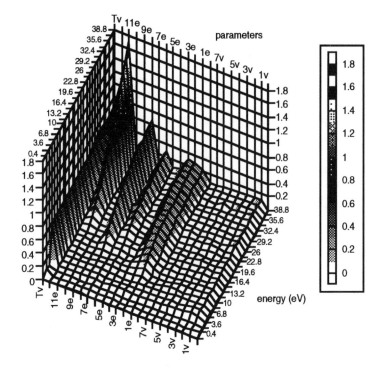

(a)

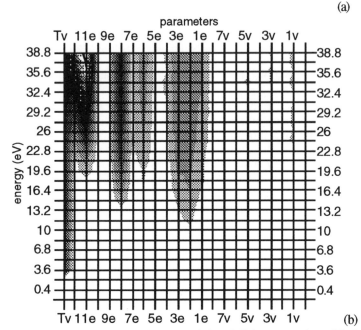

(b)

fig. 3a-b. Eedf sensitivity to simultaneous variation of all the cross sections of fig.1.

It is worth noting that the presented noise reduction technique would have a key-role in higher order sensitivity analysis, because of the huge number of normally required model calculations in this kind of analysis.

Concerning the first three points underlined in this paragraph, it is possible, by comparing fig. 3a-b with fig. 1 and separately grouping vibrational and electronic cross sections, that relevant sensitivities are generated by variation of cross sections of higher peak value, as it could be expected.

The distinction between vibrational and electronic cross sections is due to the presence of a varing vibrational temperature T_V that, when different from zero, makes superelastic processes possible, and these ones stabilize the system, that is cause very low vibrational sensitivities, much lower than expected on the basis of cross section values in comparison with electronic results.

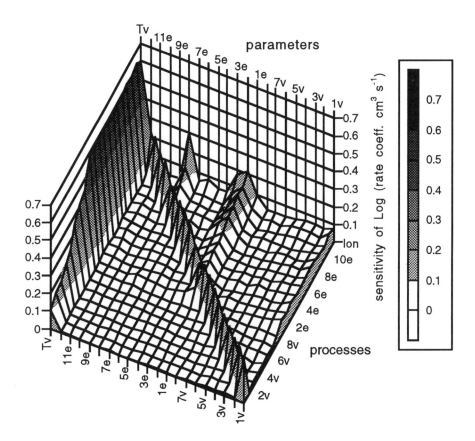

Fig.4. Sensitivities of the anelastic rate coefficients to simultaneous variation of cross sections of fig.1.

A large sensitivity in vibrational rates is on the contrary expected in the cold gas approximation (i.e. when $T_V = 0K$).

In fig. 4 results about rate coefficient sensitivity to cross section variations are presented. Here the mean error is about ± 0.02, with maximum value of ± 0.04 (that is

$10^{\pm 0.04} \approx \pm 9\%$). Obviously, there is a diagonal of constant values (about 0.35) concerning sensitivity of each rate to the cross section of the same process. These constant values are easily directly calculated, by considering that the integral (9) defining rate coefficient is, as a first approximation, linearly varied by the cross section explicitly appearing, and a variation of $10^{\pm 0.3}$ (that is an interval $\approx [0.5 \div 2]$) is in accord with the cross section interval of variation. Even in this case, sensitivity to vibrational cross sections are of the same magnitude of numerical error on results.

From the saturation trend of all eedf sensitivities, it is obvious to expect that the variation of the cross section of one process has influence on all the other processes of higher energy, that is those ones whose cross section extends mainly over high energy value. In fact, eedf sensitivity to first three electronic cross sections of tab.1, which starts at about 10 eV but reaches a constant value not before 20 eV, generates sensitivity in all the electronic processes, but in an increasing manner depending on the energy range of the considered process. This is confirmed by rate sensitivities to 6e: in this case the sensitivity threshold is very high (over 15 eV), therefore even if the eedf sensitivity value is not negligible (0.3), only ionization rate is affected by 6e cross section variation to an extent just overcoming noise. Similar conclusions are possible also for 8e, 11e and ionization cross section variations on all the other rate coefficients.

Electronic sensitivity to vibrational temperature has a constant value of 0.6; on the contrary, vibrational processes show increasing values between 0.1 and 0.45. This happens because eedf sensitivity to vibrational temperature increases very rapidly between 2 and 3.6 eV (fig.3), therefore all the electronic processes are affected in a constant way by vibrational temperature change; on the contrary, all the vibrational cross sections considered spread just in that range, with the first peak of each one of slightly larger energy value of the preceding: this explains the increasing value of sensitivity to vibrational temperature.

5. Conclusions

In this work an efficient method of global sensitivity analysis, improved with the aid of an original noise reduction technique, has been applied to the Boltzmann transport equation for electrons in a nitrogen discharge. It has been possible to obtain sensitivity of the discretized eedf and of inelastic rate coefficients to variations of the cross sections of 20 electron-molecule collision processes, and to vibrational temperature. The results are explained with a qualitative analysis based on the magnitude and position on energy axis of the varied cross sections, taking into account the different vibrational and electronic temperatures. This kind of results can be of interest in the search for better values of cross sections obtained by deconvolution of experimental data (rates and transport coefficients), because it gives a sufficiently accurate measure of the range of variation of a given model output due to a given parameter incertitude, *averaged* over the variations of all the remaining parameters (it is important to remind that considered models have generally a non-linear behaviour in the parameters, therefore the effect of variation of one parameter is strictly linked to the values simultaneously assigned to all the remaining parameters. This is the central problem of *global* sensitivity analysis).

The results has been obtained with an extremely low number of calculations on the model (1000) in comparison with the number of simultaneosly varing parameters (21), with numerical errors really small for a sensitivity analysis (10% on the average, 20% at maximum). The sensitivity method applied in this work is immediately susceptible of improvements - in particular the use of LP sequences instead of simple random sequences and higher order sensitivity analysis - and applications to much more complex and time expensive models. Some works on these topics are in progress.

References

[1]   A.V.Phelps and L.C.Pitchford, JILA report no. 26 (1985).
[2]   G.Colonna, S.Longo, F.Esposito, M.Capitelli, "Fourier Transform Sensitivity Analysis: application to XeCl self-sustained discharge-laser kinetics", *Appl.Phys.* B 59, 61-72 (1994)
[3]   G.Colonna, F.Esposito, M.Capitelli, "Sensitivity Analysis of Models in Plasma Kinetics" ICPIG XXI, 1993, Bochum, Germany.
[4]   F.Esposito, G.Colonna, M.Capitelli, "Statistical Sensitivity Analysis of Rate Coefficients for Excitation of H_2 by Electron Molecule Discharge Collisions", XII ESCAMPIG, Noordwijkerhout, The Netherlands, 1994.
[5]   F.Esposito, G.Colonna, S.Longo, M.Capitelli, "Monte Carlo Global Sensitivity Analysis of Boltzmann Equation for Electron Kinetics in H_2 Discharges", *Plasma Chem. Plasma Process.* (1996).
[6]   R.I.Cukier, C.M.Fortuin, K.E.Shuler, A.G.Petschek, J.H.Shaibly, "Study of the sensitivity of coupled reaction systems to uncertainties in rate coefficients. I" *J.Chem.Phys.* 59 (1973).
[7]   J.H.Shaibly, K.E.Shuler, "Study of the sensitivity of coupled reaction systems to uncertainties in rate coefficients. II" *J.Chem.Phys.* 59 (1973).
[8]   R.I.Cukier, J.H.Shaibly, K.E.Shuler, "Study of the sensitivity of coupled reaction systems to uncertainties in rate coefficients. III. Analysis of the approximations" *J.Chem.Phys.* 63 (1975).
[9]   R.I.Cukier, H.B.Levine, K.E.Shuler, "Nonlinear Sensitivity Analysis of Multiparameter Model Systems" *J.Comp.Phys.* 26 (1978).
[10]   I.M.Sobol, "On the sistematic search in a hypercube" SIAM J. Numer.Anal., 5, Vol.16, Oct. 1979
[11]   S.D.Rockwood, *Phys.Rev. A* 8 (1973).
[12]   G.Capriati, G.Colonna, C.Gorse, M.Capitelli, "A parametric study of electron energy distribution functions, rate and transport coefficients in non-equilibrium helium" *Plasma Chem. Plasma Proc.* 12,no.3 (1992) and G.Colonna, C.Gorse, M.Capitelli, R.Winkler and J.Wilhelm, "The influence of electron-electron collisions on electron energy distribution functions in N_2 post discharge" *Chem.Phys.Lett.*, 213, 1-2, (1993)

List of Participants

AUSTRIA

H. PAUSER — Institute für Allgemeine Physik, Technische Universitat Wien, Wiedner Haupstrasse 8-10, A1040 Wien, AUSTRIA

BELGIUM

P. ALAVILLI — VUB/ Dept. of Fluid Mechanics, 2 Pleinlaan, 1050 Brussel, BELGIUM

G.S.R. SARMA — VKI, 72 Chaussée de Waterloo, 1640 Rhod-Saint-Genese, BELGIUM

DENMARK

G. D. BILLING — Kemisk Laboratorium III, Universitetsparken 5, DK-2100 Copenhagen, DENMARK

C. NYELAND — Kemisk Laboratorium III, Universitetsparken 5, DK-2100 Copenhagen, DENMARK

FRANCE

A. BOURDON — CNRS-URA 230, CORIA, 76821 Mont Saint Aignan Cedex, FRANCE

A. BROC — ONERA / Div. de l'Aérodynamique Theorique 1, 29 Avenue de la Division Leclerc, 92320 Chatillon, FRANCE

R. BRUN — Université de Provence, Centre Saint Jerome, Case 321, 13397 Marseille 20, FRANCE

M. DE GRAAF — CNRS Laboratoire d'Aérothermique, 4 ter route des Gardes, 91290 Meudon, FRANCE

M. DUDECK — Laboratoire d'Aérothermique du CNRS, 4 ter route des gardes, 92190 Meudon, FRANCE

J. P. DUDON — I.U.S.T.I.-MHEQ, Université de Provence Case 321, 13397 Marseille Cedex 20, FRANCE

S. FAHRAT — LIMHP-CNRS, Av J.B. Clement, 93400 Villateneuse, FRANCE

K. HASSOUNI — LIMHP-CNRS, Av J.B. Clement, 93400 Villateneuse, FRANCE

783

784

A. LEROUX — Université INSA de Rouen, CNRS-URA 230 CORIA, 76821 Mont Saint Aignan Cedex, FRANCE

L. ROBIN — CNRS-URA 230 CORIA, 76821 Mont Saint Aignan Cedex, FRANCE

S. SEROR — Université de Provence, Centre Saint Jerome, Case 321, 13397 Marseille 20, FRANCE

P. VERVISCH — CNRS URA 230-Coria, Université de Rouen, Place Emile Blondel, 76821 Mont Saint Aignan, FRANCE

GERMANY

A. DAISS — IRS Stuttgart University, Pfafenvaldring 31, D-70550 Stuttgart, GERMANY

H.H. FRÜHAUF — University of Stuttgart, Institut fur Raumfahrsysteme, Pfaffennwaldring 31, 70550 Stuttgart, GERMANY

S. KANNE — University of Stuttgart, Institut fur Raumfahrsysteme, Pfaffennwaldring 31 70550 Stuttgard, GERMANY

O. KNAB — IRS Stuttgart University, Pfaffenwaldring 31, D-70550 Stuttgart, GERMANY

E. PASSOTH — FB Physik, Univeritaet Greifswald, Domstrasse 10a, D-17489 Grefswald, GERMANY

I. VINOGRADOV — Rhur Universitat Bochum, Inst. fur Experimentalphysik II, 44780 Bochum, GERMANY

M. WOLLENHAUPT — DLR, Bunsenstrasse 10 ,D-37073 Gottingen, GERMANY

INDIA

H. AGRAWAL — Department of Aerospace Engineering, Indian Institute of Technology, 208 016 Kanpur, INDIA

ITALY

I. ARMENISE — Dipartimento di Chimica, Università di Bari, via Orabona 4, 70126 Bari, ITALY

M. BARBATO — CRS4, via N. Sauro 10, 09123 Cagliari, ITALY

V. BELLUCCI — CRS4, via N. Sauro 10, 09123 Cagliari, ITALY

S. BORRELLI — CIRA, via Maiorise, 81043 Capua, ITALY

C. BRUNO — Dipartimento di Meccanica ed Aeronautica, via Eudossiana, Rome, ITALY

M. CACCIATORE — Centro Studio Chimica Plasmi CNR, Università di Bari, via Orabona 4, 70126 Bari, ITALY

M. CAMPOBASSO — Politecnico di Bari, Macchine ed Energetica, via Orabona 4, 70126 Bari, ITALY

F. CAPEZZUTO — Dipartimento di Chimica, Università di Bari, via Orabona 4, 70126 Bari, ITALY

M. CAPITELLI — Dipartimento di Chimica, Università di Bari, via Orabona 4, 70126 Bari, ITALY

R. CELIBERTO — Dipartimento di Chimica, Università di Bari, via Orabona 4, 70126 Bari, ITALY

S. CROCCHIANTI — Dipartimento di Chimica, Università di Perugia, via Elce di Sotto 8, 06100 Perugia, ITALY

S. DE BENEDICTIS — Centro Studio Chimica Plasmi CNR, Università di Bari, via Orabona 4, 70126 Bari, ITALY

F. DE FILIPPIS — CIRA, via Maiorise, 81043 Capua, ITALY

G. DILECCE — Centro Studio Chimica Plasmi CNR, Università di Bari, via Orabona 4, 70126 Bari, ITALY

F. ESPOSITO — Dipartimento di Chimica, Università di Bari, via Orabona 4, 70126 Bari, ITALY

G. FALCONE — Dipartimento di Fisica, Università di Bari, via Orabona 4, 70126 Bari, ITALY

C. GORSE — Dipartimento di Chimica, Università di Bari, via Orabona 4, 70126 Bari, ITALY

A. LAGANA — Dipartimento di Chimica, Università di Perugia, Via Elce di Sotto 1, 06100 Perugia, ITALY

S. LONGO — Dipartimento di Chimica, Università di Bari, via Orabona 4, 70126 Bari, ITALY

M. LOSURDO — Centro Studio Chimica Plasmi CNR, Università di Bari, via Orabona 4, 70126 Bari, ITALY

M. MARINI — CIRA, via Maiorise, 81043 Capua, ITALY

F. PIRANI — Dipartimento di Chimica, Università di Perugia, via Elce di Sotto 8, 06100 Perugia, ITALY

A. QUARTERONI — Dipartimento di Matematica, Politecnico di Milano, Via Bonardi 9, 20133 Milano, ITALY

S. REGGIANI — Dipartimento di Meccanica ed Aeronautica, via Eudossiana, Roma, ITALY

M. VIGLIOTTI — Dipartimento di Chimica, Università di Bari, via Orabona 4, 70126 Bari, ITALY

JAPAN

C. PARK — Department of Aeronautics and Space Engineering, Faculty of Engineering, Tohoku University, Aoba Aramaki, Aoba-Ku, 980 Sendai, JAPAN

THE NETHERLANDS

G.J.H. BRUSSAARD — Eindhoven University of Technology, Dept. of Physics, P.O.Box 513, 5600 MB Eindhoven, THE NETHERLANDS

D. GIORDANO — ESA-ESTEC, Aerothermodynamics Section, P.O.Box 299, 2200 AG Nordwjik, THE NETHERLANDS

L. MARRAFFA — ESA-ESTEC, Aerothermodynamics section, P.O. Box 299, 2200 AG Noordwijk, THE NETHERLANDS

PORTUGAL

B. GORDIETS — Centro de Electrodinamica da UTL, Instituto Superior Tecnico, Av. Rovisco Pais 1, 1096 Lisboa Codex, PORTUGAL

V. GUERRA — Centro de Electrodinamica da UTL, Instituto Superior Tecnico, Av. Rovisco Pais 1, 1096 Lisboa Codex, PORTUGAL

M.J.G. PINHEIRO — Centro de Electrodinamica da UTL, Instituto Superior Tecnico, Av. Rovisco Pais 1, 1096 Lisboa Codex, PORTUGAL

ROMANIA

E. ALDEA — Institute of Physics and Technology of Radiation Devices, Low Temperature Plasma Physics Department, P.O. Box M6-07, RO 76900 Bucharest-Magurele, ROMANIA

RUSSIA

I. BRYKINA — Institute of Mechanics, Moscow State University, Mechrinsky pr. 1, 119899 Moscow, RUSSIA

M. IVANOV — Institute of Theoretical & Appled Mech., Institutskay ul 4/1, 630090 Novosibirsk, RUSSIA

A. KOLESNIKOV — Institute for Problems in Mechanics, Russian Academy of Science Prospekt Vernadskogo IOI, 117526 Moscow, RUSSIA

S. A. LOSEV — Institute of Mechanics, Moscow State University, 1 Michurinski pr., 117192 Moscow, RUSSIA

E. A. NAGNIBEDA — Math. & Mech. Department, St Petersurg University, 198904 St Petersburg, Petrodvorets, RUSSIA

G. A. TIRSKIY — Institute of Mechanics, Moscow State University, Michurinsky 1, 119899 Moscow, RUSSIA

S. V. ZHLUKTOV — Institute for Computer Aided Design, Russian Academy of Sciences, 2-d Brestskaya 19/18, 123056 Moscow, RUSSIA

SPAIN

M. GILIBERT — Universitat de Barcelona, Department de Quimica Fisica, Marti i Franques 1, 08028 Barcelona, SPAIN

SWEDEN

T.A. GRÖNLAND — Aeronautical Research Institute of Sweden, P.O. Box 11021, S 16111 Bromma, SWEDEN

USA

I. ADAMOVICH — Ohio-State University, Colombus, OH, USA

J. N. BARDSLEY — Lawrence Livermore Laboratory, L-296, 7000 East Avenue, Livermore, CA 94551, USA

D. BOSE — University of Minnesota, 107 Akerman Hall, 110 Union St SE, Mineapolis, MN 55455, USA

G. V. CANDLER — University of Minnesota, 107 Akerman Hall, 110 Union St SE, Mineapolis, MN 55455, USA

P. CINNELLA — Engineering Research Center, P.O.Box 6176, Mississippi State, MS 39762, USA

D. GOLDSTEIN — Department Aerospace Engineering, University of Texas, Austin, TX 78712, USA

E. J. JUMPER — Hessert Center for Aerospace Research, Department of Aerospace and Mechanical Engineering, University of Notre Dame, Notre Dame, In 46556, USA

J. M. MONROE — University of Texas, 11226 Barrington Way, Austin, TX 78759, USA

J. OLEJNICZAK — University of Minnesota, 107 Akerman Hall, 110 Union St SE, Mineapolis, MN 55455, USA

H. PARTRIDGE — NASA Ames Center, Moffett Field, California, USA

C. SCOTT — NASA Johnson Space Center, Houston, TX 77058, USA

H. THUEMMEL   mail stop 230-3, NASA Ames Research
Center, Moffett Field, CA 94035-1000, USA

P. VARGHESE   Dept of Aerospace Engineering & Engineering
Mechanics, U.T. Austin, Austin, TX 78712,
USA

P. VITELLO   Lawrence Livermore Laboratory, L-296,
7000 East Avenue, Livermore, CA 94551,
USA

Analytical index